International Conference on Innovative Research (ICIR EUROINVENT 2020)

IOP Conference Series: Materials Science and Engineering
Volume 877

Iasi, Romania
21 - 23 May 2020

Editors:

Andrei Victor Sandu
Mohd Mustafa Al Bakri Abdullah
Petrica Vizureanu
Che Mohd Ruzaidi
Ion Sandu

ISBN: 978-1-7138-3620-9
ISSN: 1757-8981

Printed from e-media with permission by:

Curran Associates, Inc.
57 Morehouse Lane
Red Hook, NY 12571

Some format issues inherent in the e-media version may also appear in this print version.

Institute of Physics
Dirac House, Temple Back
Bristol BS1 6BE UK

Phone: 44 1 17 929 7481
Fax: 44 1 17 920 0979

techtracking@iop.org

Additional copies of this publication are available from:

Curran Associates, Inc.
57 Morehouse Lane
Red Hook, NY 12571 USA
Phone: 845-758-0400
Fax: 845-758-2633
Email: curran@proceedings.com
Web: www.proceedings.com

TABLE OF CONTENTS

PROCEDURES AND TECHNOLOGIES FOR MATERIALS ENGINEERING

MATERIALS APPLICATION

IOP Conf. Series: Materials Science and Engineering **877** (2020) 011001 doi:10.1088/1757-899X/877/1/011001

ICIR Euroinvent 2020

International Conference on Innovative Research

in the field of Materials Science and Engineering

May 21st to 22nd, 2020
Iasi – Romania

Organized by:

- **Romanian Inventors Forum**
- **Faculty of Materials Science and Engineering, The "Gheorghe Asachi" Technical University of Iasi, Romania**
- **ARHEOINVEST Platform, Alexandru Ioan Cuza University of Iasi**
- **Centre of Excellence Geopolymer and Green Technology CEGeoGTech), Universiti Malaysia Perlis (UniMAP)**

With support of:

- **Universiti Malaysia Terengganu**
- **International Federation of Inventors' Associations – IFIA**
- **World Invention Intellectual Property Associations – WIIPA**

Editors:

Andrei Victor SANDU, Mohd Mustafa Al Bakri ABDULLAH,

Petrică VIZUREANU, Che Mohd Ruzaidi GHAZALI, Ion SANDU

IOP Conf. Series: Materials Science and Engineering **877** (2020) 011001 doi:10.1088/1757-899X/877/1/011001

EUROINVENT
ICIR 2020

International Conference on Innovative Research
Iasi, 21st–22nd of May 2020

Foreword

This volume contains selected peer-reviewed articles presented at ICIR Euroinvent ICIR Euroinvent 2020 Conference. The event was held *on-line* in Iași, România from 21st to 22nd of May 2020.

The ICIR Conference is organized under the auspices of EUROINVENT. Euroinvent is a joint event promoting creativity in a European context, by displaying the contributions of consecrated schools from higher education and academic research and also of individual inventors and researchers.

The EUROINVENT International Conference on Innovative Research (ICIR) brings together leading researchers, engineers and scientists who will present actual research results in the field of Materials Science and Engineering.

The conference aims to provide a high level international forum for researchers, engineers and scientists to present their new advances and research results in the field of materials science and engineering.

The volume covers all the aspects of materials science, from synthesis and characterization of materials to procedures and technologies for materials engineering, as well as materials application and their involvement in the life sciences.

All the papers have been reviewed by two expert referees in their relevant topic disciplines. The papers selected for the volume depended on their quality and relevancy to the conference.

According to the World Health Organisation recommendations due to COVID-19 pandemic, all the works had to be carried online, on the same dates as the conference was planned, in order to avoid overlapping on other postponed events. The online presentation was carried on ZOOM video conferencing application, offering the option to record on cloud all the sessions and to moderate at high quality all the presentations and interventions. The entire conference is available as recording – on 3 plenary sessions, next to the program on: http://www.euroinvent.org/conference/ .

IOP Conf. Series: Materials Science and Engineering **877** (2020) 011001 doi:10.1088/1757-899X/877/1/011001

The structure of the conference was kept similar to the previous ones, having 2 Keynote Speakers (renowned scientists from Israel – Prof. Joseph KOST and Malaysia – Dr. Arif SALLEH) who had 30 minutes to present their research and to reply to questions, also 3 Invited Speakers (Hungary – Dr. Tunde KOVACS, Morocco – Dr. Hana HACHIMI and from Malaysia – Dr. Shayfull RAHIM), who had 20 minutes to present their research and to reply to questions. To the regular presenters were given 15 minutes for presentation and Q&A. In total were 31 plennary presentation next to 43 posters.

The conference was very dynamic with many questions and replies from the participants. During the 3 on-line sessions more than 200 participants joined the presentations from Bulgaria, France, Hungary, Israel, Japan, Malaysia, Moldova, Morroco, Poland, Romania, Russia, Thailand, Ukraine and United States of America.

At the end of the conference there was a small ceremony where on the decision of the Scientific Board, The Best Oral Presentation Award was presented, next to 3 Best Poster Awards.

The editors hope that this volume will provide the reader a broad overview of the latest advances in the field of materials science and engineering and that they will be a valuable references source for further research.

The editors would like to express their sincere appreciations and thanks to all the committee members of the ICIR 2020 for their tremendous efforts. Thanks also to IOP Conference Series for producing the volume with full articles.

Finally, the editors would like to thank all the authors for their contribution to this valuable volume.

Editors

IOP Conf. Series: Materials Science and Engineering **877** (2020) 011001 doi:10.1088/1757-899X/877/1/011001

Scientific Advisory Board

Chairman

Event Coordinator

IOP Conf. Series: Materials Science and Engineering **877** (2020) 011001 doi:10.1088/1757-899X/877/1/011001

EUROINVENT ICIR 2020 — International Conference on Innovative Research, Iasi, 21st–22nd of May 2020

Technical and Organizing Committee

IOP Conf. Series: Materials Science and Engineering **877** (2020) 011002 doi:10.1088/1757-899X/877/1/011002

Peer review statement

All papers published in this volume of *IOP Conference Series: Materials Science and Engineering* have been peer reviewed through processes administered by the proceedings Editors. Reviews were conducted by expert referees to the professional and scientific standards expected of a proceedings journal published by IOP Publishing.

IOP Conf. Series: Materials Science and Engineering **877** (2020) 012001 doi:10.1088/1757-899X/877/1/012001

The Relationship Between the Cutting Process Parameters and the Surface Roughness during the Aluminum Machining

A B Pop [1] and M A Țîțu [2*]
[1] SC TECHNOCAD SA, 72, Vasile Alecsandri Street, Baia Mare, România
[2] Lucian Blaga University of Sibiu, 10, Victoriei Street, Sibiu, România, The Academy of Romanian Scientists, 54, Splaiul Independenței, Sector 5, Bucharest, Romania

*E-mail: mihail.titu@ulbsibiu.ro

Abstract. The purpose of this paper is to identify based on the composite central factorial experiment, the mathematical model between the arithmetic mean deviation of the surface profile and the cutting process parameters (cutting speed, cutting depth and feed per tooth), which describes the studied system in any point of the chosen experimental domain. The coefficients of these equations represent the influence of the variables on the response. The analysis of the variance ANOVA is focused on estimating the different types of variability of the response and the estimations made with the Fisher test. This test analyses the coefficients significance of the regression equation, by comparing the ratio of two variants. The Fisher test indicate the probability to be a statistical difference between them or not. In any case, the experimental results are associated with experimental errors. It is essential to find the prediction accuracy of the coefficient's values, and not only their absolute value. In the end, will be determined the arithmetic mean deviation of the surface profile R_a, based on the mathematical model, and will be compared with the experimental results.

1. Introduction

Aluminum alloys are widely utilized as the principal workpiece material in automotive and aeronautical sectors. The surface quality, more exactly the surface roughness - represent an essential feature of product quality, as it has a significant influence on the superior capacities of the mechanical parts but also on the cost of production. Aluminum alloys have a higher workability index.

Among the most used processes is the milling one, in the aluminum alloys machining. Nowadays there are various researches that have been carried out to evaluate the influence and also the interaction of the cutting process parameters, respectively the cutting speed, the feed rate and the cutting depth exerted on the surface roughness in milling processes, using statistical techniques like Taguchi's method [1-3]. Dobrotă D et al [4] addressed the issue of chemical composition in the castings process of aluminum alloys. Țîțu A M et al [5] debate the problem of an experience plan using Taguchi's method to track the different cutting parameters influence to be able to optimize the performance of a process or system. Vamsi I and Gopinath V [6] addressed the issue of cutting forces and surface quality in the machining of aluminum alloys, using it as a technique for optimizing the Response Surface Methodology process. On the other hand, [7] studied in the cutting process, the fact that almost all the mechanical energy that is used to remove the material, is converted into heat and in this way generates a high temperature in the cutting area. As the cutting speed increases, so does the heat, which generates higher temperatures. For these authors, the newest machining challenge is to use the cutting speed with an increased value which can lead to the productivity increasing and to achieve a better surface finish. They found that this generated heat will lead to the faster tool wear and surface roughness. Other authors have studied the effect of the cutting process factors on the surface quality using the design of experiments method [8-10] like the factorial experiment, Taguchi's method, ANOVA. These methods were used to investigate the main effect of the cutting parameters, on the surface roughness in the end milling process of the Al 7136 and it results that the cutting speed, cutting depth and feed per tooth are influencing the surface quality.

In this research paper, it will be pursued to obtain a quantitative relation of the cutting process variables and the R_a (the arithmetic mean deviation of the surface profile) based on the method of the central composite experiment.

2. Research methodology

2.1. The workpiece material

The material chosen to conduct the research is the 7136 - aluminum alloy. The workpiece dimensions are 500 x 110 x 24.50 [mm]. The chemical composition of this Al7136 alloy is shown in Table 1 [9].

Table 1. The chemical composition of Al7136.

Element	Min	Max	Element	Min	Max
Silicon	-	0.12	Zinc	8.40	9.40
Iron	-	0.15	Titanium	-	0.10
Copper	1.90	2.50	Zirconium	0.10	0.20
Manganese	-	0.05	Other Elements, each	-	0.05
Magnesium	1.80	2.50	Other Elements, total	-	0.15
Chromium	-	0.05	Aluminum	remainder	

2.2. Cutting tool

SECO R217.69-1616.0-09-2AN is the cutting tool chosen to carry out the research. The related two cutting inserts are XOEX090308FR-E05, H15. The tool is a standard one in aluminum machining.

2.3. Coolant

The coolant is a mineral oil - Blasocut BC 35 Kombi SW.

2.4. Experiment design

The end milling of Al7136 was carried out on the CNC HAAS VF-YT2.

The design of experiment (DOE) implies a central composite factorial design, which is presented in Table 2. STATISTICA Software was chosen for this research.

Table 2. The cutting process parameters and their levels.

Name		Units	Type	Low (-1)	Center (0)	High(1)
A	Cutting speed	[m/min]	Numeric	495	570	660
B	Cutting depth	[mm]	Numeric	2	3	4
C	Feed per tooth	[mm/tooth]	Numeric	0.04	0.06	0.08

For this study, the design of a central composite experiment (cube plus star) was chosen with 3 factors having 16 experimental tests and 1 single block.

In the STATISTICA Software there are the following options for selecting options:

- For the Compute / use alpha for rotatability option - Alpha axial distance will be calculated by STATISTICA Software. Thus, the axial distance in this case does not depend on the number of center points in the DOE.
- When the Compute / use alpha for orthogonality option is selected, then STATISTICS calculates the alpha axial distance that depends on the number of center points in the DOE.
- In the current research the interest is the Face center star points option(for example, ± 1), to set the values of the points as being equal to the minimum and maximum of the respective factor.
- Add center points (for orthogonality & rotatability) is indicated to add center points to the design to obtain (approximately) orthogonality and rotatability. Specifically, STATISTICA Software adds the central design points.

The points location determines whether the design is orthogonal (that is, if the effects of the factor are orthogonal) or rotary. The concept of DOE rotatability refers to the characteristics by which DOE can extract the same amount of information (i.e., make predictions with the same precision) in all the analyzed directions.

In this case Alpha for rotability calculated by STATISTICA Software is 1.6818 and Alpha for orthogonality is 1.2872.

In the construction of the central composed design of three factors, the experimental domain is a cube and the 8 experimental points (nc = 8) represented by the corners of the cube, are added 6 experimental points on the three perpendicular axes (ns = 6) and two replication experiments in the center of the experimental domain (ns = 0), experiments 15 (C) and 16 (C).

The design array developed to accomplish the 16 studies runs the central composite factorial design, as presented in Table 3.

Table 3. Design matrix.

Standard Run	Control factors: A – Cutting speed [m/min]		B – Cutting depth [mm]		C – Feed per tooth [mm/dinte]		Response: R_a – Surface roughness [μm]
1	-1	495	-1	2	-1	0.04	0.235
2	-1	495	-1	2	1	0.14	0.250
3	-1	495	1	4	-1	0.04	0.342
4	-1	495	1	4	1	0.14	0.270
5	1	660	-1	2	-1	0.04	0.549
6	1	660	-1	2	1	0.14	0.723
7	1	660	1	4	-1	0.04	0.565
8	1	660	1	4	1	0.14	0.561
9	-1	495	0	3	0	0.08	0.232
10	1	660	0	3	0	0.08	0.604
11	0	570	-1	2	0	0.08	0.450
12	0	570	1	4	0	0.08	0.777
13	0	570	0	3	-1	0.04	1.133
14	0	570	0	3	1	0.14	0.424
15 (C)	0	570	0	3	0	0.08	0.872
16 (C)	0	570	0	3	0	0.08	0.872

3. Analysis of variance ANOVA

The variance analysis ANOVA uses the variance ratio tests to determine whether or not exist some very large differences between the averages of several data groups and each data group has a normal distribution or not. The analysis of variance is an elementary instrument, used to evaluate the degree of matching of a regression analysis and of the generated model following the regression analysis. The variance analysis considers the estimation of various types of response variability and the F-test estimates [11].

The first step assumes to use the F test where the variance of the modeled results it is compared to the variance of the unmodified results. Then it is considered significant if p is less than 0.05. The next phase consists of the model comparation with the replication error than the results are significant if the p value has a greater value than 0.05.

Specifically, the objective of ANOVA technique is to evaluate the influence and the interaction of machining parameters that significantly affect the surface quality. Therefore, the proper combinations of the process variables are predicted. Table 4 presents the results obtained from this analysis carried out using STATISTICA Software.

Table 4. The variance analysis of the surface roughness of Al7136.

Source of variation	Sum of square (SS)	Degree of freedom (d.o.f)	Mean Square = SS/d.o.f	F Value	P Value	
A	0.246332	1	0.246332	15.23480	0.003602	significant
B	0.043963	1	0.043963	2.71899	0.133561	
C	0.010977	1	0.010977	0.67891	0.431256	
AB	0.023891	1	0.023891	1.47759	0.255075	
AC	0.041271	1	0.041271	2.55246	0.144587	
BC	0.007539	1	0.007539	0.46628	0.511896	
Error	0.145521	9	0.016169			
Total SS	0.514799	15				
R-squared = 0.71732; Pred R-squared = 0.8542, Adj. R-squared = 0.52887						

By this analysis results that the cutting speed is the variable with the greatest effect on the surface quality. Statistically, the model value F of 15.23480 assumes that the model is significant. Moreover, the P value less than 0.05 indicates that factor A - cutting speed is important.

Next, the standardized estimation of the effects of the factors and the interactions between them will be made using the Pareto diagram (Figure 1).

As can be seen, a vertical line is displayed indicating the minimum size of statistically significant effects, taking into account the current model and choosing the error term using the statistical significance criterion. This graph shows very clearly the influence of factor A - the cutting speed on the roughness of the machined surface of the aluminum alloy.

Another important aspect is that the objective function can be undergone to a three-dimensional analysis, to follow their variation under the influence of controllable variables (cutting process parameters). By plotting the response surfaces, an equation of the regression model is obtained, which can be successfully used to estimate the values of the objective function, throughout the experimental field. The response surface corresponding to the regression equation of the model with the estimated coefficients, belongs to a multifactorial space located between the extreme levels of the influence factors, so to the interpolation of the values of this objective function (that is, with a dimension with one larger than the number of influence factors included in the model) and as such, in general (for more than two influencing factors), it can only be described analytically. However, often, in order to form an image that is more intuitive than that provided by the regression equation, one uses the graphical representation of this response surface, depending on two influencing factors [11-13]. However, it is important to note that the interpretation of experimental results, solely on the basis of these graphical representations, can lead to serious errors, since they refer only to a particular ("frozen") situation of the parameterized influence factors (which do not take representation party). As such, it provides only a partial picture of the response in the investigated multifactorial space (if the objective function depends on more than two influencing factors). For this reason, always these

IOP Conf. Series: Materials Science and Engineering **877** (2020) 012001 doi:10.1088/1757-899X/877/1/012001

graphical representations must be associated with the regression equation obtained and interpreted in the context offered by the respective equation [11].

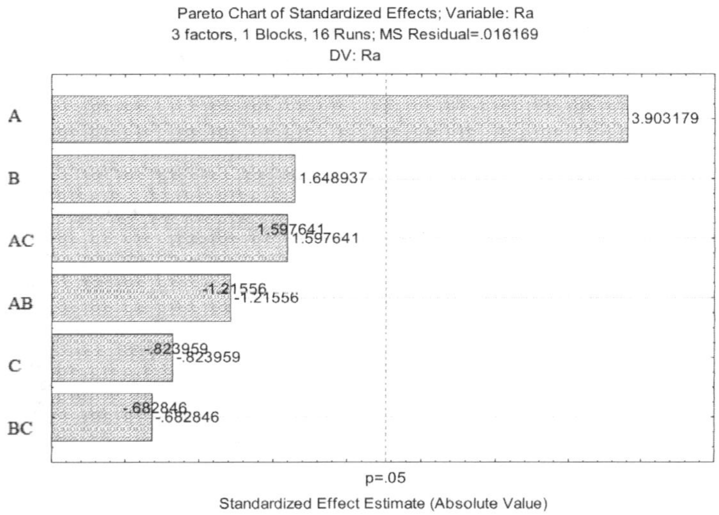

Figure 1. The Pareto Chart of the Standardized Effect.

Analyzing the graph in Figure 1, it is observed that the interactions that has the greatest effect on the roughness is the AC - the cutting speed and the feed per tooth. Therefore, the effect of this interaction on the surface roughness will be presented in Figure 2.

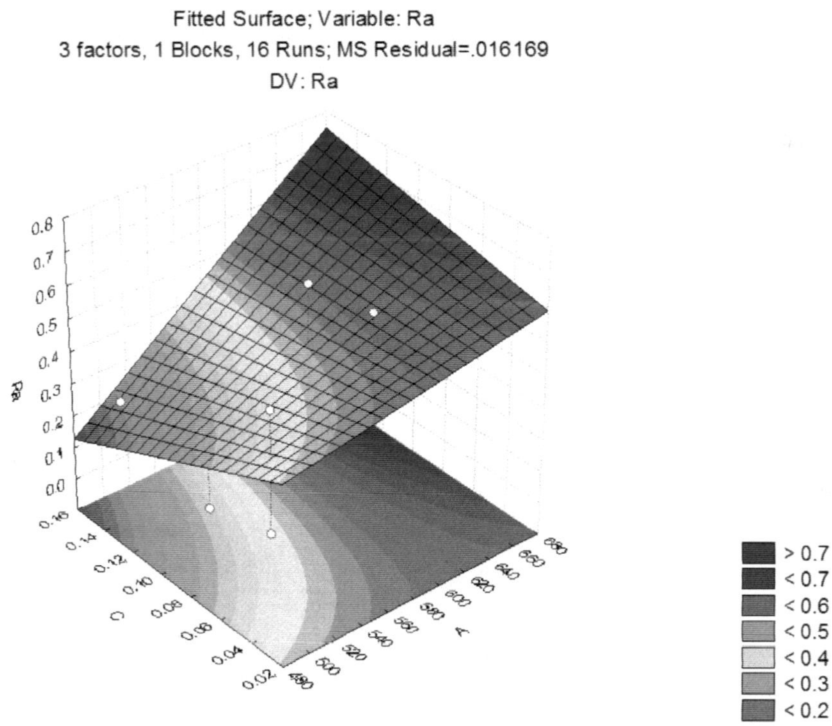

Figure 2. Fitted surface –R_a related to AC variation.

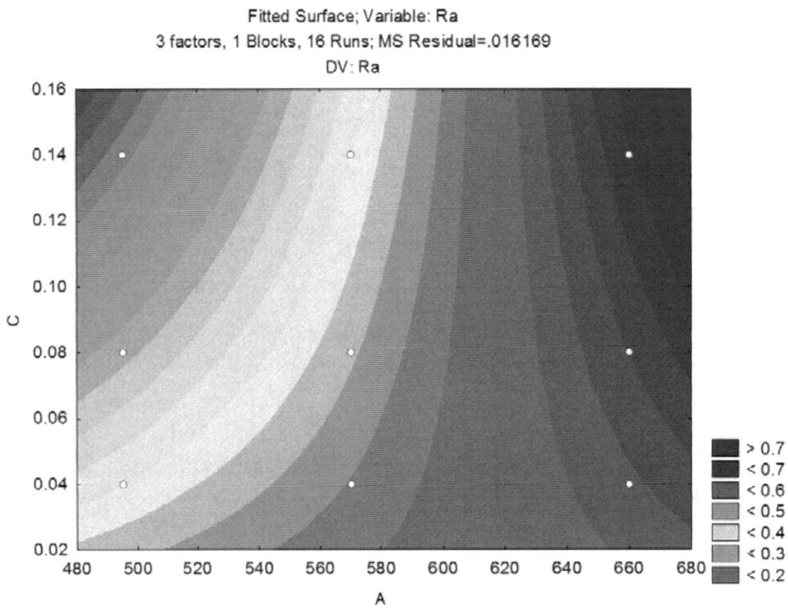

Figure 3. Fitted surface –Ra related to AC variation.

By Figures 2 and 3 it follows that with the increase of the two cutting parameters, the cutting speed and the feed rate, the value of the surface roughness increases.

The regression equation for the graph has the following form:

$$z = -1.1725290986626 + 0.0023274260714661 \cdot x - 8.8302925001351 \cdot y - 0.00066185580512082 \cdot 3 \cdot x + 0.017328190259503 \cdot x \cdot y - 0.61153657258854 \cdot 3.\cdot y + 1.51093249 \quad (1)$$

4. Statistical analysis of the coefficients of the regression equations for the surface roughness

R^2 is a statistical parameter that evaluates the fit degree of the regression model.

Table 5. Regression coefficinets of R_a Al7136.

Factor	Regression Coeff.	Std. Err.	t (9)	p	-95% Cnf.Limt.	+95% Cnf.Limt.
Mean/Interc.	-1.17253	1.157552	-1.01294	0.337546	-3.7911	1.446036
A	0.00233	0.001953	1.19147	0.263939	-0.0021	0.006746
B	0.50364	0.325591	1.54686	0.156300	-0.2329	1.240181
C	-8.83029	6.846863	-1.28968	0.229314	-24.3190	6.658388
AB	-0.00066	0.000544	-1.21556	0.255075	-0.0019	0.000570
AC	-.01733	0.010846	1.59764	0.144587	-0.0072	0.041864
BC	-0.61154	0.895570	-0.68285	0.511896	-2.6375	1.414383
R-squared = 0.71732; Pred R-squared = 0.52887, MS Residual = 0.16169						

The closer the values for R^2 are to 1, the better the experimental data are matching with the chosen model, and the ability to predict the experimental plan is better [12]. R^2 is the coefficient of multiple regression (or the coefficient of determination) and represents the response variation that can be explained by the model. R^2 take values between 0 and 1:

- 0 means that the experiments data do not match the chosen regression model at all;

- 1 is a perfect model.

The fit degree of the regression equation obtained is 71.7% (Rsqr=0.71732) (Table 5).

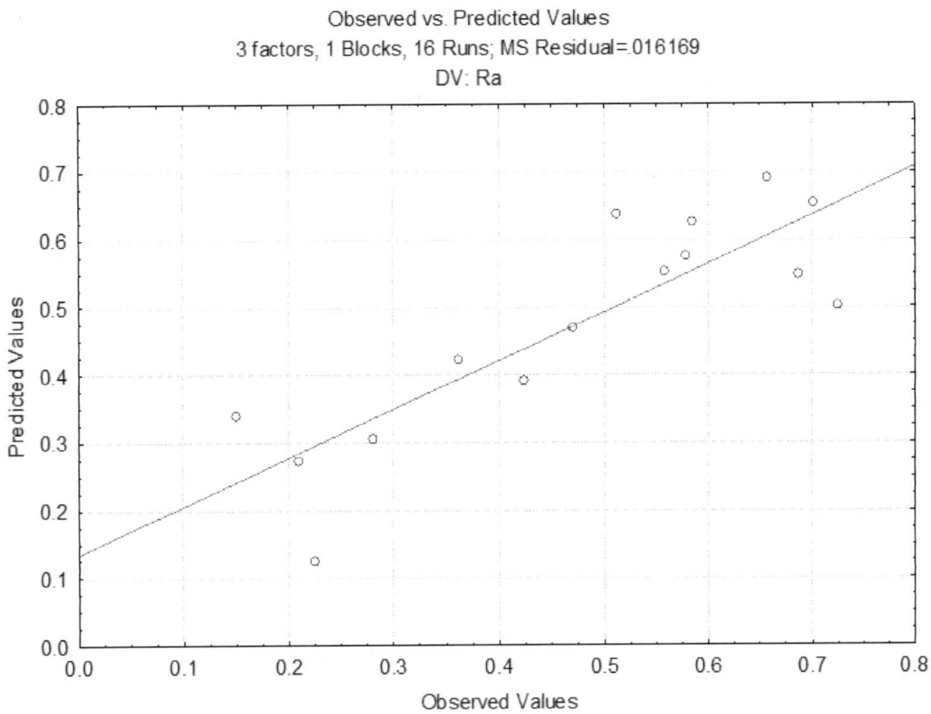

Figure 4. Observed versus predicted values of Ra.

The final conclusion that can be drawn based on the obtained data (Figure 5), is that the regression equation for R_a predicts the chosen experimental plan, in every point in the experimental domain, with a probability degree of 72%.

5. Conclusions

The surface quality and, implicitly its roughness, is an essential indicator of the product quality as it affects the product performance and also the production cost.

The reason for which the aluminum was chosen for this research is that its alloys have a higher index of workability than other materials. The milling process was chosen considering that it the most widely used process for the aluminum machining.

Starting from the principal purpose of this research, it was identified based on the method of the central composite experiment, the quantitative relation between the parameters of the cutting process parameters: cutting speed, cutting depth and feed per tooth and the arithmetic mean deviation of the surface profile, to describe the system studied at any point of the chosen experimental field.

We chose to use the ANOVA analysis to estimate the different types of response variability to determine whether or not are significant differences between the averages of the analyzed data groups and whether they have a normal distribution. This analysis followed the influence and interaction of the cutting process parameters that significantly affect the surface quality and predicted the optimal combinations of the end milling process.

At the same time, the fit degree of a regression analysis and the generated model was evaluated. The coefficients significance which belong to the regression equation was tested using the Fisher test, and the probability of a statistical difference was shown. From this analysis it resulted that the cutting speed has the greatest effect on the roughness. Statistically, the model value is F = 15.23480 which assumes that the model is significant. Moreover, the P value which is less than 0.05 indicates that the factor A - cutting speed is important.

Using the Pareto diagram, a standardized estimation of the effects of the factors and of the interactions between them was performed. It results that the highest influence on the surface roughness of the 7136-aluminum alloy, is exerted by the factor A - the cutting speed, followed by the interaction between the cutting depth, the cutting speed and the feed rate.

Analyzing the variation of the objective function under the effect of the cutting factors, the evolution of the surface roughness was subjected to a three-dimensional analysis. Thus, by plotting the response surfaces, an equation of the regression model was obtained, which can be used to approximate the values of the objective function, throughout the experimental domain. The graphical representation was associated with the obtained regression equation and interpreted in this context. The interaction which has the highest effect on the surface roughness was found to be the AC - the cutting speed and the feed per tooth. It was also observed that with the increase of the two cutting parameters, the value of the surface roughness increases.

It was also taken into account that the experimental results must be associated with experimental errors. Therefore, the coefficients values accuracy was predicted and not only their absolute value was determined but also a regression model was identified that predicts the chosen experimental plan, at any point of the experimental domain with an accuracy of 72%.

6. References

[1] Selvam M D, Shaik Dawood A K and Karuppusami G 2012 *Int. Journal* **2** (4) 544-547
[2] Krishnaraj V, Zitoune R and Collombet F 2012 *J of Mat Sc* **2** 95-109
[3] Surendra K S and Sharad K P 2014 *Procedia Engineering* **97** 141-149
[4] Dobrotă D, Țîțu A M, Dobrița F and Petrescu V 2016 *Revista de Chimie* **67** (3) 520-523
[5] Țîțu A M, Sandu A V, Pop A B, Țîțu Ș and Ciungu T C 2018 *Materials Science and Engineering* **374** 012054
[6] Vamsi I and Gopinath V 2014 *Int. J. of Eng. Research & Tech.* **3** (9) 875-878
[7] Venkatesan K, Ramanujam R, Joel J, Jeyapandiarajan P, Vignesh M, Darsh J T and Venkata K R 2014 *Procedia Engineering* **97** 677-686
[8] Pop A B and Țîțu A M 2019 *MATEC Web of Conferences* **290** 02010
[9] Pop A B, Țîțu A M, Pop G I and Țîțu Ș 2019 *Materials Science and Engineering* **572** 012043
[10] Țîțu A M, Pop A B, Țîțu Ș and Pop G I 2019 *Materials Science and Engineering* **572** 012042
[11] Burduhos-Nergis D -P, Cimpoesu N, Vizureanu P, Baciu C and Bejinariu C 2019 *Materials Today: Proceedings* 19 969–78
[12] Țîțu A M, Oprean C and Boroiu A 2011 *Cercetarea experimentală aplicată în creşterea calităţii produselor şi serviciilor* (Bucureşti: Editura AGIR)
[13] Cazac AM, Bejinariu C, Baciu C, Toma SL and Florea CD 2014 *Applied Mechanics and Materials* 657 137-141

IOP Conf. Series: Materials Science and Engineering **877** (2020) 012002 doi:10.1088/1757-899X/877/1/012002

Chemical Sensitization for Electric Properties Improvement of PBDB-T-SF Polymer for Solar Cells Application

L Hrostea[1,2*], L Leontie[1] and M Girtan[2]
[1]Faculty of Physics, Alexandru Ioan Cuza University of Iasi, Bd. Carol I, nr.11, 700506, Iasi, Romania
[2]Photonics Laboratory, (LPhiA) E.A. 4464, SFR Matrix, Angers University, Faculty of Sciences, 2 Bd Lavoisier, 49000 Angers, France

*E-mail: laura.hrostea@yahoo.ro

Abstract. This paper reports the effect of ferric chloride ($FeCl_3$) sensitization on poly [(2,6 (4,8 bis (5(2ethylhexylthio)4fluorothiophen2yl)benzo[1,2b:4,5b']dithiophene))-alt-(5,5-(1',3'-di-2-thienyl-5',7'-bis(2-ethylhexyl)benzo[1',2'c:4',5'c']dithiophene4,8 dione)] (known as PBDB-T-SF) in thin films (93 nm thickness), spin coated onto glass and ITO substrates. The optical properties do not undergo significant changes after $FeCl_3$ immersion. Instead, $FeCl_3$ sensitization leads to a dramatic decrease in resistivity of seven orders of magnitude comparing to the pristine thin films. This change is not stable over time, depending on $FeCl_3$ solution concentration and polymer thin-film thickness, but remains remarkable after a new immersion.

1. Introduction

Organic solar cells have recently become of great interest for both scientific research (academia) and industry owing to their multiple advantages. In this context, organic polymer materials have gained significant attention for photovoltaic applications, considering their attractive characteristics: conductive properties, lightweight, corrosion resistance [1–4], mechanical flexibility, etc., properties that are absent in most inorganic semiconductors [3, 5, 6]. In view of envisaged applications, both optical and electric properties of polymer materials are of great importance, but they are strictly related to the structural and morphological characteristics, varying with the size range of investigations. In the case of charge carrier transport through the polymer bulk material, from global point of view, the interconnections between polymeric chain domains are important for charge passage on the way to the electrodes, while, from local point of view, the links between the chains have to be taken into account, as much as the π-conjugation along the chains [1, 7, 8].

There are many ways to change the physical properties of polymer materials by adding a small amount of a different chemical element, through several techniques and methods; this will produce a disruption of crystal structure of the polymer, affecting, in this way, all its physico-chemical properties [9]. One method is polymer synthesis via fluorination, representing the insertion of fluorine atoms into the polymeric backbones. This is the reason of studying poly[(2,6-(4,8-bis(5-(2-ethylhexylthio)-4-fluorothiophen-2yl)benzo[1,2b:4,5b']dithiophene))-alt-(5,5-(1',3'-di-2-thienyl-5',7'-bis(2thylhexyl) benzo[1',2'c:4',5'c'] dithiophene 4,8 dione)], known as PBDB-T-SF or PCE13 [10–13], which contains two fluorine atoms. Being the most electronegative chemical element, fluorine brings many benefits to the host material. It avoids the steric hindrance of molecules, while it is down shifting the highest occupied molecular orbital (HOMO) and lowest unoccupied molecular orbital (LUMO) levels

[7, 10, 11, 14–17]. Besides, it improves the interactions between the molecules and between the polymeric chains, enabling the charge carrier transport, also facilitating exciton (electron-hole pair) separation by reducing the Coulomb potential [5, 9–11, 18, 19].

A second method, applied to enhance electric properties of this fluorinated polymer, is ferric chloride ($FeCl_3$) sensitization. The $FeCl_3$ is a cheap chemical agent, behaving as a *p*-type material and its molecules are capable of moving freely between polymer molecules, being assimilated by the polymer matrix [20, 21]. Because of its nature, depending on the involved amount, $FeCl_3$ is able to favour free electrons transport, improving the electric conductivity, or it can become an obstacle or a trap (if it is in excess) for charge carrier transport, but without damaging the polymer structure [20]. The temporal stability of the effect of ferric chloride on the polymer can be also lighter or stronger, depending on the thin-film thickness [21].

The present paper investigates the effect of $FeCl_3$ sensitization on the optical and electric properties of PBDB-T-SF thin films.

2. Materials and methods

The samples have been prepared using PBDB-T-SF (Figure 1), having an average molecular weight of 50,100, purchased from Ossila. The polymer solution in chlorobenzene (16 $g \cdot L^{-1}$) was deposited by spin coating (600 – 800 rpm) as thin films (Figure 2) onto glass substrates and indium tin oxide (ITO) coated substrates (15 Ω/□), purchased from Kintec. After drying for 1 h at 100 °C, samples were submerged for 1 h in a $FeCl_3$ solution in nitromethane (5 $g \cdot L^{-1}$).

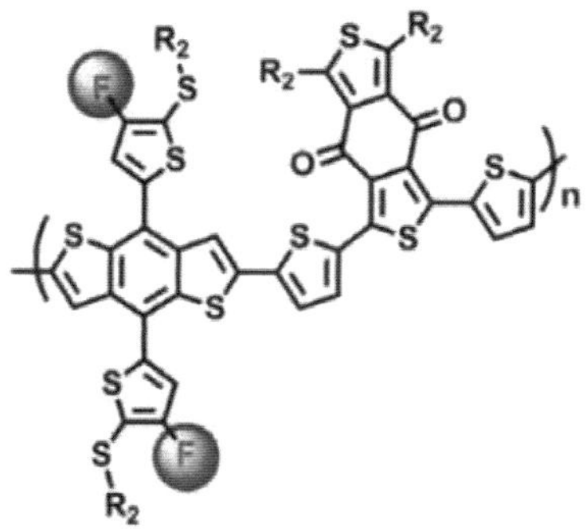

Figure 1. Molecular structure of PBDB-T-SF.

The thickness of the as-prepared films was measured with a Veeco Dektak 6M Stylus Profilometer. The optical properties have been measured at room temperature; for the spectral range of 250 – 1100 nm, a two-beam spectrophotometer UV/VIS S9000 was used for absorption spectra recording, while determination of the refractive index, extinction coefficient and dielectric constant was performed with an UVVISEL ellipsometer (Horiba Jobin Yvon).

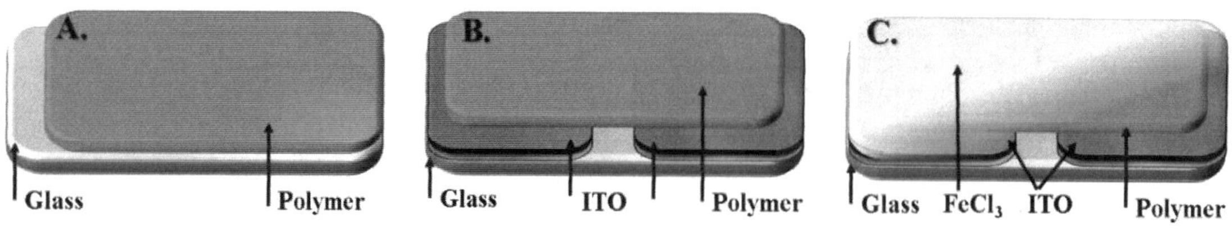

Figure 2. Thin-film patterns: A. polymer deposited on glass substrate; B. polymer deposited on ITO coated glass substrate; C. polymer sample immersed in $FeCl_3$ solution.

The electric properties were measured using a two-point technique during several successive cooling and heating cycles in temperature range of 30 – 90°C, under dark and white light illumination conditions.

3. Results and discussions

The thickness of as-prepared PBDB-T-SF thin films was about 93 nm, value which is increasing with almost 20 nm after first immersion in $FeCl_3$ solution. For the second immersion, this amount is decreasing to 15 nm, after which it remains constant (Figure 3). Therefore, it may exist a limit of the adherence tendency of $FeCl_3$ solution to PBDB-T-SF thin-film surface, fact which has been confirmed by the electric behaviour of samples.

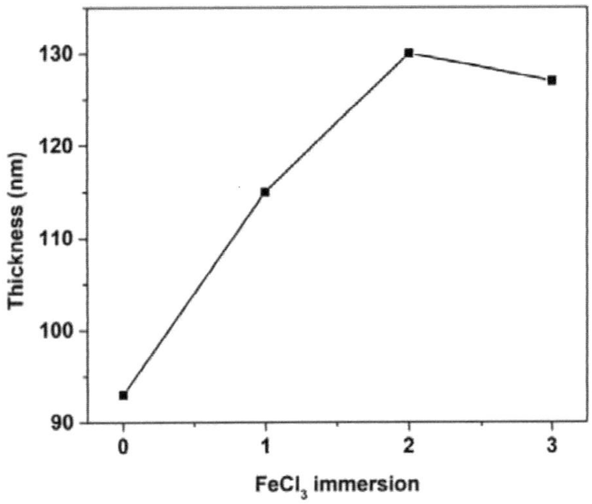

Figure 3. Thickness values for the $FeCl_3$ sensitized PBDB-T-SF thin film: 0 – no immersion; 1 – 1st immersion; 2 – 2nd immersion; 3 – 3rd immersion.

The absorption spectrum of PBDB-T-SF exhibits (Figure 4) two absorption peaks located at 576 nm and at 621 nm, notifying strong $\pi - \pi$ intermolecular interactions [10].

An important decrease in the absorption spectrum intensity of $FeCl_3$ sensitized polymer thin film is emphasized mostly in the visible wavelength range, because of the presence of $FeCl_3$, which is slightly assimilated by polymer chains. Besides, an enhanced absorption in the Near Infrared (NIR) domain can be noticed.

Moreover, it has to be taken into account that the absorption spectrum of as-prepared $FeCl_3$ layer displays low intensity bands. A chemical change that may occur in polymer matrix due to action of ferric chloride is anticipating, also responsible for the colour change, from light blue to dark blue.

The refractive index of pristine thin film, at 650 nm, was found to be $n_{max}^{PBDB-T-SF} = 2.99$ (Figure 5). After immersion, only some changes in the magnitude of the refractive index can be emphasized, but no significant changes in its behaviour: $n_{max}^{FeCl_3}$ remains constant, the initial maximum refractive index shows a decrease to $n_{max}^{PBDB-T-SF:FeCl_3} = 2.47$, maintaining its wavelength position at 650 nm; at the same time, a second peak, located at 930 nm, occurs, $n_{2max}^{PBDB-T-SF:FeCl_3} = 2.05$. Figure 6 illustrates the spectral dependence of the extinction coefficient, the features of which are similar to those of the absorption spectrum.

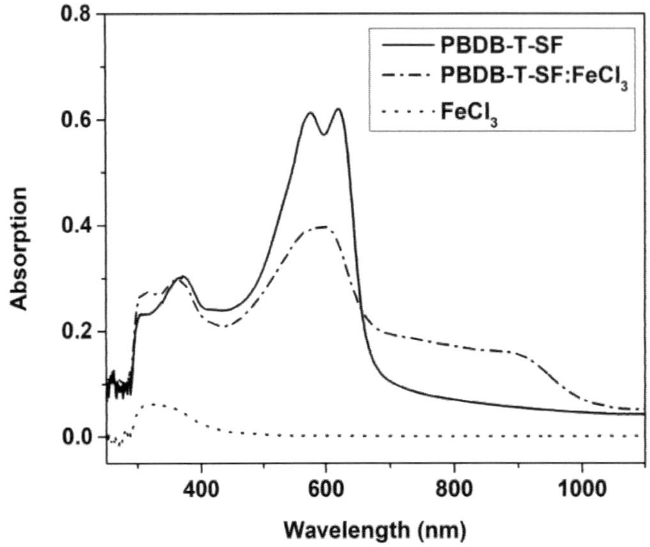

Figure 4. Absorption spectra of polymer thin film (continued line), $FeCl_3$ film (dotted line) and $FeCl_3$ sensitized polymer thin film (dashed dotted line).

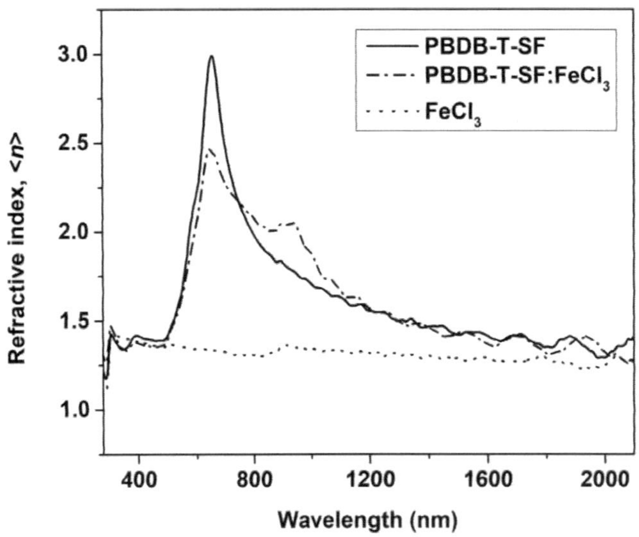

Figure 5. Pseudo-refractive index for polymer thin film (continued line), $FeCl_3$ film (dotted line), $FeCl_3$ sensitized polymer thin film (dashed dotted line).

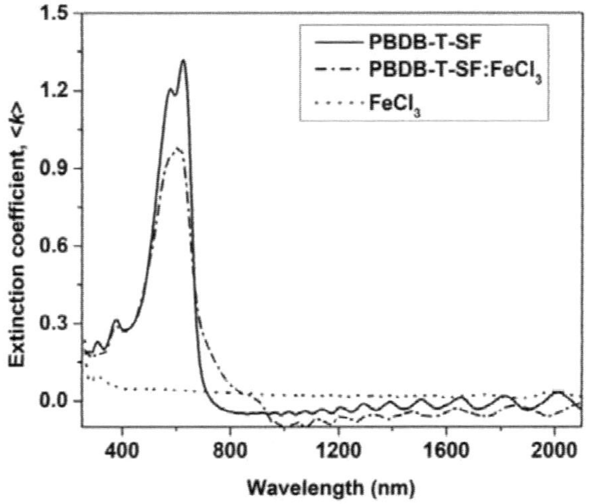

Figure 6. Pseudo-extinction coefficient for polymer thin film (continued line), $FeCl_3$ film (dotted line), $FeCl_3$ sensitized polymer thin film (dashed dotted line).

The effects of the $FeCl_3$ sensitization on polymer thin film describe a strong decrease in terms of electric resistivity, of six orders of magnitude (Figure 7), after the first immersion. It defines an unstable behaviour: after 24 hours, its resistivity increases by about three orders of magnitude, while after 48 hours, it returns to the initial resistivity of as-prepared polymer thin film. After 6 days, a new immersion proceeds and similar effects were registered, but their amplitude was less with respect to the first immersion: only three orders of magnitude decrease and a recovery time of 24 h have been noticed.

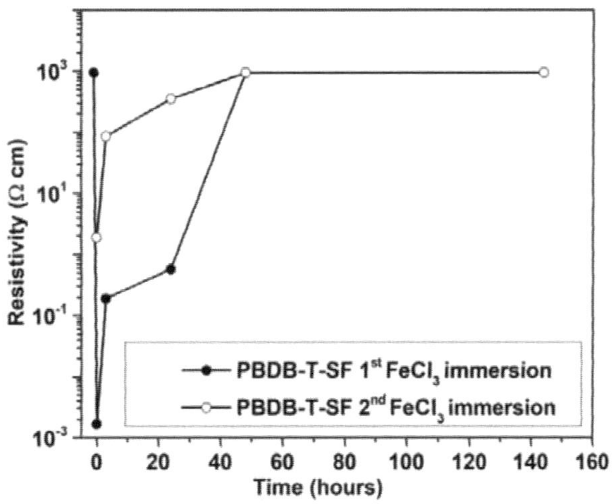

Figure 7. Time dependent resistivity with respect to $FeCl_3$ sensitization.

Concerning the temperature effect on resistivity (Figure 8), an expected behaviour is emphasized, consisting in a decrease in resistivity with increasing temperature [22].

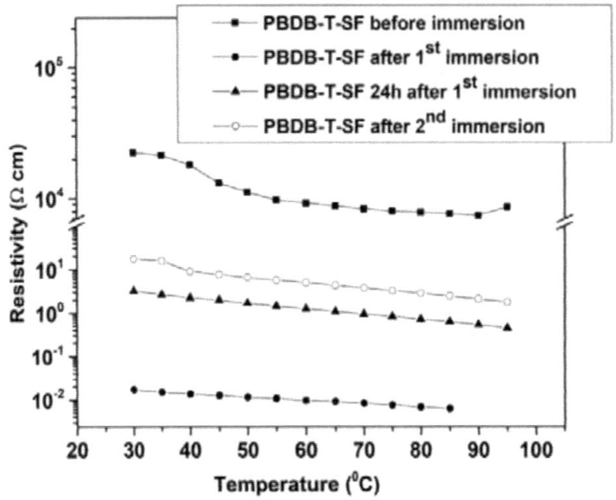

Figure 8. Temperature dependence of $FeCl_3$ sensitized PBDB-T-SF thin films under dark condition measurements.

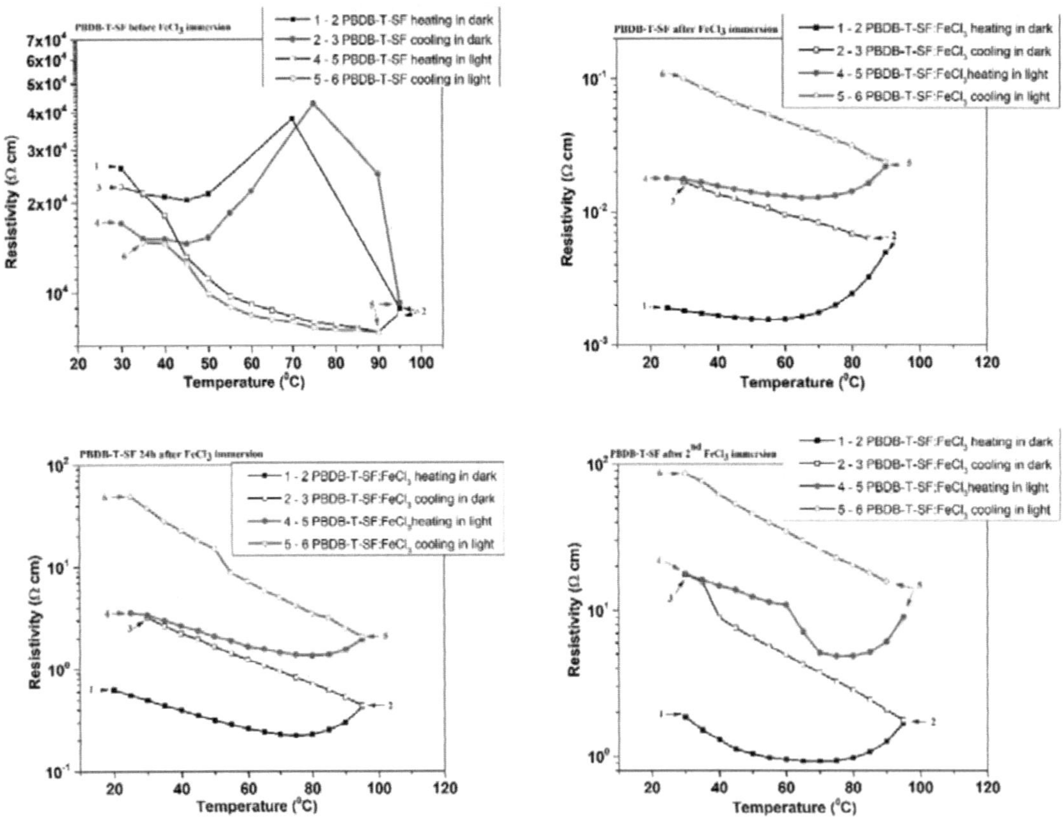

Figure 9. Temperature dependence of resistivity during several heating and cooling cycles under dark and illumination conditions (for as-prepared polymer thin film – top left graph; for $FeCl_3$ sensitized polymer thin film – top right graph; for after 24 h of $FeCl_3$ sensitization – bottom left graph; for a second immersion in $FeCl_3$ of polymer thin film – bottom right graph).

Comparing to the as-prepared thin-film resistivity, the sensitized thin films exhibit a slight decrease. This fact is presented in detail in Figure 9, during several heating and cooling cycles under dark and light conditions. In Figure 9, the top left graph describes as-prepared thin film, from which it can be easily inferred the effect of illumination-a resistivity decrease. This statement is not confirmed in the case of sensitized films, in which the presence of light increases the resistivity; this fact needs further investigations, in order to be fully understood.

4. Conclusion and perspectives

In conclusion, this paper presents a solution for enhancement of electric conductivity of polymer thin films used in solar cells, by ferric chloride ($FeCl_3$) sensitization. The PBDB-T-SF thin films (93 nm thick) exhibit a good absorption in visible range, which is slightly decreased after $FeCl_3$ sensitization. From ellipsometric measurements of refractive index and extinction coefficient, the mentioned behaviour is confirmed. Besides, the electric resistivity decreases dramatically after first immersion, from $>10^3$ Ω·cm (pristine thin films) to 10^{-3} Ω·cm, effect which lasts one day, with a slow recovery. Instead, the resistivity variation with temperature for sensitized films is slower comparing with that of pristine films, while the effect of light is different, an increased resistivity.

5. References

[1] Ludwings S 2014 vol Advances in Polymer Science 265 (Springer)
[2] Almosni S, Delamarre A, Jehl Z, Suchet D, Cojocaru L, Giteau M, Behaghel B, Julian A, Ibrahim C, Tatry L, Wang H, Kubo T, Uchida S, Segawa H, Miyashita N, Tamaki R, Shoji Y, Yoshida K, Ahsan N, Watanabe K, Inoue T, Sugiyama M, Nakano Y, Hamamura T, Toupance T, Olivier C, Chambon S, Vignau L, Geffroy C, Cloutet E, Hadziioannou G, Cavassilas N, Rale P, Cattoni A, Collin S, Gibelli F, Paire M, Lombez L, Aureau D, Bouttemy M, Etcheberry A, Okada Y and Guillemoles J-F 2018 *Science and Technology of Advanced Materials* **19** 336–69
[3] Berger P R and Kim M 2018 *Journal of Renewable and Sustainable Energy* **10**
[4] Gasparini N, Wadsworth A, Moser M, Baran D, McCulloch I and Brabec C J 2018 *Advanced Energy Materials* **8** 1703298
[5] Heeger A J 2001 *Current Applied Physics* **1** 247–67
[6] Etxebarria I, Ajuria J and Pacios R 2015 *Journal of Photonics for Energy* **5** 05721401-20
[7] Chang Y, Lau T-K, Pan M-A, Lu X, Yan H and Zhan C 2019 T *Materials Horizons* **6** 2094–102
[8] Ye L, Zhao W, Li S, Mukherjee S, Carpenter J H, Awartani O, Jiao X, Hou J and Ade H 2017 *Advanced Energy Materials* **7** 1602000-10
[9] Michaelson L 2015 Advances in conducting polymers research (New York: Nova Publisher)
[10] Zhao W, Li S, Yao H, Zhang S, Zhang Y, Yang B and Hou J 2017 *Journal of the American Chemical Society* **139** 7148–51
[11] Li S, Ye L, Zhao W, Zhang S, Mukherjee S, Ade H and Hou J 2016 *Advanced Materials* **28** 9423–9
[12] Lu L, Chen W, Xu T and Yu L 2015 *Nature Communications* **6**, 7327-34
[13] Zhang S, Qin Y, Zhu J and Hou J 2018 *Advanced Materials* **30** 1800868-75
[14] Fan Q, Su W, Wang Y, Guo B, Jiang Y, Guo X, Liu F, Russell T P, Zhang M and Li Y 2018 *Science China-Chemistry* **61** 531–7.
[15] Kan B, Feng H, Yao H, Chang M, Wan X, Li C, Hou J and Chen Y 2018 *Science China-Chemistry* **61** 1307–13
[16] Fei Z, Eisner F D, Jiao X, Azzouzi M, Rohr J A, Han Y, Shahid M, Chesman A S R, Easton C D, McNeill C R, Anthopoulos T D, Nelson J and Heeney M 2018 *Advanced Materials* **30** 1705209-16
[17] Chen T-W, Chang C-C, Hsiao Y-T, Chan C, Hong L, Zhong L, Chuang W-T, Hou J, Li Y and Hsu C-S 2019 *ACS Applied Materials & Interfaces* **11** 31069–77

[18] Jung J W, Jo J W, Chueh C-C, Liu F, Jo W H, Russell T P and Jen A K-Y 2015 F*Advanced Materials* **27** 3310–7
[19] Cui Y, Yao H, Gao B, Qin Y, Zhang S, Yang B, He C, Xu B and Hou J 2017 *Journal of the American Chemical Society* **139** 7302–9
[20] Hong C T, Lee W, Kang Y H, Yoo Y, Ryu J, Cho S Y and Jang K-S 2015 *Journal of Materials Chemistry A* **3** 12314–9
[21] Bounioux C, Díaz-Chao P, Campoy-Quiles M, Martín-González M S, Goñi A R, Yerushalmi-Rozen R and Müller C 2013 *Energy and Environmental Science* **6** 918–25
[22] Girtan M 2013 *Organic Electronics* **14** 200–5

IOP Conf. Series: Materials Science and Engineering **877** (2020) 012003 doi:10.1088/1757-899X/877/1/012003

The Influence of Thermal Factors on the Protection Characteristics of Electro-Insulating Materials

G Buică[1*], A E Antonov[1], C Beiu[1], R Dobra[2] and M Risteiu[2]
[1] National Research and Development Institute for Labor Protection "Al. Darabont", Bucharest, Blvd. Ghencea no. 35A, sect. 6, Romania
[2] "1 Decembrie 1918", Gabriel Bethlen str. no. 5, Alba Iulia, Romania

*Email: gbuica@protectiamuncii.ro

Abstract. The use of electrical insulating materials in the manufacture of electrical insulating equipment is the primary prevention and protection measure against electrocution. It must ensure the highest protection of the workers. The study aims to present the results of studies over behaviour of e electrical insulating materials under the influence of the thermal risk factors, to verify the functionality and to ensure the safety function throughout their use. Electrical insulating materials and equipment intended for use as protection mean in low and high voltage electrical installations have been tested. The verifications aimed at identifying the technical and safety parameters of the electrical insulating materials and protective equipment, in case of an event generated by one or more thermal factors or in combination with other professional risk factors, to test and guarantee the applicable technical and safety requirements.

1. Introduction

The evolution of electrical power systems, work equipment and protective equipment involves a review of continued electrical research. A particularity of using electrical power in industrial facilities is represented by the electrocution hazard [1].

The role of protective equipment is to protect workers against identified, assessed, verified and certified occupational hazards, which by design and manufacture must not cause any danger to the user and, where appropriate, to persons and animals nearby [2, 3, 5, 6].

Over time, the protective equipment diminishes its technical and protective characteristics, due to wear, aging of materials, non-compliant use. Studies on the behaviour in time of protective equipment in use revealed that during use they lose their characteristic Ile safety dielectric due to damage nature heat generated by accidental exposure to arc, exposing it to high temperatures combined with mechanical stress and non-compliant use over time.

Analysing statistically the share of factors that influence the safety feature over time, it is found that a fairly large share of thermal factors during use [2, 3].

From the categories of monitored work equipment, it was found that a share of 34% of them show traces of electric arc (the highest share being recorded by electrical insulating tools and fuse handling devices), 7% show material damage due to exposure to fire or heat generated by defects in installations such as hot spots .

Microclimate conditions can decisively influence the safety of insulating working equipment, in terms of insulation degradation, which leads to a high risk for workers [6].

The thermal stresses to which the protective equipment is subjected during use require the choice, from the design, of materials that are resistant both from an electrical point of view, as well as from a thermal and mechanical point of view, and these technical and safety characteristics must to guarantee the observance of the safety and health requirements established by the legislation and in the specific Roman standards, applicable to the respective product categories [3, 7, 8].

In this sense, when designing and manufacturing, as well as when choosing a protective equipment, the properties of electrical insulating materials from the components and subassemblies of protective equipment must be known, taking into account both safety and ergonomic requirements and technical-economic criteria.

For example fibreglass, a material widely used especially in hot sticks (electrical insulation equipment consisting of one or more electrical insulating tubes provided with coupling systems to each other, and at one end for the purpose of attaching accessories), rigid protectors (shutters, electrical insulating platforms) has high fire resistance, high heat resistance, low thermal conductivity.

Fiberglass resists the action of chemicals and biological agents, but mechanically has a weak structural strength with low resistance to compression and torsion. The tests performed in the study on electrical insulating rods made of fiberglass tubes showed the reduced resistance to compression and torsion, for this reason, this type of product was put into use in electrical installations with limited use, being restricted to mounting. with the help of these types of rods of earthing and short-circuiting devices or performing other works involving the mechanical stress of the rod such as removing the victims from the electrical installations by means of an attached accessory, hook type or in case of cleaning works (tree grooming) of the corridors of the high voltage overhead power lines. Another disadvantage of fiberglass is its low moisture permeability, thus limiting the use of fiberglass electrical insulating rods only in dry environmental conditions.

In general, the conductivity of such mixtures increases drastically at a certain concentration of the conductive component, the so-called percolation concentration. Among the parameters influencing the percolation concentration, the filler distribution, filler shape, filler/matrix interactions and the processing technique are the most important ones. On the basis of these parameters, different models have been proposed aimed at the prediction of the conductivity or the percolation concentration. It will be shown here that statistical, geometric or thermodynamic models explain the conductivity behaviour of specific mixtures on the basis of insufficient assumptions. However, the conductivity seems to be predictable with the help of structure-oriented models [9].

The purpose of the analysis of the safety level assessment is to provide the opportunity to know the real situation from the point of view of safety and to take the proper prevention methods for the given situation [10].

2. Technique identification

In order to identify the specific technical and technical safety requirements that the electrical insulating materials must meet, which are part of the electrical insulating work equipment, the following was performed:

- analysis of the types of protective equipment in use, the fields of use, the types and technical characteristics of the electrical installations in which these protective equipment are used, the works for which they are used and the environmental conditions are used.
- analysis of the new types of protective equipment, of the people of use, of the types and technical characteristics of the electrical installations in which these protective equipment will be used, of the works to be used and of the environmental conditions [11].
- the analysis performed was performed corroborating conditions of compliance established in the Law no.245 / 2004 with health and safety provisions in use stipulated in GD no.1146 / 2006 on the minimum safety and health requirements for the use of work equipment by workers work and observance of the technical conditions established in the specific Romanian standards, applicable to the respective product categories [3, 7, 8].

The electrical insulating materials of the protective equipment must be selected in accordance with the electrical, mechanical, thermal and chemical stresses to which they are subjected during use.

This material is must have adequate resistance to aging and must not propagate the flame.

Electrical insulating coatings consisting of one or more layers of electrical insulating material must be selected in accordance with the same safety requirements and technical criteria.

The electrical insulating materials in the protection equipment component must retain their characteristics and technical properties in case of thermal instability, especially in the hand-held protection equipment. For example, electrical hand tools must be used without restrictions in the temperature range of -20 ^{0}C and +70 ^{0}C. The electrical insulating materials applied on the hand tools used in live electrical installations must maintain their adhesion on the conductive side in the above-mentioned range [11, 12].

It is important that these materials do not have a high heat transfer in the metal components of the protective equipment. Heat can be transferred by conduction, convection or radiation or by a combination of all three.

A defect, pre-existing or created by the in-service stresses, gives rise to an ionization process in a volume bounded by insulating walls. With time the process does not remain identical due to both changes in the material and gas evolution, which makes it difficult to interpret the detected electrical signal. The end of life will occur through microcracks in which new discharges are initiated, activating the propagation of these structural failures [13].

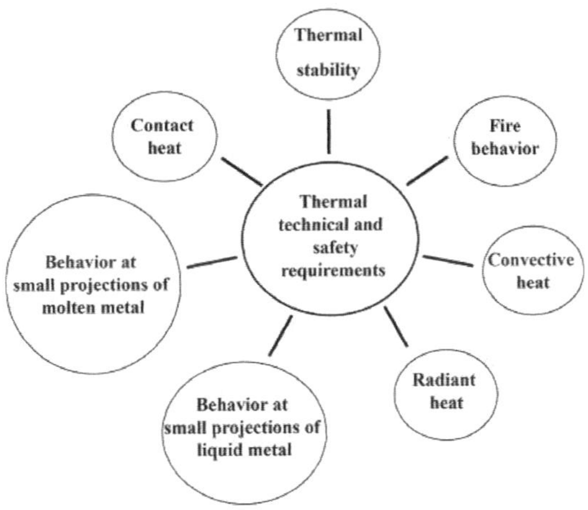

Figure 1. Technical and thermal safety requirements applicable to component electrical insulating materials of a protective equipment.

Figure 1 identifies technical requirements and safety nature of heat for material insulating components of equipment protection or for protective equipment as a whole.

3. Experimental procedures

Studies on the behaviour characteristics of protective thermal insulating materials that goes into protective equipment to u made on different categories such as: tools electrical, hot sticks, protective rigid means (plates, insulating platforms) protective flexible means ((sheaths and electrical insulating foils), devices for handling fuses with high breaking capacity. Their analysis, diagnosis and evaluation was performed on both new and in-use protective equipment.

The tests were performed on samples of protective equipment and electrical insulation materials taken from the production lines of some manufacturers who have electrical insulation products /

equipment under certification in the ICSPM - CS Certification Body, within INCDPM "Alexandru Darabont" Bucharest.

The tests were performed according to the verification methods specified in the applicable Romanian standards, the procedures of the Certification Body and the testing laboratories "Electrical and Mechanical Risks" and "Individual Means of Protection" within INCDPM.

In order to assess the quality of safety against the risks of thermal nature of the equipment protection must be identified and determined the best method is by testing. Ideally, the test methods should accurately reproduce the conditions that occur in practice, which is not possible and for practical reasons, tests on these risks are performed by accurately simulating the real effects that occur in practice [11, 14].

3.1. Flame propagation

The purpose of the test is to determine the properties of the electrical insulating materials from which the protective equipment is made or their flame-retardant coating. For the electrically insulating hand tools, the method provided in the standard SR EN 60900:2013 was used.

The test must be performed on fully assembled protective equipment (in the case of assembling hand tools); however, if this is not possible, the test must be performed on the relevant parts. The points required for the flame must be the points considered to be the weakest and are the working ends - which could be exposed to thermal stresses during use in electrical installations [15].

The outer parts of the tool head insulation were positioned horizontally above a blue flame 20 mm ± 2 mm high, generated by a methane gas burner. The exposure time of the specimens to the flame was 10 seconds, after which they were removed and for 20 seconds the behaviour of the material or coating was followed.

Three groups were tested man tools complicated to work in electrical installations, consisting of five tools / screwdrivers and pliers. Group A is made up of five tools man signed for work in electrical installations. They have electrical insulation consisting of two layers of electrical insulation material, different colours (red-orange), and the working tip has a single red coating. The internal electrical insulating coating is a rigid coating with very high insulating characteristics, and the outer one is supple and protects the inner coating for shock absorption, improving comfort in use. Hand tools in group B have an electrical insulation coating consisting of two layers of electrical insulation material, different colours (red-yellow), both layers of material are rigid. The working tip has a single red coating. Hand tools from group C have an electrically insulating coating made of a single layer of electrical insulating material, the second layer of material being applied only on the side of the tool. The working tip has a single red coating.

All specimens passed the test, during the 20 seconds of observation, the flame did not exceed 120 mm in height, according to the method from the above mentioned standard.

During the test, in addition to the height of the flame emitted by the test piece, the duration until ignition, flame propagation, heat release and drops or incandescent particles were monitored (dropping of burning particles). Rapid flame propagation was found on the rods of small diameter screwdrivers and on the working ends of tools covered with PVC electrical insulation material and also incandescent particles were found during combustion. In two of the specimens (screwdrivers) tested, the flame reached near (between 45 and 50 mm) the handles / handling area. Also, during the tests performed on the handles of the specimens with two coats of material, it was observed in the specimens from group A, that on the outer casing of electrical insulating material the flame spread more slowly, while on the base coating the flame spread faster. It has been assumed that the rapid spread of the flame on the coating of the basic electrical insulating material is due to the increase in temperature at its surface. For clarification, the tests were performed on specimens that had only the base material coating. It was found that the flame spreads faster than the flame spreads on the outer shell, but this is also influenced by the high temperature on the surface of the electrical insulating material, the manufacturer choosing, for technical and economic reasons, a material that has a lower flame propagation speed for outer coating and a higher flame propagation speed material for inner coating.

The second set of tests was performed on electrical insulating shutters. In order to test the properties of the electrical insulating materials, from which the electrical insulating plates are made, related to the flame propagation, the test method was assimilated, also used for testing the electrical insulated foils.

Electrical insulated shutters are protective equipment that apply between the disconnector contacts (the disconnector in the open position) and the fixed part of the disconnector. Nine different model samples and different materials were taken from the production lines. Three specimens (specimen code RE 0434) are made of electrical insulating material obtained by an extrusion process from non-plasticized mixtures based on polypropylene (PP), three other specimens (specimen code RE 0435) are made of material obtained by extrusion from a mixture non-plasticized PVC and three (test code RE 0436) are obtained from polyester resin coated with fiberglass (FGRPR) [2, 3, 11].

The specimens were positioned horizontally above a blue flame 20 mm ± 2 mm high. The exposure time of the specimens to the flame was 10 seconds, after which they were removed. The propagation (action) of the flame on the test sample must be observed for 60 seconds after the flame has been extinguished. The test is considered successful if during the observation period the flame does not propagate to any point at a distance of 50 mm from the centre of the specimen (the place where the flame acted). All specimens passed the test, during the 60 seconds of observation no drops or incandescent particles were recorded.

Traces of carbonization are present in the area of contact with the flame. Table 1 shows the results of flame propagation tests for polypropylene (PP) specimens.

Table 1. Flame propagation test results for polypropylene (PP) specimens.

Test sample	Flame exposure time [sec]	Observation time after flame exposure [sec]	Maximum diameter of damaged area [mm]
RE 0434-1	10	60	44
RE 0434-2	10	60	45
RE 0434-3	10	60	50

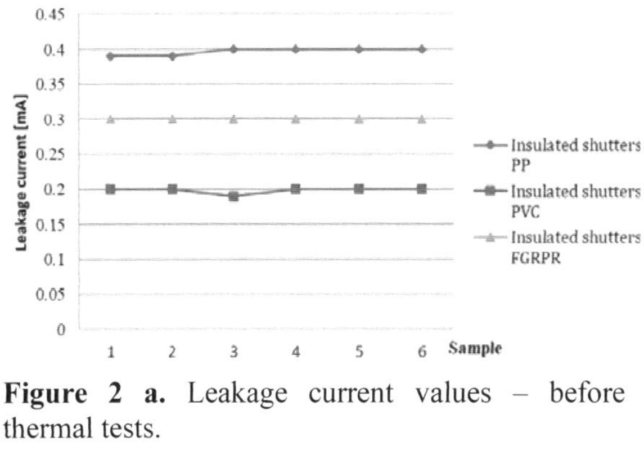

Figure 2 a. Leakage current values – before thermal tests.

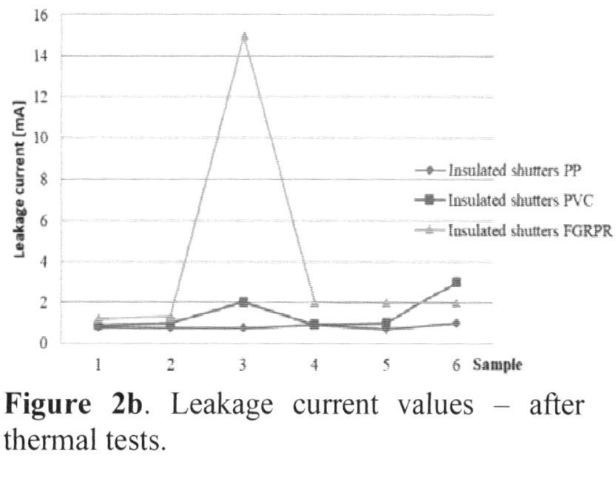

Figure 2b. Leakage current values – after thermal tests.

Result tests on specimens of PVC and FGRPR are inferior to those registered in the PP specimens, the maximum values of the damaged area being included in the limit of safety requirement.

After the flame propagation test, samples were tested at baseline and after dielectric application of heat. The dielectric test consisted of the application of a 50 kV in three points, progressively raised and maintained for 3 minutes. The electrodes were in the form of cylinders with a diameter of 25 mm and a height of 25 mm, respectively a diameter of 75 mm and a height of 25 mm [2, 3, 11].

During the tests, the leakage current, the flashovers and the local heating were monitored. It has been found that specimens that have lost the dielectric characteristic also show visible local [16].

Figure 2 shows the values of the leakage current before and after the flame propagation test. The decrease of the dielectric characteristics is found.

3.2. Thermal stability

Thermal stability tests were performed on hand tools for work in new electrical installations of different categories (pliers, screwdrivers, wrenches). Three batches of hand tools were tested for work in electrical installations. Lot A consists of 10 hand tools for working in electrical installations. They have electrical insulation consisting of two layers of electrical insulation material, different colours (red-orange).

Polymer insulating materials, are expected to be used under a variety of ambient conditions. Under severe ambient conditions, surface ageing of polymeric materials is closely related to environmental stresses. These environmental stresses affect tracking resistance in two ways. One is a direct effect on the change in the chemical structure of the surface. The other is an indirect effect on physical behaviour, such as variations of surface wetting by contamination and changes of discharge behaviour because of it [17].

The internal electrical insulating coating is a rigid coating with very high insulating characteristics, and the outer one is soft and protects the inner coating for shock absorption, improving comfort in use. Hand tools from group B have electrical insulation coating consisting of two layers of electrical insulation material, different colours (red-yellow), both layers of material are rigid.

Hand tools from group C have an electrical insulating coating made of a single layer of electrically insulating material, the second layer of material being applied only on the side of the tool.

Figure 3 shows the values of the electric leakage current for the 3 batches of electrical insulating tools from different manufacturers, before the thermal stresses are performed.

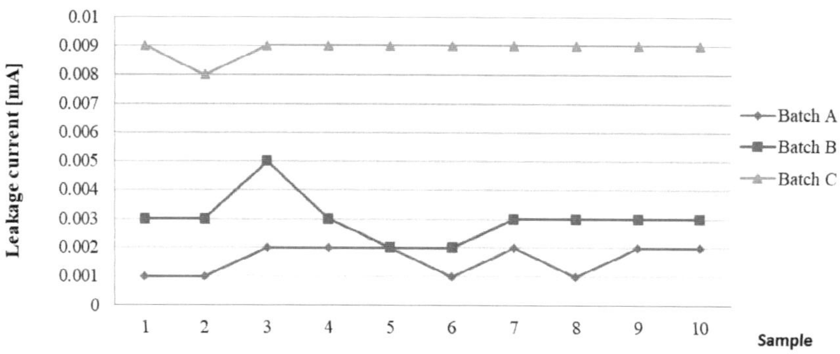

Figure 3. Leakage current values, in normal condition, before verification.

The specimens were subjected to climatic conditions at a temperature of -20 ^{0}C for 168 hours and +70 ^{0}C for 168 hours. Immediately, after each conditioning, a visual inspection of the test specimens and verification of the adhesion of the layers of electrical insulating material was performed both between them and against the metal parts. The adhesion test of the electro-insulating coating was performed according to the methods established by the standard SR EN 60900: 2013. It was found that 1% of the test specimens did not pass the test.

After checking the adhesion, the dielectric strength of the specimens was checked. Figure 4 shows the values of the electric leakage current for the three batches of electrical insulating tools, checking the thermal stability combined with the shock given by the free fall.

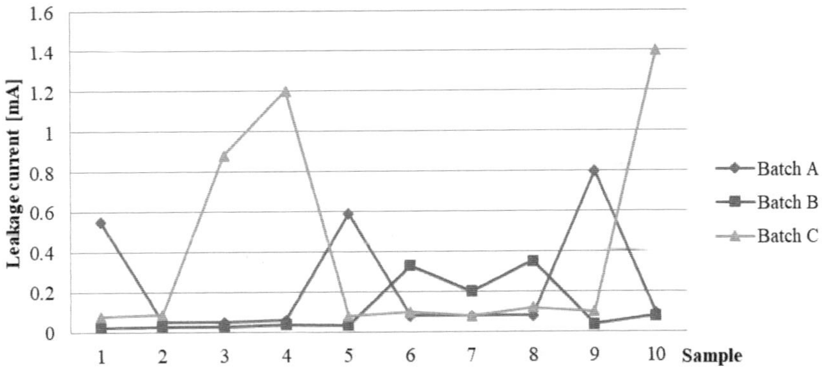

Figure 4. Leakage current values.

From the performed tests it resulted that the parameters of the dielectric characteristic of the equipment are influenced both by the thermal stability of the electro-insulating material coating, and by the way of realizing in the manufacturing procedure of their coating.

4. Results and discussions

The results of the study show that the parameters of the dielectric characteristics of protective materials are influenced by the thermal characteristics of electrical insulating materials in the protection equipment, this information allowing efficient selection and acquisition of component materials, better management of occupational hazards and ensuring a proper maintenance of the protective equipment in use, in order to establish the most appropriate measures for technical diagnosis and inspection / control of the safety level of a protective equipment [3, 11, 18].

In this regard, the appropriate choice of protective materials, which are part of a protective equipment, must take into account the behaviour of electrical insulating materials under the influence of thermal risk factors, including mechanical, in order to ensure functionality and ensure the safety function on the entire duration of their use.

This is ensured by an adequate risk management within the companies that allows an efficient management of the professional risks and guarantee of the placing on the market of some safe and efficient protective equipment.

Awareness and involvement of top management is essential in the development of organizational risk management and to increase risk awareness at different levels of the organization [20].

5. Conclusion and perspectives

The safety checks performed in the research study aimed to identify the technical and safety parameters of electrical insulation materials and protective equipment, in case of an event generated by one or more thematic factors or in combination with other occupational risk factors, in in order to test and guarantee the technical and safety requirements applicable to them.

The research study is the basis for the development of a methodology for verifying electrical insulation materials that are part of electrical equipment and protective equipment, given that during use protective equipment is subject to various risk factors generated by the use, maintenance , storage and working environment, which can change both the protection characteristics against electric shock and the protection characteristics against other risk factors, such as thermal, mechanical, which can cause a low reliability of these products.

The electrical insulating characteristics of a protection material are determined by the influence of thermal and mechanical risk factors, the environment in which the electrical insulation equipment is intended to be used, safety checks, assessment and attestation of conformity following a certification process. Ensuring the introduction on the market of compliant and safe products for users and guaranteeing a safe and healthy work environment at the workplaces in the industrial activity sectors.

6. References

[1] Pasculescu D, Lupu L, Pasculescu V, Inisconi I and Suvar M 2012 *Environmental Engineering and Management Journal* **11** (7) 1267-1271

[2] Buica G, Antonov A E, Beiu C, Dobra R and Risteiu M 2018 *IOP Conf. Series: Materials Science and Engineering* **374** 012008

[3] Buică G 2010 *Contributions of diagnostic methods of safty and health at work in electrical installations*, PhD Thesis, Universitas, Petroşani, p 70 – 149

[4] GD nr. 409/2016 - Laying down the conditions for making available on the market of low-voltage electrical equipment

[5] INCDPM, 2013 Guide for the application of GD No.1146/2006 regarding the minimum safety and health requirements for the use of work equipment by workers (in Romanian), National Institute of Research and Development for Occupational Safety Alexandru Darabont, Bucharest, Romania, Online at: http://www.inpm.ro/files/publicatii/2013-05.01-ghid-t.pdf

[6] Pasculescu V M, Vlasin N I, et. al. 2016 Proceedings of the16th International Multidisciplinary Scientific Geoconference *SGEM 2016*, Volume II, p 181-188 (Bulgaria)

[7] Antonov A, Buica G, Beiu C, Darabont D, Risteiu M, Dobra R and Ayaz Ahmad M 2019 The Principle of Compliance with Requirements of Directive 2009/104/Ec: Risk Assessment of Work Equipment in General Use, *East African Scholars Journal of Engineering and Computer Sciences,* p 103-109

[8] Buica G, Antonov A E, Beiu C, Dobra R and Risteiu M 2017 *IOP Conf. Series: Materials Science and Engineering* **209** 012094

[9] Lux F 1993 *Journal of materials science* **28** 285–301

[10] Radu S M and Pasculescu V M 2013, Safety level assessment in potentially explosive atmospheres - *Recent Advances in Industrial and Manufacturing Technologies*, p. 98-103

[11] INCDPM, 2018 Establishing the requirements and methods necessary for technical and safety verifications of protective equipment for work in live electrical installations - premise for ensuring the competitiveness of the management of companies in the energy sector, research study

[12] SR EN 60900:2013 Live working - Hand tools for use up to 1 000 V a.c. and 1 500 V d.c.

[13] Mayoux C 2000 *IEEE Transactions on Dielectrics and Electrical Insulation* **7** (5) 590-601

[14] SR EN 60695-11-5:2017 Fire hazard testing - Part 11-5: Test flames - Needle-flame test method - Apparatus, confirmatory test arrangement and guidance

[15] Pasculescu V M, Vlasin N I, Florea G D, Suvar M C and Colda I C 2017, Advances in impact resistance testing for explosion-proof electrical equipment. MATEC Web of Conferencesmatecconf/20121

[16] Dašić P, Hutanu C, Jevremović V, Dobra R, Risteiu M, Ileana I 2017 *IOP Conference Series: Materials Science and Engineering*, **209** 012068

[17] Yoshimura N, Kumagai S and Du B 1997 *IEEE Electrical Insulation Magazine* **13** (5) 8-19

[18] Buica G, Antonov A E, Beiu C, Dobra R and Risteiu M 2019, *Quality - Access to Success* **20** 159-164

[19] Dobra R, Pasculescu, Marc G., Risteiu M and Antonov A 2017 *IOP Conference Series: Materials Science and Engineering* **209** 012069

[20] Babut G and Moraru R 2006 General framework for environmental risk management, *Agir Bulletin 3*

IOP Conf. Series: Materials Science and Engineering **877** (2020) 012004 doi:10.1088/1757-899X/877/1/012004

Research Regarding the Influence of TiO_2 Nanoparticles on the Performance of Cementitious Materials

E Grebenisan[1*], A Hegyi[1] and A Lăzărescu[1]
[1]NIRD URBAN-INCERC Cluj-Napoca Branch, Romania

*E-mail: elvira.grebenisan@incerc-cluj.ro

Abstract. Urban buildings are subject to deterioration and degradation due to the action of external factors, air pollutants, water, compounds resulting from fuelcombustion used for heating and transportation, etc. Research on self-cleaning cementitious materials is developing at fast pace and contributes to the high interest in the industry in the decrease of evnironmental pollution, which is of real interest. The aim of this paper is to highlight the influence of the addition of TiO_2 nanoparticles on the self-cleaning ability of cementitious materials, under staining with an aqueous solution of a rodamina B, followed by exposures to the action of UV rays, followed by artificial rain cycles and to determine the physico-mechanical properties of the material, as well as the white dregree of the samples and their self-cleaning. Based on experimental research results it can be said that the white degree of the samples increased with the increase of the TiO_2 amount in the mixtures. Several other factors such as the intensity of the UV light also affected the self-cleaning capacity of the samples, better results being recorded for higher UVA intensity. The TiO_2 amount in the mixtures also influences fresh and hardened state properties of the material such as setting time, apparent density, water absorbtion and porosity of the material. It can therefore be said that it is necessary to identify an optimal range of TiO_2 nanopraticles concentration in the cementitious matrix, in order to obtain the maximum cumulative benefits in terms of self-cleaning capacity, physico-mechanical properties and costs.

1. Introduction

Due to the action of external factors, pollutants from air, water, compounds resulting from the combustion of fuels used for heating and transport, urban constructions are subject to deterioration and degradation. Thus, research on self-cleaning materials and contributing to pollution reduction is of real interest. Currently, worldwide, the possibility of producing cementitious composite materials with self-cleaning properties is widely reported due to the photocatalytic capacity of titanium dioxide nanoparticles (TiO_2), used as an addition or as a substitute for a part of cement. It also raised awareness of the importance of building sustainability and thus the need for a material with self-cleaning properties mainly used in urban areas to ensure a cleaner environment and reduce maintenance costs.

Accidentally, in 1995 one of the most important properties was discovered, namely the superhydrophility effect of the surface of a composition of titanium dioxide-silicon dioxide, TiO_2-SiO_2, in the presence of UV rays. This effect occurs as a result of the oxido-reduction reaction of H_2O, forming OH^- groups. Due to the multitude of OH^- formed groups, the water that reaches the surface of the material containing TiO_2-SiO_2 forms more flattened, lamellar droplets, which take the impurities from the surface of the material. Unlike TiO_2 where the photocatalytic effect ceases in the absence of

UV radiation, in the case of TiO_2-SiO_2 mixtures, the photocatalytic effect continues for hours even after the removal of the source of UV radiation [1-3]. The first official publication regarding the production of cementitious materials with self-cleaning properties was registered in 1997 by Luigi Cassar [4-6]. Subsequently, it was proved that the self-cleaning effect is actually the sum of two mechanisms that occur simultaneously: increasing the hydrophilicity of the surface of cementitious composites containing TiO_2 nanoparticles, and catalysing redox reactions, under the action of UV radiation, which allow oxidation and decomposition of organic particles on the cementitious surface (particles that produce the effect of soiling) into smaller particles, molecules with simpler structure, which can either be easily taken over by rainwater), either continue to decompose until their final reduction according (Figure 1) [7-9]. This combination of mechanisms, which creates the self-cleaning property of cementitious composites containing TiO_2 nanoparticles is also schematically represented in Figure 2.

$$\text{Organic compound} + O_2 \xrightarrow{TiO_2,\ h\upsilon} CO_2 + H_2O + \text{Residual inorganic compound}$$

Figure 1. Organic compound reduction mechanism.

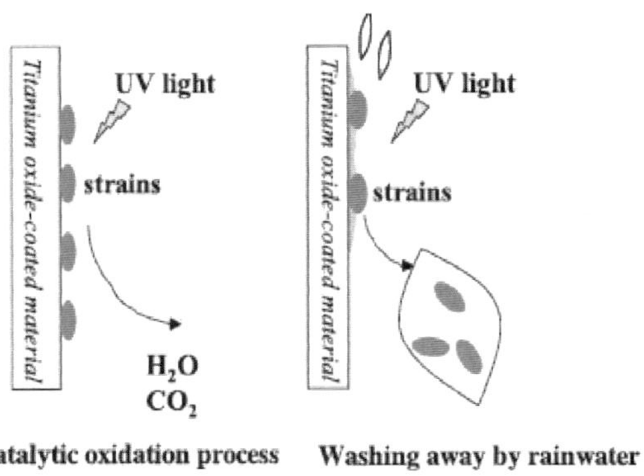

Figure 2. Schematic representation of the self-cleaning capacity by the cumulation of the two mechanisms: the process of photocatalytic oxidation of organic substances (left) and easing the removal of foreign substances by increasing surface hydrophilicity [7].

To highlight the ability of self-cleaning, research in the literature indicates several test methods, among which the method of staining with rhodamine B is the most common. The principle of this method is relatively simple: the surface of the test specimen is stained with a generally aqueous solution of rhodamine B, after which it is exposed to the action of UV rays, by combining or not with additional actions, for example artificial rain. Measurable indicators expressing the degree of staining / cleaning of the surface are identified, which are compared [10-12]. UNI 11259 standard regulates this test indicating the amount of aqueous solution of Rhodamine B applied to the surface (0,05 g / dm^2), and as a measurable indicator, the measurement of the staining degree [12]. Literature reports that comply with this standard and reports of tests performed with some deviations from the standard especially regarding the concentration of the solution of Rhodamine B (0.5 g/l and 1 g/l, respectively) [12].

Literature experimental results led to several conclusions:

- The report of the Institute of Chemical Technology, Prague, indicates a 20% discoloration after 1 minute exposure and 75-95% after 4 hours exposure to UV rays [12];

- The photocatalytic efficiency of TiO_2 nanoparticles is not altered by accelerated ageing cycles on cementitious composites [13];
- Folli indicates the dependence of self-cleaning performance on the granulometric size and distribution of TiO_2 nanoparticles. At the same time it is indicated to achieve a significant effect after only 30 minutes of exposure to UV radiation of a cementitious composite sample characterized by water/cement ratio of 0.4 and TiO_2/cement ratio 3,5:96,5, stained with Rhodamine B 1g/l solution, exposed to UV radiation with wavelength in UVA II domain. On the other hand, the exposure of the samples to accelerated ageing cycles simulating a period of 5 years under climatic conditions typical of northern Denmark, a good odelling and durability of this property was observed [10, 14];
- Khataee et al. Report the increase in self-cleaning performance with the increase in TiO_2 nanoparticles in the range 0% -3% [11];
- Odedra et al. Report the existence of self-cleaning capacity in the case of an existing structure, Church "Dives in Misericordia", Rome, made of concrete with a ratio of 3.5: 96.5, TiO_2: cement [15];
- Bianchi et al. On the one hand, confirm the effect of the self-cleaning capacity of TiO_2 nanoparticles addition in the cementitious composites, and, on the other hand, shows the influence of the type of particle, the structure of the crystal size, particle size distribution, the ratio of the three structures (rutile – tetragonal crystalline structure, anatase peaks – tetragonal crystalline structure and brukit – odelling ic crystalline structure) in the performance of these materials [16];
- Zhang et al. It indicate on one hand the reduction of the self-cleaning effect with the reduction of the time of action of UV rays, the long-term efficiency of this performance, for cementitious composites containing up to 6% TiO_2, however recommend keeping the dosage of TiO_2 to a maximum of 2%. The authors note a possible reduction in performance with increasing irradiation duration, performances that are subsequently recovered as soon as the specimen is subjected to the action of wet-dry cycles, artificial rain, so washing removal of reaction products from the surface [17].

Research in the literature has shown that adding TiO_2 nanoparticles also influences the odell-mechanical properties of cement paste, cement-based mortars and concrete. Thus, when introducing quantities of TiO_2 nanoparticles in the range 0.5% - 10% in the cementitious matrix, it was observed that the initial and final setting time decreases as the content of TiO_2 nanoparticles increases [18].
The decrease of the setting time as the amount of TiO_2 nanoparticles introduced into the cement paste increases, can be explained by the catalyst effect that nanoparticles have on the cement hydration reaction and by the fact that they can function as potential nuclei of accumulation of hydration products [11, 19, 20].

The literature indicates a reduction in porosity as the amount of nanoparticles in the cementitious mixture increases, the change in the size and orientation of the crystals of cement hydrating products and the formation of a greater amount of calcium hydrosilicate gel C-H-S [23-27].

As the amount of nanoparticles in the mixture increases, water absorption decreases, with 0.5% being the optimal amount of nanoparticles added if water absorption is assessed at the age of 28 days and 90 days, as reported by Rashad [18]. If water absorption is assessed 7 days after casting, the optimal amount of TiO_2 nanoparticles addition is 4% [18]. However, if water absorption is assessed at the age of 2 days after casting, the addition of TiO_2 nanoparticles increases this parameter [18].
The aim of this paper is to highlight the influence of the addition of TiO_2 nanoparticles on the self-cleaning capacity of cementitious composites, under staining with an aqueous solution of a Rhodamine B, followed by exposure to the action of UV rays (in 2 cases of UV radiation incidency on the surface of the composite material), followed by artificial rain cycles and to determine the odell-mechanical properties of the material, as well as the white dregree of the samples and the end of the self-cleaning procedure.

2. Materials and methods

The analysis of the influence of the addition of TiO_2 nanoparticles on the performance of the cementitious matrix, has been achieved producing samples using white Portland cement CEM I 52,5 R, with the addition of Degussa P25 TiO_2nanoparticles, while maintaining a constant ratio of water:dry mixture = 0,5, where the dry mix was considered to be the total amount of white Portland cement and TiO_2 nanoparticles. The amount of TiO_2 nanoparticles added in the preparation of cementitous composites was: 0% (control sample), 1%, 2%, 3%, 4% and 5% (percent relative to the amount of dry cement). Beginning and end of setting time was determined on the fresh mixtures, according to EN 196-3. At the same time, specimens with dimensions of 130x85x10mm were produced. The test specimens were conditioned for 24 hours in molds, at 20°C temperature and 90% relative humidity, without any light. After demolding, the specimens were completely immersed in water at a temperature of 20^0C, without light, for 27 days. After maturation, apparent density of the test specimens was determined according to EN 12390-7, absolute density according to EN 1907-6, porosity (as ratio between absolute and apparent density difference and absolute density), water absorption (EN 1015-18), and the white degree by direct measurement with portable WSB-1 leucometer, on samples exposed to natural light, under laboratory conditions, as absorbed fraction of light directed.

The measurement of the white degree was initially performed without exposure of the specimens to the action of UV radiation, after which they were exposed 24 hours to the action of UV rays and the measurement was repeated. Exposure to UV action was achieved in a closed enclosure with 4 UVA light bulbs with wavelength in the characteristic range of 315-400 nm, providing an intensity of 3.8mW/cm^2. A set of test specimens was exposed to a total luminous flux intensity of 860 lux, and the second set of test specimens, similar to the first, was exposed to a total luminous flux intensity of 405 lux. The odelling of the total luminous flux was measured at the exposed surface of the samples. The variance of this odelling of the luminous flux was achieved by exposing the specimens in the first case at a distance of 10 cm and in the second case at a distance of 20 cm from the UV bulbs.

Experimental research on the self-cleaning capacity of cementitious composites containing TiO_2 nanoparticles consisted of staining, by the method of dripping the surface of the specimens with aqueous solution of rhodamine B (1g/l), then subjecting them to UV rays cycles, artificial rain, respectively drying (Figure 3). After each step of the diagram the white degree (WD) was measured and at established intervals the surface of the specimen was microscopically analysed (MA) in the stained area. Exposure to artificial rain and drying were carried out in the absence of any light source.

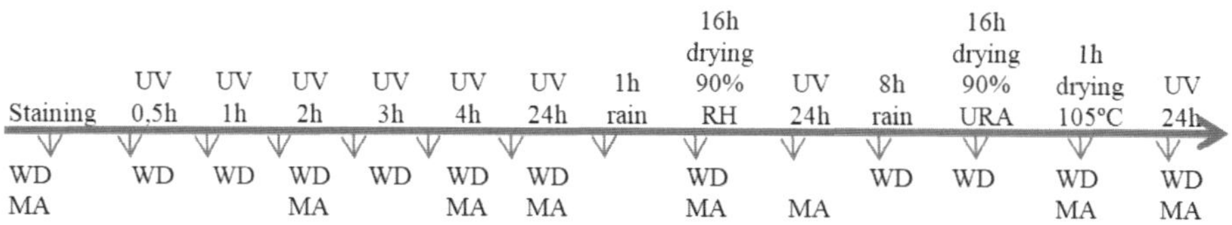

Figure 3. Test diagram.

The efficiency of the addition of TiO_2 nanoparticles on the self-cleaning character of the cementitious composites was visually analyzed by means of two measurable parameters:

- The white degree of the sample, determined in the stained area;
- Cleaning degree-representing the variation of the white degree of the sample after the test diagram, in the stained area, relative to the white degree of the stained sample but not yet exposed to UV light activation, washing under the action of artificial rain or drying, as in equation (1).

$$GC = \frac{GA_f - GA_0}{GA_0} * 100\ (\%) \tag{1}$$

where:
GC = cleaning degree
GA_f = white degree of the specimen in the tarnished area after the test diagram has been completed
GA_0 = the degree of initial whiteness of the stained sample but not yet exposed to UV light activation, washing under the action of artificial rain or drying

3. Results and discussions

The main aim of this study was to analyze the influence of the presence of titanium dioxide nanoparticles (TiO_2) on the following characteristics of cementitious composites: setting time, apparent density, absolute density, porosity, water absorption, white degree and self-cleaning capacity.

From the point of view of the influence of TiO_2 nanoparticles on the setting time, it can be said that with the increase in the amount of TiO_2 nanoparticles in the cementitious mixture, the setting time, both initial and final, decreases. The same evolution of recorded setting times can be defined by grade 2 polynomial functions (Figure 4), with a sufficiently high coefficient of accuracy (R^2 = 0,9789, respectively 0,9774) in order to be used for the assessment of the initial time, respectively the final setting time for cementitious mixtures made with another content of TiO_2 nanoparticles, contained in the range 0% - 5%, under the conditions of preservation of the previously presented preparation conditions.

On the other hand, it is noted that the greater the addition of nanoparticles, the smaller the difference between the beginning and the end of the setting time, which means a strong influence of TiO_2 nanoparticles on the setting of cementitious composites. The influence of the addition of nanoparticles on the setting times is consistent with some specifications in the literature [18] and it probably occurs as a result of the acceleration of hydration-hydrolysis processes and the operation of nanoparticles as nuclei of formation and growth of the network of hydration – hydrolysis products specific to the cement setting.

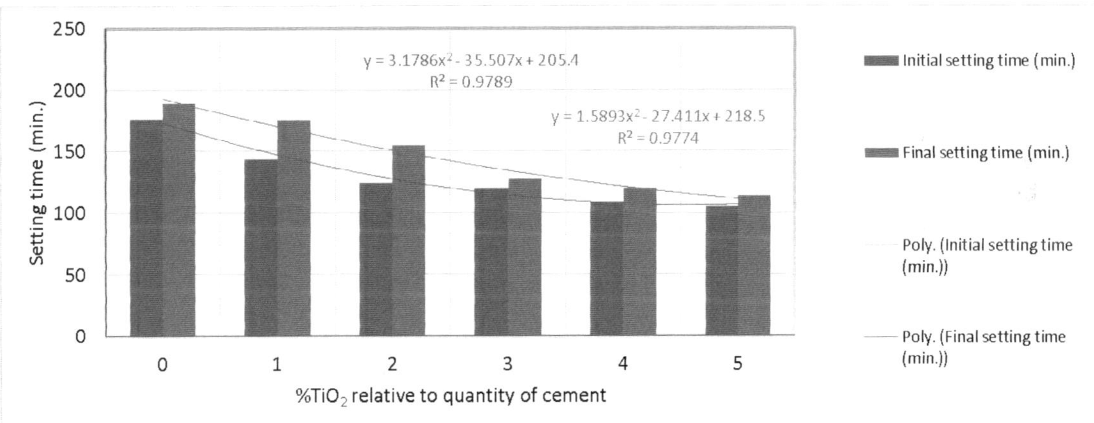

Figure 4. Identification of polynomial functions for odelling the evolution of the initial/final outlet time, in relation to the content of TiO_2 nanoparticles in the cementitious matrix.

As for the apparent density in the hardened state of cementitious composites, it is noted that it does not vary in proportion to the content of TiO_2 nanoparticles. However, a constant range of high values is recorded in the concentration range of nanoparticles (4% - 5%). In the concentration range of TiO_2 nanoparticles added to the cementitious matrix, 0% - 5% relative to the dry cement quantity, apparent density evolution in relation to this variable can be expressed mathematically by a polynomial function of order 4 (Figure 5).

Experimental results regarding the absolute density of cementitious composites with the addition of TiO_2 nanoparticles also indicates that this parameter does not vary in proportion to the content of nanoparticles, nor has it been possible to identify with sufficient precision a representative mathematical function (Figure 6).

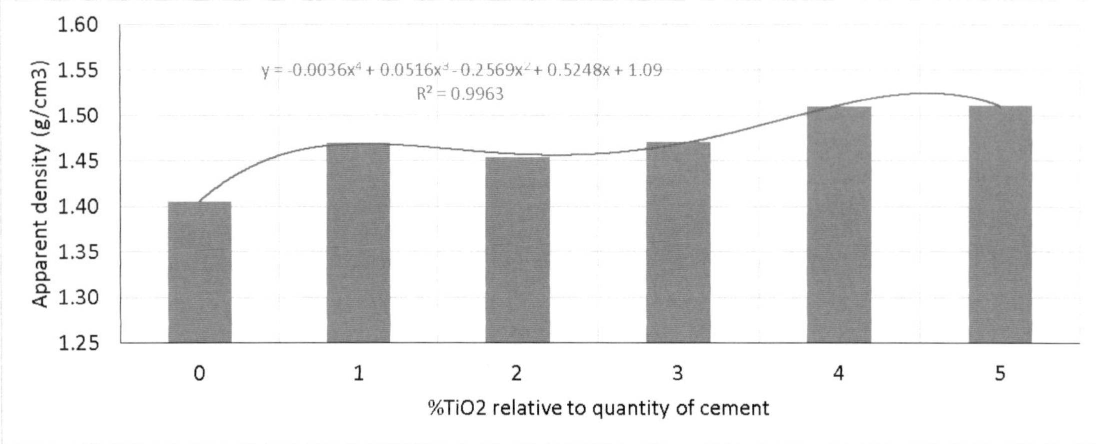

Figure 5. Polynomial function of evolution of apparent density in relation to the content of TiO_2 nanoparticles in the cementitious matrix.

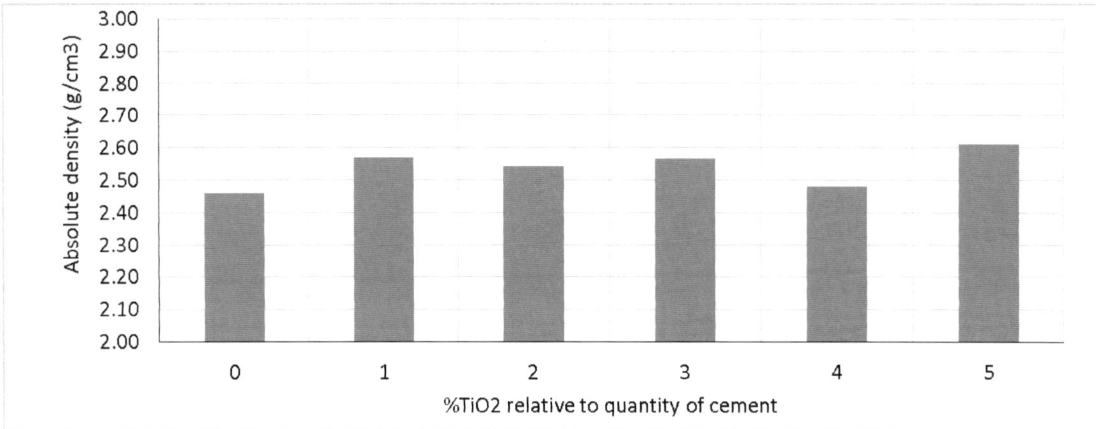

Figure 6. Absolute density of cementitious composites with the addition of 0% - 5% TiO_2 nanoparticles.

Porosity was determined by calculation based on experimentally obtained values for apparent density and absolute density, for each mixture. Thus, it was observed that the lowest value of porosity was obtained for a content of TiO_2 nanoparticles of 4% relative to the quantity of cement. In the case of cementitious materials with content of 1%, 2%, 3% and 5%, an almost constant porosity was noted. These results cannot be considered relevant for an analysis of the evolution of the parameter, considering that additional tests are necessary, possibly for a series of composites to which the addition of nanoparticles varies by one step less than 1% and over a concentration range extending beyond 5% (Figure 7). A similar situation, inconclusive and also requiring the development of experimental research, was also recorded in terms of water absorption (Figure 8).

Regarding the white degree, it was observed that for the same cementitious mixtures with the addition of TiO_2 nanoparticles, comparing the white degree initially recorded with the one recorded after 24 hours UV rays exposure, increased for all analyzed cases (Figure 9). It was noted that a maximum increase in white was recorded, both as difference between the initial and final values and as percentage change relative to the initial value, for the mixture containing 2% TiO_2.

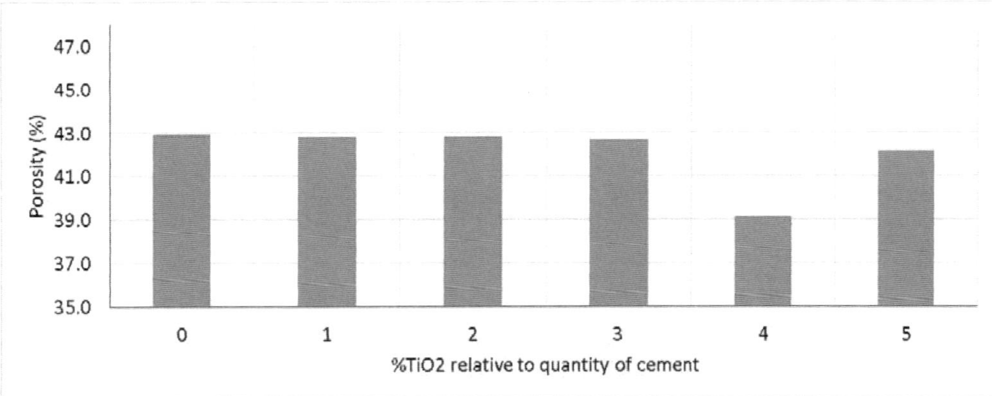

Figure 7. Porosity of cementitious composites with the addition of 0% - 5% TiO_2.

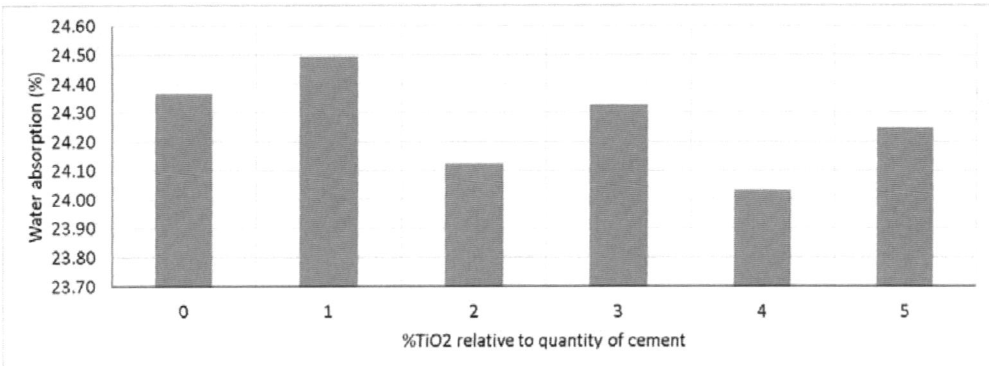

Figure 8. Water absorption of cementous composites with the addition of 0% - 5% TiO_2 nanoparticles.

Mixtures with content 1%, 3% and 4% TiO_2 nanoparticles addition had increases in the white degree, with differences between the final values and the initial values in the range of 0.4 units, as opposed to the cementitious matrix with content of 5% nanoparticles, for which there was an increase in the parameter by only 0.31 units from the initial value. This increase can be considered a first image of the photoactivation activity, respectively, the mixtures with the greater difference in the white degree with and without UV exposure are activated more strongly. The percentage increase in the degree of white after exposure to UV radiation for 24 hours relative to the initial value can be considered a measure of the effectiveness of the addition of nanoparticles on this parameter (Figure 10). It can be seen that additions of 1 – 4% TiO_2 provide a minimum 50% increase in white degree by photoactivation, and larger additions of nanoparticles, not only are not more effective, but even are unfavorable to the evolution of this parameter. Therefore, it can be said that a large amount of nanoparticles in the cementitious matrix does not guarantee a higher white degree of hardened material and no better photoactivation activity. It is essential to identify the optimal concentration range of nanoparticles, so as to achieve an optimal benefit-cost ratio, namely maximum efficiency photoactivation, while maintaining the other physical-mechanical performances of the cementitios composite, at the same time with a reasonable cost in relation to the benefits. This conclusion is consistent with some specifications in the literature [24]. The evolution of white variation after exposure to UV radiation for 24 hours relative to baseline can be mathematically described by a grade 3 polynomial function (Figure 11).

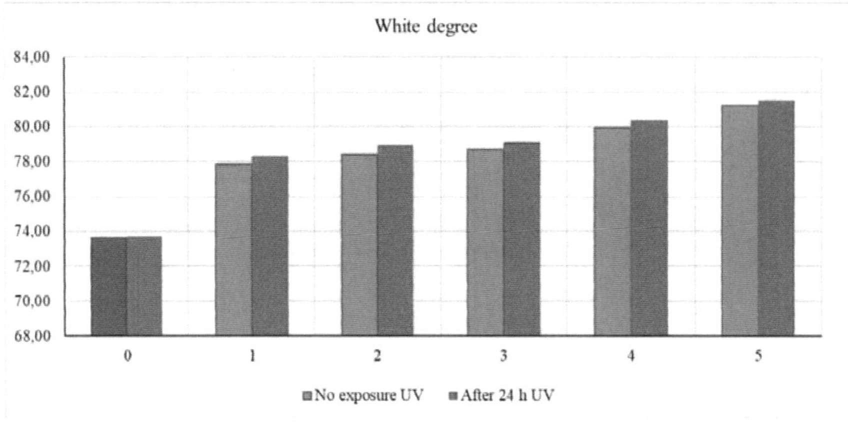

Figure 9. White degree before and after exposure to UV 24h of cementitious composites with the addition of 0% - 5% TiO_2 nanoparticles

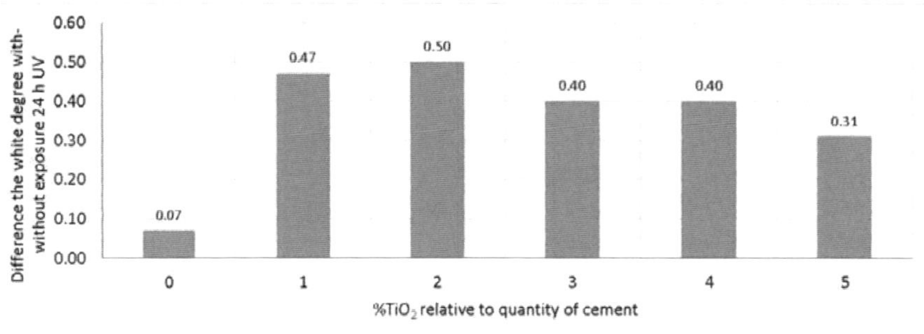

Figure 10. Difference between white degree of recorded for cementitious composites with the addition of 0% - 5% TiO2 nanoparticles exposed and, respectively, not exposed to UV action.

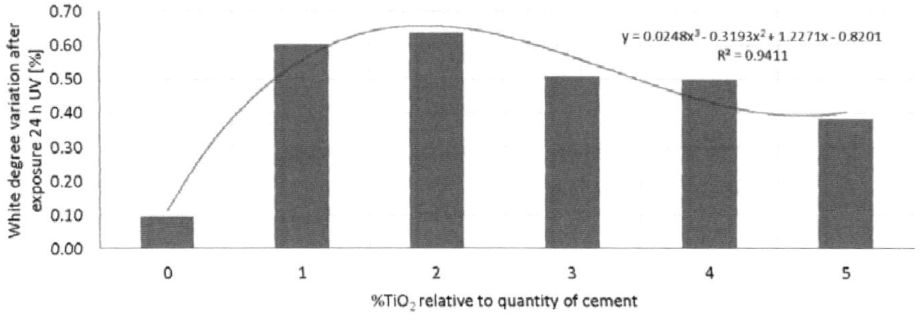

Figure 11. Polynomial function of determining the percentage increase in the white degree after exposure to UV 24h.

From the point of view of the self-cleaning capacity of cementitious composites, visual, microscopic analysis and assessment of the white degree of test specimens according to the test schedule presented in the test diagram indicated the following:

- Immediately after staining, the white degree of test specimens is significantly reduced, as expected, from values in the range (73-82) to values in the range (53 – 67), for test specimens exposed at a distance of 10 cm from UVA bulbs (860 lux) (Fig. 11) and from values in the range (73-82) to values in the range (57 – 67), for specimens exposed at a distance of 20 cm from UVA bulbs (405 lux) (Figure 13);
- After staining, the variation curve of the white degree of the specimens retains a trend similar to the variation curve of the white degree of the unpaired specimens, respectively, the higher the content of TiO_2 nanoparticles, the higher white degree;
- By exposing the tarnished specimens to the photocatalysing action of UV, the degree of white increases constantly, at each assessment after the predetermined exposure ranges (0,5 h, 1h, 2h, 3h, 4h, 24h). Visual discoloration of the spots is observed, and microscopic analysis notes that there is a discoloration especially of the surface in the immediate vicinity of micro-cracks occurring during the maturation process. The exception is the blank specimen, made without the addition of TiO_2 nanoparticles, in which only a slight change in the degree of staining is observed. This behavior of the blank specimen can be attributed both to the degradation of a small amount of organic molecules of Rhodamine B under the effect of ultraviolet radiation, and to a low but existing content of titanium dioxide in the cement composition, characteristic especially of White Cement, which was activated in the presence of UVA radiation;
- The tests carried out by the Rhodamine B staining method provide evidence of the self-cleaning capacity of cementitious matrices enriched with TiO_2 nanoparticles by photoactivation under the influence of UVA rays. However, the experimental results obtained cannot provide the basis for a documented selection of the optimal content of nanoparticles, so that the cost – benefit balance is the most favorable;
- The degree of cleaning of the test specimens obtained after completing the test diagram indicates on the one hand the influence of the concentration of TiO_2 nanoparticles in the cementitious matrix, and on the other hand the influence of the intensity of the incident ultraviolet radiation on the surface, on the self-cleaning capacity.

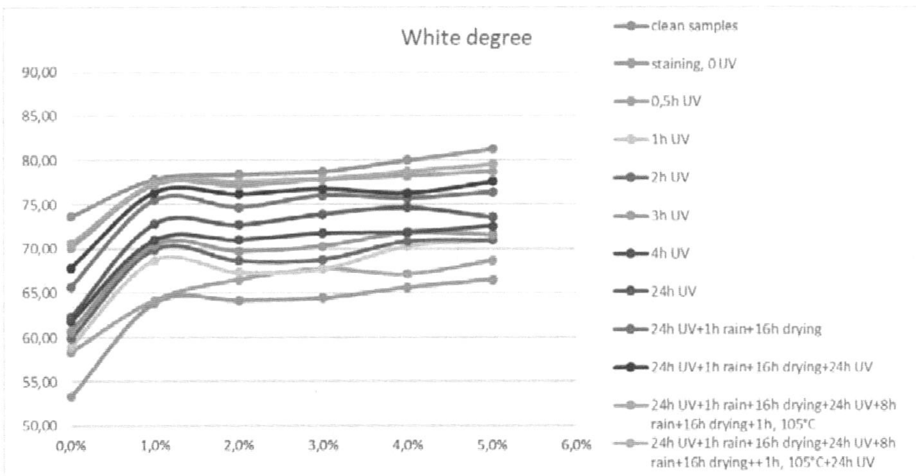

Figure 12. White degree of cementitious composites s exposed to a UVA radiation intensity of 860 lux.

As seen, at an incidence UVA intensity of 860 lux, additions of 1% - 3% TiO_2 nanoparticles (quantity related to cement quantity) provide cementitious composite matrices a better self-cleaning capacity, expressed by the quantifiable parameter "cleaning degree (CD)", compared to samples in which the addition of nanoparticles was higher (4%, 5%). However, this is not the case when the UVA light intensity is reduced (405 lux). In this situation, two phenomena aere observed. On one hand, the degree of cleaning decreases with an increase in the percentage of TiO_2 used up to 3%, after which an increase was observed again for the test specimens produced with 4% and 5%, respectively.

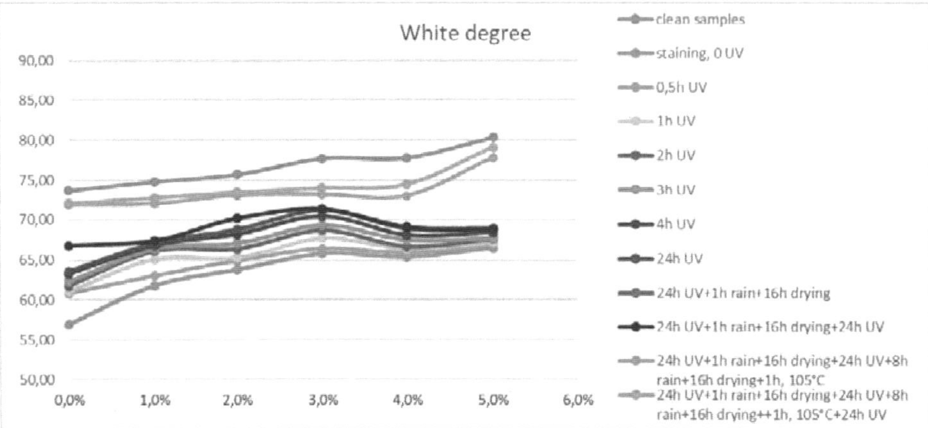

Figure 13. White degree of cementitious composites exposed to a UVA radiation intensity of 405 lux.

On the other hand, comparing the cleaning degree obtained for the composite matrix with 5% TiO_2 nanoparticles, it is observed that the difference between the values of this parameter obtained for the situation of UVA intensity of 860 lux and UVA intensity of 405 lux was very small, unlike the cases of samples with lower TiO_2 content where this difference is much greater (Figure 14). It can therefore be said that, on a case-by-case basis, in terms of the degree of cleaning, it may be more advantageous to use a lower or higher amount of TiO_2 nanoparticles. In Figure 15 and Figure 16 illustrative and demonstrative, images taken at microscopic analysis of the stained surface in various situations are presented.

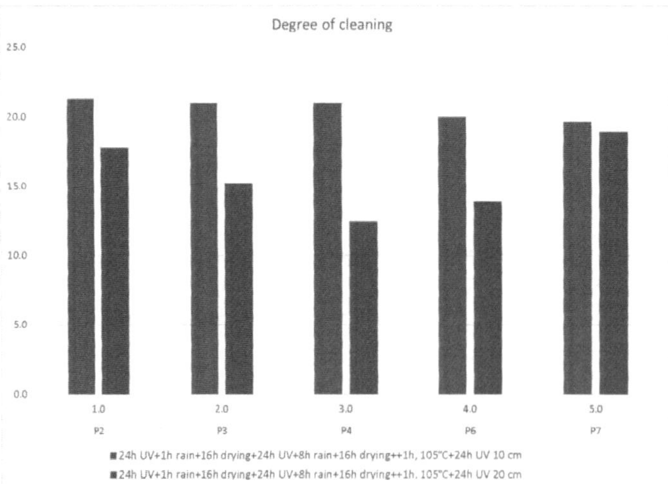

Figure 14. Degree of cleaning of cementous composites.

IOP Conf. Series: Materials Science and Engineering **877** (2020) 012004 doi:10.1088/1757-899X/877/1/012004

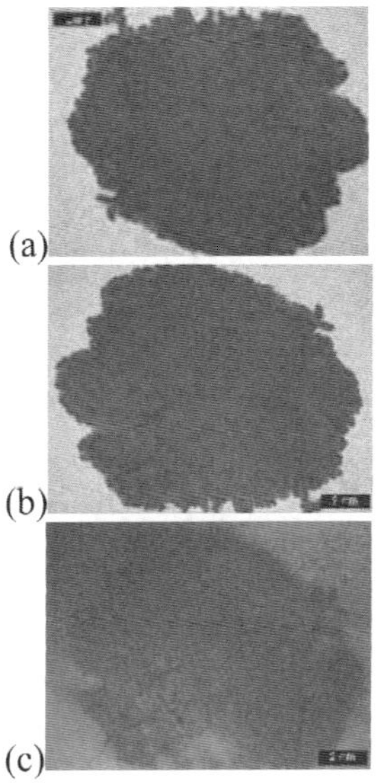

Figure 15a. UVA 860 lux intensity test control sample: (a) initial; (b) 24 h UVA exposure; (c) after completing the test diagram.

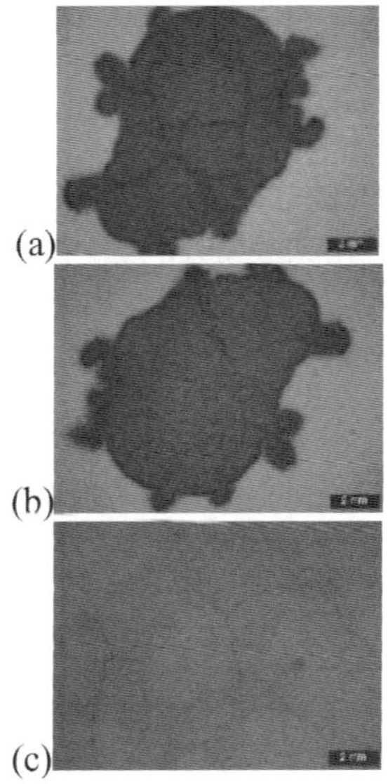

Figure 15b. UVA 860 lux intensity test 3% TiO_2: (a) initial; (b) 24 h UVA exposure; (c) after completing the test diagram.

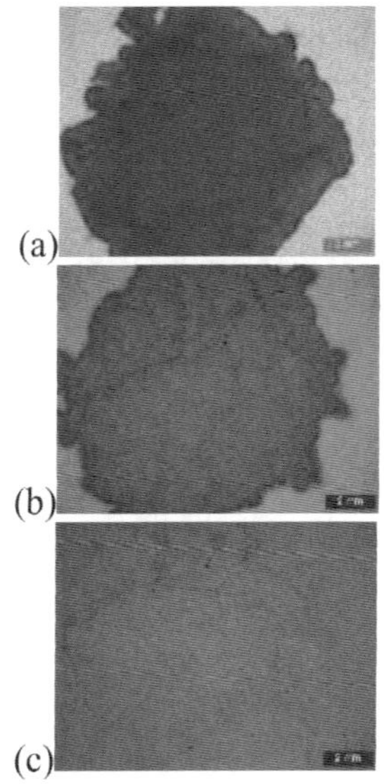

Figure 15b. UVA 860 lux intensity test 5% TiO_2: (a) initial; (b) 24 h UVA exposure; (c) after completing the test diagram.

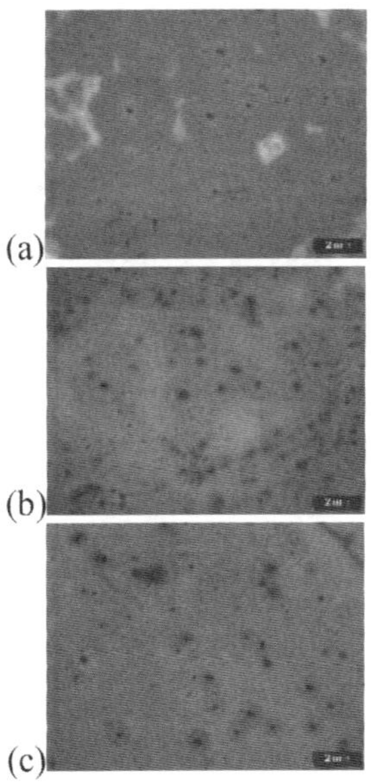

Figure 16a. UVA 405 lux intensity test control sample: (a) initial; (b) 24 h UVA exposure; (c) after completing the test diagram.

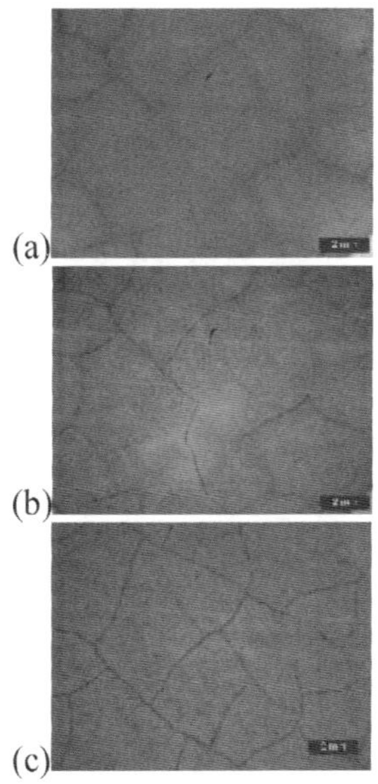

Figure 16b. UVA 405 lux intensity test 3% TiO_2: (a) initial; (b) 24 h UVA exposure; (c) after completing the test diagram.

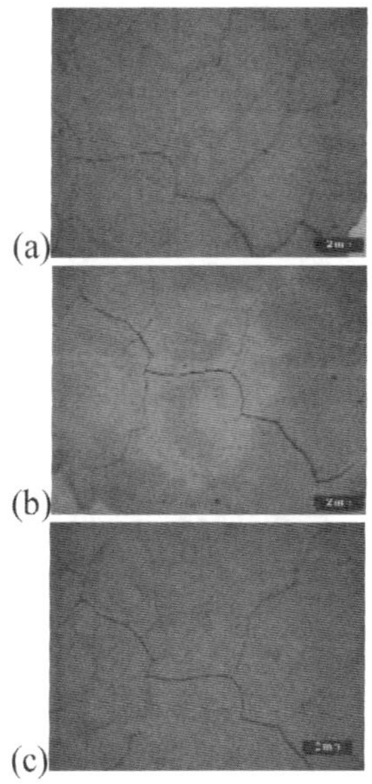

Figure 16b. UVA 450 lux intensity test 5% TiO_2: (a) initial; (b) 24 h UVA exposure; (c) after completing the test diagram.

4. Conclusions

Based on experimental research it can be said that with increasing the amount of TiO_2 in the cementitious matrix, the white degree increases, but also the capacity of self-cleaning of test specimens. The self-cleaning capacity is also influenced by the intensity of UV light flux on the surface of the sample, better results being recorded for a higher UVA intensity. In terms of setting time, it decreases with increasing concentration of TiO_2 in the cementitious matrix, which we cannot say about the apparent density, absolute density, water absorption and porosity, whose values do not vary proportionally with the content of TiO_2 nanoparticles, for the last three parameters it is not possible to identify a function of modeling their evolution in relation to the TiO_2 content.

The experimental results showed the following:

- The setting time, both initial and final, decreased, with the increase in the amount of TiO_2 nanoparticles in the cementitious mixture, conclusion in accordance with the literature;
- Regarding the apparent density of cementitious composites in hardened state, it is noted that a constant interval of high values is recorded in the concentration range of nanoparticles (4% - 5%);
- The absolute density, porosity and water absorption of cementitious composites with added TiO_2 nanoparticles tested do not vary in proportion to the nanoparticle content and functions could not be identified to express the evolution of these parameters in relation to the TiO_2 content;
- The lowest porosity value was obtained for a 4% TiO_2 nanoparticle content relative to the cement quantity. In the case of cementitious matrices with content of 1%, 2%, 3% and 5%, an almost constant porosity was noted;
- Comparing the white degree recorded initially with that recorded after 24 hours exposure to UV rays, it increased for all analyzed cases;
- With the increase in the quantity of TiO_2 introduced into the cementitious matrix, the white degree increased, but also the self-cleaning capacity of the specimens.
- Tests carried out by the method of staining with rhodamine B provide evidence of the self-cleaning ability of cementitious matrices enriched with TiO_2 nanoparticles by photoactivation under the influence of UVA rays;
- The degree of cleaning analyzed as a measurable indicator of the self-cleaning capacity was higher at a lower exposure distance from the UV source. However, in the case of cementitious composition with 5% TiO_2 following exposure to a lower UVA intensity (405 lux), the degree of cleaning obtained is almost equal to that obtained for exposure to a higher UVA intensity (860 lux). This observation indicates the need for in-depth analysis to establish the optimal addition of TiO_2 nanoparticles, depending on several factors, among which very important, the degree of sunshine, the incident light intensity on the surface;
- Similarly, the analysis of the types of functions identified as representative for the evolution of settings times, density, white degree, being functions of order greater than 1, therefore nonlinear, indicates the influence of several factors on these performances, the factor representing the TiO_2 content being only one of them;
- Achievement of an optimum ratio between benefit and cost, respectively, of photoactivation with maximum efficiency, while maintaining the other properties, physical and mechanical properties of the cementitious materials, along with a reasonable cost in relation to benefits, it is essential that the necessary conclusion is that it is consistent with the literature from the literature.

5. References

[1] Chen J and Poon C S 2009 *Build. Environ.* **44** 1899–1906
[2] Eshaghi A, Dashti A, Eshaghi A and Mozaffarinia R 2011 *Mater. Sci.* **29** 22–8
[3] Chen Y, Yang S, Wang K and Lou L 2005 *J. Photoch. Photobio. C.* **172** 47–52

[4] Caasar L 2004 *Mater. Res. Soc. Bull.* **29** 328–31
[5] Diamanti M V, Ormellese M and Pedeferri M P 2008 *Cem. Concr. Res.* **38** 1349–53
[6] Smits M, Chan C K, Tytgat T, Craeye B, Costarramone N, Lacombe S and Lenaerts S 2013 *Chem. Eng.* **222** 411–18
[7] Hashimoto K, Irie H and Fijishima A 2005 *J Appl. Phys.* **44**(12) 8269–85
[8] Fujishima A, Rao T N and Tryk D A 2000 *J. Photoch. Photobio. C.* **1** 1–21
[9] Quagliarini E, Bondioli F, Goffredo G B, Cordoni C and Munafò P 2012 *Constr. Build. Mater.* **37** 51–57
[10] Folli A, Jakobsen U H, Guerrini G L and Macphee D E 2009 *J. Adv. Oxid. Technol.* **12(**1) 126-33
[11] Khataee R, Heydari V, Moradkhannejhad L, Safarpour M and Joo S W 2013 J. Nanosci. **13** 5109–14
[12] Kaviyarasu K, Mariappan A, Neyvasagam K, Ayeshamariam A, Pandi P, Rajeshwara Palanichamy R, Gopinathan C, Mola G T and Maaza M 2017 *Surf. Interface Anal.* **6** 247–55
[13] Carmona-Quiroga P M and Martínez-Ramírezb S 2018 Appl. Surf. Sci. **433** 312–20
[14] Folli A 2010 *TiO_2 photocatalysis in Portland cement systems: fundamentals of self-cleaning effect and air pollution mitigation* (Milan: University of Milan, Italy)
[15] Ranjit K, Odedra Parmar K A and Arora N K 2014 *IJSRD* **1**(11) 2321–43
[16] Bianchi C L, Gatto S, Nucci S, Cerrato G and Capucci V 2013 *Adv. Mater. Sci. Eng.* **2**(1) 1–20
[17] Irie H, Sunada K and Hashimoti K 2006 *Optronics* **98** 102–11
[18] Rashad A M 2015 *Rev. Adv. Mater. Sci.* **40** 72–88
[19] Burduhos Nergis D D, Abdullah M M A B, Sandu A V and Vizureanu P 2020 *Materials* **13(2)** 343
[20] Baoguo M A, Hainan L I, Junpeng M and Pei O 2015 *Mater. Sci. Ed.* **30**(4) 768–73
[21] Dalton J S, Janes P A and Jones N G 2002 *Environ. Pollut.* **120** 415–22
[22] Sorathiya J V, Shah S G and Kacha S M 2017 *ICRISET2017* **1** 219–25
[23] Burduhos Nergis D D, Vizureanu P and Corbu O 2019 *Revista de chimie* **70(4)** 1262-1267
[24] Nedeff V, Lazar G, Agop M, Eva L, Ochiuz L, Dimitriu D, Vrajitoriu L and Popa C 2015 *Powder Technology* **284** 170-186
[25] Shen W, Zhang Q L, Zhang W, Cao L and Ye J 2015 *J. Clean Prod.* **87** 762–65
[26] Choy W 2006 *Catal. Surv. Asia* **10**(1) 16–28
[27] Lazăr M, Fiat D and Hubcă G 2015 *Rev. Rom. Mater.* **40**(2) 178–87

IOP Conf. Series: Materials Science and Engineering **877** (2020) 012005 doi:10.1088/1757-899X/877/1/012005

Copper Sulfate Pentahydrate Target Behavior During Pulsed Laser Deposition to Produce Dichroic Coatings for Beam Splitters

A Cocean[1], I Cocean[1], C Postolachi[1], N Cimpoesu[2], F Husanu[1], B Munteanu[1] and S Gurlui[1*]

[1]Alexandru Ioan Cuza University of Iasi, Faculty of Physics, Atmosphere Optics, Spectroscopy and Laser Laboratory (LOASL), 11 Carol I Bld. 700506 Iasi, Romania
[2]Gheorghe Asachi Technical University of Iasi, Department of Physics, 59A Mangeron Bld., Iasi, Romania

*E-mail: sgurlui@uaic.ro

Abstract. Copper sulfate pentahydrate (CuSO4·5H2O) used as target in the pulsed laser deposition (PLD) process led to producing thin films with dichroic properties. The coatings were applied on glass slab and hemp fabrics. Investigation of CuSO4·5H2O transformation during ablation and deposition based on FTIR analysis showed that the thin film chemical composition is a mixture of different products resulted from CuSO4·5H2O dehydration and decomposition, followed by melting and vaporization of the CuO and recombination of the different species in the plasma of ablation. Simulation in COMSOL of the CuSO4·5H2O laser heating anticipates the processes and phenomena to be expected during PLD. Optical properties of the thin films are investigated in UV-Vis as reflecting properties and by laser induced fluorescence (LIF). Applications are considered for dichroic plate beamsplitters, as well as others. Dichroic beamsplitters have a large range of applications among which they are useful for combining / splitting laser beams of different color. When the chemical composition provides also electrical conductivity specific properties, the coating may bring more value to the new device or material. In order to ascertain whether a dichroic system can be produced one way or the other achieving at the same time specific characteristics, it is important to understand the physico – chemical processes behind that. With this study, we determine the method to produce dichroic coatings on glass and on hemp fabric, looking at the same time into the mechanism that leads to the result.

1. Introduction

The aim of this study is to investigate pentahydrate copper sulfate target behavior under laser irradiation expected being that chemical reactions that may occur, should lead to a dichroic beam splitter based on coatings of copper derivative compounds. COMSOL simulation was performed to provide information on thermal conditions developed on the target for preliminary estimations.

Previously reported depositions of copper derivative compounds have started from copper metallic target or copper oxide, being conducted in specific gaseous atmosphere to provide chemical elements or groups for the desired end product [1-5]. Different from that, herein we consider that the target itself can supply the elements and groups for the product without gaseous atmosphere in the deposition chamber and without further preparation and processing needed neither before, nor during or after deposition.

Dichroic beamsplitters have a large range of applications among which they are useful for combining / splitting laser beams of different colors [6-8]. When the chemical composition provides also electrical conductivity specific properties, the coating may bring more value to the new device or material [9, 10] and for its possible applications in optoelectronics [5].

Recent researches have reported copper effects to inactivate coronavirus to less than 60 minutes [11], including new coronavirus SARS-CoV-2 responsible for Covid 19 disease [12], while the virus time life on other materials counts days. With regards to these studies, pulsed laser deposition of copper and copper derivatives may be used for producing coatings on materials for protection masks and other cloths used in medical and for day to day activities.

In order to ascertain whether a dichroic system or medical device can be produced one way or the other achieving at the same time specific characteristics, it is important to understand the physico – chemical processes behind that. With this study we determine the method to produce coatings on glass and on hemp fabric by PLD method that uses copper sulfate as target, looking at the same time into the mechanism that leads to the final result. Dichroic properties will be studied based on the UV-Vis properties of reflection and transmission.

2. Methods and materials

A target of pentahydrate copper sulfate was used for film deposition on substrates of glass and of hemp twill fabric. The deposition was perform by Pulsed Laser Deposition (PLD) with YG 981E/IR-10 laser system using the following parameters: τ=10 ns pulse width, λ=532 nm wavelength, α=45° incident angle, ν = 10 Hz pulse repetition time, r = 336 μm spot radius, F = 25 J/cm2 fluence, d = 3.5 cm distance between target and support, pressure in the deposition chamber (Figure 1) $3 \cdot 10^{-2}$ Torr.

3. Results and discussions

3.1. COMSOL simulation

COMSOL simulation was conducted using the numerical model reported in [13, 14] to anticipate deposition results and to evaluate chemical species that are expected to form during PLD process.

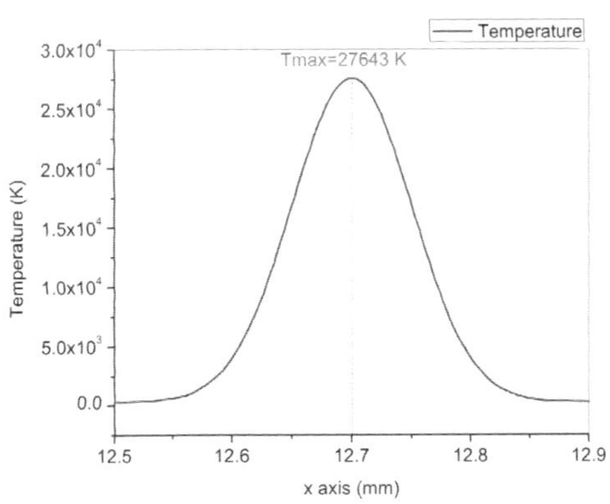

(a)

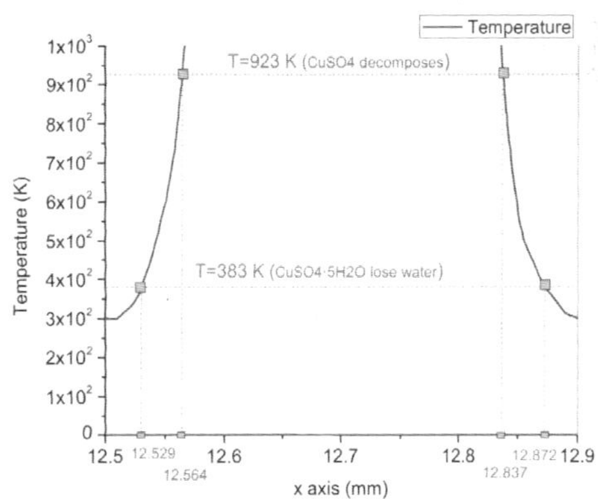

(b)

Figure 1. Target phase change T(x): Temperature diagram on target irradiated spot area with maximum value (a); phase change based on $CuSO_4 \cdot 5H_2O$ dehydration and $CuSO_4$ decomposition (b).

Based on COMSOL simulation results (Figure 1), decomposition steps of pentahydrated copper sulfate are noticed as being possible. In a first step, 383K < T < 923K, $CuSO_4 \cdot 5H_2O$ loses water molecules resulting into $CuSO_4$ anhydrous. Following that, in a second step $CuSO_4$ decomposes based on the chemical reaction described by equation (1).

$$CuSO_4 \bullet 5H_2O \rightarrow CuO + SO_2 + O_2 \quad (1)$$

The thin layer is expected to contain mainly CuO, but also $CuSO_4$ from ablation as clusters sourced in the dehydrated area, but also from recombination of the gas phase and ionic phase in the plasma plume of ablation.

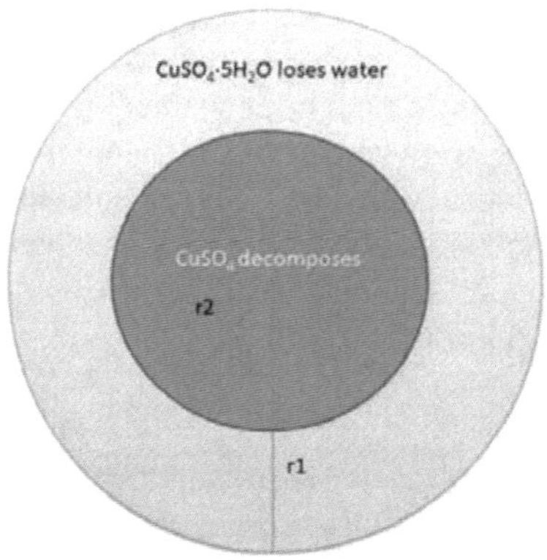

Figure 2. Schematic representation of the target spot area affected by the laser irradiation during one pulse width.

$r_1 = 0.343$ mm; $r_2 = 0.273$ mm.

$$A_2 = \frac{\pi \cdot r2^2}{2} = \frac{\pi \cdot (0.273)^2}{2} = 0.117069879\ [mm^2]$$
$$A_1 = \frac{\pi \cdot r1^2}{2} = \frac{\pi \cdot (0.273)^2}{2} = 0.184802617\ [mm^2] \quad (2)$$
$$A_{ring} = A1 - A2 = \frac{\pi}{2}(r1^2 - r2^2) = \frac{\pi}{2}((0.343)^2 - (0.273)^2) = 0.067732738\ [mm^2]$$

CuO (copper oxide) resulted in the thermal effect during laser beam interaction with the $CuSO_4 \cdot 5H_2O$ target, melts (M.P. = 1326^{0}C = 1599K) and vaporizes (B.P. = 2000^{0}C = 2273K). Plasma formation temperatures are even greater, 10^4K order, leading to further breaking and recombination.

3.2. Fourier transform infrared spectroscopy (FTIR)

The chemical composition was analyzed with Fourier transform spectroscopy (FTIR) for both the target and the pulsed laser deposited film. The results evidenced by the changes in the spectra of Figure 3, show that different chemical species were obtained in the film, including copper sulfate, but also copper oxide and copper in metallic state as hydrogen sulfide adsorbed on the film indicates.

Based on the spectra of Figure 3, the chemical groups and vibrations for the target and thin film are identified and presented in Table 1 and Table 2, respectively.

Due to the conditions during ablation, the decomposition of pentahydrate copper sulfate takes place as per equation (1), followed by other interactions and recombination resulting in a mixed chemical composition of the thin film. FTIR analyses show difference between the target and thin film. The presence of the copper sulfate is shown on FTIR bands with wavenumbers of 1109 cm^{-1}; 874 cm^{-1}; 692 cm^{-1}; 983 cm^{-1} for $(SO_4)^{2-}$. Other compounds than those of the target are formed presumably due to recombination, such as free OH and H bonded noticed at 3435 cm^{-1} wavenumber and H_2O recombined as 1648 cm^{-1} band indicates.

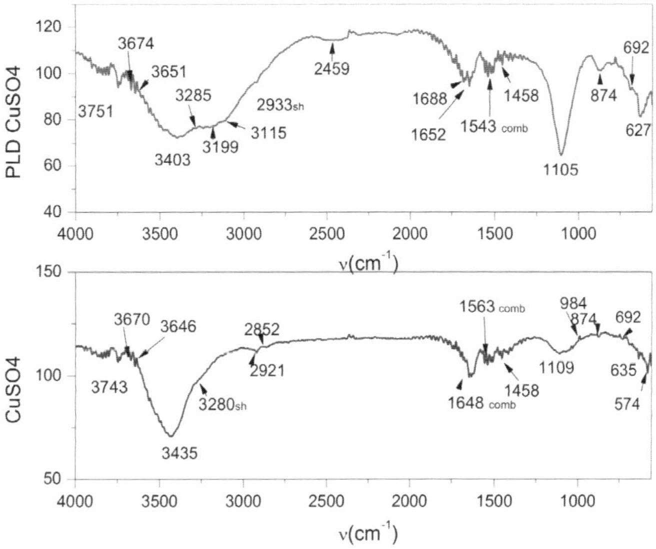

Figure 3. FTIR compared spectra of copper sulfate target and thin films obtained by PLD method.

Table 1. Target of pentahydrate copper sulfate groups and IR vibrations.

Wavenumber (cm^{-1})	Vibration and groups	References
3435	broad band – OH free and H bonded	[15]
3300-3280	Cu-OH (OH from water)	[16, 17, 18]
1648	H_2O bending vibrations (δ H_2O within 1800-1300 cm-1 spectral region)	[15]
1109	$(SO_4)^{2-}$ antisymmetric stretching modes	[15, 17, 18]
874; 692	$(SO_4)^{2-}$ bending modes	[15, 17, 18]
983	$(SO_4)^{2-}$ symmetric stretching modes	[15, 17, 18]
3743; 3670; 3649	OH free	[15, 18]

Table 2. The thin film groups and IR vibrations.

Wavenumber (cm^{-1})	Vibration and groups	References
3435	broad band – OH free and H bonded	[15]
3300-3280	Cu-OH (OH from water)	[16, 17]
1648	H_2O bending vibrations (δ H_2O within 1800-1300 cm-1 spectral region)	[15]
1109	$(SO_4)^{2-}$ antisymmetric stretching modes	[15, 17, 18]
874; 692	$(SO_4)^{2-}$ bending modes	[15, 17, 18]
983	$(SO_4)^{2-}$ symmetric stretching modes	[15, 17, 18]
3743; 3670; 3649	OH free	[15, 18]
2459	H_2S adsorbed on copper (2500-2600)	[15, 16]

Also the H_2S adsorbed on copper is indicated by the 2459 cm^{-1} band which confirms the existence of copper in metallic state. Copper oxide is confirmed by OO stretching vibration at 1105 cm^{-1} [19].

3.3. UV-Vis thin films spectra

For the same thin film deposited on glass, three UV-Vis measurements in three different spots of the same film were performed, namely PLD CuSO4/glass, PLD CuSO4/glass -edge 1, PLD CuSO4/glass -edge 2, and for the film deposited on hemp, two measurements in two different spots of the same film were performed, namely PLD CuSO4/hemp 1, PLD CuSO4/hemp 2. The UV-Vis spectra in Figure 4 (a) show reflection at the same wavelengths of 480 nm, 575 nm, 633 nm, 665 nm, 696 nm in visible and 930 nm, 967 nm in near infrared (NIR), which shows same chemical species have been deposited, but different rates of reflected wavelengths prove different quantities of each species. Transmittance spectra performed on PLD CuSO4/glass, PLD CuSO4/glass -edge 1, show maximum in NIR region at 1037 nm and 1021 nm, respectively. If deconvolution is analyzed, two spectra could be components of the main registered spectrum, with peaks at (727 nm, 996 nm) and (710 nm, 981 nm) for each of the two spots (Figure 4 b). The thin films show dichroic properties as evidenced by UV-Vis measurements (Figure 4 and Table 3), properties that are induced by the film chemical composition as a mixture of different chemical species, as it had already been evidenced by the FTIR spectra (Figure 3) and SEM-EDS mapping elemental distribution (Figure 7 a-f), as well as elemental composition of the thin films (Table 4). Another characteristic of the thin film that determine dichroic property is the granular morphology of the thin film as SEM images of Figure 6 a, b show.

Table 3. Reflection intensities (%).

λ (nm)	PLD CuSO4/ glass -edge 1	PLD CuSO4/ glass -edge 2	PLD CuSO4/ glass	PLD CuSO4/ hemp 1	PLD CuSO4/ hemp 2
480	17.77	16.08	14.85	17.63	20.15
575	38.85	40.17	39.65	42.81	49.29
633	38.37	39.88	38.3	42.53	48.2
665	46.42	37.43	36.96	46.76	54.47
696	42.94	28.71	32.98	46.76	53.93
930	7.26	3.57	5.85	8.01	8.97
967	3.99	1.46	3.74	4.53	5.55

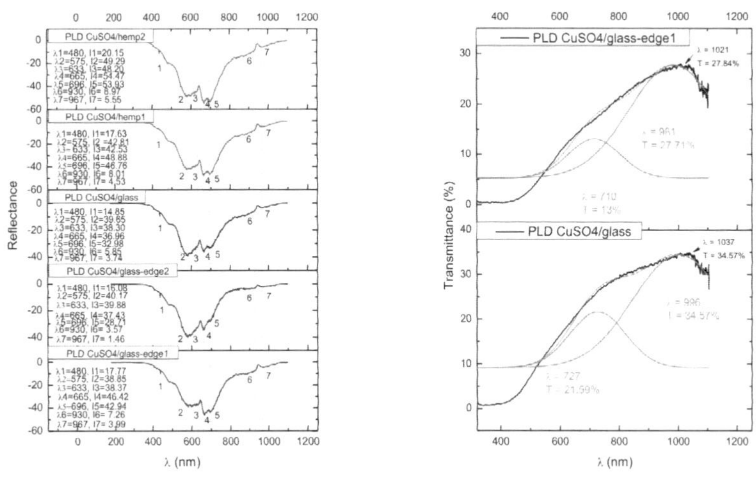

(a) (b)

Figure 4. Optical properties of thin films obtained from $CuSO_4$ using PLD method: Reflection spectra (a); Transmittance spectra of thin film PLD $CuSO_4$/glass (b).

3.4. Laser induced fluorescence (LIF)

LIF spectroscopy was conducted with YG 981E/IR-10 laser system as for PLD process, but using 355 nm wavelength laser irradiation, 10 ns pulse width, 10 Hz repetition and 150 mJ energy. The LIF spectra of Figure 5 show the presence of both copper oxide and copper sulfate. The curvature going up in 400 nm region could be assessed to the emission of sulfur dioxide (from copper sulfate), reported by Syty et al. as fluorescence emission in the range 290 - 310 nm [20-24]. The emission band at 699 nm may be assessed to the peak that goes up to 750 nm representing copper fluorescence [22, 27, 28]. Tower bands in the ranges of 482 nm – 508 nm and 587 nm – 627 nm of the spectrum are consistent with hydroxyl radical formation, according to Cocean et al. [25, 26]. The hydroxyl groups denote either molecular water in the thin film or copper hydroxyl produced under PLD process.

Table 4. Elemental composition of the thin films.

	PLD $CuSO_4$/glass		*PLD $CuSO_4$/hemp*	
Element	Mass [%]	Atom [%]	Mass [%]	Atom [%]
Oxygen			51.3498	58.17554
Carbon	66.56417	86.41754	21.43479	32.34785
Copper	25.16761	8.22657	21.09623	6.017582
Sulfur	8.268215	5.355894	6.119176	3.459024
	100	100	100	100

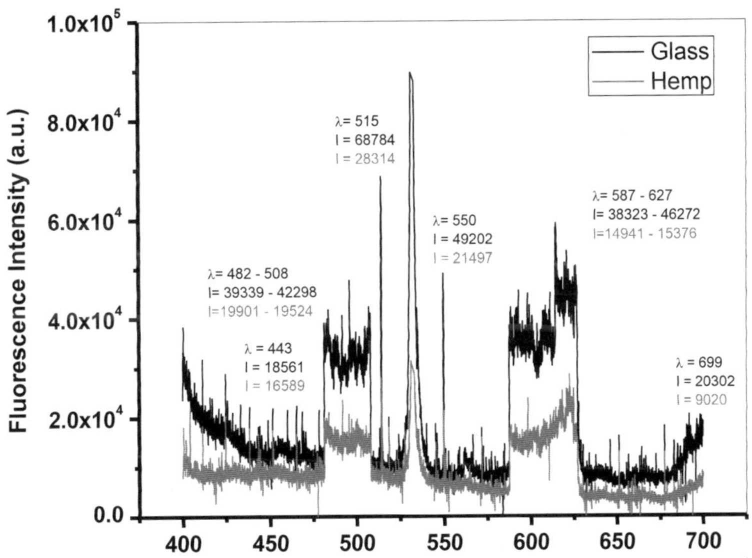

Figure 5. Laser Induced Fluorescence spectra of the pulsed laser deposited thin films.

The mitigation of intensity for the deposition on hemp fabric compared with the deposition on silica glass slab is being considered due to water sorption phenomena into the fibers, and due to the diffusion of the deposited particles among the hemp fibers from the structure of the fabric.

IOP Conf. Series: Materials Science and Engineering **877** (2020) 012005 doi:10.1088/1757-899X/877/1/012005

3.5. Scanning electron microscopy coupled with energy dispersive spectroscopy (SEM-EDS)

SEM images show the granularly structure of the film. This structure is present on both the glass substrate and the hemp substrate with large dispersion of the grains. Such morphological aspect of the thin film has an important role in their reflection properties, as already iterated herein.

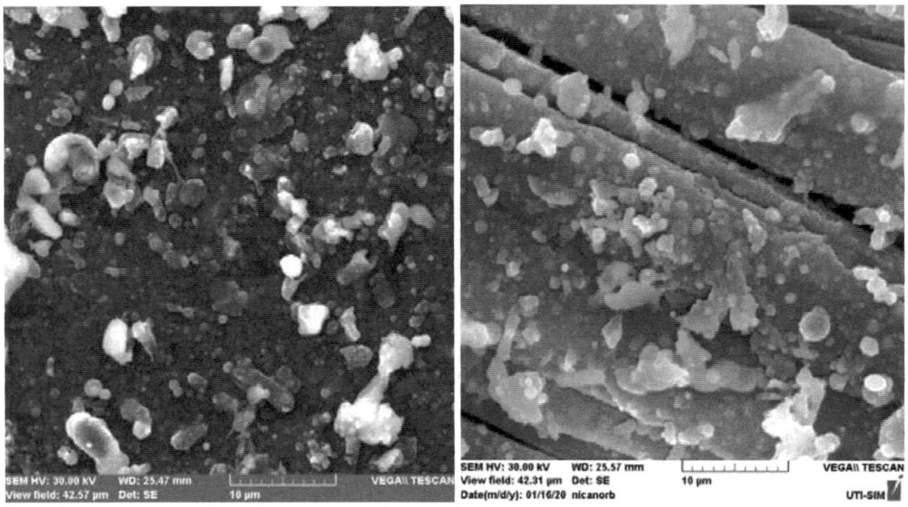

(a) (b)

Figure 6. SEM images of thin films PLD $CuSO_4$/glass (a) and PLD $CuSO_4$/hemp (b).

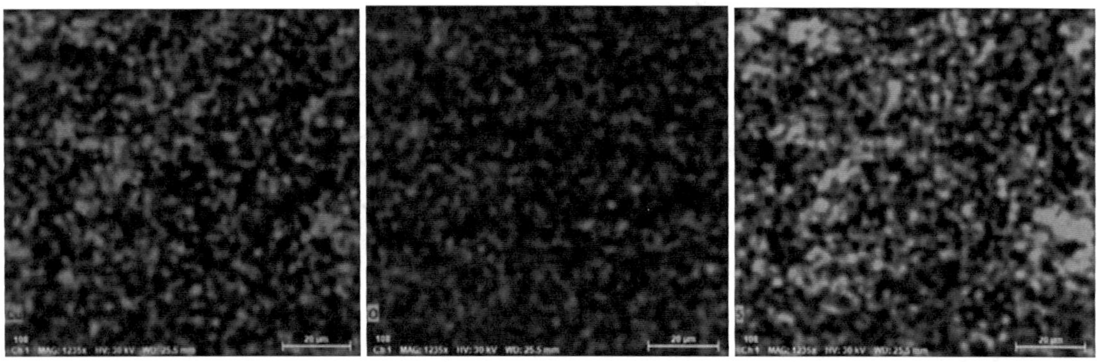

(a) (b) (c)

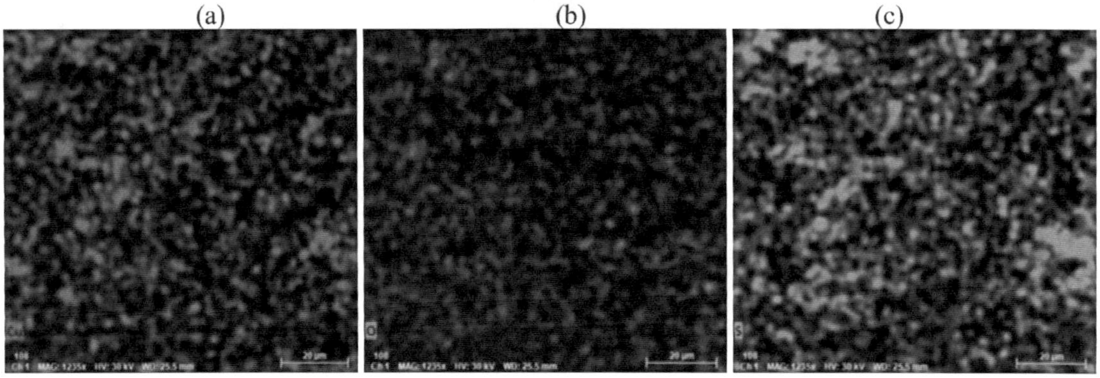

(d) (e) (f)

Figure 7. Elemental composition map for PLD-$CuSO_4$/glass: copper (a); oxygen (b); sulfur (c) and for PLD$CuSO_4$/hemp: copper (d); oxygen (e); sulfur (f).

Deposited particles variety in dimension, noticed in the SEM images, contributes to the optical properties as well. In addition to the optical properties of the deposited thin films, their important content in copper (25% in the film PLD $CuSO_4$/glass and 21% in the film PLD $CuSO_4$/hemp) makes them suitable for uses where copper is required to be added on different materials, such as textiles, in producing composites designed for masks and other equipment-ware for medicine, copper virucide properties being already proven [11, 12].

A uniform distribution appears on the elemental map as the elements superimpose over each other proving that most of them are combined in either by chemical bonds or in physical interactions, such as Van der Waals. The elements shown on map are present in the compounds found in FTIR and UV-Vis, with the mention that carbon is from the hemp substrate.

4. Conclusion and perspectives

A dichroic beam splitter has been achieved using copper sulfate. The film structure and different species deposited on substrate give it those properties. In addition, the possibility to deposit complex chemical species using PLD without special gas atmosphere in the deposition chamber is evidenced. The thermal decomposition under laser irradiation provides chemical components that can be anticipated when simulated the process in COMSOL. With careful changes of laser parameters, films can be tuned to specific needs. Such coatings find a large number of applications from optics, electronics, to medical and composite materials for protection equipment. With further studies, films with more complex behavior can be achieved

5. References

[1] Kassim A, Min H S, Haron M J and Nagalingam S 2011 *Int. J. of Pharm. & Life Sci. (IJPLS)* **2**(11) 1190-1194

[2] Aravinda C L, Mayanna S M and Muralidharan V S 2000 *Proc. Indian Acad. Sci. (Chem. Sci.)* **112**(5) 543–550

[3] Zhang Y, Aslan K and Previte M J R 2007 *Appl. Phys. Lett.* **90** 173116

[4] Seiler W, Millon E, Perriere J, Benzerga R and Boulmer-Leborgne C 2009 *Journal of Crystal Growth* **311** 3352–3358

[5] Umesh Chandra Bind U C, Dutta R K, Sekhon G K, Yadav K L, Krishna J B M, Menon R and Nabhiraj P Y 2015 *Superlattices and Microstructures* Volume **84** 24-35

[6] Xu Q, Tao J, Sun C, Zhao J, Wang Z, Du L, Niu C, Li X, Huang W 2020 *Optics Communications*, **454** 124424

[7] Yang H, Ou K, Cao G, Shang X, Liu Y and Deng Y 2019 *Optics Communications* **443** 104 – 109

[8] Fedaouchea A, Badaouia H A and Abri M 2018 *Optik* **157** 1300-1305

[9] Zhou R, Huang T, Chen L, Chen S, Lin S and Zhuo Y 2017 JLMN-*Journal of Laser Micro/Nanoengineering* **12**(2) 2017

[10] Floegel-Delor U, Riedel T, Wippich D, Goebel B, Rothfeld R, Schirrmeister P, Werfel F N, Usoskin A and Rutt A 2011 *IEEE Transactions on Applied Superconductivity* **21**(3) 2984-2987

[11] Warnes S L, Little Z R and Keevil C W 2015 *mBio* **6**(6) e01697-15

[12] van Doremalen N, et al. 2020 *The New England Journal of Medicine* 2004973

[13] Cocean A, Cocean I, Gurlui S, Iacomi F 2017 *U.P.B. Sci. Bull., Series A* **79**(2)

[14] Cocean A, Pelin V, M M, Cocean I, Sandu I, Gurlui S and Iacomi F 2017 *Applied Surface Science* **424** 324–329

[15] Pretch E, Bülmann P and Badertscher M 2009 Structure Determination Of Organic Compounds.Tables Of Spectral Data, Fourth, Revised and Enlarged Edition, *Springer – Verlag Berlin Heidelberg* 2009, ISBN 978 – 3 – 540 – 93810 – 1, doi 10.1007/978-3-540-93810-1

[16] Miller F A and Wilkins C H 1952 *Anal. Chem.* **24** (8) 1253–1294

[17] Frost R L, Jagannadha R B and Keeffe E C 2010 *Journal of Molecular Structure* **977**(1-3) 90-99
[18] Etalo A S 1988 *CAN. I. CHEM.* **66**
[19] Chertihin GV andrews L and Bauschlicher CW Jr. 1997 *J. Phys. Chem. A* **101**(22) 4026
[20] Augusta Syty 1973 *Anal. Chem.* **45**(9) 1744-1747
[21] Burduhos-Nergis D -P, Vizureanu P, Sandu A V and Bejinariu C 2020 *Applied Sciences* **10(8)** 2753 doi:10.3390/app10082753
[22] Zohora N, Kandjani A E, Orth A, Brown H M, Hutchinson M R and Gibson B C 2017 *Sci Rep* **7** 16905
[23] Vizureanu P, Perju MC, Galusca DG, Nejneru C and Agop M 2010 *Metalurgia international* **15(12)** 59-64
[24] Sharma S and Uttam K N 2017 *Vibrational Spectroscopy* **92** 135-150
[25] Cocean I, Cocean A, Postolachi C, Pohoata V, Cimpoesu N, Bulai G, Iacomi F and Gurlui S 2019 *Applied Surface Science* **488** 418–426
[26] Tsuboi Y, Kimoto N, Kabeshita M and Itaya A 2001 *J. Photochem. Photobiol. A Chem.* **145** 209–214
[27] Ionescu D, Mătăsaru D, Radu V 2013 *UPB Scientific Bulletin Series A-Applied Mathematics and Physics* **75(4)** 265-274
[28] Irimiciuc S A, Agop M, Nica P, Gurlui S, Mihaileanu D, Toma S and Focsa C 2014 *Japanese Journal Of Applied Physics* **53(11)** 173-179

IOP Conf. Series: Materials Science and Engineering **877** (2020) 012006 doi:10.1088/1757-899X/877/1/012006

Stability of Emulsions Prepared by Vapor Condensation

E T Iacob Tudose[1*]

[1]"Gheorghe Asachi" Technical University of Iasi, Faculty of Chemical Engineering and Environmental Protection, 73 Prof.Dr.Docent D.Mangeron Blvd, 700050, Iasi, Romania

*E-mail: etudose@ch.tuiasi.ro or et_tudose@yahoo.com

Abstract. Emulsions, consisting of a mixture of two immiscible liquids, one dispersed in another, with an added amphiphilic surfactant, have been extensively used in different areas such as cosmetics, pharmaceutics, material synthesis for new applications, for example thermal energy storage, new cooling systems etc.The current study focuses on emulsion preparation using a novel technique[1], namely water vapor condensation on an oil-surfactant layer containing one or two surfactants. The method was incompletely investigated previosly, thus experiments were initiated in order to assess the emulsion stability for different working conditions. Two distinctive configurations were experimentally studied, one, with an adiabatic, constant room temperature bottom and another, with a high, constant temperature bottom. A mixture of two surfactants, Tween 80 and Span 80, was used in different ratios, also, the condensation time and the vapor flowrate were varied. The obtained emulsions stabilities were assessed visually and also, by pH measurements, over couple of week time period. The estimated average emulsion particle size was within the microscale range.

1. Introduction

Emulsions are defined as dispersed systems consisting of two immiscible liquids, one reffered to as the dispersed or discontinuous phase, spreaded within another, named the continuous phase, stabilized by an amphiphilic surfactant [2]. Based on their droplet size, they can be classified in macroemulsions (mean diameter rangind from 0.1 to 100 μm) [3], microemulsions (100Å - 100 nm) [4] and nanoemulsions (20 - 200 nm) [5]. The various tunable properties of the emulsions make them suitable for application in different fields such as: cosmetics, pharmaceutics, material synthesis for new purposes such as thermal energy storage, new cooling systems etc.

One of the major characteristics of an emulsion is its physical stability consisting of a uniform distribution of the dispersed particles throughout the continuous phase. The stability is visually assessed by simple storage at different temperatures (usually 0-40°C) for 24 h up to 15 days. It can be also evaluated by accelerated aging methods, for example repeated hot-cold cycles or even, centrifugation to establish the emulsion stability index (ESI), many times reffered as creaming index (CI), known to have higher values for less stable emulsions. In an emulsion three types of physical instabilities can be encountered: creaming, flocculation and breaking.The creaming consists of two layer separation, one richer in the disperse phase than the other. It can be reduced by decreasing the droplet size, increasing the droplet concentration and even, the system viscosity by addition of different substances such as methylcellulose or paraffin. Flocculation consists of cluster formation due to the aggregation of dispersed phase particles and can be followed by sedimentation. It can be prevented by coating particles with a high charge density surfactant that keeps them separately. Breaking implies phase separation and can be prevented by controlling the factors responsible for creaming and flocullation.

A stable emulsion can be prepaired by an appropriate selection of the components and the corresponding optimum quantities, the attainment of the best hydrophilic-lipophilic balance (HLB), appropriate storage conditions, addition of some ingredients that can increase the emulsion stability etc.

The main emulsion preparation techniques are classified in low and high energy methods and have been extensively used to obtain all types of emulsions. Recently, a new technique has been proposed, namely the condensation method consisting in water vapor condensation onto a subcooled oil-surfactant solution [1]. Water droplets nucleate at the oil/air interface and spontaneously disperse within the oil, due to the spreading dynamics. Oil-soluble surfactants stabilize the resulting emulsions. The oil-surfactant concentration controls the spreading behavior as well as the droplet size, the polydispersity and the stability of the resulting emulsions [1].

Despite the discussions related to the emulsion formation mechanism, the major parameters thought to influence the particle size and ultimately, the emulsion stability, have not been investigated. Thus, this study focuses on some preliminary experiments meant to identify these aspects. Commercial sunflower oil, with an added mixture of Tween 80 and Span 80, were used as a base layer for water vapor condensation, in order to obtain emulsions. Among the parameters assumed to influence the emulsion particle size and thus, the emulsion stability, the following were investigated: the vapor flowrate, the condensation time, the surfactant mixture-oil ratio, the emulgator final HLB_f. Also, two settings with different bottom temperatures were studied.

These data can be used as a starting basis to investigate the vapor condensation technique to obtain high stability emulsions.

2. Experimental setup and measurement procedure. Data analysis

2.1. Materials and sample preparation

Tween 80 (polyoxyethylene sorbitan monooleate) and Span 80 (sorbitan monooleate), two high and respectively low hidrophilic-lipophilic emulsifiers, (with HLB=15.0 and respectively, HLB=4.3) from Sigma Aldrich, sunflower oil (SO) bought from a local store and double distilled water were used in this study.

Sunflower oil and a mixture of Tween 80 and Span 80, both measured with a pipette or a graded cylinder, depending on the needed volume, were introduced in a Berzelius glasses and homogenized for couple of minutes, at 500 rpm, using a magnetic stirrer. Afterwards, the glasses were placed in the lab apparatus described below, to enable the water vapour condensation and finally, the emulsion formation. Two types of emulsions were prepaired by varying the sunflower oil : water ratios to be either 1:4 or 4:1. Once the assessed time for the desired vapour condensation elapsed, the samples were removed from the experimental apparatus and let to cool, at room temperature. The prepaired emulsions were observed right away and also, after 24 h. If no phase separation, creaming or sedimentation occurred within first 24 h, the samples were used to be further analysed. The literature indicates that an emulsion stable for one day, remains usually stable for several days [6].

2.2. Experimental laboratory setup

The set-up used to perform all experiments is presented in Figure 1. A special chamber 1, with a heating jacket 2 (filled with water), was used to store a beaker 3, containing a sunflower oil (SO) layer where, previously, a mixture of the two above mentioned emulgators was stirred. The chamber was sealed with a flange lid 4, fitted with an inlet tube 5, for vapor admission. The whole system was placed on a heating plate 6. A steam generator 7 provided with a pressure gauge 8 and an admission valve 9 was used to provide vapour with an adjustable flowrate measured using a flowmeter 10. Digital thermometers (TR) were used to verify the temperatures within the system.

Two different configurations related to the bottom surface temperature were investigated, while the other conditions were maintained the same, in order to assess the emulsion stability. The first one consisted of an adiabatic, constant room temperature bottom (T_b= 20°C) and the second one, of a high, constant temperature bottom (T_b=90°C). The latter was achieved by placing the chamber onto the heating surface, maintained at a constant temperature, until the steady state regime was attained.

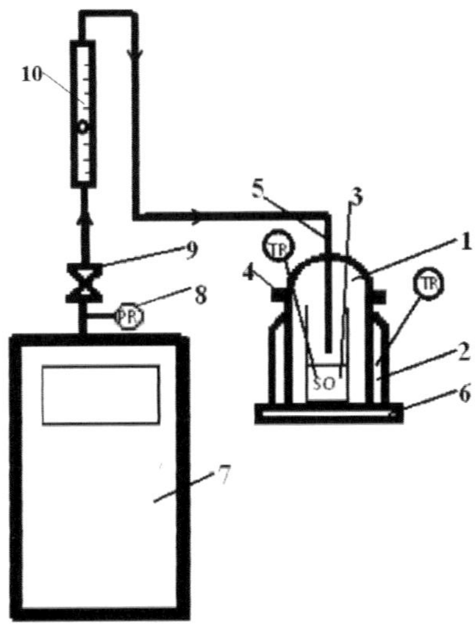

Figure 1. Experimental set-up for emulsions obtained by vapor condensation.

A total of seventeen samples, for different conditions, were prepared and their stability was visually assessed within first 24 h and over couple of days, or even weeks, time period. Also, the emulsion pHs were monitored as a measure of their stability and for the creaming emulsions, the creaming index (CI) was reported.

2.3. Preliminary calculations and data analysis

The emulsifiers were prepaired using the surfactants Span 80 and Tween 80 in different weight proportions to have a resulting HLB value of 10[7,8] or 11.6. In order to calculate the needed Span 80 mass concentration, c_{sp} (%), to achieve a certain HLB value, the following equation was used:

$$c_{sp} = \frac{HLB_{tw} - HLB_f}{HLB_{tw} - HLB_{sp}} \cdot 100 \quad (1)$$

where the Tween 80 surfactant has a HLB_{tw} =15, Span 80 HLB_{sp}=4.3 and the final HLB of the emulsifier obtained by mixing Span 80 with Tween 80, HLB_f =10[8]. Thus, the calculated mass percentage of Span 80 and respectively, Tween 80 are presented in Table 1:

Table 1. The surfactant mixture composition.

Det. No.	HLB_f	Span 80 mass concentration (%)	Tween 80 mass concentration (%)	Span 80: Tween80
1	10	46.73	53.27	8:9
3. 2	4. 11.6	5. 31.78	6. 68.22	7. 1:2

The unstable emulsions were characterized by the creaming index defined using the following equation:

$$CI = \frac{H_C}{H_E} \cdot 100\,,\%\qquad(2)$$

with H_C – the height of cream or aqueous layer and H_E the total height of the emulsion layer. It is known that the higher the CI value, the lower the system stability [9,10].

3. Results and discussions

The investigated values of the emulsifier mixture : oil ratio, the final hidrophilic-lipophilic balance (HLB_f), the water vapour flowrate and the condensation time are presented in Table 2. Also , the last column of the table 2 indicates the final characteristics of the obtained emulsions.

Table 2. Emulsion composition (S1* prepaired at T_b=20°C; S2-S17 prepaired at T_b=90°C).

Sample number	Surfactant mixture:oil ratio(w)	HLB_f	Condensation conditions		Emulsion stability
			Water vapour flowrate (m^3/h)	Time (min)	
S1*	1:1	10	0.3	5	creaming
S2	1:1	10	0.3	5	*stable*
S3	1:1	11.6	0.5	5	creaming
S4	1:1	11.6	0.3	5	creaming
S5	5:4	11.6	0.5	5	creaming
S6	5:4	10	0.3	5	creaming
S7	5:4	10	0.5	3	creaming
S8	5:4	10	0.7	2.1	*stable*
S9	3:2	11.6	0.5	5	creaming
S10	3:2	11.6	0.3	5	*stable*
S11	3:2	10	0.5	3	creaming
S12	3:2	10	0.3	5	creaming
S13	3:2	10	0.7	2.1	creaming
S14	2:1	11.6	0.5	5	creaming
S15	1:2	10	0.7	2.1	*stable*
S16	1:2	10	0.5	3	*stable*
S17	1:2	10	0.3	5	*stable*

3.1. The bottom surface temperature

Two identical samples consisting of 10 mL sunflower oil, a mixture of Tween 80 and Span 80, with a resulting HLB_f of 10, were used to establish what surface condition is more suitable to obtain a stable emulsion.

The two emulsions, marked S1, and respectively, S2, obtained by vapour condensation with a constant bottom surface temperature, T_b, of 20°C (room temperature) and, respectively, 90°C (heating plate), are presented in Figure 2 (a) and (b).

A couple of other similar samples were prepaired and checked for the bottom temperature influence on the emulsion stability within the first 24 h and yielded the same results. In the first setup, there is a larger temperature difference between the vapour and the liquid layer which would be expected to increase the mass of the condensed vapours on the subcooled substrate, according to a simple heat transfer balance.

(a)
(b)

Figure 2(a-b) Emulsions obtained by vapor condensation (a) S1 and S2 immediatelly after preparation; S1(left) –T_b =20°C; S2(right) – T_b =90°C; (b) S1 and S2 after 24 h; Emulsifier mixture:Sunflower oil=1:1; HLB_f=10; Steam flowrate = 0.3 m^3/h; t=5 min.

However, the condensed vapour mass was similar in the two setups. This suggests that on a cooler substrate, the vapour convection is more enhanced and probably the local rate of condensation becomes larger, thus, more unstable emulsions are obtained.

Thus, once the favorable condition was established, the rest of the prepaired emulsions were obtained by vapour condensation on a high temperature substrate.

Pictures of all the other prepaired oil-in-water samples are presented in Figure 3 (a1) and (b1) up to (a5) and (b5). The letters (a) and respectively, (b), stand for freshly prepaired emulsions and respectively, 24 h after emulsions preparation.

3.2. The vapor flowrate

The vapour flowrate influence on the oil/water emulsion stability was investigated for the same emulgator mixture: oil of 2:3, namely at 0.3 and 0.5 m^3/h, for the same time period of 5 minutes (samples S9 and S10, presented in Figure 3 (a2), (b2)).

It was found that the lowest flowrate produced a stable emulsion, while the other one did not render a stable system.

This probably can be explained as follows: a lower vapour flowrate condensing on the liquid layer will induce a more gradual formation and stabilization of the droplets within the liquid layer.

Some of the samples prepaired using the lowest vapor flowrate value did not render stable emulsions. (sample S6, as seen in Figure 3(a4), (b4) and sample S12, as seen in Figure 3(a5), (b5)).

The obtained water/oil emulsions (samples S15, S16, S17, as seen in Figure 4), no matter the vapour flowrate, of 0.3,0.5,0.7 m^3/h, have a relatively good stability and are characterized by a much higher viscosity (mayonnaise consistency) in comparison to the oil/water emulsions.

(a1)S3-stable, S4-creaming, S5-creaming (b1) S3-creaming,S4-creaming,S5 -creaming

(a2)S9-creaming, S10-stable (b2)S9-creaming,S10-stable (a3)S14-creaming (b3)S14-creaming

(a4)S6-creaming, S7 - stable, S8-stable (b4) S6-creaming, S7 – creaming, S8 stable

(a5)S11-creaming, S12-creaming, S13-creaming (b5) S11-creaming, S12-creaming, S13-creaming

Figure 3. Oil/water emulsions, samples S3 up to S13 (a) freshly prepaired; (b) 24h after preparation.

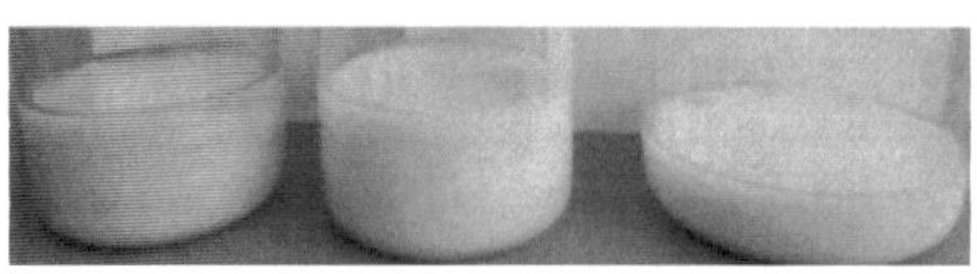

Figure 4. Water/oil emulsions, samples S15, S16, S17, 5 days after preparation.

3.3. The condensation time period
As the condensation time period increases, larger amounts of steam can be incorporated in the liquid substrate. However, for the emulsion stability, the data seem to indicate that the condensation time is a less important factor in comparison to the vapour flowrate.

3.4. Surfactant mixture:oil ratio
It is expected that a higher amount of surfactant mixture in the emulsion, provided the other experimental conditions are the same, would increase the emulsion stability due to the increase of the interfacial area (smaller diameter particles) and decrease of the interfacial tension. In Table 2, the investigated samples S1 to S14 (w/o emulsions) are ordered in ascending order of the surfactant:oil ratio values. Stable emulsions were obtained at all investigated ratios, however, one can observe that these values correspond to lower vapor flowrates, suggesting that the latter is a key parameter for stability.

3.5. Hidrophilic-lipophilic balance
It is known that a mixture of low and high HLB surfactants (hidrophilic- lipophilic balance) determines more stable emulsions since choosing different surfactant ratios allows for HLB tuning. As stated before, a HLB value around 10 was reported as suitable for a stable oil-water emulsion, however some of the prepaired emulsions using vapour condensation did not exhibit good stability at this HLB value. Another HLB_f value of 11.6 was investigated for different parameter values, as seen in Table 2. In this case the Span80:Tween80 ratio was 1:2. However, stable or unstable emulsions were obtained for both HLB_f values, thus a clear conclusion cannot be drawn for now.

3.6. Emulsion stability
The emulsion stability was assessed visually right after their preparation, after the first 24 h and also, for the next couple of weeks.

An accelerated thermal stability test consisting of repeated increasing-decreasing temperature cycles was applied to the stable emulsions in an Eccocell 55 BMT oven.The temperatures were increased from 20°C to 50°C followed by cooling at the same rate. Five cycles were applied in each case and the emulsion stability was confirmed for samples S2, S8, S10, S15, S16, S17.

For the unstable emulsions, the creaming index, calculated using equation (2), and measured after several days when its value maintained constant, is presented in Table 3:

Table 3. Creaming index for unstable emulsions.

Sample number	S1	S3	S4	S5	S6	S7	S9	S11	S12	S13	S14
CI(%)	27.57	30.00	28.57	20.00	18.60	65.12	20.00	16.22	31.82	46.34	20.00

These data seem to indicate that higher creaming index values, of 65.12% and respectively, 46.34%, are obtained in case of larger condensing vapour flowrates of 0.5 and respectively, 0.7 m^3/h.

3.7. Emulsion pH
The pH values of the obtained emulsions were monitored from the initial moment for several days. It was found that the stable emulsions are characterized by an almost constant pH (graphs S8, S10, S11 in Figure 5), while the unstable emulsions are characterized by a decreasing pH value (graphs S2, S6, S7, S12, S13) especially in the first days after preparation.

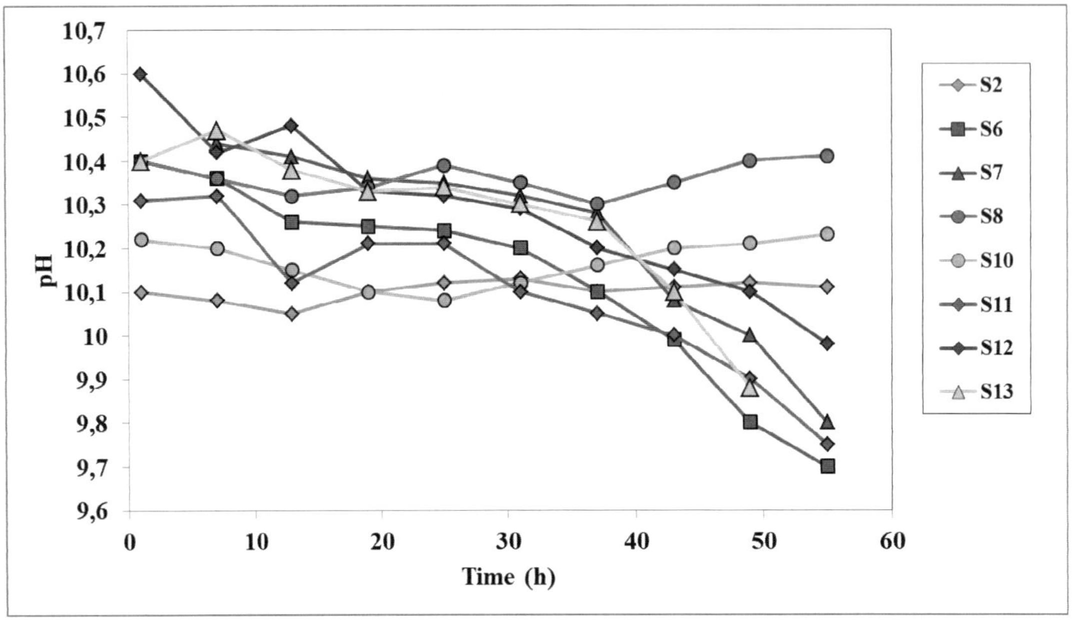

Figure 5. Emulsion pH variation in time, in the first 2-3 days after preparation.

4. Conclusions

The stability of the emulsions prepaired by condensation of water vapor on an oil liquid layer, containing a mixture of two surfactants, was investigated. The experimental study examined the influence of some of the parameters thought to be important in this technique. The obtained data seem to indicate that a lower condensing vapor flowrate renders, in most cases, more stable emulsions, even at lower surfactant mixture:oil ratio. Values of 10 and 11.6 of the hidrophilic-lipophilic balance, coresponding to a Span80: Tween 80 ratio of 8:9 and respectively, 1:2 did not seem to indicate any major difference in the obtained emulsions stabilities. Repeated heating-cooling tests operated on stable emulsions did not change their appearance and structure. Measured pH values for stable emulsions within 2-3 days after preparation remain almost constant while for the unstable emulsions, they tend to decrease.

5. References

[1] Guha I F, Anand S, Varanasi K K 2017 *Nat. Commun.* **8** 1-7

[2] Tadros T F, Vincent B 1983 *Emulsion stability*, In: Becher P (Ed.), *Encyclopedia of Emulsion* Technology, 1, Marcel Dekker, New York

[3] Schramm L L 2005 *Emulsions, foams and suspensions: fundamentals and applications*, Wiley-VCH Verlag GmbH&Co.KgaA, Weinheim

[4] Friberg S E, Venable R L 1983 *Microemulsions* In: Becher P. (Ed.) *Encyclopedia of Emulsion Technology*, 1, Marcel Dekker, New York

[5] Mason T G, Wilking J N, Meleson K, Chang C B, Graves SM 2006 *J.Phys.-Condens.Mat.* **18** 635-666

[6] Meisen L 2006 *Sc.Techn. Food Ind.* **2** 213-225

[7] ICI Americas Inc. 1976 *Communique* **37(10)** 2-22

[8] Porras M, Solans C, Gonzales C, Martinez A, Guinart A, Gutierrez J M 2004 *Colloids Surfaces A Physicochem. Eng. Asp.* **249(1-3)** 115-118

[9] Meher J G, Yadav N P, Sahu J J, Sinha P 2013 *Drug Dev. Ind. Pharm.* **39** 1540-1546

[10] Raghavendra S N, Raghavarao K S M S 2010 *J. Food Eng.* **97** 341–347

IOP Conf. Series: Materials Science and Engineering **877** (2020) 012007 doi:10.1088/1757-899X/877/1/012007

Novel Trip Steel – Characterization, Impact Test and Comparison with Existing Trip Steels

M C Pantilimon[1], C I Tarcea[1*], C S Gradinaru[1], G Coman[1], S Ciuca[1], M Sohaciu[1] and C Predescu[1]
[1]University Politehnica of Bucharest, Faculty of Material Science and Engineering, Splaiul Independentei nr. 313, RO-060042, Bucharest, Romania

*E-mail: tarceaclaudia@yahoo.com

Abstract. 4 TRIP steels were produced by use of an induction furnace with controlled atmosphere and vacuum, one of which having a completely new chemical composition with the purpose of obtaining an increased reaction during impact when compared to the other, already documented, steels. This study follows the heat treatment used in order to obtain the TRIP effect in the produced steels and the reaction that the steels exhibit when submitted to an impact test used to simulate a car crash at approximately 60 km/h. The preparation of the samples for both characterization and testing follow standard procedures in terms of analysis. The crash-like tests were performed by using an INSTRON 9340 Ceast which generated an impact energy of 18J. The comparison between the different types of TRIP steels helps to determine the best application of the purposed steel in the automotive industry.

1. Introduction

The selection of steel for critical safety applications follows two general guidelines. These guidelines are based on the principle that, in a sudden deceleration resulting from an accident, energy must be dissipated in a controlled manner so that the deceleration of the occupants does not exceed certain thresholds, maximizing survival and minimizing the chances of injury [1]. The amount of energy absorbed by the material is strongly influenced by the stress behavior of the material, which is controlled by the microstructure of the material [2]. The acronym TRIP for transformation-induced plasticity was originally invented by Zackay et al. [3] to classify all alloys with a unique combination of high strength and improved ductility imposed by the deformation-induced transformation of austenite into martensite. The phenomenological characteristics that distinguish martensitic transformations have been reviewed by Cahn et. al. [4]. Martensitic transformation is characterized by high transformation velocities in which atoms move at distances smaller than the atomic distance.

The microstructure of TRIP assisted steels is obtained by performing a heat treatment in two stages after cold rolling. The first stage of the heat treatment is performed at a slightly higher temperature in the bidirectional region $\alpha + \gamma$, leading to a microstructure of approximately 50% austenite and 50% ferrite. A cooling with isothermal holding time after partial austenitization is used to have a controlled structure formed initially in ferrite and austenite and finally the austenite will be transformed isothermally in upper bainite. During bainite formation, a little quantity of carbon respectively alloying elements retrodiffuse into the not transformed yet austenitic islands, [5] enriching them and modifying the critical points of transformation austenite-martensite (permutes to lower values the temperatures of MS and MF). According to the specific phenomena of bainitic transformation, the respective areas remain in the form of retained austenite. Usually undesirable, this will represent micro-regions in which in case of shock, they will turn into martensite, increasing the mechanical resistance to impact.

Most published damage data refer to the reliability of a steel with its energy absorption during a dynamic test; Typically, the tests performed refer to traction [6-8] or column-compression tests [9].

The elaboration process of the studied TRIP steels was detailed in a previous study which included both obtaining and rolling processes. The purpose was to change the dimensions in order to perform impact tests on them [10].

Upgrading TRIP steels using new compositions is a necessity in the industry especially due to the fact that cars have increasing speed capabilities that can involve in the event of an accident and more damaging impact forces, and the protection and safety of the passengers must be a factor of interest and research.

2. Experimental Procedure and Discussions

The heat treatments used involved rapid heating to a bidirectional temperature $\alpha + \gamma$, leading to a microstructure of about 50% austenite and 50% ferrite with maintenance at this temperature, then rapid cooling in the bainite formation range, leading to a diffusion of carbon into austenitic islands to form residual austenite, followed by cooling in air. In order to obtain this thermal path, a bath of Barium Chloride ($BaCl_2$) salt was used. The principle of heat treatment involves heating the alloy to a temperature close to the critical upper A_3, followed by cooling close to the bainite formation temperature. These temperatures vary depending on the composition of the alloy and involve a mathematical calculation to determine the critical temperatures A_1 and A_3 using the formulas:

$$\mathrm{A_1} = 727 - 14[\%\mathrm{Mn} + \%\mathrm{Ni}] + 22[\%\mathrm{Si} + \%\mathrm{Cr} + \%\mathrm{Al}] \quad (1)$$

$$A_3 = 855 - 180[\%C] - 14[\%Mn] - 18[\%Ni] = 45[\%Si] + 1.7[\%Cr + \%Al] + 45 \quad (2)$$

Following the determination of the values TA_1 and TA_3, the temperature of the heat treatment to be applied to each steel can be calculated using the formula:

$$T_T = \frac{A_1 + A_3}{2} \quad (3)$$

Using these formulas depending on the composition of each TRIP alloy, it was possible to determine the temperature required to obtain the optimal amount of residual austenite for each steel produced.

Table 1. Temperatures for TA_1, TA_3 and heat treatment (T_T) for processed steels.

Sample	*TA_1 (°C)*	*TA_3(°C)*	*T_T(°C)*
TRIP 1	750	813	781.5
TRIP 2	737	800	768.5
TRIP 3	659	773	716
TRIP 4	661	766	713.5

The compositional calculations led to the determination of a unique heat treatment regime for each alloy. These regimes can be seen in Figure 1.

As can be seen the same bainitic isothermal parameters were applied for all steels. A Nabertherm oven with a power of 13kW and a maximum operating temperature of 1600°C was used to heat the samples at the T_T heat treatment temperature, and the salt bath was heated in parallel in a Caloris sintering furnace at temperature of 400°C in order to facilitate the formation of lower bainite. This purpose was based on the reasoning that only lower bainite tends to retain in structure a small percentage of untransformed (retained) austenite.

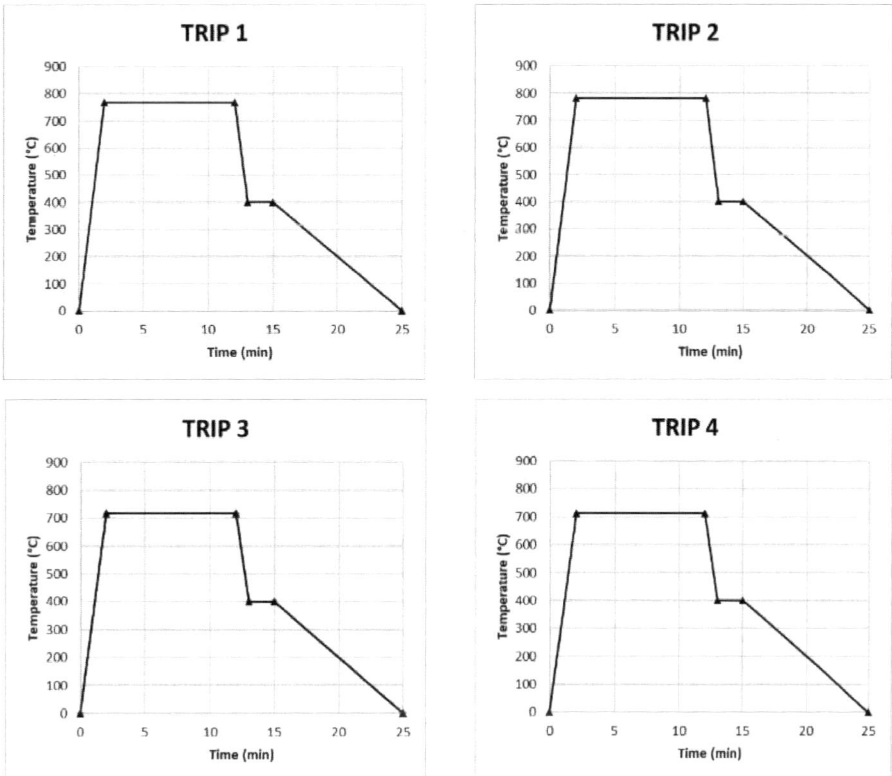

Figure 1. Heat treatment regimes applied to TRIP steels.

The steels were then analyzed using Optical Microscopy. About 4 mm was cut from each steel. These were subsequently embedded in the resin and sanded to the level of metallic luster. For the purpose of metallographic analysis, the samples were attacked with 2% NITAL reagent to highlight the grains and structure, then they were examined using an Olympus type metallographic microscope (BX 51 M), equipped with the possibility of light or dark field investigations and a magnification field. up to 1000x.

The hardness tests were performed on an Innovatest Falcon 500 microdurimeter, with an intelligent load application system, a load ranging between 1g and 31 kgf and the possibility of performing Vickers, Brinel and Knoop hardnesses. The steels were tested at a load of 5kgf and the results presented in table 8 are the average values of 3 measurements.

Table 2. Hardness values of TRIP steels after heat treatment.

Sample	Hardness (HV)	*Load* (Kgf)
TRIP 1	221	5
TRIP 2	200	5
TRIP 3	341	5
TRIP 4	316	5

Optical microscopy images can be seen in Figure 2.

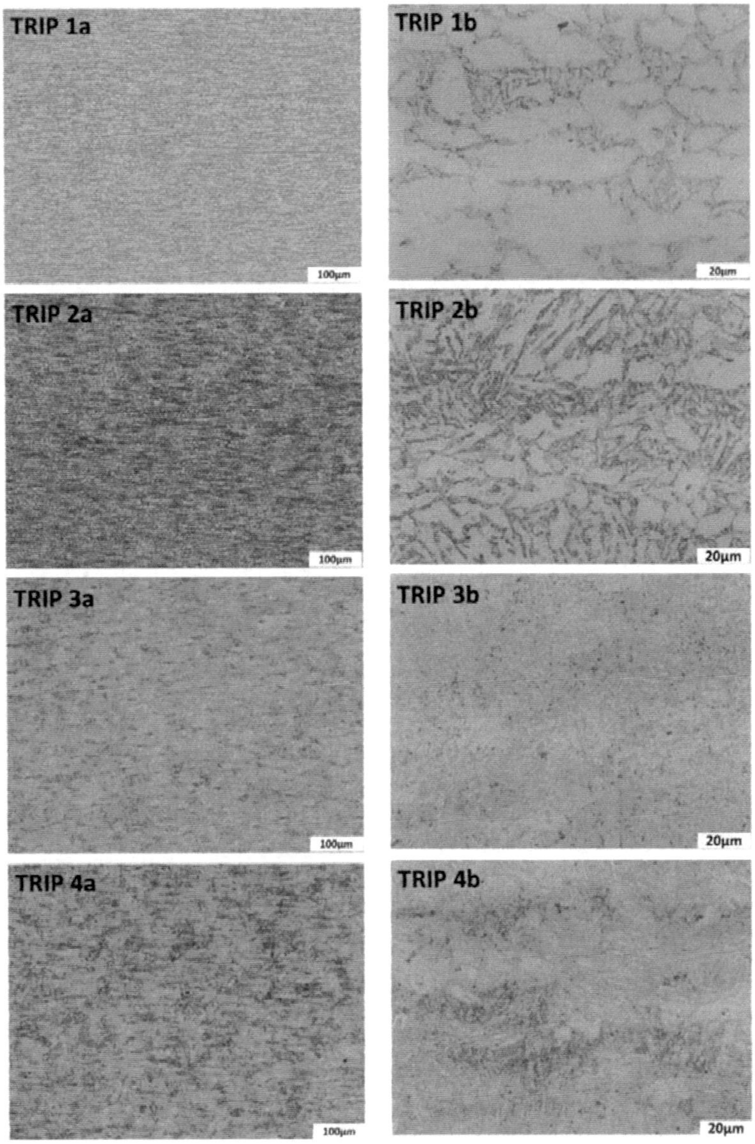

Figure 2. Optical microscopy images of TRIP steels after heat treatment: (a)-100X; (b) 1000x.

In all cases the structure consists of ferrite and lower bainite. As the alloying degree increases, the hardenability of the steel increases too, leading to a higher amount of bainite. Small bright, insular areas accompany the bainitic regions. It's easy to observe that the density and size of these areas increase with increasing of % Mn, known like an element that on the one hand increases the hardenability and on the other hand induces the tendency to retain untransformed austenite. In conclusion, the mentioned areas are clearly retained austenite. Saying this, the main objective of the purposed heat treatments is considered to be achieved. Regarding the proportion of ferrite, which intentionally exists in the structure, it is appreciated that it does not contribute to an important increase of the mechanical resistance. It should not be omitted the fact that the TRIP steel sheets, just like any other sheet used in the manufacture of the car body must first of all be processed by specific deformation techniques, and an advanced machinability is primarily ensured by the ferrite.

It must to be underlined that the main purpose of TRIP steels is to provide protection to the passengers of a car in the event of an accident. One of the most dangerous impacts is the frontal one because the energy is concentrated on a smaller surface and the direction of impact is perpendicular to the position of the car occupants. Furthermore, the research will be directed to the testing in simulated impact conditions of the 4 heat treated steels, especially having the structure adapted to withstand the impact in an optimized manner. The test results will prove which steel had the most appropriate behavior

Concretly, the impact test was designed with the idea of a frontal impact, at a speed of about 60km / h, and the samples were tested under the same conditions in order to observe the energy to which they are subjected at the time of impact. INSTRON Ceast 9340 equipment was used to perform the impact tests on the samples, which has a maximum impact energy of 405 J, a maximum force of 90kN, an impact speed of up to 4.65 m / s and the possibility to operate in an environment controlled by temperature between 50 and 100 ° C. Data is purchased electronically and displayed using VisualIMPACT V.6 and CeastVIEW software.

For testing the heat-treated samples had the surfaces cleaned of the salt film resulting from the isothermal holding period and then positioned on the impact support. The parameters used in the process are:

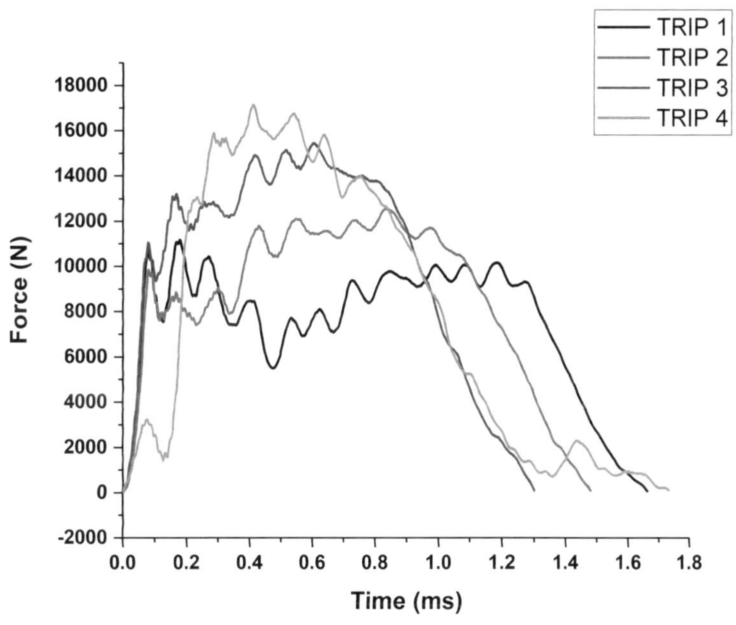

Figure 3. Comparison of the forces with which each type of steel responds at the moment of impact.

In Figure 3, the equipment registered the reaction of each steel and the absorption of energy. What is clearly visible is that in the case of TRIP 4 it presence a higher and faster energy absorption than it's counterparts which are already documented TRIP steel compositions that were reproduced for this experiment. Each of the reproduced steels was studied as follows: TRIP 1 by Krujiver et al., [11], TRIP 2 by Srivastava et al., [5] and TRIP 3 by Merwin [12].

After impact, TRIP steels have been tested for hardness changes, which should be visible when obtaining the desired transformations in the structure.

Table 3. Hardness values of TRIP steels after impact.

Sample	Hardness (HV)	*Load* (Kgf)
TRIP 1	228	5
TRIP 2	240	5
TRIP 3	348	5
TRIP 4	349	5

In all steels we can observe by comparing the data from Table 2 and Table 3 that the hardness has increased after the impact tests. The result is dual: under impact inherently the ferrite was cold hardened determining a certain increase in hardness, but the stronger effect is concentrated on the retained austenite, which transforms into martensite, This transformation is stimulated by the plastic deformation which acts like a motrice force, inducing the mentioned phenomenon. This is one of the clear indications that the transformation induced plasticity effect has been developed in the steels via the heat-treatment applied and that it varies greatly based on the composition of the steel.

3. Conclusion and perspectives

Four steels from the TRIP category have been proposed for elaboration and research, their chemical composition has been specially adapted in order to obtain a controlled structure: the carbon content was relatively low in order to obtain a majority ferritic structure which admits a corresponding plastic deformation, the presence of some alloying elements able to inhibit the carbides formation and some alloying elements having the property to keep in structure a small amount of retained austenite

A special structure according to the specific solicitations in automotive industry has imposed an unconventional heat treatment to be proposed in order to fix a required amount of ferrite and to obtain lower bainite always accompanied by a small quantity of retained austenite. Optical microscopy investigations justified the correctness of the applied heat treatment.

An impact stress test has simulated as eloquently as possible the shock in case of a road accident The behavior of the investigated steels has proved to be appropriate. These aspects have been very well justified by the choice of the structure. The presence of a small amount of retained austenite in the structure, usually unwanted, has been shown to be beneficial. During the shock it has transformed into martensite increasing the mechanical resistance and diminishing the effects of the impact. This is one of the clear indications that the transformation induced plasticity effect has been developed in the steels.

Acknowledgement

The work has been funded by the Operational Programme Human Capital of the Ministry of European Funds through the Financial Agreement 51668/09.07.2019, SMIS code 124705.

4. References

[1] Urbina P, Orta P and Ahuett-Garza H 2014 *International Journal of Automotive Technology* **15(6)** 909-17
[2] Rana R and Singh S B 2016 *Automotive Steels: Design, Metallurgy, Processing and Applications* Woodhead Publishing
[3] Zackay VF, Parker ER, Fahr D and Busch R 1967 *ASM Trans Quart* **60(2)** 252-9
[4] Cahn R and Haasen P 1983 *Elsevier Science Publishers* **3** 987-90.
[5] Srivastava A K, Jha G, Gope N and Singh S 2006 *Materials Characterization* **57(2)** 127-35
[6] Kang W, Cho S, Huh H and Chung D 1998 *SAE Technical Paper* 0148-7191
[7] Nakanishi E, Tateno H, Hishida Y and Shibata K 1998 *SAE Technical Paper* 0148-7191
[8] Bleck W and Schael I 2000 *Steel research* **71(5)** 173-8

[9] Mizui N, Fukui K, Kojima N, Yamamoto M, Kawaguchi Y, Okamoto A and Nakazawa Y 1997 *SAE transactions* 205-10

[10] Pantilimon M C, Berbecaru A, Ciuca S, Sohaciu M, Reumont G, Coman G and Predescu C 2019 *UPB Sci Bull, Series B*, **81**(4)

[11] Kruijver S, Zhao L, Sietsma J, Offerman S, Van Dijk N, Lauridsen E Margulies L, Grigull S Poulsen H F and van der Zwaag S 2003 *In situ observations on the mechanical stability of austenite in TRIP-steel. Journal de Physique IV (Proc.)* EDP sciences

[12] Merwin M 2007 *Low-carbon manganese TRIP steels. Materials science forum:* Trans Tech Publ.

IOP Conf. Series: Materials Science and Engineering **877** (2020) 012008 doi:10.1088/1757-899X/877/1/012008

Photocatalytic Degradation of Methylene Blue Dye Using TiO_2 and $Fe_3O_4/SiO_2/TiO_2$ as Photocatalysts

C I Tarcea[1], C M Pantilimon[1*], E Matei[1], A M Predescu[1], A C Berbecaru[1], M Rapa[2], A Turcanu[2] and C Predescu[1]
[1]University Politehnica of Bucharest, Faculty of Materials Science and Engineering, Bucharest, Romania.
[2]University Politehnica of Bucharest, Center for Research and Eco-Metallurgical Expertise (UPB_ECOMET)

*E-mail: cristi_pantilimon@yahoo.com

Abstract. This paper is focused on obtained two catalysts such as TiO_2 nanoparticles and $Fe_3O_4/SiO_2/TiO_2$ nanocomposite for adsorption and photocatalytic degradation of methylene blue (MB) dyes from aqueous solution. The morphology, structure and chemical proprieties of synthesized materials were investigated by X-ray diffraction (XRD) analysis, scanning electron microscopy (SEM), N_2 adsorption-desorption isotherms and Zeta potential. The photocatalytic degradation of methylene blue under UV light in the presence of different synthesized catalysts was analyzed with Spectrometer UV-Vis. The photocatalytic degradation of methylene blue was studies by focusing of photoactivity performance of $Fe_3O_4/SiO_2/TiO_2$ in comparation with TiO_2. An attempt has been made to study the effect of process parameters through amount of the catalysts and initial concentrations of methylene blue. In all cases was found that the kinetics of the MB photocatalytic degradation under UV light was fitted to the Langmuir–Hinshelwood. Even if the photocatalytic degradation study revealed that $Fe_3O_4/SiO_2/TiO_2$ and TiO_2 degraded about 90 % of methylene blue within 60 min, the magnetic nanocomposite $Fe_3O_4/SiO_2/TiO_2$ serves as better catalyst compared with TiO_2 nanoparticles. An important role in the photocatalytic degradation of MB is adsorption characteristic of TiO_2 and $Fe_3O_4/SiO_2/TiO_2$ surface. The photocatalytic performance of $Fe_3O_4/SiO_2/TiO_2$ remained greater than TiO_2 after 4 cycles of use.

1. Introduction

The accelerated growth of the different branches of consumption has developed the appearance of some pollutants that by their effect become more and more a threat to the quality of life. Nowadays there are several pollutants carcinogenic, inert and non-biodegradable such as rhodamine B, congo red, methylene blue, basic yellow 28 which can be remove using the conventional methods [1-4]. The disadvantages of the conventional methods involve low performance regarding color removal, consumption of high energy and high quantity of wastes [5]. For this reason, the researchers have developed new advanced materials with nanostructures architecture integrated as nano-catalysts in advanced oxidation processes [6]. These catalysts offer an easy and reliable way to degradation the carcinogenic and persistent pollutants. In this way, the new technologies combined with the production of materials, generate in the last decade another series of advanced materials with unique physical-chemical properties, addressed to different applications. Thus, in the field of advanced oxidation process, titanium dioxide (TiO_2) and its derivatives offer solutions in the field of environmental problems due to the specific and well defined catalytic and structural properties [7]. For such applications, the size, morphology, photocatalytic properties and specific surface disclose important characteristic of materials. Compared to the bulk material, the materials with nanostructures prove with unique chemical and physical properties. Depending on the synthesis method used these properties differ [8]. TiO_2 is well known as a catalyst having the best catalytic characteristics, chemically stable, good oxidation resistance, non-toxic and low cost of use [9]. The most used forms are the rutile (stable phase) and anatase (metastable phase). Although the anatase form has a bandwidth (Eg ~ 3.2eV) larger than the rutile (~ 3.0eV), the photocatalytic activity between the two phases highlighted the performances of the anatase phase [10]. The use of TiO_2 nanoparticles for adsorption and photocatalytic degradation of dyes in industrial waters can be considered a viable alternative based on the combination of optical, chemical and electrical properties. In most cases, in the photocatalytic applications TiO_2 is used in the form of suspension obtained by various methods such as co-precipitation method, sol-gel method, hydrothermal method etc [11]. Although it is noted by its properties that are well known, there are certain restrictive features such as band gap and difficult separation and recovery of nanoparticles [12].

To implement the advanced oxidation process in practical water treatment applications, the overall cost of the process should be kept to a minimum, so the study focused on improving the removal of material from the system and its reuse. The strategy approached in order to improve the efficiency of the recovery of the used material was highlighted by obtained a nanocomposite with magnetic core of $Fe_3O_4/SiO_2/TiO_2$. The advantage of $Fe_3O_4/SiO_2/TiO_2$ nanocomposite as photocatalyst is very efficient because it can be recovered with a magnetic from the suspension system [13, 14]. Another challenge arises when iron oxide (Fe_3O_4) interacts directly with titanium dioxide (TiO_2), a phenomenon that will occur due to the interaction of forbidden darkness and reduce photocatalytic activity [15, 16]. The silica coating between the core of Fe_3O_4 and TiO_2 to bond the two layers. Thus, the nanocomposites formed of metal oxide material incorporated in an amorphous matrix have more interesting properties [17]. These properties depend significantly on the particle size and distribution, as well as on morphology and porosity. Silica dioxide was chosen to serve as a barrier in separating the magnetic core from the TiO_2 catalyst, due to its compatibility with both oxide materials as well as to prevent the oxidation of the iron oxide core by titanium dioxide reducing its magnetism [18]. The role of silica is not only to act as a bond, but also to ensure a larger surface area, improving the photocatalytic performance and also helping to separate and recovery the catalyst. The presence of a layer of silica also increases the lifetime of generating orifices which in turn increase the photoreactivity of the magnetic core [19].

In the present paper, TiO_2 nanoparticles and $Fe_3O_4/SiO_2/TiO_2$ nanocomposite were synthesized and characterized by advanced techniques such as X-ray diffractions (XRD), scanning electron microscope (SEM), zeta potential and N_2 isotherm. Furthermore, a study of the adsorption and photocatalytic degradation of a cationic dye, methylene blue (MB), under dark adsorption and under irradiation with UV light was performed to support the validity of the Langmuir–Hinshelwood model.

2. Experimental section

2.1. Synthesis of TiO_2 nanoparticles

TiO_2 nanoparticles was obtained using a sol-gel method, by the alkoxide route, as was described in our previous work [20]. All reagents were analytical grade.

The chemical reaction is carried out for 30 minutes under magnetic stirring using titanium isopropoxide ($C_{12}H_{28}O_4Ti$) in a mixture of acetic acid and ethyl alcohol. To obtain amorphous titanium dioxide, the mixture was treated with aqueous ammonia (NaOH) until the pH reached 10. In the end, the suspension was dried and heated at 600°C. A white power of TiO_2 was obtained.

This material was subjected of XRD, SEM, Zeta potential and adsorption-desorption isotherm investigation in order to prove the crystalline phase, size, stability of nanoparticles and the specific surface.

2.2. Synthesis of $Fe_3O_4/SiO_2/TiO_2$ as nanocomposite

$Fe_3O_4/SiO_2/TiO_2$ nanocomposite was obtained in three steps as was presented in our previous work [19]. In the first stage, Fe_3O_4 nanoparticles were obtained by the method of co-precipitation by mixing of $FeCl_2$ (1M) and $FeCl_3$ (2M) in alkaline solution of NaOH (0.7 M). Then, the magnetic nanoparticles encapsulation was performed using a mixture of aqueous ammonia solution and tetraetylortosilicate (TEOS). Fe_3O_4/SiO_2 nanocomposite was put into contact with diethylenetriamine (EDTA), isopropanol and titanium isopropoxide ($C_{12}H_{28}O_4Ti$) and treated in an autoclave at 200°C. In the end, the suspension was dried was dried and heated at 600℃. A gray nanocomposite of TiO_2 was obtained. This material was subjected of XRD, SEM, Zeta potential and adsorption-desorption isotherm investigation in order to prove the crystalline phase, size, stability of nanoparticles and the specific surface.

2.3. Photocatalytic degradation

Photocatalytic degradation and adsorption experiments of MB using TiO2 nanoparticles and $Fe_3O_4/SiO_2/TiO_2$ nanocomposite were performed at the laboratory scale, at room temperature, 50 mL reaction volume and unadjusted pH conditions. In the context of ensuring a good reaction and to improve the dissolved oxygen concentration in the solution, the suspension was subjected to a compressed air filter.

Prior to the irradiation, the suspensions were shaken in the dark for 30 minutes to achieve the adsorption equilibrium of MB on the catalysts surface. In the case of the photocatalytic degradation experiments, a UV lamp equipped with 4 tubes of 15W each Philips (UV-C, λ_{max} = 365 nm) was used as the light source. The samples were taken at set time intervals, filtered through a 0.45 μm pore membrane and subsequently analyzed by UV-Vis spectrometer. The absorbances were measured at the wavelength specific to MB respectively 665 nm. The percentage of MB removal was calculated using the equation:

$$\eta = \frac{C_0 - C}{C_0} \cdot 100\ [\%] \qquad (1)$$

where: C_0 is the initial concentration [mg/L] and C is the concentration at time t, [mg/L].
The amount of MB adsorbed at equilibrium was calculated using the following equation

$$q_e = \frac{(C_0 - C)V}{w} \qquad (2)$$

where: C_0 is the initial concentration [mg/L], C is the concentration at time t [mg/L], V is the volume of the solution [mL] and w represents the amount of catalyst used [g].

3. Results

3.1. Characterization of TiO_2 nanoparticles and $Fe_3O_4/SiO_2/TiO_2$ nanocomposite

The crystalline phases of the obtained TiO_2 based nanoparticles and nanocomposites as well as the size of the crystallites were analyzed by a Panalytical X'Pert PRO MPD X-ray diffractometer with Cu-Kα radiation (wavelength of 1.54 Å). The average size of the crystallites was calculated using the Debye-Scherrer (3) equation:

$$D = \frac{K\lambda_{Cu-K\alpha}}{Cos\theta FWHM} \qquad (3)$$

where: D is the average crystallite size, K is a coefficient (0.89), $\lambda_{Cu-K\alpha}$ is the wavelength of the radiation from the diffraction tube, FWHM is the full width at half maximum of diffraction and θ is the diffraction Bragg angle.

To establish a comparison between TiO_2 nanoparticles and $Fe_3O_4/SiO_2/TiO_2$ nanocomposite a diffraction analysis was done as it is presented in Figure 1. The characteristic peaks of TiO_2 nanoparticles indicated a mixture of phases, anatase (PDF no. 01-085-5943) and rutile (PDF no.03-065-0191), the predominant phase being anatase while the characteristic peaks of $Fe_3O_4/SiO_2/TiO_2$ nanocomposite indicate anatase phase. XRD pattern indicated higher intensities for the peaks attributed to $Fe_3O_4/SiO_2/TiO_2$ nanocomposite which denotes a more crystalline material compare with TiO_2 nanoparticles. The crystallite size of particles could be evaluated as 20 nm for $Fe_3O_4/SiO_2/TiO_2$ nanocomposite and 9 nm for TiO_2 nanoparticles.

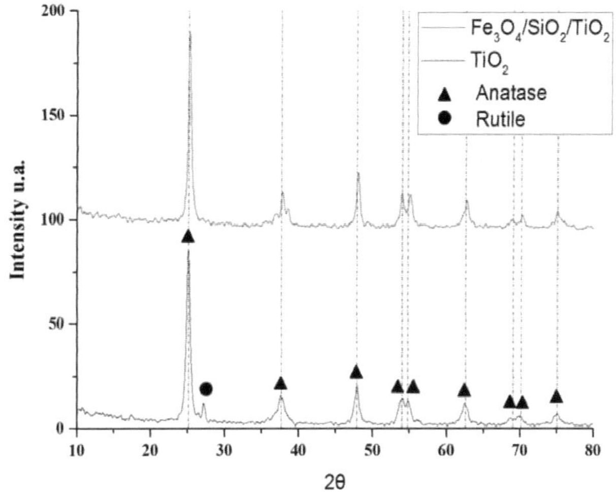

Figure 1. XRD of $Fe_3O_4/SiO_2/TiO_2$ nanocomposite and TiO_2 nanoparticles.

The morphology and structure of TiO_2 nanoparticles and $Fe_3O_4/SiO_2/TiO_2$ nanocomposite obtained were evaluated by Quanta 450 FEG scanning electron microscopy equipped with a field emission gun and an X-ray scattering spectrometer as shown in Figure 2. It can be seen that the appearance of nanomaterials based on TiO_2 have the same smooth structure. The difference is visible in terms of nanoparticle size in the case of $Fe_3O_4/SiO_2/TiO_2$ is increased due to multilayers up to a diameter of about 120 nm. In contrast, in the case of TiO_2 nanoparticles, the particle size is estimated to be 45 nm, the result is consistent with the XRD result.

Figure 2. SEM imagines of $Fe_3O_4/SiO_2/TiO_2$ nanocomposite and TiO_2 nanoparticles.

The specific surface of TiO_2 nanoparticles and $Fe_3O_4/SiO_2/TiO_2$ nanocomposites were investigated by using the adsorption-desorption isotherm of nitrogen at the liquid nitrogen temperature 77 K, as shown in Figure 3. Adsorption curve - N_2 desorption for the obtained nanomaterials suggests as a type IV isotherm, which is characteristic of mesoporous material due to high pressure values $P/P_0 > 0.8$. Under these conditions both materials obtained present the same type IV isotherm characteristic of a mesoporous material. The pore size was calculated from the desorption branch of the isotherm using the Barrett Joyner Halenda (BJH) method [21]. These results indicate that materials obtained with such physical properties, high specific surface area and high crystallinity make them good candidates for photocatalytic activity. In this context, the specific surface of TiO_2 nanoparticles is 79.219 cm^2/g with a pore volume of 0.1505 cm^3g^{-1} while the specific surface of $Fe_3O_4/SiO_2/TiO_2$ nanocomposite is 62.183 cm^2/g with a pore volume of 0.1241 cm^3g^{-1}.

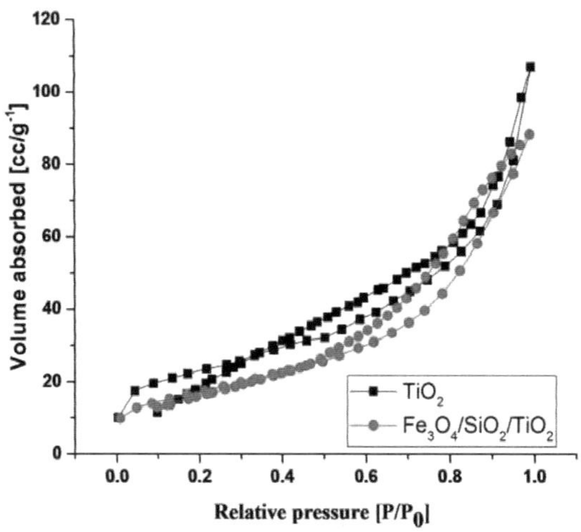

Figure 3. Adsorption-desorption isotherm of $Fe_3O_4/SiO_2/TiO_2$ nanocomposite and TiO_2 nanoparticles.

Zeta potential values of solutions were measured by a Zetasizer Nano ZSP type equipment (Malvern) equipped with a glass electrode, being calibrated with different buffer solutions. After the suspensions were allowed to equilibrate for 120 sec, the pH values of the final solutions were represented according to the initial pH values. In the case of the Zeta potential for both materials, there is a tendency for agglomeration, because the Zeta potential does not greatly exceed the range of - 30 mV to 30 mV. $Fe_3O_4/SiO_2/TiO_2$ composite shows a brief pass beyond this range, at pH values from 8.2 to 11 and 3 to 4, which corresponds to an increased stability of the material.

According to Figure 4, the value of the isoelectric point of TiO_2 nanoparticles is 6.4, being in agreement with other studies reported by K. Suttiponparnit, while the value of the isoelectric point $Fe_3O_4/SiO_2/TiO_2$ nanocomposite is 4.6. This decrease of the value of the isoelectric point in the case of the $Fe_3O_4/SiO_2/TiO_2$ nanocomposite occurs due to the increase of the particle size, result which is in accordance with the study reported by K. Suttiponparnit [22]. In this context, TiO_2 based nanomaterials are strongly positively charged under acidic conditions (below 6.4, respectively 4.6) compared to the hydroxylated surface (over 6.4, respectively 4.6) where the surface is strongly negatively charged. Therefore, under basic conditions the catalyst absorbs the positively charged molecules based on the electrostatic attraction while in acidic conditions the catalyst will have the effect of repulsion against the molecules with positive charge.

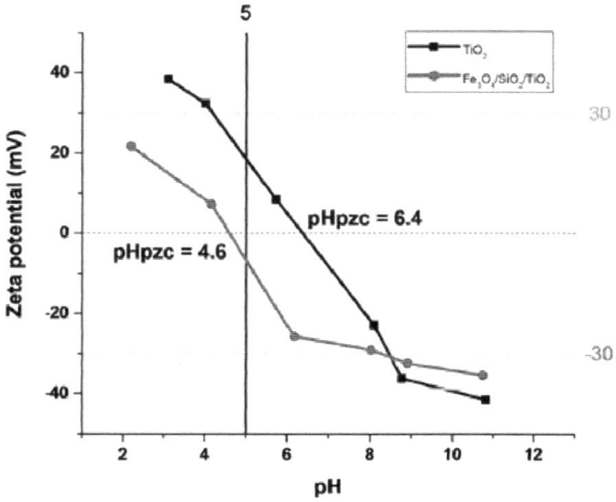

Figure 4. Zeta potential for $Fe_3O_4/SiO_2/TiO_2$ nanocomposite and TiO_2 nanoparticles.

3.2. Photolysis, adsorption and photocatalytic degradation of MB

The preliminary studies performed to determine the effect of UV radiation on the photocatalytic degradation process are presented in Figure 5. Before beginning the process of photocatalytic degradation (in the presence of catalyst and UV radiation), it is useful to study the effect of the following processes (which may affect the removal of pollutants), such as adsorption in the dark (in the presence of the catalyst without UV light) and photolysis (in the presence of UV radiation, without catalyst) under the following conditions: 50 mg of each catalyst (TiO_2 and $Fe_3O_4/SiO_2/TiO_2$), 4 mg/L methylene blue (MB) as pollutant, pH 5, 25℃ and 765 mmHg. According to Figure 5 the results indicated that the photolysis was very slow with a rate of MB efficiency by 23% and 43% respectively for the TiO_2 nanoparticles and the $Fe_3O_4/SiO_2/TiO_2$ nanocomposite after 60 min.

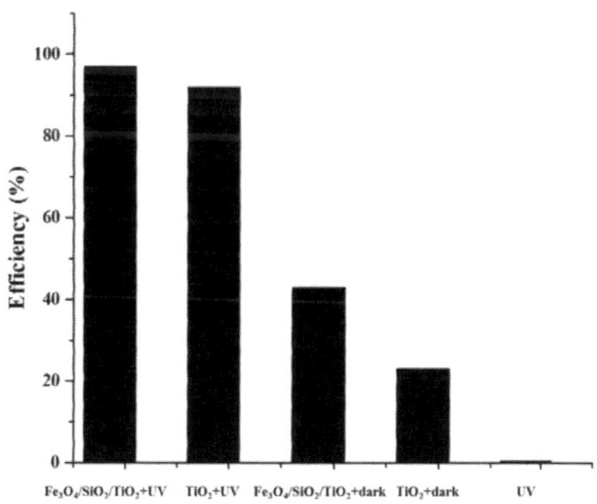

Figure 5. Comparative study of the efficiency of MB degradation using TiO_2 and $Fe_3O_4/SiO_2/TiO_2$ in photolysis, adsorption and photocatalytic degradation processes.

The efficiency of MB was significantly improved when the TiO_2 nanoparticles and $Fe_3O_4/SiO_2/TiO_2$ nanocomposite were used in the presence of UV light, showing the importance of UV light for the photocatalytic degradation process Therefore, the photocatalytic degradation of the MB in the presence of UV light was about 97% using $Fe_3O_4/SiO_2/TiO_2$ and 92% using TiO_2.

In this context, in the heterogeneous catalytic processes the reaction take place on the surface of catalyst following the preliminary adsorption of MB.

3.3. Photocatalytic degradation of methylene blue

The photocatalytic degradation of MB has been investigated using TiO_2 nanoparticles and $Fe_3O_4/SiO_2/TiO_2$ nanocomposite at constant volume and catalysts dosage, but varying the concentrations of MB in the range l-15 mg/L. The optimum dosage of each catalyst was 50 mg as found in our previous investigations [20].

The photocatalytic degradation of MB can be described by the pseudo-first-order kinetics [23].

$$r = -\frac{dC}{dt} = k_{obs}C \tag{4}$$

the integration of this equation will lead to the expected relationship:

$$\ln\left(\frac{C_0}{C}\right) = k_{obs}C \tag{5}$$

where k_{obs} is the apparent first-order rate constant, C is the concentration of MB time t, C_0 is initial concentration of MB.

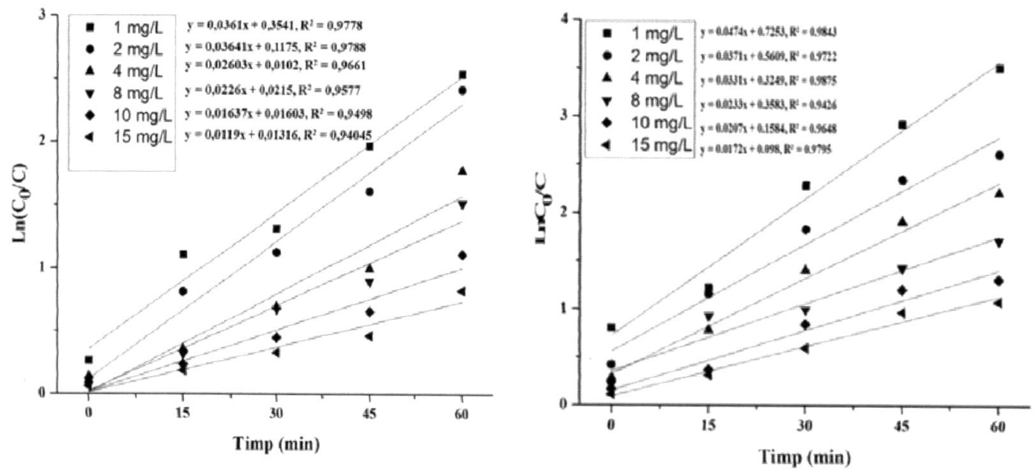

Figure 6. Linear variation of Ln (C0/C) versus time for MB degradation, using TiO2 nanoparticles and $Fe_3O_4/SiO_2/TiO_2$ nanocomposite.

According to Figure 6, the photocatalytic degradation of MB strongly depends on the initial MB concentrations. The kinetic parameters obtained directly from the regression analysis of the linear curves together with the regression coefficients are listed in Table 1.

In this context, in both cases the apparent first-order rates constant was found to decrease with increasing initial MB concentration while the values of v_0 obtained by using the equation $v_0 = k_{obs}C_0$ increased according to Figure 7.

Table 1. Pseudo-first order constant values for the different concentration of MB.

Catalyst Concentrations mg/L	TiO_2				$Fe_3O_4/SiO_2/TiO_2$			
	Experimental results		Pseudo-first-order		Experimental results		Pseudo-first-order	
	η (%)	η' (%)	k_{obs}	R^2	η (%)	η' (%)	k_{obs}	R^2
1	23	92	0.0361	0.98	41	97	0.0474	0.98
2	18	91	0.0364	0.98	35	93	0.0371	0.97
4	13	83	0.026	0.97	25	89	0.0331	0.98
8	11	78	0.0221	0.96	20	82	0.0233	0.94
10	8	67	0.0163	0.95	16	73	0.0207	0.96
15	6	56	0.012	0.94	11	66	0.0172	0.97

η (%) - adsorption efficiency
η' (%) – photocatalytic degradation efficiency

This can be explained as Ashok V. Borhade suggests that at low concentrations, the molecules are adsorbed on the active surface of the catalyst and reacts with the photogenerated holes and the hydroxyl radicals on the surface of the catalyst. In contrast, at high concentrations, the passage of UV radiation through the solution would lead to the weak interaction with the catalyst and, finally, a failure will generate electron-hole pairs [24]. This causes an insufficient hydroxyl radical to remove a high concentration of dye.

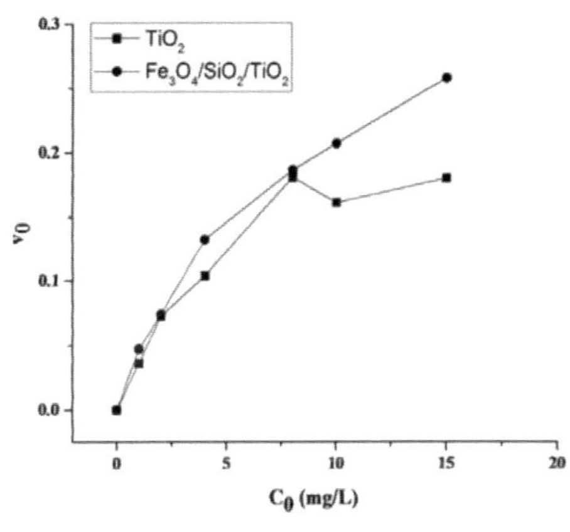

Figure 7. The variation of v_0 of MB degradation versus different initial concentrations.

The influence of the initial concentration of MB on the photocatalytic degradation rate is usually rationalized employing a Langmuir– Hinshelwood type model.

$$r = -\frac{dC}{dt} = k_r \frac{K_{LH}C_0}{1+K_{LH}C_0} \tag{6}$$

where k_r is the rate constant of surface reaction (L mg^{-1}) and K_{LH} the L-H adsorption equilibrium constant (L mg^{-1}) [25, 26]. Inserting eq. 4 and 6 with $C=C_0$ at t=0.

$$v_0 = k_{obs}C_0 = k_r \frac{K_{LH}C_0}{1+K_{LH}C_0} \tag{7}$$

rearranging the equation results.

$$\frac{1}{k_{obs}} = \frac{1}{k_r K_{LH}} + \frac{C_0}{k_r} \quad (8)$$

thus the parameters k_r and K_{LH} were obtained from the slope and the interception of the graph $1/k_{obs}$ vs C_0. (Figure 8).

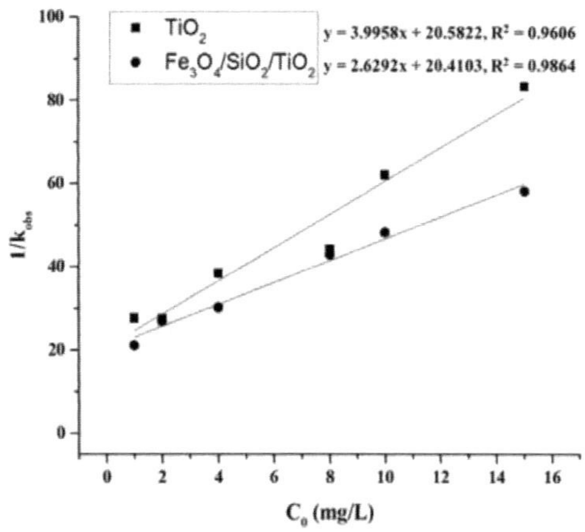

Figure 8. The variation of reciprocal of constant rate versus different initial MB concentrations.

When $k_{abs}C \ll 1$, mathematically the relationship can be written like this $k_r \frac{K_{LH}C_0}{1+K_{LH}C_0} = k_r K_{LH} C_0$. According to this condition the Langmuir–Hinshelwood type model (6) can be approximated by the pseudo-first-order kinetics as it is given in (4).

The linearization of the graphs in Figure 8 over the chosen concentrations range describes a linear behavior especially using $Fe_3O_4/SiO_2/TiO_2$ having the regression coefficient $R^2 = 0.98$. The kinetic parameters k_r and K_{LH} for TiO_2 nanoparticles and $Fe_3O_4/SiO_2/TiO_2$ nanocomposite obtained from this graph are given in Table 2.

Table 2. Langmuir–Hinshelwood fitting parameters.

Catalyst	TiO_2	$Fe_3O_4/SiO_2/TiO_2$
k_r mg L^{-1}	0.25	0.38
K_{LH} L mg^{-1}	0.19	0.12
R^2	0.96	0.98

3.4. Adsorption isotherm

From the beginning, according to preliminary studies it has been observed that the adsorption process is important for initiating the photocatalytic degradation reaction. Because the adsorption is an important process in the evolution of the photocatalytic degradation process, the adsorption isotherms have been investigated at optimum catalyst dosage 50 mg varying the initial MB concentrations. Langmuir - Hinshelwood model used in the process of photocatalytic degradation, is based on the idea that the absorption of MB can be described by the Langmuir adsorption isotherm (Figure 9).

$$q_e = q_m \frac{K_L C_e}{1+K_L C_e} \quad (9)$$

where q_m is the maximum adsorption capacity and K_L is the equilibrium constant for adsorption. Rearrangement of (9) yields.

$$\frac{C_e}{q_e} = \frac{1}{K_L q_m} + \frac{C_e}{q_m} \quad (10)$$

thus the parameters q_m and K_L were obtained from the slope and the interception of the graph C_e versus C_e/q_e Figure 9.

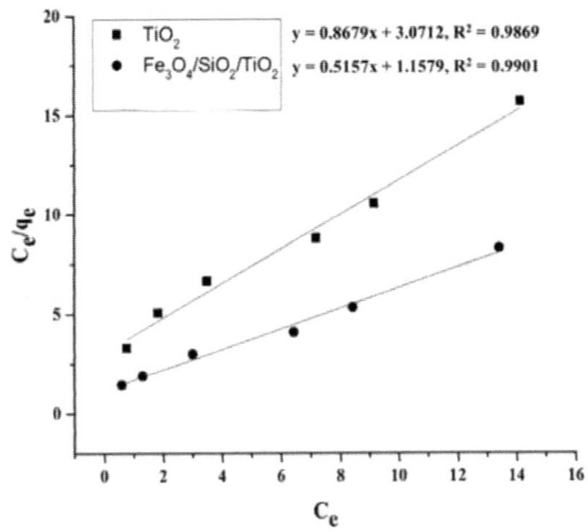

Figure 9. Adsorption isotherm of MB onto TiO2 nanoparticles and $Fe_3O_4/SiO_2/TiO_2$ nanocomposite Experimental conditions C_0 (1-15 mg/L), Catalyst dosage 50 mg, V = 50 mL.

According to Alkaim A, the constant K_L represents the affinity between adsorbate and adsorbent [27]. The values of K_L obtained can concluded that the adsorption was higher using $Fe_3O_4/SiO_2/TiO_2$ nanocomposite (0.44 L mg^{-1}) compare to TiO_2 nanoparticles (0.29 L mg^{-1}). Comparing the values K_L with the values K_{LH} (Table 1) indicate that the Langmuir isotherm is a faster process compared to the Langmuir Hinshelwood model. As a result, the process can be described by two successive reactions: adsorption followed by photocatalytic degradation in which the slow step is the rate determination step.

The difference between the maximum capacities of MB adsorption between onto TiO_2 nanoparticles (1.15 mg g^{-1}) and $Fe_3O_4/SiO_2/TiO_2$ nanocomposite (1.93 mg g^{-1}) can be explained by the nature of MB (cationic dye), the nature of the catalysts and the pH of the solutions. It is known that the surface charge of the catalyst is mainly affected by the value of its zero charge point (pH_{pzc}). This is due to the fact that TiO_2 has an amphoteric character and the photocatalytic activity depends on the pH of the solution, the zero pH_{pzc} loading point of the super phase and the dye structure as we discussed above. Under these conditions, the photocatalytic degradation of MB is higher using $Fe_3O_4/SiO_2/TiO_2$ nanocomposite compared to TiO_2 nanoparticles, a result which is due to the fact that the surface of $Fe_3O_4/SiO_2/TiO_2$ nanocomposite is negatively charged at values higher than 4.6 and as the pH of the solutions is 5, the positive structure of MB represents the key to the degradation of this dye.

According to the analysis of adsorption-desorption isotherm of nitrogen even if the total pore volume and the surface area are high for TiO_2 nanoparticles, $Fe_3O_4/SiO_2/TiO_2$ nanocomposite showed higher photocatalytic activity indicating that the photocatalytic surface is only an index that

IOP Conf. Series: Materials Science and Engineering **877** (2020) 012008 doi:10.1088/1757-899X/877/1/012008

characterizes the physical-chemical properties, not the decisive index to ensure the photocatalytic activity.

3.5. Reuse studies

To evaluate the performances of the TiO_2 based nanostructures used, their stability tests were also performed in case of reuse in several photocatalytic cycles. The reuse tests were performed under the following conditions: 1 mg/L MB, natural pH, volume of solution 50 mL, 50 mg dosage, irradiation time 60 min. After each experiment, both TiO_2 nanoparticles and $Fe_3O_4/SiO_2/TiO_2$ nanocomposite were centrifuged, washed and dried at 60°C.

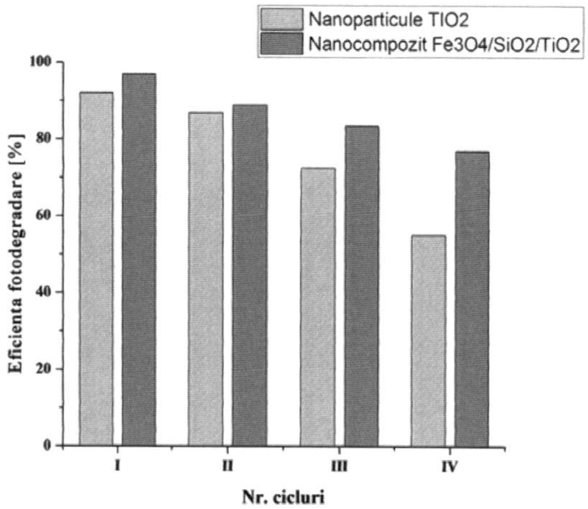

Figure 10. Adsorption isotherm of MB onto TiO_2 nanoparticles and $Fe_3O_4/SiO_2/TiO_2$ nanocomposite. Experimental conditions C_0 (1-15 mg/L), Catalyst dosage 50 mg, V = 50 mL.

From the result presented in Figure 10 after the 4 cycles of reuse, it observed a different behavior regarding the cycles of reuse of these two materials for the photocatalytic degradation of MB. In the case of the $Fe_3O_4/SiO_2/TiO_2$ nanocomposite, the efficiency ranged between 97% and 77% while for the TiO_2 nanoparticles, the efficiency varied between 92% and 54%. This decreasing tendency can be explained by the deposition of organic species on the active sites of the photocatalysts while the TiO_2 nanoparticles are lost in the filtration process.

According to these data, it seems that the $Fe_3O_4/SiO_2/TiO_2$ nanocomposite has a good stability of the photocatalytic properties, at least 4 times used repeatedly. This is also stated in the reports of M. Fanun stating that the maximum efficiency of the use of the catalyst is 4 uses [28].

4. Conclusion and perspectives

In the current context of the implementation of heterogeneous photocatalysis in practical water treatment applications, the strategy approached to improve the recovery efficiency of the material used was highlighted by obtaining a nanocomposite with $Fe_3O_4/SiO_2/TiO_2$ magnetic core. To highlight the performance of $Fe_3O_4/SiO_2/TiO_2$ nanocomposite, the results were compared with the performance of TiO_2 nanoparticles. MB adsorption was better described by $Fe_3O_4/SiO_2/TiO_2$ nanocomposite due to the regression coefficient (R^2 = 0.99). The rate of photocatalytic degradation of MB in the presence of the two TiO_2-based materials proved to be high in the case of $Fe_3O_4/SiO_2/TiO_2$ nanocomposite. The photocatalytic degradation decrease in MB concentrations during irradiation can be described by first-order kinetics. Under the experimental conditions of this study the rate of the adsorption of MB in the dark were found to be always higher than the rate of the photocatalytic reactions. In other words, the

process can be described by two successive reactions: adsorption followed by photocatalytic degradation in which the slow step is the rate determination step.
Although the photocatalytic degradation efficiencies of MB showed values of over 90% for both TiO_2-based materials in the first cycle, $Fe_3O_4/SiO_2/TiO_2$ nanocomposite was distinguished by obtaining high values of photocatalytic degradation efficiency after 4 consecutive cycles of reuse. Under these conditions, $Fe_3O_4/SiO_2/TiO_2$ nanocomposite compared to TiO_2 nanoparticles is a better catalyst which is distinguished by its excellent performance as a viable alternative in water treatment processes.

Acknowledgement
This work was supported by the Operational Programme Human Capital of the Ministry of European Funds through the Financial Agreement 51668/09.07.2019, SMIS code 124705 and a grant of the Romanian Ministry of Research and Innovation, CCCDI-UEFISCDI, project number 26PCCDI/01.03.2018, "Integrated and sustainable processes for environmental clean-up, wastewater reuse and waste valorization" (SUSTENVPRO), within PNCDI III.

The pHpzc analysis (Zetasizer Nano ZSP, Malvern UK) was possible due to European Regional Development Fund through Competitiveness Operational Program 2014–2020, Priority axis 1, Project No. P_36_611, MySMIS code 107066, Innovative Technologies for Materials Quality Assurance in Health, Energy and Environmental - Center for Innovative Manufacturing Solutions of Smart Biomaterials and Biomedical Surfaces-INOVABIOMED.

5. References

[1] Kasinathan K, Kennedy J, Elayaperumal M, Henini M, and Malik 2016 *Scientific reports* **6** 1-12
[2] Nadjia L, Abdelkader E and Ahmed B 2011 *J. Chem. Eng. Process. Technol* **2** 1-9
[3] Zuo R, Du G, Zhang W, Liu L, Liu Y, Mei L 2014 Photocatalytic degradation of methylene blue using TiO2 impregnated diatomite, *Advances in Materials Science and Engineering*.
[4]Köneçoğlu G, Toygun Ş, Kalpaklı Y and Akgün M 2015 *Advances in environmental research* **4** 25-38
[5] Guo H, Ke Y, Wang D, Lin K, Shen R, Chen J 2013 *Journal of nanoparticle research* **15** 1475
[6] Buthiyappan A, Aziz A and Daud W M A W 2016 *Reviews in Chemical Engineering* **32** 1-47
[7] Cuerda-Correa E M, Alexandre-Franco M F and Fernández-González C 2020 *An Overview Water* **12**, 102
[8] Cheng H, Wang J, Zhao Y and Han X 2014 *Rsc Advances* **4** 47031-47038
[9] Bagheri S, Muhd Julkapli N and Bee Abd Hamid S 2014 *The Scientific World Journal* **2014**
[10] Yamada Y and Kanemitsu Y 2012 *Applied Physics Letters* **101** 133907
[11] Macwan D, Dave P N and Chaturvedi S 2011 *Journal of materials science* **46** 3669-3686
[12] Jiang W F, Jiang J Y, Zhang W Z and Wang H L 2014 *Journal of Chemical Technology & Biotechnology* **89** 1910-1917
[13] Wang D, Yang J, Li X, Wang J, Zhai H, Lang,J 2017 *Physica status solidi (a)* **214** 1600665
[14] Xin T, Ma M, Zhang H, Gu J, Wang S, Liu M 2014 *Applied Surface Science* **288** 51-59
[15] Belessi V, Lambropoulou D, Konstantinou I, Zboril R, Tucek J, Jancik D 2009 *Applied Catalysis B: Environmental* **87** 181-189
[16] Beydoun D, Amal R, Low G and McEvoy S 2002 *Journal of Molecular Catalysis A: Chemical* **180** 193-200
[17] Hüsing N and Schubert U 1998 *Angewandte Chemie International Edition* **37** 22-45
[18] Ali A, Hira Zafar M Z, Haq I, Phull A R, Ali J S and Hussain A 2016 *Nanotechnology, science and applications* **9** 49

[19] Matei E, Covaliu C I, Predescu A, Coman G, Dragan C, Nitu C 2019 *Revista romana de material* **49** 475-482

[20] Tarcea C I, Predesdu C, Matei E, Predescu A M, Rapa M and Turcanu A 2019 *UPB Sci. Bull. Ser. B* **81** 169-178

[21] Tan Y H, Davis J A, Fujikawa K, Ganesh N V, Demchenko A V and Stine K J 2012 *Journal of materials chemistry* **22** 6733-6745

[22] Suttiponparnit K., Jiang J, Sahu M, Suvachittanont S, Charinpanitkul and P. Biswas 2011 *Nanoscale Res Lett* **6** 27

[23] Jawad A H, Mubarak N S A, Ishak M A M, Ismail K and Nawawi W 2016 *Journal of Taibah University for Science* **10** 352-362

[24] Borhade A V, Tope D R and Uphade B K 2012 *Journal of Chemistry* **9** 705-715

[25] Turchi C S and Ollis D F 1990 *Journal of catalysis* **122** 178-192

[26] Al-Ekabi H and Serpone N 1988 *The Journal of Physical Chemistry* **92** 5726-5731

[27] Alkaim A, Kandiel T A, Hussein F, Dillert R and Bahnemann D 2013 *Catalysis Science & Technology* **3** 3216-3222

[28] Fanun M 2014 *The role of colloidal systems in environmental protection: Elsevier*

IOP Conf. Series: Materials Science and Engineering **877** (2020) 012009 doi:10.1088/1757-899X/877/1/012009

Corrosion Protection of Metallic Surfaces in Peroxide – Surfactant Formulations

I E Chican[1], D Vărășteanu[1*], I Fierăscu[1], R C Fierăscu[1], I-L Calancea[2] and R Petre[3]
[1]National Institute for Research and Development in Chemistry and Petrochemistry ICECHIM-Bucharest, 202 Splaiul Independentei, 060021, Bucharest, Romania
[2] Research Center for Navy Constanta, 4 Stefanita Voda, 900402, Romania
[3]Scientific Research Center for CBRN Defence and Ecology – SRCCBRNDE, 225 Oltenitei Road, 041309, Bucharest, Romania

*E-mail: dvarasteanu@yahoo.com

Abstract. Formulations containing surfactants and oxidizing agents can be used to neutralize hazardous material from different surfaces contaminated in cases of chemical aggressions. The formulations act chemically through different mechanisms on hazardous materials, transforming them into inactive agents. The surfactants are used to solubilise the slightly soluble chemical warfare agents and catalyse their decontamination. In this work we evaluated the stability of surfactants in hydrogen peroxide solutions, the surface active properties of surfactants system and also performed a study of the different types of surfactants, including amino acid-based surfactants, and corrosion inhibitors in an aqueous solution of hydrogen peroxide to identify the mixture that better protect the surfaces against corrosion. Corrosion tests were performed on cast iron and aluminum surface, evaluating the pinching, staining as well as staining color intensity of the surface at 24 hours after the solution is deposited on the surface. Based on the results obtained, formulations containing a system of three surfactants and a corrosion inhibitor in an aqueous solution of hydrogen peroxide were selected.

1. Introduction

In case of CBRN (chemical, biological, radiological or nuclear) incidents, an entire perimeter that encompasses people, land, water sources, equipment and buildings is contaminated with hazardous materials. Decontamination consists in the removal of these materials from the affected area [1]. The main priority is to carry out the decontamination actions on the people, followed by the decontamination of the entire affected area, in order to avoid the subsequent danger of contamination. Surfactants were initially used in decontamination operations that involved physical processes to remove chemical or biological warfare agents or radiological agents. During the First World War, when mustard was first used as a chemical warfare agent, water and soap were used to remove the chemical from the skin [2]. Afterwards the surfactants were included in chemical decontamination compositions. The development of decontamination systems based on surfactants [3-9] could be realized due to versatility of these compounds, in terms of structure and properties. The surfactant molecules consists in a hydrophobic portion, usually a long alkyl chain and a polar group, ionisable or non-ionisable (hydrophilic portion). In solutions the adsorption of surfactants at the interfaces creates an adsorption layer which serves as a link between phases and determines the modification of the surface properties of solutions. A more or less condensed state of this layer determines the specific action of the surfactant (washing agent, emulsifier, foaming agent, wetting agent, dispersant, etc.). In the formulations of CBRN decontamination agents, surfactants are used to improve the solubility of sparingly soluble chemical contaminants in aqueous medium, allowing better contact between the contaminant and the reactive decontamination component. The most used classes of surfactants are cationic surfactants (quaternary ammonium salts), anionic surfactants (alkyl sulphates, alkylarylsulfonates, alpha olefin sulfonates) or nonionic surfactants (polyethoxylated derivatives, amine oxides). The neutralizing reagents (detoxification) of the contaminants are compounds with active oxygen or active chlorine, which pose corrosion problems on metals.

In this work we propose a ternary surfactant system including an amino acid based surfactant (sodium lauroyl sarcosinate) that can be used in decontamination systems. It was studied the synergism of nonionic surfactants (polyethoxylated derivatives, amine oxides) to sodium lauroyl sarcosinate (NaLSar) regarding the behavior of mixed micelles and the stability of surfactants in hydrogen peroxide solutions. Corrosion behaviour of surfactants in hydrogen peroxide solutions with various corrosion inhibitors on cast iron and aluminum surface were investigated.

2. Materials and Methods

2.1. Materials

The following materials were used in the experiments:

Surfactants: Sodium lauroyl sarcosinate (NaLSar) from Merck, active matter 95%, molecular weight 293.40; lauryl dimethyl amine oxide (AO) from Enaspol, R = C_{12}-C_{14} (even numbered), active matter 30%, molecular weight 229.40; C_{12}-C_{14} alchool ethoxylated 7EO (L7) from PCC Exol, active matter 90%, HLB 12.9, molecular weight 506.

Glycols: Polyethylene glycol 2000 (Merck) density 1.21 g/cm^3, propylene glycol (Merck) density 1.04 g/cm^3.

Oxidizing agent: H_2O_2 30% solution (Scharlau), density 1.11 g/cm^3.

Corrosion inhibitors: triethanolamine (Merck) 98.5% purity; sodium silicate (Scharlau) density 1.37 g/cm^3; benzotriazole (Merck) g/cm^3.

Distilled water was used for the preparation of surfactant solutions and formulations.

2.2. Methods

2.2.1. Determination of critical micelle concentration. The critical micelle concentration (CMC) of each surfactant NaLSar, AO, L7 and for ternary surfactant system (NaLSar:L7:AO with mass ratio 1:2:3) was determined from surface tension measurements carried out on a KSV automatic tensiometer Sigma 700, by the Du Nouy ring technique. After preparing the physical setup, the experiment proceeds in stages. Precisely known amounts of stock fluid were added in a vessel filled with 40 mL distilled water with the help of an automatic dispenser connected to the instrument. The instrument stirred the solution and the surface tension values were measured after the required stability of the solution is accomplished. The critical micelle concentration value is determined from the graph of surface tension in function of logarithmic concentration from the intersection of slope and baseline.

2.2.2. Determination of surfactant stability in hydrogen peroxide. The stability of surfactants in hydrogen peroxide was evaluated by determining the active oxygen content of solutions every two weeks. Active oxygen was determined by permanganometry, using potassium permanganate to titrate an acidified test solution. Blank solutions of hydrogen peroxide were used in order to evaluate the effect of surfactants on the stability of surfactant formulations.

2.2.3. Corrosion testing. The corrosion behavior of metallic surfaces when exposed to surfactant formulations containing oxidizing agent and corrosion inhibitors on the metallic was examined. The metallic plates used in experiments consisted in grey cast iron with tensile strength 200 MPa and aluminum plate with tensile strength 90 MPa. In case of cast iron plate, the corrosion experiment is done by placing 2g of steel chips (OL50 steel chips) on the clean surface of the cast iron plate. 2 mL solution is dripped onto the chips, in order to cover the entire amount of chips. The cast iron plate was placed into desiccator for 24 hours, than the chips are removed and the cast iron plate examined. In the case of aluminum plate, the corrosion experiment is performed by dripping onto the plate 2 mL solution, and after 24 hours in which the plate was held in a desiccator, the surface of aluminum plate was examined.

3. Results and discussions

3.1. Determination of critical micelle concentration

Surfactants solutions act in accordance with the laws of ideal dilute solutions up to the reach of a particular concentration named critical micelle concentration, above which the surfactant molecules are associated in spherical aggregates called micelles, having the hydrophilic regions in contact with surrounding water molecules and the lipophilic tail sequestered in the micelle centre, to have the lowest contact surface towards the aqueous phase. At the critical micelle concentration the surface tension of the solution is minimal, and although the surfactant concentration increases the surface tension no longer decreases, remaining constant. Critical micelle concentration is of great importance for the use of surfactants, since it specifies the optimal concentration for meaningful use.

For the examined surfactants the critical micelle concentration was determined. The surface tension variation with concentration for individual and ternary surfactant system is shown in Figure 1.

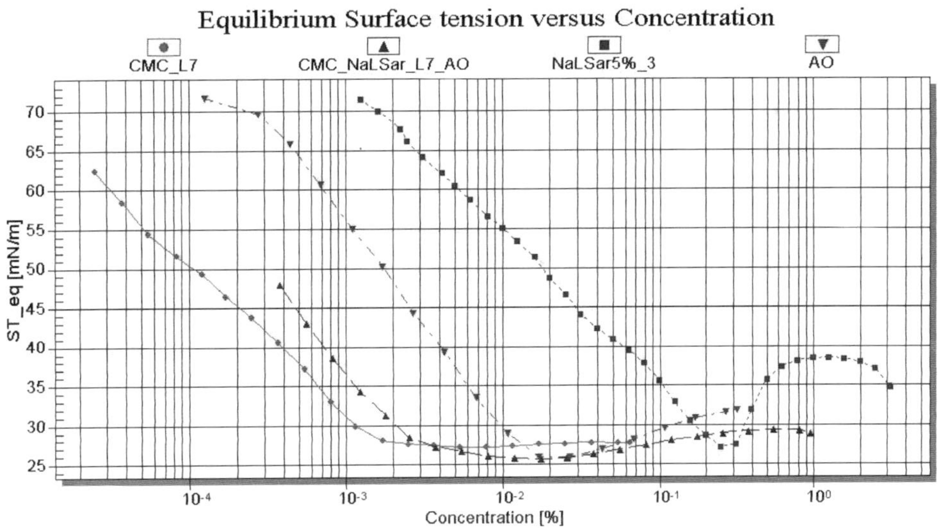

Figure 1. Plot of surface tension versus concentration for single surfactants and ternary surfactant system.

The works are a continuation of the studies performed for binary systems of surfactants with sarcosinated sodium lauroyl or sodium lauroyl glycinate content [10]. Besides the critical micelle concentration, the data supplied by the software of automatic tensiometer are used in the calculation of the maximum surface excess concentration at the air–liquid interface (Γ_{max}) and the minimum area per surfactant molecule at the air–liquid interface (A_{min}). Γ_{max} is evaluated with Gibbs adsorbtion equation:

$$\Gamma_{max} = -\frac{1}{2.303nRT}\left(\frac{d\gamma}{d\log c}\right) \quad (1)$$

Where: R is gas constant, dγ/dlogc is the surface activity, T is temperature in K and n is a factor dependent on the type of surfactant (n=2 for ionic surfactants).

For the maximum surface excess concentration, the minimum area per molecule A_{min} is calculated according to relation:

$$A_{min} = \frac{1}{N\Gamma_{max}} \quad (2)$$

N representing Avogadro's number.

Surfactant efficiency can be calculated according to equation 3, as a reduction from the surface tension value γ_0 of distilled water of 71 mN/m and surface tension of aqueous surfactant solution γ at critical micelle concentration.

$$\pi = \gamma_0 - \gamma \quad (3)$$

In Table 1 are summarized the surface parameters of surfactant aqueous solutions.

Table 1. Surface properties of single surfactants and ternary surfactant system.

Sample	Surfactants and mixtures	CMC (%)	Surface tension at CMC (mN/m)	Efficiency	dγ/dlogc (mN/m /decade)	10^{10}x Γ_{max} (mol/cm^2)	10^2xA_{min} (nm^2)
1	NaLSar	2.405x10^{-1}	27.33	43.67	-19.87	1.59	104
2	AO	1.496x10^{-2}	25.94	45.06	-25.31	4.04	41
3	L7	1.726x10^{-3}	27.57	43.43	-18.49	2.96	56
4	NaLSar+L7 +AO	3.155x10^{-3}	25.93	45.07	-22.12	3.54	47

As shown in Figure 1 and Table 1, adding non-ionic surfactants to the anionic surfactant NaLSar improves the surface parameters of mixed surfactant system.

The CMC value of the mixture NaLSar:L7:AO with mass ratio 1:2:3 is lower than the CMC values for NaLSar and AO aqueous solutions. The surface tension at critical micelle concentration of the mixture NaLSar+L7+AO is decreasing to the value of 25.93 mN/m, which is more efficient than the surface tension at CMC of NaLSar and L7 solutions and practically equal with the surface tension at CMC of AO solution. The minimum area per molecule for mixture NaLSar+L7+AO also shows a decreased value, especially compared to minimum area per molecule for NaLSar solution.

The lower values for surface parameters obtained for the aqueous solution of mixture NaLSar+L7+AO compared with individual surfactants suggests a synergistic effect might occur by the interaction of surfactants.

The evaluation of the interactions between NaLSar and non-ionic surfactants can be determined from the deviation of experimental CMC values form the ideal CMC values. In ideal solutions, CMC obeys a linear behaviour consistent with micellar composition. However, many surfactant systems do not respect a linear relation toward surfactant composition.

For surfactant systems the ideal CMC, CMC_{ideal}, can be calculated using Clint's equation [11]:

$$\frac{1}{CMC_{ideal}} = \frac{\alpha_1}{CMC_1} + \frac{\alpha_2}{CMC_2} + \frac{\alpha_3}{CMC_3} \quad (4)$$

where CMC_1, CMC_2, CMC_3 represent CMC for single surfactants and α_1, α_2, α_3 the mass fractions of individual surfactants.

The ideal CMC value obtained with the above formula for mixture NaLSar+L7+AO, 4.4x10^{-3} %, is lower than the experimental CMC value, resulting in a negative deviation of the ideal behavior. The synergism of mixtures of ionic and nonionic surfactants is given by the higher entropy of the double layer of counterions far from the micelle interface and lower chemical potential of aggregated surfactant [12].

3.2. Evaluation of surfactant stability in hydrogen peroxide

Hydrogen peroxide is an oxidizing agent widely used due to the low price, easy handling and its decomposition products consisting in water and oxygen [13]. A destabilizing effect on the hydrogen peroxide may occur by adding organic components such as surfactants [14]. The need to use stable surfactants in hydrogen severely limits choices. The most common choice is amine oxide type surfactants [15]. In order to diversify surfactant classes used in decontamination we propose a mixed surfactant system containing an amino acid based surfactant (NaLSar), used in combination with AO and L7. Evaluation of active oxygen content in function of time for H_2O_2 – surfactant combinations are presented in Figure 2.

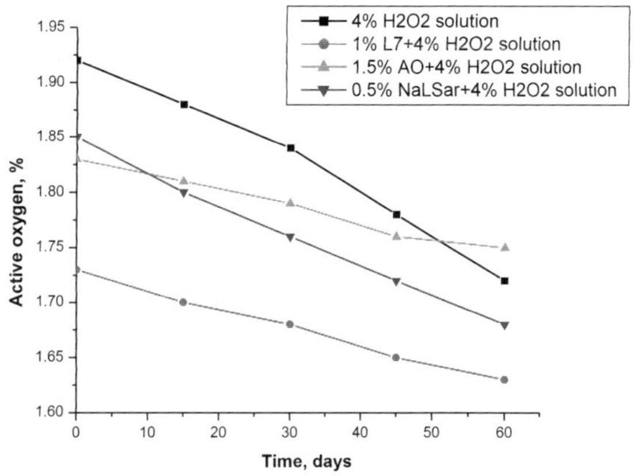

Figure 2. Active oxygen content variation in function of time for hydrogen peroxide-based formulations.

The slow decrease of active oxygen content for all the formulations during the 60 days of observation shows a good stability, so that the surfactant system can be used in formulating decontamination agents. The tests performed with GC-MS technique on the NaLSar + L7 + AO surfactant system in 4% hydrogen peroxide to neutralize the dimethyl methylphosphonate (DMMP) used as a neurotoxic warfare chemical simulant resulted in the total neutralization of the simulant used.

3.3. Corrosion testing and evaluation

Surfactant-based decontamination systems must take into account corrosive effects on the metal surfaces of the oxidizing component, hydrogen peroxide, which ensures the decontamination effect of some chemical contaminants. Hydrogen peroxide is an oxidizing agent with no negative effects on the environment used in industrial applications and as antimicrobial agent for medical applications. The oxidizing character of hydrogen peroxide causes an inherent potential for metal corrosion. Therefore, in the corrosion tests several corrosion inhibitors were used. Triethanolamine (TEA) helps prevent corrosion of zinc and steel, is an organic amine corrosion inhibitor widely used for corrosion protection of metals in metal working fluids or water systems. Organic amines inhibit corrosion on metals by a mechanism of adsorption on metal surface and the subsequent formation of a passivated protective film [16-18]. Sodium silicate is inhibitor for iron and steel in cooling water systems. Soluble sodium silicate is an industrial chemical compound used extensively in corrosion inhibition. Silicates form protective films of silica and iron oxides [19, 20]. The deposition of thin silicate coatings on aluminum and its alloys have been investigated [21]. Benzotriazole (BTA) and its derivatives are organic corrosion inhibitors for copper alloys used in aqueous environment. BTA is also used to protect against corrosion the steel and aluminum alloys [22].

The use of the above corrosion inhibitors (triethanolamine, sodium silicate and benzotriazole) on corrosion protection on metal surfaces was evaluated in this study. Mixtures of NaLSar+L7+AO with mass ratio 1:2:3, 4% hydrogen peroxide and 12% mixture of propylene glycol and polyethylene glycol were prepared, and different corrosion inhibitors were added. Sample D3 contains 5% triethanolamine, sample D4 contains 1.5% sodium silicate and sample D5 contains 0.2% benzotriazole.

The results obtained by applying the test method presented above on the gray cast iron plate and aluminum is presented in Figures 3 and 4.

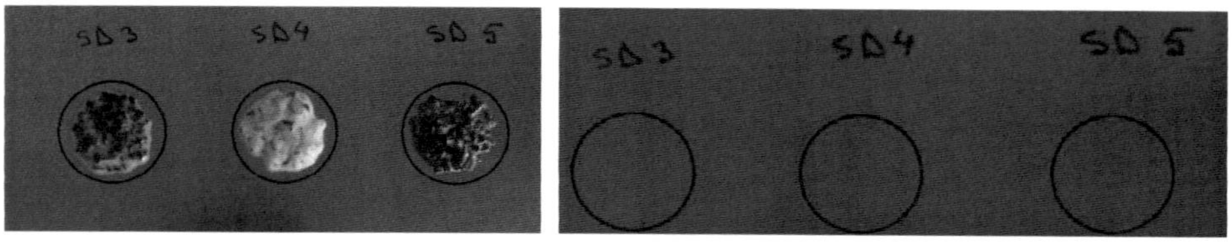

(a) (b)

Figure 3. Initial appearance (a) of the surface of cast iron plate and after 24 hours (b) for samples SD3, SD4 and SD5.

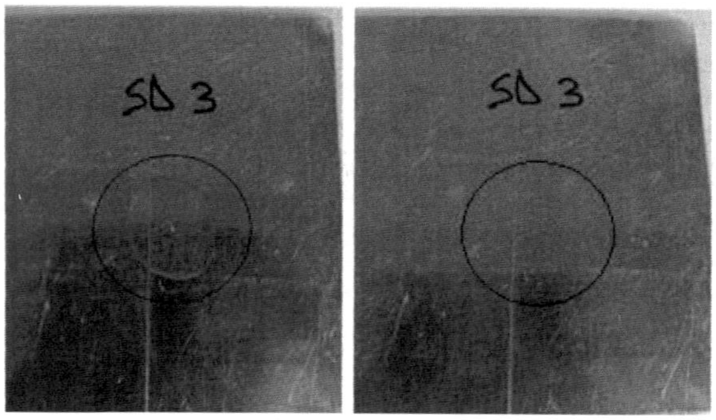

(a) (b)

Figure 4. Initial appearance (a) of the surface of aluminum plate and after 24 hours (b) for sample SD3.

The examination of the cast iron plate takes into account the proportion of stained surface and the colour intensity of the stains as well as the number of pinches. Corrosion tests performed on grey cast iron plate with formulations containing different corrosion inhibitors led to the conclusion that the most efficient corrosion inhibitor is triethanolamine (sample SD3), the metallic surface is free of stains or pinches. Sodium silicate is not as effective as triethanolamine on preventing corrosion of cast iron, it can be observed a stained surface hardly noticeable on an area of less than 10% of the test surface (sample SD4), while in the case or benzotriazole used as corrosion inhibitor pinches can be observed, as well as a stained surface on an area of less than 10% of the test surface (sample SD4). For sample SD3, the aluminum plate showed no signs of corrosion.

4. Conclusions

The use of an amino acid-based surfactant, namely sodium lauroyl sarcosinate, in a ternary surfactant system intended to be used in formulations with oxidizing agents for the neutralization of hazardous materials proved to be efficient in terms of surface properties and also in terms of stability in hydrogen peroxide solutions. The experimental critical micelle concentration value for the mixture of sodium lauroyl sarcosinate with lauryl dimethyl amine oxide and C_{12}-C_{14} alchool ethoxylated 7EO are lower than the ideal critical micelle concentration, demonstrating the synergistic action of the mixture, due to the greater interaction between the chemical species involved than the individual surfactants. Stability in hydrogen peroxide solutions are high for all the surfactants included in the study. In the presence of an appropriate corrosion inhibitor, the formulations containing surfactants and hydrogen peroxide present no signs of corrosion on grey cast iron or aluminum surfaces. The results obtained lead us to the conclusion that it can be used in such formulations amino-acid-based surfactants, due to their stability in hydrogen peroxide and the compatibility and synergism with nonionic surfactants, but also due to their low toxicity and high biodegradability.

Acknowledgments
This work was financially supported by the Romanian National Authority for Scientific Research and Innovation –UEFISCDI, Contract No. 70PCCDI/2018.

5. References

[1] Kumar V, Goel R, Chawla R, Silambarasan M and Sharma R K 2010 *Journal of Pharmacy And Bioallied Sciences* *2*(3) 220-238
[2] Fitzgerald G J 2008 *American journal of public health* **98(4)** 611-625
[3] Purdon J G, Claude L C and Burczyk A F H 1999 Broad spectrum decontamination formulation and method of use WO2000048684A1
[4] Cronce D T 1998 Chemical warfare agent decontaminant solution using quaternary ammonium complexes US Patent No. 5760089
[5] Maciejewski P, Zielonka Z and Wrzesiński J 2011 *Bezpieczeństwo i Technika Pożarnicza* 51-58
[6] Khan A W, Kotta S, Ansari S H, Ali J and Sharma R K 2013 *Defence Science Journal* **63(5)** 487-496
[7] Brown J S, Hodge R C and Schilling A S 2003 NAVAL SURFACE WARFARE CENTER DAHLGREN DIV VA
[8] Brame J A, Medina V F, Smith I and Procell L 2016 (No. ERDC/EL-SR-16-2) Army Engineer Research and Development Center Vicksburg, United States
[9] Singh B, Prasad G K, Pandey K S, Danikhel R K and Vijayaraghavan R 2010 *Defence science journal* **60(4)** 428-441
[10] Chican I E, Vărăşteanu D, Fierăscu I, Fierăscu R C and Deaconu M 2019 In *IOP Conference Series: Materials Science and Engineering* **572(1)** 012009
[11] Akisada H, Kuwahara J, Noyori K, Kuba R, Shimooka T and Yamada A 2005 *Journal of colloid and interface science* **288(1)** 238-246
[12] Bergström M and Eriksson J C 2004 *Trends in Colloid and Interface Science XVI* (pp. 16-22). Springer, Berlin, Heidelberg
[13] Lesage G, Penate I Q, Cognet P and Poux M 2012 *International Journal of Chemical Reactor Engineering* **10(1)** 1-18
[14] Bonislawski D J, Lovetro D C and Schneidewind L 2014 U.S. Patent Application No. 20140005090
[15] Johansson I and Somasundaran P (Eds.). 2007 *Handbook for Cleaning/decontamination of Surfaces*. Elsevier
[16] Choi H, Song Y K, Kim K Y and Park J M 2012 *Surface and Coatings Technology* **206(8-9)** 2354-2362
[17] Tang Z 2019 *Current Opinion in Solid State and Materials Science* **23** 100759
[18] Shang W, He C, Wen Y, Wang Y and Zhang Z 2016 *RSC advances*, **6(115)** 113967-113980
[19] Rushing J C, McNeill L S and Edwards M 2003 *Water Research* **37(5)** 1080-1090
[20] Gao H, Li Q, Chen F N, Dai Y, Luo F and Li L Q 2011 *Corrosion Science* **53(4)** 1401-1407
[21] Hamdy A S 2006 *Surface and Coatings Technology* **200(12-13)** 3786-3792
[22] Bokati K S and Dehghanian C 2018 *Journal of Environmental Chemical Engineering* **6(2)** 1613-1624

Synthesis of Carbon Quantum Dots from Food Products by Hydrothermal Carbonization Method

Z Iwińska[1], and P Pietrusiewicz[1]
[1] Department of Physics, Faculty of Production Engineering and Materials Technology, Czestochowa University of Technology, 19 Armii Krajowej Str., 42-200 Częstochowa, Poland

*E-mail: pawel.pietrusiewicz@pcz.pl

Abstract. The article describes the preparation and filtration of a solution containing carbon quantum dots. In this paper carbon quantum dots were obtained from food products, cheap and environmentally friendly hydrothermal carbonization method, without using toxic chemicals reagents. Moreover, to study the optical properties of the carbon quantum dots UV-VIS absorbtion were carried out. The maximum absorbance was determined in the range of λ ~377 to ~408nm. Furthermore, carbon quantum dots were found to emit green light.Introduction

1. Introduction

Carbon Quantum Dots (CQDs) are a new class of carbon nanomaterials, typically less than 10nm in size. Moreover, it is considered that CQDs structure consists of a graphite core with surface functional groups. Their discovery was documented in 2004, until 2006 they received the name "carbon quantum dots". Since then, CQDs have attracted significant and growing interest due to their unique properties that enable their use in many branches of science. The most important of them are photoluminescence, biocompatibility, good water solubility, photostability and low toxicity. Therefore, research is being carried out on the use of CQDs in, among others, bioimaging, drug delivery, photocatalysis and photovoltaic devices. Furthermore, they are also used as biosensors (e.g. for visual glucose monitoring) [1-3]. The first methods of synthesis used the strong influence of energy on the carbon source, which resulted in the formation of fluorescent carbon nanoparticles. An example of this approach is laser ablation [4-6]. Nowadays, CQDs synthesis methods are divided into two main groups: bottom-up and top-down. Bottom-up methods rely on the synthesis of CQDs from molecular precursors, whereas top-down methods rely on breaking down larger carbon structures [1]. During the synthesis of CQDs, there are three basic problems to consider: carbonaceous aggregation during carbonization, control of size and homogeneity, surface properties that can be modified during preparation and after processing [7]. Top-down processing methods include molecular beam epitaxy (MBE), electron lithography and X-ray lithography. Bottom-up methods include into wet-chemical and vapor-phase methods. Moreover, bottom-up methods also include hydrothermal methods consisting in the synthesis in an aqueous solution, at high temperature and under high pressure of steam[8-10]. The rapid development of nanomaterials synthesis techniques has brought many benefits in various fields of science and engineering, but also introduced several new problems. One of them is the synthesis of nanomaterials, which usually requires high energy consumption and the use of toxic chemical reagents. It is generally thought that synthesis methods increase production costs and aren't environmentally friendly. Therefore, research is ongoing on the use of environmentally friendly and renewable raw materials [11-12]. Through, hydrothermal carbonization method, carbon quantum dots can be obtained from any biomass. It is a cheap, eco friendly and non-toxic way to produce CQDs from many carbon precursors. Typically, the carbon precursor solution is sealed and reacts in a hydrothermal reactor at high temperature [9,13-14].

In this paper carbon quantum dots were synthesised with hydrothermal carbonization method from gelatin and a hypoallergenic milk for the babies. Gelatin and a hypoallergenic milk were used for the synthesis due to the high content of amino acids, which has a beneficial effect on the preparation of CQDs with high photoluminescence quantum yield. Toxic chemicals weren't used to obtain CQDs [15-20].

2. Experimental

The precursor of carbon to obtain carbon quantum dots was commercial gelatin, spirit vinegar (10% acidity) and a hypoallergenic milk. Sample 1 was prepared from 13.2g of gelatin dissolved in 40ml distilled water with the addition of 2.5ml spirit vinegar. Then the solution was poured into a stainless steel autoclave with a teflon liner and heated for 6h at 180°C. Figure 1 presents picture of autoclave. Subsequently, the autoclave was cooled to room temperature.

Figure 1. Autoclave with a teflon liner.

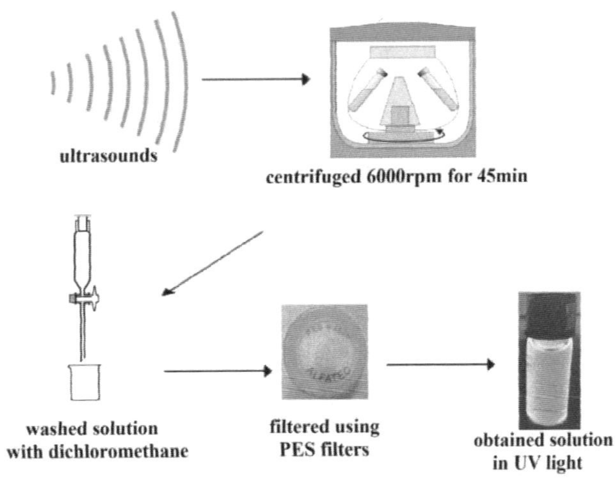

Figure 2. Scheme of production and filtration.

The resulting solution was diluted three times, five times and ten times with distilled water. Sample 2 was prepared from 1g of gelatin dissolved in 40ml distilled water with the addition of 1ml spirit vinegar. Next, the solution was placed in an autoclave and heated for 3h at 200°C. Sample 3 was prepared and filtered similary to Sample 2, without the addition of spirit vinegar. Sample 4 was prepared with 4.5g of hypoallergenic milk and 30ml of distilled water, then the solution was placed in an autoclave and heated for 3h at 180 ° C. All the resulting solutions were ultrasonicated for 10min, centrifuged for 45min at 6000 rpm, washed with dichloromethane to remove the unreacted organic moieties and then filtered using PES filters. Figure 2 illustrated scheme of production and filtration solution of CQDs.

The UV-VIS cary 100 Bio spectrophotometer from Varian and the Al-Prof SM1 Hm spectrometer were used to characterize the optical properties of carbon quantum dots.

3. Results and discussion

Using UV-VIS spectrophotometry, the maximum absorbance of carbon quantum dots obtained from food products was determined. The absorption specrum for sample 1, diluted three times is illustrated in Figure 3.

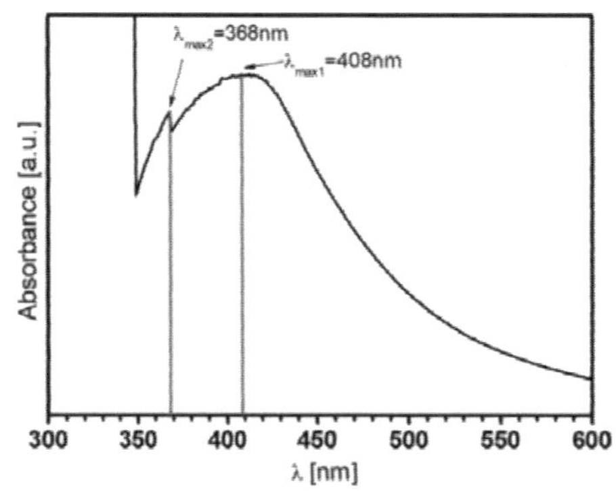

Figure 3. UV–VIS absorption spectrum.

Figure 3 shows the UV - VIS spectrum for an aqueous CQDs solution obtained from gelatin, diluted three times. As shown in Figure 3, the absorbance spectrum shows a broad peak in the range of ~ 368nm to ~ 440nm, with a maximum at 408nm. Based on the range of absorbance spectra and its maximum values, it can be assumed that there was n–π* transition of CQO bonds or other connected groups in all samples.

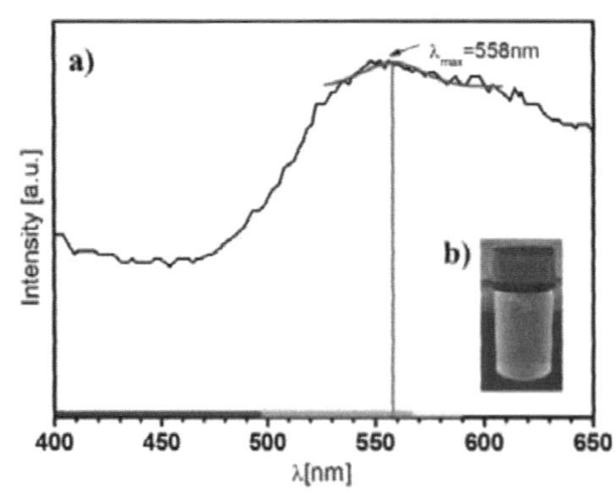

Figure 4. Sample 1: (a) photoluminescence emission spectrum, (b) sample under UV light.

The maximum emission of CQDs was determined using the Al-Prof SM1 Hm spectrometer. The results are presented in Figure 4 for sample 1, diluted three times.

Figure 4 presents the photoluminescence spectrum of sample 1 diluted three times. The maximum emission of excited carbon quantum dots was ~558nm. Determination of the maximum corresponds to the wavelength for green emission.

4. Conclusions

Commercial gelatin and hypoallergenic milk for babies was used for the research. Based on the literature data, this food products were selected due to the high content of amino acids in the composition, which in turn affects the quantum yield and fluorescence properties of the obtained quantum carbon dots. Optical properties were tested using a UV - VIS spectrophotometer. All of the tested samples show broad absorbance peaks which may indicate molecular excitation in the wavelength range corresponding to ultraviolet and violet light. Based on UV-VIS spectra, it can be assumed that the concentration of the solution affects its optical properties. The excitation of the obtained carbon quantum dots was induced by a UV lamp emitting light in the range from 360 to 370 nm. All soulutions were found to emit green light. Based on the UV-VIS spectrophotometric and spectrometric analysis and based on the literature data, it can be assumed that the obtained solutions contain carbon quantum dots. This article proves that it is possible to obtain CQDs from food products, cheap and environmentally friendly hydrothermal carbonization method. Toxic chemicals weren't used to produce CQDs.

5. References

[1] Yuan T, Meng T, He P, Shi Y, Li Y, Li X, Fan L and Yang S 2019 *Journal of Materials Chemistry C* **7** 6820
[2] Goncalves G and Tobias G 2018 Nanotechnology: Engineering nanomaterials for cancer therapy and diagnosis, *Springer*
[3] Xu X, Ray R, Gu Y, Ploehn J H, Gearheart L, Raker K and Scrivens A W 2004 *Journal of the American Chemical Society* **126** 12736-12737
[4] Jelinek R 2017 Carbon Quantum Dots Synthesis, Properties and Applications, *Springer*
[5] Hola K, Zhang Y, Wang Y, Giannelis P E, Zboril R and Rogach R A 2014 *Nanotoday* **9** 590-603
[6] Sun Y, Zhou B, Wang W, Fernando S, Pathak P, Meziani J M, Haruff A B, Wang X, Luo G P, Yang H, Kose E M, Chen B, Veca M and Xie S 2006 *Journal of the American Chemical Society* **128** 7756-7757
[7] Wang Y and Hu A 2014 *Journal of Materials Chemistry C* **2** 6921-6939
[8] Valizadeh A, Mikaeili H, Samiei M, Farkhani S, Zarghami N, Kouhi M, Akbarzadeh A and Davaran S 2012 *Nanoscale Research Letters* **7**
[9] Mattoussi H, Palui G and Na H B 2012 *Advanced Drug Delivery Reviews* **64** 138–166
[10] Birudavolu S, Nuntawong N, Balakrishnan G, Xin Y C, Huang S, Lee S C, Brueck S R J , Hains C P and Huffakera D L 2004 *Applied physics letters* **85** 2337–2339
[11] Liu M, Chen B, Li Ch and Huang Ch 2019 *Green Chemistry* **21** 449–471
[12] Sangam S, Gupta A, Shakeel A, Bhattacharya R, Sharma A K, Suhag D, Chakrabarti S, Garg S K, Chattopadhyay S, Basu B, Kumar V, Rajput S K, Dutta M K, and Mukherjee M 2018 *Green Chemistry* **20** 4245-4259
[13] Hu B, Wang K, Wu L, Yu S H, Antonietti M, Titirici M M 2010 *Advanced Materials* **22** 813-828
[14] Titirici M M, Antonietti M 2010 *Chemical Society Reviews* **39** 103-116
[15] Liang Q, Ma W, Shi Y, Li Z, Yang X 2013 *Carbon* **60** 421-428
[16] Hsu P C and Chang H T 2012 *Chemical Communications* **48** 3984–3986
[17] Tarnita D, Catana M, Dumitru N and Tarnita D N 2016 *New Trends in Medical and Service Robots, Springer Publishing House* 61-77.

[18] Kasa SN, Omar MF, Abdullah MMA, Ismail IN, Ting SS, Vac SC and Vizureanu P 2017 *Materiale plastice* **54(1)** 91-97
[19] Vizureanu P, Cimpoesu N, Radu V and Agop M 2015 *Experimental heat transfer* **28(1)** 37-57
[20] Duceac L D, Marcu C, Ichim D L, Ciomaga I M, Tarca E, Iordache A C, Ciuhodaru M I, Florescu L, Tutunaru D, Luca A C and Stafie L 2019 *Rev. Chim.* **70(7)** 2622-2626

Influence of the Imposed Parameters on XRD Patterns of Electrochemically Top-Down Obtained Nanoporous Al_2O_3 on Al1050 Alloy

V Basliu[1], L Benea[2*], and N Simionescu[2]
[1] Dunarea de Jos University of Galati, Cross-Border Faculty, Domneasca Street 47, 800201 Galati, Romania
[2] Competences Center: Interfaces-Tribocorrosion-Electrochemical Systems (CC-ITES) Faculty of Engineering, Dunarea de Jos University of Galati, 47 Domneasca street, Galati, Romania

*E-mail: lidia.benea@ugal.ro

Abstract. 1050 Aluminum alloy samples polished electrochemically in solution of 15% Na_2CO_3 and 5% Na_3PO_4 were subjected to the anodic oxidation process in an acid electrolyte, of 1 M H_2SO_4 to which 1 g / L $Al_2(SO_4)_3$ x 18 H_2O was added for the formation on their active surfaces of aluminum oxide nanoporous layers by top-down nanotechnology method. The applied parameters during the anodic oxidation processes were varied in order to obtain the most uniform aluminum oxide nanoporous layers. Thus the potential was varied between 1 V and 25 V, the duration of the oxidation processes was between 10 minutes and 48 minutes and the stirring rate of the electrolyte was between 0 and 700 rotations per minute. The formation of nanoporous layers of aluminum oxide was observed when the applied potential was between 15 V and 21 V, the duration of the anodic oxidation process was varied between 25 and 45 minutes and the stirring rate of the electrolyte was between 0 and 500 rpm. Too low duration and / or too low potential caused an electrochemical polishing phenomenon to occur and the imposition of too high potential caused dissolution of the alloy substrate. Also, the use of a dynamic electrolyte regime produced a homogenization of the electrolyte temperature during the anodic oxidation processes and the growth of the aluminum oxide nanoporous film is slowed down. XRD diffractogram analysis reveals an enhancement in peak intensity correspondent to aluminum oxide concomitantly with a reduction in peak intensity of metallic aluminum for aluminum nanoporous oxide layers in comparison to the surface of polished Al1050 electrochemically.

1. Introduction

Aluminum and its alloys always have a fine layer of natural oxide (alumina) on their surface. This film is topic to deterioration due to its fragility, which is due to its reduced thickness and its inhomogeneity [1-2].

The anodizing methods afford the thickness check of this oxide layer, obtaining a homogeneous and resistant layer. Using special dyes it is possible to color the anodized (eloxated) layer. Coloring of the deposited layer is an additional and optional operation. The only way to discolor the piece is to destroy the anodized layer. The completion of the anodization (eloxation) treatment is done by clogging the pores by hydrating the oxide layer formed. At once the pores are clogged; best corrosion behaviour is obtained [1-3].

The deposited aluminum oxide layer also has dielectric properties. The presence of anodized (eloxated) film on the surface of an aluminium part prevents the flow of electricity. The porosity of the anodized surfaces permits a really great attachment to the painting in electrostatic field or in the situation in which two surfaces are bonded [1, 4-6].

In the last decades, the researches in the areas of functional areas have concerned a significant number of researchers and their studies determine to obtain a diversified range of functional surfaces with unique properties. The manufacture of functional surfaces on aluminum and its alloys can be accomplished by a diversity of process, but the most commonly used is the electrochemical anodic oxidation method [7-16].

The anodic oxidation method implies the imposition of external parameters during the electrochemical process, and their variation allows a rigorous control of the morphological and functional structure of the oxide layers formed on the surface of the aluminum and its alloys. The functional surfaces obtained by anodic oxidation offer improved physical-chemical and mechanical properties to the aluminum substrate on which they are formed [10, 11, 14, 15].

Due to unique combinations of properties, aluminum and its alloys are some of the most versatile and economically attractive materials, being used in a wide range of fields. 1050 aluminum alloy is part of group 1 of aluminum alloys, which means a purity of at least 99.5%, also known as pure commercial aluminum. Pure commercial aluminum (Al1050, 99.5% purity) is malleable, ductile and the main alloying elements or impurities are iron and silicon.

This research work is dealing with anodic oxidation of 1050 Aluminum alloy in order to grown a nanoporous layer or film on its surface by top-down nanotechnology. The obtained nanoporous aluminum oxide layers are analyzed by XRD diffraction in order to evaluate the influence of applied potential and time duration of oxidizing process on structural aspects of obtained nanoporous aluminum oxide layers.

2. Experimental procedures

2.1. Materials and methods

Aluminum alloy type 1050 (Alro Slatina, Romania) was used as a substrate for the growth of nanoporous aluminum oxide layers by electrochemical methods.

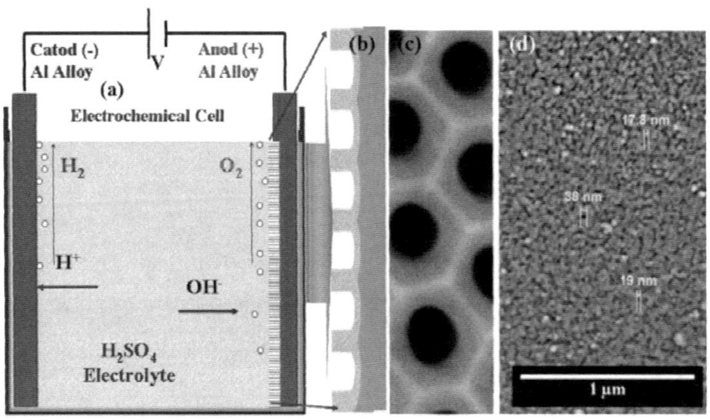

Figure 1. Schematic drown of controlled growth of nanoporous aluminium oxide film by electrochemical oxidation: (a) electrochemical cell with anode , cathode and electrolyte, (b) schematic cross section of porous oxide film, (c) schematic view of surface porous aluminum oxide film, (d) SEM micrograph of obtained nanoporous aluminum oxide film with measured nanopores (17-38 nm).

Figure 1 shows schematically the steps taken to carry out the anodic oxidation procedure and to obtain the aluminum oxide nanoporous films.

The anodic oxidation process involves the forming of an aluminum oxide coating on the surface of the 1050 aluminum alloy used as substrate. In order to obtain nanoporous layers of aluminum oxide on the surface of aluminum alloy, different electrolytes based on sulfuric acid were tested.

The most reproducible results were obtained for 1 M H_2SO_4 electrolyte to which 1 g / L Al_2 $(SO_4)_3$x18 H_2O was added. Due to the fact that the anodic oxidation process causes an elimination of aluminum ions, which contaminates the electrolyte, the optimum solution identified for the electrolyte reuse was to add an aluminum salt, so that an aged electrolyte (reused) was simulated from the first use of the electrolyte in the electrochemical process.

The anodic oxidation process took place in a conventional electrochemical cell, shown in Figure 1 (a), where the electropolished Al1050 plate with an active surface of 4 cm^2 was used as an anode and for the cathode an Al1050 plate was used with an active surface of 12 cm^2.

The two Al1050 aluminum plates were immersed in 200 mL of electrolyte, at a distance of 5 cm, and then with the help of two pliers were connected to a great potential voltage source (TDK LAMBDA GEN 300-8).

Electrochemically polished Al1050 aluminum samples were subjected to an anodic oxidation process, under potentiostatic regime, in 1 M H_2SO_4 electrolyte to which 1 g / L $Al_2(SO_4)_3$x18 H_2O was added., and the physical-chemical parameters of the electrolyte is indicating Table 1.

Table 1. Physico-chemical parameters of oxidation solution.

pH	Salinity [‰]	Conductivity [mS/cm]
0.08	68.9	97.1

Table 2. Variation of imposed parameters to anodic oxidation process.

Substrate	Stirring speed [rpm]	Potential [V]	Duration [min]
Al1050	0	15	25
Al1050	0	15	35
Al1050	0	15	45
Al1050	0	18	25
Al1050	0	18	35
Al1050	0	18	45
Al1050	0	21	25
Al1050	0	21	35
Al1050	0	21	45
Al1050	500	15	25
Al1050	500	15	35
Al1050	500	15	45
Al1050	500	18	25
Al1050	500	18	35
Al1050	500	18	45
Al1050	500	21	25
Al1050	500	21	35
Al1050	500	21	45

In order to obtain nanoporous layers of aluminum oxide with a uniform cell structure, the imposed parameters in the anodic oxidation procedure were varied as follows: the imposed potential in the anodic oxidation procedure was alternated between 1 and 25 V, the time of the anodic oxidation procedure was between 10 and 48 minutes, and the stirring speed of the electrolyte was between 0 and 700 rotations per minute.

After identifying the values of the optimal parameters for growing the aluminum oxide nanoporous layers, they were varied to observe their influence on the morphological, compositional and structural characteristics like the chemical, mechanical, physical properties of the aluminum oxide nanoporous layers.

Table 2 shows the variations of the values of the parameters implicated in the anodic oxidation procedure. Also, the anodic oxidation processes were repeated at least 20 times, for each set of parameters, to obtain reproducible results.

2.2. Preparation of Al alloy for oxidation process

The aluminum sheet was cut to the size of 2 x 30 x 35 mm, the obtained plates were mechanically glazed with emery paper with a granulation of #1500 and #2000, in order to remove both macroscopic defects and the native oxide layer formed on aluminum alloy.

The polished samples they were rinsed with distilled water and then chemically purify by submerged in 5 M NaOH for 30 seconds, washed under running water for 60 seconds, rinsed with distilled water for 30 seconds and dried in an oven under hot air (50°C) for 60 minutes, after which they were stored in a desiccators.

Finally, they were embedded in insulating resin, with an active surface of 6.5 cm^2 being delimited. Before being used in the anodic oxidation procedure, the aluminium plates were subjected to electrochemical polishing treatment in a solution of 15% Na_2CO_3 and 5% Na_3PO_4 (Brytal), to obtain a surface with a small roughness and with a low number of macroscopic defects.

After the electrochemical grinding process, the surfaces of the Al1050 samples were cleaned under distilled water and after that dried in the oven under hot air (50°C) for 50 minutes.

The samples of electrochemically polished aluminum 1050 subjected to anodic oxidation treatment, at various parameters, were isolated with insulating resin in order to achieve a very properly defined surface area of 4 cm^2.

The anodically oxidized aluminum alloy surfaces were extracted from the resin and then prepared to be structurally characterized. Structural characterization of the aluminium oxide nanoporous layers was performed using the X-ray diffraction (XRD) method to observe the phase changes of the surfaces of the aluminum oxide nanoporous film obtained by the anodic oxidation method, depending on the parameters imposed in within the electrochemical process.

Structural characterization of aluminium oxide nanoporous films achieve by the electrochemical anodic oxidation method was performed using the ray diffraction method X (XRD) using Dron-3 equipment. XRD diffractograms were recorded using a Molybdenum anode (Mo, λKa = 0.71073 Å), and then they were analyzed using MATCH 3 software connected to Free Crystallography Open Database (COD) database.

3. Results and discussions

3.1. Structural investigations of 1050 aluminium substrate electrochemically polished

Figure 2 exhibit the XRD diffraction patterns recorded for the surface of Al1050 having the surface only electrochemically polished.

From the XRD diffractogram analysis shown in Figure 2, two peaks (maximum) of high intensity diffraction were identified corresponding to the metallic aluminum (2θ = 28.55° and 2θ = 33.65°), according to the COD pdf standard. No. 96-431-3207.

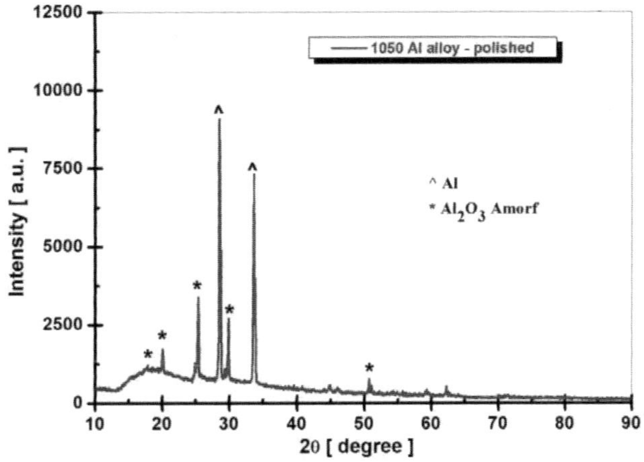

Figure 2. XRD patterns of electrochemically polished Al1050 alloy.

3.2. Influence of the potential applied in the anodic oxidation procedure on the structure of aluminium oxide layers

The XRD diffraction patterns register for the aluminium oxide nanoporous layers attained at anodic oxidation duration of 45 minutes and varying the imposed anodic potential are shown in Figure 3 (a-c) for the static regime of the electrolyte and respectively in Figure 4 (a-c) for its dynamic regime.

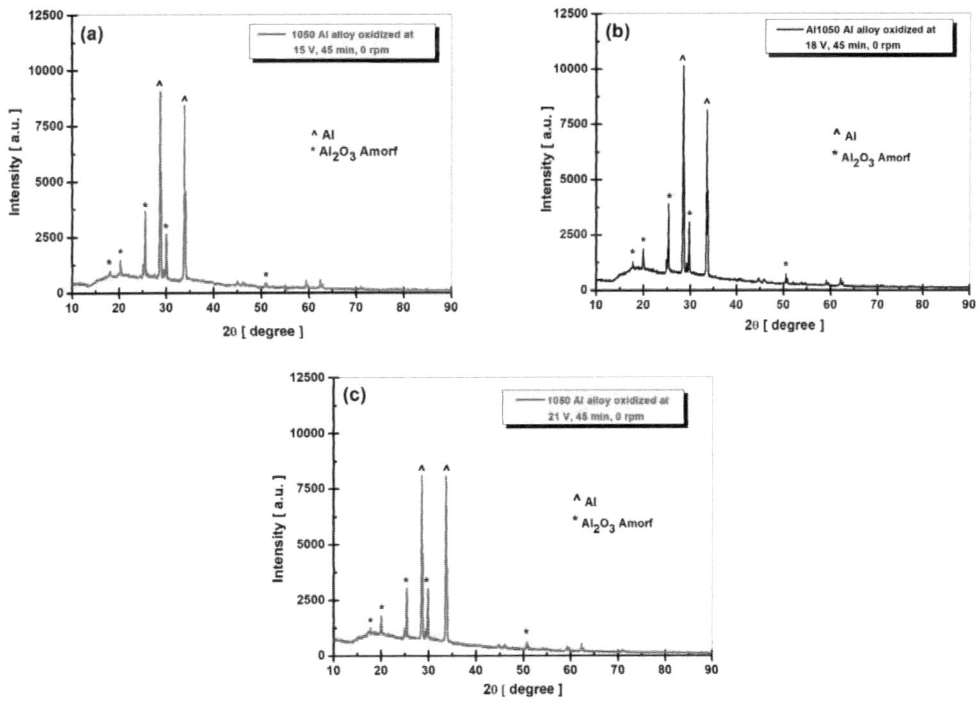

Figure 3. Influence of imposed oxidation potential on XRD patterns corresponding to the obtained aluminum oxide films during static regime of electrolyte: (a) 15 V, (b) 18 V and (c) 21 V.

From the analysis of the diffraction patterns presented in Figures 3 and 4 it is established that for the anodic oxidized surfaces, peaks to the metallic aluminium were identified in similar positions ($2\theta = 28.55^{o}$ and $2\theta = 33.65^{o}$) to those identified for the electrochemically polished surface of 1050 Al alloy, but the intensity of the diffraction peaks related to the amorphous aluminum oxide increased together with the value of the imposed oxidized potential during the anodic oxidation procedure while the intensity of the peaks corresponding to the metallic aluminum shows a downward trend. The obtaining of nanoporous layers of aluminum oxide with an amorphous structure has been observed by other researchers also [17-19].

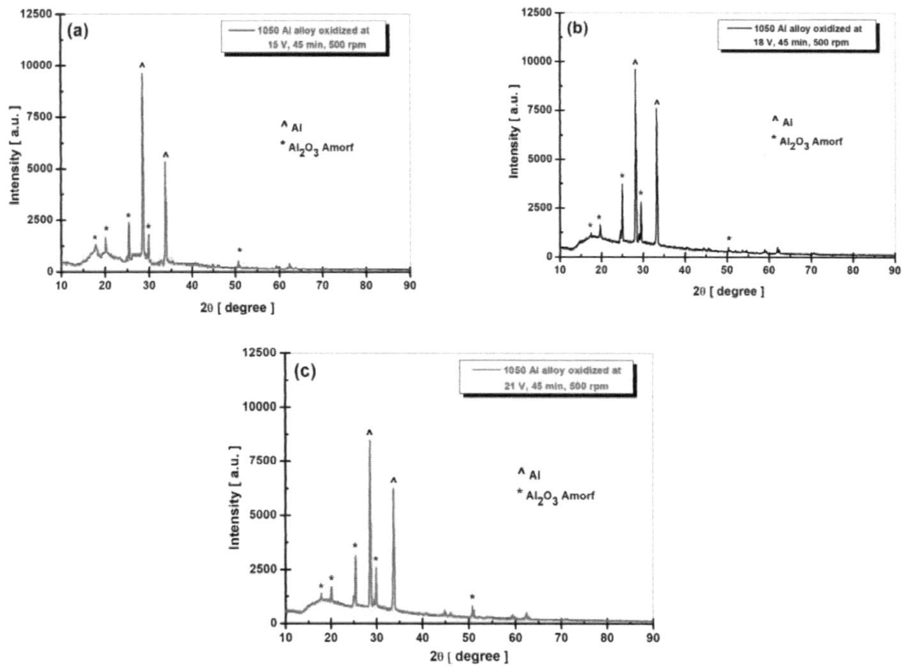

Figure 4. Influence of imposed oxidation potential on XRD patterns corresponding to the obtained aluminum oxide films during dinamic regime of electrolyte: (a) 15 V, (b) 18 V and (c) 21 V.

Also, from the XRD diffraction patterns presented in Figures 3 and 4, it is very clear that the intensities of the peaks related to the aluminum metal show the highest intensities within the XRD diffraction patterns, regardless of the duration or potential imposed during the anodic oxidation processes. The same growth trend of the peaks related to the amorphous aluminum oxide is manifested no matters if it is a static or dynamic regime of the electrolyte during oxidation process.

R K Choudhary et al. [19] Analyzed the XRD diffraction patterns recorded for the aluminum surfaces (96% purity) anodically oxidized at 20 V and 30 V in 10% $C_2H_2O_4$, and the peaks with the highest intensity correspond also to the metallic aluminum.

3.3. The influence of the duration of the anodic oxidation procedure on the structure of the aluminium oxide layers

The XRD diffraction spectra recorded for the anodic oxidized surfaces at 21 V for duration of 25 minutes, 35 minutes and 45 minutes are shown in Figure 5 (a-c) for the static regime of the electrolyte and in Figure 6 (a-c) for its dynamic regime.

From the XRD patterns corresponding to the oxidized surfaces at 21 V for different durations of oxidizing process and presented in Figures 5 and 6, the corresponding peaks to the metallic aluminum were identified in similar positions (2θ = 28.55° and 2θ = 33.65°) to those recorded for the surface of 1050 aluminum alloy electrochemically polished (Figure 2).

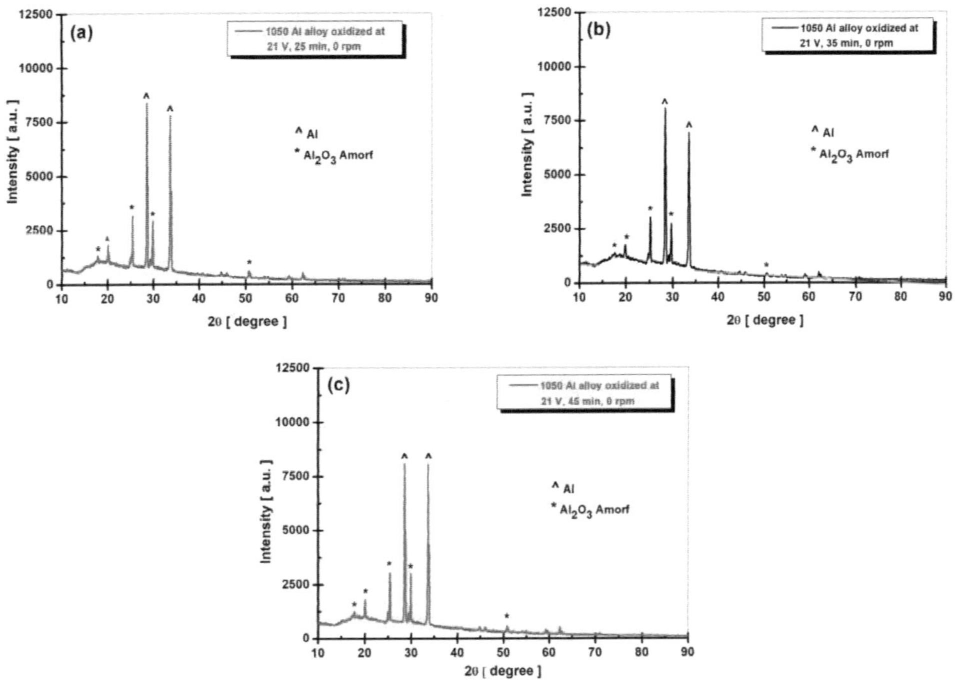

Figure 5. Effect of time period of anodic oxidation process on XRD diffraction patterns corresponding to aluminium oxide surfaces obtained in static regime of electrolyte: (a) 25 minute, (b) 35 minute and (c) 45 minute.

The intensity peaks related to metallic aluminum show a decrease in intensity as can be seen in Figures 5 and 6, and the intensity peaks related to aluminum oxide have a tendency to increase in intensity with the duration of the anodic oxidation process.

The most important factor on which the production of a nanoporous oxide film depends is the ability of the electrolyte to produce a continuous flow of Al^{3+} ions from the substrate to the solution. The passage of Al^{3+} ions from the substrate to the electrolyte is achieved by two mechanisms. The first mechanism is the direct rejection of the ions by the applied electric area and the second mechanism is the dissolving of the oxide film already formed [3]. The source of the pore nuclei as well as the mechanism of pore nuclei is greatly unknown; in this sense several models are proposed.

A model uses as a source for pore nucleation the electric field that assists local chemical dissolving both at the level of the solution / oxide interface and at the level of the oxide / metal interface.

Therefore the applied potential is a very important parameter in order to obtain a homogenous nanoporous aluminum oxide film.

This model explains how the metal Al^{3+} ions migrate into the oxide film when using an electrolyte with a pH lower than 5. The electrolyte with a pH below 5 produces a continuous flow of Al^{3+} ions, which is incorporated at the layer of oxide, thus protecting regions with an unstable oxide layer at the oxide / solution interface. These instable oxide regions cause variations in the applied electric area which manifests in increasing dissolving rate [9].

This mechanism take place at the base of metal / oxide and oxide / solution interfaces and manifests itself in a large number of hemispheric depressions / cm^2, a number that corresponds to the pore density. In these depressions, due to the hemispherical shapes, the electric area lead to be more crowed, thus increasing the dissolution rate. These hemispheric depressions form the basis of the porous structure.

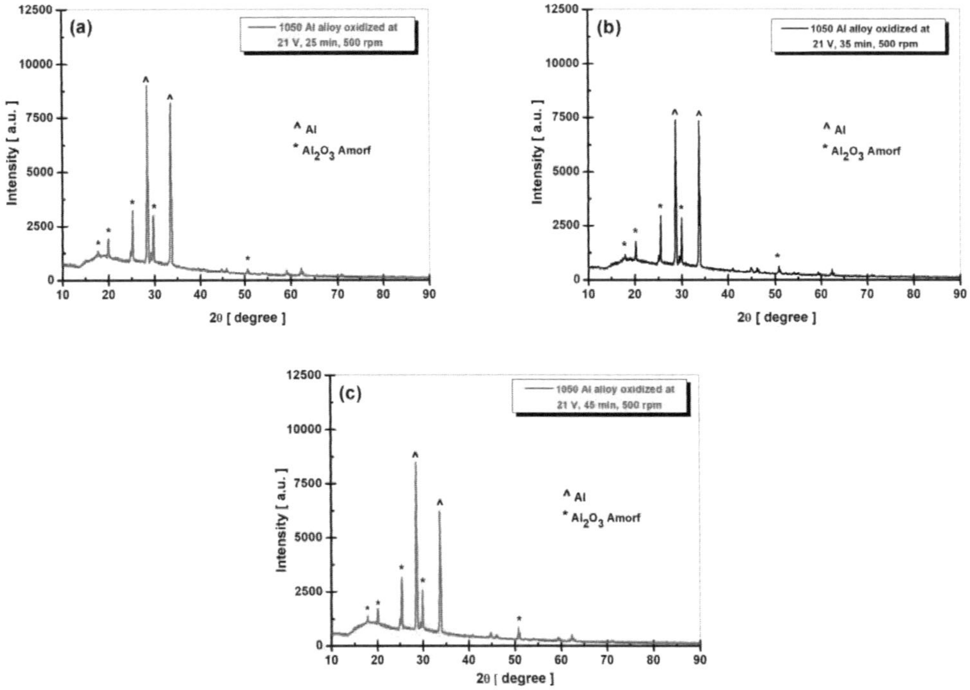

Figure 6. Effect of time period of anodic oxidation process on XRD diffraction patterns corresponding to aluminium oxide surfaces obtained in dynamic regime of the electrolyte at: (a) 25 minute, (b) 35 minute and (c) 45 minute.

The location of these valleys is too affected of the surface imperfections like impurities holes, scratches, grain boundaries, surface topography or treatments to which the surface was subjected prior to anodization.

In the first phases of the anodization procedure, the Al^{3+} ions migrate by the aluminium substrate through the metal / oxide interface forming an oxide film. Meantime, ions O_2- formed from water at the oxide / solution interface migrate to the oxide film. Through these stages, almost 70% of the Al^{3+} and O_2- ions redound to the formation of an oxide film [7] while the ions remain are dissolved in the solution. Over the oxide formation, the barrier film is constantly regenerated by increasing the oxide thickness and transforming in a hemispherical oxide layer, with a constant thickness and which will form the pore base [3].

Through the formation of the nanoporous oxide film, at the anode the aluminium dissolves according to the reaction below:

$$2Al \rightarrow 2Al^{3+} + 6e^{-} \tag{1}$$

And at the cathode the following reaction occurs in which hydrogen is produced:

$$6H^{+} + 6e^{-} \rightarrow 3H_2 \tag{2}$$

Reaction from anode it takes spot at the interface among metal and oxide (oxygen anions interact with aluminum):

$$2Al + 3O^{2-} \rightarrow Al_2O_3 + 6e^- \quad (3)$$

The reaction from the anode it takes spot at the interface among the oxide and the electrolyte (aluminum cations react with water molecules):

$$2Al^{3+} + 3H_2O \rightarrow Al_2O_3 + 6H^+ \quad (4)$$

The total reaction that takes place at the electrode (the total reaction of anodizing of aluminum):

$$2Al + 3H_2O \rightarrow Al_2O_3 + 3H_2 \quad (5)$$

This depict mechanism clarifies the dependence of the size of the pore diameters on the electric field produced by the anodizing voltage. Can be remark that the power of the electric domain inside the pore walls is very poor to be able to influence the ion course in any way [2, 5]. Further investigations must be mode to correlate the applied potential for anodic oxidation process with the dimensions of pores from obtained nanoporous oxide layers.

4. Conclusions

Taking the anodic oxidation process on the surfaces of the 1050 Al alloy, a nanoporous film of aluminium oxide could be increased which would radically change the structural aspects in comparison with polished surface samples

Considering that all the parameters implicated in the anodic oxidation process can affect the compositional and structural characteristics of the aluminium oxide nanoporous film, the present work sought to highlight the influences produced by the variation of the imposed potential and the length. of the anodic oxidation process, like properly the role of the dynamic or static regime of the solution on the characteristics of the aluminium oxide nanoporous film obtained in 1 M H_2SO_4 solution to which 1 g / L $Al_2 (SO_4)_3$x 18 H_2O were added.

XRD diffraction patterns analysis reveals an enhancement in the peak intensity corresponding to aluminium oxide concomitantly with a reduction in peak intensity of metallic aluminum for nanoporous aluminum oxide layers in comparison with the surface electrochemically polished of Al1050.

The optimum parameters as applied potential and time duration of anodizing process could be revealed by analyzing the XRD diffraction patterns of obtained nanoporous aluminum oxide layers.

5. References

[1] Vargel C 1999 *Corrosion de l'aluminium* Paris Edition Dunod, ISBN: 2-10-004191-6
[2] Benea L 2017 *Coroziune şi Protecţii Anticorozive - De la Teorie la Practică* Galati Editura Academica ISBN: 978-973-8937-99-4
[3] Poinern G E J, Ali N and Fawcett D 2011 *Materials* **4(3)** 487
[4] Diggle J W, Downie T C and Goulding C W 1969 Chemical Reviews **69(3)** 365
[5] Kalska-Szostko B 2012 *Electrochemical methods in nanomaterials preparation, Recent trend in electrochemical science and technology*, Edited by U.K. Sur InTech ISBN: 978-953-307-830-4
[6] Kelly R G, Scully J R, Shoesmith D W and Buchheit R G 2003 *Electrochemical techniques in corrosion science and engineering*, Ed. Marcel Dekker, New York, ISBN: 0-8247-9917-8
[7] Palibora E 1995 *Electrochim Acta* **40** 1051
[8] Masuda H, Yada K and Osaka A 1998 *Jpn J Appl Phys* **37** 1340
[9] Poinern G E J, Le X T, Hager M, Becker T and Fawcett D 2013 *Am J Biomed Eng* **3(6)** 119
[10] Dumitrascu V M, Benea L and Mardare L 2016 *Influence of Anodizing Voltage on the Morphology and Corrosion Resistance of 1050 Aluminum Alloy* Conf Proce Book 6 – Nano,

Bio and Green – Technologies for a Sustainable Future, Vol. 1 – Micro and Nano Technologies, Advances in Biotechnology, ISBN 978-619-7105-68-1 pp.167-174, DOI: 10.5593/SGEM2016/B61/S24.022

[11] Dumitraşcu V and Benea L 2015 *The Annals of Dunărea de Jos University of Galati. Fascicle IX Metallurgy and Materials Science*, **3** 10

[12] Dumitraşcu V and Benea L 2017 *Rev Chimie* **68(1)** 77

[13] Madakson P B, Malik I A, Laminu S K and Bashir I G 2012 Int J Mater Eng **2(4)** 38

[14] Benea L and Dumitraşcu V 2019 *RSC Advances* **9** 25056

[15] Benea L, Mardare-Danaila E and Celis J P 2014 *Tribol Int* **78** 168

[16] Veys-Renaux D, Chahboun N and Rocca E 2016 *Electrochim. Acta* **211** 1056

[17] Bocchetta P, Sunseri C, Chiavarotti G and Di Quarto F 2003 *Electrochim Acta*, **48** 3175

[18] Vazquez A L, Carrera R, Arce E, Castillo N, Castillo S and Moran-Pineda M 2009 *J Alloys Compd* **483** 418

[19] Choudhary R K, Mishra P, Kain V, Singh K, Kumar S and Chakravartty J K 2015 *Surf Coat Tech* **283** 135

IOP Conf. Series: Materials Science and Engineering **877** (2020) 012012 doi:10.1088/1757-899X/877/1/012012

Characterization of Zinc and Manganese Phosphate Layers Deposited on the Carbon Steel Surface

D P Burduhos-Nergis[1], A M Cazac[1], A Corabieru[1], F Matcovschi[1] and C Bejinariu[1*]

[1] "Gheorghe Asachi" Technical University of Iasi, Faculty of Materials Science and Engineering, 41 "D. Mangeron" Street, 700050, Iasi, Romania

*E-mail: costica.bejinariu@yahoo.com

Abstract. The phosphating process is usually used to improve metals corrosion resistance properties. In this article, the carbon steel surface was coated with a phosphate layer based on zinc or manganese. Three types of phosphating solutions with different substances and concentrations were used for this process. The deposited layers were structurally characterized by Fourier-transform infrared spectroscopy (FTIR) in order to highlight the main compounds specific to the phosphate layers. In the same time, by means of this analysis it is possible to confirm and evaluate the formation of coating layer and its degree of hydration.

1. Introduction

Phosphate conversion coating is one of the most important and widely used metal coating techniques. Phosphating is a special surface treatment of metals which, by spraying or immersion in a phosphate bath and phosphoric acid, produces the chemical transformation of the metal surface. This coating technique is commonly used for steel. The metal surface transformation consists of the creation by conversion of phosphate crystals or compounds that improve the material corrosion resistance, facilitating the subsequent painting process [1].

By immersing the parts in acidic solutions based on zinc, iron or manganese, a solid, difficult to dissolve phosphate layer is created, by a chemical reaction, with the base material [2].

Therefore, coating the steel with phosphate layers has some advantages, including high protection against corrosion, reducing friction and gliding or electrical insulation properties improving, also, subsequent coating by paints or oils can be considered [3].

The deposition of phosphates on the surface of the metal is used in the car construction industry, in cars, shipbuilding, aircraft, in the oil industry and the refrigeration industry. Due to their insulating properties, phosphate coatings can also be used in the electrical industry [4, 5].

The metal coatings by phosphates deposition are, mainly, used in the car construction industry, shipbuilding, aircraft industry, oil industry and refrigeration industry. Due to their insulating properties, phosphate coatings can additionally be used in the electrical industry [6, 7].

Depending on the environment and the process conditions, the phosphate layers are of two types: crystalline or amorphous. Generally, the crystalline layers are preferred which are deposited on a substrate of ferrous materials such as carbon steel, weak alloy steel and cast iron. The phosphate layers deposited on non-ferrous materials, as well as on stainless steels, have an amorphous structure, the deposition process and the parameters used are different in each case [8, 9].

The crystalline phosphating of metallic materials is obtained by spraying or immersing with an acidic solution that contains metallic phosphates of the bivalent metals zinc, iron and manganese. The layers deposited on the metal surface do not change the properties of the base metal, such as hardness, elasticity or magnetic properties [10-13]. This paper aims to analyze the zinc, zinc-iron and manganese phosphate layers deposited on the surface of carbon steel from a structural point of view using Fourier-transform infrared spectroscopy (FTIR).

2. Materials and methods

The base material used in this work is a carbon steel utilized in the manufacture of personal protective equipment components, work equipment parts etc. [7]. The main chemical elements composition, except iron, are carbon (0.45%), manganese (0.98%), silicon (0.22%), chromium (0.17%) and copper (0.15%).

Three phosphate layers of different compositions (1-zinc-based solution; 2-iron/zinc-based solution and 3-manganese-based solution) were obtained and structurally tested.

In order to obtain a homogeneous and uniform deposited layer on the carbon steel surface, the sample was prior prepared by two steps. These two stages with the role of cleaning the sample surface, consist of:

- the first stage, degreasing, has the role of removing from the steel surface any fat particles (oils or greases) prior used against atmospheric corrosion.
- the second stage, pickling, has the role of removing from the metal surface most of the iron oxides appeared due to the prior corrosion.

After each stage, the sample is rinsed in water, by immersion, in order to remove the chemical compounds that have appeared as a result of metal surface degreasing and pickling.

Following these two stages, the samples were phosphate conversion coated by immersion in three types of phosphating solutions, with different compositions. The technological parameters of the phosphating process, the temperature and the pH of the solution, as well as the samples maintaining time in the solution, have been strictly controlled. After phosphating, the samples were dried at room temperature.

All three types of phosphate layers deposited on the carbon steel surface were chemical structurally analyzed using a Bruker Hyperion 1000 FTIR spectrometer in a wavelength range of between 4000 cm^{-1} and 600 cm^{-1} at a spectral resolution of 4 cm^{-1} and 64 scans for each surface analyzed., the interferograms obtained being analyzed using the OPUS 65 Bruker software.

3. Results and discussion

Structural characterization by Fourier-transform infrared spectroscopy was performed in order to highlight the main compounds specific to phosphate layers [14]. Furthermore, by means of this analysis, it is possible to confirm the formation of the coating layer and its degree of hydration.

The FTIR spectrum of the sample coated with a zinc phosphate layer graphically represented by the number of the wave on the abscissa and the intensity of the infrared radiation in absorbance per ordinate is shown in Figure 1.

As can be seen in Figure 1, the vibration band between 1100 and 850 cm^{-1} has several peaks of significant intensity. Their positioning at 871, 976 and 1055 cm^{-1} can be attributed to the stretching asymmetric valence bonds of the reaction products between phosphate and hydroxide in the form of PO or POH groups of complex compounds of $HOPO_3^{3-}$, PO_4^{3-} and $H_2PO_4^-$ being specific to phosphate layers deposited by conversion [15].

The vibration band between 3100 cm^{-1} and 3600 cm^{-1} with a significant peak at 3278 cm^{-1} is attributed to the hydroxide groups. Their presence may be due to the valence vibration of the adsorbed water, but also to the hydroxide group stretching from the hopeite structure (specific to zinc phosphate) [16].

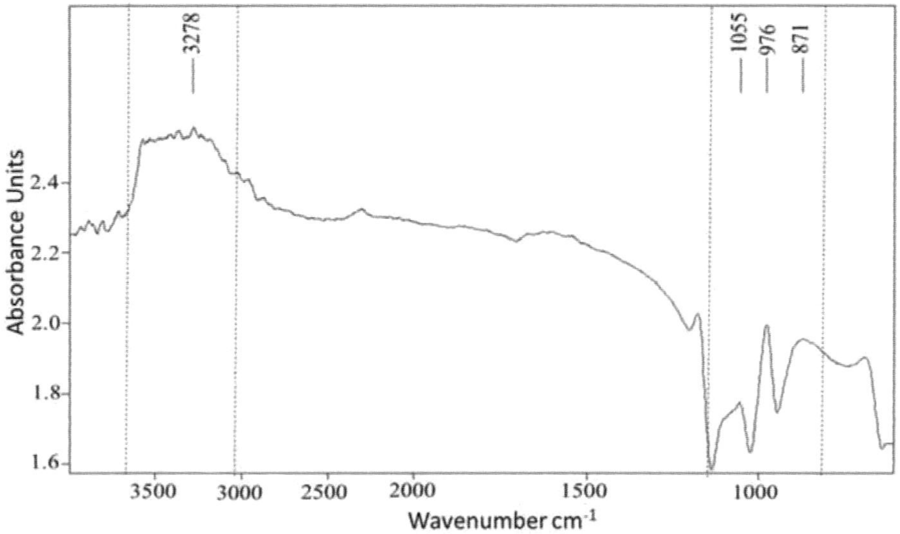

Figure 1. FTIR spectrum of zinc phosphate layer.

The FTIR spectrum of the sample coated with zinc phosphate and iron presents three vibration bands, as can be seen in Figure 2.

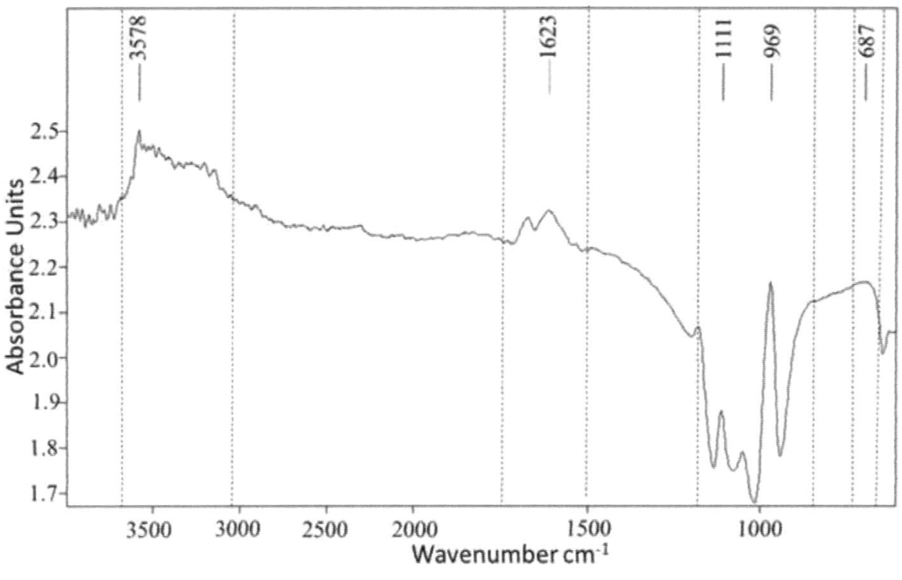

Figure 2. FTIR spectrum of zinc-iron phosphate layer.

Due to the similarity between the chemical composition of these two zinc phosphating solutions, the FTIR spectra present a large common proportion. The vibration band between 710 cm^{-1} and 680 cm^{-1} shows a peak at the wave number 687 cm^{-1} that can be attributed to the asymmetric valence vibrations of the metaphosphate groups. The appearance of the peaks in the vibration band between 1700 cm^{-1} and 1500 cm^{-1} corresponds to the adsorbed water molecules, which highlights the hydration degree of the zinc-iron phosphate layer [17]. Also, the vibration band between 3600 cm^{-1} and

3100 cm^{-1} which corresponds to the valence vibrations of the hydroxide groups is present, but also a band between 1100 cm^{-1} and 850 cm^{-1} which is attributed to the vibrations of the phosphate ions [18].

The large width of the vibration band between 1100 and 850 cm^{-1} is due to the different phosphate groups in the structure, the high number of vibrations in this area appears due to the bonds between these groups [16]. The peak positioned at 1111 cm^{-1} can be attributed to the asymmetric valence vibrations of PO_4^{3-} from the phosphophyllite structure, and the peak positioned at 969 cm^{-1} can be caused by the stretching and deformation vibrations of the P-O groups [19].

In the case of the samples coated with a layer of manganese phosphate, the FTIR spectrum shows major changes (Figure 3).

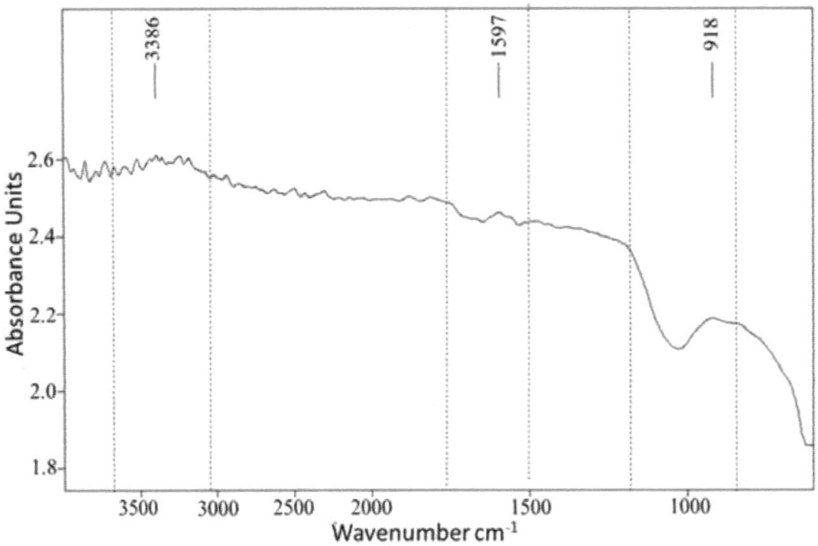

Figure 3. FTIR spectrum of manganese phosphate layer.

The peak positioned at 918 cm^{-1} in the vibration band between 1100 and 850 cm^{-1} confirms the formation of manganese phosphate by the presence of P-O and P-O-H groups in $Mn_5H_2\ (PO_4)_4.4H_2O$ [17]. Moreover, the peak positioned at 3386 cm^{-1} is attributed to the hydroxide groups of complex chemical compounds [16]. While the band between 1700 and 1500 cm^{-1} is attributed to the deformation vibrations of the H-O-H groups in the adsorbed water molecules [20-22].

4. Conclusion

The FTIR spectra of the phosphate layers deposited by conversion present multiple similar vibrations bands that correspond to the hydroxide groups, but also different positioned bands that highlight the particularities of the formed compounds. In both the zinc phosphate and manganese layers, the spectrum presents a vibration band between 1100 cm^{-1} and 850 cm^{-1}, which is assigned to the P-O and P-O-H groups. On this vibration band, sample 1 shows a high-intensity peak positioned at 976 cm^{-1}, which confirms the formation of the hopeite, but after the introduction of Fe in the phosphating solution the main compound resulted on the surface is phosphophyllite, resulting in a high porosity layer with low crystallinity.

5. References

[1] Hafiz M H, Kashan J S and Kani A S 2008 *Eng. & Technology* **26(5)** 501-511
[2] Burduhos-Nergis D P, Nejneru C, Burduhos-Nergis D D, Savin C, Sandu A V, Toma S L and Bejinariu C 2019 *Rev Chim* **70(1)** 215-219

[3] Jafari H, Danaee I, Eskandari H and Rashvandavei M 2013 *J Environ. Sci Health A Tox Hazard Subst Environ Eng.* **48(13)** 1628-1641
[4] Burduhos-Nergis D P, Nejneru C, Cimpoesu R, Cazac A M, Baciu C, Darabont D C and Bejinariu C 2019 *Qual -Access to Success* **20(1)** 77-82
[5] Darabont D C Antonov A E and Bejinariu C 2017 *E D P Sciences: Cedex A* **121** 11007
[6] Bejinariu C, Burduhos-Nergis D P, Cimpoesu N, Bernevig-Sava M A, Toma S L, Darabont D C and Baciu C 2019 *Qual-Access to Success* **20(1)** 71-76
[7] Tomachuk C R, Elsner C I and Di Sarli A R 2013 *Mat. Res.* **17(1)** 61-68
[8] Cunha C A, Lima N B, Martinelli J R, Bressiani A H A, Padial A G F and Ramanathan L V 2008 *Mat. Res.* **11(2)** 137-143
[9] Darabont D C Moraru R I Antonov A E and Bejinariu C 2017 *Qual.-Access Success* **18** 11–14
[10] Abdalla K, Rahmat A and Azizan A 2014 *Materials and Corrosion* **65** 977-981
[11] Forero A B, Nunez Milagros M G and Bott I S 2013 *Mat. Res.* **17(2)** 461-471
[12] Bejinariu C, Darabont D C, Baciu E R, Georgescu I -S, Bernevig-Sava M A, and Baciu C 2017 *Sustainability* **9** 1263
[13] Bejinariu C, Darabont D C, Baciu E R, Ionita I, Sava M A B and Baciu C 2017 *Environ. Eng. Manag. J.* **16(6)** 1395–1400
[14] Sankara Narayanan T S N 2005 *Rev.Adv.Mater.Sci.* **9** 130-177
[15] Ghoneim N A, Abdelghany A M, Abo-Naf S M, Moustafa F A and ElBadry Kh M 2013 *J.Mol.Struct.* **1035** 209-217
[16] Fernandes K S, Alvarenga E A, Brandao P R G and Lins V F C 2011 *R.Esc.Minas* **64(1)** 45-49
[17] Moustafa Y M and El-Egili K 1998 *JNCS* **240** 144-153
[18] Otsuka Y, Takeuchi M, Otsuka M, Ben-Nissan B, Grossin D and Tanaka H 2015 *Colloid Polyn Sci.* **293** 2781-2788
[19] Wang J D, Li D, Liu J K, Yang X H, He J L and Lu Y 2011 *SNL* **1** 81-85
[20] Alibakhshi E, Ghasemi E, Mahdavian M and Ramezanzadeh B 2016 *Prog. Color Colorants Coat.* **9** 233-248
[21] Ionescu D, Mătăsaru D, Radu V 2013 *UPB Scientific Bulletin Series A-Applied Mathematics and Physics* **75(4)** 265-274
[22] Toma SL, Bejinariu C, Gheorghiu DA and Baciu C 2013 *Advanced Materials Research* **814** 173-179

The Effects of Solid to Liquid Ratio on Fly Ash Based Lightweight Geopolymer

W M W Ibrahim[1,2*], R Ahmad[1,2], B T Coman[3], M M A B Abdullah[1], A Puskas[4] and V S Jaganathan[2]

[1] Center of Excellence Geopolymer and Green Technology, School of Materials Engineering, Universiti Malaysia Perlis (UniMAP), P.O. Box 77, d/a Pejabat Pos Besar, 01007, Kangar, Perlis, Malaysia.

[2] Faculty of Engineering Technology, Uniciti Alam Campus, Universiti Malaysia Perlis (UniMAP), 02100, Sungai Chuchuh, Padang Besar, Perlis, Malaysia.

[3]Gheorghe Asachi Technical University, Faculty of Machine Manufacturing and Industrial Management, Department of Physics, 67 Prof. Dimitrie Mangeron Blvd., 700050, Iasi, Romania.

[4]Technical University of Cluj-Napoca, Department of Structures, Baritiu Street no. 25, Cluj Napoca, Romania.

*E-mail: wanmastura@unimap.edu.my

Abstract. Geopolymer material was used as the raw material because it promotes the green technology. In this study, lightweight geopolymer was produced using fly ash as raw material with the addition of alkali activation which is mixture of sodium silicate and sodium hydroxide, foaming agent that gives lightweight properties and finally, underwent curing process. The molarity of sodium hydroxide (NaOH) used was fixed at 12 M while the ratio of fly ash to alkali activator (solid to liquid) used were varied in the range of 2.0, 2.5, 3.0 and 3.5, by mass. Besides that, foaming agent (Polyoxyethylene alkyether Sulfate) was added to the geopolymer sample to give the lightweight properties. The samples were cured at 80 °C for 24 hours in the oven for curing process and left at room temperature prior for testing for 14 days. The testing of sample was conducted in this study which includes density test, compression strength test, water absorption test and scanning electron microstructure (SEM) test. The results obtained for optimum solid to liquid ratio is 2.5, by mass with the optimum value of compressive strength density value. The mechanical and physical properties of lightweight geopolymer were based on the ASTM International Standard.

1. Introduction

Geopolymer is relatively a new class of construction materials. Its materials are usually made of an aluminosilicate that acts as precursors. It will be incorporated with an activating liquid which consist of sodium hydroxide and sodium silicate solution [1]. Thereupon, reducing the emission of carbon dioxide (CO_2) which can be represented as an environmentally friendly innovative for cement. Moreover, this technology is getting interests in various application fields like fire resistance materials, low cost ceramics, refractory filters and lightweight panels for thermal and acoustic isolation [2].

Fly ash is commonly used and competent waste material in geopolymerization. This is because huge amount of fly ash produced worldwide. It is approximated to be around 780 million tons annually [3]. Currently, lightweight concrete is widely well-known as it caters additional advantages to the concrete production industry. Lightweight concrete can be exemplified as a type of concrete which involves an expanding agent. This agent surge the capacity of the mixture. Hence, this gives extra properties such as workability and lower dead weight [4].

The dead load for normal concrete is higher which is approximately 2400 kg/m^3. The density of conventional lightweight concrete is higher which is 1440 kg/m^3 to 1840 kg/m^3 [5]. However, the density of lightweight concrete varies between 300 kg/m^3 to 1800 kg/m^3 depending on the materials used [6]. Lightweight concrete has high flow ability, less uses of aggregates, optimum low strength, low weight and excellent thermal-insulation properties. Moreover, it also speed up the building rates in construction and lower the haulage and handling costs [7].

This research proposes to study the strength properties by the effects of fly ash to alkaline activator ratio on lightweight geopolymer. The foaming agent which is Polyoxyethylene Alkylether Sulfate is incorporated into geopolymer paste to produce a lightweight geopolymer. It is a synthetic foaming agent. This type of inorganic geopolymer has highly potential to be used as thermal insulator material and lightweight concrete in construction field.

2. Experimental

2.1. Material

The raw materials that are involved in this research are fly ash, alkaline activator and foaming agent. Sodium hydroxide (NaOH) and sodium silicate (Na_2SiO_3) solution are combined together to be act as alkali activator solution. Fly ash is a by-product of coal combustion.

The type of fly ash used in this research is Class F Fly Ash. The sodium hydroxide in pallet form was diluted in distilled water to produce 12 M concentration. Sodium silicate are in the form of aqueous solution. The foaming agent used in this research was Polyoxyethylene Alkylether Sulfate. It is a synthetic foaming agent. It has high pH as it is alkali substance and has the appearance of a brown liquid and manufactured from coconut oil. This foaming agent is cheap, foams quickly and can be cleansed effectively.

2.2. Mixing Process with Different Solid to Liquid Ratio

The effect of solid to liquid ratio is investigated by fixing the ratio of sodium hydroxide (NaOH) to sodium silicate (Na_2SiO_3) at 2.5, by mass [8].

The solid to liquid ratio was fixed at 2.0, 2.5, 3.0 and 3.5, by mass. However, the ratios of 1.0 and 1.5, by mass were unable to use because of high workability of the geopolymer paste. The ratio of foaming agent to water and foam to geopolymer paste were fixed at 1:10, by volume and 1:1, by volume. The details of mix design are stated in Table 1.

The geopolymer sample was poured in (50 x 50 x 50) mm cubic moulds for all ratios. Then, the samples were cured in the oven for 24 hours at 80°C. After 24 hours, the geopolymer samples were left at room temperature for 1 day before demoulded and cured in room temperature for another 14 days before testing process.

Table 1. Mixing design proportions.

Ratio of fly ash to alkaline activator	Ratio of sodium silicate to sodium hydroxide	Ratio of foam to paste	Fly ash (g)	Sodium silicate (g)	Sodium hydroxide (g)
2.0	2.5	1.0	468.7	143.49	67.0
2.5	2.5	1.0	502.21	143.49	57.4
3.0	2.5	1.0	527.30	140.6	35.2
3.5	2.5	1.0	546.86	111.6	44.6

IOP Conf. Series: Materials Science and Engineering **877** (2020) 012013 doi:10.1088/1757-899X/877/1/012013

2.3. Curing Process
During curing process, the geopolymerization process starts to build up. Due to the increasing of temperature, polymerization become quicker and the strength can increase 70% within 3 hours to 4 hours of curing. The samples were dried by using an oven at temperature about 80 °C for 24 hours.

2.4. Mechanical and Physical Analysis
There are several physical and mechanical tests that will be done in this research which are compressive strength, density and water absorption test to determine the properties of lightweight geopolymer. These test were done after curing process.

2.4.1. Compression Strength Test. The compressive strength test for lightweight geopolymer was carried out to measure the maximum resistance of a sample specimen to the axial loading. The compressive strength was measured by using a Universal Testing Machine (UTM) in compliance with ASTM C 109-16a on the (50×50×50) mm specimens. The dry density of lightweight geopolymer can be calculated by the formula as shown in equation (1) below.

$$\rho = \frac{M}{V} \quad (1)$$

Where:
ρ = Dry density;
M = Mass of sample;
V = Volume of sample.

*2.4.2. Water Absorption Test.*The water absorption test was conducted accordance to ASTM C140. After the curing process, the samples are weighed and will be immersed in water for 24 hours. Then, samples were removed from the water, pattered dry, blotted with cloth to remove excess water before it will be weighed again. After weighing the saturated sample, the sample is cured in the oven at 105 °C for 24 hours.

$$Water\ absorption = \left[\frac{(Wet\ weight - Dry\ weight)}{Dry\ weight} \times 100\right] \quad (2)$$

The water absorption was calculated based on dried weight by using equation (2) below.

2.4.3. Density Analysis. The density of lightweight geopolymer samples were calculated based on the value of mass of sample and the volume of sample. The densities were measured after fulfilling the aging period of 14 days. Equation (3) is used to measure the density.

$$Density = \frac{mass\ of\ the\ sample\ (kg)}{volume\ of\ the\ sample\ (m^3)} \quad (3)$$

Average of three samples were tested for each ratios.

3. Results and Discussions
3.1. Compressive Strength of Lightweight Geopolymer
The result of compressive strength for fly ash based lightweight geopolymer with different solid to liquid ratio was illustrated in Figure 1. The highest strength is detected at solid to liquid ratio 2.5, by mass with value of 19.3 MPa while lowest strength is observed at solid to liquid ratio 3.5, by mass with 12.1 MPa. Highest compressive strength is found to be at solid to liquid ratio 2.5 because it has the optimum amount of fly ash and alkaline activator that gives a good workability and activate the fly ash at the highest rate of the geopolymerisation process before it mixes with foam. Hence, this provided sufficient Si species and raises the ratio of SiO_2/Al_2O_3. Besides, this can be related to the less of water available as the solid to liquid ratio increased and consequence in faster hardening of gel

phase and rapidly bond together the geopolymer framework [9]. During geopolymerization process, the SiO_2 and Al_2O_3 presented in fly ash need sufficient alkalis to allow the glassy structure to dissolve in alkaline medium to form the geopolymer gel structure [10].

The change in the alkaline activator content affects the (Si) species content, the sodium (Na^+) ions content and base water content. All these factors may affect during geopolymerization process and on the strength of the resulting geopolymer.

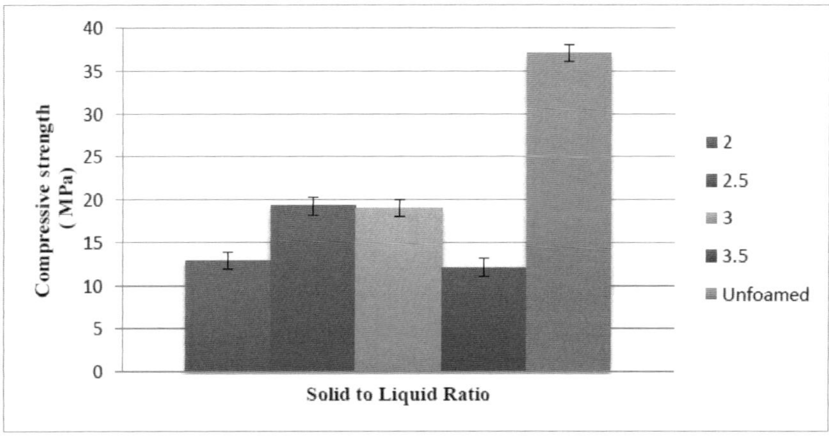

Figure 1. The compressive strength of lightweight geopolymer with different solid to liquid ratio.

A decrease in the solid to liquid ratio increases the Si species content, because the alkaline activator containing sodium silicate which contain high amount of Si species, increased the SiO_2/Al_2O_3 ratio and also increased the strength of geopolymer samples because of more Si-O-Si bonds produced which are stronger compared with Si-O-Al [11-13].

3.2. Water Absorption of Lightweight Geopolymer

The results of water absorption of lightweight geopolymer were shown in Figure 2. Water absorption is an important factor for determining the durability of concrete.

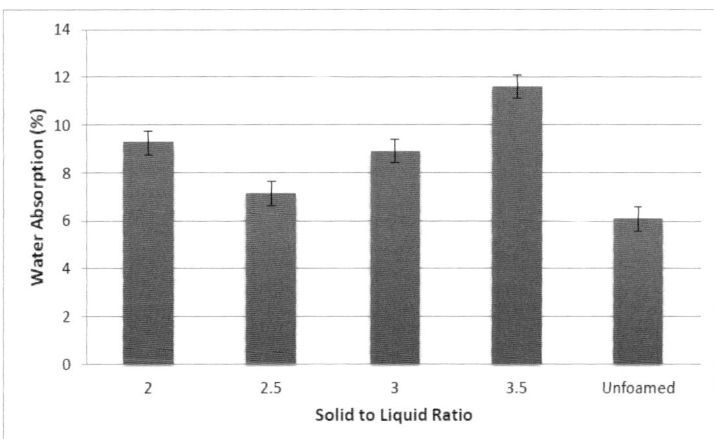

Figure 2. The water absorption of lightweight geopolymer with different solid to liquid ratio.

The less water infiltrates into concrete, the more durability of the concrete and resistance to the natural environment are anticipated. The lowest water absorption is found at solid to liquid ratio 2.5, by mass with 6.08 % while highest water absorption is found to be at solid to liquid ratio 3.5, by mass with 11.61 %. An increment in water absorption from 7.14 % to 11.61 % was observed with the increase of solid to liquid ratio from 2.5 to 3.5, by mass.

The results found that the water absorption increases with increasing of solid to liquid ratio. The previously reported by Zhao et al, [10], with the increasing volumes of foam, more and more interconnected pores which are advantage for flow of water are performed. Nevertheless, the increase of water absorption could decrease compressive strength. Based on the results, the porosity also affects the water absorption. When the porosity decreases, the pore size also decreases [14].

3.3. Density of Lightweight Geopolymer

The densities of lightweight geopolymer with different solid to liquid ratio are shown in Figure 3. The densities vary in the range of 1307 kg/m^3 to 1800 kg/m^3. The highest density is found at solid to liquid ratio 2.5, by mass with 1375 kg/m^3. The lowest density is observed at solid to liquid ratio 3.5, by mass with 1267 kg/m^3. The density of lightweight geopolymer has slightly decreases from solid to liquid ratio 2.5 to 3.5, by mass ranging from 1373 kg/m^3 to 1267 kg/m^3.

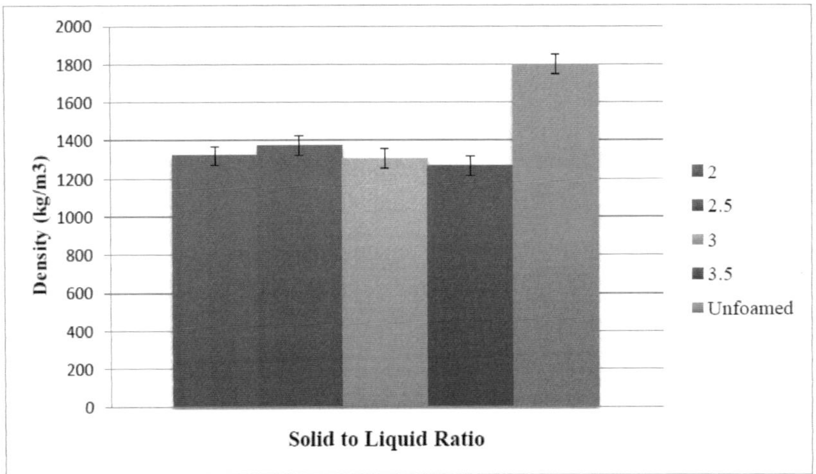

Figure 3. The density of lightweight geopolymer with different solid to liquid ratio.

Density at solid to liquid ratio 3.5, by mass is the lowest because more volume of foams has been used in the paste. Mass of fly ash is more higher compared to alkaline activator which means less liquid is inadequate in the paste. Therefore, to improve the workability, more volume of foams is added in the paste. The increases of voids in the geopolymer samples due to the foaming agent will reduce the density of the sample. Thus, the compressive strength will also decrease with the increment of those voids [15].

3.4. Microstructure of Lightweight Geopolymer

Figure 4 shows the microstructure of lightweight geopolymer with different solid to liquid ratio. The microstructure of lightweight geopolymer showed some unreacted or partially reacted fly ash particles and formed of a continuous alumina-silicate gel. For solid to liquid ratio 2.0, by mass unreacted fly ash particles are found on the microstructure. Pores also formed. This is because there is too much water from alkaline activator and foams. Therefore, the concentration of sodium hydroxide decreases resulting in lower compressive strength and pore formation on the microstructure. Cracks also observed on the surface can lead to the reduced mechanical strength observed at these activation conditions [7].

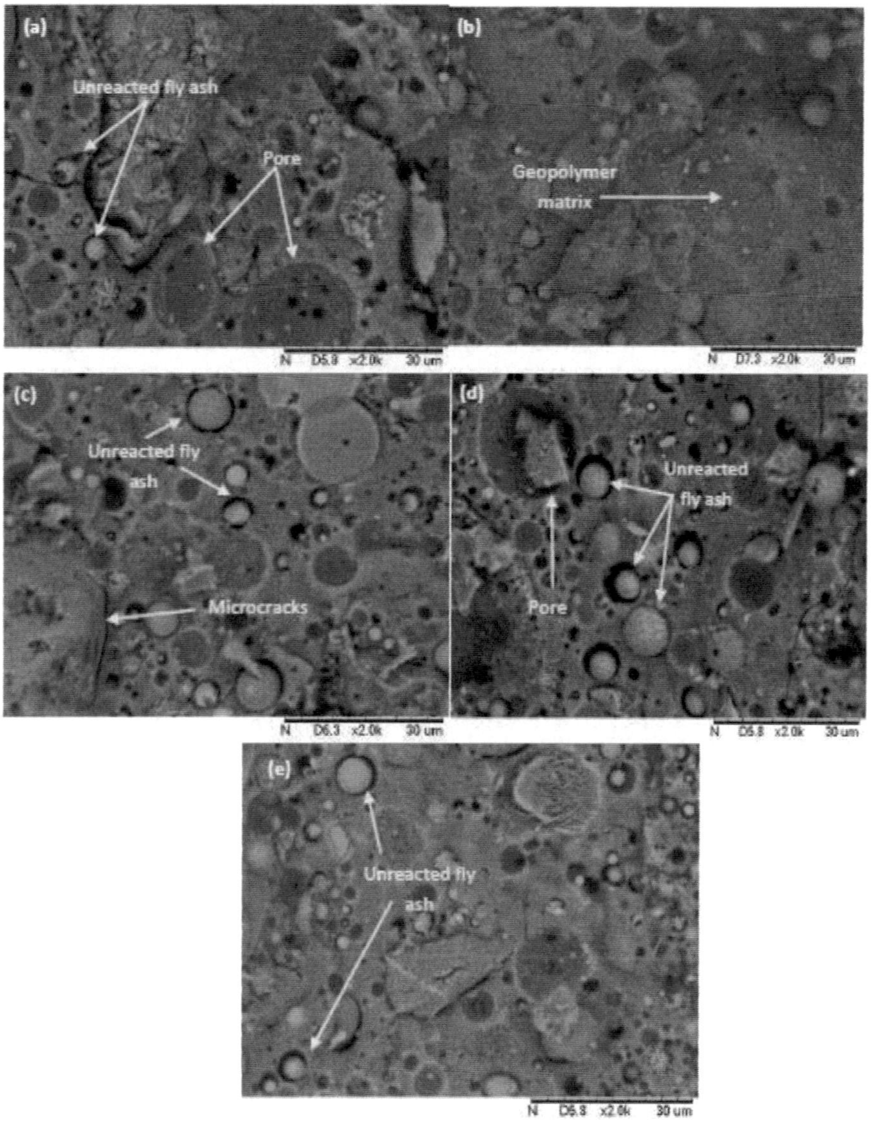

Figure 4. Microstructure of lightweight geopolymer with different solid to liquid ratio; (a) 2.0, (b) 2.5, (c) 3.0, (d) 3.5 and (e) Unfoamed.

Therefore, the concentration of sodium hydroxide decreases resulting in lower compressive strength and pore formation on the microstructure. Cracks also observed on the surface can lead to the reduced mechanical strength observed at these activation conditions [10]. For solid to liquid ratio 3.5, more unreacted particles are found on the microstructure compared to the other ratios. There were small and large pores due to incomplete reaction with fly ash, alkaline activator and foam. Microcracks were also observed which contributed to their lower strength by increasing their water absorption and porosity. The lower values of compressive strengths of lightweight geopolymer attributed to the presence of voids happen when cured in the oven. This might have resulted due to poor surface structure of the lightweight geopolymer as shown in Figure 4(d) [16-17]. For unfoamed geopolymer, there were minimal numbers of unreacted fly ash particles found on the microstructure. This is because there were adequate amount of fly ash and alkaline activator to react with each other.

In conclusion, solid to liquid ratio 2.5, by mass has the better microstructure compared to other ratios. It has less unreacted fly ash, more completely reacted fly ash, no microcracks and tiny pores as shown in figure 4. Hence, contributing to higher strength and lower water absorption.

4. Conclusions

An investigation of the effect of different solid to liquid ratio on fly ash-based lightweight geopolymer was conducted. The effects of solid to liquid ratio were studied. From the research, the solid to liquid ratio that gives the best strength is 2.5, by mass. This ratio caters the highest compressive strength of 19.3 MPa, lowest percentage of water absorption at 7.14% and density at 1373 kg/m^3 of lightweight geopolymer. The microstructure of the concrete is was smoother and denser with less unreacted fly ash particles and pores. It was shown that the content of alkaline activator solution is the most crucial parameter affecting the strength of lightweight geopolymers.

Acknowledgments

The author would like to acknowledge the support from the Fundamental Research Grant Scheme (FRGS) under a grant number of FRGS/1/2019/TK10/UNIMAP/02/21 from the ministry of Education Malaysia.

5. References

[1] Adesanya E, Ohenoja K, Luukkonen T, Kinnunen P and Illikainen M 2018 *Journal of Cleaner Production* **185** 168-175
[2] Toniolo N and Boccaccini A R 2017 *Ceramics International* **43(17)** 14545-14551
[3] Kaur M, Singh J and Kaur M 2018 *Ceramics International* **44(2)** 1534-1537
[4] Posi P, Thongjapo P, Thamultree N, Boontee P, Kasemsiri P and Chindaprasirt P 2016 *Construction and building materials* **127** 450-456
[5] Ghazali N, Muthusamy K and Ahmad S W 2019 *IOP Conference Series: Materials Science and Engineering* **601(1)** 012023
[6] Abdullah M M A B, Hussin K, Bnhussain M, Ismail K N, Yahya Z and Abdul Razak R 2012 *International journal of molecular sciences* **13(6)** 7186-7198
[7] Ibrahim W M W, Hussin K, Abdullah M M A B and Kadir A A 2017 Geopolymer lightweight bricks manufactured from fly ash and foaming agent *AIP Conference Proceedings* **1835(1)** 020048
[8] Abdullah M M A, Hussin K, Bnhussain M, Ismail K N and Ibrahim W M W 2011 *Int. J. Pure Appl. Sci. Technol* **6(1)** 35-44
[9] Rahim A, Hanim R, Azizli K A, Man Z, Rahmiati T and Ismail L 2014 *Applied Mechanics and Materials* **625** 46-49
[10] Zhou W, Yan C, Duan P, Liu Y, Zhang Z, Qiu X and Li D 2016 *Materials & Design* **95** 63-74
[11] Sanjayan J G, Nazari A, Chen L and Nguyen G H 2015 *Construction and Building Materials* **79** 236-244
[12] Burduhos Nergis D D, Abdullah M M A B, Sandu A V and Vizureanu P 2020 *Materials* **13(2)** 343
[13] Burduhos Nergis D D, Vizureanu P and Corbu O 2019 *Revista de chimie* **70(4)** 1262-1267
[14] Abdullah A, Abdullah M M A B, Hussin K and Tahir M F M 2017 Effect of different sintering temperature on fly ash based geopolymer artificial aggregate *AIP Conference Proceedings* **1835(1)** 020050
[15] Farahani J N, Shafigh P and Mahmud H B 2017 *Procedia engineering* **184** 778-783
[16] Barbosa T R, Foletto E L, Dotto G L and Jahn S L 2018 *Ceramics International* **44(1)** 416-423
[17] Scripcariu L, Mătăsaru D, Diaconescu V 2017 UPB Scientific Bulletin Series C-Engineering and Computer Science **73**(3) 83-92

IOP Conf. Series: Materials Science and Engineering **877** (2020) 012014 doi:10.1088/1757-899X/877/1/012014

Void Distributions in Sn-3.0Ag-0.5Cu (SAC305) Composite Lead Free Solder Subjected to Thermal Ageing Using Acoustic Micro Imaging Technique

N S M Zaimi[1*], M A A M Salleh[1], M M A B Abdullah[1], M Mostapha[1], M S Baltatu[2] and R Ahmad[3]
[1]Centre of Excellence Geopolymer and Green Technology, School of Materials Engineering, Universiti Malaysia Perlis (UniMAP), Taman Muhibbah 02600, Jejawi, Arau, Perlis, Malaysia.
[2]Gheorghe Asachi Technical University, Faculty of Materials Science and Engineering, Department of Physics, 67 Prof. Dimitrie Mangeron Blvd., 700050, Iasi, Romania.
[3]Faculty of Engineering Technology, Uniciti Alam Campus, Universiti Malaysia Perlis (UniMAP), Perlis, Malaysia.

*E-mail: syahirahzaimi25@gmail.com

Abstract. The formations of the voids in SAC305 lead free solder and SAC305 with additions of kaolin geopolymer ceramics were studied. The composite solders were fabricated by using powder metallurgy with microwave sintering method. The samples were sandwiched between two copper substrates and reflowed in a reflow oven and aged at 125°C for 0 and 7 days. The acoustic micro imaging was used to analyse the distributions of voids in the solder joints of SAC305 and SAC305 with additions of kaolin geopolymer ceramics. Results shows that, rhe void in SAC305 are larger in size and numbers as compared to SAC305 with additions of kaolin geopolymer ceramics for both reflowed and aged conditions.

1. Introduction

Nowadays, the trend in electronic products towards higher speed, lightweight, multifunctional and continual miniaturization has led to the urgent need for a development of very reliable solder material. Solder play a very important role in electronic packaging where it mechanically and electronically connects the electronic components to the printed circuit board (PCB). For more than 50 years ago, tin-lead solder alloy was successfully and widely used in the electronic industry as the main solder materials due to the outstanding properties served by this alloy [1]. However, the toxicity of lead towards environments and peoples, causes the usage of lead is strictly banned by Restriction of Hazardous Substance (RoHS) and the European Waste Electrical and Electronic Equipment (WEEE) directives. Therefore, many researches have been done in order to replace the conventional lead-based solder which could perform same or better than tin-lead solder alloy. Several lead-free solder alloy have been developed including tin-silver-copper (Sn-Ag-Cu) solder alloy. Sn-Ag-Cu solder alloy has been proposed to be the most promising solder alloy to replace lead-based solder especially in surface mount technology (SMT) due to good properties such as better solderability and well compatibility with the current electronic components [2]. Even so, Sn-Ag-Cu solder alloy need to further improved in the reliability as to fulfil the current demand of higher performance solder alloy. To date, development of composite solder has took placed among many researchers as to increase the reliability of lead free solder alloy [3-5]. Incorporation of reinforcements either metallic or non-metallic particles believed could give better results in terms of the microstructure formation, solderability and strength of the solder joints [6-8].

In this study, Sn-3.0Ag-0.5Cu (SAC305) lead free solder will be incorporated with kaolin geopolymer ceramics particles that are expected to suppressed the formation of IMCs and improve the performance of existing SAC305 lead free solder. Apart from that, the formations of voids in solder could not be avoided. Thus, in this study an acoustic micro imaging will be used as one of a technique in analysing void distributions in the developed composite solders. The acoustic micro imaging is one of non-destructive technique that widely used to inspect the defects such as cracks, voids and delamination in the electronic components and assemblies [9]. Besides that, acoustic micro imaging utilises acoustic microscope that is equipped with transducers. The transducers emits high frequency of ultrasound in the range of 5MHz to more than 400MHz and when the ultrasound waves penetrates the samples it may scattered, reflected or absorbed by the internal features of the material itself [9,10]. The acoustic images were generated as the ultrasound wave is reflected from an internal part of the sample which had been travelled through the entire thickness of the samples itself. Besides that, Zhang et.al. [11], in his study had used acoustic micro imaging in analysing the micro cracks propagation in the microelectronic packages. Semmens et al. [12], also utilised acoustic micro imaging to determine the failure modes of the flip chip interconnections. Therefore, in this research an attempt was made by using acoustic micro imaging as one of the techniques in analysing the void distributions in SAC305 lead free solder incorporated with kaolin geopolymer ceramics which are also subjected to thermal ageing process.

2. Methodology

In this study, the samples of Sn-3.0Ag-0.5Cu (SAC305) with the addition of 1.0 wt.% kaolin geopolymer ceramic was fabricated by using a technique of powder metallurgy microwave sintering. The base material which is SAC305 lead free solder with the particle size in the range of 25-45μm. Firstly, SAC305 powders with 1.0 wt.% kaolin geopolymer ceramic were mixed by using a planetary mill machine at a rotating speed of 200 rpm for an hour in an artight container. Then, the mixture of SAC305 powders with kaolin geopolymer ceramics were uniaxially compacted by using a Specac 15-ton Manual Hydraulic Press at a pressure of 4.5 tonnes. The compacted pellets were then microwave sintered for about 3 minutes at a temperature of 185°C under ambient conditions by using a 800 W, 50 Hz domestic microwave oven.

Then, the sintered samples were cold rolled by using a rolling machine. The solder sheets obtained from the rolling process were then sandwiched between two copper substrates as in Figure 1 and then reflowed by using a F4N reflow oven with the aid of rosin mildly activated flux (RMA). The reflowed samples were then undergo thermal ageing at a temperature of 100°C for 7 days.

Acoustic micro imaging technique was used to analyse the internal features of the thermal aged samples and it was carried out at Liverpool John Mores University, United Kingdom. The acoustic microscope of the Sonoscan Gen6™ with the 100Mhz 0.5inch transducer was used in this study.

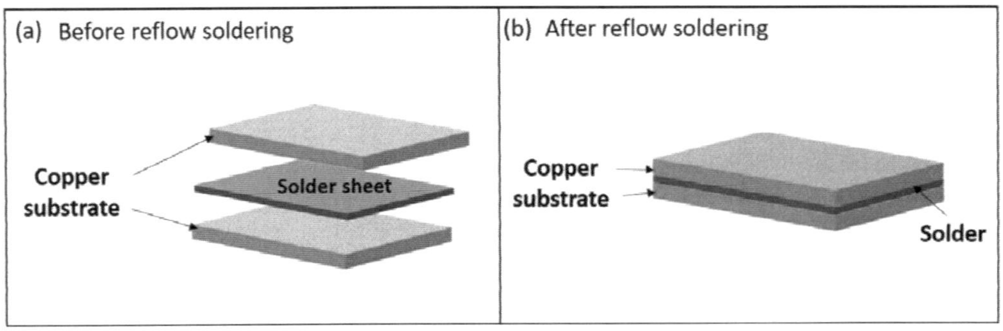

Figure 1. Sample preparation for the reflow soldering of sandwiched samples.

3. Results and Discussions

Figure 2 shows the result of scanning from acoustic micro imaging for the samples of SAC305 and SAC305 with additions of kaolin geopolymer ceramics, subjected to thermal ageing at 125 °C for 0 and 7 days. Based on Figure 2, the brighter circle regions of lower density corresponds to the solder voids. While, the darker regions represents higher density of solder.

The size of the voids in the pure SAC305 were bigger and with larger numbers than the sample of SAC305 with additions of kaolin geopolymer ceramics as revealed in Figure 2(a), 2(b) and Figure 3. It shows that the numbers of solder voids were lesser in the SAC305 with additions of kaolin geopolymer ceramics in as-reflowed and aged samples.

The formation of the voids in the solder was due to the outgassing flux that were entrapped in the solder during soldering. The sources of outgassing flux was due to the evaporation and thermal decomposition of flux constituents that were entrapped in the solder joint. Moreover, the formation of the voids in the solder joint could affect the reliability and the performance of the solder joints. Size of the voids will influence the robustness of solder joint in where larger size of voids increases the tendency of solder joint cracks compared to smaller voids [13, 14]. Besides that, IPC-7095 standard had specified the maximum allowable of void area [14, 15].

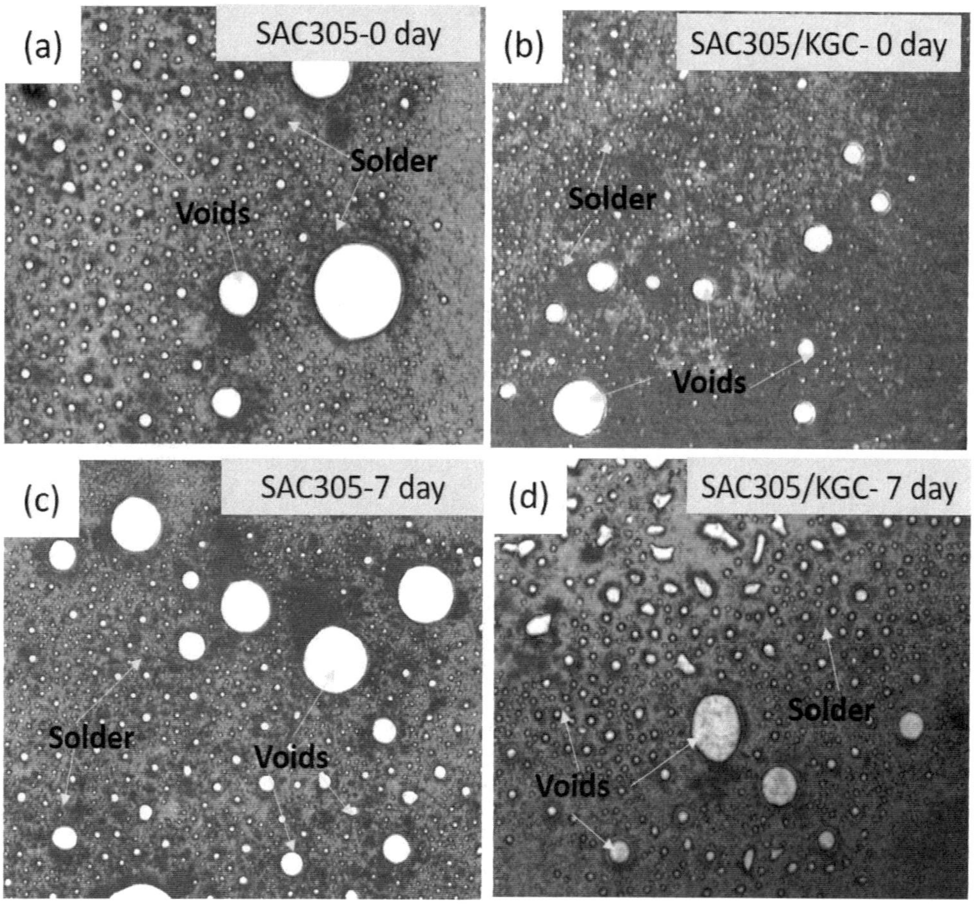

Figure 2. Distribution of voids in SAC305 as a function of thermal ageing temperature at 125°C for (a) 0 day, (c) 7 days and SAC305 with addition of kaolin geopolymer ceramics as a function of thermal ageing temperature at 100°C for (b) 0 day, (d) 7 days.

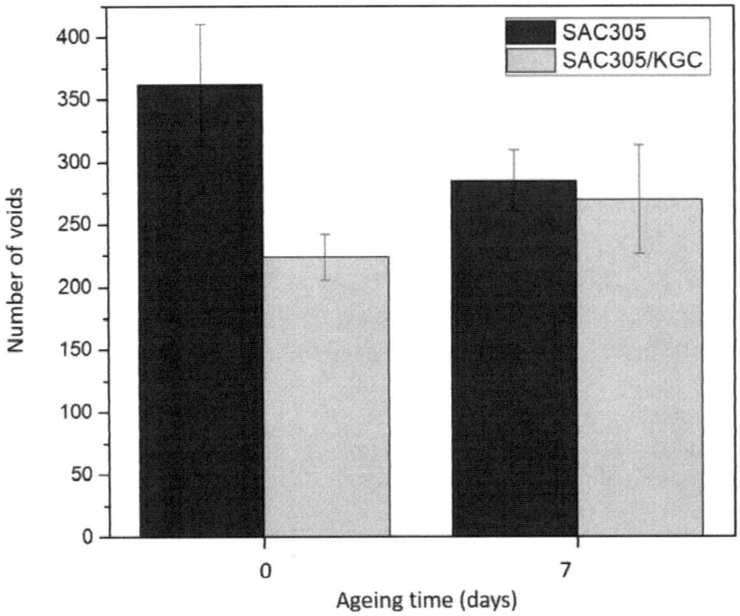

Figure 3. Number of voids as a function of thermal ageing temperature at 125°C for 0 and 7 days in SAC305 and SAC305 with addition of kaolin geopolymer ceramics.

Apart from that, these results also proved that the utilization of acoustic micro imaging in analysing the internal features of solder joint could give new insight in this study area. It was also proved that, the use of high frequency transducer which is 100MHz 0.5inch in this study were able to provide better acoustic image with high resolution since the resulting acoustic image depends on the material itself [9]. However, the results obtained from the acoustic micro imaging could be supported by other characterisation techniques.

4. Summary

A new lead-free composite solder system with the addition of kaolin geopolymer ceramics that are subjected to thermal ageing were sucessfully fabricated. The acoustic micro imaging technique was used to abalyse the distributions of voids in the solder joints. It was clearly observed that the formations of void in SAC305 with additions of kaolin geopolymer ceramics were reduced as compared to pure SAC305 solder joints either on as-reflowed or aged conditions.

The authors acknowledge the financial support through the Ministry of Higher Education, Malaysia, under the Fundamental Research Grant Scheme (FRGS/1/2017/TK10/UNIMAP/03/2) (9003-00612). This work was also supported by an Institutional Links grant, ID 332397914, under the Newton-Ungku Omar Fund partnership. The grant is funded by the UK Department of Business, Energy and Industrial Strategy (BEIS) and Malaysia and delivered by the British Council. The authors also acknowledge Center of Excellence Geopolymer & Green Technology (CEGeoGTech) and School of Materials Engineering, Universiti Malaysia Perlis.

References

[1] Cheng S, Huang C, and Pecht M 2017 *Microelectronics Reliability* **75** 77-95
[2] Gain, A K, Chan Y C, and Yung W K C 2011 *Microelectronics Reliability* 2011 **51(12)** 2306-2313

[3] Zaimi N S M et al., 2019 *Influence of kaolin geopolymer ceramic additions to the wettability and electrical properties of Sn-3.0Ag-0.5Cu (SAC305) lead free solder.* IOP Conference Series: Materials Science and Engineering **701** 012033
[4] Sukpimai K, Suwannakrue W, and Kanlayasiri K 2019 *Wettability and printability of SAC305-xTiO 2 Pb-free solder paste on Cu substrate.* IOP Conference Series: Materials Science and Engineering **635** 012009
[5] Qu M, et al. 2019 *Journal of Materials Science: Materials in Electronics* **30(21)** 19214-19226
[6] Cheani F, et al., 2018 *The International Journal of Advanced Manufacturing Technology* **96**
[7] Gain A K and Zhang L 2018 *Materialia* **3** 64-73
[8] Mohd Salleh M A A, McDonald S D and Nogita K 2017 *Journal of Materials Processing Technology* **242** 235-245
[9] Achitei DC, Vizureanu P, Minciuna MG, Sandu AV, Buzaianu A and Dana DI 2015 *Materiale Plastice* **52(2)** 165-167
[10] Ghaffarian R 2016 *Packages and Assemblies* **31** 88-107
[11] Lee C, et al., 2015 *NDT & E International* **79**
[12] Semmens J E and Kessler L 1997 *Characterization of flip chip interconnect failure modes using high frequency acoustic micro imaging with correlative analysis* 141-148
[13] Yunus M, et al., 2000 *Effect of voids on the reliability of BGA/CSP solder joints* **43** 207-213
[14] Bušek D, et al., 2016 *Microelectronics Reliability* **60** 135-140
[15] Vizureanu P, Perju MC, Galusca DG, Nejneru C and Agop M 2010 *Metalurgia international* **15(12)** 59-64

Comparison Study on Microstructure Properties of Kaolin Based Geopolymer Ceramics with Addition of UHMWPE under Different Sintering Condition

R Ahmad[1,2*], M M A B Abdullah[2], W M W Ibrahim[1,2], A V Sandu[3], P Vizureanu[3] and M S Tengah[1]

[1] Faculty of Engineering Technology, Universiti Malaysia Perlis (UniMAP), P.O Box 77, D/A Pejabat Pos Besar, 01000 Kangar, Perlis, Malaysia

[2] Center of Excellence Geopolymer and Green Technology (CEGeoGTech), School of Materials Engineering, Universiti Malaysia Perlis (UniMAP), P.O Box 77, D/A Pejabat Pos Besar, 01000 Kangar, Perlis, Malaysia

[3] Gheorghe Asachi Technical University of lasi, Faculty of Materials Science and Engineering, Blvd. D. Mangeron 71, 700050, lasi, Romania

*E-mail: romisuhani@unimap.edu.my

Abstract. To better understand the structure-mechanical properties relation of additively fabrication of kaolin geopolymer ceramics with sintering method, a comparison study was performed. Kaolin based geopolymer ceramics were synthesized starting from the powders of kaolin based geopolymer, using powder metallurgy method. Typically, the sintering method used are one step (1200 °C – 5 min) and novel two step sintering processes (600 °C-5 min and 1200 °C – 5 min) for both kaolin based geopolymer ceramics with and without ultra high molecular weight polyethylene were applied and compared. The outcome revealed that there is no phase changes on the both sintering method pattern and the two step sintering method giving a smooth surface owing to the densification process during the preheat treatment.

1. Introduction

The combination of compacting and sintering process in the fabrication of ceramic materials has become increasingly important in modern industry due to superior mechanical and physical properties. A crystalline or partly semi-crystalline structure of ceramic are fabricated by a great number of small crystals, or grains separated from one another by grain boundaries [1]. Green properties of ceramic components are depending on the powder packing while the additives such as binder, surfactants, plasticizers, dispersants and lubricants also plays a major role, irrespective of the various processing technique [2]. The binder were used to plastically distort the ceramic particles in order to convey sufficient strength to the green compact, subsequently sustaining the integrity of the green part throughout ejection and presintering process. Although, the present of the binder also can affect the relative [3]. Thus, the usage of ultra high molecular weight polyethylene (UHMWPE) perhaps can develop the strength of the green bodies by binding the geopolymer ceramic particles due to their characteristic features includes the dimensional sheets covalently bond of network, the functional group types, and the length of molecular weight, chain and distribution [4].

Inorganic polymers called geopolymer harden at ambient temperature, having great potential applications as construction materials, matrix in ceramics other composite materials and coatings [5]. Geopolymer are synthesized by the polycondensation of silica-alumina structure. The reaction between solid aluminosilicate with high concentrated of aqueous alkali hydroxide or silicate solution can produces a synthetic alkali aluminosilicate material [6]. Geopolymer that having amorphous to semi-crystalline phases which contain random tetrahedral network of silicon and aluminium atom will transform into crystalline phase upon heating [7]. Among the unique feature of geopolymer is the high temperature in ceramic materials fabrication is no longer needed. With the help of geopolymerization, the homogeneous of the geopolymer system will influence the structural arrangement during the phase changes thus promoting the nucleation and densification process.

In general, the engineering properties of ceramic materials are measured by the microstructure such as grain size and density. In order to enhance this microstructural, the sintering mechanism is one of the factors that should be discovered due to the material densification will be promoted during this process [8]. Different methodologies have been made with a view to develop the mechanical properties through powder consolidation that appear to offer major property benefits.The influence on the sintering method in the microstructure of ultrafine alumina were studied by Chinelatto et al., (2012) [9]. By heating the samples to 1050 °C, smaller final grain were observed due to the removal of finest particles and narrow particles size distribution. Vogiatzis et al., (2015) found that the two step sintering can avoid any thermal shock that lead to cracking and deteriorating the structural of the samples [10]. Thus, this study was to investigate the effect of sintering method on the microstructural properties of kaolin based geopolymer ceramics with and without addition of UHMWPE as binder.

2. Experimental Work

2.1. Materials

Kaolin is a clay mineral composed of $Al_2Si_2O_5(OH)_4$ chemical composition (Table 1) with good characteristics such as natural whiteness, fine particle size, non-abrasive stability and chemical stability. Kaolin that a layer of silicate mineral with one sheet of tetrahedral silica (SiO_4), linked through oxygen atoms to one sheet of octahedral alumina (AlO_6) octahedral was bought from the Kaolin Industries Associated Sdn. Bhd. Malaysia in powder form.

Table 1. Composition of kaolin.

Composition	SiO_2	Al_2O_3	Fe_2O_3	K_2O	TiO_2	MnO_2	ZrO_2	LOI
Percentage (%)	54.0	31.7	4.89	6.05	1.41	0.11	0.10	1.74

Sodium silicate (Na_2SiO_3) solutions were supplied by South Pacific Chemicals Industries Sdn. Bhd. Malaysia. The chemical compositions of of the solution given by the supplier are 30.1% SiO_2, 9.4% Na_2O and 60.5% H_2O. While for the sodium hydroxide (NaOH) is supplied by Formosa Plastic Corporation, Taiwan.with a brand name of Formosoda-P were used in this study.

Ultra high molecular weight polyethylene (UHMWPE) was used as binder with a molecular weight of 5 X 106 g/mol and density of 0.94 g/ml, in powder form was purchased from Ticona Engineering Polymer, China.

2.2. Fabrication of Kaolin Based Geopolymer Ceramics

Alkali activator was prepared by mixing 12M of NaOH solution with Na_2SiO_3 at a ratio of 0.24. The solutions were then allowed to mature for 24 h permitted to dissolve the silica homogeneously. The geopolymer paste was prepared by mixing the kaolin powder with alkali activator solution at solid-to-liquid ratio of 1.0 and was stirred by using mechanical mixer and cured at 80 °C for 24 h. The ratio was chosen based on our previous work since it allowable for optimal physical, mechanical and thermal properties.

The kaolin geopolymer were then crushed and sieved manually by using 150μ siever. The kaolin based geopolymer powder was then mixed with 4 wt.% of UHMWPE by using planetary mill in dry condition method for 4 minutes at 100 rpm with reverse direction. Then, the samples through compaction process by using a 12-mm-diameter cylindrical stainless steel die at 5 ton for 2 min and sintered at two different sintering condition which are one step and two steps method with a cooling and heating rate of 5 °C/ min. For the two steps method, the samples were heated at lower temperature of 600 °C with 60 min of soaking and proceed to second steps of heating up to 1200 °C with 120 mi of soaking.

2.3. Microstructure Testing
XRD 6000, SHIMADZU diffractometer were used to characterize the phase involved in the samples. Samples were prepared by classic preparation technique based on the pressing of powder samples into aluminium holders with the operating settings at 40 kV and 30 mA using Cu-Kα radiation. The peak were identified by using X'Pert High Score Plus. The microstructure of the sample was revealed by using JSM-6460LA model Scanning electron microscope (JEOL). The specimens were prepared and coated by using Auto Fine Coater; model JEOL JFC 1600.

3. Results and Discussion
3.1. Microstructure
The XRD patterns of kaolin based geopolymer ceramics with and without addition of UHMWPE at different methods of sintering are shown in Figure 1 and Figure 2. Both samples with and without addiotion of the binder shows the same pattern. Nepheline ($NaAlSiO_4$) phase appeared as major phase in both XRD patterns. Crystallization of nepheline ($NaAlSiO_4$) from kaolin geopolymer gradually occurred on heating [11] and its proof that there is no phase change occurred due to the same sintering temperature used. However, in the XRD pattern kaolin geopolymer ceramics with addition of various content of UHMWPE, appearance of extra peak of carbon (C) were observed at 2θ values of 17.7 °, 31.0 °, 35.9 °, 37.5 °, 41.7 °, 45.7 ° and 56.2 ° credited to the decomposition of UHMWPE while sintering at high temperature [12]. The samples sintered with two step method illustrate a sharper peak compared to samples sintered with conventional method. In addition, the intensities increased in the XRD pattern of samples that sintered using two step method in which consistence with the high value of strength [13, 14].

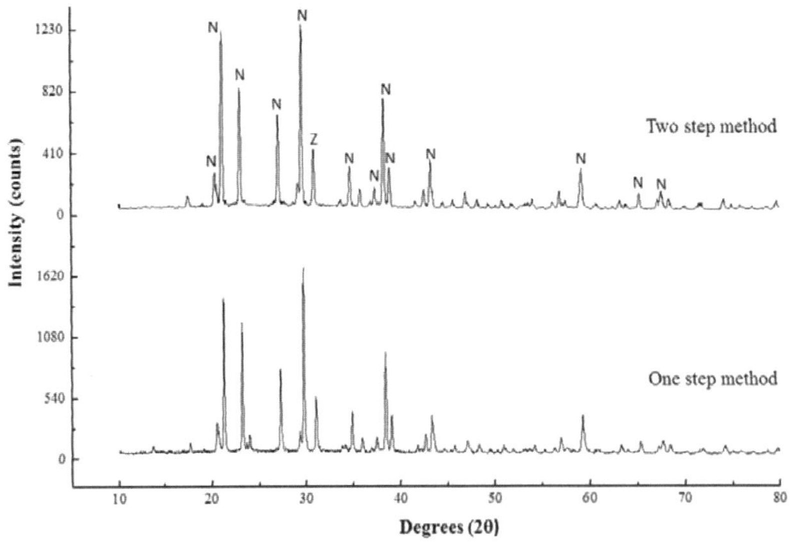

Figure 1. XRD pattern for kaolin based geopolymer ceramics at different sintering method (N= Nephaline, Z= Zeolite).

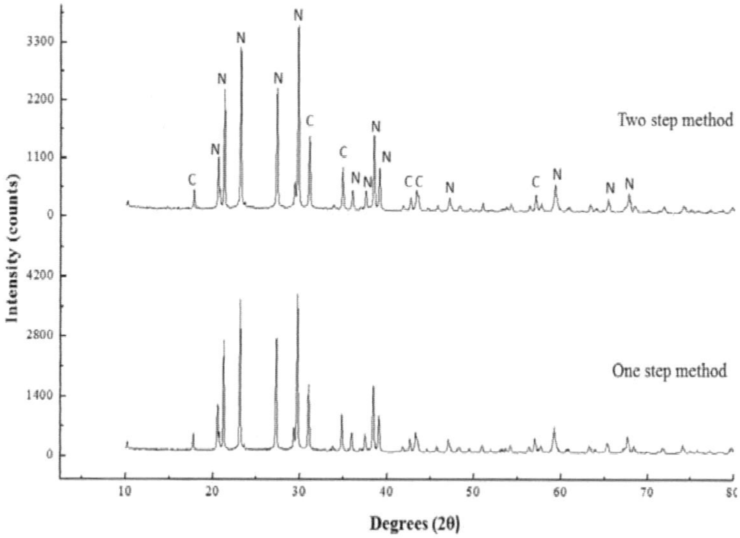

Figure 2. XRD pattern for kaolin based geopolymer ceramics with addition of UHMWPE at different sintering method (N= Nephaline, C= Carbon).

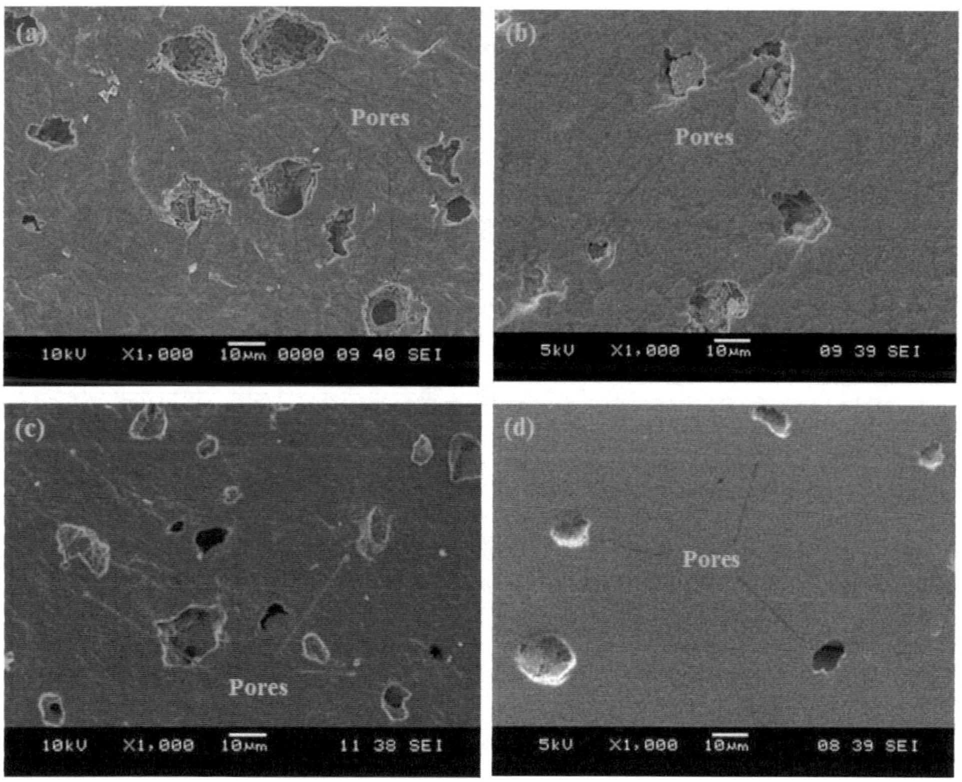

Figure 3. SEM micrograph of kaolin based geopolymer ceramics sintered using (a) one step method, (b) two step method and with addition of UHMWPE, (c) one step method, (d) two step method.

The SEM micrographs of the fractured surface of kaolin based geopolymer ceramics with and without addition of UHMWPE at different sintering method were compared in Figure 3. Both samples with and without the addition of UHMWPE that sintered using two step method shows a more uniform and homogeneous microstructure compared with the samples using one step sintering method. This is due to the preheating phase in two step method that induced densification process by eliminating the finest particles of the samples [9].The fractured surface of both samples with and without addition of UHMWPE shows formation of pores for both method. However, the pore size become irregular and some of the pores confined within the grains during the faster densification of finest particles in two step sintering method [15-18]. In the study on the effect of the pre-sintering on microstructural evolution of alumina carried out by Lin et al. (2005), the average of pore size of a pre-sintered compact were found to increases after the preheat treatment and the pore-size distribution becomes narrower [13].

In general, the microstructure of the ceramic bodies showed a network of small dense zones interconnected by a porous phase. However a noticeable improvement is shown in the microstructure of the samples that heated with two steps method in which would improve the physical and mechanical properties of the ceramic materials.

4. Conclusion

The use of two steps sintering methods shows a noticeable improvement in microstructural characteristics properties for both samples with and without the addition of UHMWPE. XRD results of all the samples show significant similarities in term of peaks appearance of nepheline phase which used to occur at the same sintering temperature. While the microstructure analysis demonstrated that the two step method allowing an improvement of the microstructure, thus can improves the mechanical properties of the samples.

5. References

[1] Fragassa C 2015 *Int. J. Qual. Res.* **9(2)** 279–298
[2] Rahaman M N *Ceramic processing* CRC/Taylor & Francis 2007.
[3] Taktak R, Baklouti S and Bouaziz J 2011 *Mater. Charact.* **62**(9) 912–916
[4] Pizette P, Martin C L, Delette G and Geneves S F T 2013 *J. Eur. Ceram. Soc.* **33**(5) 975–984
[5] Daud Y M, Hussin K, C M R Ghazali, Osman A F, Abdullah M M A B and Bin Hussain M 2016 *Key Eng. Mater.* **673** 55–63
[6] Shahedan N F, Abdullah M M A B, Mahmed N, Kusbiantoro A and Bouissi A 2018 *AIP Conf. Proc.* **2030** 2018.
[7] Duxson P, Lukey G C and Van Deventer J S J 20017 *J. Mater. Sci.* **42**(9) 3044–3054
[8] Lóh N J, Simão L, Faller C A, De Noni A and Montedo O R K 2016 *Ceram. Int.* **42**(11) 12556–12572
[9] Chinelatto A and Pallone E 2012 *Sintering of Ceramics - New Emerging Techniques* 401–422
[10] Vogiatzis C A, Tsouknidas A, Kountouras D T and Skolianos S 2015 *Mater. Des.* **85** 444–454
[11] Kuenzel C, Grover L M, Vandeperre L, Boccaccini A R and Cheeseman C R 2013 *J. Eur. Ceram. Soc.* **33** 251–258
[12] Mera G, Riedel R, Poli F and Müller K 2009 *J. Eur. Ceram. Soc.* **29** 2873–2883
[13] Romisuhani A, Abdullah M M A, Kamarudin H and Andrei S V 2017 *IOP Conf. Ser. Mater. Sci. Eng.* **267** 2017
[14] Acosta-Humánez F, Almanza and Vargas-Hernández O C 2014 *Superf. Vacío* **27** 43–48
[15] Burduhos Nergis D D, Abdullah M M A B, Sandu A V and Vizureanu P 2020 *Materials* **13(2)** 343
[16] Calin MA, Curteza A, Toma S and Agop M 2013 *Metalurgia International* **18** 19-22
[17] Nedeff V, Lazar G, Agop M, Mosnegutu E, Ristea M, Ochiuz L, Eva L and Popa C 2015 *Powder Technology* **269** 452-460
[18] Lin F J T, Jonghe L C and Rahaman M N 2005 *J. Am. Ceram. Soc.* **80** 2269–2277

IOP Conf. Series: Materials Science and Engineering **877** (2020) 012016 doi:10.1088/1757-899X/877/1/012016

Experimental Study of Thermochemical Treated Biomedical 3D Printed Titanium Alloy Implant

J Kónya[1*] and K Kulcsár[1]
[1]Dent-Art-Technik Kft., Csokonai 10. Győr 9024, Hungary

*E-mail: janos@dentarttechnik.hu

Abstract. For medical and dental individual implants are made by additive technology. The patient-specific usable implant material is the Ti alloy (Grade 23). The used 3D printer do LMF (Laser Metal Fusion) process and melts the metal particle by the laser layer by layer. Surface treatment of implants affects mechanical strength, aesthetic and osseointegration properties [1]. Of these, our study investigates the mechanical strength properties of plasma nitride surface treatment. The plasma nitrided 3D printed biomedical Ti alloy (Ti6Al4V, Grade 23) mechanical properties. The influence of the nitrided layer on tribological behaviour of plasma nitrided film was studied using microhardness tester. The microhardness results pointed out that the surface treatment increased the surface hardness and reduced the plastic deformation of the alloy.

1. Introduction

In the medical field, custom-made implants are used for individual cases. These implants are manufactured with additive technologies thanks to 21st-century innovations in manufacturing technology. In this study, custom-made patient-specific subperiosteal dental implants were investigated. These are called cortically supported implants in current practice. They have substantially changed in recent years considering the framework, perforation and fixation of their supporting base and the detachable joint of the denture. Fixing method to the cortical ridge, surgical insertion strategy, implant material and postoperative care have all transformed significantly. In the case of the medical implant required the corrosion resistance, the suitable mechanical properties and the biocompatibility and high corrosion resistance.

1.1. 3D printing of patient-specific dental implants

Nowadays, the development of 3D printing technologies has made ever-increasing demands on perfect and precise products. During product processing, the design is carried out in a virtual environment, the product is printed out additively, its geometric sizes are then checked, and the processed blank implant part inspected after burr-drilling. Metal additive manufacturing technology provides a more homogeneous material structure for the manufactured part in comparison with the previously used casting procedure. Manufacturing productivity is also higher, and it is easier to control the final product quality.

In current practice, implants, which fit patients' bone anatomy, are designed in a virtual environment based on CBCT images. Afterwards, an STL file is created from DICOM format. Manufacturing is carried out with a metal 3D printer. Our 3D printer uses the energy of laser photons to fuse metal particles layer-by-layer with LMF (Laser Metal Fusion) technology [2-4]. The material of patient-specific implants is Grade 23 titanium alloy.

1.2. Surface treatment by nitridation
It was treated the 3D printed test samples by nitridation. The nitriding efficiency depends on the chemical composition of the treated sample, it needs to contain some high nitrogen affinity elements. By the way of diffusion, the nitrogen establishes in the surface layer some nitride compound and increase the surface hardness [5]. The formed compound volume depends on the nitridation treatment temperature and time, and also the media nitrogen concentration and the test samples chemical composition. The used alloy base metal is the titanium, which has a big affinity to join by nitrogen to form TiN or TiN2 compounds in the surface layer.

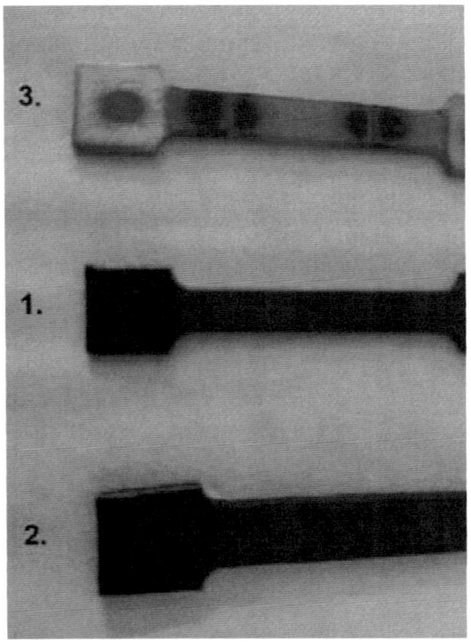

Figure 1. The nitrided test samples.

The nitrided microstructure also consists of a region of nitrogen-stabilized α-titanium, that is, α-case, and a nitrogen diffusion zone (typically 15–25 μm deep) underneath the compound layer. The plasma-nitrided microstructure is strongly correlated with the composition of titanium substrate and the nitriding process parameters such as duration, temperature, pressure, and composition of the nitriding medium, among which temperature has the most significant influence [6, 7]. The nitrided test samples are shown in Figure 1.

2. Used material and heat-treating processes

2.1. 3D printed test samples
The test samples made by additive technology (3D printing). LMF process was performed by using a Sisma Mysint 100 3D printer. The parameters for manufacturing Ti6Al4V was next: laser power 125 W, scanning speed 1000m/s, layer thickness 20 μm, laser beam spot size 50 μm. Argon atmosphere was used to sheld the melted pool during the additive process.

Table 1. Chemical composition (content in wt. %) of Ti alloy, Grade 23 remain is Ti.

V	Al	C	Fe	N	H
3.5-4.5	5.5-6.5	Max. 0.08	Max. 0.25	Max. 0.03t	Max. 0.0125

Chemical composition of the test samples showed in Table 1. The Grade 23 alloy is a special high purity version of the Ti 6Al 4V alloy. This kind of titanium alloys suitable for biomedical applications [6]. It has high corrosion resistance, it is lightweight and biocompatible also the mechanical properties are suitable for several applications [7, 8]. Density is 4,43 g/cm3, the melting point is between 1604-1660°C. Mechanical properties are the next: tensile strength is 860 MPa, yield strength 790 MPa, hardness 341 HV, elastic modulus 113,8 GPa.

2.2. Annealing of samples
The test samples were annealed after the printing process. The heat-treating is useful for homogenise the printed microstructure and decrease the heat affected stress conditions [9-11]. The heat-treating process parameters shown in Table 2. The heat-treating T-t diagram shown in Figure 2.

Table 2. Heat treating of the samples.

Treatments	Parameters
Annealing temperature	600 °C
Cooling media	Water and air

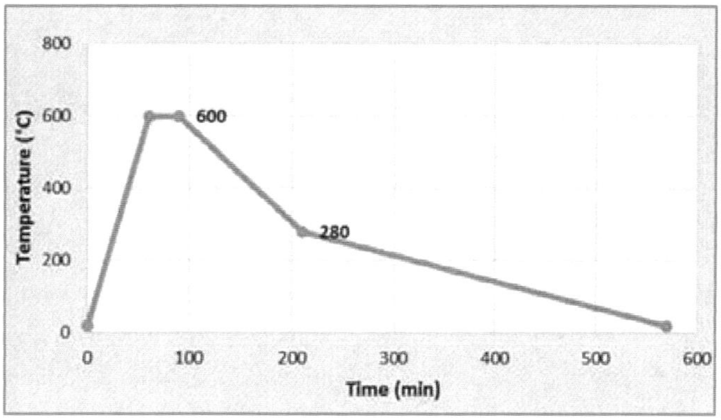

Figure 2. Annealing T-t diagram of the 3D printed samples.

2.3. Surface thermochemical treating processes
The plasma nitridation carried out at 525 °C for 24 hours in a nitrogen-containing medium. The other group of the test samples was nitrided by gas nitridation carried out at 570 °C for 2 hours and other ones for 12 hours. All samples surface was prepared by polishing (Ra=0,01 μm) before the nitridation.

Table 3. Nitridation parameters.

Number of the tested samples	Thermochemical treating parameters	Nitridation Temperature	Nitridation time (h)
1	Gas nitridation	570 °C	2 h
2	Gas nitridation	570 °C	12 h
3	Plazma nitridation	525 °C	24 h

The nitridation process parameters summarized in the Table 3. The used temperature much lower than the typical nitridation temperature (700-1100 °C) in the case of the same Grade 23 Ti alloys, but the high temperature can involve a microstructural changing (grain growing, overaging, embrittlement, unwanted phase precipitations, etc.), what we wanted to avoid [12]. The surface treating little bit modified the surface roughness (after nitridation Ra=0,05 μm). The average surface roughness measured by GD120 MarSurf surface roughness tester. The surface roughness did not change significantly after the nitridation process. The suitable surface roughness is important about the wear and corrosion resistance [12, 13].

3. Results and conclusions

3.1. Results

The original hardness of the test samples was 341 HV. The nitridation aimed to increase surface hardness and wear resistance. The hardness tests made by microhardness tester (CSM Instruments MHT), using the diamond Vickers indenter, the maximum load was 50 g.

Table 4. The hardness of the tested samples.

Number of the tested samples	Hardness $HV_{0,05}$
1.	386 $HV_{0,05}$
2.	412 $HV_{0,05}$
3.	423 $HV_{0,05}$

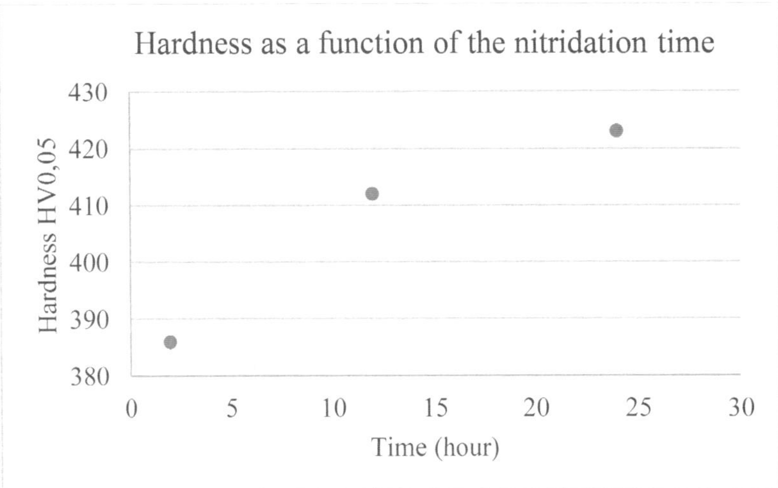

Figure 3. The nitrided surface layer hardness as a function of the nitridation time.

The hardness of the nitrided sample shown in Table 4 and the relationship between the nitriding time and the earn hardness shown in Figure 3.

4. Conclusion

The nitridation significantly modified the hardness of the tested Grade 23 titanium alloys surface. The increasing of the surface hardness shows a strong relationship with the nitridation time. The used nitridation temperature was different than the typical nitridation temperature to avoid the unsuitable microstructural changing. The high temperature can increase the grain growth of the titanium grains and establish precipitation.

It can conclude, that even that the used nitridation temperature was much lower than the typical temperature, as a function of the nitridation time the surface thermochemical process was successful. It needs some microscopy test to compare the high temperature (700-1100°C) and the used (525-570°C) temperature effect for the surface microstructure and the nitridation affected differences in the grain size and mechanical properties. The used surface nitridation process not affected significant modification of the surface roughness.

It can conclude that the used nitridation parameters are suitable to increase the Grade 23 titan alloy surface hardness.

5. References

[1] Katić J, Šarić A, Despotović I, Matijaković N, Petković M and Petrović Ž 2019 *Coatings* **9** 612
[2] Cerea M and Dolcini G A 2018 *Hindawi BioMed Research International* 2018 1-11
[3] Kulcsár K and Kónya J 2018 *Műszaki Tudományos Közlemények* **8** 51–60
[4] Lauer G, Wiedmann-Al-Ahmad M J, Eotten U, Hübner U, Schmel-Zeisen R and Schilli W 2020 *Biomaterials* **22(20)** 2799-2809
[5] Edrisy A and Farokhzadeh K 2016 *Plasma Nitriding of Titanium Alloys* IntechOpen (2016) http://dx doi org/10 5772/61937
[6] Hassanen J and Tünde K 2019 *Materials Research Express* **6(8)** 082003
[7] Čaplovič P P and Ľ Kovács T 2007 *Materials World* **2(1)** 1-6
[8] Tünde K 2010 *Materials Science Forum* **659** 191-196
[9] W Niu et al. 2013 **46** 640-644
[10] Tóth L 2018 *European Journal of Materials Science and Engineering* **3(3)** 165-170
[11] J Huand et al. 2020 *Materials Today Communications* **24** 101171
[12] Tóth T, Haraszti F and Kovács T 2018 *Journal of Materials Science and Engineering* **3(2)** 38-42
[13] Tóth T, Haraszti F and Kovács T 2018 *Acta Materialia Transylvanica* **1(1)** 53-56

IOP Conf. Series: Materials Science and Engineering **877** (2020) 012017 doi:10.1088/1757-899X/877/1/012017

Effect of Iron Oxide (Fe_2O_3) on the Properties of Fly Ash Based Geopolymer

W W A Zailani[1,2*], M M A Abdullah[2], M F Arshad[1], D D Burduhos-Nergis[3] and M F M Tahir[2]

[1]Faculty of Civil Engineering, Universiti Teknologi MARA, Shah Alam, Selangor 40450, Malaysia.

[2]Center of Excellence Geopolymer& Green Technology (CEGeoGTech), School of Materials Engineering, Universiti Malaysia Perlis, 01000 Kangar, Perlis, Malaysia, India.

[3]Gheorghe Asachi Technical University, Faculty of Materials Science and Engineering, Department of Physics, 67 Prof. Dimitrie Mangeron Blvd., 700050, Iasi, Romania.

*E-mail: warid.wazien92@yahoo.com

Abstract. Geopolymer is an attractive construction binder owing to its ability to improve the properties of the concrete and preserves the environment from the high CO_2 emission. Geopolymer technology will convert the potential hazardous industrial waste such as fly ash into valuable construction materials. However, there is a need of studying the properties of iron-based geopolymer in order to enhance the fundamental and knowledge of the geopolymer research also development in this study area. Fly ash which contains a significant amount of iron oxide (Fe_2O_3) was used as a precursor and tested at different curing duration (1, 3, 7, 14 and 28 days). Crystallization of iron oxide (Fe_2O_3) contained in the fly ash under geopolymerization process will be able to turn waste fly ash into a strong concrete materials, simultaneously creating a waste-to-wealth economy. Furthermore, the formation of fayalite detected from the microstructure characterization is mainly contribute to the strength development of the fly ash after 28 days curing.

1. Introduction

Geopolymer is an inorganic polymer which made up of two main constituent namely aluminosilicate-source materials and alkaline activator solutions [1]. Generally, aluminosilicate source materials that contain high silica, alumina, and calcium oxide minerals are commonly used in the geopolymer production. The combination of sodium silicate (Na_2SiO_3) and sodium hydroxide (NaOH) has been used widely as the alkaline activator in liquid form [2].

Fly ash, slag, kaolin and metakaolin are examples of aluminosilicate source materials which gives different behavior, properties and specialities depends on many factors [3-5]. One of the factors affecting the properties of geopolymer final product is the composition of aluminosilicate source materials. Fly ash is by far the most commonly used as a precursor for geopolymer concrete due to its commercial and performance reasons. It is an industrial by product which residue resulting from the combustion of coal [6]. Fly ash generally can be categorized into two classes which are class F and class C and the content of calcium in the chemical composition is one of the indicator to make the classification. Silicon oxide (SiO_2) and aluminium oxide (Al_2O_3)

can found mostly in fly ash thus make it the suitable source of aluminum and silicon for geopolymerization [7].

Geopolymer becomes an attractive research due to its ability to improve the concrete properties and preserves the environment from the high CO_2 emission from the concrete production. In addition, the geopolymer technology will convert the potential hazardous industrial waste such as fly ash, bottom ash, slag and etc. into the valuable construction materials [8]. However, it is very limited study that reported on the performance of high iron based geopolymer. The current research and development of geopolymer usually focussing on high silica and alumina based geopolymer due to its high resistance against temperature and high durability.

Over the past few years, iron based geopolymer have attracted increasing attention and have been considered for applications such as repair materials and etc. Dimitros et al. (2009) produces geopolymer from red mud and achieved very high strength and high durability when exposed to high temperature [9] and also excellent applied as flue gas desulfurization sorbent [10]. While Pacheco et al. (2008) explored tungsten mine waste based geopolymer and obtained very good mechanical properties of geopolymer [11]. However, the effect of iron oxide is not thoroughly discussed and reported. Hence, there is a clear need of searching and studying the properties of high iron based geopolymer to enhance the current state of the geopolymer research and development in this study area. Thus, this study focuses on the producing a geopolymer by using high iron content in the selected fly ash and mixed with alkali activator and left hardened under geopolymerization process at room temperature. The effects of the iron oxide to the microstructure and mechanical properties of geopolymer were assessed.

2. Experimental details

Fly ash has been used as the main aluminosilicate source materials and mixed with the alkali activator solution to produce geopolymer. The combination of sodium silicate (Na_2SiO_3) solution and 12 Molar sodium hydroxide (NaOH) solution at ratio 2.5 was used as the alkali activator solution [12]. The physical and chemical characteristics of the fly ash have a significant role in determining the properties of the fly ash based geopolymer. Thus, compressive strength and microstructure characteristics were carried out to explore the properties of high iron based geopolymer.

2.1 Chemical Composition Characterization

The chemical composition characterization of fly ash was determined by using a bench top X-ray fluorescence (XRF) spectrometer with a brand name PAN analytic PW4030, MiniPAL 4, X-Ray Flourescence (XRF) spectrometer. The chemical composition was reported in the percentage by weight of the sample.

2.2 Microstructure Characterization

The microstructure of the fly ash and fly ash based geopolymer were obtained using JSM-6460LA model Scanning Electron Microscope (JEOL) utilizing the secondary electron detectors. All specimens were coated with palladium by using Auto Fine Coater JEOL JFC 1600 model before test. At the same time, the surface elemental composition was taken by Energy Dispersive X-ray (EDX) spectroscopy using the same procedure.

2.3 Compression Test

The compressive strength was evaluated using Instron machine series 5569 Mechanical Tester with a loading capacity of 3000 kN using a loading rate of 0.5 MPa/s. Compressive strength was performed on cubes with the dimensions of 50 mm × 50 mm × 50 mm according to ASTM C109 [13]. The average of three specimens was tested to determine the compressive strength of geopolymer mortar and the mean value of three samples was reported.

3. Physical and chemical composition analysis of fly ash

3.1. Physical analysis of fly ash

Morphological studies of fly ash and sand have been conducted using Scanning Electron Microscope (SEM) as shown in Figure 1. The spherical particles structure of raw fly ash can be seen clearly in the Figure 1. Existing of the ferrosialate and cennosphere with small particle fly ash is from the special source of coal which able to improve geopolymerisation rate and shows high reactivity. Thus, it is believed that different structure of fly ash will have impact to the mechanical properties.

The ferrosphere (dendritic type) as shown in the figure was identified. Ferrosphere contains very high iron oxide and this kind of particles structure was rarely observed in the fly ash. The ferrosphere has a significant impact in the formation of geopolymer gel. Needle crystallites on the surface of ferrosphere indicate that crystallizes from amorphous aluminosilicate melt during combustion of fly ash but not having transformed directly from fly ash [14].

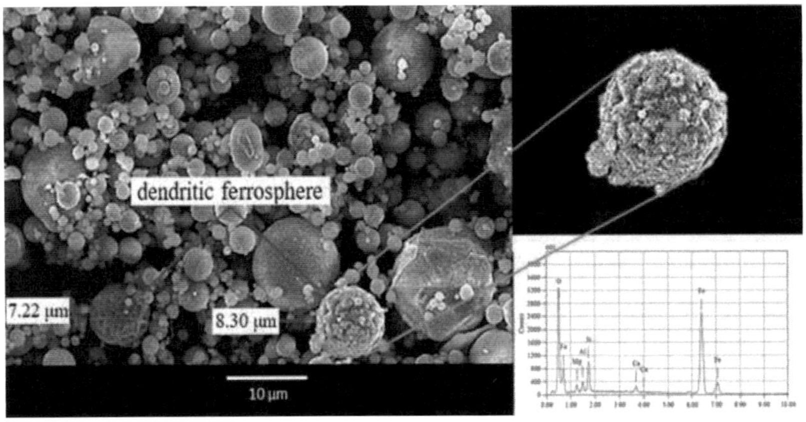

Figure 1. Ferrosphere shape of fly ash.

Needle crystallites of ferrosphere were capable of raising the strength of fly ash based geopolymer which was good for the long term durability.

3.2. Chemical composition analysis of fly ash

X-ray fluorescence (XRF) was used to identify and measure the chemical composition of fly ash as shown in Table 1. The main chemical composition of fly ash is silica oxide SiO_2, followed by Fe_2O_3, Al_2O_3, and CaO. The content of SiO_2, Fe_2O_3 and Al_2O_3 are 40%, 23% and 14% and sum of these oxide minerals are more than 70 %. Fly ash is classified as Class F and suitable to be used as raw material for geopolymer as referred to American Society of Testing and Materials (ASTM C618) as shown in Table 2 [15]. It also indicates that this fly ash classified as pozzolanic material and has cementitious properties [16]. Those SiO_2 and Al_2O_3 will undergoes geopolymerization to form aluminosilicate gel or known as (N-A-S-H) gels.

Table 1. Chemical composition of fly ash from different sources.

Chemical	SiO_2	Al_2O_3	Fe_2O_3	TiO_2	CaO	MnO	CuO	K_2O	Na_2O	SO_3	SrO
Fly ash (wt.%)	38.8	14.7	19.48	1.02	18.1	0.16	0.04	1.79	-	1.50	0.11

Table 2. Fly ash classification.

Chemical	wt.%	ASTM C618 (%)
SiO_2+ Al_2O_3+ Fe_2O_3	72.98	70.00

Based on the chemical composition testing, iron (III) oxide content (Fe_2O_3) was the second highest mineral oxide that exist in fly ash which was 19.57 %. The content of Fe_2O_3 from the source material was probably have a positive impact on the durability in term of final compressive strength of geopolymer. The atomic diameter of iron was higher than silica and alumina, thus the high content of iron in the raw materials may longer the geopolymerization reaction with alkali solution to form iron silicate binder. Thus, the longer aging duration of the geopolymer sample left at room temperature help to enhance the strength development of geopolymer. The significant amount of iron oxide content in the raw materials gives a good indication to improve the overall properties of geopolymer [17-20].

3.3. Compressive strength of fly ash based geopolymer at different curing duration

Trends for compressive strengths against different curing duration were observed. The purpose of curing is to identify the appropriate period of sample to gain the desired strength and allow for the further development of its optimum. Figure 2 shows that the compressive strength of geopolymer strength increase as the curing time increased. Geopolymerisation is a time dependent process that may continuously.

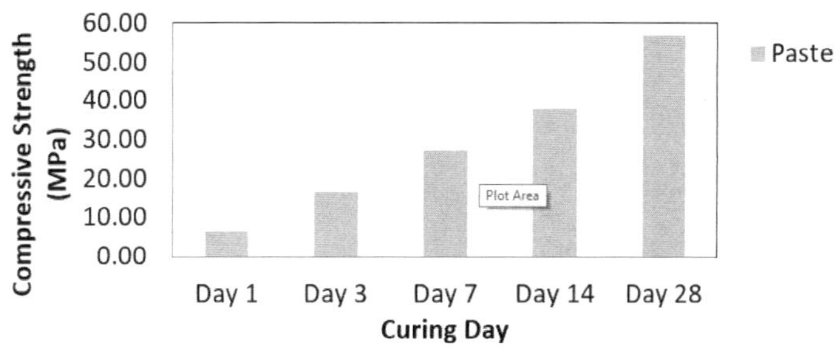

Figure 2. Compressive strength of geopolymer with different curing time.

The increment of the compressive strength between 14 days and 28 days of curing of fly ash based geopolymer was about 38 %. It was also noticed that the strength development of geopolymer differs significantly at others curing duration. This can be explained by the high percentage of iron (Fe_2O_3) available in the fly ash which is about 19.57 %.

Fly ash with high content of iron (19.57 %) takes time to react with the alkali solution in the geopolymer system in order to form the iron silicate binder gel. This was due to the high atomic mass and large atomic diameter of the iron compared to silica and alumina. Cannio et al. (2018) identified the ferrosialate in addition to the sodium aluminosilicate gel in the geopolymer which contribute to the durability of the geopolymer [20-22]. Thus, the longer curing duration led to increase the compressive strength which was contributed by the iron oxide.

3.4. Morphology of fly ash based geopolymer

The microstructure analysis for fly ash based geopolymer paste cured at different curing days are presented in Figure 3. The microstructure of selected samples were cured at different aging duration which presented by the 1 day and 14 days aging as shown in Figure 3 (a) and Figure 3 (b), respectively. It was found that a denser matrix has been obtained at 14 days of curing time compared to that obtained at 1 day. When compared between at 1 day with 14 days ageing period, compressive strength increased with sample age.

Ferrosphere are well known iron rich spheres with a chemical composition that rich in iron oxide (Fe_2O_3). The EDX analysis for the early curing ages (at 1 day) marked with 'X' at Figure 3 (a) has found that this partially reacted fly ash is rich in iron. This kind of fly ash microstructure is formed by the conglutination after the oxidation of iron bearing minerals contains high amount of iron with more than 80 %. The white needle like structures shows the reaction on the surface and formed from the crystallization of iron oxide and from the geopolymerization between iron and alkali activator solution but requires time (Xue et al., 2008).

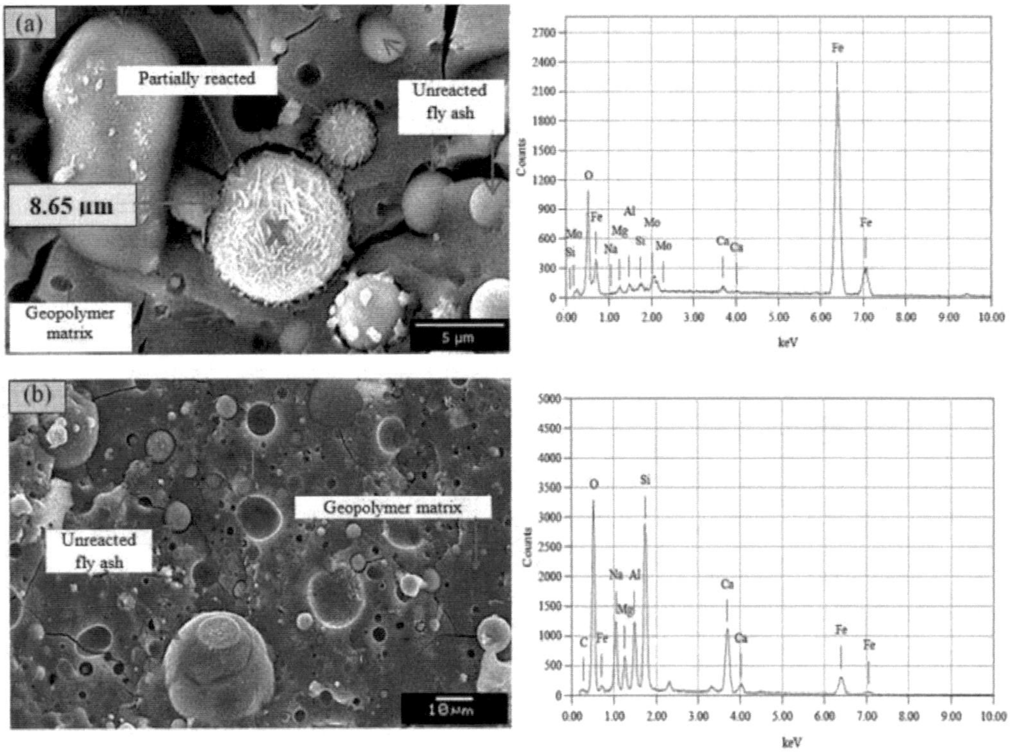

Figure 3. Microstructure of geopolymer samples (a) 1 day curing and (b) 14 days curing.

This EDX results confirmed that this shape is one of the ferrosphere particles which is partially reacted at early age according to the high percentage of iron element content.

4. Conclusion and perspectives

The properties requirement as grouting material led us to propose the optimum grout geopolymer. From the behaviour of the geopolymer, this clearly highlighted that the use of geopolymer is potentially to provide and increased stiffness under services condition and the high tensile strength of geopolymer. Thus, the main purpose of this paper is to give an overview on the most recent studies using polymer based materials and geopolymer with different raw material as an excellent repair material to developed and applied to solve civil engineering problems indeed. Geopolymer is compatible with the standard requirements and shows the potential to become as an excellent repair material. Geopolymer are good in durability and high mechanical strength thus they can be an alternative as an excellent repair materials.

5. References

[1] Davidovits J 1994 *First Intern. Conf. Alkaline Cem. Conc.* **1(10)** 131–149
[2] Zainal F F, Fazill M F, Hussin K, Rahmat A, Abdullah M M A B and Wazien W 2018 *Solid State Phenom.* **273** 175-180
[3] Ahmad R, Mustafa A B A M, Mastura W I W, Sandu A V, Wazien A Z W and Nizam S N S 2019 *IOP Conf. Ser. Mater. Sci. Eng.* **551(1)** 012083.

[4] Fang G, Ho W K, Tu W and Zhang M 2018 *Constr. Build. Mater.* **172** 476–487
[5] Fauzi A, Fadhil M, Malkawi A B and Mustafa M B A 2016 *Procedia Eng.* **148** 487–493
[6] Assi L N, Eddie E, Elbatanouny M K and Ziehl P 2016 *Constr. Build. Mater.* **112** 807–815
[7] Yuvaraj D and Manu S 2020 *Cem. Concr. Res.* **129** 105959
[8] Mohamed R, Abd R R A, Mustafa A B A Khimi S R, Aida M M and Wazien A Z W 2019 *IOP Conf. Ser. Mater. Sci. Eng.* **551** 012093
[9] Dimitrios D D, Ionna P G and Dimitrios P 2009 *Min. Proc. Ext. Met. Rev.* **30** 211-239
[10] Nie Q, Hu W, Ai T, Huang B, Shu X and He Q 2016 *Constr. Build. Mater.* **125** 905–911
[11] Pacheco-Torgal F, Castro-Gomes J P and Jalali S 2008 *Constr. Build. Mater.* **22(3)** 154–161
[12] Mustafa A B A M, Binhussain M, Zarina Y and Rafiza A R 2012 *ACI Mater. J.* **109(5)** 503-508
[13] ASTM-C109/C109M 2013 *American Society for Testing and Materials*, West Conshohocken, PA 19428-2959, United States
[14] Ahmad Z W W, Bouaissi A, Abdullah M M A B, Abd Razak R, Yoriya S, Mohd Salleh M A A, Rozainy M A Z M and Fansuri H 2020 *Appl. Sci.* **10(9)** 3321
[15] ASTM-C618 2012 *American Society for Testing and Materials*, West Conshohocken, PA 19428-2959, United States
[16] Mustafa A B A M, Kamarudin H, Nizar I K, Sandu A V, Binhussain M and Rafiza A R 2013 *Rev. Chim-Bucharest* **64(4)** 382-387
[17] Davidovits J 1991 *J. Therm. Anal.* **37(8)** 1633–1656
[18] Mustafa A B A M, Kamarudin H, Bnhussain M, Khairul N I, Rafiza A R and Zarina Y 2011 *Adv. Mater. Res.* **328-330** 1475–1482
[19] Burduhos Nergis D D, Abdullah M M A B, Sandu A V and Vizureanu P 2020 *Materials* **13(2)** 343
[20] Burduhos Nergis D D, Vizureanu P and Corbu O 2019 *Revista de chimie* **70(4)** 1262-1267
[21] Diaconescu V D, Scripcariu L, Matasaru P D, Diaconescu M R and Ignat C A 2018 *EUROINVENT International Conference on Innovative Research (ICIR)* 1757-8981
[22] Cannio M, Cyriaque R, Myllyam L, Rosa R, Kamseu E, Chinje U and Leonelli C 2018 *J. Clean. Prod.* **199** 849–859

IOP Conf. Series: Materials Science and Engineering **877** (2020) 012018 doi:10.1088/1757-899X/877/1/012018

Recycling of Plastic Throughout Pyrolysis and Distillation Process to Recover an Alternative Fuel Sources

S Muzarpar[1], A M Leman[1*], K A Rahman[2] and M T R M Foudzi[3]
[1] Faculty of Engineering Technology, Universiti Tun Hussein Onn Malaysia (UTHM), Pagoh Higher Education Hub, KM 1, 84600, Muar, Johor, Malaysia
[2] Department of Pertochemical Engineering, Politeknik Tun Syed Nasir (PTSN), Pagoh Higher Education Hub, KM 1, 84600, Muar, Johor, Malaysia
[3] Department of Mecanical Engineering, Politeknik Nilai, Kompleks Pendidikan Nilai, Bandar Enstek, 71760 Labu, Negeri Sembilan, Malaysia

*E-mail: mutalib@uthm.edu.my / mutalib1480@gmail.com

Abstract. Pyrolysis process is a process whereby plastic undergo the heating process and transform into another type of product either liquid form or waxy form. The basic purposed of the study was to determine the type of product formed from different type of plastic used. From the product collected in pyrolysis process it will undergo a further distillation process to increase the quality of the product. The methods that was involved in the making of the product is by using pyrolysis process, distillation process whereby a few product are derived. For pyrolysis process, plastic is heated until vaporized and the vapor is collected and condense to collect the product, while for the distillation process the separation process is done to separate the product according to its boiling point. For data collection, every sample is tested by using the Fourier Transform Infrared Spectroscopy (FTIR) and the bomb calorimeter for its calorific value. From three types of plastic used which is polystyrene, polyethylene, and polyethylene tetraphlate the most suitable type of plastic is polystyrene due to its product which is in liquid form. The distillation process will give three end product which is similar to commercial petrol, diesel and high density oil.

1. Introduction

Plastic consumption has increased from a last few decades. Approximately is about 60 million tons annual consumption of plastics in occidental Europe. From the annual capacity, 23 million tons is plastics waste. Packing and packaging is the main product of plastic waste approximately 63 wt% [1]. Agriculture sector is the main user of the plastic materials since they need to cover the vegetables, fruit, some of the farm and etc. Since they need to cover the vegetables, fruit, some of the farm and etc. On the other hand the plastic packaging also need for other industries which is high-volume market and they directly look into the advantages of plastic if compared with the traditional materials. Nowadays the plastic that normally use for packing and packaging is the main contributors to the most visible in the waste stream. A lot of debate and a public criticism on solid materials. It have look into the component and it comparatively short life-cycles and usually are non-degradable. But it have received a great deal of public since the price and maybe their durability is the characteristic that are prefer by the consumer [2]. Basically, the common ways to treat waste plastics is via landfill and incineration. The municipal solid waste is increase year by year and plastic is one of the waste stream.

Normally, the waste plastic that dump into the landfill is not converted to the energy recovery. Incinerator is the solution recovering energy from the plastic waste but this process give an impact to atmosphere since the carbon content in the plastics being mostly converted into CO_2 and released. To reduce the plastic fraction of municipal solid waste which is recycling process. But so far the rate below 15%, representing a waste of a resource [3].

In the modern era, pyrolysis may be an attractive ideas. At the same time it will increase the chemical recycling percentage, and to obtain valuable liquid and gas fuels from plastic wastes [4]. Pyrolysis process normally produce three phases such as solid (char), liquid (tar), and gas that it is spontaneously and immiscible. Immiscible means that those phases cannot mixed together after pyrolysis process. Immiscible means that. However, three phases mixture material could be produced by temperature performed [5-9]. Thermal phenomena may include in this process due to degradation process of large molecules may occur caused by temperature, catalyst presence and residence time. However, large molecules degradation occurred with or without the catalyst presence [10]. Degradation of these large molecules depends on a number of different conditions. Basically the temperature, residence time, and the presence of catalysts is the factor that should be considered when dealing with pyrolysis process. And at the same time, the pyrolysis reaction can be carried out with or without the presence of catalyst. Thermal and catalytic pyrolysis reaction according to the setting for the process.

2. Materials

More than hundreds of commercial plastic materials available. Only the reliable and quality material will be user preference. High volume and relatively low price is the factors that will be consider for a thermoplastics commodity. Figure 1 below show the consumption on a global basis according to the plastics and their fractional. Highest consumption plastic types is Low-density polyethylene (LDPE), high- density PE (HDPE), polypropylene (PP) and PVC approximately 61% while the other rest is PVC, PS, EPS, PET and PUR is around 20% and Others is 19% [11].

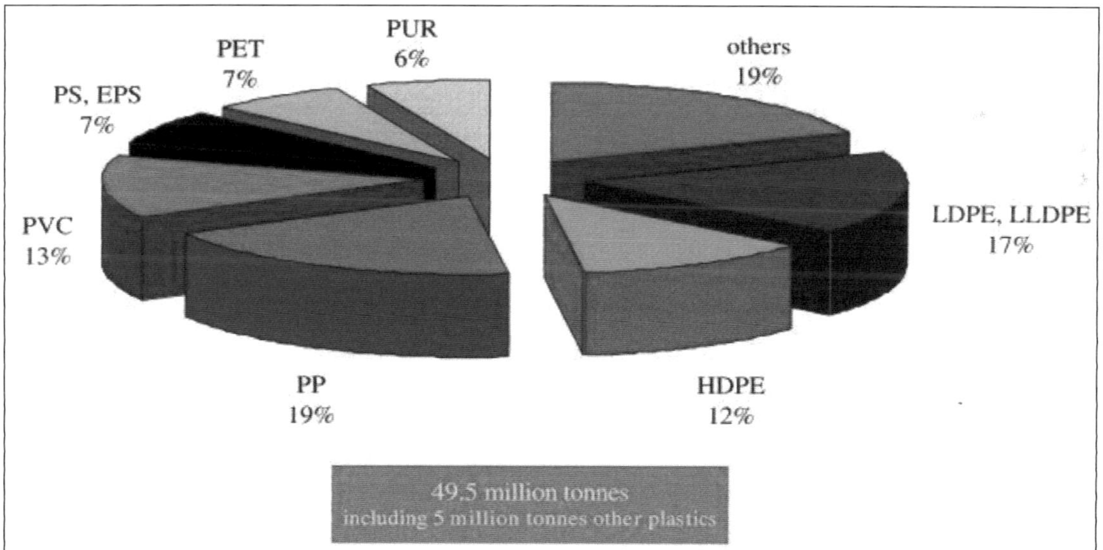

Figure 1. World plastic demand by resin type [12].

For the raw materials, the focus is on polystyrene (PS), polyethylene (PE) which include low and high density of polyethylene (LDPE, HDPE), and Polyethylene tetraphlate (PET). Meanwhile Figure 2 shows the monomer and polymer of certain plastic. For polyethylene its monomer is ethylene which is derived from ethane. Polypropylene is long chain of propylene which is derived from propene.

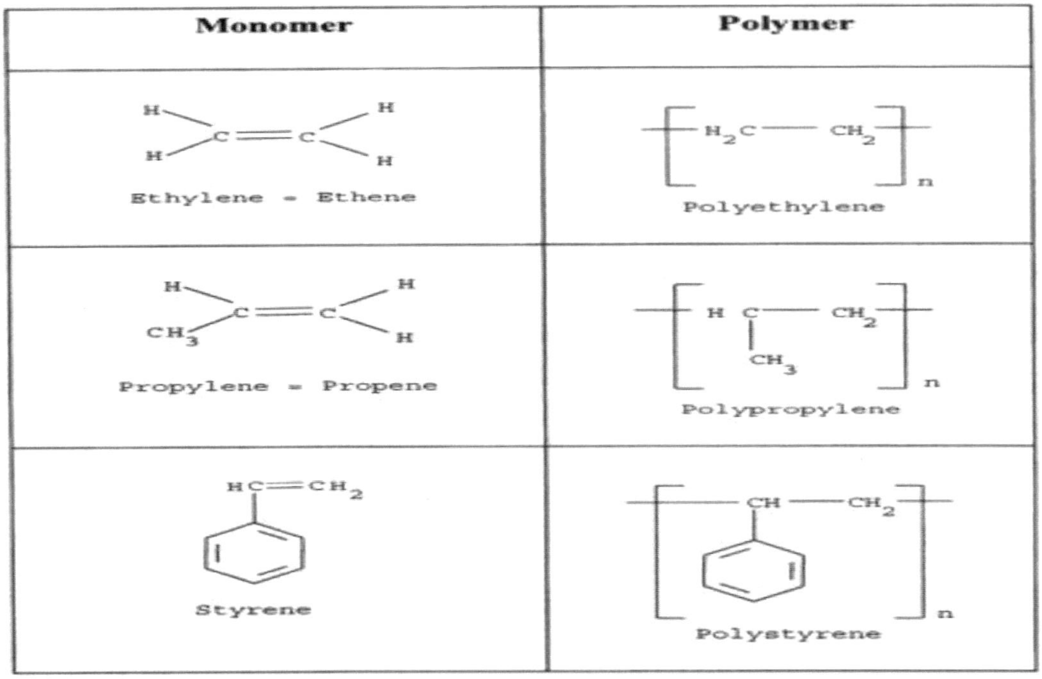

Figure 2. Representative of plastic monomer and its polymers [13].

Styrene monomer contain of benzene ring and long monomer chain will be called as polystyrene [13].

3. Process Involved

3.1. Pyrolysis process

Common pyrolysis system is shown in Figure 3. It consists of reactor, 2 heat exchanger system which called by first and second heat exchanger, 2 bottles as condenser system and gas storage. They conduct the pyrolysis process at temperature of 500 ^{0}C, and various temperature increasing rates between 6 and 14 ^{0}C/min [14].

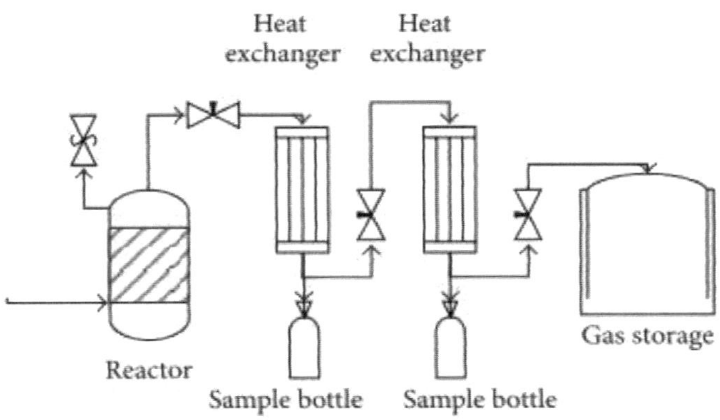

Figure 3. Schematic diagram of pyrolysis process [14].

Figure 3 shows the schematic diagram of pyrolysis. Meanwhile, in this study the similar system was applied. However, it was conducted by different temperature and heating rate in order to eliminating oxidation product which may developed during the pyrolysis process. The process is started with inserting the plastic according to its type into the combustion chamber. Next the lid of combustion chamber is closed tightly to prevent the vapour from escaping. The condenser is filled with ice and the collection jar is placed beneath it. After completing the setting up procedure, heating process takes place. Turn on the gas and ignite the stove, be very careful in this process because it could cause some serious damage. The process is continued until all the vapour from the plastic is converted into fluid / oil, next repeated the process by using different type of plastic. The product of each type of plastic is collected and recorded the date, time and materials used. Based on the pyrolysis process that we have done, two types of product is in liquid form for polystyrene and waxy form for PE and PET. Both products were stored for analysis and for further process.

3.2. Distillation process

After pyrolysis process was complete, the collected product will be differentiated into liquids form and wax form. In this stage, the oil or product collected cannot be used as fuel to run the diesel or gasoline engine. Thus to make the oil possible to use as alternative fuel sources, some process should be done in order to convert the oil into petrol and diesel. The process is a distillation process. The petrol and diesel can be extracted from the oil collected by separating them according to their boiling point. The distillation process was conducted in the laboratory under controlled environment.

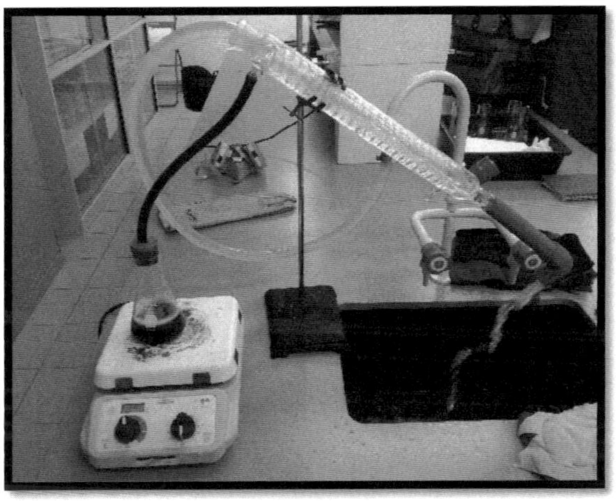

Figure 4. Experimetal setup for Distillation Process.

The most important equipment that's needed in this process is a hot plate, beaker, condenser, and product to be distiller and setting temperature are between 100 ^{0}C and 300 ^{0}C. Figure 4 shows the apparatus setup for distillation process.

4. Result and Discussion

For the pyrolysis end product the result is tested by using two different analyzer / machine which are FTIR machine and Bomb Calorie Meter.

4.1. FTIR Result for Pyrolysis of Polystyrene

Figure 5 shows the FTIR result for the pyrolysis process of polystyrene. The red fluctuation line indicate the major group of chemical composition that present in the product, while the blue color graph indicate the exact position of the compound related to their wavenumbers. Theoretically, the first blue graph which has the highest peak around 2900 cm^{-1} to 3000 cm^{-1} indicates that the product is hydrocarbon of CH_3 and has asymmetrical stretch. For the blue graph which has the highest peak around 700 cm^{-1} indicated that there is aromatic group present in the product.

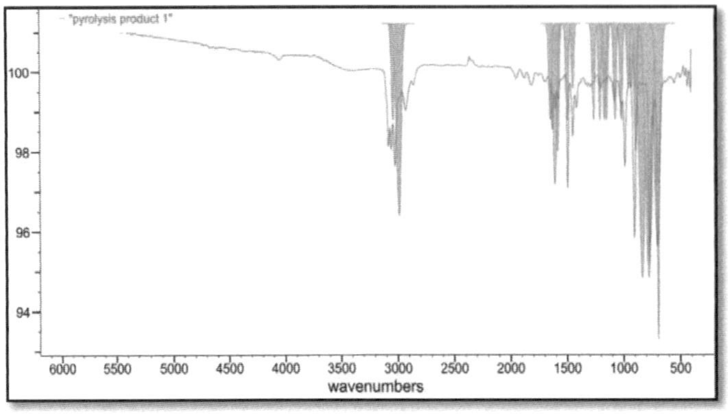

Figure 5. Graph of spectrometer for Polystyrene.

The strong odor that produce within the product is the proved of the aromatic group present in the product.

4.2. FTIR Result for Pyrolysis of Polyethylene

Figure 6 shows the FTIR result for pyrolysis of polyethylene. In general the red line can indicate the composition of the product is aliphatic hydrocarbon, due to lack of fluctuation of the red line at the back of the graph. The blue graph which has highest peak, about 3000 cm-1 shows that the product contained hydrocarbon. To be specified the hydrocarbon is CH3 with asymmetrical stretch. The blue graph that spike at wavenumbers of 1470 cm-1 indicates the product contain alkanes group which is methyl group and has asymmetrical C-H bending.

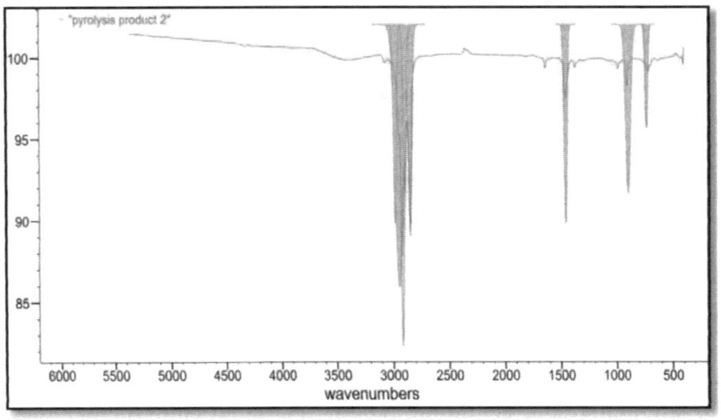

Figure 6. Graph of spectrometer for pyrolysis of polyethylene.

The highest peak that aligns with wavenumbers of 700 cm-1 can be categorized as aromatic group. The aromatic group is very small in amount thus produce less odor compared to other product.

4.3. FTIR Result for Pyrolysis of Polyethylene Terephthalate

Figure 7, shows the FTIR result for pyrolysis of polyethylene terephthalate. The highest hydrocarbon peak is at wavenumber of 3000 cm^{-1}. These indicate that the product has CH_3 type of bonding. The fluctuation at the back side of the graph is refer to the aromatic compound that present inside the product, and the highest peak was at wavenumbers of 700 cm^{-1}. Next there was also benzene ring structure that can be detected by the machine which ranges about 1500 cm^{-1} to 2000 cm^{-1}. High intensity of the benzene ring will produce a compact structure of the product thus gave the structure the solid waxy formed. These phenomenon cause the product hard to distillate and required higher temperature in order to break the chain reaction. There was lack of light absorption in between the range of 3600 cm^{-1} to 4000 cm^{-1} which indicate that the component is lack of alcohol group inside the product.

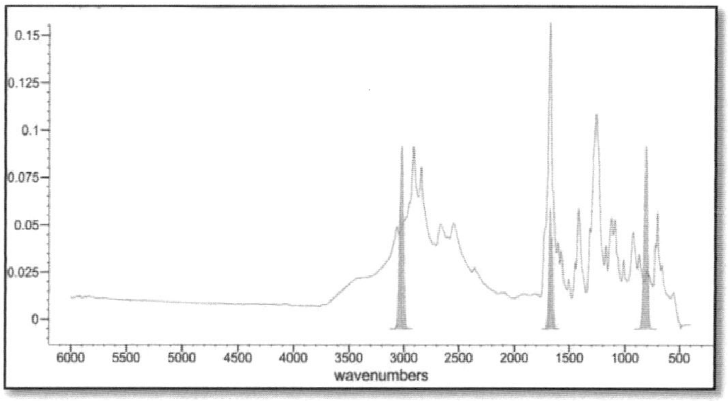

Figure 7. Graph of spectrometer for pyrolysis of polyethylene terepthalate.

From the Pyrolysis process the product collected can be either in liquids form or in waxy formed. The liquids product is derived from the pyrolysis process of polystyrene (PS) while the waxy product is derived from the pyrolysis of polyethylene (PE) or polyethylene terephlate (PET). The Distillate product undergo burning test to determine the soot produce, and duration of combustion. The burning test result is shown in Table 1 and Figure 8 shows the distillation end product.

Table 1. The burning test for distillate product.

Product	The temperature required	Viscosity	Color of product	Soot produce
Product 1	100 °C	Low	Light yellow	Low
Product 2	300 °C	Medium	Light brownish	High
Bottom product	More than 500 °C	High	Dark brown	Very high

Figure 8. Product form distillation process.

Figure 8 shows the product end product of distillation process form the pyrolysis product. The distillation process is aimed to retrieve thee product which is similar to gasoline, diesel and high density oil.

4.4. Calorific Value for Each End Product

The calorific value for each product that collected was shown in Table 2 below.

Table 2. Calorific value data.

Product	Calorific value (MJ/kg)
Pyrolysis Product In Liquids Form (PS)	41.10
Pyrolysis Product In Wax Form (PE and PET)	45.50
Distillate Product At 100 °C	41.00
Distillate Product At 300 °C	41.50
Bottom Product	40.60

From this test all the products were confirmed to be fuel base composition. Beside that the product also can be used as fuel source according to their function. The calorific value for the petroleum base product was usually in the range in between 40 MJ/kg to 50 MJ/kg. Due to the product testing, the calorific value was in the range of the commercial petrol and diesel. From the data, this similarity means that the product also can run the vehicle that used the commercial petrol and diesel. The high value of this calorific value also indicated that it stored a lots amount of the energy. It is better because it's can run the petrol or diesel transportation in much higher duration due to their higher energy stored. In comparison with the commercial petrol and diesel, the product also can run the transportation as good as the commercial product. As for the distillation process, the liquid product from the pyrolysis process can be separated into several products according to its hydrocarbon chain and boiling point. From the distillation process, three more products can be extracted which are petrol, diesel and high viscosity oil. All the product were essential to human daily life, such as powering transportation and also act as an industrial fuel source.

5. Conclusion

Plastic could be recycled into alternative sources through series of process. It could be generate as an alternative fuel such as petrol or diesel, wax or other alternative fuel source. Base on liquids collected it will undergo further process which is a distillation process to extract raw material. While the wax could be instantaneously used in daily life, such as for candle and act as surviving kits when required to start a fire. In comparison with the commercial product, testing is vital in order to test the oil either it could be performed as well as a commercial product. From the FTIR test, the entire product has exactly the same compound as a commercial product while the calorific value of the entire product is in the same range as the calorific value of commercial product. This testing shows that the amount of energy that could be generated by the product as similar with the commercial product.

Acknowledgement

The authors would like to thank the Ministry of Higher Education Malaysia (MOHE), Universiti Tun Hussein Onn Malaysia (UTHM) and Centre for Graduate Studies – UTHM. Thanks to Politeknik Tun Syed Nasir (PTSN) and Politeknik Nilai (PNS) for the involvement of this project.

6. References

[1] Cimadevila F 2018 *Plastics Europe* Exporecicla Technical Conferences. Zaragoza, Spain
[2] Kumar S P, Achyut K and Singh R K 2011 *Resources Conservation and Recycling* **55** 893-910
[3] Williams P T 2005 *Waste treatment and disposal*. Chichester. John Wiley & Sons

[4] López A, Marco I D, Caballero B M, Laresgoiti M F and Adrados A 2011 *Chemical Engineering Journal* **173**(1) 62-71
[5] Hall W J and Williams P T 2007 *Journal of Analytical and Applied Pyrolysis* **79**(1) 375–386
[6] Kaminsky W and Kim J S 1999 *Journal of Analytical and Applied Pyrolysis* **51**(1) 127–134
[7] L´opez A, de Marco I, Caballero B M, Laresgoiti M F and Adrados A 2010 *Waste Management* **30**(4) 620–627
[8] Williams P T and Slaney E 2007 *Resources, Conservation and Recycling* **51**(4) 754–769
[9] Jung S, Cho M, Kang B and Kim J 2010 *Fuel Processing Technology* **91** 277–284
[10] Scheirs J and Kaminsky W 2006 *Feedstock Recycling and Pyrolysis of Waste Plastic*. Converting Waste Plastics into Diesel and Other Fuels. Wiley Series in Polymer Science
[11] Andrady L A and Neal A M 2009 *Philos Trans R Soc Lond B Biol Sci* **364** 1977-1984
[12] Plastics Europe 2008 *The Compelling Facts About Plastics*, Analysis of Plastics Production, Demand and Recovery for 2006 In Europe. Belgium, Plastics Europe
[13] Pinto F, Costa P, Gulyurtlu I, Cabrita I 1999 *Journal of Analytical and Applied Pyrolysis* **51**(1) 239-55
[14] Hoseini S M F and Dastanian M 2013 *Journal of Chemistry* **5** 487676

Microstructural Analysis of Multiple Layer Depositions on Cast Iron Using the Electrospark Deposition Method

M C Perju[1,2], C Nejneru[1*], P Vizureanu[1,2], D D Burduhos Nergis[1] and A V Sandu[1,2,3]

[1]"Gheorghe Asachi" Technical University of Iasi, Blvd. Mangeron, No. 51, 700050, Iasi, Romania

[2]Center of Excellence Geopolymer and Green Technology, School of MaterialsEngineering, Universiti Malaysia Perlis (UniMAP), P.O. Box 77, d/a Pejabat PosBesar, 01007, Kangar, Perlis, Malaysia

[3]Romanian Inventors Forum, St.Sf. P. Movila 3, Iasi, 700089, Romania

*E-mail: nejnerucarmen@yahoo.com

Abstract. The present article is focused on the obtaining of a thin layer system on Fe-C alloys by deposition of 3 successive layers with several electrodes. These layers offer to the material a broad range of enhanced properties as: mechanical resistance, surface hardness, compression behavior and a martensitic structure of the carbides, which reduce the effect of the wear, having positive results on the life-cycle of the product. The principle of the hardening consists in an electrical discharge by spark on which the action of the rectified current impulse transfers material from the electrode, considered cathode to the surface of the material which is the anode. This material reacts chemically with the nitrogen, from the atmosphere, with carbon and with support material, forming a diffusion layer, resistant to wear.

1. Introduction

Thin-film coatings are generally used to improve hardness, wear and oxidation resistance. Along with traditional technologies of coatings achievement we assist to the development, improvement and extension of some modern coating methods, through physical and physical-chemical methods, which assure high purity and adherence through a very large variety of coating realizing procedures [1].

Increasingly, nowadays, electro-spark deposition (ESD) is used for reconditioning the dimensions of used parts, because this method allows the obtaining of high wear resistance coatings [2-5].

The proposal to use the method consists in prolonging the functioning period of Fe-C alloys pieces that work under the intense regime (ex. piston rings for Diesel engines, cutting tools, dies for heat and cold deformation, medical instruments, different mechanical pieces) [6, 7].

Procedure's advantage is that the heat density of the piece is minimum maintaining chemical composition and the properties of the basic material.

The thin layers system hardened through impulse electric discharges method is divided into an exterior layer with a strongly modified structure at the surface and an interior layer (diffusion layer) with the properties corresponding to the basic material and added material. The hardened layer presents cracks that are advantageous to the lubrication process (during exploitation) by protecting the basic material of excessive wear [8-10].

2. Materials and methods

In this study, gray cast iron with the chemical composition shown in Table 1 was used as base material. The elemental chemical composition was analysed using a Foundry-Master Spectrometer [11].

Table 1. Chemical composition of ferritic-pearlitic cast iron.

Element	Fe	C	Si	Mn	P	S	Cr	Ni	Cu	Mo
%, wt.	92.23	3.89	2.84	0.27	0.05	0.06	0.29	0.14	0.19	0.04

As consumable materials, three types of electrodes were used: tungsten carbide (WC), titanium carbide (TiC) and tungsten (W).

Furthermore, the depositions were obtained through the electrospark deposition method. The process of hardening the metal surfaces by this method consists in performing electrical discharges between the tool-electrode, designed in the shape of an electrode, and the surface of the processed workpiece. Both, the electrode and the workpiece are connected to the discharge circuit of the pulse generator. In other words, in the case of electric spark discharge, in a gaseous environment, the essence of the process, mainly, consist of the erosion of the electrode material (the anode) and the transfer of the erosion products from the electrode to the surface of the workpiece (the cathode).

In this study, an ELITRON 22A equipment was used to obtain the thin layer depositions through impulse discharge method.

3. Results and discussion

The micromorphology analysis of the triple surface deposition (WC/TiC//W) shows a relatively smooth surface [12-15], with partial stripping of graphite, cracks, splashes and micro-adhesions zones specific to the tungsten layer which was the last deposited layer, i.e. the outer layer, as can be seen in Figure 1.

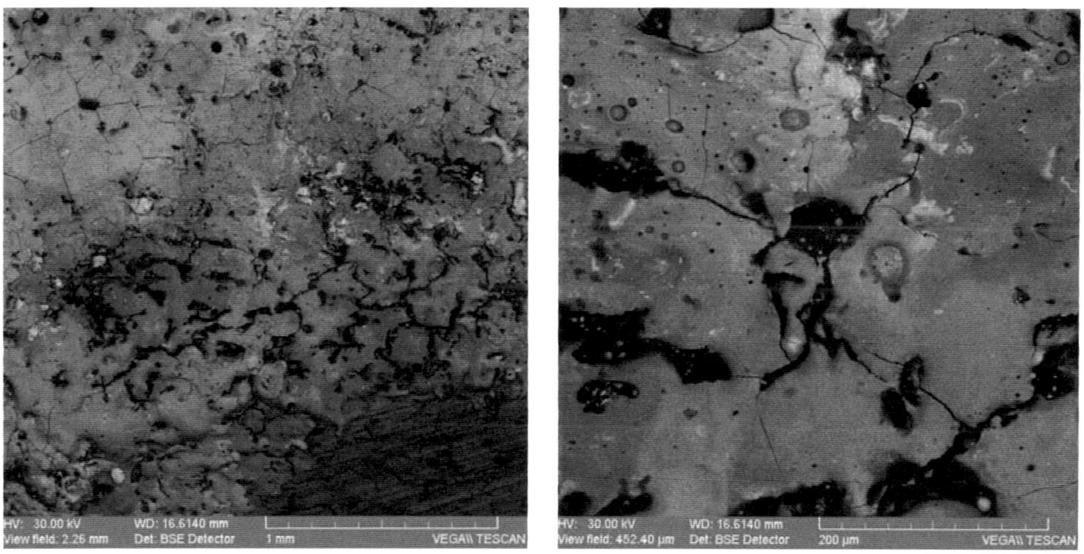

(a) (b)

Figure 1. Micromorphology of the triple-layer (WC/TiC/W) coating: (a) BSE micrography, 1 mm; (b) BSE micrography, 200 μm.

The high surface tension of the TiC melt results in the appearance of pinches or micro-holes in the deposited layer, however, in this case, the concentration of defects is much lower than in case of deposition with only TiC electrode. Tungsten carbide layers have good adhesion but important non-

IOP Conf. Series: Materials Science and Engineering **877** (2020) 012019 doi:10.1088/1757-899X/877/1/012019

uniformities in both structure and layer thickness. The wolfram-carbide (WC) layer is relatively hard and it has a moderate ability to take over the apparent splashing defects from subsequent deposits. The complex tungsten and iron carbides from its composition facilitate the formation of compact and uniform micro-alloy layers with tungsten and titanium carbides. The EDX analysis shows a high percentage of tungsten of 27.35%, a percentage of carbon of 10.08% and a percentage of titanium of 1.54% in the third deposited layer (Table 2 and Figure 2). However, some of the titanium disappeared either by combustion, either by spraying.

Table 2. Chemical composition of the WC/TiC/W deposited layer.

Element	Fe	C	W	Ti
%, wt.	61.03	10.08	27.35	1.54

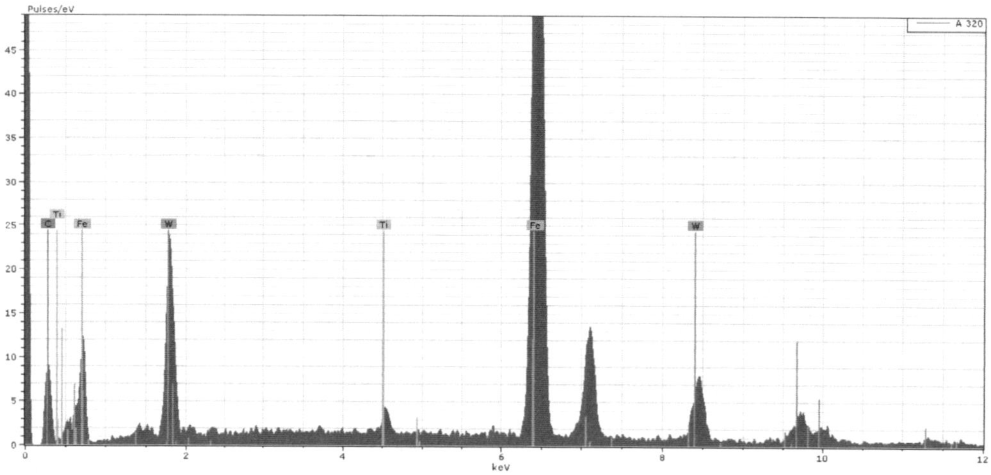

Figure 2. EDX spectra of the triple-layer coating (WC/TiC/W).

According to the chemical mapping (Figure 3) the distribution of the elements on the coating surface shows a relatively uniform dispersion of titanium and tungsten even on the areas with carbon lamellae.

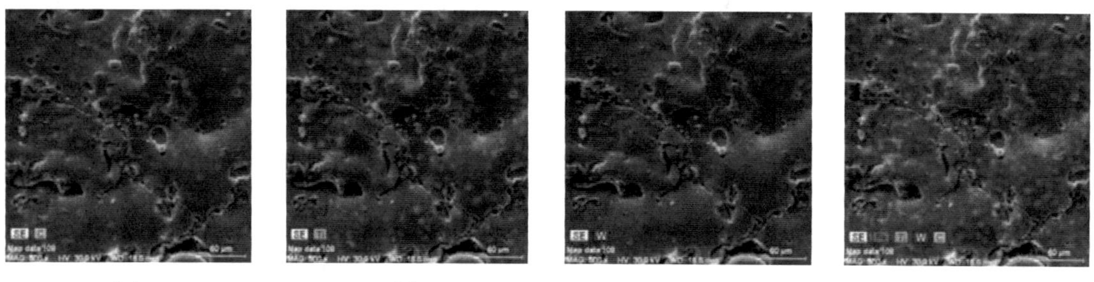

(a) (b) (c) (d)

Figure 3. Elemental mapping of the triple-layer coating (WC/TiC/W): (a) Fe distribution; (b) Ti distribution; (c) W distribution; (d) Fe, Ti, W, C distribution.

The triple-layer coating shows a relatively compact surface, with few adhesions and exfoliations, with not too high unevenness. The morphology characteristics are related to the high concentration of tungsten from the deposition, both in the most basic form as tungsten carbide or in the unalloyed state. Usually, in deposited layers, tungsten gives very good properties of hardness and wear resistance, leading to quenching in liquid media (glassy layers are obtained).

Because the titanium carbide deposition shows higher compactness than tungsten and tungsten carbide deposits, it can cover the surface unevenness and the tearings resulting in a homogeneous layer with constant thickness and high quality (lack of defects).

The outer layer (the third) made by tungsten deposition leads to a very hard surface, on which many oxides, burns and cracks are created. However, due to the presence of titanium carbide, the cracks are not so obvious, meaning that the intermediate layer has changed the coefficient of expansion and contraction of the workpiece-deposited layer assembly resulting in a layer with high compactness.

4. Conclusion

The heterogeneous layers deposited have, as the main advantage, the obtaining of a coating with appropriate roughness and hardness of the surface and good adhesion and compactness of the substrate. These characteristics are obtained by combining the advantages of each type of electrode.

Tungsten carbide deposition has good adhesion characteristics to the grey cast iron substrate, therefore, those have good compatibility and doesn't react with the surface in a normal atmospheric condition which can result in areas with pores and absorbed gases.

The titanium carbide layer increases the fluidity of the complex metal bath that is formed during the second deposition. Therefore, the areas uneven coated after the tungsten deposition, because of the high surface tension of the W drops, are uniformized by the second deposition.

The third layer, made with tungsten, generates the final properties of the workpiece surface, namely high hardness and good corrosion resistance.

Due to the good compatibility between the consumable electrodes materials and the base material, ie lamellar grey cast iron, an adhesive coating with good mechanical, thermal and chemical resistance was obtained.

5. References

[1] Jilani A, Abdel-wahab M S and Hammad A H 2017 Advance Deposition Techniques for Thin Film and Coating, *Modern Technologies for Creating the Thin-film Systems and Coatings*, Nikolay N. Nikitenkov, Intech Open, DOI: 10.5772/65702

[2] Frangini S and Masci A 2010 *Surface and Coatings Technology* **204**(16-17) 2613–2623

[3] Aghajani H, Hadavand E, Peighambardoust N S and Khameneh-asl S 2020 *Surfaces and Interfaces* **18** 100392

[4] Pliszka I and Radek N 2017 *Procedia Engineering* **192** 707-712

[5] Jiao Z, Peterkin S, Felix L, Liang R, Oliveira J P, Schell N, Scotchmer N, Toyserkani E and Zhou Y 2018 *Journal of Materials Engineering and Performance* **27**(9) 4799-4809

[6] Wang J, Zhang R, Yan 2011 *Advanced Materials Research* **146-147** 1601-1604

[7] Liu D, Gao W, Li Z, Zhang H, Hu Z 2007 *Materials Letters* **61** 165-167

[8] Chen Z and Zhou Y 2006 Surf. Coat. Technol. **201** 1503-1510

[9] Padgurskas J, Kreivaitis R, Rukuiža R, Mihailov V, Agafii V, Kriukiene R, Baltušnikas A 2017 *Surf. Coat. Technol.* **311** 90–97

[10] Miller T, Pirolli L, Deng F, Ni C and Teplyakov A V 2014 *Surf. Coat. Technol.* **258** 814-821

[11] Perju M C and Vizureanu P 2014 *Revista de Chimie (Bucharest)* **65**(6) 694-696

[12] Perju M C, Nejneru C, Vizureanu P and Stefănică R G 2012 XPS chemical analysis for the multilayer deposition WC/TiC/W on gray cast iron using electric impulse discharge method, *ModTech International Conference-New face of TMCR, Proceedings of The 16th International Conference, Modern Technologies, Quality and Innovation*, vol II, p 737-740

[13] Dandu-Bibire L, Borsos Z, Matasaru D, Casian-Botez I, Niculescu A and Agop M 2011 *University Politehnica of Bucharest Scientific Bulletin-Series a-Applied Mathematics and Physics* **73**(2) 175-184

[14] Buzea CG, Agop M, Galusca G, Vizureanu P and Ionita I 2007 *Chaos solitons & fractals* **34(4)** 1060-1074
[15] Baltatu M S, Tugui C A, Perju M C, Benchea M, Spataru M C, Sandu A V and Vizureanu P 2019 *Revista de Chimie* **70(4)** 1302-1306

Influence of the Stand-off Distance and of the Layers Thickness on the Adhesion and Porosity of the 97MXC Deposits Obtained by Arc Spraying Process

R A Haraga[1], D L Chicet[1], N Cimpoiesu[1], S L Toma[1*] and C Bejinariu[1]
[1] "Gheorghe Asachi" Technical University of Iasi, Materials Engineering and Industrial Safety Department, No. 67 Prof. D. Mangeron Blvd., 700050, Iasi, Romania.

*E-mail: stl_toma@yahoo.com

Abstract. In this paper, the surfaces of some weakly alloyed steel specimens were covered, by arc spraying process, with a layer of high alloyed steel: 97MXC, using core wires. The technological parameters, used to spray the high alloy steel, were kept constant except for the spray stand-off distance (SOD) which varied on three levels. The research was performed on deposits with a thickness between 0.2-2.4 mm. The microstructure of the deposits, the chemical composition, the physical properties of the layers were investigated by SEM analyzes, X-ray diffraction, microhardness tests and adhesion tests. The porosity of the deposits was determined by optical microscopy. The adhesion of the layer to the substrate was researched by the tensile test. Studies have shown the fact that SOD significantly influences the adhesion and porosity of the deposits. Thus, the increase of SOD determines the decrease of the adhesion of the layer to the substrate by approximately 21.4%, the increase of the porosity of the deposits by approximately 14.2% and variations up to 5% of the microhardness.

1. Introduction

The versatility and economic efficiency of the arc spray process (ASP) make it often to be used in obtaining surfaces with new physical and mechanical properties, or in restoring the surfaces of used parts [1-4].

In the ASP, the raw material - in the form of wire, is melted in the electric arc developed at the contact between the tops of the wires [5-7]. A jet of compressed air divides the drop formed into fine particles, which it projects onto the surface of a substrate, [8-12]. The sprayed particles collide with the surface of the substrate and flatten on it [13-17], finally resulting in a layer that presents a lamellar structure and contains various chemical compounds, oxides, pores or unmelted particles [18-23]. It is known that the physical, chemical and mechanical properties of the deposits are strongly influenced both by the intrinsic properties of the adition material [24-27] and by the technological parameters of the ASP, [28, 29]. The latter are defined as being the sum of the parameters that directly or indirectly influence the melting process, the flight characteristics (speed, temperature and particle size) [30] and the kinematic spraying parameters, that influence the process of layer formation. The spraying distance, generically defined spray stand-off distance (SOD) - represents the distance between the nozzle of the spray device and the surface of the substrate [31-34]. It belongs to the category of kinematic parameters that dictate the state of the particles before impact, the heat and mass transfer at the substrate surface being specified to each sprayed material [35-38].

The purpose of this study is the influence of the spray stand-off distance and the thickness of the deposits on the physical-mechanical characteristics of the 97MXC layer obtained by arc spray process.

2. Experimental procedure

2.1. Materials and equipment

Substrates made of low alloy steel C15 –EN10083, with dimensions of 40x40x10 were covered by ASP using the 97MXC filler material - in the form of cored wires, a product of the company Praxair-Tafa, USA. The chemical composition of the sample material, as well as that of the additive material are presented in Table 1.

Table 1. Chemical composition of the materials used.

Materials	Elements (wt%)										
	C	Si	Cr	Ni	Mn	B	WC	TiC	Fe	P	S
C15	0.14	0.15	0.3	0.3	0.43	-	-	-	-	0.04	0.04
97MXC	-	1.25	14.0	4.5	0.55	1.87	26.0	6.0	balance	-	-

For the carrying out the deposits, we used a Tafa CoArc System electric spray installation (manufactured by Praxair-Tafa, USA), equipped with a 9935-spraying gun, before spraying.

Before deposits carrying out, the surface of the samples was cleaned by sandblasting using alumina powders (alumina Al_2O_3) with an average diameter of 536 ± 24μm, sprayed with the pressure of 4 bar at the distance of 30 mm. To remove residues left on the surface, the specimens were ultrasonically cleaned by immersion in an ethanol bath (C_2H_6O) for 10 minutes and dried under compressed air. All the samples examined were deposited at a constant speed, at the same value of pressure and current intensity.

Table 2. Technical parameters of magnetron sputtering.

Parameters	**Value**
Current intensity (A)	220
Voltage (U)	32
Air pressure (bar)	6.0
Movement speed of the gun (m/s)	0.14
Coating thickness (mm)	0.53/1.24/2.22
Stand-off distance (mm)	100/125/150
Spray angle	90^0

During the experiments the spray distance varied on three levels. In Table 2 are presented the parameters of the arc spraying.

2.2. Deposits characterization

The investigations on the qualitative elemental chemical analysis and on the morphology of the 97MXC layers were performed using the scanning electron microscope (SEM): FEI - Quanta 200 SEM-EGF. The crystalline spectra of the deposits were highlighted by X-ray diffraction, using X-ray diffractometer - type X'PERT PRO MRD made by PANalytical Holland, using Cu Kα radiation (λ = 0.15406 nm), with an emission current of 40mA and acceleration voltages of 40kV. The tensile test of the flat samples was used to determine the adhesion of the layer to the substrate, in accordance with EN 582. With the help of the CV Instruments 400 DM micro-hardness tester - by using a load of 300g for 10s, we determined the Vickers micro hardness of the deposits. We performed a number of 15 investigations in the cross section of the layers and reported the average values of the measurements. With the help of the micrometre we determined the deposits thickness and we reported its average values - obtained after performing a number of 5 determinations on the surface of each sample. The layers porosity was determined in the cross section of the test samples, by image analysis, using the IQ Materials program (Japan).

3. Results and discussions

Microstructures of the 97MXC deposits - made by thermal spraying in electric arc, at different values of the spraying distance, are presented in Figure 1. It is observed that the deposits obtained at short SOD presents a homogeneous structure, the tungsten and titanium carbides are evenly distributed and they are well anchored in the metal matrix.

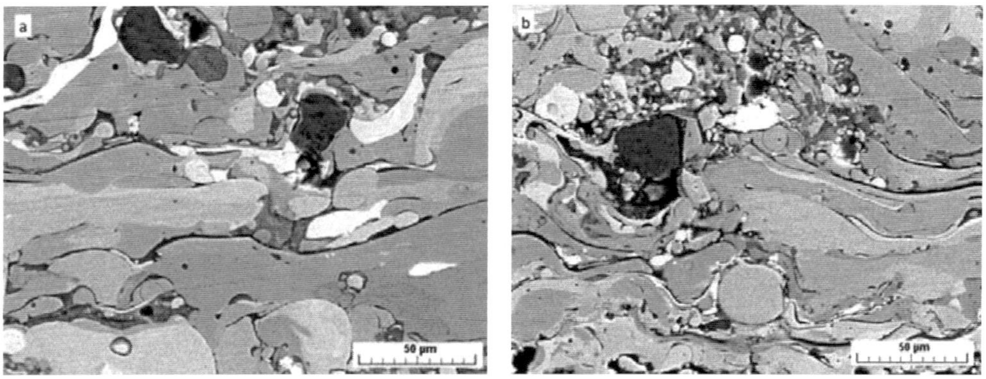

Figure 1. SEM micrographs showing the cross-section of 97MXC deposits obtained at p = 6.0 bar, I = 220 A and under the following conditions: (a) SOD = 100 mm; (b) SOD = 150 mm.

By increasing the spraying distance, the structure of the deposits becomes inhomogeneous and in the structure of the deposits appear particles of unmelted material. The electric arc formed at the contact between the electrodes melts, by Joule Lentz effect, the metallic material outside the wires and the powder inside it. The jet of compressed air sprays the particles on the surface of the substrate interacting with the outer surface of the particles. Their flight time and implicitly the contact time between the particles and the jet of compressed air, is shorter when the SOD is shorter, reason for which the particles arrive on the layer surface in a liquid phase.

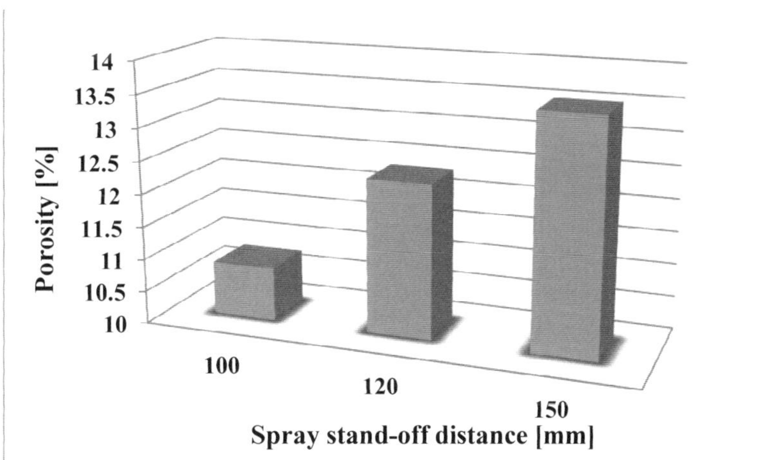

Figure 2. Variation of the porosity of the 97MXC layers with the SOD.

The increase of SOD determines the increase of the flight time of the metallic particles, respectively of the particle-compressed air interaction time, reason for which the particles reach the value of the solidification temperature and reach the surface of the substrate in solid, semi-molten or molten state.

The variation of the porosity of the 97MXC deposits with the spray distance (SOD) is presented in Figure 2. It can be seen that denser coatings are obtained when the spray distance is shorter. The increase by up to 14.2% of the porosity of the 97MXC deposits obtained at SOD = 150mm, the presence of less flattened (thick) lamellae and of unmelted particles inside the deposition presents a clear proof of the fact that at high SOD, the pulverized particles cool and reach the surface of the substrate in a predominantly solid state.

X-Ray diffraction investigations carried out on 97MXC layers, deposited by ASP in demonstrate that their structure consists of hard compounds such as WC and TiC, ferroborides, ferrite alloyed with Ti or ferrite alloyed with tungsten (FeW) or titanium alloy ferrite, as well as complex compounds of the type FeNiCrW, FeNiCrTi - ferrite alloyed with Ti or ferrite alloyed with tungsten (FeW) or titanium alloy ferrite, as well as complex compounds of the type FeNiCrW, FeNiCrTi - independent of the SOD, reason for which only one was exposed - Figure 3.

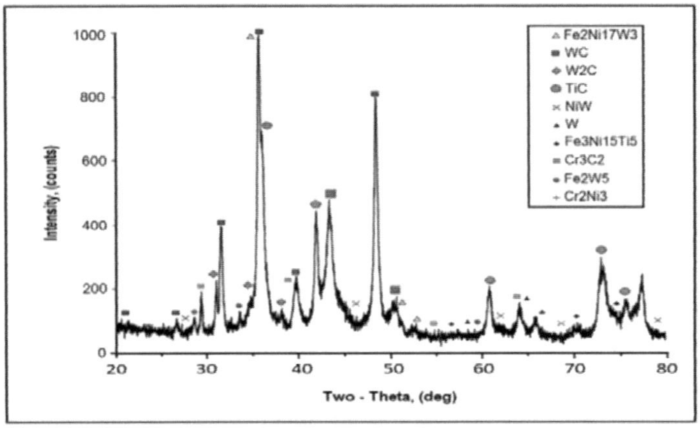

Figure 3. The XRD analysis of the 97MXC coatings obtained at p = 6 bar, I = 220 A, and SOD = 100 mm.

Thus, the solidification mechanism is complex, depending on the concentration and the number of component elements of the filler material. During the flight, inside the pulverized particles diffusion phenomena appear. Some of the dissolved tungsten and titanium is alloying with carbon, forming WC or TiC - an aspect which is in concordance with the reports of Tillman et al. [35]. As the temperature decreases, in the layer are formed substitution solid solutions as type of ferrite alloyed with tungsten or ferrite alloyed with titanium.

In Figure 4 is presented the variation of the HV micro hardness of the 97 MXC deposit with SOD. It is remarked that the HV microhardness decreases with the increasing of SOD value, due to both the phenomenon of rapid solidification of particles before hitting the surface of the substrate and the inhomogeneous and porous structure of the deposit.

The variation of the deposition adhesion of 97MXC with the layer thickness, for the three considered values of the spraying distance is presented in Figure 5. It is observed that the adhesion of the carried-out deposits decreases with the spraying distance, on average by 21.4%. This aspect is due to the existence of pores, the high speed of the particles when the spraying distance is small, as well as the presence of solidified particles in the layer. The increasing of the deposits thickness determines the adhesion decreasing, on average by up to 11.5%.

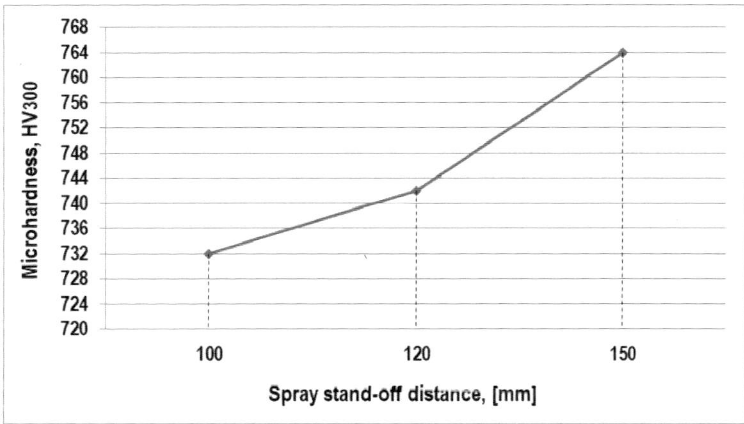

Figure 4. The effect of SOD on the microhardness of 97MXC coatings obtained by ASP.

This aspect is due to the fact that the adhesion of 97MXC deposits - obtained by the electric ASP with two wires, is mainly mechanical and the presence of hard carbides and complex solid solutions stiffens it - detailed aspect by Haraga et colab., [17].

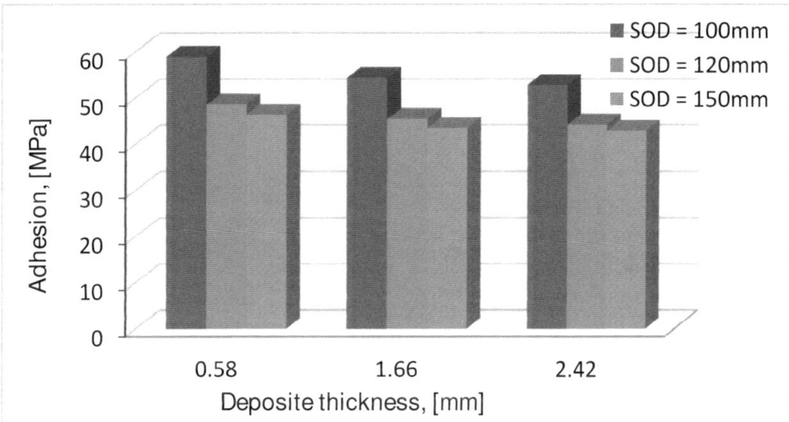

Figure 5. Variation of adhesion of 97MXC coatings with deposits thickness and SOD.

It can be suggested that by reducing the spraying distance the particles speed is high, the particles are in a liquid state, collide the substrate and fix in its asperities - aspect in concordance with the reports of Toma et colab. [9].

4. Conclusion and perspectives

The study performed on the influence of the spraying distance (SOD) and of the thickness of the deposits on the physical-mechanical properties of the 97MXC deposits, obtained by arc thermal spraying highlights the following aspects:

– at small values of the spray distance, uniform, homogeneous deposits are obtained in which the carbides are uniformly embedded in the metal matrix. The deposits obtained at higher values of the spraying distance (SOD) have inhomogeneous structures, present unmelted pores and particles and the tungsten and titanium carbides are unevenly distributed inside the layer - see Figure 1. Regardless of the SOD value, the layers contain ferrite alloyed with W or Ti, WC, CTi, complex compounds such as FeNiCrW, FeNiCrTi;

– the porosity of the deposits increases by approximately 14.2%, when the SOD increases from the value of 100mm to the value of 150mm - see Figure 2. Increasing the porosity of 97MXC deposits has a negative effect both on the micro hardness - see Figure 4 and on the adhesion of the layer to the substrate - see Figure 5;

– the presence of hard carbides in the layer influences its elastic properties, which is why the increase in the thickness of the deposits determines the decreasing of the layer adhesion to the substrate.

All these aspects suggest the fact that in the moment of increases the spraying distance, the interaction time between the compressed air jet and the molten particles increases, the latter cool intensely and hit the substrate surface in a predominantly solid or semi-molten state having negative effects on mechanical properties.

5. References

[1] Sacriste D, Goubot N, Dhers J, Ducos M and Vardelle A 2001 *J. Therm. Spray Technol.* **352** 35-358

[2] Pawlowski L 2008 in: 2nd ed., *The Science and Engineering of Thermal Spray Coatings*, Ed. Wiley, Chichester, England

[3] Toma B F, Baciu R E, Bejinariu C, Cimpoieşu N, Ciuntu B M, Toma S L, Burduhos-Nergis D P and Timofte D 2018 *IOP Conf. Series: Mat. Science and Engineering* **374** 012017

[4] Cazac A M, Bejinariu C, Ionita I, Toma S L and Rodu C 2014 *Applied Mechanics and Materials* **657** 193-197

[5] Tillmann W, Walther F, Luo W F, Haack M, Nellesen J and Knyazeva M 2018 *J. of Therm. Spray Technol.* **27** 50-58

[6] Popovici R, Scripcariu I S, Himiniuc I, Murarasu M and Grigore M 2019 *Med. Surg. J.-Rev. Med. Chir. Soc. Med. Nat.* **123**(1) 97-101

[7] Irimiciuc S A, Agop M, Nica P, Gurlui S, Mihaileanu D, Toma S and Focsa C 2014 *Japanese Journal Of Applied Physics* **53**(11) 173-179

[8] Panadda S and Hathaipat K 2014 *Wear* **317** 194–200

[9] Toma S L 2013 *Surf & Coat. Technol.* **220** 261–265

[10] Tîrnovanu M, Onofriescu M, Tîrnovanu S, Tîrnovanu V, Ciuntu B M, Timofte V, Himiniuc L and Toma B 2020 *Med. Surg. J.-Rev. Med. Chir. Soc. Med. Nat.* **124**(1)107-112

[11] Matz M M, Aumiller M P 2014 *Journ. of Thermal Spray Technology* **23**(8) 1470-1477

[12] Bloch K, Titu M A and Sandu A V 2017 *Rev. de Chimie* **68** 2162-2165

[13] Guilemany J M, de Paco J M, Nutting J and Miguel J R 1999 *Metall. Mater. Trans.* A **30** 1913

[14] Cazac A M, Bejinariu C, Baciu C, Toma S L and Florea C D 2014 *Applied Mechanics and Materials* **657** 137-141

[15] Tugui C A, Vizureanu P, Perju M C, Savin C, Nejneru C, Baltatu S M, Bejinariu C and Benchea M 2018 *IOP Conference Series-Materials Science And Engineering* **374** 012029

[16] Toma S L, Bejinariu C, Eva L, Sandu I G and Toma B F 2015 *Key Engineering Materials* **660** 86-92

[17] Haraga R A, Bejinariu C, Cazac A, Toma B F, Baciu C and Toma S L 2019 *IOP Conference Series: Materials Science and Engineering* **572** 012056

[18] Arif Z U, Shah M, Rehman E and Tariq A 2020 *International Journal of Advanced and Applied Sciences* **7**(7) 25-39

[19] Tănase A E, Toma B, Himiniuc L, Petrica M, Hutanu D and Onofriescu M 2020 *Med. Surg. J. Rev. Med. Chir. Soc. Med. Nat.* **124** (1) 86-93

[20] Baltatu M S, Tugui C A, Perju M C, Benchea M, Spataru M C, Sandu A V and Vizureanu P 2019 *Rev. Chim* **70**(4) 1302

[21] Li S, Wang Y, Xiang D et al 2016 *Rare Metal Materials And Eng.* **45** 2555-2560

[22] Nanu C, Poeata I, Popescu C, Eva L, Toma B F and Toma S L 2018 *Materiale Plastice* **55** 85-90

[23] Berger L M, Saaro S, Naumann T, Wiener M, Weihnacht V, Thiele S and Suchánek J 2008 *Surf. Coat. Technol.* **202** 4417

[24] Nurisna Z T, Muhayat N and Wijayanta A T 2016 *Book Series AIP Conference Proceedings* **1717** 040012

[25] Toma S L, Badescu M, Ionita I, Ciocoiu M and Eva L 2014 *Applied Mechanics and Materials* **657** 296-300

[26] Nedeff V, Mosnegutu E, Panainte M, Ristea M, Toma S and Agop M 2012 *Powder Technology* **221** 312–317

[27] Calin M A, Curteza A, Toma S and Agop M 2013 *Metalurgia International* **18** 19-22

[28] Toma S L, Bejinariu C, Gheorghiu D A and Baciu C 2013 *Advanced Materials Research* **814** 173-179

[29] Sroka M, Nabialek M, Szota M et al 2017 *Revista de Chimie* **68** 737-741

[30] Tillmann W, Hagen L and Kokalj D 2017 *Journal of Thermal Spray Technology* **26** 1685-1700

[31] Katranidis V, Kamnis S, Allcock B and Gu S 2019 *J Therm Spray Tech* **28** 514–534

[32] Masoumeh G, Shahrooz S, Mahmood G and Ahmad S E 2018 *J. Theor Appl Phys* **12** 85–91

[33] Tanase A E, Tibeica A, Tanase G V, Toma B, Nhambasora F, Nemescu D and Onofriescu M 2018 *Filodiritto Publisher* 799-803

[34] Przybyl A, Wnuk I and Wyslocki J J 2019 *Rev. Chim.* **70**(11) 4086-4088

[35] Nejneru C, Cimpoesu N, Stanciu S, Vizureanu P and Sandu AV 2009 *Metalurgia international* **14(7)** 95-98

[36] Burduhos-Nergis D -P, Vizureanu P, Sandu A V and Bejinariu C 2020 *Applied Sciences* **10(8)** 2753 doi:10.3390/app10082753

[37] Achitei DC, Vizureanu P, Minciuna MG, Sandu AV, Buzaianu A and Dana DI 2015 *Materiale Plastice* **52(2)** 165-167

[38] Tillmann W and Abdulgader 2012 *Journal of Thermal Spray Technology* **22** 352-362

CNC End Milling Process Parameters Optimization on Aluminium Alloy by Using Taguchi and ANOVA Methods

M A Țîțu[1*] and A B Pop[2]
[1] Lucian Blaga University of Sibiu, 10, Victoriei Street, Sibiu, România, The Academy of Romanian Scientists, 54, Splaiul Independenței, Sector 5, Bucharest, Romania
[2] SC TECHNOCAD SA, 72, Vasile Alecsandri Street, Baia Mare, România

*E-mail: mihail.titu@ulbsibiu.ro

Abstract. This paper is focused on the experimental research which is carried out on the machining of the aluminum alloy. Its main objective is to find the optimum cutting parameters used in the end-milling process. An experimental plan was established based on the L27 orthogonal matrix. The experiments were performed using the recommended values for the cutting parameters by the cutting tools manufacturer on the aluminum machining. The design of experiment methods that were used, are the Taguchi method, the variance analysis and the regression analysis. The controllable factors were the cutting speed, the cutting depth and the feed per tooth. The surface roughness represents the objective function and this was obtained by the studied cutting process. The surface roughness was measured longitudinally and transversely in the cutting direction.

1. Introduction

Surface roughness is an essential indicator of surface quality. This indicator significantly influences the technological time and costs, in other words the productivity.

We have to define the surface quality starting with the first phase of the product development – the design one, considering some aspects like functionality, costs, appearance, etc. Therefore, numerous scientific papers address the roughness modeling and cutting process optimization to achieve the tracked level of surface quality. For this purpose, the correct selection of the cutting process parameters is essential in order be able to obtain the minimum value of the surface roughness and, at the same time to extend the tool life.

The experimental design is widely used in the cutting process in order to predict it and optimize it to obtain a given surface roughness. Rashid M F F et a. [1] has developed a mathematical model because he wants to predict the surface roughness during the front milling process. Barkallah M et al. [2] have create a statistical model of the surface roughness in the front milling at high cutting speeds with cooling, by changing the rotations number, and also the cutting speed and the cutting depth. By his research, [3] analyzes the cutting depth influence, the rotation speed and the cutting feed exerted on the surface roughness.

Due to the primary impact on the production, but also on the cost and productivity, the investigations of the surface quality represent the main subject in many another research [4, 5].

Modeling the quality of the machined surface of an aluminum alloy using the active type experiment is approached by [6] and [7-8] and the objective function optimization - the surface quality was approached in [9] based on Montgomery principles [10] (the principle of random character, the principle of replication and the one of working in "blocks").

Taguchi method is a technique frequently utilized to study the process robustness; and was addressed by [11].

In this scientific work, the main objective was to analyze the influence exerted by the cutting parameters on the surface roughness during the end milling process of an aluminum alloy. Then, will be followed by the elaboration of a statistical regression model using the design of experiment methodology in order to predict the surface roughness.

2. Methods and methodology

The aluminum-alloy 7136 end milling experiments were performed using SECO R217.69-1616.0-09-2AN cutting tool, having two teeth, diameter is 16 mm, and the tool engagement consist of 100% (16 mm), and the coding of the corresponding cutting inserts, is XOEX090308FR-E05, H15.

The process parameters with the three levels are listed in Table 1. HAAS VF-YT2 is the CNC machine tool used to carry out the experiments.

Table 1. Cutting process parameters and their levels.

Factors	Process parameters	Unit	Level 1	Level 2	Level 3
A	Cutting speed (v)	m/min	610	660	710
B	Cutting depth (ap)	mm	3	3.5	4
C	Feed per tooth (fz)	mm/tooth	0.4	0.6	0.8

Table 2. L27 orthogonal array to conduct the experiments and the obtained measurements of surface roughness.

No. Exp.	A - Cutting speed (v) [m/min]	B - Cutting depth (ap) [mm]	C - Feed per tooth (fz) [mm/tooth]	R_z [μm]	R_{zT} [μm]
1	610	3	0.4	3.167	2.301
2	610	3	0.6	5.978	5.390
3	610	3	0.8	3.850	4.498
4	610	3.5	0.4	4.778	4.177
5	610	3.5	0.6	5.768	3.027
6	610	3.5	0.8	6.667	4.150
7	610	4	0.4	4.394	10.234
8	610	4	0.6	8.017	6.780
9	610	4	0.8	6.867	5.455
10	660	3	0.4	3.167	2.042
11	660	3	0.6	4.141	1.213
12	660	3	0.8	4.114	1.492
13	660	3.5	0.4	3.620	1.592
14	660	3.5	0.6	4.261	1.302
15	660	3.5	0.8	3.901	1.724
16	660	4	0.4	4.299	1.650
17	660	4	0.6	3.528	1.549
18	660	4	0.8	3.177	2.433
19	710	3	0.4	4.349	3.620
20	710	3	0.6	4.104	2.923
21	710	3	0.8	4.913	3.078
22	710	3.5	0.4	4.780	2.345

No. Exp.	A - Cutting speed (v) [m/min]	B - Cutting depth (ap) [mm]	C - Feed per tooth (fz) [mm/tooth]	R_z [µm]	R_{zT} [µm]
23	710	3.5	0.6	3.850	4.261
24	710	3.5	0.8	3.189	3.826
25	710	4	0.4	3.756	4.006
26	710	4	0.6	4.647	2.722
27	710	4	0.8	3.743	5.253

The end milling operations were carried out in accordance with the experiment array L27 presented in Table 2. With the portable Mitutoyo SURFTEST SJ-210 were obtained the experimental data. In Table 2 are also presented the roughness values measurements obtained longitudinally (R_z) and transversely (R_{zT}) in the milling direction.

3. Results and discussion

For the above experimental conditions, the results were explained using the figures, diagrams and graphs presented below, using Minitab Statistical Software.

Using Taguchi design the Orthogonal Array Design L27 (3^3) was obtained. The 27 experimental tests were obtained based on the 3 factors with 3 levels each. The influence of the surface roughness on each individual factor (A, B, C) and on the interaction between them (AB, AC, CA and ABC) was followed.

The percentage of influence of the factors and their interactions on the R_z and R_{zT} measured on the milling direction was determined. These percentages are shown in Table 3.

Table 3. Determining the percentage influence of the factors and interactions on the surface roughness.

Source	R_z Contribution	R_{zT} Contribution
Regression	49.13%	37.21%
A	21.35%	10.36%
B	3.12%	9.70%
C	2.44%	0.00%
A*B	12.20%	4.99%
A*C	8.03%	1.83%
B*C	0.16%	1.18%
A*B*C	1.82%	9.16%
Error	50.87%	62.79%
Total	100%	100%

As it results from Table 3 the cutting process variable with the highest influence exerted on the surface roughness is the cutting speed (A) in both situations for the R_z longitudinally (R_z)- the percentage being 21.35%; as well as for the R_z transversely (R_{zT})- the percentage being 10.36%. Regarding the interaction of the parameters, the greatest influence has the cutting speed associated with the cutting depth (A*B) being a percentage of 12.2% for the R_a longitudinally and 4.99% for the R_z transversely. A noteworthy aspect is that on the R_z transversely a significant influence has the interaction of the 3rd degree of the process parameters (A*B*C), which has a percentage of 9.16%. This influence is not felt on R_z longitudinally.

The determined regression equations that describe the surface roughness according to the factors chosen for the established experimental field are the following:

$$R_z = 14 - 0.020 \cdot A + 6.0 \cdot B - 100 \cdot C + 0.0105 \cdot A \cdot B + 0.156 \cdot A \cdot C + 38.3 \cdot B \cdot C - 0.0592 \cdot A \cdot B \cdot C \quad (1)$$

$$R_{zT} = -347 + 0.513 \cdot A + 107.6 \cdot B + 491 \cdot C - 0.1579 \cdot A \cdot B - 0.727 \cdot A \cdot C - 147.8 \cdot B \cdot C + 0.219 \cdot A \cdot B \cdot C \quad (2)$$

Figures 1 and 2 show the graphs of the main effects of the roughness related to the cutting process parameters: A (cutting speed), B (cutting depth) and C (feed rate).

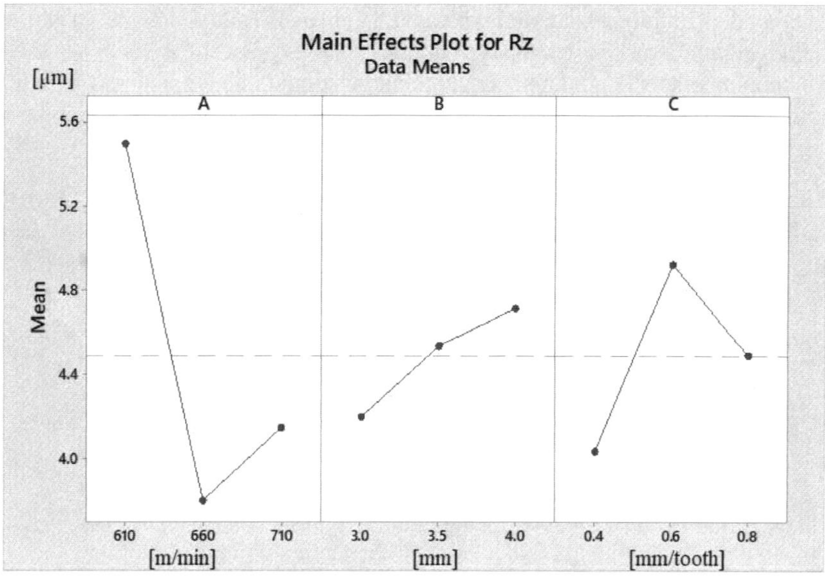

Figure 1. The main effect plot for R_z.

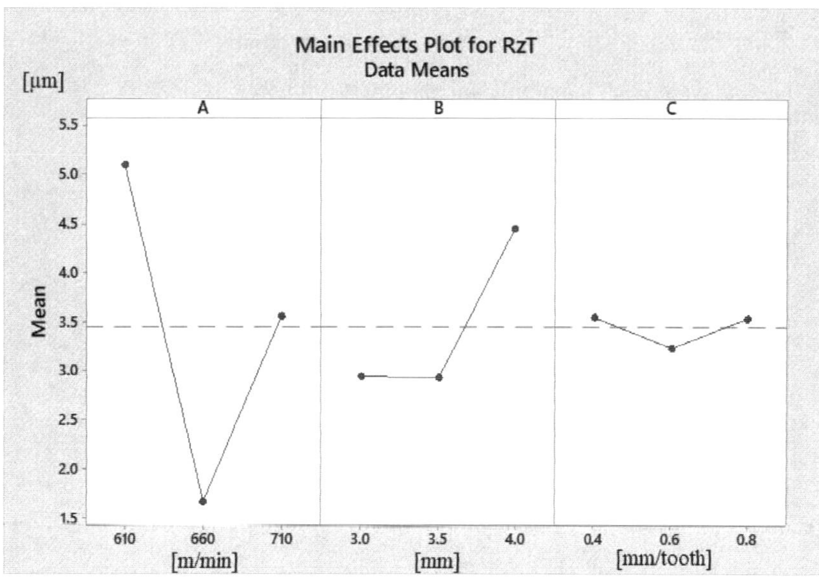

Figure 2. The main effect plot for R_{zT}.

As can be seen from these graphs, the maximum value of the R_z of 8,017 [μm] is obtained at the A = 610 [m/min], B = 4 [mm] and C = 0.6 [mm/tooth]. Regarding the R_{zT}, the maximum value of 10,234 [μm] is obtained at A = 610 [m/min], B = 4 [mm] and C = 0.4 [mm/tooth]. The only difference between the two measurements is the feed per tooth parameter.

Figures 3 and 4 show the influences of the interactions of the cutting parameters exerted on the surface roughness.

The interactions graphs of the process parameters show that both surface roughness R_{zT} and R_z measured in the milling direction are largely influenced by the interaction of A and B process parameters. Specifically, from Figures 3 and 4 it results that that the greatest roughness values is recorded when A = 610 [m/min] associated with B = 4 [mm]. If it is desired to have the best surface quality, in other words, the surface roughness measured longitudinally R_z as small as possible, it is recommended - according to the obtained data - to adopt a cutting regime in which all the analyzed process parameters take the minimum values. Regarding R_{zT}, the recommended cutting regime is that A = 660 [m/min], B = 3 [mm] and C = 0.6 [mm/tooth].

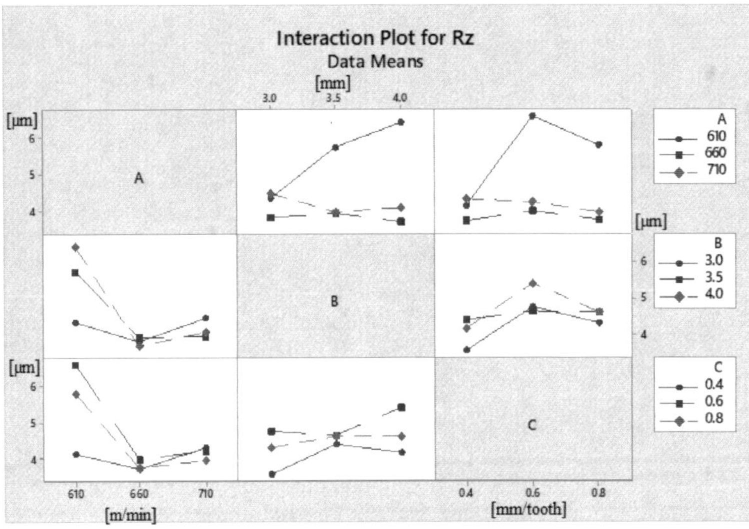

Figure 3. Interaction plot for R_z.

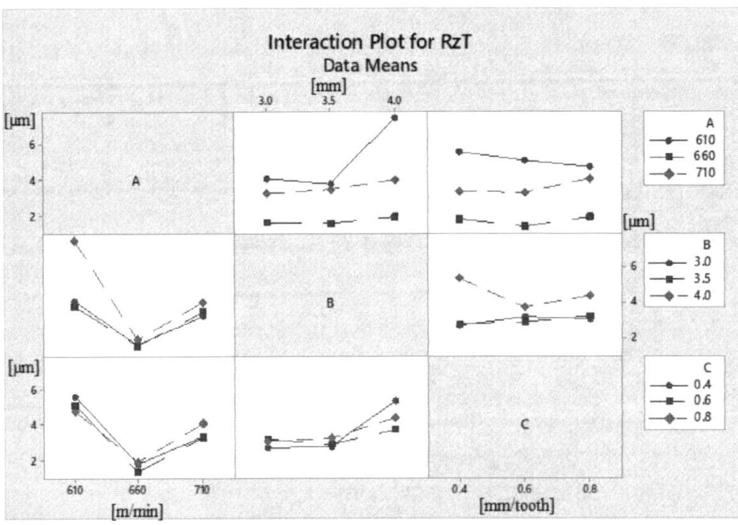

Figure 4. Interaction plot for R_{zT}.

In the Figures 5 and 6 are presented the residual diagrams related to the regression equations obtained for the two objective functions analyzed.

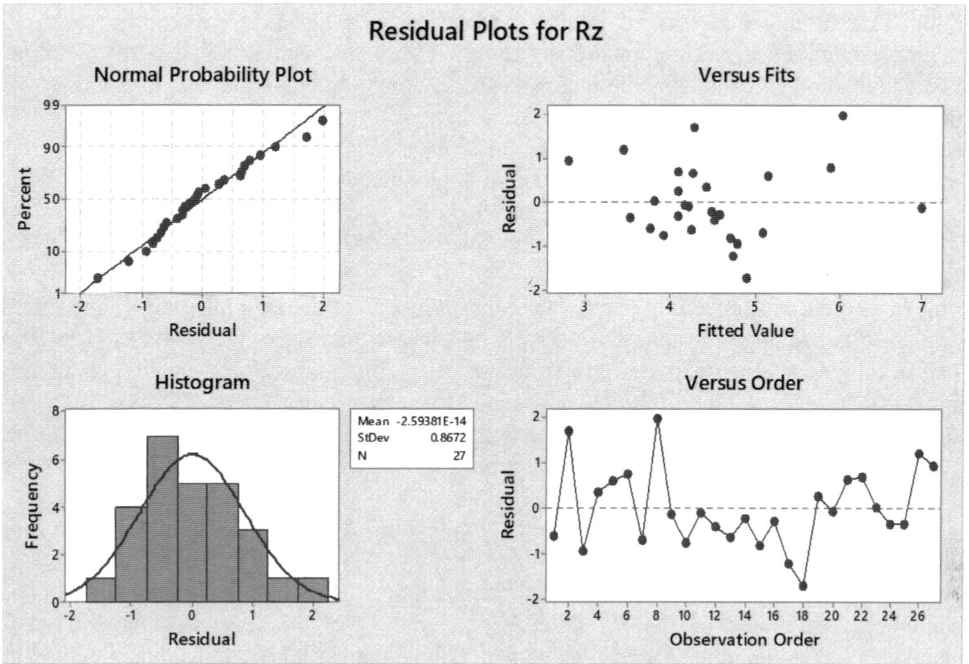

Figure 5. The residual plots for R_z.

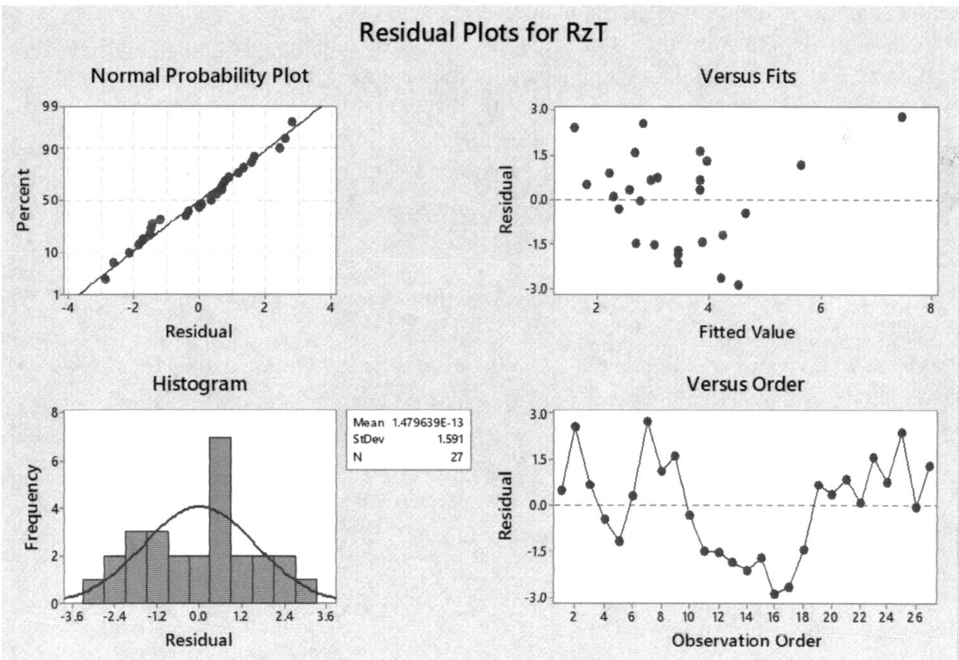

Figure 6. The residual plots for R_{zT}.

The first graph of these figures presents the probability of the normal distribution. From here it can be seen that the residuals are to a certain extent approximating a straight line and the histogram of the residual distribution shows the residual values which exceed the Gauss's distribution curve.

Regarding the fitted values diagrams and the one of their observation orders, results that the residuals distribution is not uniform respecting the zero line and, sometimes, they show large jumps value from one to another, caused by the recorded values at cutting speeds of 610 [m/min].

4. Conclusions

The surface quality resulted by the end milling process is particularly important because it significantly influences the productivity.

To obtain an optimal roughness the base is the correct selection of the cutting parameters from the design phase of the product development.

Starting from the main objective of analyzing the influence of the cutting process variables which is manifested on the surface roughness machined by the end milling process of the 7136-aluminum alloy, was developed a statistical regression model to predict the surface quality using the DOE methodology.

The influence percentage of the factors and their interactions on the surface roughness measured longitudinally (R_z) and transversely (R_{zT}) on the milling direction was determined.

The conclusions which were drawn punctually, are:

- the cutting process parameter which have the greatest influence exerted on the surface roughness is the cutting speed (A);
- regarding the interaction of the parameters, the greatest influence is the cutting speed associated with the cutting depth (A*B);
- on the surface roughness measured transversely (R_{zT})a significant influence has the interaction of 3rd degree of the process parameters (A*B*C);
- the regression equations describing the surface roughness were determined according to the factors chosen for the established experimental field;
- the maximum R_z value measured of 8.017 [μm], is obtained when A (cutting speed) is 610 [m/min], B (cutting depth) is 4 [mm] and C (feed per tooth) is 0.6 [mm/tooth];
- the maximum value of the R_{zT} of 10.234 [μm] is obtained when the A (cutting speed) is 610 [m/min], B (cutting depth) is 4 [mm] B (cutting depth) is 0.4 [mm/tooth];

To obtain the lowest roughness value, it is recommended to adopt a cutting regime in which all the analyzed process parameters take the minimum values.

5. References

[1] Rashid M F F, Gan S Y and Muhammad N Y 2009 *World Academy of Science - Engineering and Technology* **53** 393-396

[2] Barkallah M, Louati J and Haddar M 2012 *Int. J. of Simulation Modeling* **11(1)** 5-16

[3] Svalina I, Sabo K and Simunovic G 2011 *Int. J. Adv. Manuf. Tech.* **57(9-12)** 1099-1106

[4] Cosic P, Lisjak D and Antolic D 2011 *Tehnicki vjesnik / Technical Gazette* **18(4)** 479-484

[5] Pop A B and Țîțu A M 2019 *MATEC Web of Conferences* **290** 02010

[6] Pop A B, Țîțu A M, Pop G I and Țîțu Ș 2019 *Materials Science and Engineering* **572** 012043

[7] Țîțu A M, Sandu A V, Pop A B Ceocea C and Țîțu Ș 2018 *Materials Science and Engineering* **374** 012034

[8] Focsaneanu S C, Vizureanu P, Sandu A V, Ciobanu G, Baltatu S M and Vlad D 2019 *Revista de Chimie* **70(6)** 2273-2275

[9] Țîțu A M, Pop A B, Țîțu Ș and Pop G I 2019 *Materials Science and Engineering* **572** 012042

[10] Montgomery D C 2009 *Design and Analysis of Experiments* (7th ed.) (Hoboken, John Wiley and Sons)

[11] Asilturk I and Neseli S 2012 *Measurement* **45** 785-794

Laser Induced Dyeing (LID) with Reactive Blue 21 on Hemp Fibers

A Cocean[1], I Cocean[1], C Postolachi[1], D Pricop[1], F Husanu[1] and S Gurlui[1*]
[1] Alexandru Ioan Cuza University of Iasi, Faculty of Physics, Atmosphere Optics, Spectroscopy and Laser Laboratory (LOASL), 11 Carol I Bld. 700506 Iasi, Romania

*E-mail: sgurlui@uaic.ro

Abstract. Effects of 532 nm and 320 mJ laser beam on aqueous solution of Reactive Blue 21 (RB21), a copper based on reactive dye, have been studied in both presence and absence of garnet gemstone with the aim to receive a proper coating of the hemp fibers contained in the yarns of a woven texture while reducing the dyeing time compared to the classic exhaustion dyeing method. Results show that both laser beam and garnet influenced the dyeing process with RB21, enhancing the absorption and reaction of RB21 with the composite structured hemp fibers, effect observed in the color intensities for all samples measured with the grayscale. The best results are noticed for laser effects in presence of garnet gemstone when highest coating is achieved among all samples, and the process is 11 times shorter than the classical exhausting method. Compared FTIR and UV-VIS spectra of initial dyestuff and residual dyestuff isolated from the wastewaters offered information about dyestuff exhaustion process under garnet and/or laser influence. Laser Induced Fluorescence completed the study with the information about RB21 behavior under UV laser beam of 362 nm.

1. Introduction

The dyeing with reactive dyes of cellulose fibers included either in yarns or in fabrics is a technological process that can last between 60 minutes and 110 minutes, at high temperatures (over 60 degrees Celsius) and with the use of auxiliaries according to the manufacturers' recommendations. Most of the time, dyeing is followed by various fixing processes, from those with specific chemical compounds to thermal ones. Also, because reactive dyes have affinity for both the OH group in the cello-fiber structure and the OH group in the water, some of the dye will hydrolyze, becoming inactive for reaction with the cellulosic fiber. The hydrolyzed dye is deposited on the cellulosic fiber, without fixing and will have to be removed by washing following dyeing [1, 2]. This produces wastewater both in the dyeing process itself and in the next stages of post-dyeing treatments that involve water. This shortcoming has led to numerous studies on wastewater treatment, which have led to observations on the stability of dyes reactive to various physical and chemical exposures [3- 8]. Some of the wastewater treatments resulting from reactive dyeing have even focused on their reuse in the technological dyeing process [9- 12].

The present study aims to investigate a new method of dyeing using the reactive dye contained in the Color Index (C.I.) with the code Reactive Blue 21. The proposed method is an absolutely new one, namely laser induced painting (LID). Both the dyeing effect on the hemp fabric (cellulose fiber) and the changes occurred on the dye structure during the process are studied. The method aims at reducing the dyeing process time, the energy consumption, and the consumption of chemical auxiliaries commonly used in dyeing processes with reactive dyes.

2. Method and materials used for laser induced dyeing (LID)

The laser induced dyeing (LID) with C.I. Reactive Blue 21 (RB21) was tested on 100% hemp fabric (noted herein as HMP). The procedure consists in irradiating the dyeing system in free atmosphere with pulsed laser beam of 532 nm wavelength, 10 ns pulse width, 10 Hz repetition rate and 320 mJ energy, using YG 981E/IR-10 laser system. A dye bath using 5 ml Reactive Blue 21 aqueous solution 1g/l was used for the first sample P1 (RB21+HMP+LSR) and for the second sample P2(RB21+GRNT+HMP+LSR) a garnet gemstone granule was added. The dye baths were used for 0.131 g hemp fabric (HMP) each. The dyeing process was conducted in test tubes and laser irradiated through the walls of the glass tubes. Irradiation time was of 10 minutes for each sample.

3. Results, spectroscopy analyses and discussions

3.1. Gray Scale dyeing results evaluation

The dyeing results are evaluated in dyeing intensity on the 10 tones gray scale and a difference from 8 for classical dyeing (CLSS) to 6 for sample P1 and 7 for sample P2, as seen in Figure 1, is noticed. As 1 tone on the 10 tones gray scale is the equivalent of ½ tones in the 5 tone gray scale used in textile finishing, it turns out that that the dyeing intensity is higher by ½ tone on the textile gray scale as it follows: P1 (RB21+HMP+LSR) < P2 (RB21+GRNT+HMP+LSR) < CLSS. It turns out that the garnet stone produces an intensification of the laser-induced dyeing (LID) process. Also, the two dyes obtained by the LID method are very close in intensity to the dyeing by the classical process.

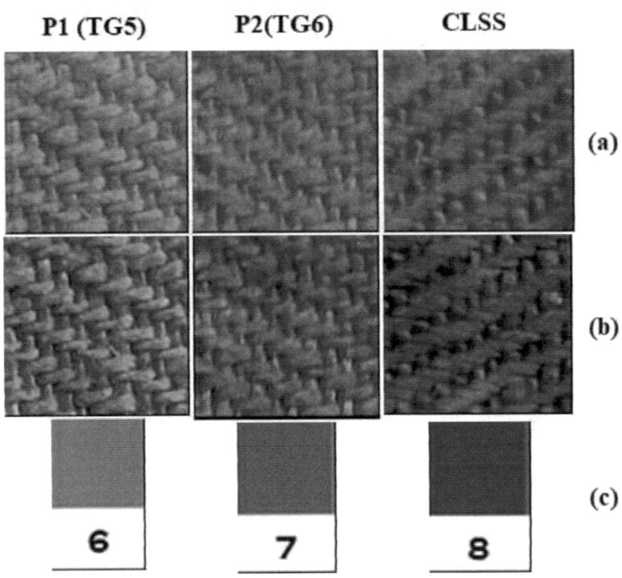

Figure 1. Dyeing intensity compared evaluation of dyed samples P1 (RB21+HMP+LSR), P2 (RB21+GRNT+HMP+LSR) and P3 (RB21+AUX+HMP+LSR) with classical procedure (a,b) based on the 10 tones gray scale (c).

Note: (The classical dyeing involved the same concentration of 1g/l RB21 in the dye bath, 10g/l Na_2CO_3, 50g/l NaCl, 80^0C temperature and 110 minutes was the time for the dyeing process).

3.2. Fourier transform infrared spectroscopy (FTIR)

The FTIR compared spectra (Figure 2) of the residual dyestuff obtained after drying out the exhausted baths of the two dyeing processes (P1: RB21+HMP+LSR and P2: RB21+GRNT+HMP+LSR) show changes in the chemical structure of the dyestuff during the pulsed laser induced dyeing (LID) process. Specific functional groups in accordance with the chemical structure of the C.I. Reactive Blue 21 are identified in Table 1 and their changes during LID process are noticed.

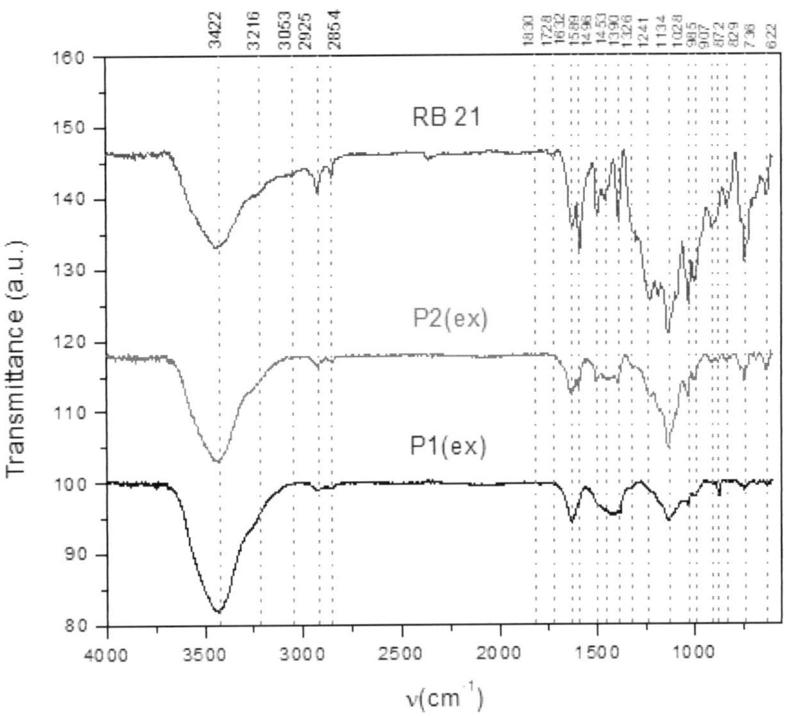

Figure 2. FTIR compared analysis of Reactive Blue 21 (RB21) and residual dye extracted from exhausted dye baths P1ex (RB21+ HMP+LSR); P2ex: RB21+GRNT+HMP+LSR.

In FTIR spectra (Figure 2) and Table 1, most of the vibrations of the specific functional groups of C.I. Reactive Blue 21 dyestuff are also identified for the exhausted dye baths resulted from the laser induced dyeing (LID) process, exception those related to the side chain. The groups in the main chromogen structure (the chromofore and auxochromes) are proved to be stable during the two processes P1(ex) and P2(ex). That is evidenced by the vibrational bands of 3422 cm^{-1} (N – H and O – H stretching), 3216 cm^{-1} (chelates and N – H stretching), 3053 cm^{-1} (aromatic C – H stretching), 1632 cm^{-1} (C = C aromatic stretching, C – N stretching; N – H deformation), 1589 cm^{-1} (C – C aromatic stretching, N – H deformation and ring skeleton in aromatics and heteroaromatics), 1496 cm^{-1} (Cu – N complexation, C – C ring modes), 1453 cm^{-1} (O – H deformation), 1390 cm^{-1} and 1326 cm^{-1} (- SO_2 asymmetric stretching), 1241 cm^{-1} (C – N stretching), 1134 cm^{-1} (- SO_2 symmetric stretching), 1028 cm^{-1} and 985 cm^{-1} (aromatic C – H deformation), 872 cm^{-1} (S – O stretching, = C – H deformation), 736 cm^{-1} (o-di-substitution in 6-membered aromatic rings, C – H deformation), 622 cm^{-1} (C – S stretching). Thus, the very weak intensity of the 907 cm^{-1} band of S – N vibrations in the P2(ex) sample and missing in the P1(ex) sample denotes formation of the components CHR (chromogen) in Figure 3 b and SCH (side chain) in Figure 3 c, after S – N bond is broken. The missing 829 cm^{-1} band in the P1(ex) spectrum, assessed to p-di-substitution in 6-membered aromatic rings, indicates an advanced decomposition of SCH. The component SCH (Figure 3 c) may decompose into amino-benzene (phenylamine or aniline), vinyl sulfone or into amino-benzene sulfonic acid and sulfur dioxide. Formation of aniline could explain the more greenish color received on the P1(ex) and P2(ex) samples, the green shift being more advanced for the P1(ex) sample. That is in accordance with the FTIR spectra of P1(ex) where the 907 cm^{-1} and 829 cm^{-1} bands are missing. As 907 cm^{-1} is assessed to S – N group, it means that in a first decomposition of the dyestuff, the component SCH in Figure 3(c) was formed when the S – N bond was broken. Also, as the 829 cm^{-1} band is assessed to p-di-substitution in 6-membered aromatic rings, it means that a second decomposition of the SCH component into alanine, vinyl sulfone and sulfur dioxide followed the first one.

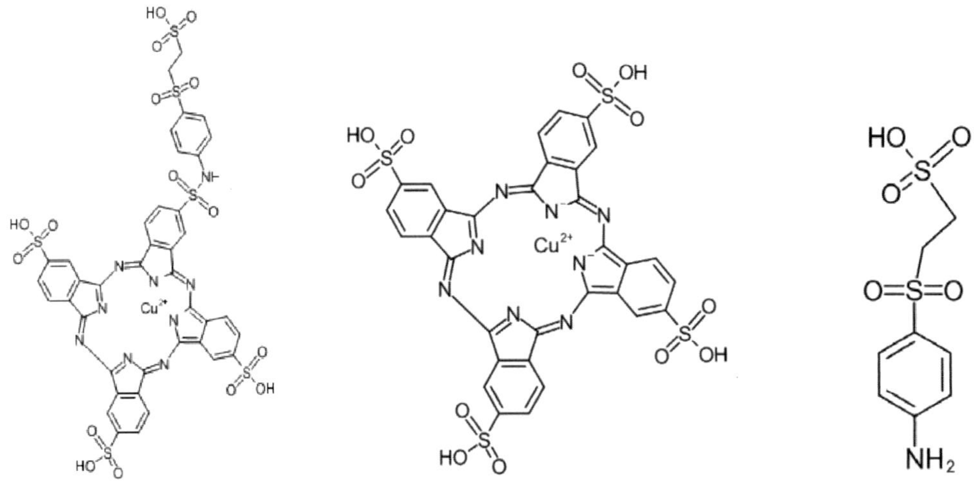

(a) (b) (c)

Figure 3. Chemical structure of the C.I. Reactive Blue 21 (RB21): a) RB21 derivatives resulted after the pulsed laser induced dyeing (LID); b) process on hemp CHR; c) the SCH component.

It is important to notice that the component denoted as CHR (chromogen) in Figure 3 (b) represents a copper phthalocyanine structure with sulfone groups linked to the benzene groups of the molecule. Such structure is very similar to the C.I. Direct Blue 86 and it will be still able to dye the cellulosic fibers, but following a mechanism specific to direct dyestuffs, different from the dyeing mechanism specific to reactive dyestuffs. Also, the FTIR analysis shows the effect induced in P2 sample by the garnet gemstone during laser induced dyeing (LID) process, in the way of preserving the dyestuff structure compared to the P1 process.

Tabel 1. Correlation between vibration bands and the specific functional groups.

IR bands (cm^{-1})	Functional groups [13, 14, 15]	RB21	P1 (HMP+RB21+ LSR)	P2 (HMP+RB21+ GRNT+LSR)
3422	$-NH$ stretching free; $-O-H$ stretching H - bonded	Strong and broad	Stronger and broad	Strong and broad
3216	Chelates: 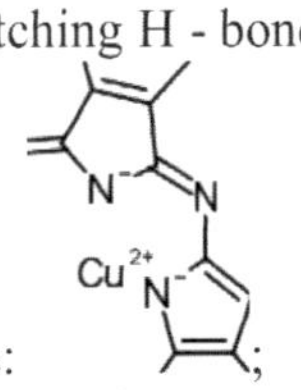; $-N-H$ stretching H-bonded	Overlaps with 3422 – 3216 broad band	Overlaps with 3422 – 3216 broad band	Overlaps with 3422 – 3216 broad band

IR bands (cm^{-1})	Functional groups [13, 14, 15]	RB21	P1 (HMP+RB21+LSR)	P2 (HMP+RB21+GRNT+LSR)
3053	$Aromatic - C - H$ stretching	Very weak	Very weak	Very weak
2925	$Aliphatic - C - H$ stretching; $-O - H$ stretching, H – bonded	Weak	Very weak	Very weak
2854	$Aliphatic - C - H$ stretching	Weak	Very weak	Very weak
1830 – 1728 comb	Substitution in 6-membered aromatic rings	Very weak	Very weak	Very weak
1632	$-C = C -$ aromatic; $-C = N -$; $-N - H$ deformation	Medium	Weaker	Weaker
1589	$-C - C -$ aromatic stretching; $-N - H$ deformation; Ring skeleton in heteroaromatics	Medium	Weak	Weak
1496	Complexation $Cu \leftarrow N$; $-C - C -$ ring modes	Medium	Weak	Weak
1453	$-O - H$ in plane deformation	Weak	Weak, overlaps on 1390 band, denoting H-bonding and Van der Waals interactions	Weak, broader and overlapped on 1390 band, denoting H-bonding and Van der Waals interactions
1390	$O = S = O$ stretching asymmetric in $R - SO_2 - OR$	Medium	Weak, broad, denoting H-bonding and Van der Waals interactions	Weak, broad, denoting H-bonding and Van der Waals interactions
1326	$O = S = O$ stretching asymmetric in $R - SO_2 - OH$ and $R - SO_2 - N$	Medium	Very weak	Weak
1241	$-C - N$ stretching;	Strong	Very weak	Weak
1134	$O = S = O$ stretching symmetric in $R - SO_2 - OH$, $R - SO_2 - OR$ and $R - SO_2 - N$	Strong	Weak broad	Medium to strong
1028	$Aromatic - C - H$ in plane deformation	Strong	Weak	Weak
985	$Aromatic - C - H$ in plane deformation	Medium	Weak	Weak
907	$-S - N$ stretching;	Medium	Missing	Very weak

IR bands (cm^{-1})	Functional groups [13, 14, 15]	RB21	P1 (HMP+RB21+ LSR)	P2 (HMP+RB21+ GRNT+LSR)
872	$-S-O$ stretching; $Aromatic - C - H$ out of plane deformation; $= C - H$ out of plane deformation; Mono-substitution in 6-membered aromatic rings	Missing	Weak	Weak
829	p-di-substitution in 6-membered aromatic rings	Weak	Missing	Very weak
736	o-di-substitution in 6-membered aromatic rings; $-C-H$ out of plane deformation	Strong	Very weak	Weak
622	$-C-S$ stretching vibrations	Weak	Weak	Very weak

3.3. UV-Vis spectroscopy

In UV-Vis spectroscopy, the analysis was conducted on the waste waters (exhausted dye baths P1ex, resulted in RB21+HMP+LSR procedure and P2ex, resulted in RB21+GRNT+HMP+LSR procedure) and compared with RB21 aqueous spectrum. In UV-Vis spectroscopy, the solutions were analyzed in quartz tubes. UV-Vis absorption spectroscopy completes the information about the Reactive Blue 21 aqueous solution (RB21aq) during P1 and P2 processes, namely laser induced dyeing (LID) without and with garnet gemstone.

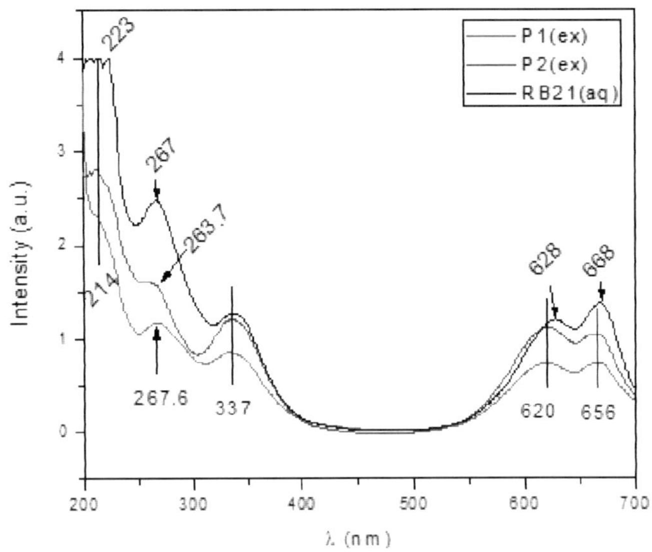

Figure 4. UV-Vis compared spectra of Reactive Blue 21 aqueous solution and exhausted dye baths resulted in laser induced dyeing (LID), namely P1(ex): RB21aq+HMP+LSR and P2(ex): RB21aq+GRNT+HMP+LSR.

The band at 337 in the UV-Vis spectra denotes Cu^{2+} ions complexation assessed to copper phthalocyanines and on visible spectra, copper complexation with phtalocyanine group is noticed in the peak at 668 nm for RB21aq and 656 nm for the samples P1(ex) and P2(ex). Absorption spectra in visible denote hypsochromic shift from 628 nm and 668 nm in the RB21 aqueous solution to 620 nm and 665 nm, respectively, in the exhausted dye baths of the P1 and P2 samples (Figure 4). That indicates changes in the functional groups playing the role of auxochromes. It is thus evidenced that some of the auxochromes have been removed from the dyestuff molecule under the laser irradiation. The auxochrome groups influence the dyeing color, exerting its bathochrome or hypsochrome effect. The auxochromes in p- π conjugation with the chromophore increase the wavelength at which the light is absorbed and the intensity of absorption. Once such groups are removed, the opposite effect of shifting light absorption to lower wavelengths and decreasing the intensity of absorption occurs. This is the case in the present study when the wavelength of the absorbed light decreases compared to the initial dye (blue shift), by losing one or more auxochromes which are situated in the side chain. Literature reports absorbance wavelength in the same range for copper binding with proteins and other organic compounds as ligands [16, 17].

However, the large chromogene (phthalocyanine with sulfone groups) involved in a complexation with copper ions remains stable during laser irradiation, and despite there are changes in the side chain which is a vinyl sulfone structure that is responsible for binding with the cellulose and in the bridging aniline structure, there is still possible that the component CHR (Figure 3b) binds with the cellulose through the sulfonic groups attached to the chromofore benzenic groups, under a mechanism specific to the direct dyestuffs, as iterated before.

3.4. Laser induced fluorescence (LIF)

Laser induced fluorescence spectroscopy (LIF) analysis performed on Reactive Blue 21 aqueous solution (RB21aq) and on the exhausted dye baths P1(ex): RB21+ HMP+LSR and P2(ex): HMP+RB21+GRNT+ HMP+LSR, in quartz tubes, at 30 ns, 40 ns and 50 ns delay using the same YG 981E/IR-10 laser system as for the laser induced dyeing, this time with 355 nm beam wavelength, 10 ns pulse width, 10 Hz repetition and 150 mJ energy.

In the LIF spectra (Figures 5, 6), the copper fluorescence is indicated by the peak contouring towards 699 nm and that could go up to 750 nm [18]. Sulfure dioxide generation due to UV laser beam interaction with sulfone groups is indicated by the spectra curves going up towards 400 nm and that could form a peak in the 300 nm wavelength region [19, 20]. The radicals formation, mainly ·OH radicals, is indicated by the tower shaped bands in the 481 nm – 508 nm and 588 nm – 627 nm as noticed before in the literature [21, 22-25]. The two emission lines at 514 nm and 550 nm may be assessed to a mirrorless laser phenomena induced during the interaction of UV laser beam and the copper phtalocyanine structure in the dyestuff. The RB21aq presents a strong 30 ns delay emission in the green region (524 nm – 550 nm) which mitigates for the exhausted dye baths P1(ex) and P2(ex), indicating that the emission is related to the initial chemical structure of the dyestuff.

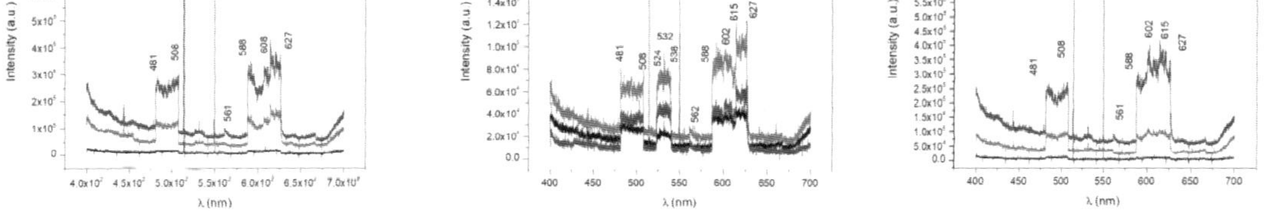

(a) (b) (c)

Figure 5. LIF spectra of RB21aq: Reactiv Blue 21aqueous solution and exhausted dye baths P1(ex): RB21+ HMP+LSR and P2(ex): HMP+RB21+GRNT+ HMP+LSR at: (a) 30 ns; (b) 40 ns; (c) 50 ns delay.

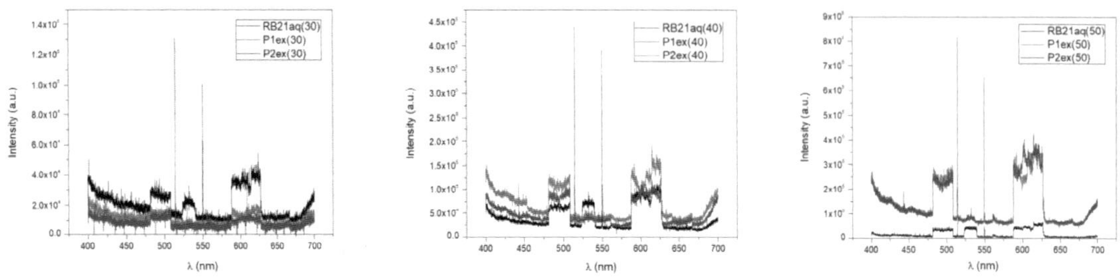

(a) (b) (c)

Figure 6. LIF compared spectra at: (a) 30 ns, (b) 40 ns and (c) 50 ns delay of the RB21aq: Reactive Blue 21 aqueous solution versus exhausted dye baths resulted in laser induced dyeing (LID) of P1ex: RB21+ HMP+LSR and P2ex: HMP+RB21+GRNT+ HMP+LSR samples.

Table 2. LIF spectra emission intensities (x10^5).

Sample	RB21			P1(ex)			P2(ex)		
Wavelength (nm)	30 (ns)	40 (ns)	50 (ns)	30 (ns)	40 (ns)	50 (ns)	30 (ns)	40 (ns)	50 (ns)
481	0.3	0.8	0.4	0.1	1.1	2.6	0.1	0.8	2.4
508	0.3	0.8	0.4	0.2	1.3	3	0.2	1	2.5
514	1.3	2	0.6	0.8	2	8.2			
515							0.4	3.5	7.9
524	0.2	0.7	0.4						
538	0.2	0.7	0.4						
550	1	1.6	0.5	0.7	1.6	6.5	0.4	3.4	6.3
561				0.1	0.5	1	0.06	0.4	0.9
562	0.16	0.3	1.5						
588	0.18	8	0.45	0.2	1.4	3.1	0.1	0.9	2.8
602	0.4	1	0.5				0.1	0.9	4
608				0.2	1.9	4.3			
615	0.4	1.2	0.7				0.1	1	4.1
626	0.4	1.2	0.7						
627				0.2	1.6	3.3	0.07	1	3.4

Compared spectra in Figure 6 show a longer fluorescent time life of the exhausted dye baths P1(ex) and P2(ex) compared to the Reactive Blue 21 aqueous solution (RB21aq). That indicates intermolecular interactions during UV laser irradiation, as well as different chemical structures of the three analyzed solutions. The level of copper complexation with the phthalocyanine may influence the variation in emission intensity because of the energy transfer from the excited state of the ligand to the metal ions causing decrease in intensity [17]. Table 2 includes the emission intensities in the three fluorescence spectra.

4. Conclusion and perspectives

The laser induced dyeing method purposed and studied herein proves to be suitable, with very good results regarding fiber coating when compared with the classical dyeing. Considering the substantial reduction of the dyeing process time, namely 11 times compared to classical process, the fact that no auxiliary substances were used, as well as the very good dyeing/coloring results, the method proves useful and deserves further investigation and possibly improved.

It is not clear yet how the garnet gemstone influenced the physico – chemical process induced by laser interaction with the dyeing system, but the results of the FTIR analysis proved that the structure of the RB21 dyestuff was preserved in P2(ex) when the garnet gemstone was used, compared with the P1(ex) when decomposition at bridging group level (S – N) and furthermore is clearly evidenced. Further investigation is needed to explain the phenomena.

It is very important that the LID method can be applied for dyeing small amounts of textile materials. Currently, the dyeing of small amounts of textile material is a problem primarily in terms of optimizing energy consumption, as well as materials and chemicals. Applications of the laser induced dyeing (LID) method could be used especially for dyeing materials with special destination (medical, cosmetics, etc.).

5. References

[1] Grindea M, Forst T and Hanganu A 1983 *Dyeing and Printing Technology of Textiles* (Bucharest: Technical Publishing)
[2] Ojstršek A, Doliška A and Fakin D 2008 *Analytical Sciences* **24** 1581-1587
[3] Wang P, Ma Q, Hu D and Wang L 2015 *Reactive & Functional Polymers* **91–92** 43–50
[4] Srivastava P, Goyal S and Tayade R 2014 *The Canadian Journal of Chemical Engineering* **92** 41-51
[5] Brijpuriya D, Jamdarkar M, Agrawal P and Nandy T 2013 *International Journal of Scientific & Engineering Research* **4(11)** 1893-1897
[6] Imen K, Benoît M and Raja Ben A 2010 *Chemical Engineering Journal* **156** 121–133
[7] Amani-Ghadima A R, Aber S, Olad A and Ashassi-Sorkhabi H 2013 *Chemical Engineering and Processing* **64** 68– 78
[8] Alinsafi A, Khemis M, Pons M N, Leclerc J P, Yaacoubi A, Benhammou A and Nejmeddine A 2005 *Chemical Engineering and Processing* **44** 461–470
[9] Ma Q and Wang L 2015 *Journal of the Taiwan Institute of Chemical Engineers* **50** 229–235
[10] Sanz J F, Monllor P, Vicente R, Amat AM, Arques A and Bonet M 2013 *Textile Research Journal* **83(13)** 1327–1334
[11] Cleto M, Ribeiro M, Clara M, Starling V M, Maria M, Leão D and de Amorim C C 2017 *Environ Sci Pollut Res* **24**:6165–6175
[12] Shin C-H, Bae J-S and Rudolph V 2014 *Journal of Industrial and Engineering Chemistry* **20** 710–716
[13] Miller FA and Wilkins C H 1952 *Anal. Chem.* **24 (8)** 1253–1294
[14] Pretch E, Bülmann P and Badertscher M 2009 Structure Determination Of Organic Compounds.Tables Of Spectral Data, Fourth, Revised and Enlarged Edition (Berlin: Springer – Verlag)
[15] Frost RL, Reddy B J and Keeffe E C 2010 *Journal of Molecular Structure* **977(1-3)** 90-99
[16] Girotto S, Cendron L, Bisaglia M, Tessari I, Mammi S, Zanotti G and Bubacco L 2014 *J. Biol. Chem.* **289** 10887-10899
[17] Hazra M, Dolai T, Pandey A, Dey S K and Patra A 2017 *Journal of Saudi Chemical Society* **21(1)** S240-S247
[18] Sharma S and Uttam K N 2017 *Vibrational Spectroscopy* **92** 135-150
[19] Syty A 1973 *Anal. Chem.* **45(9)** 1744-1747
[20] Zohora N, Kandjani A E, Orth A, Brown H M, Hutchinson M R and Gibson B C 2017 *Scientific Reports* **7(1)** 16905
[21] Cocean I, Cocean A, Postolachi C, Pohoata V, Cimpoesu N, Bulai G, Iacomi F and Gurlui S 2019 *Applied Surface Science* **488** 418–426
[22] Tsuboi Y, Kimoto N, Kabeshita M and Itaya A 2001 *J. Photochem. Photobiol. A Chem.* **145** 209-214

[23] Achitei DC, Vizureanu P, Minciuna MG, Sandu AV, Buzaianu A and Dana DI 2015 *Materiale Plastice* **52(2)** 165-167
[24] Tarnita D, Berceanu C and Tarnita C 2010 *Materiale plastice* **47(3)** 328-334
[25] Irimiciuc S A, Agop M, Nica P, Gurlui S, Mihaileanu D, Toma S and Focsa C 2014 *Japanese Journal Of Applied Physics* **53(11)** 173-179

IOP Conf. Series: Materials Science and Engineering **877** (2020) 012023 doi:10.1088/1757-899X/877/1/012023

Influencing Factors on Premature Fracture of Cast Iron Support of Electric Motor in Aluminium Rolling Line

F Preda[1*], I Riposan[1] and S Stan[1]
[1] Materials Processing & Ecometallurgy, Politehnica University of Bucharest, 313, Independentei Spl., 6th District, 060042 Bucharest, Romania

*E-mail: preda.florin89@yahoo.com

Abstract. There is a limited increase in the temperature of the cast iron supports during rolling (up to 30...50^{0}C), but there is a wide range of vibration level (2.6 - 4.4 mm/s) representing 58-98% (75-80% as medium values) from the maximum admitted level (4.5 mm/s, ISO 10816). The maximum level of vibration is identified in the area where a fan is installed to cool the electric motor. In all electric motor types it appears a difference between several measurements, which highlights the appearance of various conditions in the rolling process, or in the quality of the cast iron support. The fracture of the support takes place predominantly in the area of fixing it with a screw. The cracks connect firstly the areas with casting defects (inclusions, pinholes, shrinkage and micro-shrinkages), but also coarse graphite particles (Kish or Type C- ASTM, spiky graphite) or undercooled graphite morphologies (Type-D ASTM). Lamellar graphite cast irons (traditionally used in this domain) are characterized by a low level of fracture resistance but high vibration damping capacity due to which it is necessary to make parts without casting defects and with a controlled graphite phase (high amount of graphite, graphite Type-A ASTM, without undercooled graphite), usually obtained through high performance inoculation [1,2].

1. Introduction

The objective of this study case is represented by the analysis of the causes leading to the premature fracture of the cast iron support of the electric motor from the rolling-line components and the rolling bridges in the aluminium industry and identifying ways to prevent this fracture.

The rolling line consists of three important machines, the principal machine being the hot rolling which has the role to processing by plastic deformation using two work cylinders and two reversible support cylinders. They reduce the size of the aluminium products in the direction of pressing from raw thickness of 530 mm until 7 mm.

The next two components of the rolling-line are two cutting to length machines located at the line extremities, namely: one cutting to length machine of 3430 kN (F350) which cuts the material up to a maximum thickness 40 mm and one cutting to length machine of 7355 kN (F750) which cuts the material with a minimum thickness of 40 mm up to 180 mm. Regarding the material transport from the hot rolling to the cutting machine and back, it is used a transporting roller including an electric motor type : TML 100A B35, ASU - IM 3001, B30 – KL 160 with a power of 11 kW.

The rolling bridges have the role of lifting and carrying in the horizontal direction the load (coils, rolling cylinders, plates, mechanical parts with oversize) and have a maximum lifting weight of 79 kN, 118 kN, 618 kN or 1177 kN. The rolling bridges have the following components parts:

a) A crane trolley, which is composed of the lifting mechanism of load and the movement mechanism of the crane trolley which are both made from an electric action motor type: ASU 225M-8, CEI34-1 or K500 with a power of 25kW, a coupling, an industrial gear reducer and an electro-brake;

b) The movement mechanism of the bridge which is composed of an action group at the end of the line made of an electric action motor type: TML 100A B25, CEI 34-1, K500 or ASU 200L-8 with a power of 35kW, a coupling, a gear reducer and a electro-brake which drives the rolling wheels.

The electric motor includes a cast iron support (Figure 1) with the role of fixing the motor in the transporting roller and rolling bridges area. Due to the mechanical shocks and the continuous vibrations during the use, these pieces/parts could present fractures prematurely and after that to block the transporting roller and damage the rolling material. If a motor does not work, the rolling bridge cannot be used, this delaying the production process.

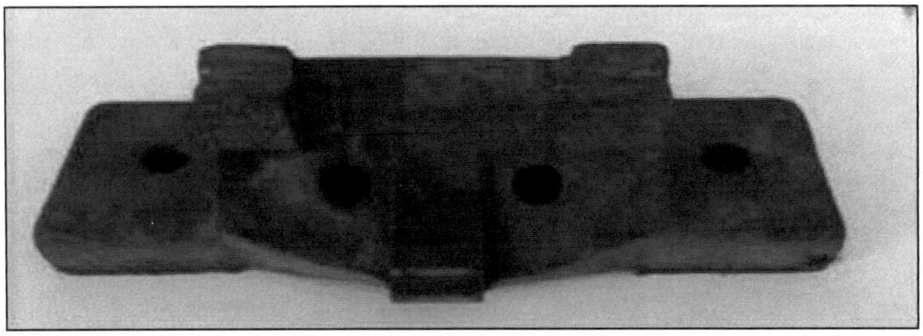

Figure 1. Iron cast support from the electric motor ensemble.

2. Experimental Procedure

In this study there are analyzed 20 grey iron castings (with lamellar graphite) used as supports for the electric motor from the rolling assembly component (Figure 2), and which have been subjected to vibrations and mechanical shocks in different circumstances.

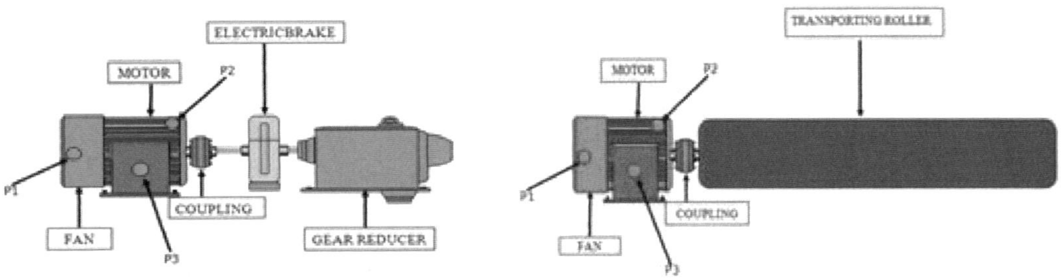

(a) (b)

Figure 2. General scheme of the ensemble and vibration measurement points (P1, P2, P3): (a) motor scheme mounted on the crane bridge; (b) motor scheme mounted at the transporting roller.

There is a number of 6 cast iron supports at 3 motors from the transport roller that are positioned in 3 points on the rolling line namely: one is positioned near the cutting to length machine of F350, one near rolling and one near the cutting to length machine of F750. The last one is placed as close as possible of the main machine of the rolling line because the vibrations make their presence felt.

The remaining 14 supporting cast iron parts from 7 engines are analyzed in the component of the rolling bridges besides the vibration chock that comes from the impact of the bridges at the end of the line and also the crane trolley. It is also taken into consideration the possibility of thermal demand because of comparing to the ambient temperature (for example, in winter can be -5°C and in summer up to 30°C) in some cases will be added also the heat generated by electric furnace.

Table 1. Parameters of the experiment conditions.

Electric Motor Type	Motor work area	Tested Supports	Moment of vibration measurement
TML 100A B35	Transporting roller (rolling area)	2	When transporting raw material, during the rolling process; when transporting thin but long material (coil)
ASU - IM 3001	Transporting roller (F350)	2	When transporting the material before to be cut, after cutting and during the cutting process
B30 – KL 160	Transporting roller (F750)	2	When transporting the material before to be cut, after cutting process, during cutting and when the scrap drops
ASU 225M-8	Rolling bridges (travel load)	2	Without load, with the load allowed 12t, during load lifting
CEI 34-1	Rolling bridges (travel crane trolley)	2	During transport with the load allowed 12t, without load
K500	Rolling bridges 177kN (travel crane bridge)	2	During transport with the load allowed 12t, without load, during travel the crane trolley
TML 100A B25	Rolling bridges 1177kN (travel load)	2	Without load, with load of 12t, during load lifting
ASU 200L-8	Rolling bridges 1177kN (travel crane trolley)	2	When transporting with the load of 12t, without load
CEI 32-1	Rolling bridges 1177kN (travel crane bridge)	2	When transporting with the load of 12t, without load, during to travel the crane trolley
ASU 200L-8	Rolling bridges 1177kN (travel crane bridge)	2	When transporting the work cylinders (98t) for change

The equipments used to realize this study case are: a) SKF microlog analyser used to measure and analyze the vibrations of the motor and also its temperature; b) a digital thermometer with laser Trotec for measurement of temperature; c) microscopes Olympus and 3D Hirox for performing the structural analyze to the cast iron support, broken premature.

In Table 1 is presented the analysis procedure for 20 cast iron supports, the type of motor and the moments of vibrations measurement.

3. Results and Discussion

3.1. Variation of the temperature

Figure 3 includes the measured value regarding to temperature in the case of the TML 100A B35 motor (Table 1), mounted on the transporting roller (rolling area). It could be observed an increase and a decrease of the temperature of the cast iron support of the electric motor, during its using: in the moment when the material for rolling (to 500°C) is transported on the rollers; when it is rolled; where it is transported to cutting. The time for the entire cycle of rolling is around 20min/rolled product.

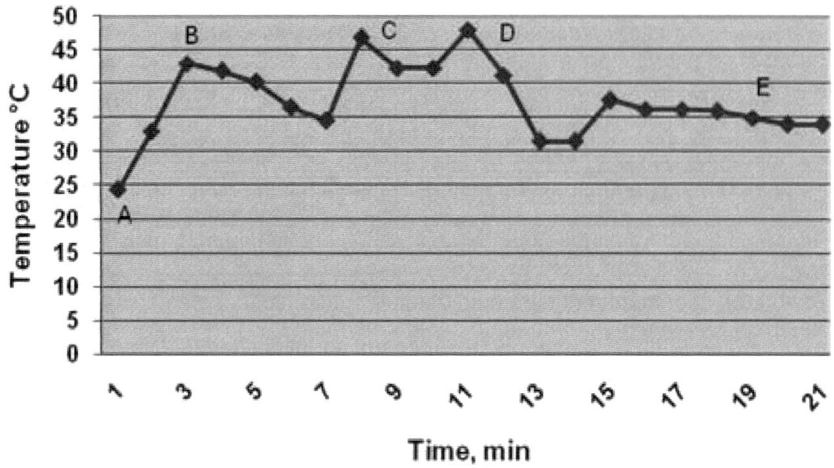

Figure 3. Temperature variation of the electric motor during the rolling process (the motor TML 100A B35, Tab. 1) (A) - start of measurement; (B) - transport of material on the roller; (C) - after the 10 pass of rolling; (D) - after the last pass of rolling; (E) - go to the cutting.

It is appreciated that the heating of the electric motor ensemble is limited (30....50^{0}C), and as a result, the cast iron support shouldn't be affected concerning its behaviour in the exploitation.

3.2. Vibration level evaluation

Vibrations are measured throughout the rolling process from putting the aluminium product on the transporting rollers to the last rolling pass. Figure 4 illustrates a typical evolution of the level of vibrations measured in the case of the motor TML 100A B35 (transporting roller).

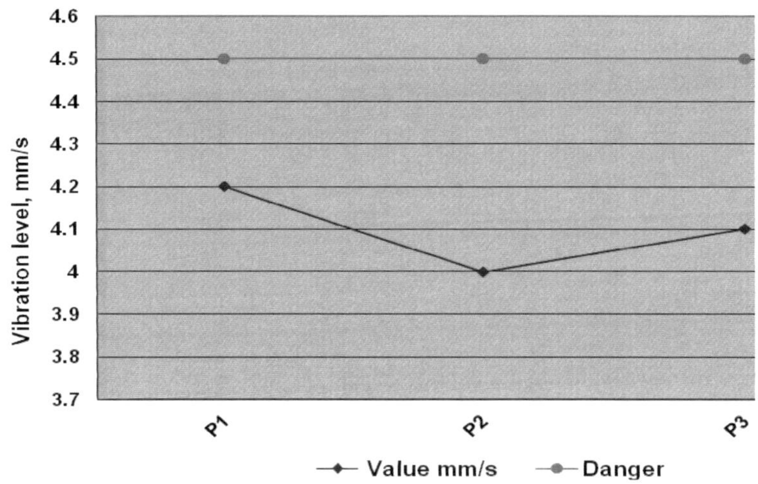

Figure 4. Typical example of the vibration level of the electric motor (P1, P2, P3 - measurement points, see Figure 2).

The Table 2 and the Figures 5 and 6 include the level of vibrations measured in the case of the 20 tests performed, two for each motor considered. P1, P2 and P3 represent the vibration measurement points of the electric motor assembly, highlighted in Figure 2.

The measurements performed showed that there is a wide range of framing of vibration level of electric motors, respectively 2.6 - 4.4 mm/s. At the same type of motor, it appears differences, sometimes important, between two measurements, which highlights the appearance of various conditions in the rolling process. There are not obviously different behaviours between the electric motors used to transporting roller and respectively the rolling lines of aluminium products.

Regarding the measuring points, the maximum level vibration is at P1, in the area where a fan for cooling the electric motor is mounted. In case of this fan has been turned off, as at the motor CEI 34-1 (see Table 2), the measured value of the vibrations decreased substantially. Then follows the measuring point P3, and, at the lowest level, the measuring point P2.

Table 2. The measured values of the electric motor vibrations.

Motor Electric Position	Electric motor Type (are working)	Measured value of vibrations, mm/s, in points: P1	P2	P3
Transporting Roller	TML 100A B35 (rolling)	3.8	4	4.1
		3.0	2.6	3.3
	ASU - IM 3001 (F350)	3.5	3.0	3.4
		4.0	3.9	4.2
	B30 – KL 160 (F750)	3.8	3.6	3.2
		3.7	3.5	3.7
	Average	3.63	3.43	3.65
Rolling Bridges	ASU 225M-8 (travel load)	3.7	3.5	3.6
		3.2	2.8	3.3
	CEI 34-1* (travel crane trolley)	3.1	2.9	3.4
		3.2	2.8	3.3
	K500 - 1177kN (travel crane bridge)	4.4	4.0	4.3
		3.4	3.1	3.2
	TML 100A B25 - 1177kN (travel load)	4.0	3.8	3.9
		3.2	2.8	3.0
	ASU 200L-8 - 1177kN (travel crane trolley)	3.5	3.4	3.3
		3.9	3.2	3.4
	CEI 32-1 - 1177kN (travel crane bridge)	4.2	3.8	3.9
		4.0	3.7	3.7
	ASU 200L-8 - 1177kN (travel crane bridge)	3.8	3.6	3.2
		3.8	4.0	4.1
	Average	3.67	3.39	3.54
	General Range	3.0-4.4	2.6-4.0	3.0-4.3
	General Average	3.66	3.40	3.58

* - without fan (without electric motor cooling)

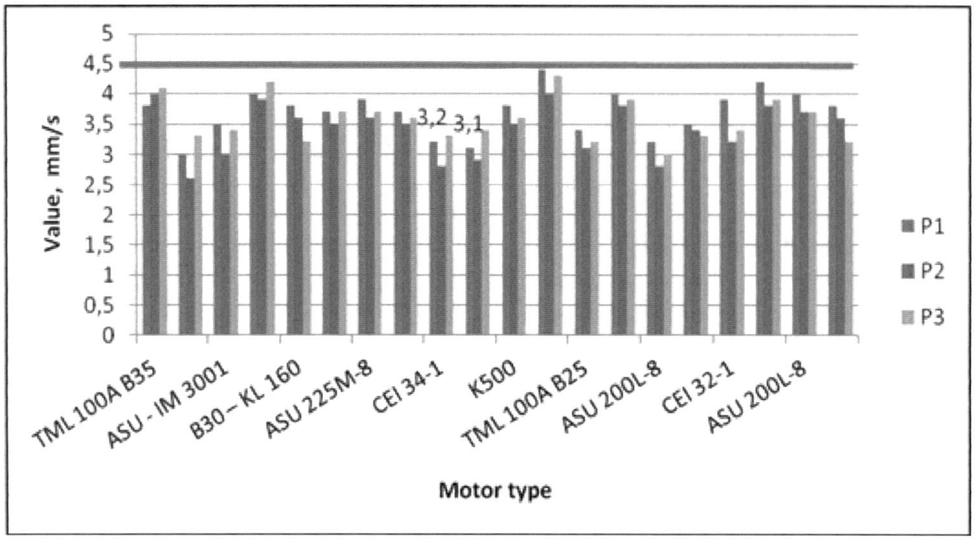

Figure 5. The vibration level of the electric motors that operate in the presented conditions in Table 1. (P1, P2 and P3 – points of measuring, Figure 2) (2 measurements for each type of electric motor).

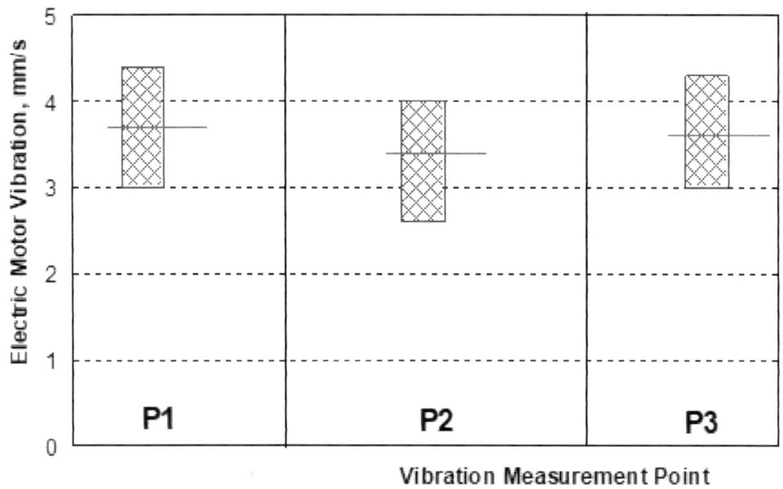

Figure 6. Framing range and average vibration values, in the measurement points P1, P2 si P3 (see Figure 2).

In accordance with the international standard ISO 10816, the maximum allowed value of the vibration level of electric motor is 4.5mm/s. In the analyzed case, the vibration level represents 58 - 98% of the maximum accepted level, respectively at level 75-80% of their environments. It is appreciated that in the analyzed case, this level is high, and that measures are needed to reduce vibration. In this case, the iron cast support of electric motor can play an important role.

In case of cast iron parts, graphite has a positive important role in vibration damping, favouring the dissipation of the energy appeared such a request. As a result, the amount of graphite, the graphite morphology and the degree of compactness of the graphite particles play an important role. Suggestive in this sense are the data presented in Figure 7, obtained by processing the data from the paper [3] that highlight the damping capacity of the cast iron with lamellar graphite (grey cast iron), graphite compact/vermicular (compacted graphite cast iron) and with nodular graphite (ductile iron). In this case, damping capacity is expressed by cyclic toughness, as a ratio between hysteresis curve area (□W) and area under force-strain curve (W). The larger it is the higher value is, with that damping capacity is bigger. In case of lamellar graphite (LG) cast iron, the chemical composition is varied to a large extent expressed by the equivalent carbon value (CE = 3.5 - 5%). For compacted graphite cast iron and nodular graphite cast iron, chemical composition was used of the typical values for each cast iron type, respectively CE = 4.2 - 4.5%.

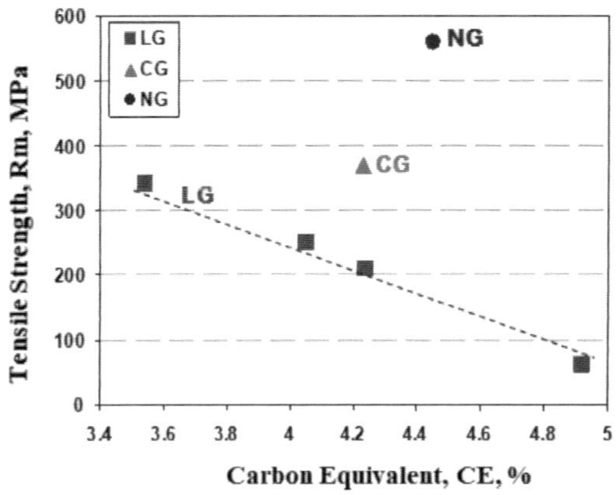

(a)

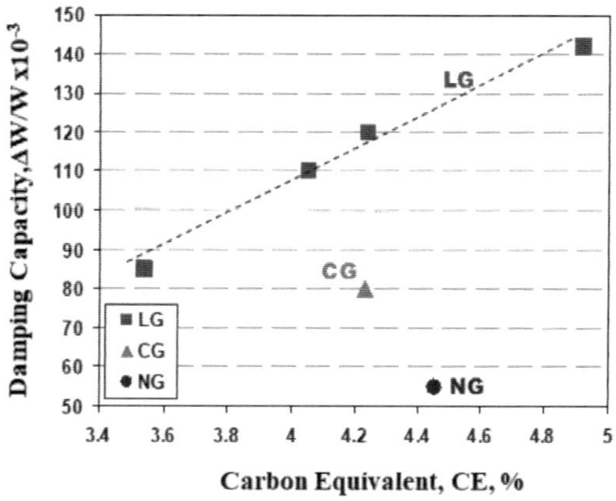

(b)

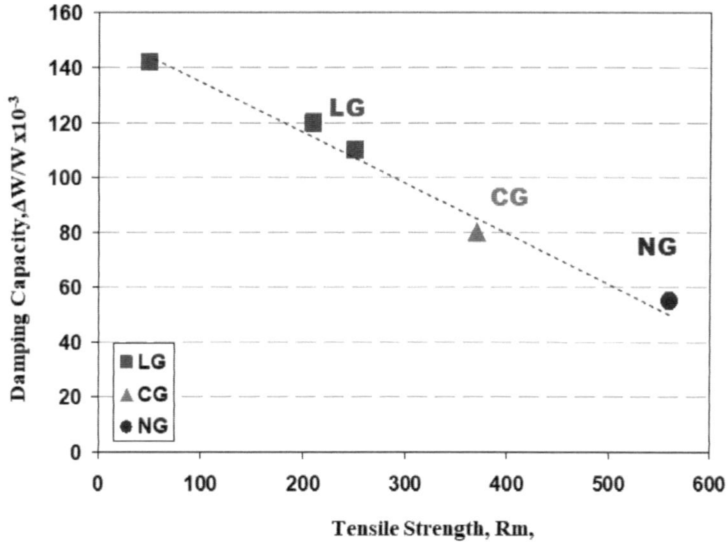

(c)

Figure 7. Tensile strength (a) and damping capacity (b), (c) of cast irons with lamellar graphite (LG), compacted/vermicular graphite (CG) and nodular graphite (NG).

Increasing the value of the equivalent carbon (CE) means the transition from the hypo-eutectic domain, through the eutectic domain up to the hyper-eutectic domain, action accompanied by the increase significant amount of graphite in the cast iron structure. Consequently, tensile strength (Rm) decreases significantly in the case of lamellar graphite cast iron (Figure 7 a)). Intermediate compacting of graphite until to morphology of compacted graphite (CG) substantially increases the tensile strength, with the maximum value reached the maximum degree of compactness of the graphite,

IOP Conf. Series: Materials Science and Engineering **877** (2020) 012023 doi:10.1088/1757-899X/877/1/012023

respectively the nodular graphite (NG), despite the fact that these castings are usually characterized by high values of carbon equivalent.

A variation into reverse shall be recorded in damping capacity (Figure 7 b)): increasing the amount of graphite with the increase of the equivalent carbon leads to the increase of the vibration damping capacity to castings iron pieces. The graphite compacting decreases the vibration damping capacity, which is why in case compacted graphite (CG) and especially nodular graphite (NG), lower values of this parameter are reached, although these cast irons have a large amount of graphite.

Figure 7 c) illustrates the strong connection between damping capacity and tensile strength, with graphite positioning as morphology (LG, CG, NG). It is found that in the case of compacted graphite cast iron (with a large amount of graphite in structure), the vibration damping capacity is of the same order of size with that one of lamellar graphite cast iron, with a low carbon saturation degree (small amount of graphite), respectively high tensile strength.

It is known that in the case of grey cast irons, used in the automotive industry, in production for engine block, lamellar graphite morphology is also important: the presence of undercooled graphite (D-type Graphite, ASTM) determines the decrease not only of tensile strength but also of the damping capacity. As a result, metallurgical inoculation treatment of grey cast irons appears to be useful for achieving simultaneous improvement of tensile strength and damping capacity, both of them very important to avoid premature fracture of the cast iron support of electric motors.

3.3. Characterization of fracture areas of cast iron supports

It is identified the area where it was produced the fracture of the grey cast iron (lamellar graphite) support of the electric motor (Figure 1), mounted in the conditions presented in Figure 8. Several broken parts were analyzed (Figure 9), regard the initiation and development for the cracks (Figures 10 and 11).

(a) (b) (c)

Figure 8. Identify the fracture areas of cast iron supports of an electric motor: (a) motor K500; (b) motor ASU 200L-8 (crack thickness of 18mm for motors of 35KW); (c) motor B30 – KL 160 (crack thickness of 9mm for motors of 11KW).

(a) (b)

Figure 9. Example of fracture of cast iron support, for electric motor in aluminium rolling lines.

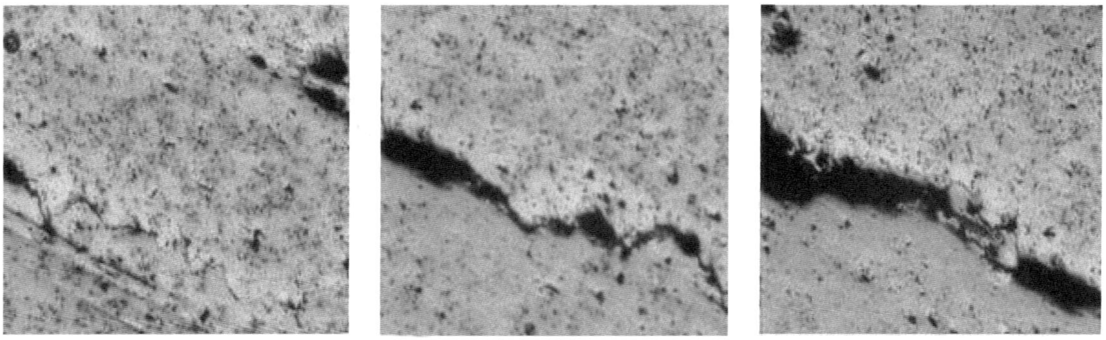

(a) (b) (c)

Figure 10. Evolution of the crack of the cast iron support: (a) micro-crack initiation; (b) crack development; (c) cracking before the piece fractured (Olympus Microscope analysis).

The analysis of a large number of cast iron supports, used to fixing the electric motors on the aluminium rolling line, leads to conclusion that fracture of these parts occurs predominantly in the screw area. Usually, micro-cracks are initiated on the surface of the support from contact with the screw, then developed into cracks in the mass of the cast iron part. Usually, the braking is favoured by the lack of material addition for avoiding the occurrence of defects and lack fall of pass from thin-walled to the thick-walled.

Following the route of the cracks, it is found that they unite in the first time areas with castings defects, of the type pinholes (gaps due to gases) and voids/micro-voids (gaps due to contraction phenomena). In the case of these grey cast iron pieces (lamellar graphite), it was also found that graphite morphology (type) play an important role. In this regard the coarse forms of graphite such as the type C-ASTM (specific to primary graphite, Kish graphite) as well as those specific to degenerated graphite (example Spiky graphite) favour the accelerated development of cracks. Simultaneously with the preferential development of cracks through the mentioned areas, there are not found cracks in the structural areas where appears lamellar graphite of type A-ASTM. This graphite morphology, that maximizes the achievement of maximum strength characteristics, usually includes relatively evenly distributed blades in the structure, to medium size (50 - 250 μm), uncut tips. This goal is usually through inoculation treatment, based on treating liquid cast iron with active substances, including inoculating elements [4]. These elements are able to positive influence quality of graphitization sites and thereby reducing the degree of undercooling at the eutectic solidification, necessary condition to form Type-A graphite [3].

4. Conclusions

The analysis of the premature fracture of cast iron parts, used as supports for electric motors included in the rolling lines of aluminium products, in terms of the use of these motors in industrial conditions, leads to the following conclusions:

a) It is a limited increase in the temperature of these parts during rolling (up to 30 ....50^{0}C), which should not particularly affect their behaviour in plant conditions.

b) The performed measurements highlight the existence of a wide range of framing of the vibration level of electric motors, 2.6 - 4.4 mm/s, constituting 58 - 98% (75-80% as medium values) from the maximum allowed level (4.5 mm/s, ISO 10816).

c) The vibrations level appears to be maximum in the area where a fan is installed for cooling the electric motor (obviously decreases when the fan is stopped).

d) At the same type of motor there are sometimes important differences, between several measurements, which highlights the appearance of various conditions in the rolling process, or in the quality of the cast iron support.

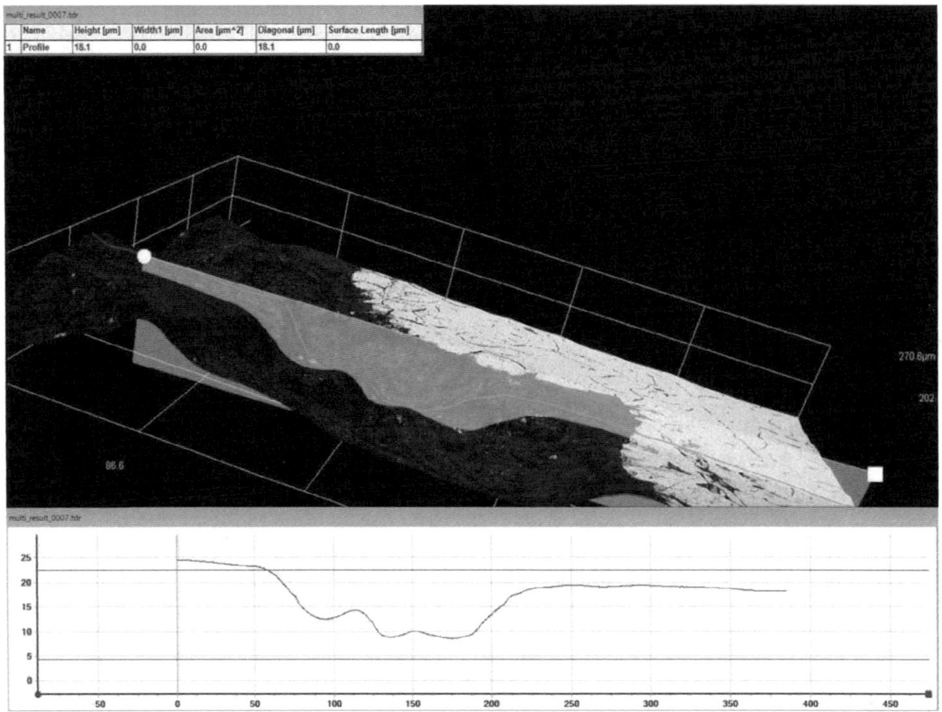

a)

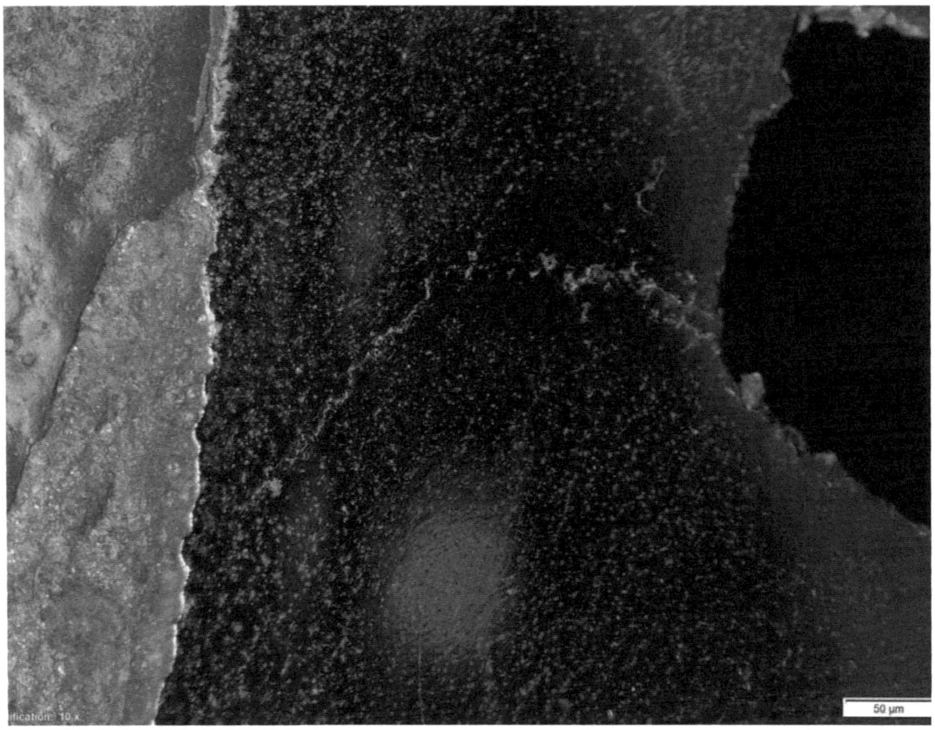

b)

Figure 11. Cracking analysis before fracture using 3D Hirox Microscope: (a), depth of the beginning of the crack 18 μm); (b) the starting area of the crack in the piece of cast iron with lamellar graphite.

e) It is appreciated that the vibration level is high which means that measures are required to reduce it, including by improving the quality of cast iron support.

f) The broken of these parts takes place predominantly in the grip area with screwing the micro-cracks initiated at contact with the screw developing cracks in the mass of the cast iron piece.

g) The cracks primarily unite areas with casting defects (inclusions, pinholes, shrinkage and micro-shrinkages), but also coarse graphite particles (Kish or Type C- ASTM, spiky graphite) or undercooled graphite morphologies, such as Type-D ASTM.

h) Lamellar graphite cast irons, traditionally used in this field are characterized by low resistance of fracture level but high vibration damping, the reason which is why it is necessary to make parts without casting defects and with a controlled graphite phase (large amount of graphite, Type-A ASTM), obtained by specific metallurgical treatment, inoculation.

Acknowledgements
The work has been funded by the Operational Programme Human Capital of the Ministry of European Funds through the Financial Agreement 51668/09.07.2019, SMIS code 124705.

5. References

[1] Chisamera M, Riposan I, Sofroni L and Stan S 2010 *Fabricarea Pieselor din Fonta Prin Turnare. Capitol in: Tratat de Stiinta si Ingineria Materialelor Metalice* Vol. 4 (Bucharest: ASTR AGIR) 534-717, ISBN: 978-973-720-313-7

[2] Riposan I, Chisamera M, Sofroni L, Stan S 2009 *Fonte Chapter in Tratat de Stiinta si Ingineria Materialelor Metalice*, Vol. 3, 4 (Bucharest: ASTR AGIR) pp. 463-580, ISBN: 978-973-720-261-1

[3] Riposan I, Chisamera M and Sofroni L 1985 *AFS Trans.* **94** 35-48

[4] Riposan I and Skaland T 2017 *Modification and inoculation of cast iron. ASM Handbook on the Cast Iron Science and* Technology **1A**, ed D.M. Stefanescu (American Society of Materials) 160-176

Transfer Processes at the Manufacture of Metal Matrix Composite Materials

R M Florea[1*]
[1] "Gheorghe Asachi" Technical University of Iasi-Romania, Dept. of Materials Science, Blvd. Mangeron, No. 61A, 700050, Iasi, Romania.

*E-mail: raluca.m.florea@gmail.com

Abstract. The main elements of composite materials technological design consist of choosing the appropriate matrix/reinforcement systems and optimal thermodynamic conditions for ensuring compatibility during processing and operation. Composite material properties depend on the volumetric characteristics of components as well on the intensity of links between these components. Initially, the mass transfer takes place under unstable conditions in the liquid matrix and at the solid/liquid interface, and finally, after casting or infiltration, stationary. Modelling of reinforcement elements transfer from gas to liquid is based on total variation of transfer energy or on the variation of the amount of the forces acting on the reinforcement. Once the basis of the models investigated and of the actual model proposed are analyzed, the densities influences, particles granulations, liquid viscosity, thermal conductivities, the energies at liquid/reinforcement interface and critical speeds of reinforcement elements, transfer from liquid to solid.

1. Generalities

Composites are multiphase materials with distinct and well defined interfaces between the constituent phases which are strongly interrelated and it presents exceptional properties in well-defined directions and plans.

An important direction of research in these areas is to provide models to successfully predict if interactions of physical or chemical nature occur between matrix and reinforcement, as well as for the thermodynamic evaluation of the interfaces during processing and exploitation.

2. Physico-chemical processes at the matrix-reinforcement interface

Composites processing and its characteristic depend on the components density and the intensity of connections which are formed between them. According to Dupré's relation, the adhesion energy, W_a, between two different phases can be evaluated in terms of the interfacial tension and contact angle:

$$W_a = \sigma_{SG} + \sigma_{LG} - \sigma_{SL} = \sigma_{LG}(1 + \cos\theta) \tag{1}$$

Where: σ_{SG} is the solid-gas interfacial tension; σ_{LG} is the liquid-gas interfacial tension; σ_{SL} is the solid-liquid interfacial tension; θ is the contact angle.

Chidanibaran et. al (1992), [1], consider that ceramics wettability by metals is achieved only if the chemical bond are formed between atoms of the two phases, totally ignoring the contribution of physical interactions. Chemical forces are much stronger and manifests themselves when the chemical reaction between the atoms of liquid and non-metallic component, respectively, shows a negative variation of the free enthalpy. Type of chemical bond between metallic and non-metallic component structures has a particular importance because the transition layer's structure and its mechanism of formation determine the crystallisation and hence the structure of the composite material.

The first model that tried to explain the metal-ceramic wettability was proposed by Mc Donald and Eberhart in 1965. It claims that between the chemical component of adhesion energy and the variation of metal-ceramics chemical reaction free energy a linear relation can be established in the form [2]:

$$W_{a(ch)} = a - b\,\Delta G^{\circ} \tag{2}$$

Where a is the contribution of the physical interactions, while b is a constant.

Experimentally it has been shown [3] that adhesion energy W_a does not depend on ΔG° when $\Delta G^{\circ} > 300$ kJ/mol. For $\Delta G^{\circ} < 300$ kJ/mol, the value of W_a increases with decreasing of ΔG°.

Because the chemical reaction between some metals with high reactivity (such as Al) and ceramics is associated with a variation of the free energy smaller than 100 kJ/mol, we might conclude that they'll wet any solid material.

In reality, even in the most severe conditions of achievable depression an oxide layer is forming which will affect the wettability conditions. During the production of composite materials there are mechanisms of irreversible loss of energy during the processes of interface creation. These losses are sometimes very important and dictate the process parameters such as the applied pressure or preheating temperature of reinforcement elements. Under these circumstances, it is possible that a contact angle less than $\pi/2$ not be a sufficient condition for spontaneous wetting of ceramics during composite processing.

2.1. Physico-chemical processes at metal-oxide interface

In the case of non-metallic components consisting of oxides, adherence to melt improves with increasing the metal affinity to oxygen and oxygen concentration, respectively. If it is noted with Me the matrix metal and with M the oxide-bounded metal, then at the interface the next reactions occur:

$$y\,Me + M_xO \leftrightarrow Me_yO + x\,M \qquad \Delta G_1^{\circ} \tag{3}$$

$$M_xO \leftrightarrow xM + 1/2\,O_2 \qquad \Delta G_2^{\circ} \tag{4}$$

$$Me_yO \leftrightarrow y\,Me + 1/2\,O_2 \qquad \Delta G_3^{\circ} \tag{5}$$

$$y\,Me + 1/2\,O_2 \leftrightarrow Me_yO \qquad \Delta G_4^{\circ} \tag{6}$$

As well as dissolutions and mass transfer between the two phases:

$$M \rightarrow [\,M\,]_{Me} \qquad \Delta G_5^{\circ} \tag{7}$$

If the free enthalpy variation is negative, the reactions arise in the normal direction, and adhesion energy will increase with the increasing of the absolute value of this variation. The sign of free enthalpy variation of the chemical Equation (3) is a criterion for assessing the metal-ceramic reactivity or non-reactivity. A more rigorous criterion for assessing interface chemistry takes into account that the dissolution of M in Me will always takes place and $\Delta G_{1\text{-}5}^{\circ}$ is calculated with the Equation:

$$\Delta G_{1\text{-}5}^{\circ} = \Delta G_1^{\circ} + \overline{\Delta H_{M(Me)}}^{\infty} \tag{8}$$

Where: $\overline{\Delta H_{M(Me)}}^{\infty}$ is the partial entropy at infinite dilution of M into Me.

In metal-oxide systems, frequently used for the production of composite materials, the maximum dissolved amount is of the order of few parts per million. Such systems with $\Delta G_{1\text{-}5}^{\circ} > 0$ are known as non-reactive systems and are characterized by a weak and limited wettability. In this case the adhesion energy, W_a, represents only 15-40% of the metal cohesion energy [4], which is considered to be the $W_c \approx 2\ \sigma_{Lv}$. On the basis of the thermodynamic considerations, in 1987 Chatain et. al show that even for nonreactive metal-oxide systems, chemical interactions occurs at the interface and they propose an expression for the adhesion work [5]:

$$W_a = -\frac{C}{N_0^{1/3} V_{Me}^{2/3}}\left[\overline{\Delta H_{0(Me)}}^{\infty} + \frac{1}{S}\overline{\Delta H_{M(Me)}}^{\infty}\right] \tag{9}$$

Where: $\overline{\Delta H_{0(Me)}}^{\infty}$ and $\overline{\Delta H_{M(Me)}}^{\infty}$ are mixture partial enthalpies at infinite dilution of oxygen and oxide metal in the metal matrix; N_0 – Avogadro's number; V_{Me} – molar volume of liquid metal; C and S – are constants that depend only on the nature of the oxide substrate (for alumina C = 0,2 and S = 1,5 while for silica C = 0,2 and S = 2).

The product $N_0^{1/2}\ V_{Me}^{2/3} = \Omega_{Me}$ represents the area occupied by a monolayer of The product $N_0^{1/2}\ V_{Me}^{2/3} = \Omega_{Me}$ represents the area occupied by a monolayer of a mole of matrix metal atoms.

According to Equation (9) dissolution of oxygen in the matrix metal impoves its adhesion to ceramics. On consider that the dissolved oxygen combines with the metal atoms and MeO clusters are forming, clusters with partially ionic character as a result of charge transfer from the metal to oxygen. Nowadays it is widely accepted the model proposed by Naidich [6] in 1981, which assumes that dissolved oxygen forms with nearby metal dipole $Me^{2+}\ O^{2-}$ which is adsorbed to the interface due to electrostatic attraction forces exerted on the Me^{2+} cation by the anions existing on the ceramic surface.

Improving adhesion in non-reactive conditions can be achieved by using immiscible alloying elements into the base metal, which reduce its surface tension. Thus, in the Al alloys, In, Bi, Cd, Na can be added, which are adsorbed at the surface, diffuse together with Al atoms in the ceramic surface and favour the adhesion.

2.2. Physico-chemical processes at the metal-carbon or metal-carbides interfaces

Carbon and carbides wettability by liquid metals can be analyzed by considering three chemical contributions [2]: Me metal affinity to carbon; formed Me_xC carbide stability; mixing enthalpy of M in Me.

Liquid metal adhesion to carbon is strongly dependent on the reactivity between the two phases and it is slightly influenced by the crystallographic shape of the solid. An interdependence between metal's position in Mendeleev's table and their reactivity in respect to carbon was found. Metals from groups I-b, III-a, IV-a and V-a and periods 4, 5 and 6, like: Au, Ag, In, Ga, Ge, Sn, Pb, Sb, Bi are inert to carbon [7], have low adhesion energy, only 100 ÷ 300 mJ/m^2 and the contact angle greater than $\pi/2$.

Elements: Al, Si, B, react and form carbides that wet the carbon, adhesion energy being 1000 ÷ 1500 mJ/m^2. Even in this case, wettability will practically occur only at temperatures above 900°C because the oxide films prevent direct contact between the two components.

Transition metals forming stable carbides are: Ti, V, Cr, Mn, Fe, Co, Ni from period 4, Zr, Nb, Mo, Pd from 5th period and Ta, W, Re, Pt from period 6. In their case the adhesion energy to carbon is 2000 ÷ 3000 mJ/m^2, increasing with the decrease of the number of electrons on the d layer. Adhesion energy between the metal matrix containing such elements and carbon is more than 90% due to the chemical interaction.

Alkaline, alkaline-earth and some rare metals, although diffuse in carbon components and form carbides, not always wet such particles or fibres [15-16].

Experimentally it was proved that Li and Na wet the graphite [17-18].

Through its influence on the nature and kinetics of metal-carbon interfacial reactions, temperature is a control parameter for wettability.

Nayeb-Hashemi and Seyyedi [8], in 1989, support the fact that the formation of rhombohedral crystalized aluminium carbide Al_4C_3 can occur in long time even at temperatures around 500°C.

Wang et. al [9] in 2015 after determining the activities and the contact angle, had been calculated the free enthalpy of reaction:

$$4[Al] + 3\ C_{(s)} = Al_4C_{3(s)} \tag{10}$$

Using Equation:

$$\Delta G_a = \Delta G_a^\theta + RT\ ln \frac{a(Al_4C_3)}{a^4_{([Al])} a^3_{(C)}} \tag{11}$$

They have obtained a free enthalpy significant variation of aluminium carbide formation function by content of Mg from matrix Al alloy.

In Figure 1 it can be seen that up to approx. 10% Mg, the free enthalpy of reaction (10) is negative, so Al_4C_3 can form, but at higher concentrations occurs the decomposition of carbide.

To improve adhesion between the graphite's fibres or particles and Al alloys based metal matrix, Ti or B coatings on carbon solid components were made. In this case, special measures are required to prevent titanium or boron oxidation, as well as to limit the interaction because the strong reactions increase the fragility and lead to the degradation of reinforcement.

Another measure to improve the adhesion of the aluminium metal matrix alloys and carbon or graphite reinforcement is microalloying with superficially active elements such as Mg, Li, Na or Sr.

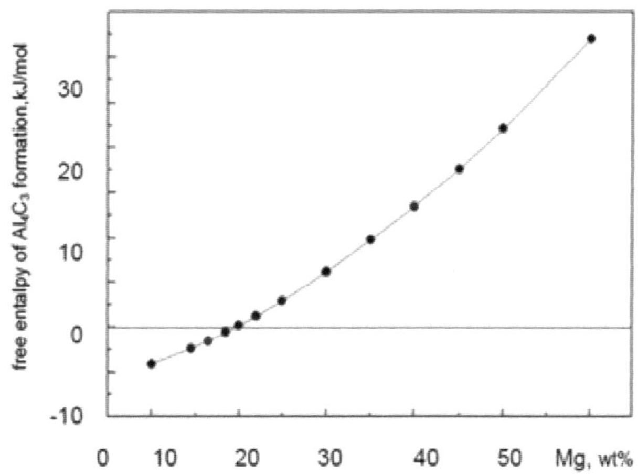

Figure 1. Variation of formation Al_4C_3 free enthalpy depending on the Mg content [9].

In Figure 2, SEM and EDS investigations conducted by Anna J. Dolata et. al [10] in 2016 in $AlSi7Mg2Sr0.03/SiC_p+C_{gp}$ composites highlighted a good bond between matrix and reinforcement, as well as an increase in the concentration of the strontium and magnesium concentrations at the interface between matrix and glassy carbon.

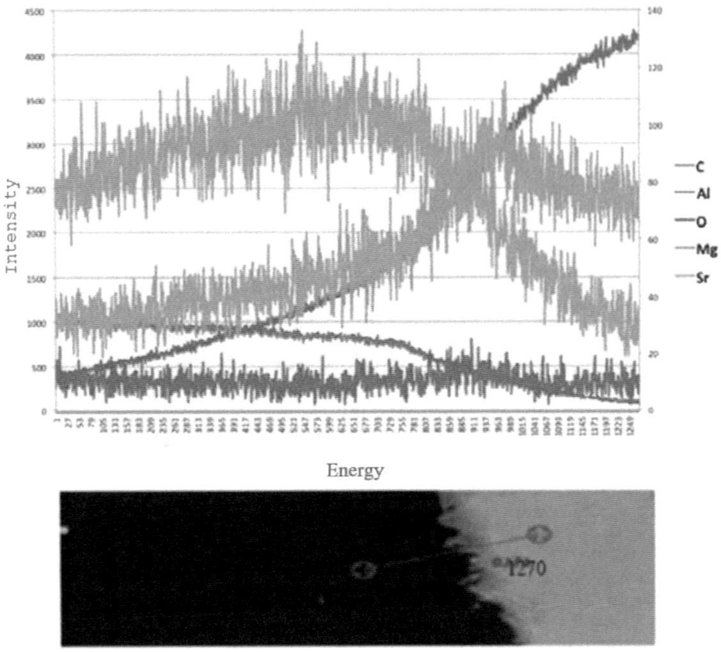

Figure 2. The AlSi7Mg2Sr0.03/SiCp+Cgp composite: linear distribution of elements at the interface between Al matrix alloy and Cgp, EDS [10].

While Nayeb-Hashemi and Seyyedi [8] noted the formation of chemical compounds at interface: Al_4O_4C, Al_4C_3, other, Siemens [11], in 1989, highlights the dependence existing between carbide concentration and temperature, as it can be observed in Figure 3. Research has revealed that the interaction between metal matrix with reinforcement elements containing carbides is much weaker because of the high energy of carbides formation.

Wettability of solid carbides like: SiC, B_4C, etc. is similar as in the case of carbon because the metals generally have similar behaviour against these elements which forms the carbides.

Estimations made on the processes at the SiC-metal interface or on the link between the two phases are not yet theoretically and experimentally grounded. First, the nature and composition of the interface are not precisely known.

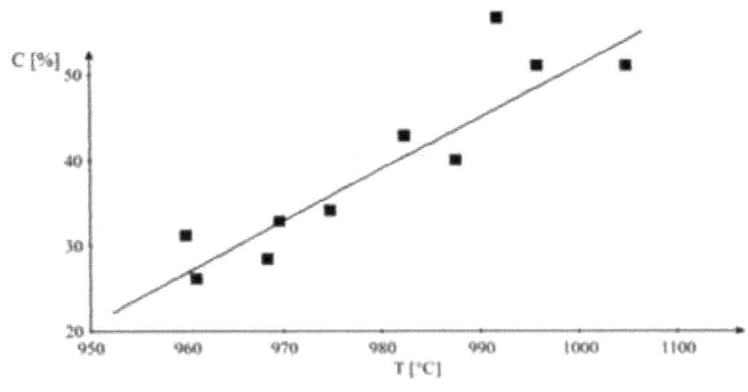

Figure 3. Variation of carbides concentation at the interface with metallic melt temperature [7].

Thus, in the case of Al-SiC interfaces, three hypotheses exist:
- a SiO_2 film forms at the interface [12];
- no interactions exist at interface, distinct areas being one of SiC and another of Al [13];
- an Al_4C_3 film forms at the interface [4];

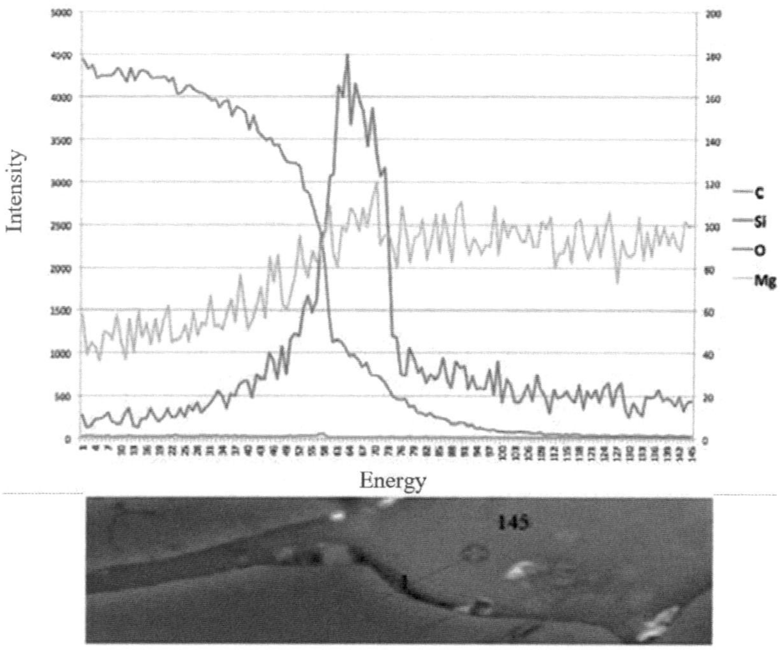

Figure 4. The AlSi7Mg2Sr0.03/SiC_p+C_{gp} composite: linear distribution of elements at the interface between Al matrix alloy and SiC_p, EDS [10].

In the case of Al-Mg alloys an MgO film may form at the interface, as it is deduced from Figure 4.

3. Conclusion

Theoretical study at atomic level of the interface is too complex due to reduced symmetry, high number of atoms in the elementary cell and complications that occur when trying to use interatomic potentials along the interface.

Li and Arsenault [14] in 1988, claim that for a good understanding of the bond at interface it is necessary to take into consideration the following:

a. interface may consist of areas with different aluminium and silicon carbide crystallographic orientations;

b. SiC has two different structures: α and β; whiskers have β structure, and particles have α structure which may be in six polymorphic forms;

c. interface can be form between carbon and aluminium films or between Si and Al;

d. at the interfaces formed with different orientations between SiC and Al, the atoms cannot form a commensurable structure, i.e. it is a network mismatch at the interface;

e. it is expected that at the interface atoms move in positions of minimum energy, but there is no experimental or theoretical studies confirming this.

Eustathopoulos and Mortensen [4] in 1993 supports the idea of slow decomposition of SiC, the enrichment of aluminium in Si and C at the interface and formation of fine and flat particles of Al_4C_3. To verify this hypothesis, large additions of Si and Al have been made, finding an increase of SiC thermodynamic stability and a reduction of Al_4C_3 formation. Wettability kinetics is limited by the dissociation velocity of SiC and the stationary value of contact angle is reached when Al surface layers are saturated in Si and C. The initial angle of 160° is due to Al_2O_3 film on the Al surface.

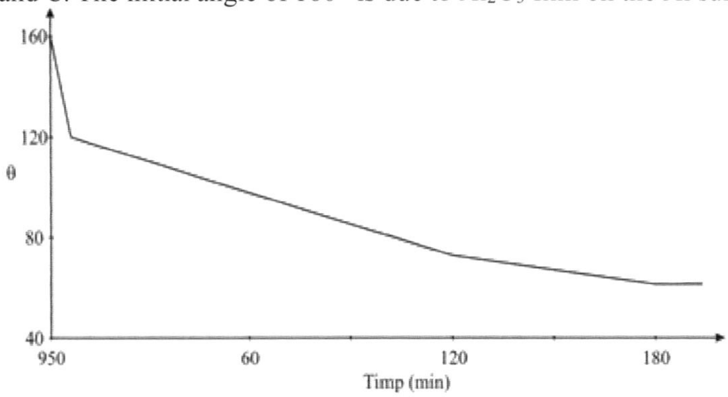

Figure 5. Variation of contact angle θ with time t, for Al-18%Si alloy with SiC, at 800°C under vacuum of 10-5Pa [7].

The real value of θ for Al-SiC is 120°, and after a three-hour contact angle θ decreases to 60°. Improvement of silicon carbide wettability by the Al is obtained by Li addition, which is a superficially active element, and Cu addition, which easily dissolves Si. Much stronger the interatomic links of carbides are, much higher the interphase energy is and a lower degree of wetting is obtained. Since the formation heat of B_4C is higher than that for SiC forming, the contact angle formed with an Al melt is 120° and 106.3°, respectively, at a temperature of 700°C.

4. References

[1] Chidanibaran P R, Edwards G R and Olson D L 1992 *Metall. Trans. B* **23(B)** 215-222
[2] Delannay F, Froyen L and Demyttere A 1988 *Cast Reinforced Metal US* 81-85
[3] Kaptay G 1991 *Solidif. and Micrograv. Mater. Sci. Forum* **77** 315-330
[4] Eustathopoulos N and Mortensen A 1993 *Fund. Met.-Matr. Comp. (*Oxford: Butterworth-Heinemann) pp 42-58
[5] Chatain D, Rivollet I and Eustathopoulos N 1987 *J. Chim. Phys.* **84** 201-203
[6] Naidich MV 1981 *Prog. in Surf. and Membr. Sci. NY* **14** 353-484
[7] Florea RM, Peter I, Rosso M, Mitrica D and Carcea I 2013 *JOAM* **15(7-8)** 833-840
[8] Nayeb-Hashemi H and Seyyedi J 1989 *Metall. Trans. A* **20(A)** 727-739
[9] Wang X, Wang C, Jiang D and Zhang Z 2015 *Rare Met. Mater. and Eng.* **44(4)** 830-833
[10] Dolata A, Dyzia M and Boczal S 2016 *Mater. Today: Proceed.* **3** 235-239
[11] Siemensen CJ 1989 *Metall. Trans. A* **20(A)** 194
[12] Carcea I and Florea R M 2017 *Transfer Processes at the Manufacture of Metal Matrix Composite Materials (Advances in Materials Science and Research* vol 30) ed Maryann C Wythers (New York: Nova Publishing) chapter 1 pp 1-81
[13] Monteiro AP and Barbedo A 1988 *Cast Reinf. Met. Comp. US* 361-367
[14] Li S and Arsenault R J 1988 *Cast Reinf. Met. Comp. USA* 33-39
[15] Baltatu M S, Vizureanu P, Balan T, Lohan M and Tugui C A *Book Series: IOP Conference Series-Materials Science and Engineering* **374** 012023
[16] Cazac AM, Bejinariu C, Ionita I, Toma SL and Rodu C 2014 *Applied Mechanics and Materials* **657** 193-197
[17] Vizureanu P, Cimpoesu N, Radu V and Agop M 2015 *Experimental heat transfer* **28(1)** 37-57
[18] Vizureanu P, Perju MC, Galusca DG, Nejneru C and Agop M 2010 *Metalurgia international* **15(12)** 59-64

IOP Conf. Series: Materials Science and Engineering **877** (2020) 012025 doi:10.1088/1757-899X/877/1/012025

Methods for Assessing the Freeze-Thaw Resistance of Road Concrete Used in our Country and at European Level

L M Nicula[1], O Corbu[1,2,3*] and M Iliescu[1]
[1] Technical University of Cluj-Napoca, Faculty of Civil Engineering, Romania
[2] Research Institute for Construction Equipment and Technology, Bucuresti
[3] Center of Excellence Geopolymer& Green Technology (CEGeoGTech), School of Materials Engineering, University Malaysia Perlis, 01000 Kangar, Perlis, Malaysia, India

*E-mail: Liliana.nicula@infra.utcluj.ro; *ofelia.corbu@staff.utcluj.ro; iliescu.mihai@staff.utcluj.ro

Abstract. The paper addresses three methods of assessing the freeze-thaw resistance of road concretes prepared with conventional and artificial materials. The blast furnace slag was used, as an artificial material, in granulated and milled form and in the form of crushed aggregates from the steel industry. The most common method used in our country to determine the freeze-thaw resistance is the destructive method by measuring the variation of the compressive strength of the samples subjected to repeated freeze-thaw cycles. The most severe method of testing the freeze resistance considered at the European level, is based on the calculation of the amount of exfoliated material in the presence of defrosting agents. Another method used is to determine the values of the dynamic modulus of elasticity after repeated cycles of freeze-thaw. The results obtained were interpreted according to different evaluation criteria and compared with limit values proposed by the standardized methods.

1. Introduction

For the beginning it is presented an general analysis of the requirements and recommendations related to the freeze-thaw exposure of the concrete structures at international level, but the purpose of the present paper is to select the testing methods for freeze-thaw used in our country, used at European level, with applicability for the on-going experimental study. This study is related to the usage in the road concrete mixtures of blast furnace slag as mineral addition from the ground-granulated blast furnace slag (GGBS) and as partial substitution of natural aggregates with crushed, non-granulated air-cooled furnace blast slag (ABS). The concept of equivalent durability defined in the paper "Advanced solutions for durable concrete", 2019 [1] in compliance with CEN/TR16563 [2], has as objective the correlation of the ground experience with the results obtained in the laboratory by developing a data base with different concrete mixtures as reference and with exposures to different exploitation environments. In the European countries the concept of concrete equivalent performance can be found in the Directive EN 206 [1, 3], but it's application in most of the countries is not implemented because in requires a national system renowned with clear information about the reference concrete. By definition, the concrete freeze durability represents the capacity of resisting in repeated freeze-thaw cycles during over the life span without deteriorating caused by material loss of the concrete surface or by internal cracks, [1]. The general parameters established by projection of the freeze-thaw resistant concrete are: the water/cement (w/c) ratio or the water/binder (w/l) ratio, the compressive strength and the content of occluded air, but besides these, there would be other requirement that are researched with a lot of interest: the factor of pore spacing, the proportion of supplementary constituents with cement characteristics (CSCC) defined as in the paper with reference [4], the limiting of the permeability and the freeze-thaw resistance of aggregates and concrete. Most of these parameters are tested according to the European standards, exception is the factor of pore spacing that can be found in the durability requirements of Norway, Denmark, Canada and USA, [1,4]. The standard PN-EN 480-11 [5] describes the structure of the air entrained in the concrete using the following parameters: the total air content A [%], the specific surface α [mm^2], the spacing factor L [mm] and the content of micropores A_{300} [%]. The spacing factor L of the air pores can be determined after the simplified model of Powers, which supposes that all the air bubbles have the same diameter and are distributed in the corners of a cube. Another approach for determining the spacing factor (L) of the air pores is based on the Pilleo concept, which takes into consideration both the air gap system and the characteristics of the aggregate particles. The principles of the calculus methods of the spacing factor was described by J. Wawrzeńczyka in the paper [6]. As example, the Canadian standard CSA A23.1-09 A23.2-09 [7] establishes the requirements for the occluded air spacing factor of 230 µm, [1]. In Russia, the standard GOST 26633-2015 [1,8] limits the (w/c) ratio at 0.45 and the content of occluded air at 5÷7% for road structures. In our country, by applying the standard SR EN 206+A1:2017 [9] and the normative NE 014:2002 [10] is previewed for freeze resistant concrete, the usage of air-entraining agents with minimum values established depending on the maximum dimension of the aggregate, then the limiting of the (w/c) ratio and minimum values for the compressive strength. For the road concrete are stipulated supplementary requirements regarding the minimum values for the flexural strength and the freeze-thaw resistance.

As for the use of supplementary constituents with cement characteristics (CSCC) in concrete mixtures exposed to freeze-thaw, the German standard DIN 1045-2 [11] indicate limit values for the concrete with slow hardening, such as those with fly ash, the Canadian standard [7] suggests the use of (CSCC) for the sulphur and chlorine combined attacks, and the requirements for the materials for concrete resistant to freeze, the Russian standard GOST R 55224-2012, clause 5.11.2 [12] accepts the blast furnace slag as main mineral supplement in the cement manufacturing for road concrete, reinforced elements and bridge structures [1]. In our country, in compliance with SR EN 206 [9] it is accepted the use of type II supplements (the pozzolanic or the latent hydraulic supplements) thus, fly ash maximum 33% of the cement mass, depending on the usage domain, and the ultrafine silica in percentage of maximum11% of the cement mass. Hooton mentions that for obtaining a durable concrete for the most aggressive exposure conditions is conditioned by the materials, the projection of the mixtures and the compliance with some appropriate construction technologies [1,13]. The testing methods of the freeze resistance and the acceptance criteria differ from one area to another, depending on the experimented type, the sample conservation plan, the duration and the temperature interval in the thermostat chamber and the acceptance criteria. All these parameters are conditioned by the exposure environment of the road concrete structure. Several main observations can be noticed regarding the testing methods, such as for the structures placed in a dry environment, but made of high-performance concrete, the resistance to freeze is evaluated only by the internal cracks, through method ASTM C666, procedure A [14]. But the road authorities require first the performance of the exfoliation testing with thawing agents, [1,15]. The Russian standards require for a freeze-thaw class F a minimum compressive strength that needs to be reached before the freeze-thaw exposure. There are other methods to assess the freeze resistance, such as the method of assessing the factor of protecting the pores known as (PF), representing a criteria for the assessment of the concrete freeze-thaw resistance, in compliance with the Finish standard SFS 4475:1998, [1, 16-18], and another one by assessing the concrete permeability [1, 19]. In Europe, the assessing of the internal cracks of the structure exposed to freeze is done in compliance with CEN /TR 15177:2006 [20], by which, depending on the chosen procedure, it is measured the dynamic elasticity module, the length variation and the water uptake after 56 freeze-thaw cycles, and the material loss testing of the concrete surface is done in compliance with the Swedish standard SS 137244 [21] and CEN /TS 12390-9:2009 [22], by which it is measured the material mass loss of the concrete surface after

56 or 112 freeze-thaw cycles, [1]. In North America, the assessment of the internal cracks is done in compliance with ASTM 666/C666-15 [14], after 300 freeze-thaw cycles, and the material mass loss testing of the concrete surface is done in compliance with ASTM C672 [1,23] at 50 freeze-thaw cycles [1]. The methods adopted in China for the internal crack assessments are in compliance with the standard GB/T 50082-2009 [24] by which it is measured the mass loss and the compressive strength after minimum 25 freeze-thaw cycles [1]. This method is close to the one used in our country, in compliance with the national standard SR 3518-2009 [25], regarding the preparation of specimens, the length and the temperature of the freeze-thaw cycles and the acceptance criteria. From this paper it can be observed that the testing methods are structured in two deterioration types: first the scaling of the concrete surface and the second the internal cracks of the concrete structures. The term of scaling is currently used in international studies, but in our country the usual term is exfoliation of the concrete surface. The definition of the term scaling can be found in CEN /TS 12390-9:2009 and represented the material loss of the concrete surface caused by the freeze-thaw cycles. Regarding the acceptance criteria, the results obtained by different testing methods there are reservations regarding the preparation, preserving modes, the age of the samples compared to the real exposure of structures. Due to these disagreements it is concluded that some methods do not present explicitly the acceptance criteria, such as method CEN/TR 15177:2006 used in Europe for the internal cracks testing, and the obtained results are seldom compared with those of the reference concrete. The described testing methods are frequently used for the assessment of the freeze-thaw resistance and they are still presented in several experimental studies that approach this issue.

Peter B. tested the freeze resistance in compliance with CEN/TS 12390-9:2007 of concrete made with three types of cement CEM I, CEM II, CEM III supplied by four European factories. The surface exfoliation was increased in mixtures containing CEM II and CEM III compared to mixtures containing CEM I. By changing the preserving conditions previously to the freeze-thaw testing of the concrete containing CEM II/III the material loss of the concrete surface was within the CEM I range [26]. The technical bulletin Frostprüfung von Beton from the Federal Waterways Engineering and Research, [27] recommends a 14-day preservation instead of 7-day under water for slow hardening concrete and the freeze-thaw testing to be done after 56 or 90 days. D Kocab *et al* [28], from Czech Republic, determined the freeze-thaw resistance of concrete made with aggregates coming from different crushing quarries. The chosen testing methods were GB/T50082-2009, a destructive method used for assessing the internal cracks of the structure, and ASTM 666-03, a non-destructive method used for the measuring of the relative dynamic modulus (RDM) of elasticity. Daria J.N.[29], carried out a study, in Poland, related to the influence of the light aggregate on the deterioration of the concrete surface, in the present of defrost agents due to the freeze-thaw phenomena. She used the Borås method, in compliance with the Swedish standard SS137244. Syamak Tavasoli, Mahmoud Nili and Behrad Sepoush, [30] used the ASTM666 method for determining the relative module of elasticity, the length variation and the mass loss in the study regarding the influence of GGBS on the freeze resistance in self-compacting concrete (SCC). Marta K.K.[31], studied the freeze-thaw resistance by testing the surface scaling in compliance with the method CEN/TS 12390-9:2007 of concrete (slab test) with fly ash, undergoing 112 freeze-thaw cycles in presence of the NaCl solution with 3% concentration. Marta K.K. showed that the test results analysed statistically, in compliance with the accepted criteria from SS 137244:1995, allow the usage of up to 25% fly ash, taking into consideration the air-trainer supplement and appropriate hardening conditions, lead to obtaining a durable concrete. Jun Phil Hwang, Hyun Bo Shim, Sooyoung Lim and Ki Young Ann, [32] studied the durable properties of concrete containing recycled aggregates and pozzolanic materials such as fly ash and grounded blast furnace slag. The freeze resistance was determined using the ASTM666 method through a fast test. Huai-Shuau and Ting-Hua Yi, 2013 [33] studied the freeze-thaw durability of air-entrained concrete. They measured the relative dynamic modulus of elasticity and the mass loss up to 400 freeze-thaw cycles using the GB/T50082-2009 method. The results were compliant up to 300 freeze-thaw cycles. In our country, there were experimental researches for the assessment of the freeze-thaw resistance of concrete using the method CEN/TS 12390-9 (cube test) for the exfoliation of the surface, using the national standard SR 3518:2009 and the method CEN/TR 15177:2006 for assessing the internal cracks of the structure, [34].

In this paper there are presented three testing methods for the freeze-thaw resistance by material loss of the concrete surface (scaling) using the method CEN/TS 12390-9:2007 [22] (slab test), by assessing the internal cracks of the structure using the method CEN/TR15177:2006 [20] and using the destructive method in compliance with the national standard SR 3518:2009 [25]. The first two methods are freeze-thaw testing methods used in the European countries, and the third one is the most frequent method used in our country.

2. Materials and Methods

2.1. Prime conventional materials

The cement used in road concrete mixtures is type CEM I 42,5R, and it was supplied from the factory of Aleşd, by the company S.C. HOLCIM Romania S.A. The natural sand, 0/4 mm dimension, comes from the gravel pit from Beclănuţ, Bistriţa County, the coarse cover aggregate (gravel), 4/8 mm dimension, comes from the gravel pit from Sanicoara, Cluj County, and the coarse cover aggregate (crushed), 8/16 mm and 16/25 mm dimension, comes from the company SC Grandemar Cluj. The used aggregate proportions were 32% sand (fine aggregate) and 68% coarse aggregate. The granularity curve of the total aggregate mixture was within the granularity range in compliance with NE 014-2002 [10]. The testing procedure on aggregates was done in compliance with the normative references SR EN 12620:2003 and SR EN 12620:2003+A1:2008 [35] and SR 667:2001 [36]. The used additives were supplied by the company BASF Romania, the super-plasticizer additive MasterGlenium SKY 527 and the air-trainer additive Master Air 9060. The water from the concrete was taken from the water supply system of the city Cluj-Napoca, in compliance with SR EN 1008:2003 [37].

2.2. Prime non-conventional materials

The blast furnace slag represents a secondary product of the metallurgic industry after the melting of cast iron. This secondary product can be reused as prime material for constructions. Depending on the chosen cooling process, two types of blast furnace slag can be obtained granulated and non-granulated distributed by the company ArcelorMittal from Galaţi, (Romania). The granulated blast furnace slag was grounded at the cement factory from Aleşd, and the specific surface with the value of 5770 cm^2/g was determined by the university's laboratory. The grounded-granulated blast furnace slag called (GGBS) is part of type II supplements category due to the fact that it has hydraulic properties, in compliance with SR EN 15167-1:2007 [38], with usage in concrete mixtures for improving certain properties and to grant special properties in compliance with the provisions from SR EN 206 [9]. According to the chemical composition, the blast furnace slag from the company ArcelorMittal fits within the category of basic slags because of the basicity index, Nicula *et al* [39], because the basic oxides prevail (CaO, MgO, FeO, MnO) in proportion of 49.01%, see the Table 1.

Table 1. Chemical composition of the granulated blast furnace slag (GGBS), %.

SiO_2	Al_2O_3	MnO	MgO	CaO	Fe_2O_3	Na_2O	K_2O	TiO_2	Mn_2O_3
36.44	11.60	0.55	5.8	41.81	0.78	0.345	0.428	0.30	0.62

The non-granulated blast furnace slag is not used in cement manufacturing, it presents with a crystallin molecular structure [40], which after crushing can be used as air-cooled blast furnace slag (ABS) crushed aggregate in different field. The crushing of the non-granulated blast furnace slag was ensured by the supplier at a dimension of 0/4 mm. The characteristics of artificial blast furnace slag were tested by the supplier in the reports [41, 42] for the blast furnace slag dimension 0/4 mm.

2.3. Methods

It is known that the granulated blast furnace slag has hydraulic properties when it is activated suitable in compliance with SR EN 197-1:2011 [43] and with SR EN 15167-1:2007 [38], it was introduced in the road concrete mixture grounded at a dimension of 63 μm, as a binding agent. The non-granulated, crushed blast furnace slag with the dimension of 0/4 mm was used as substitute for the related sand that was part of the total mixture of aggregates defined within SR EN 12620:2003+A1:2008 [35]. Three mixtures made with blast furnace slag were compared with two road concrete mixtures made with conventional materials with different cement dosages. The samples were named symbolically to make the connection with the quantities of the used material, see Table 2. The quantity of 54 kg of (GGBS) was used as addition 15% to the mass of the cement, compared with the S 360c mixture, and compared with the S 414c mixture, the quantity of 54 kg of (GGBS) substituted 13% from the mass of the cement. The mixtures were projected to be used in exposure conditions class XF4 (exposure to freeze F with strong water saturation with defrost agents, with humidity level 4). In compliance with the durability requirements for the road concrete established within [9, 10, 44], the minimum conditions for the projecting were: a cement dosage of minimum 340 kg/m^3, the W/L ratio maximum 0.45, minimum resistance class C35/45 and air-entraining agent within the mixtures. All the tests were performed in the labs of Cluj-Napoca Technical University, Faculty of Constructions.

Table 2. Material quantities within mixtures.

Quantities	Mixtures [Kg/m^3]				
Mixture (Kg/m^3)	S 360 c	S 414 c	S 54/20	S 54/40	S 54/60
Cement (C)	360	414	360	360	360
Ground-granulated blast furnace slag powder (GGBS)	-	-	54	54	54
Total binding agent (L)	360	414	414	414	414
W/L, (water/binder)	0.44	0.41	0.41	0.43	0.43
Aggregate	1864	1831	1863	1810	1810
Fresh state density	2380	2415	2444	2402	2402
Super-plasticizer additive	3.60	4.14	4.14	4.97	4.97
Air-trainer additive	1.08	2.07	2.07	2.07	2.07

2.3.1. Casting specimens, hardening, maintaining and preparing samples for testing. The mixtures analysed in this study are part of the first casting series. For each mixture we casted 4 cubic samples with side of 200 mm and 12 cubic samples with side of 150 mm, kept in air for 24 h, afterwards they were removed from forms and sunk in water at a temperature of (20±2)°C. At the age of 7 days, the samples were removed from water and kept in air, in a climatic chamber, at a temperature of (20±2) °C and a humidity of (65±5) % for the freeze-thaw testing, in compliance with SR-3518:2009 [25]. At the age of 50 days started the freeze-thaw testing using the destructive method. From the 200 mm-side cubes we cut concrete stripes of 150x150x50 mm, and at 150 days we started the freeze-thaw testing with defrost agents in compliance with CEN/TS 12390-9:2009 [22] and using the method of measuring the propagation time of ultrasonic impulse (UPTT) in compliance with CEN/TR 15177:2006 [20].

2.3.2. Determining the freeze-thaw resistance using the destructive method. The freeze-thaw resistances of the concrete mixtures were determined on 12 samples with a side of 150 mm (6 control samples and 6 samples that underwent 150 and 300 repeated freeze-thaw cycles), in compliance with the national standard SR-3518:2009 [21]. The testing period of our study extended compared to the provisions from NE014-2002 [7] from 100 to 300 freeze-thaw cycles, similarly to the number of freeze-thaw cycles requested by the standard ASTM C666/C666M-03 [18], but the testing procedure for the two standards is different. The method ASTM C666 has the average temperature range between (+5 and -18)°C, the length of a cycle between (2 ÷5) h, the freeze-thaw cycle in water (in compliance with procedure A), respectively the freeze cycle in the air/the thaw cycle in the water (in compliance with procedure B, also called the fast test). According to the testing procedure mentioned in SR-3518:2009, the thermostat chamber maintains the temperature for the freeze cycle at (-17±2)°C for 4h, and for the thaw cycle up to (20±2)°C for 4h. 4 days before starting the sample testing, they were taken from the climatic chamber and inserted in the water at a temperature of (20±5)°C for saturation. The water was added gradually on ¼ of the height of the samples for 3 days, after 3 days the water was added up to 20 mm over the height of the specimens. The samples completely covered with water were kept like that for 24 h, and afterwards the control samples continued to be kept in water, while those prepared for the freeze-thaw cycles were introduced in the thermostat chamber. The admissibility condition in compliance with SR-3518:2009 is that the compressive strength losses of the samples undergoing successive freeze-thaw cycles should not decrease with more than 25% compared to those of the control samples kept in water. The value of the resistance loss after n freeze-thaw cycles is calculated using the equation 1.

$$\eta_n = \frac{f_{cmwater} - f_{cmfreez}}{f_{cmwater}} x100\ (\%) \tag{1}$$

η_n – the compressive strength loss after "n" freeze-thaw cycles,

$f_{cm\ water}$ – the average compressive strength of the samples maintained in water during "n" freeze-thaw cycles,

$f_{cm\ freeze}$ – the average compressive resistance of the samples maintained in the thermostat chamber during "n" freeze-thaw cycles,

2.3.3. Determining the freeze-thaw resistance by measuring the exfoliant mass of the concrete surface. The material loss at the hardened concrete surface after the 56 repeated freeze-thaw cycles, defined shortly as scaling of the concrete surface, in the presence of defrost agents, is determined using the alternative "slab test": method mentioned within SR CEN/TS 12390-9:2009 [22]. The reference method requires the testing to start at the age of 31 days, and the alternative method allows the beginning of the testing at a concrete age different than 31 days. The exfoliated mass measuring method on the sample surface is appreciated as the most severe method of testing the freeze resistance and it is based on the Swedish standard SS 13 72 44 using the Borås method [21]. For each mixture, we cut 4 concrete stripes of 150x150x50 mm from the 200 mm cubes that were preserved in a climatic chamber at a humidity of (65±5) % and a temperature of (20±2) °C until the testing. A 3 mm rubber sheet was first glued an all the sample surfaces, except for the testing surface, then a silicon glue cord was applied between the rubber and the concrete surface. A test was performed to check the tightness of the rubber sheet by pouring on the testing surface 67 ml of deionized water, maintained for 72 h. The water evaporation rate was measured in a container filled with water with a surface of 225 cm^2, for which it resulted, after 7 days, an evaporation of 44.59 g. After the tightness testing all the sample surfaces, except the testing surface, were isolated thermic with polystyrene of (20±1) mm. 15 minutes before introducing it in the thermostat chamber, on the testing surface it was poured 67 ml solution for testing with thaw agent with a thickness of 3 mm, consisting of 97% drinkable water and 3% sodium chloride NaCl of the total mass. Then the samples were introduced in the thermostat chamber, from the brand Controls, they were covered with a polyethylene sheet and they were tested in repeated freeze-thaw cycles. The length of a freeze-thaw cycle was 24 h, of which the thaw cycle range was between (7÷9) h, and the freeze cycle range was between (15÷17) and the temperature range minimum/maximum between (-22 and +22) °C. The temperature variation registered during the testing at the testing surface is represented in Figure 1.a), and in Figure 1.b) there is the image with the samples deposited in the thermostat chamber. The freeze-thaw resistance was assessed by measuring the exfoliated mass on the sample surface after 7, 14, 28 and 56 freeze-thaw cycles. At 7, 14, 28, 42 and 56 cycles, at the end of the thaw phase it was collected in a filter paper the exfoliated material from the testing surface and a new testing medium was added on

IOP Conf. Series: Materials Science and Engineering **877** (2020) 012025 doi:10.1088/1757-899X/877/1/012025

the surface exposed to the freeze-thaw in the climatized chamber. The exfoliated material and the used filter were dried until the constant mass of $(110\pm10)^0C$ and then weighted (images just like in Figure 2).

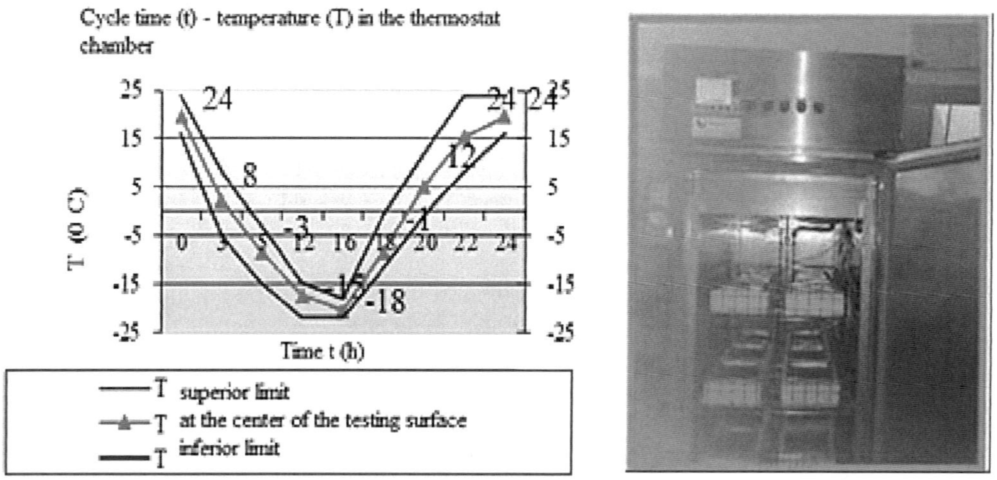

(a) (b)

Figure 1. (a) Cycle time (t) - temperature (T) in the thermostat chamber at the center of the testing surface of a specimen; (b) the thermostat chamber.

Figure 2. Collecting the exfoliated material, filter with exfoliated material after drying, replacing the testing medium with a new one (3% sodium chlorine NaCl).

The total dried exfoliated material mass after (n) freeze-thaw cycles was calculated using equation 2.

$$m_{s,n} = m_{s,inainte} + (m_{v+s+f} - m_{v+f}) \tag{2}$$

The cumulated value S_n of exfoliated material, after (n) freeze-thaw cycles in kg/m^2 is expressed using the equation 3.

$$S_n = \frac{m_{s,n}}{A} x10^3 \tag{3}$$

$m_{s,n}$ – total mass of the exfoliated and dried material after (n) freeze-thaw cycles, rounded value to the closest 0.1 g;

$m_{s,\,before}$ – total mass of the exfoliated dried material from the previous cycles, g;

$m_{v+s(+f)}$ – the mass of the container containing the exfoliated dried material with used filter, rounded value to the closest 0.1 g;

$m_{v(+f)}$ – the mass of the empty container and of the used filter, rounded value to the closest 0.1 g;

A=0,021025 m^2 surface total area, measured between the tightening cords rows, rounded to the closest 100 mm^2.

The accepted criteria used for the freeze-thaw testing with defrost agents in this experiment are compliant with the Swedish standard SS 13 72 44 (the Borås method), method included in the study [1] from 2019 performed by Andrei Shapak, and criteria adopted as well by Daria J.N. 2002 in the study [29]. The Swedish standard SS 13 72 44 (the Borås method) is based on the average exfoliated mass at 28 days (m_{28}), at 56 days (m_{56}) and, optionally, at 112 days (m_{112}), see Table 3.

Table 3. Acceptance criteria for the freeze resistance, the exfoliation testing.

Scaling resistance	Requirements
Very good	m_{56} <0.10 kg / m^2
Good	m_{56} <0.20 kg / m^2 or m_{56} <0.50 kg / m^2 and m_{56} / m_{28} <2 or m_{112} <0.50 kg / m^2
Acceptable	m_{56} <1.00 kg / m^2 and m_{56} / m_{28} <2 or m_{112} <1.00 kg / m^2
Unacceptable	m_{56} ≥1.00 kg / m^2 and m_{56} / m_{28} ≥2 or m_{112} ≥1.00 kg / m^2

2.3.4. Determining the freeze-thaw resistance by percentage decrease the dynamic module of elasticity. In compliance with the Technical Report CEN/TR 15177:2006 [20], the freeze-thaw resistance was determined through the alternative method "the slab test", as relative dynamic modulus of elasticity. The measurements were made on prismatic sample of 150x150x50 mm before test, after 28 freeze-thaw cycles and after 56 freeze-thaw cycles, on prisms of 150x150x50 mm. The method is based on measuring the propagation time of ultrasonic impulses (UPTT) between 2 transducers in compliance with SR EN 12504-4, [45]. The measuring equipment, the brand Controls, having its own frequency in the range of 40-100 KHz, was graduated before usage. The transducers were pressed on the coupling environment of the concrete surface in order to reach a minimal constant value. The transmission time was read with an approximation of 0.2 μs. The percentage decrease of the dynamic module of elasticity was determined using equation 4 for each mixture. The method CEN/TR 15177:2006 does not present a value threshold for the acceptance criteria of the relative dynamic modulus of elasticity. The GB/T 50082-2009 method used in China limits the RDM loss up to 60% [1, 24], and with the method used in Russia GOST 10060-2012 [1, 46], the module RDM variation must be smaller than 25%. Because the testing procedure is different from one method to another, for this study the acceptance criteria was based on the equivalent performance concept of mixtures with blast furnace slag compared with reference mixtures made with conventional materials.

$$RDM_n = \left(\frac{t_0}{t_n}\right)^2 x100\ (\%), \tag{4}$$

Where t_0 and t_n is the initial propagation time of ultrasounds, respectively after n cycles (μs), RDM the relative dynamic modulus of elasticity after n freeze-thaw cycles, E_{d0} and E_{dn} in (MPa) the dynamic module of elasticity at the beginning and after n freeze-thaw cycles, (MPa).

3. Results and discussions

3.1. Freeze-thaw results using the destructive method

The results obtained using the method from SR 3518:2009 are presented in Table 4, and graphically represented in Figure 3 a) and b). The obtained values represent the compressive strengths for the control samples and for the samples undergoing repeated freeze-thaw cycles. The compressive strength losses were calculated between the control samples kept in water and the samples undergoing repeated freeze-thaw cycles.

Table 4. Resistances and compressive strength losses after 150 and 300 freeze-thaw cycles.

Mixture	fcm,150 water (MPa)	fcm,150 freeze (MPa)	η150 (%)	fcm,300 water (MPa)	fcm,300 freeze (MPa)	η300 (%)
S 360 control	85.83	88.32	-2.91	87.78	83.06	5.38
S 414 control	81.46	75.64	7.14	80.96	71.15	12.10
S 54/20	87.06	85.75	1.51	84.80	77.82	8.32
S 54/40	80.85	76.29	5.63	79.38	73.06	7.96
S 54/60	76.79	67.92	11.55	74.77	60.95	18.47

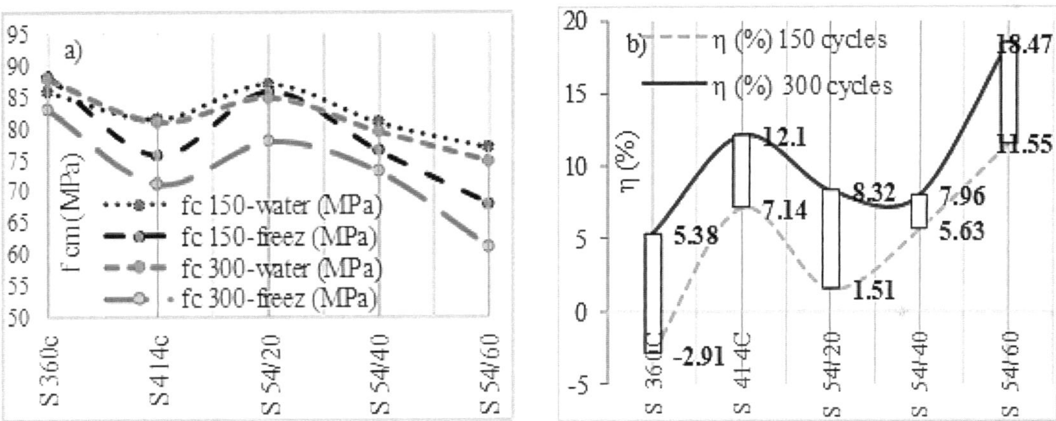

Figure 3. (a) Compressive strengths and (b) compressive strengths losses after 150 and 300 freeze-thaw cycles, (MPa).

In Figure 3 (a) can be noticed that the highest compressive strengths were obtained in the S 360c mixture, and the closest to these were the values of the S 54/20 mixture. By comparison with the resistances of the control mixture S 414c, the obtained values were superior in the S 54/20 mixture, closer in the S 54/40 mixture, but inferior in the S 54/60 mixture. Also, the resistance losses in the S 54/20 and S 54/40 mixtures are smaller than those in the control mixture S 414c. The control mixture S 360c registered the lowest resistance losses, but the S 54/60 mixture registered the highest compressive strengths losses. In all mixtures, the resistance losses are lower than the admitted limit of 25%.

3.2. Results of freeze-thaw resistance by measuring the exfoliated mass of the sample surface

Individual values were calculated for each specimen, and then the cumulated value of exfoliated material for each mixture at 7, 14, 28, 42, 56 freeze-thaw cycles, see Table 5, represented graphically as in Figure 4 a) and b).

Table 5. The cumulated quantity of exfoliated concrete S_{56} after 56 freeze-thaw cycles, in stripes of 150x150x50 mm reported to the measured surfaced of 0.02103 m^2.

Mixtures	$m_{s,7}$ [g]	$m_{s,14}$ [g]	$m_{s,28}$ [g]	$m_{s,42}$ [g]	$m_{s,56}$ [g]	$m_{s,\ cumulated}$ [g]	$S_{56,}$ cumulated [kg/m^2]
	7 cycles	14 cycles	28 cycles	42 cycles	56 cycles	56 cumulated	
S_{360} c	0.200	0.225	0.300	0.400	0.550	1.675	0.080
S_{414} c	0.225	0.250	0.350	0.425	0.550	1.800	0.086
S 15/20	0.225	0.250	0.275	0.350	0.450	1.550	0.074
S 15/40	0.250	0.275	0.325	0.400	0.475	1.725	0.082
S 15/60	0.225	0.250	0.350	0.450	0.575	1.850	0.088

The exfoliation resistance was evaluated qualitatively considering the criteria (SS 13 72 44- the Borås method) presented in Table 3. The scaled mass value S_{56} was lower than 0.1 kg/m^2 in all mixtures, criterium that has placed the concretes with very good exfoliation resistance. The usage of the blast furnace slag had a positive influence on the scaling resistance in mixtures S 54/20 and S 54/40, in which the rate of exfoliated material cumulation was lower than that in the two control mixtures.

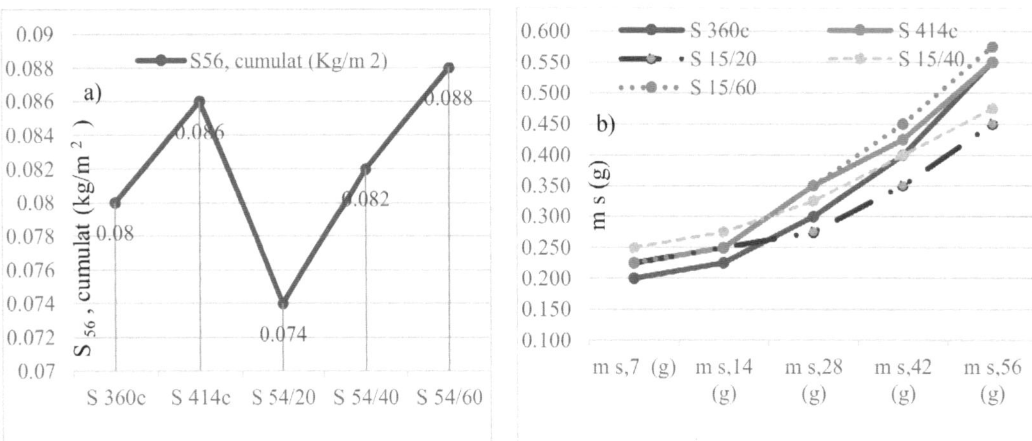

Figure 4. (a) Cumulated exfoliated concrete quantity S_{56}; (b) Exfoliated mass at 7, 14, 28, 42 and 56 cycles.

The results are similar with those obtained in the study carried out by [31], in which the exfoliation resistance after 56 freeze-thaw cycles was very good, for concretes in which fly ash was used up to 15%.

3.3 Results of the freeze-thaw resistance by percentage decrease of the relative dynamic modulus RDM of elasticity

The values registered for the propagation time of ultrasonic impulses (UPTT) and the values calculated for RDM are presented in Table 6 and images in Figure 6.

As observed in Figure 5, the relative dynamic modulus of elasticity registers a reduction of up to 2% in the following mixtures S 360c, S 54/20, S 54/40 and up to 3% in the following mixtures S 414c and S 54/60. The percentage decrease in the mixtures S 54/20 and S 54/40 is lower than that in the control mixture S 414c, but higher for the S 54/60 mixture. Compared with the S 360c mixture the percentage decrease is lower for the S 54/20 mixture, but higher for the following mixtures S 54/40 and S 54/60.

Table 6. Measuring the propagation time of ultrasounds (UPTT) before and after 56 freeze-thaw cycles and the RDM calculus.

	Mixtures				
Characteristics	S 360c	S 414c	S 54/20	S 54/40	S54/60
T_0 (μs)	31.97	32.35	31.98	32.21	32.44
T_{28} (μs)	32.19	32.63	32.15	32.47	32.79
T_{56} (μs)	32.24	32.75	32.23	32.53	32.89
RDM_{28} (%)	98.68	98.30	98.93	98.43	98.20
RDM_{56} (%)	98.37	97.59	98.49	98.09	97.65
(100-RDM_{56}) %	1.63	2.41	1.51	1.91	2.35

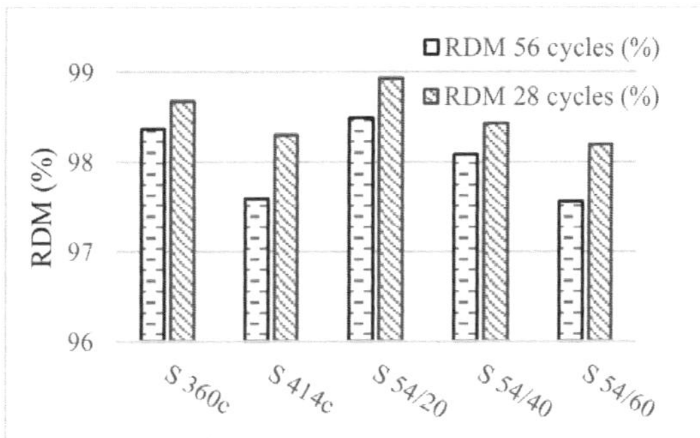

Figure 5. RDM at 28 and 56 freeze-thaw cycles.

(a) (b)

Figure 6. (a) and (b) Images for measuring (UPTT).

The ultrasonic impulse method allows the localisation of structural degrading, such as cracks caused by the freeze-thaw phenomenon through values registered during the measuring of the propagation time of the ultrasonic impulses (UPTT). If the cracks create a right angle with the propagation velocity of the impulse, the wave suffers a diffraction around the crack and leads to the increase of the duration of the wave. If the direction of the crack coincides with the direction of propagation of the impulse the wave passes through the concrete on both sides of the crack and the duration of the wave is not affected [47,48]. The lowest the travel time (UPTT), indicate the highest the relative dynamic modulus (RDM) of elasticity and the compressive strength of the concrete. If the wave travel time is low, it means that the concrete sample that was tested has fewer voids and internal cracks, which leads to the conclusion that the density of the concrete is higher [49-51]. From the analysis of the results obtained in Table 6 it can be noticed that the lowest propagation duration (UPTT) was obtained for the following mixtures S 360c and S 54/20, in which the best results were obtained in all the performed testing, because these

IOP Conf. Series: Materials Science and Engineering **877** (2020) 012025 doi:10.1088/1757-899X/877/1/012025

mixtures have lower porosity. The propagation duration (UPTT) increased, while the freeze-thaw resistance results decreased, in the order enumerated in the following mixtures S 54/40 and S 414c. The longest propagation duration (UPTT) was registered for the S 54/60 mixture, which obtained the weakest freeze-thaw resistance, because it is a mixture with porosity higher than the reference mixture S 414c.

4. Conclusions

In this paper are presented three testing methods for the freeze-thaw resistance, two for the assessment of the internal degrading and one for the material loss at the road concrete surface. Selecting a working method depends on many factors: the research target, the concrete usage field, the severity of the climatic conditions of exposure, etc. The testing represents the measuring of different concrete parameters, comparing results between methods being extremely difficult. But the testing of a mixture using several methods offers supplementary results for each tested parameter that afterwards leads to a more precise interpretation regarding the freeze-thaw behaviour of the concrete. The obtained results by using the methods in this study for testing the road concrete with blast furnace slag in repeated freeze-thaw cycles leads to the following conclusions:

The freeze-thaw resistances in the two control mixtures were higher in the S 360c mixture, compared with S 414c. These results are due to the fact that by increasing the cement dosage from 360 to 414 kg/m^3 it led to the increase of the water quantity necessary and it decreased the quantity of aggregate in the S 414c mixture;

The compressive strength losses determined in compliance with SR-3518:2009 (the destructive method for testing the internal cracks) after 150 freeze-thaw cycles, were the lowest in the S 54/20 mixture, and after 300 freeze-thaw cycles in the S 54/40 mixture. The two mixtures with blast furnace slag have the compressive strength losses lower than the control mixture S 414c, but higher than S 360c.

Applying the method SR CEN/TS 12390-9:2009 of assessing the material loss of the concrete surface (the exfoliated mass) proved that the concrete with blast furnace slag 54/20 obtained better results than the two control mixtures S 360c and S 414c, and the S 54/40 mixture situated at a superior level only compared to the S 414c mixture. Because the road concretes are exposed to the freeze-thaw phenomenon with defrost agents, we consider this method very important and compulsory for the freeze-thaw testing.

The results of the experiments show that the percentage decrease of the relative dynamic modulus RDM of elasticity determined using the method CEN/TR 15177:2006 (non-destructive method for testing the internal cracks) is situated in the 2÷3% range, in all mixtures. It can be noticed that the evolution of the percentage decrease of the RDM is similar to that of the compressive strength losses after 150 and 300 freeze-thaw cycle, but the size order is different.

An important advantage for the non-destructive method is the possibility of observing the progress of the internal damages caused by the freeze-thaw during the testing, using the same set of specimens, while for the destructive method it is a disadvantage because we need a higher number of specimens. Using the destructive method, with each age the measuring is performed on different sets of specimens, which can affect the interpretation of the results. The destructive method (resistance loss) is more severe compared to the non-destructive method (percentage decrease of RDM) as it can be noticed from this study. After 300 freeze-thaw cycles, for the same mixture S 54/60, the compressive strength losses reached the value of 18.47%, and the percentage decrease of the RDM_{300} was only 2.35%, which shows that the destructive method is more precise. But the two methods prove to be complementary because the propagation time of ultrasonic impulses (UPTT) through concrete is the direct liaison with the compressive resistance having the common factor the concrete density.

The results of the study show that by replacing the cement with 13% (GGBS) and up to 40% natural aggregate with artificial aggregate from crushed blast furnace slag (ABS) with a dimensions of 0/4 mm in the following mixtures S 54/20 and S 54/40, the durability factors are above those of the control mixture S 414c, hence the material proportions combined in these mixtures led to equivalent performances with the reference mixture S 414c made with conventional materials.

For the mixture with blast furnace slag S 54/60, the results obtained were below those of the two control mixtures, both in assessing the internal damages, and the assessment of the structure surface damages, caused by the increase of porosity in the structure of this mixture.

Acknowledgments
We would like to thank the Technical University of Cluj-Napoca for the didactic and logistic support offered during this entire experimental program.

5. References

[1] Andrei S and Stefan J 2019 *Report No.06: Requirements and recommendations for frost durable concrete, DaCS Durable advanced Concrete Solutions* (ISBN: 978-82-7482-116-3)
[2] PD CEN/TR 16563:2013 *Technical Report of Principles of the equivalent durability procedure* (European Committee for Standardization (CEN)
[3] NS-EN 206:2013+NA:2014 *Standard for Concrete-Specification, performance, production and conformity, National annex NA informative, Use of NS-EN 206:2013 in Norway 2014 (NO)* (Norvegian Standard)
[4] Corbu O Ioani A M Al Bakri M M Meiță V, Szilaghiy Henriette and Sandu A V 2015 *Key Engineering Materials* **660** 237-243
[5] PN-EN 480-11:1998 *Standard for Admixtures for concrete, mortar and grout - Test methods - Part 11: Determination of air void characteristics in hardened concrete* (Polish Committee for Standardization)
[6] Wawrzeńczyk J and Kozak W 2015 *7th Scientific-Technical Conference Material Problems in Civil Engineering (MATBUD'2015):Procedia Engineering* **108** 102- 107
[7] CSA A23.1-09-A23.2-09 2011 *Standard for Concrete materials and methods of concrete construction/Test methods and standard practices for concrete* (Canadian Standards Association)
[8] GOST 26633-2015 *Standard for Heavy-weight and sand concretes Specifications* (Russian standards)
[9] SR EN 206+A1:2017 *Standard for Concrete-Part 1: Specification, performance, production and conformity* (Bucharest ASRO- Romanian Standardization Association)
[10] NE 014-2002, *The national norm for the execution of the road clads of cement concrete in system of fixed and sliding formworks* (Bucharest:Matrix Rom, 2007 ISBN 978-973-755-185-6)
[11] DIN 1045-2:2014-08 *Standard for Concrete, reinforced and prestressed concrete structures - Part 2: Concrete - Specification, performance, production and conformity - Application rules for DIN EN 206* (German standards)
[12] GOST R 55224-2012 *Standard for Cements for transport construction Specifications* (Russian standards)
[13] Hooton, R.D. 2012 Durability Aspects of Fly ash and Slag in Concrete, *Proceedings of Workshop on Oslo Feb 2012 "Thirty Five Years Experience with Slag Cement Concrete in North America"*
[14] ASTM C666/C666M-2015 2015 *Standard Test Method for Resistance of Concrete to Rapid Freezing and Thawing, West Conshohocken, PA,* Doi: 101520/C0666_C0666M-15 (ASTM American Society for Testing and Materials, International)
[15] Kjell Tor Fosså 2017 *Kværner Concrete Solutions AS. Personal communication*
[16] Process code 2, 2015 *Standard description for bridges and quays. Main process 8. Nr.R672 in Road administration handbook series* (Norwegian Public Road Authorities)
[17] Laboratory research Nr. R210 in 2014 *Road administration handbook series* (Norwegian Public Road Authorities p 431)
[18] Mohammad H B Per J H and Stefan J, 2011 Effect of Internal Hydrophobation, Silica Fume and W/C on Water Sorption of Hardened Cement Pastes, *12th International Conference on Durability of Buiding Materials and Components, Porto -Portugal*

[19] GOST 127305-1984 *Standard for Concretes Methods for determination of water impermeability* (Russian standars)
[20] CEN/TR 15177:2006 *Technical Report of Testing the freeze-thaw resistance of concrete - Internal structural damage* (European Committee for Standardization (CEN)
[21] SS 137244:2005 *Standard Concrete testing - Hardened concrete - Scaling at freezing* (Swedish Standards Institute)
[22] CEN/TS 12390-9:2006 *Technical Specifications of Testing hardened concrete - Part 9: Freeze-thaw resistance – Scaling* (European Committee for Standardization (CEN)
[23] ASTM C672/C672M-2012 *Standard Test Method for Scaling Resistance of Concrete Surfaces Exposed to Deicing Chemicals* (ASTM American Society for Testing and Materials, International)
[24] GB/T 50082—2009, *Standard Chinese for Test Methods of Long-term Performance and Durability of Ordinary Concrete* (Ministry of Housing and Urban-Rural Development of China Beijing: China Architecture & Building Press)
[25] SR 3518:2009, *National standard Tests on concrete Determination of the freeze-tawing resistance by measuring the variations of the resistance strength and/or of the dynamic relative elastics modulus* (Bucharest ASRO- Romanian Standardization Association)
[26] Peter B and Zbigniew G, 2010 Testing the frost resistance of concrete with different cement types – experience from laboratory and practice, *The Silesian University of Technology, Arhitecture Civil Engineering Environment* **2**
[27] BAW-Merkblatt 2004 Concrete frost resistance test *National Hydraulic Institute 0 Karlsruhe, Hamburg, Ilmenau (in German)*
[28] D Kocáb B Kucharczyková, P Daněk, T Vymazal, P Hanuš and R Halamová 2018 *IOP Publishing IOP Conf Series: Materials Science and Engineering* **379** 012022
[29] Daria J-Niedźwiedzka, 2002 Influence of pre-wetted lightweight aggregate on scaling resistance of high performance concrete, *3rd European Conference on Advanced Materials and Technologies At: Bucharest, Romania* **1**
[30] Syamak T Mahmoud N and Behrad S 2018 *Construction and Building Materials* **165** 717–722
[31] Martha K-Kazberuk 2013 *Procedia Engineering* **57** 605–613
[32] Jun P H Hyun B S Sooyoung L and Ki Y A 2013 Enhancing the Durability Properties of Concrete Containing Recycled, Aggregate by the Use of Pozzolanic Materials, *KSCE Journal of Civil Engineering* **17**(1):155-163
[33] Huai-S Shang and Ting-H Y 2013 Freeze-Thaw Durability of Air-Entrained Concrete, *Hindawi Publishing Corporation The ScientificWorld Journal*, 650791
[34] Georgescu D Apostu A Miron G 2010 *Romanian Journal of Materials* **40(2)** 122-131
[35] SR EN 12620+A1 2008 *Standard for Aggregates for concrete* (Bucharest ASRO- Romanian Standardization Association)
[36] SR 667, 2001 *National standard for Natural aggregates and worked stone for roads* (Bucharest ASRO- Romanian Standardization Association)
[37] SR EN 1008, 2003 *National standard for Preparation water for concrete* (Bucharest ASRO- Romanian Standardization Association)
[38] SR EN 15167/1, 2007 *Standard Ground granulated blast furnace slag for use in concrete, mortar and grout Part 1: Definitions, specifications and conformity criteria* (Bucharest ASRO- Romanian Standardization Association)
[39] Nicula L M Corbu O and Iliescu M 2018 The effect of recycles blast furnace slag waste on the characteristic and durability of cement–based mortars, *Conference Proccedings, Albena, Bulgaria* vol 18 pp 379-390
[40] Federal Highway Administration Research and Technology *User Guidelines for Waste and Byproduct Materials in Pavement Construction*, Publication Number: FHWA-RD-97-148
[41] ArcelorMittal *Declaration of performance no 13242-A.M. for Artificial slag furnace aggregates*
[42] ArcelorMittal *Laboratory Test Report No.706/13.07.2016, for crushed aggregate 0/4 mm*

[43] SR EN 197-1, 2011 Standard Cement- Part 1: Composition, specification, and conformity criteria common cements (Bucharest ASRO- Romanian Standardization Association)
[44] Georgescu D P, 2001 *Guidance for the design of the durability of the concrete according to the National Annex of application of SR EN 206-1, Durability classes* (Bucharest ISBN 978-973-0-04914-5)
[45] SR EN 12504-4, 2004 *Standard of Research on concrete Part 4: Determining the speed of ultrasound propagation* (Bucharest ASRO- Romanian Standardization Association)
[46] GOST 10060-2012 , *Standard Freeze-thaw resistance* (Russian standars)
[47] Neville A M 1979 *Properties of concrete* (Bucharest Technical Publishing House)
[48] R Jones 1960 *RILEM Bul* **9** 110-14
[49] Burduhos Nergis D D, Abdullah M M A B, Sandu A V and Vizureanu P 2020 *Materials* **13(2)** 343
[50] Burduhos Nergis D D, Vizureanu P and Corbu O 2019 *Revista de chimie* **70(4)** 1262-1267
[51] Vizureanu P, Cimpoesu N, Radu V and Agop M 2015 *Experimental heat transfer* **28(1)** 37-57

Improvement of Microplastics Separation from Synthetic Samples – A Key Step for Their Analysis

I-E Ciobotaru[1], E Marcu[1], C Maria[1*], A A Ivanov[1], I Savin[1], A Moncea[1], C Tociu[1] and Gy Deák[1]
[1]National Institute for Research and Development in Environmental Protection, 294 Splaiul Independenței Blv, District 6, Zip code 060031, Bucharest, Romania

*E-mail: cristina.maria@incdpm.ro

Abstract. Identification of microplastics from the environment is a challenge mostly due to the complex nature and behavior of samples and lack of an efficient separation method. This paper approaches the issue of microplastics separation as determining step of their analysis. The separation of two types of commercial plastics from synthetic samples was investigated, namely polyethylene terephthalate (PET) and high density polyethylene (HDPE). Subsequent filtration through filters with different porosities was tested and improved by adding a surfactant in the extraction medium. For the qualitative analysis of microplastic, optical microscopy and laser-granulometry were employed. The results proved the separation of microplastics from the samples and showed a distribution between 10-100 μm for PET and 10-200 μm for HDPE. These results provide basis for the efficient microplastics separation and further quantitative analysis from more complex samples, such as environmental samples.

1. Introduction

The presence of pollutants in water is an issue of interest in the field of water management as new compounds included on the list of potentially toxic and hazardous substances are found to be present in water. A special attention was recently granted to emerging pollutants and scarce information is available regarding their effects on the environment or human health. The presence of micro- and nanoplastics in the aquatic environment raised the interest of environmental scientists due to recent reports regarding their presence and potential effects on aquatic organisms and humans [1-11].

Since the synthesis of the first plastic in the early 1900s, the production of these materials has increased significantly and millions of tons of polymers are produced every year. Plastics are versatile materials, with low weight, high durability and flexibility. They are cheaper than the traditional materials such as wood, metal, glass or paper which facilitated their wide use. The applications of such materials vary greatly from industrial (e.g. food packaging, electronics) to medicinal (drug delivery) and household purposes (e.g. grocery bags and storage flasks). However, their disposal after only one use causes the generation of significant volumes of plastic waste and a part of this waste ends up in the environment where under the influence of UV and environmental conditions degrade to microplastics and nanoplastics. These particles were shown to be present and have effects on the aquatic environment and food chain. The small size of microplastics enables their penetration in organisms and the main effect reported is physical damage. However, microplastics also have the ability to adsorb and release other toxic pollutants into the organisms which in conjunction with their size pose a greater risk to organisms after penetration [3-8, 10, 12-15].

One of the main issues related to the monitoring of microplastics in the environment is the difficulty of separating them from complex samples. The small size and the tendency to float increase the time required for their separation and limit the methods that may be used for their separation. Among these methods, density separation has been used intensely. Various solutions with density greater than the density of micropplastics have been employed, such as sodium chloride, sodium iodide, zinc chloride or sodium polytungstate. This method does not ensure the complete separation of microplastics and some of the employed solutions are harmful for the environment [9, 16-20].

Sequential filtration through sieves or filters with different porosities is another method currently used for the separation of microplastics after applying the density separation method. The analysis of separated particles is firstly accomplished through a visual method followed by instrumental methods that are currently employed for the analysis of nanoparticles. The main issue of filtration is the clogging of the filters and the long time required for the passing of the aqueous medium through the filter. The addition of a surfactant, such as sodium lauryl sulphate, in the filtration medium could overcome this disadvantage and allow a better and fast separation of the microplastics [9, 16-25].

This paper presents the results obtained on the improvement of microplastics separation. The experimental procedure employed the separation of investigated microplastics by subsequent filtration of a sodium lauryl sulphate-enhanced aqueous medium followed by optical microscopy and laser-granulometry analysis. The tests were conducted on synthetic samples of widely used commercial materials, namely polyethylene terephthalate (PET) and high density polyethylene (HDPE).

2. Materials and methods

The plastic materials selected for testing the separation procedure were commercial polyethylene terephthalate (PET) and high density polyethylene (HDPE). Two types of PET were employed, colorless and brown.

The experimental procedure consisted of two main steps, the obtaining of the microplastic particles and their analysis. The obtaining of microplastics particles was achieved by shredding with a metallic scissor (Figure 1) followed by grinding. A specific quantity of shredded material (5-15 g) was subjected to grinding in a planetary ball mill (PM 100, Retsch) together with 50 g of sodium chloride (NaCl, density 2.165 g/cm^3). The grinding was performed for 10 min at 550 rpm and direction reversal every 3 min followed by a 10 s pause. Afterwards, 10 mL of distilled water were added and the grinding continued for another 50 min (Figure 2). At the end, the sample was taken using distilled water and filtered coarsely through a metallic sieve with 1 mm pores followed by filtration through polycarbonate filters with 0.4 μm porosity (Figure 3). The addition of a solution of sodium lauryl sulphate 10% and 5% ($NaC_{12}H_{25}SO_4$, density ~1 g/cm^3) was tested. This was added to the filtration suspension to improve the recovery of microplastics and enable the dispersion of microparticles agglomerates. Based on the results of visual identification, the use of 5% sodium lauryl sulphate proved to be more efficient and thus was used throughout the experiments.

The clogging of the filter was prevented by adding ethanol (C_2H_5OH, density 0.7894 g/cm^3) and the removal of potential metallic particles was achieved using hydrochloric acid solution 1:1 (HCl, density 1.08 g/cm^3). The final wash of the filter was with distilled water. The samples were introduced in a Petri dish and dried at 20±4°C.

The analysis of microplastics was achieved by visual identification and laser granulometry. For the visual analysis, a specific quantity (0.04 mg) of the dried microplastics was dispersed in 500 mL distilled water and 1 mL solution of sodium lauryl sulphate 5% was added. The solution was further filtered through a vacuum filtration system provided with polycarbonate filter with 0.4 μm porosity. The visual analysis was performed using an optical microscope Carl Zeiss Jena) in trinocular version, with planachromatic objectives and magnification of up to 1000x.

The size and distribution of separated microplastics were analyzed by laser diffraction. This provides indirect size measurements of spherically equivalent particles. The size and distribution of separated microplastics were determined using a laser particle sizer (ANALYSETTE 22 NanoTec, Fritsch) using wet dispersion. The material was dispersed into a liquid that circulated across a quartz measurement cell illuminated by a laser beam. The particles with a specific size will diffract the light at a specific angle which increases logarithmically with decreasing size. This particle sizer performs measurements in the range 0.1-2100 μm. The measured data were processed and evaluated using the software provided by the manufacturer.

3. Results and discussions

Images of the shredded plastics and microplastics obtained after the milling procedure are shown in Figures 1-3.

(a) (b) (c)

Figure 1. Images of shredded plastic materials: (a) colorless PET, (b) brown PET, (c) HDPE.

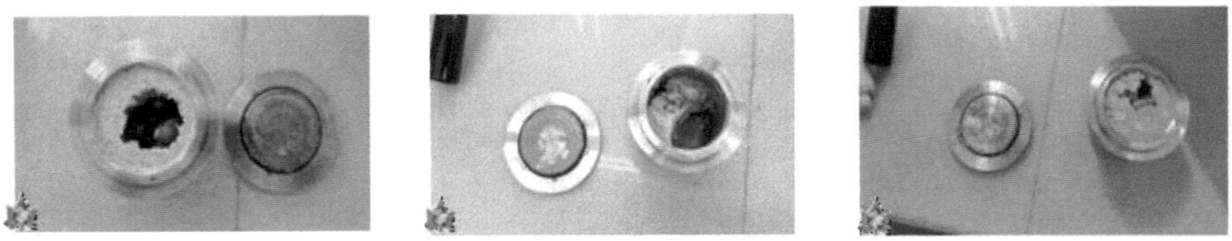

(a) (b) (c)

Figure 2. Plastic materials after grinding: a – colorless PET, b – brown PET, c – HDPE.

(a) (b) (c)

Figure 3. The microplastic particles resulting from the experimental procedure: (a) colorless PET, (b) brown PET, (c) HDPE.

The results of the visual analysis are presented in Figure 4 and show a complex and slightly irregular shape of the particles (Figure 4 a and d). The results of filtration tests conducted using solutions of sodium lauryl sulphate at two concentrations, namely 5% and 10%, are presented. For the 10% surfactant solution, a pilling tendency was observed, as depicted in Figure 4 a for colorless PET. When a 5% surfactant solution was employed, the pilling tendency was reduced and the obtained microplastics were better separated, as shown in Figure 4 b-d for all plastics.

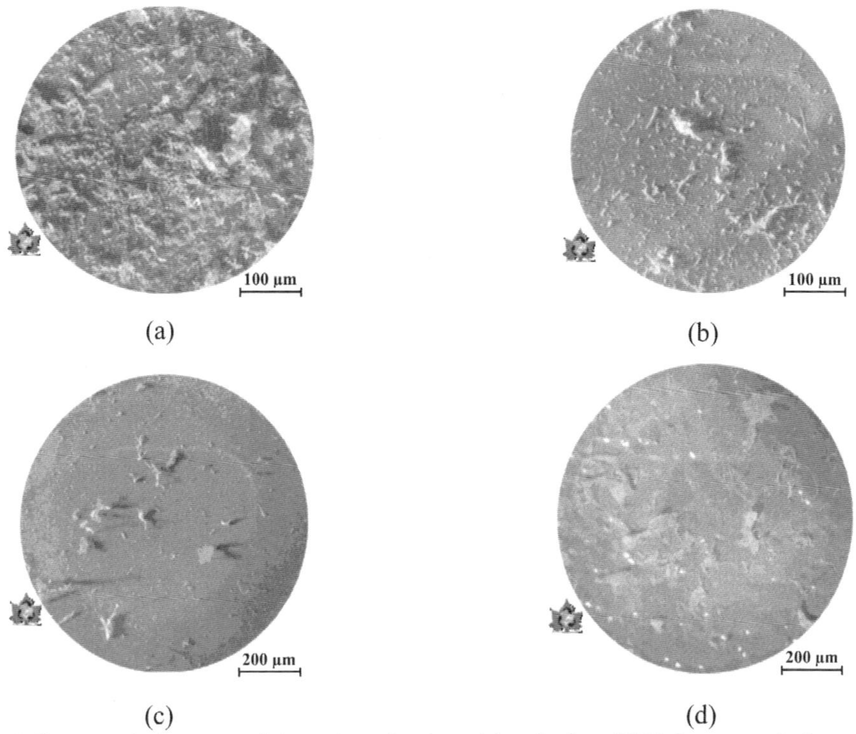

Figure 4. Microscopic images of the microplastics: (a) colorless PET from a solution containing 10% sodium lauryl sulphate, (b) colorless PET from a solution containing 5% sodium lauryl sulphate, (c) brown PET from a solution containing 5% sodium lauryl sulphate, (d) HDPE from a solution containing 5% sodium lauryl sulphate.

The dimensional characteristics of the obtained microplastics determined by laser-granulometry are shown in Figures 5-7 for PET and HDPE. The cumulative curve of distribution Q3(x) (left) indicating the standardized total quantity of all particles with equivalent diameters smaller than 100 µm for colorless PET and respectively 200 µm for cultured PET and HDPE together with the distribution density dQ3(x) (right) defined as the first derivative of Q3(x) are presented.

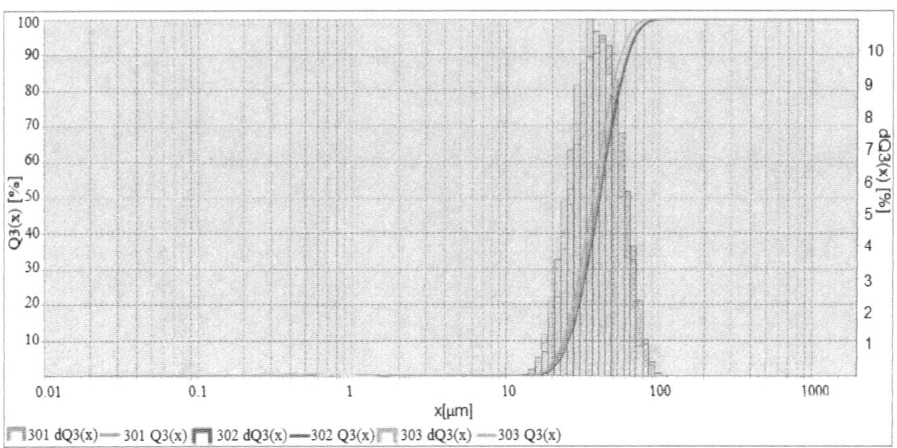

Figure 5. Granulometric distribution of colorless PET.

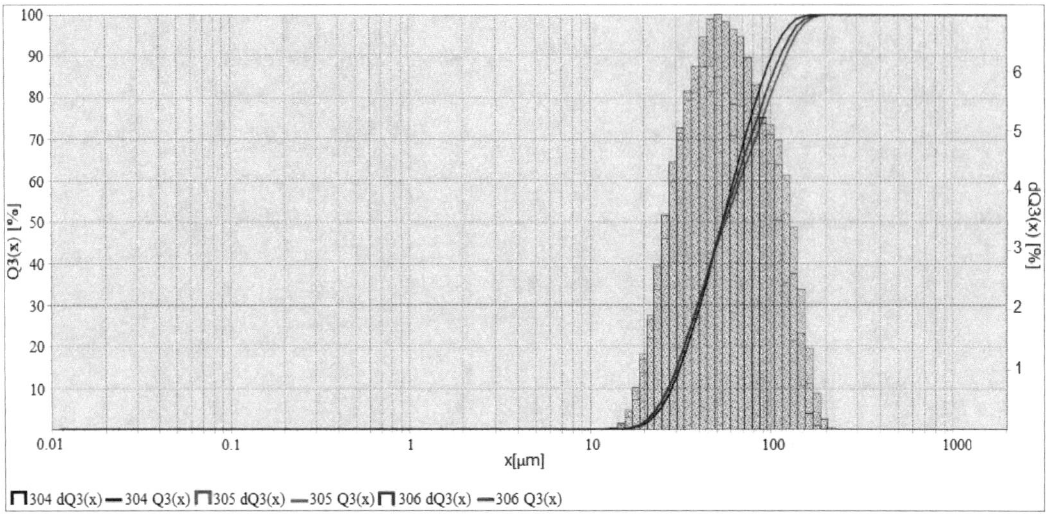

Figure 6. Granulometric distribution of brown PET.

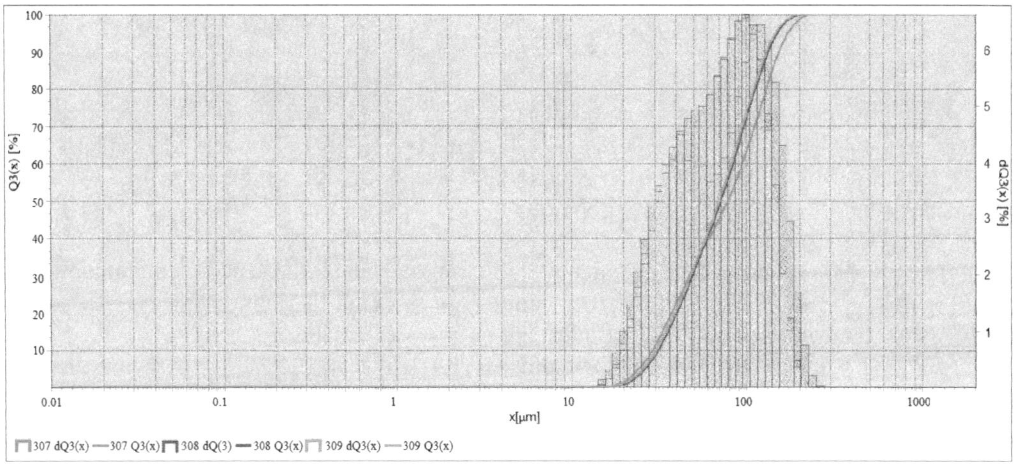

Figure 7. Granulometric distribution of HDPE.

The results showed a monomodal distribution for both PET and HDPE, with sizes between 10-100 µm for colorless PET and between 10-200 µm for brown PET and HDPE.

4. Conclusions

This paper presented an improved separation method for microplastics from synthetic samples consisting in subsequent filtration through filters with different porosities. The method employs the addition of a surfactant in the extraction medium, to facilitate the filtration and prevent the fouling of the filters.

The improved separation method was applied on two commercial materials, PET and HDPE. To evaluate the separation of the method, the determination of microplastics was performed by visual analysis and confirmed by laser-granulometry. The results proved the presence of microplastics with size between 10 and 100 µm in the case of PET and between 10 and 200 µm in the case of HDPE.

Acknowledgments
The results presented in this paper were obtained within the Nucleu Programme RESCMANS, project code PN 19 43 04 01.

5. References

[1] Ciobotaru I-E, Maria C, Marcu E, Ivanov A-A and Tociu C 2019 *Proc. RICCCE* **21** S5-381
[2] Resetar-Deac A M, Deák Gy, Marinescu P, Daescu V, Holban E, Csergo R, Tanase G S and Gaman S 2015 *J. Environ. Prot. Ecol.* **16** 183
[3] Reddy M S, Shaik B, Adimurthy S and Ramachandraiah G 2006 *Estuar. Coast. Shelf S.* **68** 656
[4] Cole M, Lindeque P, Halsband C and Galloway T S 2011 *Mar. Pollut. Bull.* **62** 2588
[5] Koelmans A A, Besseling E and Shim W J 2015 *Marine Anthropogenic Litter* eds M Bergmann, L Gutow and M Klages (Switzerland: Springer Open) pp 325-340
[6] Ravel M, Châtel and Mouneyrac C 2018 *Curr Opin Environ Sci Health* **1** 17
[7] Blair RM, Waldron S, Phoenix V and Gauchotte-Lindsay C 2017 *Springer Sci. Rev.* **5** 19
[8] Mattson K, Hansson L-A and Cedervall T 2015 *Environ. Sci.: Processes impacts* **17** 1712
[9] Van Cauwenberghe, Devriese L, Galgani F, Robbens J and Janssen CR 2015 *Mar. Environ. Res.* **111** 5
[10] Petersen A 2016 *EFSA Journal* **14(6)** 1-30
[11] Toussaint B, Raffael B, Nagers-Loustau A, Gilliland D, Kestens V, Petrillo M, Rio-Echevarria IM and Van den Eede G 2019 *Food. Addit. Contam.* Part A **36** 639
[12] Pelley J 2018 *Plastic Contamination of the Environment: Sources, Fate, Effects and Solutions*, (Washington: American Chemical Society)
[13] European Commission 2018 *Communication from the Commission to the European Parliament, the Council, the European Economic and Social Committee and the Committee of the Regions* (Brussels)
[14] Velzeboer I, Kwadijk C J A F and Koelmans A A 2014 *Environ. Sci. Technol.* **48** 4869
[15] Song J, Cui S, Li P and Li M 2018 *IOP Conf. Ser.: Earth Environ. Sci.* **186** 012027
[16] Wang W and Wang J 2018 *Trend. Anal. Chem.* 2018 **108** 195
[17] Lusher A, Brate I L N, Hurley R, Iversen K and Olsen M 2017 *Testing of methodology for measuring microplastics in blue mussels (Mytilus spp) and sediments, and recommendations for future monitoring of microplastics* (Oslo: NIVA) NIVA report 7209-2017
[18] Zobkov M B and Esiukova E E 2018 Oceanology **58** 137
[19] Löder M G J and Gerdts G 2015 *Marine Anthropogenic Litter* eds M Bergmann, L Gutow and M Klages (Switzerland: Springer Open) pp 201-227
[20] Nguyen B, Claveau-Mallet D, Hernandez L M, Xu E G, Farner JM and Tufenkji N 2019 *Acc. Chem. Res.* **52** 858
[21] Deák Gy, Dumitru F D, Moncea M A, Panait A M, Baraitaru A G, Olteanu M V, Boboc M G and Stanciu S 2019 *Int. J. Conserv. Sci.* **10** 343
[22] Yuvedha S, Yoganandhan and Nampoothiri NVN 2019 *IOP Conf. Ser.: Mater. Sci. Eng.* **561** 012026
[23] Tarnita D, Tarnita D.N, Popa D, Grecu D and Niculescu D 2010 *Romanian Journal of Morphology and embryology* **51(1)** 145-150
[24] Duceac L D, Marcu C, Ichim D L, Ciomaga I M, Tarca E, Iordache A C, Ciuhodaru M I, Florescu L, Tutunaru D, Luca A C and Stafie L 2019 *Rev. Chim.* **70(7)** 2622-2626
[25] Chiriac V, Stratulat D N, Calin G, Nichitus S, Burlui V, Stadoleanu C, Popa M and Popa I M 2016 *IOP Conference Series-Materials Science and Engineering* 133 012055

IOP Conf. Series: Materials Science and Engineering **877** (2020) 012027 doi:10.1088/1757-899X/877/1/012027

Magnetic Properties of Composites Based on Amorphous Iron Alloys Produced with the Use of a Non-Magnetic Binder and Covered with High Temperature Varnish

B Jez[1*], K Jez[1] and M Nabialek[1]
[1] Department of Physics, Faculty of Production Engineering and Materials Technology, Czestochowa University of Technology, 19 Armii Krajowej Str., 42-200 Czestochowa, Poland

*E-mail: bartek199.91@o2.pl

Abstract. The continuous increase in the demand for electricity causes the need to search for new energy sources and to look for savings during its transmission. One of the areas of interest are materials dedicated to low loss transformer cores. Due to good magnetic properties, amorphous and nanocrystalline iron-based alloys are used in this area. The limited dimensions of these materials make their application difficult. Magnetic composites based on ferromagnetic amorphous alloys are a new direction of research. Usually, composites are made using a non-magnetic binder. However, the separation of magnetic particles worsens the magnetic properties. The solution to this problem may be production of composites coated with high temperature varnish. As part of the work, the results of research on the structure and magnetic properties of composites based on amorphous iron alloys are presented. Composites were made with 0.5% resin by mass and without resin covered with high temperature varnish. Material produced without the use of resin had a lower coercive field value.

1. Introduction

The rapid economic development that has been taking place for several decades has brought a very large increase in the demand for electricity. Unfortunately, despite the development of technologies in the field of obtaining and processing energy resources, the production of electricity is becoming more and more expensive. The development of renewable energy sources alone does not cover the growing demand for energy. The development of energy technologies must be supported by progress in energy transmission and processing. One of the aspects many scientists are working on is the losses caused by transformers. Modern magnetic materials are used for their construction. Commonly used materials for transformer cores are Fe-Si sheets [1-2] and amorphous alloys based on Fe or Co [3-7]. Another object of research are materials intended for wedges in nurseries of electric motors or magnetic screens [8]. In recent years, rapid-quenched alloys with an amorphous structure and nanocrystalline alloys created by appropriate thermal treatment have enjoyed special interest [9-12]. Depending on the alloy composition used and the type of structure obtained, these materials are characterized by soft magnetic properties such as low coercive field value and high saturation magnetization [13-16]. Materials of this type are produced on an industrial scale by the melt spinning method (casting a liquid alloy on a copper rotating drum in the form of ribbons) [17]. Another way to make amorphous material is to use a method of injected a liquid alloy into a copper mold. The method is characterized by a much lower cooling rate but allows the production of alloys with amorphous structure with much larger dimensions [18].

An important disadvantage of rapid-quenched alloys is their limited geometry. One of the possible solutions to this problem are magnetic composites. Composite is a material consisting of at least two physically and chemically separate phases, which give the material new properties. In the case of magnetic composites, these phases are ferromagnetic powder obtained during crushing of an alloy characterized by the desired magnetic properties and a type of bonding binder giving forming properties, for example epoxy resins [19]. The addition of resin negatively affects the magnetic properties but significantly extends the application possibilities of the material. The challenge is to find optimal solutions during the production of composites so as to obtain the appropriate mechanical properties enabling the formation of desired shapes while maintaining good magnetic properties. Studies show that the ratio of resin to powder and the type of powder (particle size and shape) have an impact on the properties of magnetic composites [19].

However, the fact of magnetic particle separation limits the occurrence of exchange interactions, which significantly affects the deterioration of magnetic properties, increase the value of the coercive field and decrease the saturation magnetization [19-21]. A way to reduce this negative effect is to create a resin-free composite, ensuring the consistency of the material with a high-temperature varnish layer. In this way, an electrical insulator material can be obtained while maintaining good magnetic properties. The differences between the resin-containing composite and the high-temperature varnish are illustrated in the diagram in Figure 1.

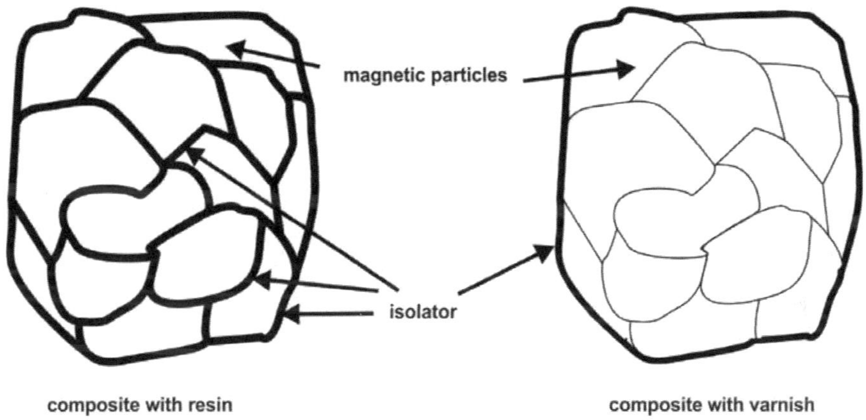

Figure 1. Scheme of composites made with resin and high temperature varnish.

As part of this work, magnetic composites based on $Fe_{70}Nb_5Y_5B_{20}$ amorphous alloy were produced. Composites were made with and without resin - these materials were covered with high-temperature varnish. The purpose of the work was to compare the magnetic properties of composites made according to two different methods.

2. Methods and materials

The starting material for making the magnetic composite is an alloy with the chemical composition $Fe_{70}Nb_5Y_5B_{20}$. To produce the alloy, components with a purity of 99.99% were used. The alloy components were measured using a laboratory balance with an accuracy of 0.001 gram. The ingredients were melted in an arc furnace to combine and mix them. The process takes place in an argon protective atmosphere after pumping out the furnace working chamber. In order to obtain the most even distribution of alloy components in the ingot volume, the melting is repeated several times. The ingot produced in this way was subjected to mechanical cleaning and using an ultrasonic cleaner. On the basis of a polycrystalline ingot, a rapid-quenched alloy was made by injection molding. The process was carried out in an argon atmosphere. The charge was melted by means of eddy currents.

The liquid alloy was injected into a water-cooled copper mold. The cooling rate achieved by this method is up to 10^3 K/s. The alloy obtained in the form of 0.5 mm thick plates was ground. The ferromagnetic powder was divided into fractions (20-50 µm, 50-100 µm, 100-200 µm) using a set of laboratory sieves with shaker. Based on the selected powder fraction, two types of magnetic composites were produced: with 0.5% resin content (by mass) and without the addition of resin with a high-temperature varnish coating. Composites were made in the form of toroids with an outer diameter of 10mm, an inner diameter of 5mm and a thickness of 2mm. The sample production scheme is shown in Figure 2.

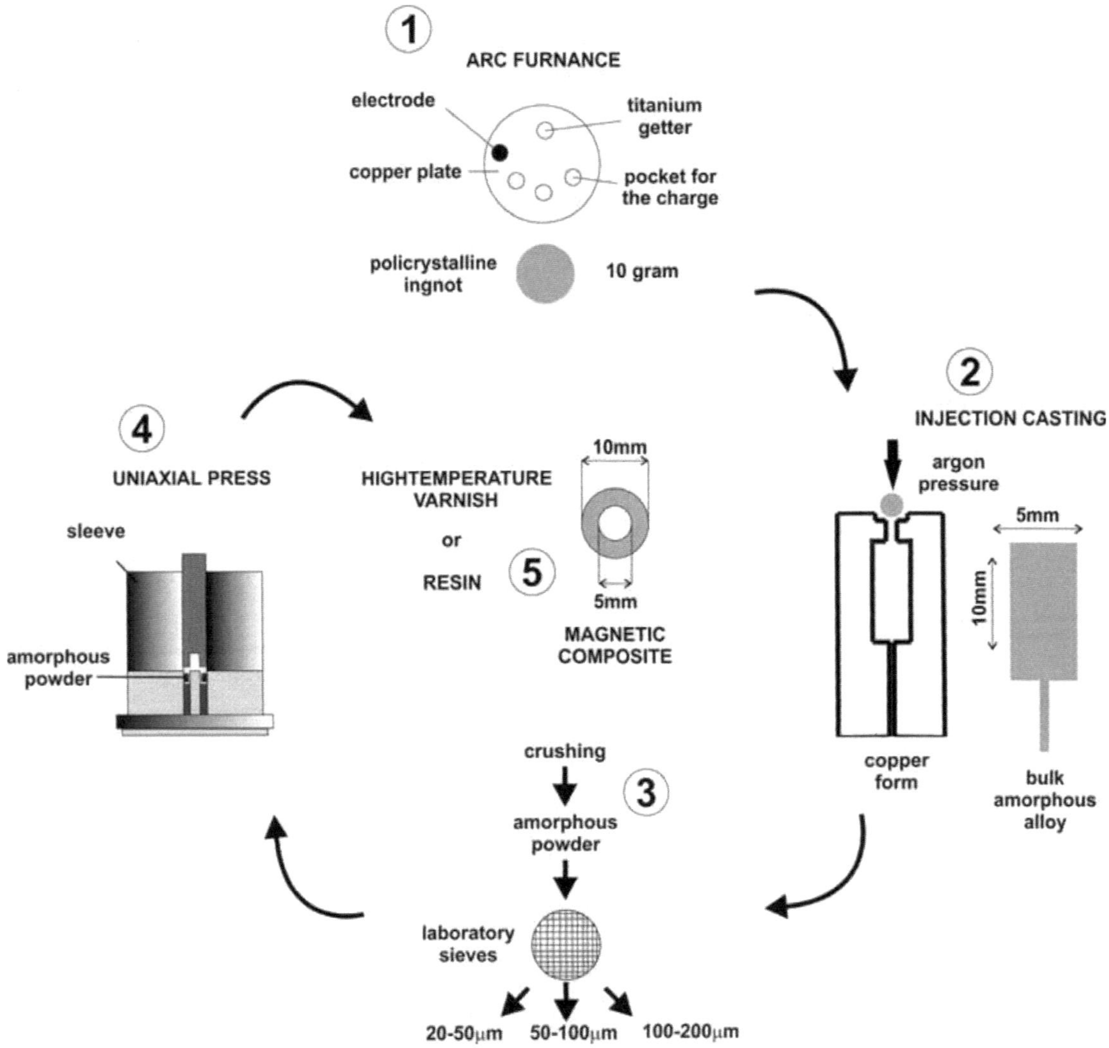

Figure 2. Scheme for producing magnetic composites.

Composites were made using a hydraulic press. The powder was pressed at 200 bar for 30 minutes at room temperature. The molding without the addition of resin during removal from the mold was covered with a layer of high temperature varnish. The resulting composites were allowed to dry at room temperature. Figure 3 contains photos of produced materials.

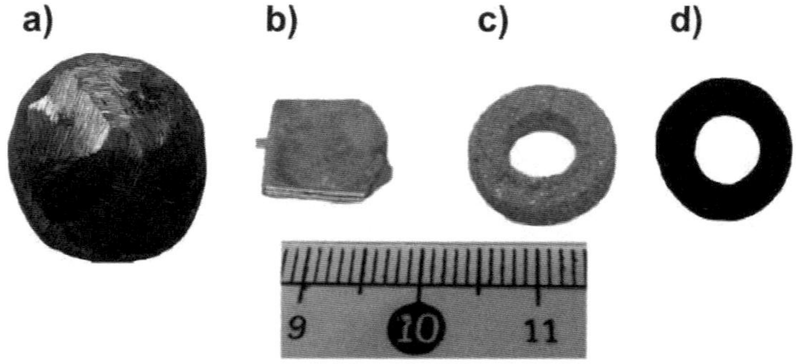

Figure 3. Photographs of produced materials: (a) polycrystalline ingot, (b) amorphous alloy in solid state, (c) composite with 0.5% resin content, (d) composite coated with high temperature varnish.

The structure of the materials produced was examined using an BRUCKER model Advanced 8 X-ray diffractometer. X-ray irradiation was performed in an angle of 2Θ with an exposure time of 5 s and a measuring step of 0.02°. Magnetic properties were tested using a LakeShore vibrating magnetometer to obtain static hysteresis loops. The measurement was carried out in the range of external magnetic field strength up to 2T.

Using the relationship (1), the spin wave stiffness parameter parameter D_{spf} was calculated (based on the primary magnetization curve).

$$b = 3{,}54 g \mu_0 \mu_B \left(\frac{1}{4\pi D_{spf}} \right)^{3/2} kT (g\mu_B)^{1/2} \tag{1}$$

Where:

b – slope of the linear fit corresponding to the thermally induced suppression of spin waves by a magnetic field of high intensity, μ_0 – magnetic permeability of vacuum, k – Boltzman's constant, μ_B – Bohr magneton, g – gyromagnetic factor, T – temperature.

3. Research results

Figure 4 contains X-ray diffraction patterns for the materials tested.

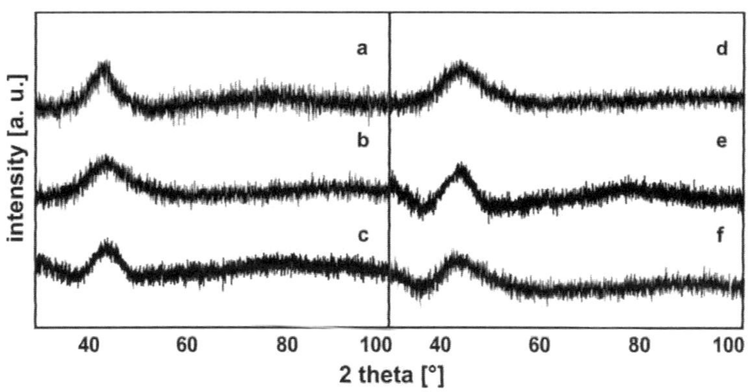

Figure 4. X-ray diffraction patterns for: (a) $Fe_{70}Nb_5Y_5B_{20}$ alloy after solidification, (b) 20-50 μm powder fraction, (c) 50-100 μm powder fraction, (d) 100-200 μm powder fraction, e) composite with resin, (f) composite coated with high temperature varnish.

Testing of the alloy after solidification was carried out for the volume of the material, i.e. for the alloy in the powder state. Diffractograms for composites were measured only for the surface of the materials. The recorded diffractograms are typical as for materials with an amorphous structure. X-ray diffraction images show only one maximum in the 40-50° two theta angle range. Figure 5 contains static magnetic hysteresis loops for the materials tested.

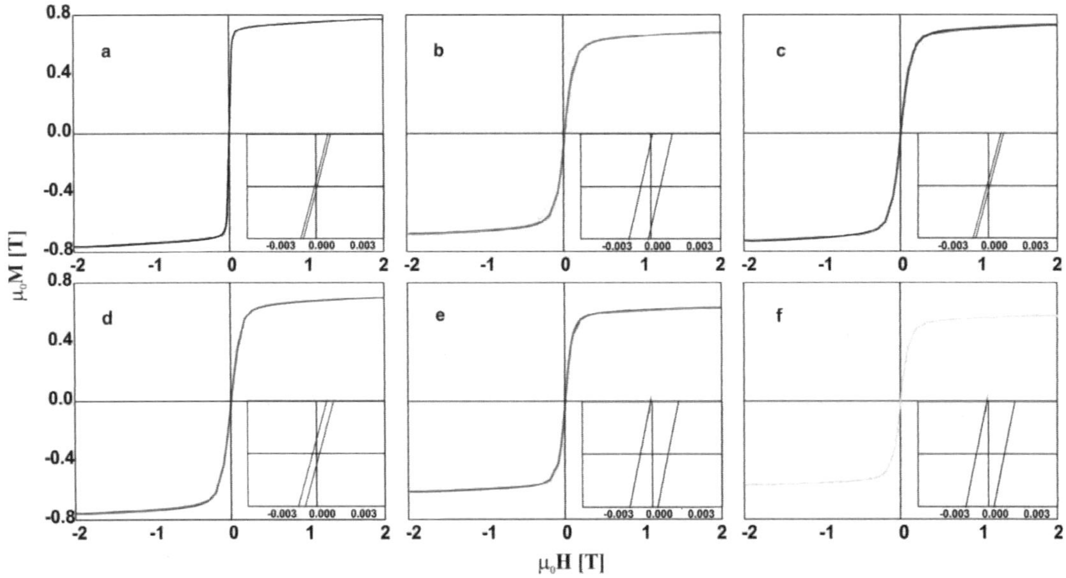

Figure 5. Static magnetic hysteresis loops for: (a) $Fe_{70}Nb_5Y_5B_{20}$ alloy in solidified state; powder fractions: (b) 20-50 μm, (c) 50-100 μm, (d) 100-200 μm; (e) composite with 0.5% resin for 50-100 μm fraction, f) composite for 50-100 μm fraction covered with high temperature varnish.

Defining the structure significantly affected the magnetic properties of the tested materials. The static magnetic hysteresis loop measured for a solid alloy has an almost rectangular pattern. This curve shape is typical for materials with so-called soft magnetic properties. Measurements carried out for various fractions of amorphous powder indicate a significant impediment to the magnetization process. The tested samples achieve saturation value at a much higher intensity of the external magnetic field. Amorphous powder with grain size 50-100 μm has the best properties among the three fractions isolated. On the basis of this fraction, composites with resin or high-temperature varnish were made. The registered loops are characterized by a similar course to the loops measured for the separated fractions. After increasing the origin of the M-H coordinate system, the coercive field value was determined. In addition, the value of saturation magnetization and the value of the external magnetic field at which the magnetization sample was associated with the damping of thermally excited spin waves was determined.

The results of magnetic properties are shown in Table 1.

Table 1. Magnetic properties of produced materials.

	H_C [A/m]	M_S [T]	$H_{H\text{-}P}$ [T]	D_{spf} [meV/nm^2]
ASQ	10	0.75	0.35	42.9
20-50 μm	565	0.68	0.7	35.4
50-100 μm	31	0.73	0.7	35.5
100-200 μm	116	0.73	0.74	35.2
composite with resin	780	0.62	0.7	42.1
composite with varnish	690	0.61	0.67	43.3

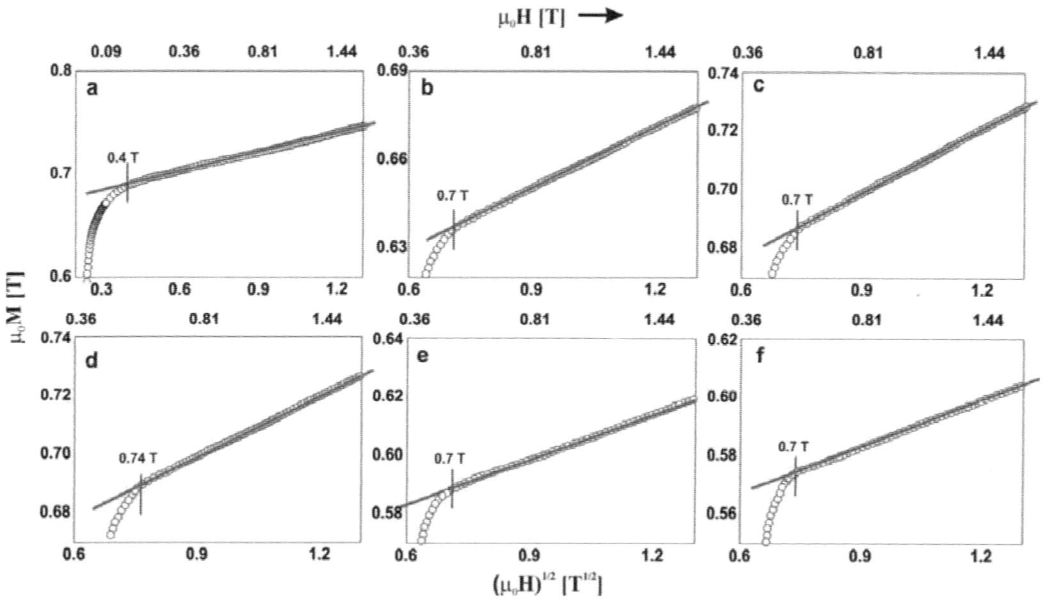

Figure 6. Magnetisation as a function of $(\mu_0H)^{1/2}$ for: a) $Fe_{70}Nb_5Y_5B_{20}$ alloy in solidified state; (a) powder fractions: (b) 20-50 μm, (c) 50-100 μm, (d) 100-200 μm; (e) composite with 0.5% resin for 50-100 μm fraction, (f) composite for 50-100 μm fraction covered with high temperature varnish.

As expected, the 20-50 μm powder has the worst magnetic properties due to the higher degree of magnetic particle separation. This fact weakens the exchange interactions. In this case, the magnetic properties are created to a greater extent by dipole interactions than exchange interactions. The composites produced have a lower saturation magnetization value, which is associated with a much lower material density. The value of the coercive field for composites increased more than 20 times compared to the value measured for the powder itself. Composite coated with high-temperature varnish is more easily re-magnetized compared to a molded with resin. This means that a smaller degree of separation of magnetic particles improves the magnetic interactions that occur in the volume of this material. In higher magnetic fields, the process of magnetizing powdered samples is similar. The tested materials achieve magnetic saturation at an external magnetic field strength of about 0.7T.

4. Conclusions

As part of this work, the impact of manufacturing technology on the soft magnetic properties of composites based on bulk amorphous alloys was investigated. The alloy was tested after solidification and after division into fractions: 20-50 μm, 50-100 μm and 100-200 μm. The 50-100 μm fraction shows the best soft magnetic properties slightly different from the values measured for the solid alloy. The composites produced had a similar saturation magnetization value of 0.61-0.62T and a much higher coercive field value compared to the solid alloy and 50-100 μm powder fraction. In a composite coated with high-temperature varnish, the arrangement and distances between magnetic particles are more favorable than for a composite with resin, which results in a lower coercive field value of almost 100 A/m. In addition, the composite coated with high-temperature varnish is characterized by a significantly higher applicability due to the lack of brittleness (which ensures the mechanical properties of the varnish) and better insulation properties (due to the "closure" of magnetic particles in the varnish matrix).

5. References

[1] Ibarrondo I and Degauque J 1999 *Vacuum* **53** 75-78

[2] Kubota T Makino A and Inoue A 2011 *Journal of Alloys and Compounds* **509S** S416–S419
[3] Gondro J 2019 Revista de Chimie **70(7)** 2699 2702
[4] Gondro J, Błoch K, Nabiałek M and Walters S 2017 *Archives of Metallurgy and Materials* **62(2)** 707-710
[5] Bloch K, Titu M A and Sandu A V 2017 *Revista de Chimie* **68(9)** 2162-2165
[6] Gondro J, Zbroszczyk J, Ciurzyńska W, Olszewski J, Nabiałek M, Sobczyk K, Świerczek J and Łukiewska A 2010 *Archives of Metallurgy and Materials* **55** 85-90
[7] Bloch K, Nabialek M and Gruszka K 2017 *Revista de Chimie* **68(4)** 698-700
[8] Azarewicz S and Węgliński B 2007 *Problem Notebooks - Electric Machines* **77** 87-90
[9] Inoue A 2000 *Acta Materialia* **48** 279-306
[10] Li X, Wang Z and Duan H 2019 *Journal of Non-Crystalline Solids* **517** 114-118
[11] Nabialek M, Jez B and Jez K 2018 *Revista de Chimie* **69(9)** 2545-2549
[12] Mchenry M E, Willard M A and Laughlin D E 1999 *Progress in Materials Science* **44** 291
[13] Błoch K 2018 *Revista de Chimie* **69(4)** 982-985
[14] Dou L, Liu H, Hou L, Xue L, Yang W, Zhao Y, Chang C and Shen B 2014 *Journal of Magnetism and Magnetic Materials* **358-359** 23–26
[15] Wang F, Inoue A, Han Y, Kong F L, Zhu S L, Shalaan E, Al-Marzouki F and Obaid A 2017 *Journal of Alloys and Compounds* **711** 132-142
[16] Li W, Yang Y Z and Xu J 2017 *Journal of Non-Crystalline Solids* **461** 93–97
[17] Duwez P, Willens R H and Crewdson R C 1956 *Journal of Applied Physics* **36(7)** 2267-2269
[18] Narayan R L, Singh P S, Hofmann D C, Hutchinson N, Flores K M and Ramamurty U 2012 *Acta Materialia* **60** 5089–5100
[19] Nabialek M 2015 *Przetwórstwo Tworzyw* **2** 139-142
[20] Grimberg R, Udpa L, Savin A, Steigmann R, Vizureanu P, Bruma A and Udpa SS 2008 *Research in nondestructive evaluation* **19(4)** 202-218
[21] Scripcariu L, Mătăsaru D, Diaconescu V 2017 *UPB Scientific Bulletin Series C-Engineering and Computer Science* **73**(3) 83-92

IOP Conf. Series: Materials Science and Engineering **877** (2020) 012028 doi:10.1088/1757-899X/877/1/012028

Development of Cement-Based Materials Enriched with Polymeric Coated Reactive Grains as Long-Term Promoter of Matrix Continuous Hydration

C Voinitchi[1], C Baeră[2]*, M Zapciu[3,4], C Matei[5] and H Szilagyi[6]
[1] Technical University of Civil Engineering, Bucharest, Romania
[2] NIRD URBAN-INCERC Timişoara Branch, Romania
[3] SC INCERTRANS SA Bucharest, Romania
[4] Politehnica University of Bucharest, Romania
[5] NIRD URBAN-INCERC Bucharest Branch, Romania
[6] NIRD URBAN-INCERC Cluj-Napoca Branch, Romania

*E-mail: cornelia.baera@incd.ro; baera.cornelia@gmail.com

Abstract. The continuing hydration of unhydrated cement grains was proven to be one of the most important processes for promoting the self-healing effect within cementitious composites, by generating the CSH gels as valuable healing products, not only sealing the microcracks but also being able to provide some mechanical recovery of the material, as well [1]. It was also concluded that the process slows down in time, being strongly connected to concrete age. In order to ensure the continuous hydration potential for the cementitious materials, also as essential self-healing (SH) promoter, the addition of reactive grains is considered. This paper presents preliminary aspects regarding the possibility of polymeric encapsulation of some reactive grains and the feasibility of the concept in terms of matrix compatibility to the addition and also their SH performance under induced, controlled cracking. The considered self-healing addition behaves intelligent as it would react with water only when the cracking occurs, creating gaps in the waterproofing coating. The object of this research is less focused on regaining the mechanical characteristics of concrete, like pre-cracking strength, but mainly on preventing aggressive agents from entering in the concrete mass and aggressing the reinforcement.

1. Introduction

The concept of Self-Healing (SH) of cementitious materials is fundamentally connected to the state of cracking - micro-cracking in the structure of the material, as a form of induction of degradation in the mass of the material. The Self-Healing effect materializes by closing the cracks, sealing them, partially or totally. Most of the time the initiated recovery is not only at the physical level, connected to the microstructure of the material, but also to its functionality, recovery of the initial mechanical characteristics [2].

The Self-Healing of cementitious materials represents a research direction gaining consistent attention among worldwide research groups and also the construction industry, taking into consideration the substantial potential for ensuring, via prevention and intelligent materials design, the durability of infrastructure and considerable reduction of repair and maintenance activities, and consequently the costs savings [3].

Autogenous self-healing processes (SH) in concrete have been noticed since ancient times; also, their efficiency in closing micro-cracks and consequently in preventing the damage of cementitious composites microstructure and preserving their durability was considered to bring effects in the survival of ancient structures, from the Middle Ages or even antiquity (Figure 1).

(a)

(b)

(c)

Figure 1. Examples of Self-Healing effect on ancient structures: (a) Roman Forum (1st century BC - I d.Hr; (b) Old bridge in Amsterdam; (c) Colosseum, Rome.

Sustained research [1, 2, 4, 5] has shown that the phenomenon of autogenic curing of cementitious composites represents the consequence of mechanical processes (particles swelling), physical processes (crack filling with fine particles usual present in the water entering the crack) and most important, the chemical processes (continuous hydration and calcium carbonate precipitation); they can occur simultaneously and interfere or they can happen independently.

The continuing hydration of unhydrated cement grains represents an important chemical process generating self-healing of cementitious composites, generating the CSH gels as crack closing valuable products, not only sealing the microcracks but also generating mechanical recovery of the material, as well [1]. The intensity of the process slows down in time, strongly connected to concrete age: In time the unhydrated clinker nuclei are consumed and the generation of further hydration products is less productive. Another factor with significant contribution to the decrease of the binder continuous reactiveness in time is the nowadays cement production and technology: the clinker is strongly grounded and the cement finesse today is superior with respect to the previous years. Consequently, cement grains are easily accessible to the fast, initial hydration, reaching their peak till 28 days. After reaching their strength class, at the age of 28 days, a small further strength growth is recorded, confirming the assumption. The continuous hydration potential of the cementitious materials for the self-healing (SH) effect inducing as well is considered to be stimulated by the means of reactive grains addition is considered. The added grains would be polymeric coated for further preservation of their reactivity for when it would be necessary: initiation of a crack and the necessity of its sealing, when also the humidity is provided.

This paper presents preliminary aspects regarding the encapsulation techniques that could be used for the considered reactive grains and their compatibility to the cementitious matrix, as well as their potential use as Self-healing effect promoter within the material.

2. Materials and methods

Developing an intelligent, self-healing addition, based on an active hydraulic material (grain), covered with a layer of waterproofing material and then testing its positive effect involves the compliance with some specific requirements: identifying active hydraulic material grains with compatible waterproofing coating, controlled cracking of the samples and accelerated aging of concrete for a fast initial evaluation, identifying an appropriate method for highlighting the self-healing inducing of the addition, correctly emphasised and distinct from the natural, autogenous self-healing capacity of concrete.

The design concept of the intelligent addition includes some specific requirements for the reactive grains and for the waterproof polymeric coating, as well. The graphical representation of the theoretical concept is presented in Figure 2.

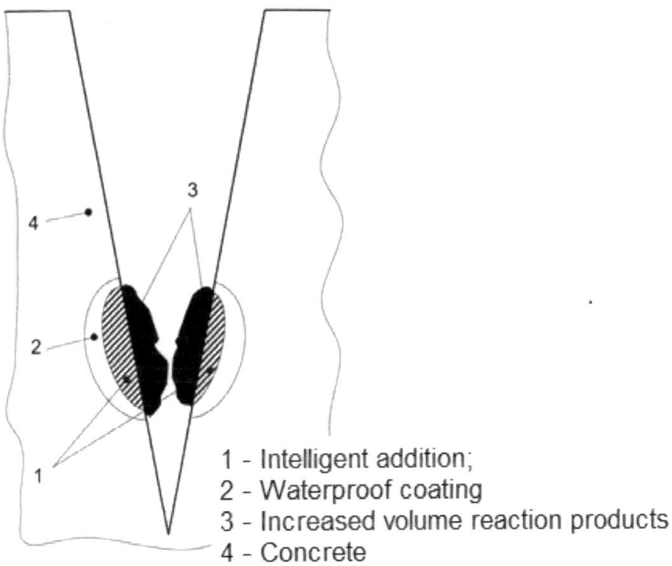

Figure 2. Graphical representation of the theoretical design for the reactive grains concept.

Regarding the hydraulic material, it must be able to generate new compounds by reaction with water or ions dissolved therein, compounds which can be deposited in the crack and diminish/cancel the access for the aggressive agents. In the same time, the reactive grain should not contain or generate sulphates, chlorides, alkalis or other aggressive concrete ions, as the presence of sulphates can lead to the formation of ettringite with a remedial action by creating a barrier to water access. There is identified the risk of an out-of-control reaction which can lead to the destruction of concrete; simultaneously, it must not negatively influence the hydration and hardening properties of Portland cement in case of accidental break of the waterproof protection.

The waterproof polymeric coating must protect the reactive material from water until the cracking would occur and also it has to be brittle and sensitive enough so it would break when necessary (when cracking happens), in order to allow the water in infiltrate and produce the healing compounds. It has to withstand efforts during the concrete preparation process, be resistant in the alkaline concrete environment, do not suffer positive volume variations in alkaline environments and not negatively influence the cement hydration processes. The adherence between the reactive granules and the coating represents an important feature, responsible to provide the required strength to withstand to mechanical shocks during mixing.

From the economic point of view, both materials, the reactive grains and the polymeric coating, should be available in large quantities and they should have the ability to be used in technological processes, in order to be attractive to the industry and produce a real effect as concrete durability enhancement, maintenance and environment pollution reduction.

In order to obtain the intelligent addition a hydraulic active material was considered. 0.09-0.16 mm particles were the main choice for the physical development of the composite capsules. Regarding the encapsulation development, a polymeric material, soluble in ordinary solvents and with a reasonably low melting point, generally complying with the previous described features, was considered.

Two different methods were considered as possible encapsulation techniques for the chosen hydraulic active grains (H):

1) Mixing the granules with the polymer (P) dissolved in compatible solvent, at various solution concentration and also various H/P ratios;

2) Mixing the granules with melted polymer at a reasonable temperature (Figure 3).

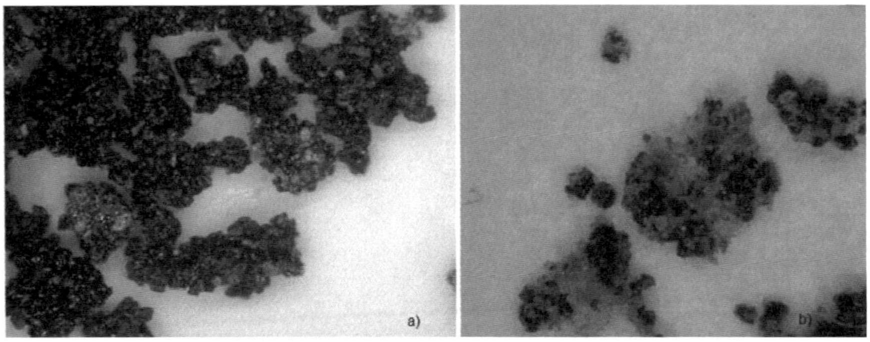

Figure 3. Active hydraulic grain (H) aspect during the coating procedures: (a) after mixing with melted polymer (Method 2); (b) before the grinding and sieving operations.

As variable parameters, the ratio hydraulic active grains (H) to Polymer (P), H/P and also the concentration of the polymer solution were considered (Method 1). An active hydraulic material with the size of granules ranging from 90 and 160 μm, wrapped in a polymer film of a thickness of 4-7 μm was obtained. Further on, the material is grinded for breaking the adhesions between the shells of the granules and then sieved through the 0.5 mm sieve, considered to generate the proper dimensions of the final coated grains.

Although mixing the reactive granules with the polymer solution or with the melted polymer generates complete coating of them, during the grinding operation, partial polymeric cover is removed and the full grain coating is lost. The coating efficiency, necessary to be evaluated, is determined by using a conductivity method, considering that the chemical reaction rate in a heterogeneous system depends on the contact surface between the two phases, the solid and the liquid. The conductivity of the distilled water, containing a certain amount of reactive addition / protected reactive addition is measured. The coating efficiency actually represents the decrease of conductivity, expressed as a percentage, and it is related to the reactive material conductivity. It was noticed that the relevant conductivity changes happened in the first 60 s after the mixing of distilled water with the considered amount of reactive grains (virgin and coated). Conductivity tests were performed for several types of coated grains, in terms of polymer concentration and H/P variation.

The Degree of coating (GA, %) was determined by using equation (1):

$$\text{GA}\ (\%) = 100 \times \frac{Cond_{hp} - Cond_h}{Cond_h} \quad (1)$$

Where:
GA (%) = Degree of coating;
$Cond_{hp}$ = conductivity of the coated grains suspension;
$Cond_h$ = conductivity of the uncoated grains suspension;
The results, presented in Table 1 show a small range of variation, proving a general reliable coating efficiency of the reactive addition and also the efficiency of the testing methodology. Seven of the tested compositions prove a GA exceeding 80%, four of them almost reach a convenient 90%, which is considered a very good performance.

Table 1. Degree of coating for relevant compositions.

Composition code	GA (%)
2	83
4	85
6	88
8	89
20	88
30	85
5	79
10	66
15	71
20	70
Melted polymer (Method 2)	87

The compatibility of the smart addition to the cementitious matrix was determined by integrating it in mortar samples produced in accordance to EN 196-1 [6]. CEM I 52.5R was used in order to obtain cement grain fast complete hydration and consequently, to reduce the natural, autogenous self-healing capacity of the mortar, for further better emphasise of the crack sealing effect of the smart addition. A small amount of polypropylene fibres was added to the mixes, including to the reference (R), which contains no other addition. The smart addition was added to the mortar samples as sand (0/0.5 mm fraction) replacement, as follows: Mix 130P40 contains 130 kg/m^3 smart addition, Mix 260P40, contains 260 kg/m^3 smart addition (developed by Method 1, polymer solution); Mix 225PT contains 225 kg/m^3 smart addition (developed by Method 2, melted polymer).

The influence of the smart addition to the mechanical resistance is observed by the means of the specific flexural (3PB) and compressive tests, by using the 40 x 40 x 160 mm^3 prismatic specimens, at the age of 28 days, in accordance to EN 196-1 [6]. It was recorded no negative influence of the smart addition to the cementitious material, on the contrary, a slight strength increase was recorded, attributed to the hydraulic active material, available from the remaining uncovered area.

The method considered reliable and conclusive for the initial evaluation of healing efficiency in terms of crack sealing and regenerating the concrete is based on a ASTM methodology [7]. The method principle is related to the specific electric load passing through the samples: cracked concrete and healed concrete. Prismatic specimens 40 x 90 x 160 mm^3 were prepared using the mixes previously mentioned (R, 130P40, 260P40 and 225PT). Accelerated aging of the samples is obtained by 2 weeks exposure to hygrothermal curing at temperature T 80°C, in humid conditions (water bath). The pre-cracking of the specimens, namely the cutting of a 5 mm thick and 15 mm deep channel, is performed by the help of a diamond disc saw, across the sample, in the median area, in order to determine the path for the further natural cracking, induced by mid span loading; the set-up is equipped with an horizontal LVDT for obtaining the desired crack width of 0.3 mm. The set-up used for the pre-cracking state of the specimens is presented in Figure 4.

Figure 4. Experimental set-up for fixed-width of 0.3 mm mid-span crack inducing into the specimens.

In order to increase the self-healing process of the specimens, they were cured by exposure in high moisture environment. After 3 days the samples were extracted from the water bath and another channel, similar to the previous one, was induced across the sample, located at 27 mm from the edge of the prismatic specimen. This way the same specimen includes both the cracking state and the totally healed cracked state, for a reliable healing evaluation via electric load passing through the two specified media. Two polyethylene tubes 50mm in diameter were stuck on the casting face using a silicone seal, one in the central area above the crack and one above the channel centred at 27mm from the prism end. The samples were dried in an oven and saturated with water in a vacuum of 20mm Hg. The samples thus prepared were introduced at the same level in a tray containing a 0,1N NaOH solution. A 3% NaCl solution is inserted into the polyethylene tubes. A difference in potential of 24V was applied for 15 minutes and the intensity of the electrical current through the cracked I_{fp} and uncracked Inp area of each sample was measured, including the I_{fr} and I_{nR} reference area. The self-healing efficiency in accordance to this procedure represents on-going research and the concluding data analysis is currently performed.

3. Results and discussions

The first results of the current research are offering some optimist conclusions related to the general viability of the concept.

The conductivity testing confirms the efficiency of the polymeric coating of the reactive grains, for both methods: Seven from the eleven tested compositions proved a coverage efficiency of over 80%, four of them approaching 90%. Figure 5 shows the conductivity development over time, for the reactive grains, without coating vs. a) the coated grains (Method 1) and b) the coated grains (Method 2).

The mechanical performance of the mortar proves not to be affected by the smart addition. Figure 6 shows the graphical comparative analyses of the 28 days results, obtained for all the mortar mixes considered. Initial specific testing performed on specimens show significant healing degrees even after short curing exposure, thus proving the research superior potential.

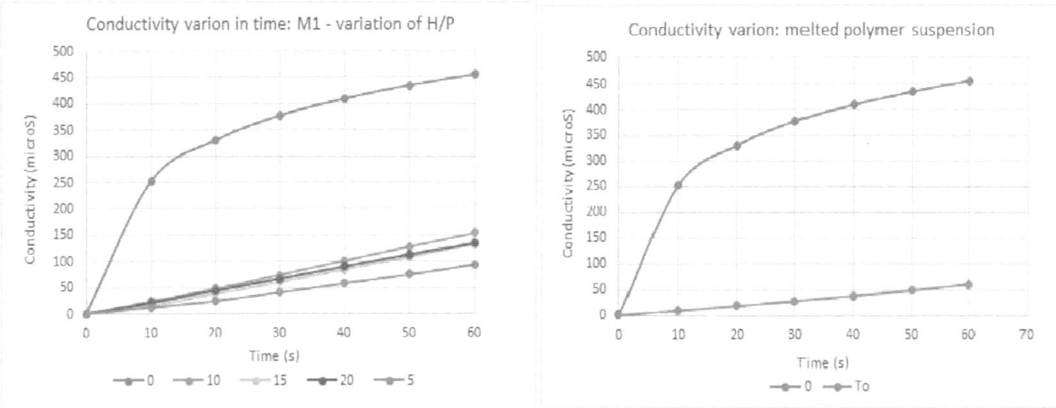

Figure 5. Graphical variation of the conductivity over time: uncoated vs. coated grains.

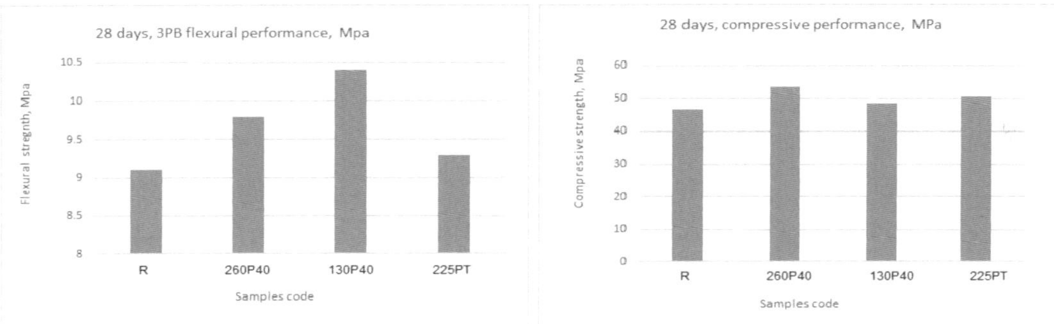

Figure 6. 28 days flexural and compressive strength for samples with or without smart addition.

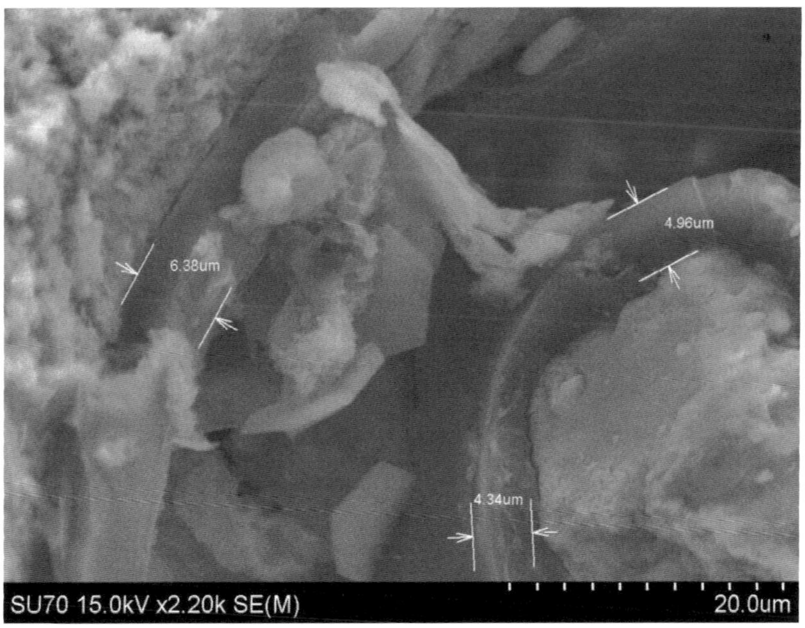

Figure 7. Graphical variation of the conductivity over time: uncoated vs. coated grains.

Simultaneously, the *visual analyses* performed by SEM microscopy coupled with EDAX (Figure 7), showed hydraulic active material granules with broken shell next to the crack and also its hydraulic active material, inside, starting to hydrate, developing new hydration, healing compounds, especially hexagonal crystals of $Ca(OH)_2$; intact coating at the edge of the crack was also noticed, offering further protection of the grain.

4. Conclusion and perspectives

The initial testing and the corresponding results are offering positive conclusions regarding the viability of the concept. A composite smart addition, consisting in hydraulic active grains, waterproof coated by a 4 - 7 μm polymer shell has been developed.

The two methods used for the encapsulation technique show good waterproof performance, in terms of conductivity evaluation, proving their efficiency.

The compatibility of the coated reactive grains to the cement based matrix was tested in terms of mechanical performance evaluation with respect to the reference matrix, with no addition; the tensile and compressive strength did not experience any decrease due to the smart addition, but a slight increased attribute to the reactive grains becoming active during mixing.

The self-repairing effect of intelligent addition is evaluated on the mortars by electro diffusion of chloride ions. Crack healing degrees of more than 70% were obtained for some of the tested compositions, according to the initial data analysis.

Optimization of the smart addition and supplementary validation performance of its repairing effect in concrete still needs to be performed.

Acknowledgments

This paper is supported by the Programme: Innovative solutions for transport infrastructure protection by the use of building materials with special self-maintenance and self-repair properties", Programme code: 6 PS / 13.09.2019, financed by the Romanian Government.

5. References

[1] De Belie N, Gruyaert E, Al-Tabbaa A, Antonaci P, Baeră C, Bajare D and Litina C 2018 *Adv. Mater. Interfaces* **5(17)** 1800074

[2] Snoeck D 2015 *Self-healing and microstructure of cementitious materials with microfibres and superabsorbent polymer* (Belgium: Ghent University)

[3] Joseph C, Lark R, Jefferson T and Gardner D 2009 *Potential application of self-healing materials in the construction industry* (Cardiff: Cardiff University)

[4] De Rooij M, Van Tittelboom K, De Belie N and Schlangen E 2013 *Self-healing phenomena in cement-Based materials: state-of-the-art report of RILEM technical committee 221-SHC: self-Healing phenomena in cement-Based materials* vol 11 (Berlin: Springer Science & Business Media)

[5] Van Tittelboom K and De Belie N 2013 *Mater.* **6(6)** 2182–217

[6] EN 196 2016 Methods of testing cement. Determination of strength, *Romanian Standard Association, Bucharest, Romania*

[7] ASTM C 1202 2005 *Standard Test Method for Electrical Indication of Concrete's Ability to Resist Chloride Ion Penetration* (Washington: ASTM International)

Optimized Conversion of Nyamplung Seeds Oil to Biodiesel Using Box-Behnken Response Surface Methodology (RSM)

Z Ilham[1*] and M R A Mansor[2]
[1] Biomass Energy Laboratory, Institute of Biological Sciences, Faculty of Science, University of Malaya, 50603 Kuala Lumpur, Malaysia
[2] Department of Mechanical and Materials Engineering, Faculty of Engineering, Universiti Kebangsaan Malaysia, 43600 Selangor, Malaysia

*E-mail: ilham@um.edu.my

Abstract. This research is carried out for optimization of transesterification reaction parameters to convert oil from Nyamplung (Calophyllum inophyllum) seeds into biodiesel. Calophyllum inophyllum seeds were normally left unused due to the toxicity of calophynic acid in the unrefined extracted oil. The optimal variables (parameters) for converting Calophyllum inophyllum seeds oil into biodiesel were found to be at a methanol: oil ratio (6 to 1), 67.5 °C temperature, 1 wt% catalyst and agitation rate of 750 (rpm). The biodiesel yield obtained (93.01 wt%). The results also indicated that Calophyllum inophyllum methyl esters (CIME) fuel properties such as kinematic viscosity 4.55 $mm^{2/s}$ acid value 0.027 mg KOH/g, cloud point 3 °C, pour point 1 °C, flash Point 161 °C, cetane index 49 min and density of 856 kg/m3 conformed with the international standard specifications except for the oxidation stability, which was at 4.38 hour. In addition, ANOVA of the developed model displayed presentation (R^2 0.99). Calophyllum inophyllum seeds oil could be a suitable non-edible feedstock for biodiesel, especially in the South East Asian region.

1. Introduction

Globally, researchers keep on looking for alternative fuels to shift the sole reliance on fossil fuel as well as ensuring energy security. Biofuel is gaining scientific community interest as it is an environmentally-friendly product obtained from agricultural derived and plant biomass, fat of animal and its other wastes [1]. As far as many countries are concern, fossil fuels are expected to be lower in production in the next 20 years with annual capacity drop rate at 14 % [2]. Many countries will be affected by this issue as they are depending mostly on fossil fuels to power their industries [3]. Various feedstocks has been used for biodiesel production from all over the world [4] such as soybean, palm kernel [5], tobacco, palm [6], cotton and canola oil [7].

In the past, biodiesel was produced mostly from edible oils and fats which creates a competition between energy and food [8]. Globally, third world countries faced severe concern as they require the edible oil for food consumption of their rising population. Edible oils commonly used for fuel is from soybean, canola, coconut and palm oil, which are a staple to the society [9]. Hence, it is need of time to focus more on non-edible seed oils for biodiesel production.

Calophyllum inophyllum (Nyamplung) belongs to Clusiaceae family and it is currently being used traditionally as topical medicine in many countries in the South East Asian region especially Malaysia and Indonesia, despite its known toxicity. Over the globe, the species can be found in coastal areas of the Indian and Pacific oceans from Australasia and the Pacific Islands throughout Asia to East Africa. It is widely distributed in tropical region and native to Peninsular Malaysia and Indonesia, commonly planted as ornamental in Florida, Hawaii and Guam. Nyamplung is normally planted to provide shade along the roadsides in many tropical municipalities. Its fruits are globose with one large seed and the flowers are in small clusters, white and fragrant. Tropical climate is needed for its cultivation (humid and warm) with 28 °C to 35 °C for normal growth. The highlights of Nyamplung seeds is the high oil content from the seeds at the range of 40-60 wt%, and the 20 tonnes seed per hectare productivity [10].

In this study utilising Nyamplung seeds oil, the optimisation on different variables and their response on biodiesel yield has been carried out using RSM (Box-Behnken) design, which has not previously reported. The outcomes could be used to validate the RSM modelling, producing an acceptable results and statistically proven technique [11].

2. Experimental

2.1. Materials and setup

Nyamplung (*Calophyllum inophyllum*) seeds used in this study (Figure 1) were collected in Jelebu District, Negeri Sembilan, Malaysia. The seeds were then dried at 60°C for 3 days before oil extraction using Soxhlet method [9]. Batch process setup used in this study includes three neck round bottom flask, heating mantle, hot plate agitator, temperature probe and digital temperature control unit [5]. In addition, pure chemicals (99% purity) and reagents were procured and used.

Figure 1. Trees, fruits and seeds of Calophyllum inophyllum (Nyamplung).

2.2. Design of experiment

In order to optimize oil extraction from Nyamplung seeds oils, DoE (RSM) was employed based on Behnken approach. Selected extraction parameters were extraction time (h): X1, hexane volume (ml): X2, and temperature (°C): X3. The un-explained variability effect minimization was done by using randomization approach in the response observed.

2.3. Oil extraction from BJO

Oil is extracted using a Soxhlet apparatus with the capacity of 500ml containing hexane [9]. Later, grinded seeds in the form of powder was placed into condenser thimble. A specified volume of solvent was stationed in the apparatus with a tight fixed condenser. Heating temperature was set at 70 °C for heating mantle. The equation for oil yield is presented below in Equation 1.

$$Oil\,Yield = \frac{Weight\ of\ oil\ produced}{Weight\ of\ oil\ used} \times 100 \quad (1)$$

2.4. Biodiesel Production

Two step biodiesel production was carried out and steps are given below.

2.4.1. Esterification. Esterification were done as a pre-treatment to minimize the free fatty acids in the oil. This process used sulphric acid (1 wt%) as a catalyst and preceded before transesterification was carried out in the second phase [9].

2.4.2. Transesterification. Transesterification is done in a bioreactor with the capacity of 500ml. A specified volume of solvent (methanol: oil) was stationed in the apparatus with a tight fixed condenser. Heating temperature was 70 °C in mantle and the oil stored in flask. Total of 29 experimental runs with optimized variables were executed [12]. Sodium hydroxide was used as catalyst.

2.5. Data Analysis

RSM technique is utilized to analyze the 29 runs of the experiment, in order to fit the quadratic function. Response of Y (oil yield) and independent variables were correlated as; A (methanol-oil ratio), B (temperature) and C (agitation rate, rpm) and D (catalyst). Further, multiple regressions were used to fit the coefficient of the polynomial model of the response. The range of optimized parameters are shown in Table 1. The quality of fit of the model was evaluated using test of significance and analysis of variance (ANOVA). The fitted polynomial equation is presented below in Equation 2.

$$"y = b_0 + \sum_{i=1}^{k} biXi + \sum_{i=1}^{k} bii\, X^2 i + \sum_{i<j}^{k} bijXiXj + e" \quad (2)$$

"y: yield
bo: value of intercept
bi: co-efficient model
bij: interaction effects
bii: co-efficients of quadratic Xi
e : "error

*2.5.1. GC-MS analysis (Gas chromatography mass spectrometry).*Gas chromatography mass spectrophotometry analysis (GC-MC) was done by using Trace GC-2000 gas chromatograph coupled to a Polaris Q Ion trap mass-spectrometer (Thermo-Finnigan, Austin, TX USA) with Zebron ZB-5ms (Phenomenex,Torrance CA USA).

2.6. Analysis of Biodiesel Properties

Karl Fischer MKC-520 (Kyoto Electronics MFG. Co. Ltd.) was used to check water content in the produced fatty acid methyl esters according to the international standards.

3. Results and Discussion

3.1. Physico-chemical characterizations of Nyamplung oil

Oil yield from this Nyamplung seeds in this study was 43 wt%. As the chemical and physical properties are very essential, Table 1 depicted the results of these properties. Nyamplung seeds oil is yellowish appearance at room temperature with refractive index (RI) of 2.51. The oil acid value is 4.243 mg KOH/g, indicating the presence of free fatty acids (FFA). Oxidation stability of this plant oil was tested at 5.48 h (Rancimat) with the kinematic viscosity (KV) of 16.613 mm^2/s.

Table 1. Physico-chemical propertied of Nyamplung seed oil.

Physico-chemical properties	Nyamplung oil
Ultrasonic speed	1.25 ms^{-1}
Refractive index (25 °C)	2.51
Colour	yellow
Oxidative stability	5.48 h

Physico-chemical properties	Nyamplung oil
Oil	43 wt%
Kinematic viscosity	16.613 mm^2/s
Acid value	4.243 mg KOH/g

3.2. Variable Optimization

The predicted and experimental values for Calophyllum inophyllum methyl esters (CIME) yield as well as responses at the design points and all variables are shown in Figure 2. The experimental and predicted CIME yield indicated that both values are very close to each other.

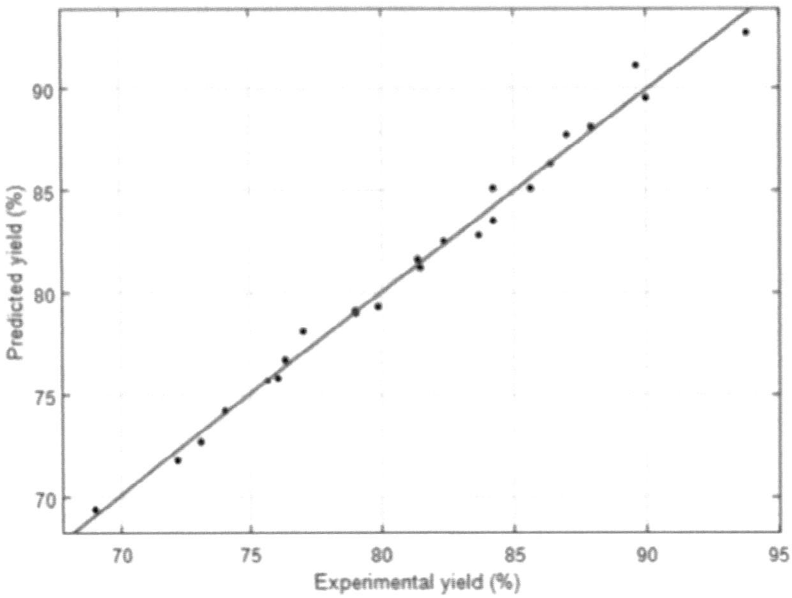

Figure 2. Experimental vs predicted yield of Calophyllum inophyllum methyl esters (CIME).

Hence, the reliability of the developed model to see the correlation between the process parameters and the CIME yield is valid. The results achieved from all four variables (methanol to oil ratio, temperature, agitation rate and catalyst) which were investigated are displayed in Table 2. The results from RSM showing multiple regressions was used to fit the coefficient of the polynomial model of the response as A, B, C and D and presented below in the equation 3.

$$\begin{aligned} \mathit{Biodiesel\ yield} &= 91.6 - 0.78 \times A - 4.79 \times B + 0.22 \times C + 1.31 \times D - 0.77 \times A \times B \\ &+ 0.68 \times A \times C + 1.51 \times A \times D + 0.26 \times B \times C - 0.38 \times B \times D \\ &+ 1.97 \times C \times D - 2.36 \times A^2 - 7.59 \times B^2 - 2.43 \times C^2 - 0.032 \times D^2 \end{aligned} \quad (3)$$

In addition, RSM Box-Behnken design with 4 levels of operating variables and 29 experimental runs ANOVA is shown in Table 3. The validity is tested by looking at model p and R-square values. The ANOVA displayed a p-value of > 0.05 and shows its significance. AB, AC, BC, A^2, B^2, C^2 and D^2 are model terms in this model. A quadratic model was used which shows p-value less than 0.05 conferring significant at 95 % confidence level. R-square (regression) with value of 0.99 and Adj r-square = 0.98. Total variability is indicated as R^2 and a value greater than 4 is desirable [13].

IOP Conf. Series: Materials Science and Engineering **877** (2020) 012029 doi:10.1088/1757-899X/877/1/012029

Table 2. CIME yield versus predicted yield.

Run	A:Methanol to oil ratio ratio	B:Temperature °C	C:rpm rpm	D:catalyst wt%	CIME yield wt%	Predicted yield %
1	6	60	750	0.5	87.24	87.34
2	6	67.5	750	1	91.93	91.63
3	6	67.5	750	1	91.21	91.63
4	6	75	750	1.5	80.45	80.74
5	6	67.5	750	1	90.97	91.63
6	9	67.5	750	0.5	85.87	85.03
7	6	75	900	1	77.01	77.89
8	9	75	750	1	75.88	75.94
9	3	67.5	750	1.5	89.22	89.24
10	6	67.5	600	1.5	88.12	88.19
11	9	67.5	600	1	85.34	85.54
12	6	67.5	900	0.5	86.44	86.05
13	3	75	750	1	78.31	78.53
14	6	60	900	1	86.31	86.12
15	9	67.5	750	1.5	90.99	91.19
16	9	60	750	1	86.76	86.21
17	3	67.5	750	0.5	90.13	90.10
18	6	75	600	1	76.01	76.37
19	9	67.5	900	1	86.24	86.37
20	6	60	600	1	86.34	86.63
21	6	60	750	1.5	90.76	89.73
22	3	67.5	900	1	87.41	87.60
23	6	67.5	750	1	93.01	91.63
24	6	75	750	0.5	78.44	77.84
25	6	67.5	900	1.5	92.34	92.03
26	3	67.5	600	1	89.22	87.43
27	6	67.5	600	0.5	89.1	88.88
28	6	67.5	750	1	91.22	91.63

Table 3. Analysis of Variance (ANOVA) to produce *Calophyllum inophyllum* methyl esters (CIME).

ANOVA for Response Surface Quadratic Model
Analysis of variance table

Source	Sum of Squares	Degree of freedom	Mean Square	F Value	p-value	
Model	736.79	14	52.63	106.2 1	< 0.0001	significant
A-Methanol to oil ratio	7.25	1	7.25	14.64	0.0019	
B-Temperature	274.85	1	274.85	554.7	< 0.0001	
C-rpm	0.57	1	0.57	1.15	0.3008	
D-catalyst	20.44	1	20.44	41.24	< 0.0001	
AB	2.36	1	2.36	4.76	0.0468	
AC	1.84	1	1.84	3.71	0.0748	
AD	9.09	1	9.09	18.35	0.0008	

ANOVA for Response Surface Quadratic Model
Analysis of variance table

Source	Sum of Squares	Degree of freedom	Mean Square	F Value	p-value	
BC	0.27	1	0.27	0.54	0.4765	
BD	0.57	1	0.57	1.15	0.3016	
CD	15.52	1	15.52	31.33	< 0.0001	
A^2	36.07	1	36.07	72.8	< 0.0001	
B^2	373.74	1	373.74	754.27	< 0.0001	
C^2	38.44	1	38.44	77.58	< 0.0001	
D^2	6.61E-03	1	6.61E-03	0.013	0.9097	
Residual	6.94	14	0.5			
Lack of Fit	4.17	10	0.42	0.6	0.765	not significant

3.3. Effect and Impact of variables on biodiesel yield

As displayed in Figure 3, methanol to oil ratio of 6:1 showed highest yield of biodiesel (96.49 wt%). As the ratio increase further, a slight decline is seen in the yield. It has been observed that at the start of the reaction, the biodiesel yield increased with increasing methanol oil molar ratio, peak point was achieved and then a decrease is seen in yield as the ratio cross a certain range. It could be observed that whenever methanol exceeds the optimal value, it will be more difficult to separate esters from glycerol. In other words, gravity decantation is hindered whenever methanol is utilized in excess [14]. Literature also reveals that no substantial effects can been seen in biodiesel production when excess methanol is used apart from increasing the processing cost. A molar ratio of less than 6:1 may lead to a dilution effect. Utilization of molar ratio at 6:1 achieved the highest biodiesel yield at 94.34 wt% [13]. Hence, it can be said that ideal molar ratio is required for a successful transesterification.

The independent variable's impact on yield was drawn with help of response curve using 3-D surface analysis. The peak point was obtained when independent variables interact perfectly and optimal yield was achieved, which lead to the conclusion that response variable depends on relationship and interaction between independents variables.

Methanol to oil ratio is considered as an important parameter in transesterification to get an optimum yield of biodiesel. Varying ratio of methanol to oil was used, ranging from 3:1 to 9:1 for the methanol to oil molar ratio in order to study the effects on Calophyllum inophyllum methyl esters yield.

The optimum biodiesel yield (93.01 wt%) was also achieved at an optimum temperature of 67.50 ^{o}C. Temperature increase can lower down the viscosity and help converting oil into methyl esters which are in agreement with previous findings [16, 17]. On the other hand, catalyst variable follows the similar pattern as molar ratio. Excessive catalyst utilization result in emulsions which eventually reveals on high viscosity. Moreover, biodiesel recovery become difficult as side products formed during the reaction [15]. It can be observed that a catalyst concentration of 1 wt% showed the highest biodiesel yield of (93.01 wt%) as displayed in Figure 3. When the catalyst amount crosses the range above 1 wt%, a declined pattern could be seen. This showed that insufficient use of catalyst may result lower biodiesel yield and offered an incomplete transesterification [16]. Both increase and decrease in amount of catalyst affect reaction and show reverse pattern [13].

As shown in Figure 3, it could be observed that the agitation speed is also an important independent variable which affect the yield. The CIME biodiesel yield somewhat increases with an increase in the agitation speed to a certain value. Mixing intensity or agitation is an important parameter which could directly affect solution homogeneity in the case of transesterification, with catalyst is present in the reaction. Increasing agitation periods facilitate transesterification and help in mixing the solution and increases the contact area between methanol, oil and CaO solution [17]. Reaction without agitation speed occurs only at the interface of two layers and this could slow down transesterification process. Previous literatures revealed that increasing the agitation speed in transesterification process could stimulates the homogeneity of reactants and hence, increase the yield of biodiesel significantly [17, 18]. Moreover, these findings showed that agitation speed of 750 rpm was enough to achieve the maximum biodiesel yield in the transesterification reaction.

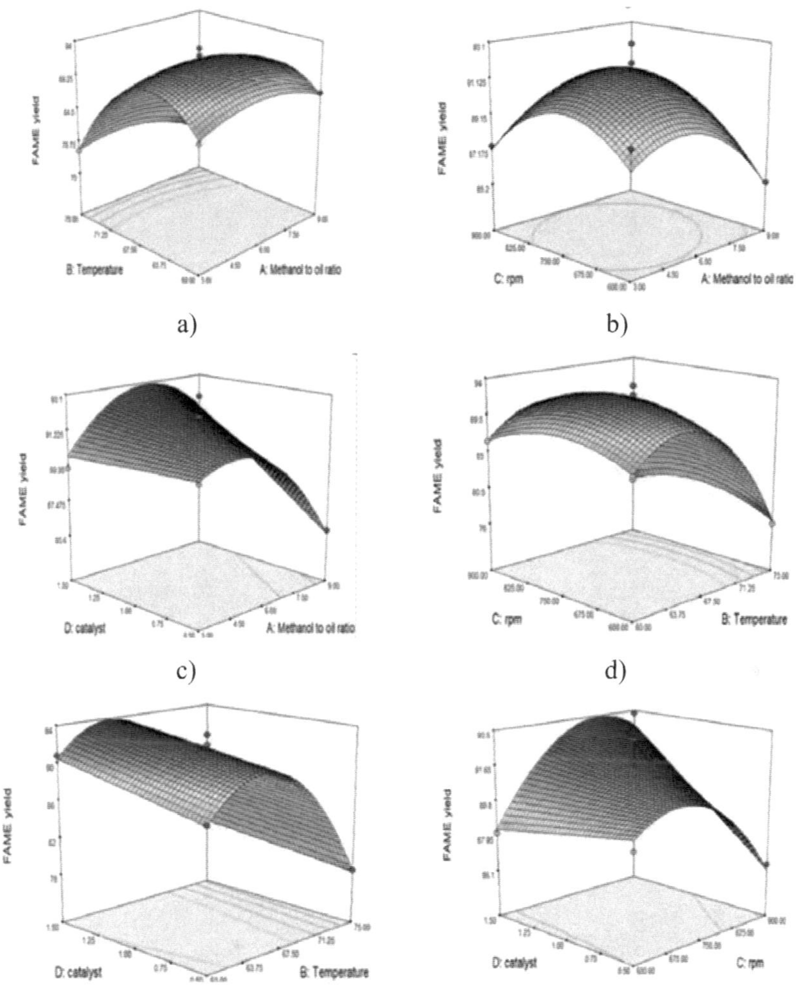

Figure 3. 3-D surface plot showing parameters effect on *Calophyllum inophyllum* methyl esters (CIME) yield: (a) methanol ratio and temperature (b) rpm and methanol ratio (c) methanol ratio and catalyst (d) temperature and rpm (e) temperature and catalyst and (f) rpm and catalyst.

3.4. Calophyllum inophyllum methyl esters (CIME)

ASTM D6751 and European Standard EN14214 are the international standards commonly used by researchers to evaluate the fuel properties of the produced fatty acid methyl esters or biodiesel. Kinematic viscosity (KV) of HSME was found to be at 4.55 mm^2/s, which falls within both standards.

The CIME density at 856 kg/m^3 also comply to the range of specifications. Increased of density in fuel could affect the fuel Injection system, where the fuel become more viscous. Acid value (AV) of CIME was detected at an acceptable value of 0.054 mg KOH equivalent [24]. High AV is not recommended due to the initiation of corrosion, disturbing the engine internal operations. The oxidative stability is the property tested to represent the potential storage time of the produced biodiesel.

Table 4. Comparison of Calophyllum inophyllum (Nyamplung) methyl esters with biodiesel produced from other non-edible plant seed oils.

Plant Seed oils	Density (mg KOH/g)	Cloud point (°C)	Flash point (°C)	Oxidative stability (hour)	Viscosity (mm^2/s)	Cetane number (min)
Nyamplung	856	3	161	4.38	4.55	49
Neem	884	14.4	--	7.1	5.21	57.83
Mahua	850	--	208	--	3.98	--
Castor	899	-13.4	--	1.1	15.25	--
Jatropha	880	2.7	135	2.3	4.80	52.31
Karanja	--	--	150	--	4.80	55.84
Malada pahit	871	2	164	3.0	3.55	51
EN14214	900	--	120	6	5	55
ASTM	--	-3 to 12	170	3	6	47

The oxidative stability of the produced HSME was at 4.38 hours, exceeding the standard 3 hours set by ASTM but less than 6 hours as required by EN 14214 standard. Although oxidation stability of this fuel was found not at par with the conventional diesel at 15.2 hours, flash point (161 °C), cloud point (3 °C) and Cetane number (49) values are all acceptable as displayed in Table 4.

4. Concluding Remarks

This study was conducted in order to optimize the parameters for biodiesel production from Nyamplung seed oils, a non-edible feedstock. The optimum variables (parameters) for conversion of Nyamplung seed oils into biodiesel were methanol to oil ratio of 6 to 1, operating temperature at 67.5 °C, catalyst amount at 1 wt% and 750 rpm of agitation rate. The CIME yield with these optimum parameters is at 93.01 wt%. The results also showed that most of Calophyllum inophyllum methyl esters (HSME) fuel properties conformed within international standards of ASTM D6751 and EN 14214 except for the oxidation stability at 4.38 hour. The results indicated in this study showed a high possibility for Calophyllum inophyllum seed oils to be converted into biodiesel to prevent the food versus fuel dilemma.

Acknowledgements

This research was funded by research grants (GPF016B-2018, ST006-2019 and IIRG017B-2019) of University of Malaya awarded to Dr. Zul Ilham as Principal Investigator, which the authors highly acknowledged.

5. References

[1] Rathour R K, Ahuja V, Bhatia R K and Bhatt A K 2018 *Int. J. of Energy Res.* **42** 4532-4545
[2] Oh T H, Pang S Y and Chua S C 2010 *Renew. Sustain. Energy Rev.* **14** 1241-1252
[3] Jayed M H, Masjuki H H, Kalam M A, Mahlia T M, Husnawan M and Liaquat A M 2011 *Renew. Sustain. Energy Rev.* **15** 220-235
[4] Ilham Z and Nimme F H 2019 *AIMS Agri. Food* **4** 303-319
[5] Zakaria M, Vijayasekaran, Ilham Z and Muhamad N A 2014 *Proc. Chem.* **13** 218-220

[6] Husin M, Li A R, Ramli N, Romli A Z, Hakimi M I and Ilham Z 2017 *Int. J. Adv. Appl. Sci.* **4** 51-58
[7] Ilham Z, Hamidon H, Rosji N A, Ramli N and Osman N 2015 *Proc. Chem.* **16** 164-170
[8] Freire da Silva R and Veras G 2019 *Int. J. of Energy Res.* **43** 5004-5012
[9] Hakimi M I, Ilham Z and Kohar R A 2019 *Waste Biomass Valor.* **10** 2681-2688
[10] Fadhlullah M, Widiyanto S N and Restiawaty E 2015 *Energy Proc.* **68** 177-185
[11] Supramani S, Rahayu A, Ilham Z, Annuar M S M, Klaus A and Wan-Mohtar W A A Q I 2019 *AIMS Microb.* **5** 19-38
[12] Taib N M, Zainuddin M S, Mansor M R and Ilham Z 2018 *J. Mech. Eng. Sci.* **12** 3689-3699
[13] Hasni K, Ilham Z, Dharma S and Varman M 2017 *Energy Conv. Mngmt.* **149** 392-400
[14] Leung D Y C and Guo Y 2006 *Fuel Process. Technol.* **87** 883-890
[15] Salehmin M N, Annuar M S M and Ilham Z 2015 *Chem. Eng. Comm.* **202** 1628-1634
[16] Sohedein M N, Wan-Mohtar W A A Q I, Ilham Z, Babadi A A, Hui-Yin Y and Siew-Moi P 2020 *J. App. Phyco.* 1-4
[17] Meher L C, Dharmagadda V S S and Naik S N 2006 *Bioresour. Technol.* **97** 1392-1397
[18] Rashid U and Anwar F 2008 *Fuel* **87** 265-273
[19] Jimoh A, Abdulkareem A, Afolabi A, Odigure J and Odili U 2012 *Energy Cons.* 197
[20] Saifuddin N and Rafel H 2014 *Res. J. App. Sci. Eng. Technol.* **8** 1149–1159
[21] Saka S, Goembira F and Ilham Z 2011 *Bioresour. Technol.* **102** 3989
[22] Mohammed I Y, Abakr Y A, Kazi F K, Yusup S, Alshareef I and Chin S A 2015 *Energies* **8** 3403-3417
[23] Shuit S H, Lee K T, Kamaruddin A H and Yusup S 2010 *Environ. Sci. Technol.* **44** 4361-4367
[24] Yang F, Hanna M A, Marx D B and Sun R 2013 Int. J. Energy Res. **37(13)** 1600-1609

Reliable Method and Multistage Process Involved in the Production of Activated Carbon Based on Raw Material- A Systematic Review

S Muzarpaar[1], A M Leman[1*], K A Rahman[2] and S Z A Rahim[3]
[1]Occupational Safety, Health and Working Environment (OSH-WE) Research Group, University of Tun Hussein Onn Malaysia (UTHM), Higher Education Hub of Pagoh, KM 1, 84600, Muar, Johor, Malaysia
[2] Department of Petro Chemical, Politeknik Tun Syed Nasir (PTSN), Higher Education Hub of Pagoh, KM 1, 84600, Muar, Johor, Malaysia
[3]Green Design and Manufacture Research Group, Centre of Excellence Geopolymer and Green Technology (CEGeoGTech), Universiti Malaysia Perlis, 01000 Kangar, Perlis, Malaysia

*E-mail: mutalib1480@gmail.com

Abstract. Activated carbon AC) has versatile usage in the modern days for water treatment, air filtration and for multipurpose adsorption. Carbonaceous material with highly developed porosity known as AC after some process took place. AC also known as artificial materials characterized by the high surface area and the extensive surface chemistry. Significant adsorptive properties of the materials according to extensive surface chemistry which are responsible for the surface properties. Nowadays, AC are widely used in a large range of applications, for example medical uses, catalysis, gas storage, removal of pollutants and odors, gas separation and purification. Important Character for based materials of AC, it must accessible, affordable price, and with valorization potential in agricultural residues and industrial base. The most important raw materials for the production of AC is lignocellulosic materials. Over the time researcher around the world conducting research to improvised and perfecting the method for producing the AC. The aim of this paper is to critical review the reliable method for AC production. The process involved in the production of activated were pre- carbonization process (PCP), carbonization process (CP), and activation process (AP). Carbonization either using furnace or microwave. For AP, there are two main activation used namely chemical activation CA) and physical activation (PA). However there also combination of chemical and physical activation which called physiochemical activation (PCA).

1. Introduction

Activated carbon (AC) has versatile usage in modern days for water treatment, air filtration and also for multipurpose adsorption applications. AC is a carbon-based material that undergoes several types of process which eventually develop high porosity that could be function as adsorbent and catalyst [1]. João M. Valente Nabais et al. stated that, AC are materials which characterized by mean of high surface area with weak van der waal's forces surrounding them [2]. Nowadays, AC become the high significant usage and widely used especially in the area of medication, pollutant and odour removal, catalysis, gas storage, separation of gas and purification. Characteristic for based materials of AC must affordable, accessible and with valorization potential, such as agricultural residues and industrial [3]. The production of 300,000 tons/year of AC were reported, as the result of corresponding usage of this material. General domestic application, metal ore processing, and removal of both organic and inorganic trace contaminants makes AC considered as one of the best available technology for this particular area [4].

2. Production flow process of activated carbon

There is a lot of technique to make an AC, but the common process is shown in Figure 1.

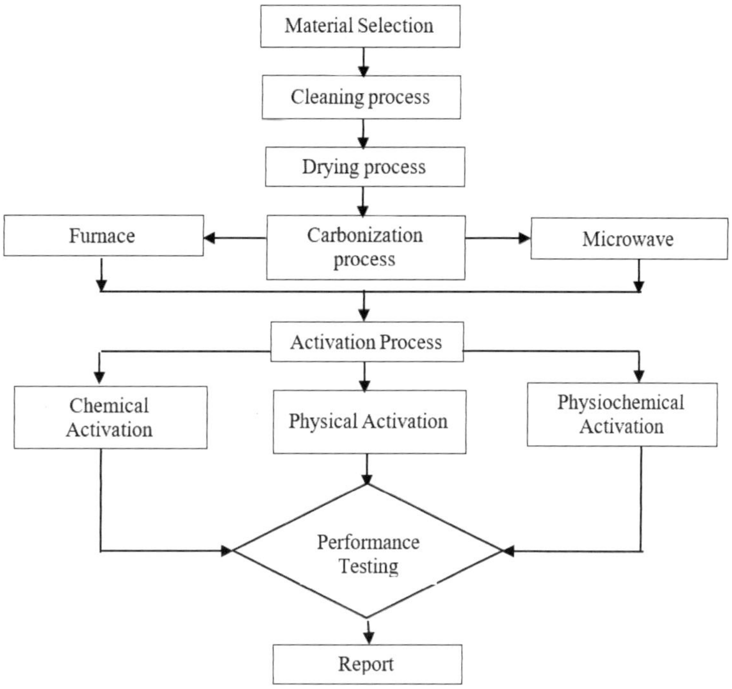

Figure 1. The general flow process of activated carbon production.

2.1. Pre- Carbonization process

Pre carbonization process (PRB)is an optional process. The process include in pre carbonization process were, raw material selection, material cleaning, and drying the material before proceed with the carbonization process. To remove the excessive dirt that stick to the material surface, the raw materials were cleaned intensively by using distilled water. The precursor then heated at 110 °C until dried, with sized of 4 mm [5] For material selection, certain criteria should be followed such as the availability of material and cost of material. Meanwhile drying of the material before carbonization process will helps in reducing the moister content [6].

2.2. Carbonization/ burning process of activated carbon

Carbonization process (CP) was describe by many researchers as the process of burning the raw materials with specific temperature to convert them into charcoal. CP could be done either using furnace, microwave or any device that generate heat. Carbonization of any biomass will produce charcoal, that later on be activate using chemical reagent to obtain AC [7]. Charcoal or the carbon based can be retrieved throughout carbonization process in electrical tubular furnace at 450 °C with a ramp temperature 10 °C/min, after that it will be cooled to room temperature, washed and dried under vacuum at 70 °C for the period of 24 hours [8].

2.2.1. Carbonization using furnace. The process of carbonization using furnace were the most dominant method for converting the raw material to charcoal. Usually this method involved in high temperature usage under controlled environment. Continuous flow of nitrogen gas with the temperature of 700°C using vertical furnace were used to carbonize the agricultural wastes [9]. Before activation process, the precursor will leaved for a while in the furnace for cooling purposed. Hence, the carbon or charcoal were soak into activated agents such as H_3PO_4, HNO_3, K_2CO_3, NaOH or KOH [10].

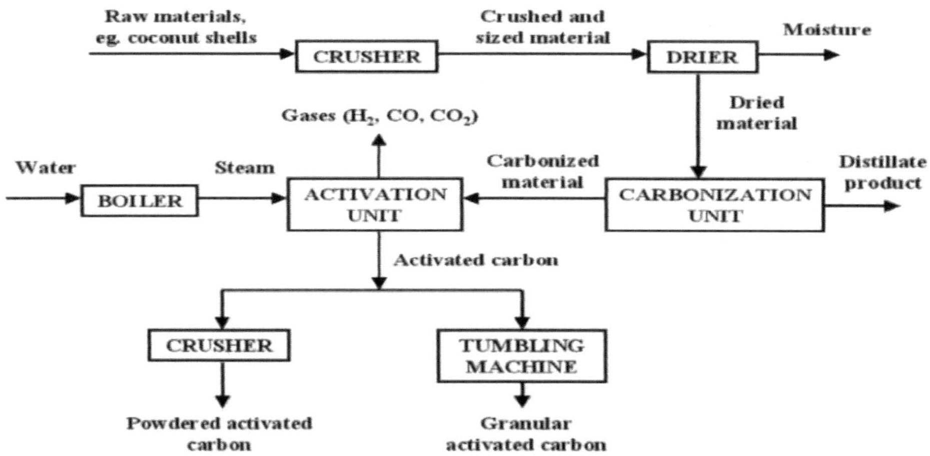

Figure 2. The process of AC using furnace with physical activation [13].

These processes were supported by other researcher which mention in production of AC the CP of the bamboo particles were heated to 500 ◦C [11]. Furthermore, to obtained large amount of charcoal, CP must be conducted with the absence of oxygen in temperature range between 600–900 °C, with the absence of oxygen. The absence of oxygen surrounding can be achieve by purging the inert gas such as argon or nitrogen inside the furnace [12]. Figure 2 show the process of AC furnace with physical activation.

2.2.2. Carbonization using microwave. Method for carbonization of carbon can be done using microwave [14]. Dipolar and interfacial polarization effect resulting from the material receiving energy. The process will determine the successful rate of dielectric mechanism. Dipolar polarization occurs when water molecule rotates and align in both permanent at certain frequency cause by the electric component of microwave. Microwave heating and heat loss were the result from molecular rotation and moving throughout the increase of friction and collision generated by molecule. Indication of volumetric characteristic and uniformity of microwave were generally describe as the higher of the induced polarity. It is the main and important characteristic for the influence of microwave [15, 16]. From the preliminary experiment the performance of different microwave heating time and temperature condition were selected. Recently, Microwave assisted pyrolysis for AC preparation commonly reported as a reliable process [17]. Shorter treatment time which is less than 10 minutes show the microwave heating advantages as compared to conventional heating, simultaneously reduce the energy consumption and expenses during the pyrolysis process [18,19]. The biomass will never have direct contact with the radiation sources using microwave heating [20]. Figure 2 show the Activated carbon Preparation conventional and microwave technique [21]. Figure 3 show the AC preparation using conventional and microwave technique.

3. Activation process

Activation of AC is a process whereby the enhancement of pore structure in AC, this pore will determine the adsorption of the AC. The higher the pore formation the higher the adsorption power. Activation process will highly determine the dependency of pore structure and adsorption properties of

AC [22]. Physical activation (PA) and chemical activation (CA) were the common method to produce AC. PA of AC consists of two steps. Elimination of oxygen and hydrogen content through carbon base pyrolysis material at high temperature with the absence of oxygen were the first step in PA. The second step involved is activation process.

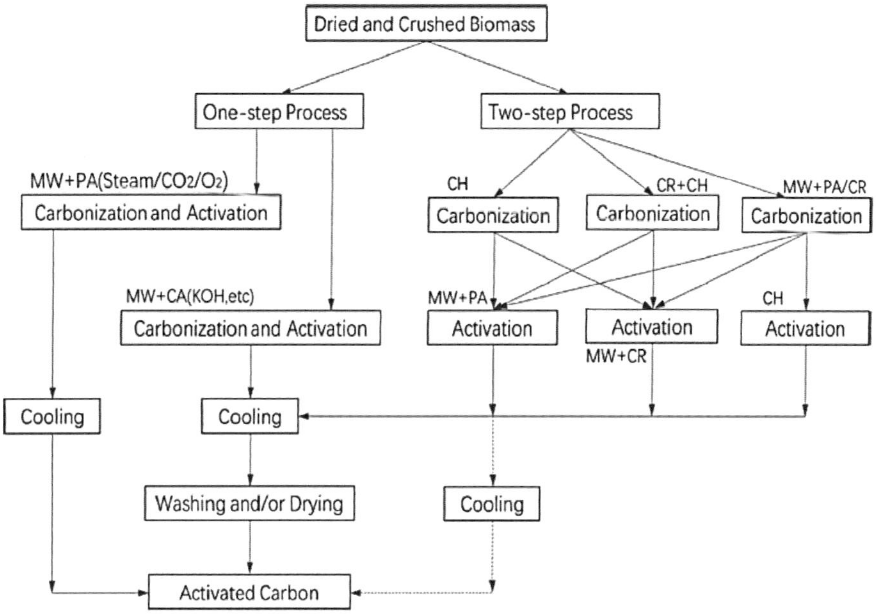

Figure 3. AC preparation using conventional and microwave technique [21].

The charcoal in the present of CO_2 or steam which will act as oxidizing agent at a certain high temperature. Throughout endothermic reaction, structure of porous of the carbon were extracted by these agents. The usage of gas phase entirely in these activation methods formally known as thermal or PA [23]. The example of PA reactions are as follow:

$$C + CO_2 \rightarrow 2CO \tag{1}$$

$$C + H_2O \rightarrow CO + H_2 \tag{2}$$

CA will generate porosity of carbon by mean of the treated of the material base with chemical reagent. It was followed by heat treatment during CP. For lignocellulosic and cellulosic ACs precursor, most common dehydrating agent were phosphoric acid and zinc chloride [24].Single step activation process at low temperature can be accomplish by carrying out thermal decomposition normally using dehydration agents. Among the necessary process for the ACs preparation were base and acid washing at first and last process respectively. Surface area and pore volume could be improved effectively by the base-leaching procedure. Residual activating agent could be cleaned using the acid-washing procedure [25]. Example of the reaction for chemical activation are as follow:

$$C + 2KOH \rightarrow 2K + H_2 + CO_2 \tag{3}$$

$$C + 2KOH \rightarrow 2K + H_2O + CO \tag{4}$$

$$CO_2 + 2KOH \rightarrow K_2CO_3 + H_2O \tag{5}$$

Physiochemical activation (PCA) were the correlation process between PA and CA. These complex processes encourage formation of AC with different chemical characteristics and textural t from single activation method [26]. Pore blockage after washing due to eliminated chemical agent can be solved through PCA process. In order to promote the pore formation, present of steam and CO_2 was important using further gasification. The combination reaction are as follow:

$$CO_2 + 2KOH \rightarrow K_2CO_3 + H_2O \tag{6}$$

$$H_2O + C + 2KOH \rightarrow K_2CO_3 + H_2 \tag{7}$$

3.1. Physical activation (PA)
PA of AC was conducted without using any chemical reaction there were many techniques for these types of activation. Due to low reactivity at high temperature PA make easier, the usage of CO_2 was preferred among researcher. $CO_2$and steam play different role in PA, CO_2 will excite the formation of microporosity whereas, the steam will encourage microporosity widening. Gasifying agent in PA plays a major role in development of the micropores structure depending on partial pressure and activation condition. For high grade of AC produced, both carbon dioxide and steam can be acting as suitable activation agent. [27]. Removing the oxygen during carbonization process at specific temperature by purging the inert gas can also be classified as PA].

3.2. Chemical activation (CA)
CA process were the most recognized among researcher in activating the carbon, this process occurs after the CP and involved the usage of chemical agent. Carbon containing materials usually activated using CA [28]. Process in CA begin with blending the precursor with chemical agent, then the material were heated to desired temperature for carbonization process with inert atmosphere. Dehydration and degradation of AC with the present of chemical agent helps to promote and develop the AC porosity. There were several types of chemical agent that can be used for activating the carbon such as KOH, NaOH, and $ZnCl_2$. Normally, considering the lower activation temperature the CA was preferable compared to PA. BET (Brunauer Emmett Teller) method determine that AC produce using CA exhibit high specific surface area with good pore formation and carbon yield [29]. Among the CA process, using KOH as chemical reagent will produce high micropore as compared to $ZnCl_2$ or H_3PO_4 activation.

3.3. Physical and chemical activation (PCA)
Combination of PCA would produce a better AC. Moreover, the complicated process makes these methods much difficult to produce the AC. However, this process proven to be the best method for producing the AC. It was reported that for methane adsorption, the efficient method was activating the palm oil shell using ZnCl2 followed by CO_2 stream for extra PA [30]. Alternatively, conventional method either PA or CA can be improvised by involving the CA raw material followed by PA process for better result. With using the horizontal furnace sample was activate under CO_2 at 800 °C in range of 1 to 5 hours and finally collected sample were cleaned with ionized water and dried at 100°C for another period of 24h.

4. Conclusion
After review from the various source in, the information of reliable method of activated carbon production has been summarized. The process involved in the production of activated were pre-carbonization process, carbonization process, and activation process. Carbonization could be either using furnace or microwave. There are two main activation process respectively, physical activation (PA) and chemical activation (CA). Other than that there also combination of physical and chemical activation (PCA) which called physiochemical activation.

Acknowledgement
The authors express thanks to Faculty of Engineering Technology, University of Tun Hussein Onn Malaysia and UTHM Center for Graduate Studies (CGS). Also to Green Design and Manufacture Research Group, Center of Excellence Geopolymer and Green Technology (CEGeoGTech), UNiMAP.

5. References

[1] Zhang X, Li Y, Li G and Hu C 2015 *RSC. Adv.* **5** 4984– 4992
[2] Valente Nabais J M, Laginhas C E C, Carrott P J M and Ribeiro Carrott M M L 2011 *Fuel Processing Technology* **92** 234–240
[3] *Activated Carbon usage retrieved from*, http://www.roskill.com/reports/activated, (2008). On 1st May 2020
[4] Sekirifa M L, Hadj-Mahammeda M, Pallier S, Baameura L, Richard D and Al-Dujaili A H 2013 *Journal of Analytical and Applied Pyrolysis* **99** 155–160
[5] Köseog˘lu E and Akmil-Basar C 2015 *Advanced Powder Technology* **26** 811–818
[6] Hunaidah M A, Undu A, Fayanto S, Sulwan K and Setiawan Y 2019 *Journal of Physics: Conference Series.* 1321
[7] Hakan D, I˙lknur D, Belgin K and Fatma T 2011 *Chemical Engineering Research and Design* **89** 206–213
[8] Akshay J, Balasubramanian R and Madapusi P S 2015 *Microporous and Mesoporous Materials* **203** 178–185
[9] Foo K Y and Hameed B H 2012 *Chemical Engineering Journal* **184** 57–65
[10] Koo W K, Gani N A, Shamsuddin M S, Subki N S and Sulaiman M A 2015 *Journal of Tropical Resources and Sustainable Science* **3** 54-60
[11] Denga S, Niea Y, Dua Z, Huanga Q, Menga P, Bin Wanga J H and Yua G 2014 *Journal of Hazardous Materials* **2014 282** 150-157
[12] Subramani T and Revathi P K 2015 *IOSR Journal of Engineering (IOSRJEN)* **5 5** 54-63
[13] Activated Carbon Manufacture: Steam Activation retrieved on 6th May 2020, https://www.acarbons.com/activated-carbon-manufacture-steam-activation/
[14] Wenya A, Jie F, Xiao M, Qinhao K, Chunmei R, Yang L, Hedong Z, Zuopeng G, Jing L, Guangqing L and Jianjun D 2018 *Renewable and Sustainable Energy Reviews* **92** 958–979
[15] Fernandez Y and Menendez J A 2011 *Journal of Analytical and Applied Pyrolysis* **91** 316–322
[16] Shoujie R, Hanwu L, Lu W, Quan B, Shulin C, Joan W, James J and Roger R 2012 *Journal of Analytical and Applied Pyrolysis* **94** 163–169
[17] Chayid M A and Ahmed M J 2015 *Journal of Environmental Chemical Engineering* **3** 1592-1601
[18] Hesas R H, Arami-Niya A, Daud W M A W, Sahu J N 2013a *Chemical Engineering Research and Design* **9** 2447-2456
[19] Hesas R H, Daud W M A W, Sahu J N, Niya A A 2013b *Journal of Analytical and Applied Pyrolysis* **100** 1–11
[20] Puchana-Rosero M J, Adebayo M A, Lima E C, Machado F M, Thue P S, Vaghetti J C P, Umpierres C S and Gutterres M 2016 *Colloids and Surfaces A: Physicochemical and Engineering Aspects* **504** 105-115
[21] dos Reis G S, Wilhelm M, de Almeida Silva T C, Rezwan K, Hoffmann Sampaio C, Lima E C and Guelli Ulson de Souza S M A 2016 *Applied Thermal Engineering* **93** 590–597
[22] Arami-Niya A, Wan Daud W M A and Mjalli F S 2011 *Chemical Engineering Research and Design* **89** 657–664
[23] Rodríguez-Reinoso M 2006 *Activated Carbon* Elsevier Science & Technology Books
[24] Nakagawa M S and Reinoso R 2007 *Microporous Mesoporous Mater.* **103** 29–34
[25] Soltani N, Bahrami A, Pech-Canul M I and González L A 2015 *Chemical Engineering Journal* **264** 899–935

[26] Bashir M J, Ibrahim N, Ismail M N and Jaya M A T *Physical treatment technologies for landfill leachate: performance and limitation, Control and Treatment of Landfill Leachate for Sanitary Waste Disposal* 2015 IGI Global, USA 250-285

[27] Chiung-Fen C, Ching-Yuan C and Wen-Tien T 2000 *Journal of Colloid and Interface Science* **232** 45–49

[28] Yorgun S and Yıldız D 2015 *Journal of the Taiwan Institute of Chemical Engineers* **100** 1–10

[29] Patil B S and Kulkarni K S 2012 *International Journal Advance Engineering Research Study* **1** 109

[30] Arami-Niya A, Wan Daud W M A and Mjalli F S 2010 *Journal of Analytical and Applied Pyrolysis* **89** 197–203

IOP Conf. Series: Materials Science and Engineering **877** (2020) 012031 doi:10.1088/1757-899X/877/1/012031

Electrochemical Behaviour of Ti-Mo Alloys for Medical Application in Biological Solution

M S Bălțatu[1], P Vizureanu[1,2*], I Bălțatu [1], D D Burduhos Nergis[1], D C Achitei[1] and M C Perju[1]
[1]"Gheorghe Asachi" Technical University of Iasi, Faculty of Materials Science and Engineering, 41 "D. Mangeron" Street, 700050, Iasi, Romania
[2]Centre of Excellence Geopolymer & Green Technology School of Materials Engineering, University Malaysia Perlis, Kompleks Pengajian Jejawi 2, 02600

*E-mail: peviz2002@yahoo.com

Abstract. Nowadays, primordial concern represent population health with all materials that are used in medical applications. It were obtained three alloys of titanium with addition of elements like molybdenum, zirconium and tantalum, in a vacuum arc melting furnace (VAR), in argon atmosphere. These alloys were investigated in biological solution, similar with to the human body by electrochemical study method (linear and cyclic polarization). Results revealed that it possesses a good behaviour which does not influence the human body, with future applications in medical field.

1. Introduction

Metallic materials are used of a few thousand years ago for the reconstruction of tissues in the human body but important progress they have achieved in the last century [1-3].

Applications of the biomaterials are found in applications such as orthopedics, cardiovascular surgery, ophthalmology, dentistry, urology, aesthetic surgery, neurology etc.

Titanium-based alloys are commonly found in medical applications for properties close to the tissue that prostheses it like mechanical properties, biocompatibility and corrosion resistance. The most used titanium alloy is the Ti4Al6V alloy, which has remarkable physico-mechanical, chemical and biocompatibility characteristics. However, vanadium being expensive and toxic, began to be replaced with various biocompatible elements: Fe, Si, Ta, Zr, Mo etc., [4-7].

As implant materials are also used precious and semi-precious metals, or alloys thereof, based on silver, platinum, zirconium or niobium, but which currently have too high a cost price and are therefore far from common use in the realization of implants.

Three TiMoZrTa (TMZT) alloys for medical applications have been designed based on the literature. Alloying elements such as Mo, Zr and Ta alloyed with titanium, have very good characteristics, close to human bone. All elements are biocompatible with human tissue, which means that the proposed alloys can be used successfully in medical applications [8-9].

This work highlights corrosion behavior of Ti15Mo7ZrxTa (where x= 5, 10, 15 wt%) alloys. Corrosion monitoring is the practice of qualitative assessment and quantitative measurement of the corrosion of an environment on a metal or alloy immersed in this environment. Monitoring tests can be performed using mechanical, electrical, electrochemical or chemical methods. Corrosion processes depend on the nature of the metal and the corrosive environment, the conditions of pressure and temperature, the static or dynamic conditions of the corrosive environment [10-15].

2. Experimental procedure

Three alloys were prepared by arc remelting method with MRF ABJ 900 equipment. The raw materials used: titanium (99, 8%), molybdenum (99, 8%), zirconium (99, 8%), tantalum (99, 5%), supplied by Alfa Aesar by Thermo Fisher Scientific. After elaboration, alloys were cuted to specific dimensions without being thermally influenced.

The corrosion behavior of three alloys of TiMoZrTa with various compositions was studied. Alloys composition are presented in Table 1.

Table 1. Elemental compositions of TMZT alloys.

Alloy	Ti [wt. %]	Mo [wt. %]	Zr [wt. %]	Ta [wt. %]
Ti15Mo7Zr5Ta	75.46	13.56	6.34	4.64
Ti15Mo7Zr10Ta	73.63	12.33	6.80	7.24
Ti15Mo7Zr15Ta	66.00	12.56	6.99	14.45

The study of corrosion behavior was done by electrochemical methods (Linear Potentiodynamic Polarization and Cyclic Potentiodynamic Polarization) in artificial (simulated) physiological environment.

Was prepared one of simulated physiological solutions, currently used in the literature for the study of alloys corrosion, respectively: Ringer's solution. This solution have the following compositions:

Ringer solution: NaCl –9.00 g/L: $NaHCO_3$ –0.20 g/L; $CaCl_2 \cdot 6H_2O$ – 0.40 g/L, KCl – 0.43g/L: pH = 6.8.

The measurements were performed at 25°C, the solution being naturally aerated, the metal surface being freshly polished. The electrochemical corrosion studies and the characterization of the corroded surfaces were performed with The VoltaLab PGP 201 Potentiostat.

The VoltaLab PGP 201 Potentiostat (VolataLab 21) (Radiometer Analytical SAS - France) is a compact potentiostat / galvanostat that can be operated either manually, when programmed from the instrument panel, or with VoltaMaster 4 acquisition and processing software. When operated manually the maximum potential scanning speed is 2.5 mV / s, and when operated with Voltamaster 4 the maximum scanning speed is 10 mV / s. The device, together with the operating system, allows the evaluation of the polarization resistance, the corrosion potential over long periods of time, the torque potential and tests to highlight the corrosion in points. Under these conditions the device is ideal for electrochemical corrosion studies.

VolataLab 21 was used for: obtaining anodic polarization curves, at low scanning speeds of the electrode potential, curves then used to evaluate the polarization resistance, the instantaneous corrosion current and the instantaneous corrosion rate; obtaining cyclic polarization curves, at scanning potential speeds of 10 mV/s based on which the type of corrosion was assessed and the corrosion rate under load was evaluated (when the alloy is polarized at potentials far from the corrosion potential); obtaining and processing potential dynamic data, both for the evaluation of the corrosion speed and for the analysis of the influence of the sweep speed on the electrochemical behavior.

A cell with three electrodes, horizontal, was used for the potentiodynamic measurements. The surface of samples exposed to corrosion media was in all cases equal to 0.283 cm^2. The samples were polished on SiC paper, degreased and washed with distilled water. A platinum electrode was used as an auxiliary electrode and a saturated calomel electrode as a reference.

3. Results and discussions

To obtain the electrochemical parameters that characterize the corrosion resistance of the developed TMZT alloys, the electrochemical study method by linear (Figure 1) and cyclic polarization (Figure 2) was used. These methods can directly and quantitatively determine the corrosion rate.

Corrosion represents the physical-chemical, spontaneous, irreversible and undesirable destruction of metals and alloys under the chemical, electrochemical or biological action of the environment.

The behavior of TMZT alloys during the electrochemical corrosion test at linear polarization is represented in Figure 1 by Tafel diagrams.

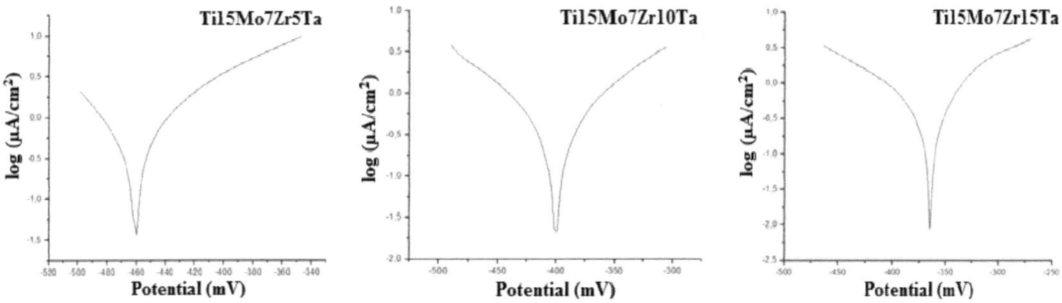

(a) (b) (c)

Figure 1. Tafel diagrams of alloys investigated by corrosion test in Ringer's solution: (a) Ti15Mo7Zr5Ta alloy, b) Ti15Mo7Zr10Ta alloy, c) Ti15Mo7Zr15Ta alloy.

The values of the electrochemical parameters for the TMZT alloys immersed in Ringer's solution are presented in Table 2.

Table 2. The main parameters of the corrosion process for TMZT alloys in Ringer's solution.

Alloy	E_0 [mV]	β_a [mV]	β_c [mV]	R_p [kΩ/cm²]	J_{cor} [µA/cm²]	V_{cor} [µm /year]
Ti15Mo7Zr5Ta	-461.20	117.90	-39.50	22.00	1.08	12.67
Ti15Mo7Zr10Ta	-400.10	92.10	-91.20	46.22	0.37	4.31
Ti15Mo7Zr15Ta	-365.40	88.30	-115.00	38.78	0.47	5.49

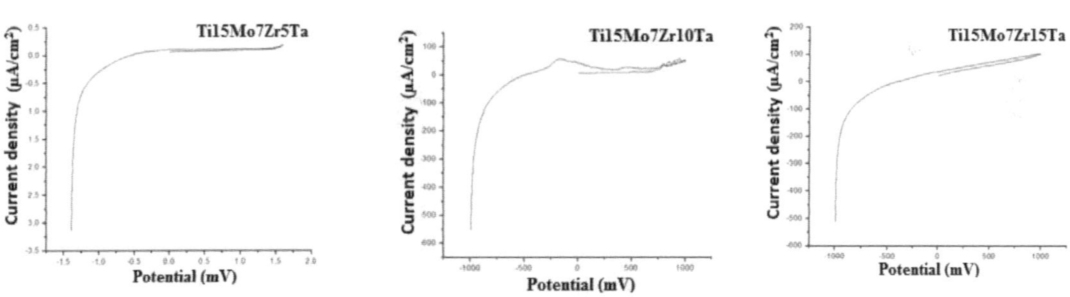

(a) (b) (c)

Figure 2. Cyclical diagrams of alloys investigated by corrosion test in Ringer's solution: (a) Ti15Mo7Zr5Ta alloy, (b) Ti15Mo7Zr10Ta alloy, (c) Ti15Mo7Zr15Ta alloy.

The corrosion potential (E_0) has values close to -400 [mV] for all TMZT samples. The best alloy, which has a much delayed corrosion priming is Ti15Mo7Zr15Ta having E_0 = - 365.4 [mV], which suggests that it has a very high surface passivation chemical stability. The alloy with the most oxidation sensitive corrosion potential is Ti20Mo7Zr5Ta, having E_0 = -461.20 [mV].

Table 1 shows that the samples with very good corrosion resistance are Ti15Mo7Zr10Ta (Rp = 46.22 [kΩ/cm²]).

For all samples the process that takes place on the surface is the oxidation of titanium with the formation of an oxide adhering to the surface, which in fact produces the passivation of the alloy.

Corrosion rates (actually passivation rates) are of the same order of magnitude, relatively small for all alloys.

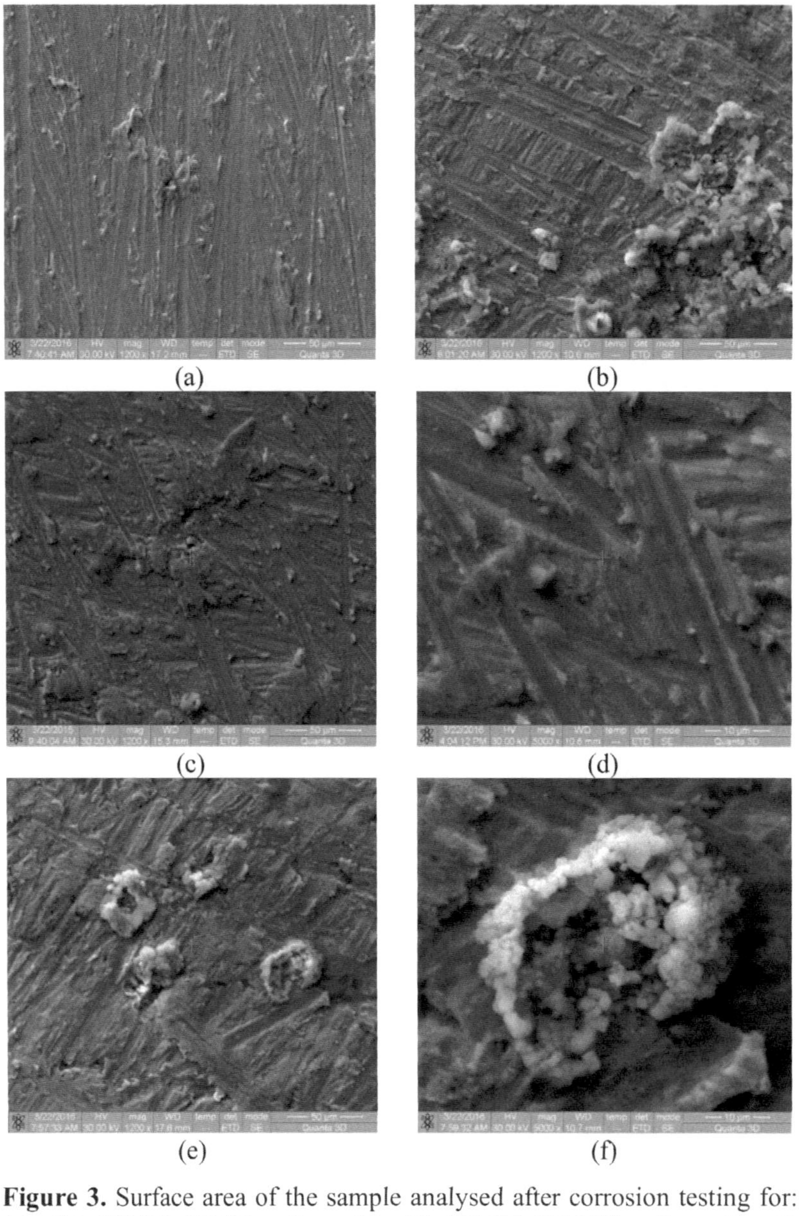

Figure 3. Surface area of the sample analysed after corrosion testing for: (a), (b) Ti15Mo7Zr5Ta alloy; (c), (d) Ti15Mo7Zr10Ta alloy; (e), (f) Ti15Mo7Zr15Ta.

Passivation of these alloys can be influenced by the presence of alloying elements as well as the presence of impurities in the material. Although titanium, in the context of the electrochemical series, is a metal that should corrode strongly, but due to the formation of a oxide layer, it remains passive in the human body, which makes it very useful in medical applications [10-13].

After investigating the alloys by linear and cyclic polarization, it is found that the elaborated TMZT alloys have surfaces slightly affected by corrosion, without traces of pitting corrosion.

SEM micrographs and surface chemical compositions highlight the formation of chemical compounds from the Ringer's solution on the surface of the alloy.

Analyzing the surface of the samples subjected to electrocorrosion in a simulated biological environment, slight unevenness is observed that follow the surface imperfections (non-formalities due to processing and deepened after corrosion), Figure 3.

4. Conclusions

The electrochemical behavior of TMZT alloys was investigated by the electrochemical study method (linear and cyclic polarization), and the selected corrosion medium was the Ringer's solution. The value of tantalum concentrations in the studied alloys is important in the case of corrosion studies. The higher percentage of tantalum show a better corrosion resistance (12.67[µm /yr] for 5%Ta, 4.31[µm /yr] for 10%Ta and 5.49 [µm /yr] for 15%Ta). The aspect of the corroded samples indicate that the passive oxide film formed on TMZT alloys is compact, has a uniform aspect with isolated corrosion compounds, specific feature of titanium alloys that resulting a good corrosion resistance.

Acknowledgements

This work was supported by a grant of the Romanian Ministry of Research and Innovation, CCCDI – UEFISCDI, project number PN-III-P1-1.2-PCCDI-2017-0239 / 60PCCDI 2018, within PNCDI III, with the support of COMPETE project nr.9PFE/2018, financed by the Romanian Government, and under accordance research grant of the TUIASI, project number GnaC2018_48 / 2019.

5. References

[1] Geetha M, Singh A K, Asokamani R and Gogia A K 2009 *Mater. Sci.* **54**(3) 397-425

[2] Minciună M G, Vizureanu P, Geantă V, Voiculescu I, Sandu A V, Achiței D C and Vitalariu A M 2015 *REV. CHIM.* **66**(6) 891-894

[3] Istrate B, Rau J V, Munteanu C, Antoniac I V and Saceleanu V 2020 *Ceramics International* **46**(10) 15897-15906

[4] Bombac D M, Brojan M, Fajfar P, Kosel F and Turk R 2007 *RMZ-M&G.* **54**(4) 471-499

[5] Chen Q and Thouas G A 2015 *Materials Science and Engineering R* **87** 1–57

[6] Niinomi M 1998 *Mater. Sci. Eng., A* **243**(1-2) 231-236

[7] Gordin D M, Gloriant T, Nemtoi G, Chelariu R, Aelenei N, Guillou A and Ansel D 2005 *Materials Letters* **59** 2936 – 2941

[8] Baltatu M S, Tugui C A, Perju M C, Benchea M, Spataru M C, Sandu A V and Vizureanu P 2019 *Revista de Chimie* **70**(4) 1302-1306

[9] Sandu A V, Baltatu M S, Nabialek M, Savin A and Vizureanu P 2019 *Materials* **12**(18) 2973

[10] Bălțatu M S, Vizureanu P, Mareci D, Burtan L C, Chiruță C and Trincă L C 2016 *Materials and Corrosion-Werkstoffe und Korrosion* **67**(12) 1314-1320

[11] Bălțatu M S, Vizureanu P, Cimpoeșu R, Abdullah M M A B and Sandu A V 2016 *Revista de Chimie* **67**(10) 2100-2102

[12] Textor M, Sittig C, Frauchiger V, Tosatti S and Brunette D M 2001 Properties and Biological Significance ofNatural Oxide Films on Titanium and Its Alloys Titanium in *Medicine: Springer; Berlin Heidelberg* 171-230

[13] Gomes C C, Moreira L M, Santos V J, Ramos A S, Lyon J P, Soares C P and Santos F V 2011 *Genetics and molecular biology* **34**(1)116-121

[14] Calin G, Mitrea G, Marcu C, Burlui A, Trusca R, Vasile E and Iacomi F 2019 *Rev. Chim.* **70**(9) 3444-3447

[15] Duceac L D, Stafie L, Banu E A, Paduraru O, Calin G and Ciuhodaru M I 2017 *Rev. Chim.* **68**(11) 2542-2545

New Materials and Processes Developed for Cranioplasty

I M Mates[1], A Semenescu[1], S I Dumitrescu[2] and R Vidu[1,3*]
[1]University POLITEHNICA of Bucharest, Faculty of Materials Science and Engineering, Splaiul Independentei nr.313, Bucharest, Romania
[2]Central University Emergency Military Hospital, Bucharest, Romania
[3]University of California Davis, Department of Electrical and Computer Engineering, Davis, CA 95616, USA

*E-mail: rvidu@ucdavis.edu

Abstract. Traumatic brain injury is the leader in the ranking of mortality and invalidity. The surgical repair of a defect of the skull by cranioplasty has been practiced since ancient times, when materials of non-biological origin were used for this purpose. New materials and processes are sought to improve osseintegration of implants. Like any surgical procedure, cranioplasty involves complications that may be related to the surgical technique and/or to the patient's tolerance to the material used. This work described a biocompatible medical device that include two supported meshes for providing mechanical strength and osseointegration properties of the implant, and a multiplayer porous material in between them that is loaded with the required bioactive antibacterial compound to promote a controlled and sustained release of the pharmaceutical agents at the site of surgical intervention. To increase osseointegration, meshes are designed with an open structure and coated with biocompatible materials such as hydroxyapatite. The composition gradient in the multilayer porous material is attained by loading successive layers of porous material with different amounts of bioactive materials and then stacking them to create a gradient of composition across the porous material.

1. Introduction

Traumatic brain injury is the leader in the ranking of mortality and invalidity. Of the total number of patients with severe head and intracranial injuries, more than 70% will remain with certain physical and mental disabilities. Also, it is recognized that every 11 minutes a child suffers from a head and brain trauma that causes significant disorders in their motor, linguistic and cognitive functioning throughout their life. In the literature, the disturbance of mnesic function is most often associated with intracranial trauma, which affects not only memory, but also other cognitive functions, such as learning.

Intracranial traumas are the most common cause of death of people between the ages of 0 and 44 years old. Currently, about 11.5 million of Europe's population that survived from brain trauma suffers from certain forms of physical disabilities or mental impairment. There are approximately 500,000 - 600,000 cases of head injury in the US annually, of which more than 10% are fatal. While intracranial traumas have various mechanisms of occurrence, the most common causes are car accidents, aggression, trauma in sports and trauma by firearms. The trauma of the skull and brain is the most common type, more serious and in continuous ascension, of injury at childhood. There has been an increased incidence of severe head and brain trauma in children requiring surgical treatment.

2. Cranioplasty and Methods to Fix Implant

The surgical repair of a defect of the skull by cranioplasty has been practiced since ancient times, when materials of non-biological origin were used for this purpose. In the literature, there are data that attest that bone defects closed with silver plate or coconut bark were found in the collection of skulls of the Incas. In the etiological structure of acute brain lesions, brain trauma is in the first place, followed by vascular accidents (ischemic, haemorrhagic strokes), brain tumours, and diffuse post-resuscitation and post-strangulation cerebral hypoxia.

The pathophysiological process secondary to a craniocerebral trauma is the occurrence and the propagation of cerebral edema, which results in increased intracranial pressure, decreased cerebral oxygenation and irreversible ischemic lesions that lead to permanent deficits and, in many cases, to death. Since the early 1970s, decompressive craniectomy is a neurosurgical method of combating refractory cerebral edema in drug therapy, due to the significant reduction of intracranial pressure and flow into the cerebral blood vessels. Although this aggressive way of lowering intracranial pressure has proven to be effective in saving the patient's life, severe deficiencies that may occur later have sparked controversy among neurosurgeons.

Like any surgical procedure, cranioplasty involves complications that may be related to the surgical technique and/or to the patient's tolerance to the plastic material used. It is known that a large variety of surgical techniques, materials and medical devices are used in performing cranioplasty procedures. With the evolution of the manufacturing systems, novel medical devices used in the reconstruction of the defects of the bone system have emerged in terms of the types of materials used and of the constructive forms, so that they are as close as possible to the anatomy of the patient.

Well-fitting prosthetic implants are required in large cranial defects to protect the patient's brain from trauma or infection, while it performs an aesthetic mission to restore the shape to the patient's head. Malleable and biocompatible materials are used to make an implant when not enough cranial bone is available for grafting. Dean et al. [1] presents a computer aided design method for producing an implant for a patient prior to operation using a non-invasive 3D (3-dimensional) scan of the patient's defect site that digitally represents the area that will receive the implant; designing and validating an implant on a computer based on digital data generated from a volume image of the patient; and fabricating the implant based solely on the implant design data generated on computer. Due to the advantages, the manufacture of personalized prostheses is currently considered the one that brings the most benefits in terms of compliance with the patient's needs. Also, a personalized prosthesis significantly reduces the duration of surgery and, implicitly, postoperative complications. However, an intracranial prosthesis that is custom made implies higher production costs compared to the standardized devices available on the market. However, drilling holes in skull is the main method used in cranioplasty to secure the cranial implant in place.

3. Surface Modification of Implants

Besides searching for methods to secure in place the cranial implant, researchers have sought different ways to improve surface properties of implants to promote adherence, osseointegration and reduce inflammation. Pacifico et al. [2] presents a composition as bone haemostat, bone adhesive, bone void filler, or bone cement. The surface of the plate has to be modified to ensure adherence. To obtain nanostructured surfaces, the prosthesis is modified by mechanical methods using abrasive materials, chemical methods using acid etching and electrochemical processes, physical methods or a combination of these methods. There are also other methods that involve different coatings on the surface of the implants. These coatings include hydroxyapatite, calcium phosphate biomimetic layers, biomolecule layers, but also coatings that have a combined synergistic effect. The main purpose of all these surface modification treatments is to improve the bioactivity, biocompatibility, wear resistance and corrosion of titanium and titanium alloys for their particular applications in medicine.

Surface modification of medical devices, such as cranial implants made of titanium mesh, can be applied to help induce the process of osseointegration. There are many clinical cases where the osseointegration process of Ti and its alloys is not enough but an osseo-inductive behaviour of the implant surface is required, especially when rapid healing or bone quality and/or bone quantity is mandatory. One of the main objectives of the current research on biomaterials is the complex system of biological inducements and surface responses, including interface processes [3]. Antibacterial surfaces capable of avoiding biofilm formation are extremely important for implants that are in direct contact with bone tissues. It was found that fewer macrophages and inflammatory reactions are reported on Ti and its alloy surfaces compared to stainless steel or polyether ketone surfaces. Titanium is well tolerated by the body, as long as the prosthesis is in perfect condition, mechanically stable and uninfected. If these conditions are not met, the prostheses may be associated with an acute or chronic inflammatory reaction, osteolysis, weakening and failure of the prosthesis. Therefore, the implant should have both osseointegration and antibacterial properties.

Human osteoclasts can corrode stainless steel, cobalt and Ti alloys, leading to the production of metal ions, which are toxic for the body and responsible for inflammatory reactions. The healing reaction in the case of a prosthesis made of Ti can occur by osseointegration, by fibrous encapsulation or by chronic inflammation. The last two reactions indicate the failure of the procedure. The events related to the inflammatory response of the body to the prosthesis is complex and include exudation, adsorption of the protein surface, development of a provisional blood-based matrix, recruitment of cells of the innate immune system (leukocytes, platelets), neutrophil migration, monocyte substitution and macrophage differentiation, foreign body reaction, production of reactive oxygen species, fusion of mono-cytes/macrophages to form giant cells or apoptosis.

Different strategies to obtain bioactive and antibacterial titanium surfaces have been also investigated. The inorganic antibacterial agents that have been considered are mainly metal ions and nanoparticles, as well as their oxides (e.g. Ag, Cu, Zn and Ce). Among organic agents, mainly antibiotics have been considered. The strategy to reduce the resistance to infections involves the modification of the biomaterial, that is, the modification of the design of the polymeric material following two main directions: 1) modification of the surface to minimize the adhesion of the bacteria; 2) incorporation of antimicrobial species with the role of destroying bacteria that contaminate the vicinity of the prosthesis (e.g. incorporation of antibiotics, silver, silver compounds etc).

The advantages of inorganic antibacterial agents are the broad spectrum of activities that allow the treatment of polymicrobial infections, the prevention of contamination with unknown bacteria and the development of resistance. Therefore, the problem of resistant bacterial strains is currently one of the most important problems regarding the use of antibiotics. However, the main disadvantages are their applications, which involve difficulties that come both from the field of bureaucracy, but especially with regard to finding the optimal therapeutic window, which will allow an effective antibacterial behaviour, without cytotoxic effects.

The placement, fixation and fastening the implants to the bone structure are generally done by suturing or screwing the implant to the skull via holes. Drilling holes is the main method used in cranioplasty to repair a defect or deformity of the skull. Implant migration and micro-movements may hinder osseointegration and even result in rejection of the implant. Bone screws, tacks and the like are destructive intrusions into the bone structure and, once applied, exert (lateral) compressive forces on the bone structure. Compressive forces in turn may cause resorption of bone and loosening of the fixing means. Micro-movements due to play between different parts of the implant or the fixing means or between the fixing means and the bone structure and/or implant may increase damage and loosening and should be prevented. Further, the implant and its fixation means form an invasive volume inside the patient's body. Such volume should be minimal, in particular for a cranial implant to prevent damage to the brain.

4. Novel Biocompatible Implant

We present a novel biocompatible implant that include two supported meshes for providing mechanical strength and osseointegration properties of the implant, and a multiplayer porous material in between them that is loaded with the required bioactive antibacterial compound to promote a controlled and sustained release of the pharmaceutical agents at the site of surgical intervention.

4.1. Porous Material with Gradient porosity

The porous multilayer is presented in Figure 1 and consists of successive layers of various porosities. The layers are arranged either in ascending or descending order of porosity. The biocompatible implant integrates the multilayer porous material in between two structural meshes (Figure 2). The thickness of the multi-layered assembly is equal or less the thickness of the skull.

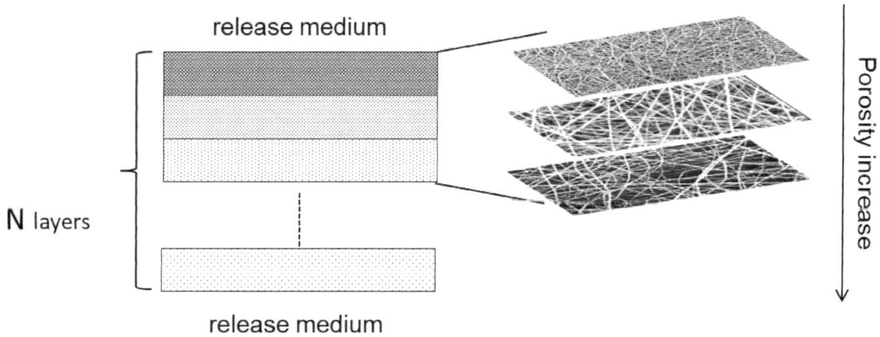

Figure 1. Illustration of the porous multilayer with simple gradient composition made of nanofibers.

The porous material includes a multitude of porous membranes that are formulated with the highest amount of bioactive material in the most porosity layer. Successive layers of polymeric nanofibers can be obtained by electrospinning. By varying the electrospinning variables conditions, various nanofibers sizes and porosities of the resulted membrane can be obtained. Nanofibers layer are obtained with progressive increased porosity.

The drug diffusion is the mechanism by which the drug is released in the body. In practice, blends consisting of a drug and polymer or blends of polymers are used. The morphology of the material and the molecular size of the drug are only a few factors that influence the diffusion of the drug into the body. The rate of releasing of a drug shows that a large amount of drug is discharged first, which indicates that the transport is nonlinear with a plateau. For the same drug, the rate of releasing is slowed down due to the resistance of polymer film [4]. The stack of ultrathin membranes of various porosities can be arranged in a simple gradient of porosity or in a double gradient of porosity as shown in Figures 1 and 2, respectively. In Figure 2, an illustration of the biocompatible implant having, in between 2 meshes, a porous biomaterial comprising of successive layers of various porosities and stacked up from high to low porosity from the inside toward outside of the material.

Each membrane of a specific porosity is loaded with bioactive agents, the amount of the drug being related to the amount of the drug retained by the membrane. Stacking up the membranes in a pre-set order controls the release the drug in a desired and precise fashion. The controlled release refers to the release of the pharmaceutic agents from the inside of a medical device to surface at a predetermined rate. The bioactive compound does not come off in an unpredictable fashion in a controlled release system, i.e. it does not discharge erratically in contact with a biological environment unless specifically intended to do so. Reducing the thickness of the layers, a continuous change in porosity can be obtained.

In certain applications, an initial burst of drug may be desirable followed by a more gradual release thereafter. The release rate may be steady state (commonly referred to as "timed release" or zero-order kinetics), that is the drug is released in even amounts over a predetermined time (with or without an initial burst phase) or may be a gradient release. A gradient release implies that the concentration of drug released from the device surface changes over time. The difference in porosity between adjacent layers can be used to control the release. In early stages of diffusion, the quantity of the bioactive compound delivered can be controlled by its distribution across the layers. Increasing the number of the layers with contrasting properties, the release profile can be engineered to better control the drug release. The effects of the relative thickness of the layers and the relative resistance to diffusion offered by each layer's composition on the drug release profile are critical. The dissimilarity in porosities of two adjacent layers can be used to control the release, while the quantity of drug delivered can be modified by varying the distribution of drug across the layers. Therefore, both microstructural and loading variances between multi-layers with porosity gradient can be used to tune the properties of the coating materials to obtain the desired drug release profile for a given application.

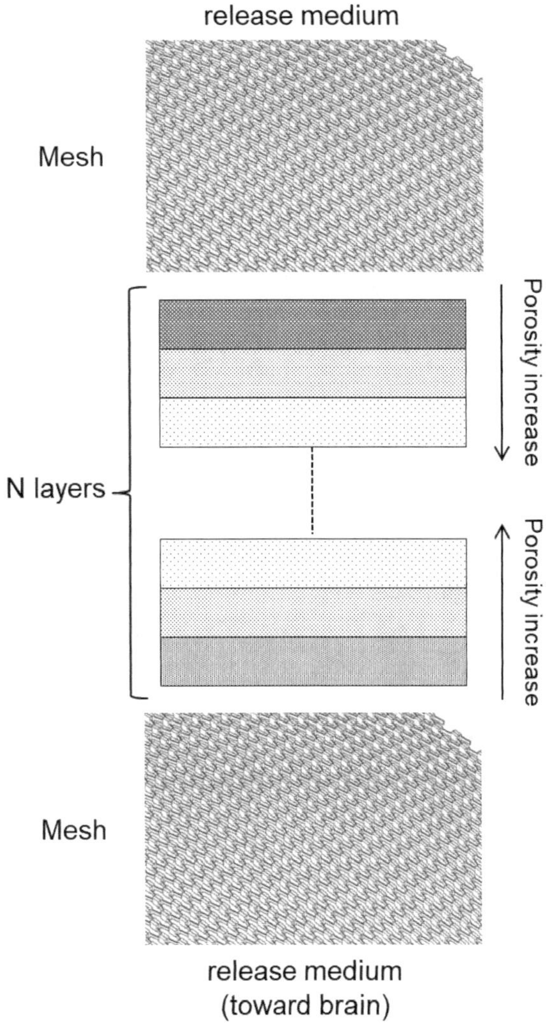

Figure 2. Illustration of the biocompatible implant showing a porous biomaterial of successive layers of various porosities and dual gradient composition in between 2 meshes.

Nanofibers used in the porous multilayer stack can be obtained by electrospinning, which is the method of choice to produce fibers. Electrospinning uses electric force to draw charged threads of polymer solutions or polymer melts up to fiber diameters in the order of nanometers. The process does not require coagulation or high temperatures to produce solid threads from solution. This makes the process particularly suited to the production of fibers using large and complex molecules. Electrospinning ensures that no solvent can be carried over into the final product. Depending on the size of the collector, large areas of membranes can be obtained. Electrospinning have been used for medical purposes. The electrospun scaffolds made for tissue engineering applications can be penetrated with cells to treat or replace biological targets. Nanofibrous wound dressings have excellent capability to isolate the wound from microbial infections. Other medical textile materials such as sutures are also attainable via electrospinning. Through the addition of a drug substance into the electrospinning solution or melt diverse fibrous drug delivery systems and transdermal patches can be prepared.

4.2. Mesh Structure and Function

To increase osseointegration, meshes are designed with an open structure and coated with biocompatible materials such as hydroxyapatite, titanium materials and compounds, synthetic resins, bone cements etc. Surface modification of medical devices, such as cranial implants made of titanium mesh, can be applied to help induce the process of osseointegration. There are many clinical cases where the osseointegration process of Ti and its alloys is not enough but an osseo-inductive behaviour of the implant surface is required, especially when rapid healing or bone quality and/or bone quantity is mandatory. Several strategies have been investigated; some of them are based on the bioactivity approach (i.e. hydroxyapatite precipitation induced “in vivo”) and other are based on osteoblast stimulation through surface roughness. For Ti and its alloys, the bioactive behaviour can be obtained by applying a layer of foreign material such as apatite or a bioactive glass using electrochemical processes such as anodic oxidation, or surface chemical treatments in acids or oxidative media. A classification of bioactive surfaces can be done according to the mechanism of bioactivity, which can be related to ion exchange process with the body fluids and/or with the effects of the topography of the surface of the prosthesis at micro and nano scale. Currently, the clinical demand has been oriented towards multifunctional surfaces, which are capable of simultaneously providing a specific response due to both colonization by different bone cells such as osteoblasts, fibroblasts, macrophages, and colonization with infectious agents such as bacteria and viruses.

Polymethyl methacrylate (PMMA) can be reinforced with titanium mesh to increase its mechanical strength but also to increase the osteo-conductive properties of the titanium mesh. PMMA is a synthetic resin produced from the polymerization of methyl methacrylate, was first used in medicine in the 1960s and is a polymerized organic combination of acrylic acid, transparent and easy to shape and. During the polymerization, the material solidifies in a short period of time, making it suitable for use in surgical procedures. The viscosity of the mixture increases rapidly, therefore the surgeon must model the cranial prosthesis in a very short time, which is why PMMA is suitable only for modeling small-sized cranial prostheses. In medical applications, however, we must take into account the chemical toxicity of the methyl methacrylate monomer but also the exothermic polymerization reaction to obtain PMMA. After polymerization, the low elasticity, but also the low resistance to mechanical stresses, makes PMMA susceptible to fracture under high mechanical pressure, which results in the production of particles by rubbing PMMA with a hard material surface. Thus, PMMA particles can provoke a cellular inflammatory response that can lead to osteolysis. This phenomenon is known as “cement disease”. In order to improve the X-ray or CT examinations, zirconium dioxide can be added to PMMA. The most important advantages of PMMA are: transparency, ease of modeling, mechanical properties and low price. PMMA is well tolerated by the body so that immediately after implantation, fibrous tissue will develop at the interface between the material and surrounding tissue, and the osseointegration process will begin with a host reaction. The body's immune response is expressed both locally and systemically by activating macrophages.

In neurosurgery, PMMA is used in stabilizations and replacements of the vertebrae, but also in cranioplasty, either in the form of cement or in the form of solid pre-molded prosthesis. Thus, for small defects in the vertebrae or in the skull, the cement is prepared and molded intraoperatively, then fixed by the surgeon. For large skull defects, the preformed solid prosthesis should be very well sized and designed based on the patient's imaging results.

In addition, the use of PMMA cement is limited by the increase in temperature during polymerization, the toxicity of the liquid monomer and the reduction of vascularization at the interface between the material and surrounding tissue after implantation. In spine surgery, these factors can promote bone resorption, followed by failure of the procedure, while in cranial neurosurgery, increased temperature can affect surrounding tissues, causing even thermal necrosis of the healthy bone, thus affecting the nervous system. When PMMA is used in cranioplasty, the cranial prosthesis can be obtained by different methods such as modeled by hand during surgery, prefabricated or obtained intraoperatively using modern design methods. In the first case, the modeling of the cranial prosthesis by hand by the surgeon, intraoperatively, is a method used only in the case of very small size defects with relatively regular margins. In this case, no special tools are needed for cranial reconstruction. The surgeon models the prosthesis, then, after it is hardened, fixes it inside the cranial defect. The second situation is specific for larger head defects. In this situation, the prosthesis is performed before surgery, based on imaging investigations. Intraoperatively, the prefabricated prosthesis can be adjusted by the surgeon and then fixed by the patient's skull. The third situation involves the use of modern computer-aided design techniques, but also the use of a 3D printer. The prosthesis thus obtained is used as a mold for the prosthesis that the surgeon will model intraoperatively. After the material hardens and the prosthesis is to be implanted, holes are made at its periphery. The prosthesis is then placed inside the intracranial defect, and then fixed by the skull through suture threads. This technique is used especially for large skull defects with irregular edges. The prosthesis made in this way, by using the modern means of design and printing, will perfectly cover the cranial defect, the aesthetic result being clearly superior to the other two mentioned methods. The mesh made of titanium has been used as a support for PMMA placement, thus reducing the fracture potential of PMMA, especially in large-scale reconstructions of the skull cap. In addition, the base network may allow the intracranial prosthesis to be embellished when using PMMA.

Hydroxyapatite (HA) can also be used to impregnate the titanium mesh, in order to increase the osteoconductive properties of the titanium mesh. Hydroxyapatite is a calcium phosphate stable in aqueous media, with the chemical formula Ca10(PO4)6(OH)2. HA contains approximately 40% calcium and 18.5% phosphorus (as a percentage by mass). The Ca/P ratio of hydroxyapatite is 1.667, which is an important indicator used to evaluate different processes for obtaining this material. It is known that about 70% of human bone is made from a non-stoichiometric form of hydroxyapatite, so that this material can be considered as an ideal bone replacement. Bone apatite contains the following elements: magnesium 0.7% wt, sodium 0.9% w; potassium 0.03% wt, chlorine 0.13% wt, fluorine 0.03% wt, trace elements: Sr^{2+}, Pb^{2+}, Zn^{2+}, Cu^{2+}, Fe^{2+}.

Hydroxyapatite has no cytotoxic or carcinogenic effect on the human body. It is characterized by high calcium content and high biocompatibility with regard to soft and hard tissues. For this reason, intracranial prostheses made of this type of ceramic material, can be used in direct contact with the bone. Also, the structure of synthetic hydroxyapatite may be similar to that of natural bones. By different manufacturing methods the size and number of pores can be controlled in the case of porous HA. The porous surface of the intracranial prosthesis can be covered by the new bone tissue that forms at the interface, which allows for a strong and lasting connection between the bone and the prosthesis. Currently, various forms of porous hydroxyapatite are available that are used either to repair bone defects or to support bone tissue regeneration in the human body. The solubility of HA is also a very important parameter, as these materials are designed to fill cavities or to be embedded in the surrounding bone tissue. This solubility depends on several factors such as i) the pH and solvent type (hydroxyapatite is insoluble in bases, but soluble in acids), ii) in potassium, sodium, magnesium and strontium salt solutions it dissolves better than in distilled water according to the following order Sr> Ba> Mg> Na> K, iii) the presence of amino acids, proteins, enzymes and other organic compounds. The solubility of HA under "in vivo" conditions is highly dependent on the degree of crystallinity, the size of the crystal, the amount of crystalline defects, stress level and porosity. Intracranial prostheses made from porous hydroxyapatite dissolve faster than densely structured prostheses due to the larger contact surface. Hydroxyapatite is manufactured and used in medicine both in dense and porous forms, as well as in granules and powders. The production process consists of several stages including obtaining powders, making prostheses, compaction and sintering, final treatment (sharpening the sharp edges), sterilization and packaging. In the process of manufacturing intracranial prostheses, antibiotics, growth factors and hormones or other cell types (drug delivery, cell cultures) may be introduced.

However, the HA geometry, the porosity and the substitution network remain the most important features involved in the healing process of bone defects. These properties have been tested on animals. In order to evaluate the biological properties of the non-stoichiometric form of hydroxyapatite, cell cultures were evaluated in terms of their chemical composition. Following the evaluation it was observed that hydroxyapatite with high carbonate content increased the activity of the osteoclasts, which suggests that bone resorption (the phenomenon underlined by the activity of the osteoclasts) is directly influenced by the functional groups incorporated in the crystalline hydroxyapatite network. Another conclusion relates to the substitutions of fluoride ions, which stimulate cell proliferation. However, there are some major disadvantages of hydroxyapatite, represented by its fragility that is specific ceramics, by the reduced resistance to stretching and by the high risk of infection after surgery. Larger bone defects can be difficult to repair with hydroxyapatite, due to the reduced osseointegration process and structural changes that occur in contact with the cerebrospinal fluid.

Titanium mesh can also be impregnated with calcium phosphate. Calcium phosphate is the main component of human bone and has been used since the 1890s to stimulate bone regeneration. However, positive results were obtained only in 1920, when Albee discovered that tricalcium phosphate stimulates bone formation. In the 70's, it was discovered that the bioglass, i.e. glass containing calcium phosphate and hydroxyapatite ceramics, are osteoconductive. From that moment, calcium phosphate was widely used in medicine, in the research of novel medical prostheses and coatings.

The most used calcium phosphate ceramics include hydroxyapatite, tricalcium phosphate (TCP) and mixtures thereof (BCP - dicalcium phosphate). The dissolution behavior is the only important thing that differentiates them and implicitly the bioresorption rates. HA ceramics obtained from coral or synthetic apatite powder dissolve extremely slowly, compared to TCP ceramics which dissolve much faster. Although the TCP resorption rate is influenced to the same extent by both macrostructure and microstructure, TCP ceramics is considered a bioresorbable ceramic. These ceramics are characterized by a low mechanical strength, which is why their use in making prostheses is limited. However, they can be used to obtain cranial prostheses, but of small size. They are widely used as coatings for prostheses used in orthopedics but also for titanium nets used in cranioplasty. The metal matrix has the role of conferring the mechanical resistance necessary for the prosthesis, and the calcium phosphate comes with the osteoconductive properties that favor the process of osseointegration. HA, calcium phosphate cranioplasty prostheses are considered to have important regeneration and reintegration properties. For instance, Engstrand *et al.* [5] has shown that an intracranial prosthesis made of calcium phosphate was manufactured by a casting technique, sterilized and then surgically implanted. After approximately 50 months after surgery, a revision of this prosthesis showed that the prosthesis was integrated and a vascularization network was developed. Also, the histological examination revealed a compact bone that was in direct contact with the remains of inert ceramic materials.

In another embodiment, natural polymeric film encapsulating propolis nano-formulation for cutaneous wound healing can be used. Cavalu *et al.* [6, 7] produced collagen films containing propolis encapsulated in chitosan nanoparticles, for biomedical applications such as cutaneous wound healing. The vibrational marker bands of propolis were well preserved in the final polymeric mixture, indicating the stability of bioactive compounds upon the encapsulation procedure. The antibacterial effect depends on the nanoparticles concentration in collagen film, the effect being more evident with respect to E. coli than S. aureus, while the antioxidant capacity indicated a synergic effect of chitosan nanoparticles matrix and propolis extract, incorporated in collagen films.

4.3. Clamping System

Part of the biocompatible implant is the clamping device that helps place, fix and fasten the implant to the bone structure. More particularly, the clamping system provides a fast attachment of the biocompatible implant to the bone structure and a rapid placement of the three layers of porous material with different properties: i.e. the two meshes that promote osseo-integration of the implant and one multilayer porous material loaded with bioactive compounds having a gradient composition to control the release of the drugs incorporated in the porous material.

The clamping system that allows fixing the biocompatible implant to the defect is presented in Figure 3. The clamps further allows fixation such that the implant is placed and remains in position when the holding force between the fastener and the implant is reduced. The implant may not be custom made pre-operatively; instead the meshes and the multilayered porous material may be cut on the spot to fit the anatomic shape of the defect. The clamp-like mounting piece is pre-assembled at the extended position in a fast and secure way to the bone, while the actual assemble of the successive layers is hold in place by the fastening mechanism. The main function of this system is to fix the medical device to the bone structure, and to support and connect the meshes used in the implant for remodelling and restoring bone tissue such as cranial implants. Mashes may have a flat or a convex/concave surface to fit the original shape of the skull. As illustrated in Figure 3, the clamping system consists of three pieces: the lower piece, the upper piece and the middle piece, as follows:

- **Lower piece** has a cylindrical part that sits on a support. The cylindrical part has a notch that corresponds to the exterior groove cut into the cylindrical part of the middle piece. The support of the middle piece has a knock on which the first mesh is placed.
- **Middle piece** helps adjust the height of the clamp to fit the thickness of the skull and supports the upper mesh. The middle piece has an external groove on the cylinder that helps adjust the height of the clamp. The bar that has the role to fasten the clamp to the bone structure.
- **Upper piece** fastens together the clamp to the bone structure and to the biocompatible medical device. The upper part has a lever and a knob that fix in place the clamp system. The upper lever comes over the upper mesh when the level is rotated to a closed position.

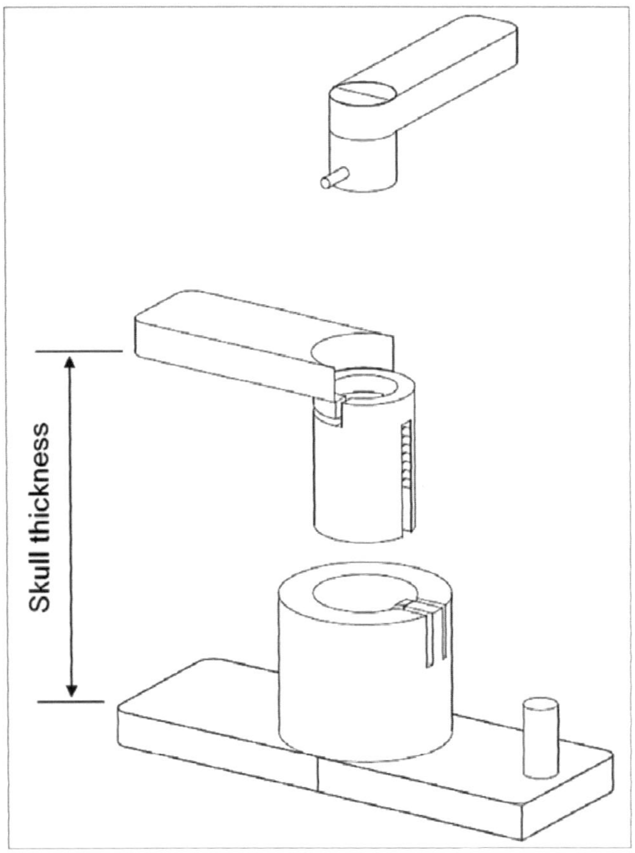

Figure 3. Illustration of the clamping system that can be used in cranioplasty containing 3 pieces that allow for placing and immobilizing the biocompatible implant.

4.4. Assembling the device

The attachment of the implant to the skull includes attaching the clamping system to the edge of the cranial orifice at different points, creating a base for attaching the lower mesh. First, the lower and the middle pieces are pre-assembled at the maximum extension length, and then mounted at the edge of the skull and then tightened up until it is secured at the edge of the skull. This is possible because both parts have a gear rack-like system that allows the two parts to be tightened up but do not allow them to move so that the assembly remains fixedly attached to the edge of the skull orifice. Then the first mesh is placed and hooked to the knock. The lower piece has a hook on the side of the support that is oriented away from the skull, which has dual roles, i.e. to fasten the clamp to the skull and to attach the first mesh. After the lower mesh is placed, the multilayered porous material can be placed on the first mesh. This layer has anti-microbial properties with slow release of the drug. Then the top mesh is placed and the upper piece is inserted and rotated to the ideal position for gripping and immobilizing the mesh.

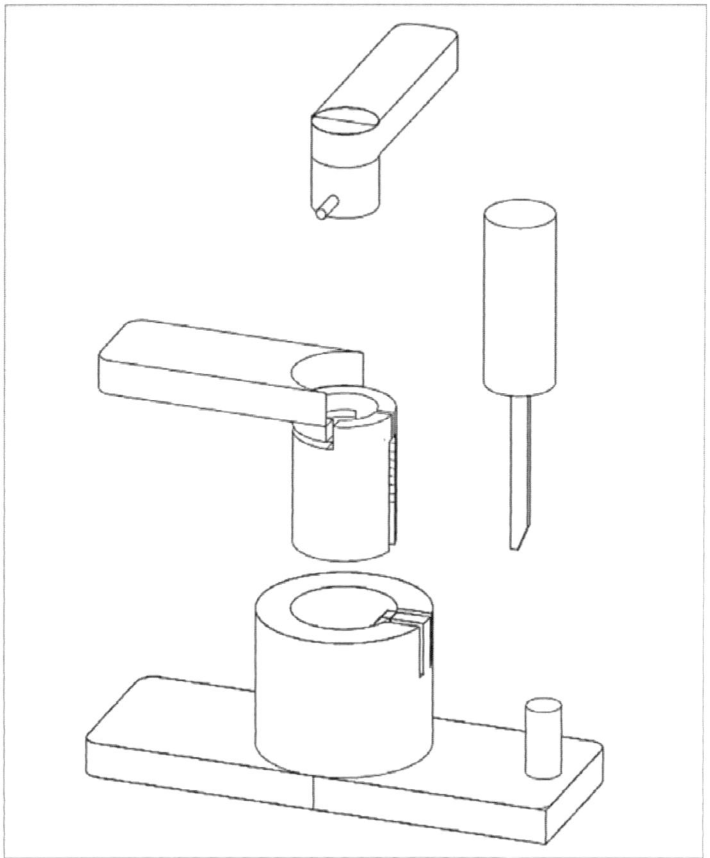

Figure 4. Illustration of the clamping system that fixes an implant in a defect in the skull containing 3 pieces that allow for placing, immobilizing, and rearrangement of the implant.

In Figure 3, the attachment of the implant is irreversible, i.e. the implant is fixed in place Figure 4 present a similar configuration for the clamping system presented in Figure 3, but the implant is reversible such that the implant may be removed if complications, such as bleeding, occur. The additional key permits to lock and unlock the clamping system.

5. Conclusions

We presented a novel biocompatible medical device that includes both an implant and a clamping system. The device presents the following advantages: i) it decreases the surgical craniotomy time by simply pressing the clips and insert the plates; ii) the two meshes doesn't have to be costumed made;

iii) the surgical procedure is more efficient due to the easiness of mounting operations including attaching the clips, laying the bottom mesh, laying the intermediate material, laying the upper mesh and fixing the upper mesh; the clamping system has a minimum height profile that helps to reduce the protrusions; iv) the clamping system does not involve any invasive procedure for clamping the implant, such as drilling and screw clamps; v) easy placement of the gradient loaded multi-layered porous material in between the meshes; vi) a good contact interface of the upper mesh at the surface of the skull.

Acknowledgments

The work was developed under the project "Work-based learning systems through entrepreneurial scholarships for doctoral and post-doctoral students (SIMBA) - MySMIS 124705" within Priority Axis 6 - Education and skills - POCU 2014-2020.

6. References

[1] Dean D, Min K-J and Ratcheson R A 2010 Computer-aided-design of skeletal implants, Case Western Reserve University (Cleveland, OH), US Patent 7,747,305 B2
[2] Pacifico J, Darr A and Kronenthal R L 2017 Settable surgical implants and their packaging, US Patent 9,827,349
[3] Cavalu S, Antoniac I V, Fritea L, Mates I M, Milea C, Laslo V, Vicas S and Mohan A 2018 *Journal of Adhesion Science and Technology* **32(22)** 2509-2522
[4] McGinty S, King D and Pontrelli G 2017 *Med Eng Phys* **45** 51-60
[5] Engstrand T, Kihlstrom L, Neovius E, Skogh A C D, Lundgren T K, Jacobsson H, Bohlin J, Aberg J and Engqvist H 2014 *Journal of Neurosurgery* **120(1)** 273-277
[6] Cavalu S, Bisboaca S, Mates I M, Pasca P M, Laslo V, Costea T, Fritea L and Vicas S 2018 *Revista de Chimie* **69**(12) 3756-3760
[7] Cavalu S, Pasca P M and Brocks M 2019 *Materiale Plastice* **56(3)** 479-483

IOP Conf. Series: Materials Science and Engineering **877** (2020) 012033 doi:10.1088/1757-899X/877/1/012033

A Review Regarding the Use of Natural and Industrial by-Products in the Production of Geopolymer Binders

B A Ionescu[1*] and A Lăzărescu[1]
[1]NIRD URBAN-INCERC Cluj-Napoca Branch, Romania

*E-mail: bradut.ionescu@incerc-cluj.ro

Abstract. Rapid growth in population has led to the generation of large amounts of waste and environmental degradation. Environmental protection for future generations and to preservation of the already limited mineral resources, could be obtained with sustainable solutions for traditional Portland Cement Concrete. In order to produce geopolymers binders, raw materials rich in aluminum and silicon are alkali-activated. Some of the raw materials, for the production of the geopolymer binders can be obtained from natural sources, such as kaolin, metakaolin (calcined kaolin), diatomite, volcanic rock, etc., or industrial waste arising from production processes, such as fly ash, iron slag, blast furnace, granulated blast furnace slag, silica fume, marble dust, etc. The aim of this paper is to present relevant data in the field of alkali-activated geopolymer materials and study the opportunities of using Romanian mineral local raw materials in order to produce these types of binders.

1. Introduction

The problems facing mankind today are related to global warming, caused by the accumulation in the atmosphere of greenhouse gases. Changes in the natural characteristics of the atmosphere, as a result of the actions of physical, chemical, biological agents that manifest a harmful effect on the environment, lead to air pollution. The main pollutants are: carbon monoxide (CO), carbon dioxide (CO_2), sulphur dioxide (SO_2), nitrogen dioxide (NO_2), particulate matter (PM), volatile organic compounds (VOC), benzene and heavy metals. The major sources of CO_2 production, one of the main pollutants, are the production of thermo-electric energy, manufacturing and metallurgical industries.

The construction industry presents the fastest development in the world. According to the statics carried out in recent years, approximately 260,000,000 tonnes of cement are needed annually, in order to satisfy the needs for development [1]. The cement industry is responsible for producing large amounts of CO_2 emissions because, for producing one tonne of Portland cement, about one tonne of CO_2 is released into the atmosphere [2] resulting in high pollution of the environment.

Rapid population growth and urbanisation in developing countries have led to large amounts of solid waste and increased environmental degradation. Around 90-95% of all waste in the world is dumped on open land or warehouses, creating considerable environmental problems. Also, given that the natural reserves are limited, alternative methods of using these types of wastes should be found, by keeping it in optimal parameters and provide the opportunity for future generations to enjoy a clean air and nature.

Managament of construction waste, recycling and re-using are key strategies adopted in order to protect natural resources and to prevent environmental pollution, by using waste materials or industrial by-products, and can also contribute to the economy having several other advantages [3, 4]:

- Reducing demand for new resources;
- Reducing energy costs for transport and production;
- Use of waste that would otherwise be lost in landfills;
- Sustainable urban development by conservation of land areas;
- Improving environmental conditions.

An alternative to traditional Portland cement concrete could be offered by Geopolymer concrete, which does not use Portland cement, and could provide significant environmental benefits. Geopolymers are a subclass of alkali-activated materials. The usual geopolymer materials mixture includes fly ash, alkaline activators, mineral additives and aggregates (Figure 1).

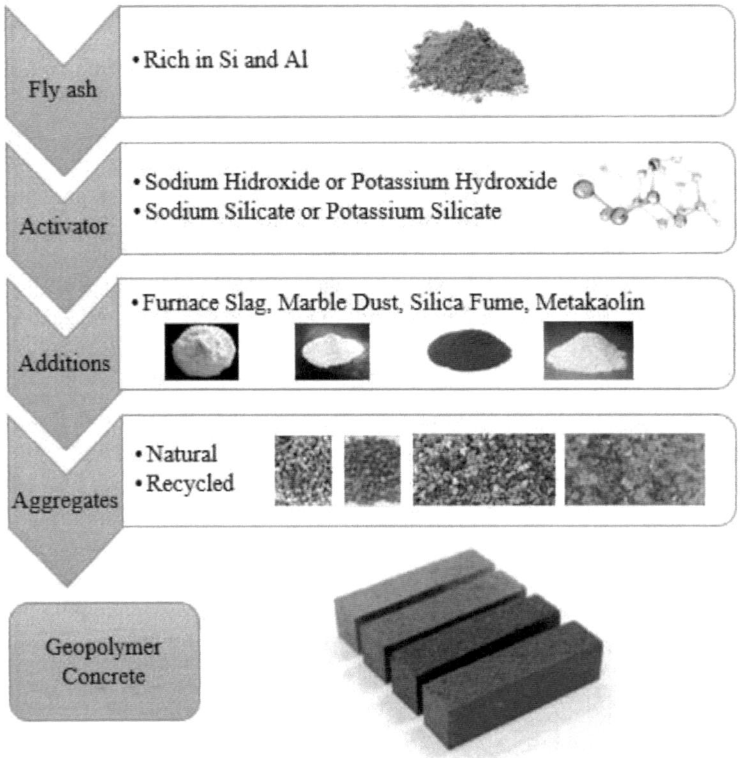

Figure 1. Schematic representation of a compositional Geopolymer Concrete.

The term geopolymer was originally introduced by Davidovits, in 1991 [5], representing a wide range of inorganic materials. According to him, among the nine different classes of geopolymers, Geopolymer concrete is of particular interest, consisting of aluminosilicate materials that could be used to completely replace traditional Portland cement concrete [6].

This class of geopolymers is also called alkali-activated cement or inorganic polymer cement, which is produced by dissolving thermally activated natural materials such as kaolinous clay, fly ash or slag in an alkaline activation solution to provide a source of silicon (and) and aluminium (Al), generating the polymerisation process. Inorganic molecular chains and networks are then created and a hardened binder is obtained [7, 8].

In short, since the geopolymerization reaction occurs when the alumino-silicates contained in the raw material react with a strong alkaline solution, any material containing silicon and aluminum can be used in the production of geopolymer materials [5-9].

IOP Conf. Series: Materials Science and Engineering **877** (2020) 012033 doi:10.1088/1757-899X/877/1/012033

2. Producing geopolymer concrete with different raw materials

Minerals containing aluminum and silicon can be found everywhere especially as by-products. Therefore, the production of geopolymer concrete is not only an alternative to recycle a large amount of waste, but also to produce concrete with special properties, such as very good compressive strength from an early age, good fire and aggressive environment resistance, etc. The multitude of raw materials that can be used in the production of alkali-activated geopolymer materials has been intensevely studied and three main categories were established based on their origin: industrial waste (IW), general waste and recycled materials (GWRM) and natural materials (NM) [10].

2.1. Industrial waste (IW) used as raw material

Worlwide huge demand for services and products leads to large quantities of industrial waste and by-products such as coal-fired combustion ash, metallurgical slag, mine waste or agricultural waste [11], most of them being stored or dumped on the ground in landfills. Partial use of the total amount of waste generated is already being used in civil engineering industry in order to improve concrete properties [12, 13], but most of it is not capitalized.

2.1.1. Fly ash (FA). Since 1930, the term fly ash was used in the industry, with reasearch on the possibility of using this type of by-product in concrete starting in 1937, where Davis and his colleagues from the University of California present this possibility. Research showed in 1956 that parts of Portland cement in concrete could be replaced by fly ash with no effects on the mechanical properties [14]. Worlwide production of fly ash is increasing (Figure 2) since energy demand for population is rising, but the quality of this type of waste should also increase, as the amount of fly ash with a low content of L.O.I. (Loss on Ignition) available on the market is continuously decreasing [15, 16].

As seen in Figure 2, fly ash production increased in countries such as India, China and the U.K. in 2013, but declined in the U.S., when compared to 2008. Simultaneously with the increase in production, in India, China and the U.K., the use of fly ash in different fields has also increased. In countries such as Denmark, Italy and the Netherlands, the re-use of ash production is 100%, which shows that a complete reintegration into the circular economy concept of this type of industrial waste can be achieved (Figure 3). Due to the total re-use there are no slag dumps that would lead to environmental pollution.

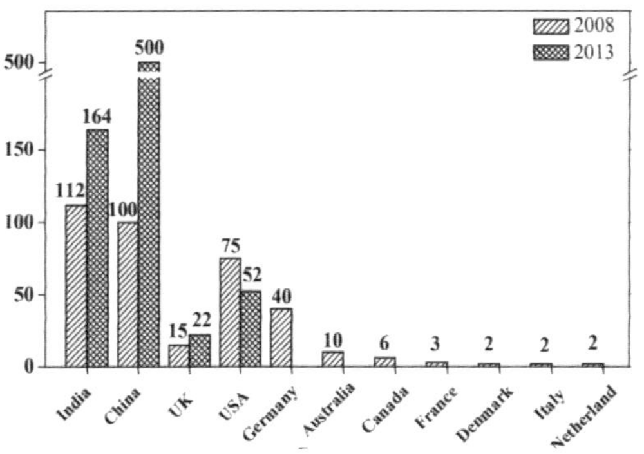

Figure 2. Production of fly ash in different countries (million tons per year) [15].

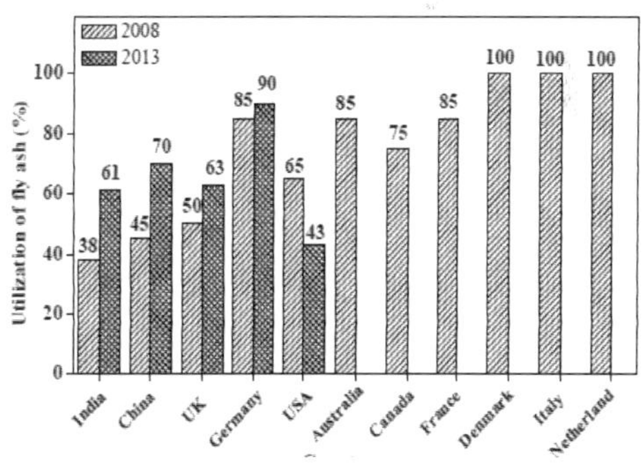

Figure 3. Percentage use of thermal power plant ash in different countries [15].

Usually, fly ash is used as main binding material for the production of alkali-activated materials. Due to its chemical composition, careful consideration is needed when using this type of waste and certain properties should be take into consideration [16]. Togheter with the alkaline activator (liquid part), fly ash (solid) represents the main constituents of the geopolymer binder [17]. Therefore, in the production of alkali-activated fly ash-based geopolymers, two types of fly ash are generally used, with specific ranges of chemical compounds (Table 1).

Table 1. Chemical composition typical fly ash [18].

Component	SiO_2	Al_2O_3	Fe_2O_3	CaO	MgO	K_2O	Na_2O	SO_3	LOI	TiO_2
Class C-Percentage wt%	15-45	20-25	4-15	15-40	3-10	0-4	0-6	0-10	0-5	<1
Class F-Percentage wt%	20-60	5-35	6-24	1-12	0-5	0-3	0-4	0-4	0-15	1-2

2.1.2. Granulated blast furnace slag (GBFS). Granulated blast furnace slag is a byproduct of the iron manufacturing process, obtained by extinguishing molten slag from the furnace with water. The removal of such large quantities of slag is a major concern and problem for steel producers. Clay is produced by the action of various flows on the materials of iron ore during the process of manufacturing iron and steel. Slag produced at the furnace (BFS) during the manufacture of iron is called slag furnace. Slag produced by steel smelting is known as steel slag. First of all, slag is formed from calcium, magnesium, manganese and aluminum silicates in various combinations together with iron oxide. The main difference between furnace slag and steel slag are the iron content. Typical mixtures of alkali-activated geopolymer materials produced using GBFS were reported by Shekhawat and the typical chemical composition of the GBFS used in his studies are presented in Table 2 [19].

Table 2. Chemical composition of GBFS- granulated blast furnace slag [19].

Component	SiO_2	Al_2O_3	Fe_2O_3	CaO	MgO	Loss on ignition
Percentage wt%	32,97	17,97	0,72	35,08	10,31	0,58

2.1.3. Marble Dust (MD). Marble dust, obtained by processing marble slabs, is a mineral addition that can be used successfully to produce environmentally friendly concrete based on mineral binders, alkali-activated. Compressive strength of geopolymers containing marble dust and granulated blast furnace slag in different compositions are shown in Figure 4 [20].

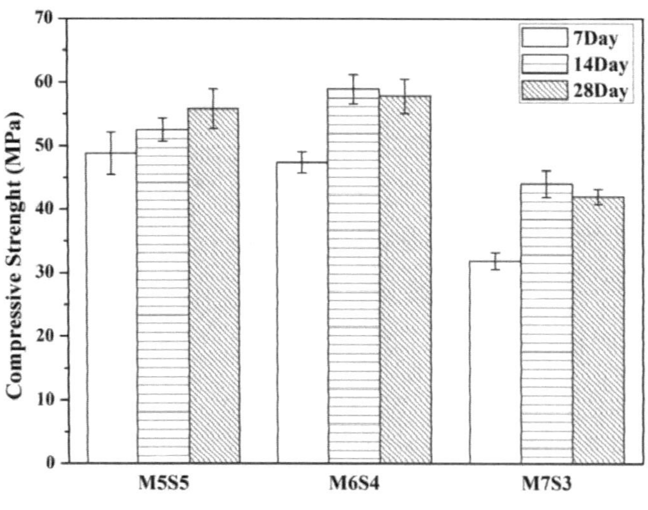

Figure 4. Effect of MD / BFS ratio on alkali-activated geopolymer materials compressive strength (M5S5=MD50%+BFS50%, M6S4=MD60%+BFS40%, M7S3= MD70%+BFS30%) [20].

As seen in Figure 4, increasing the percentage of marble dust (MD), at an early age of 7 days, compressive strength decreases. Therefore, if it is desired to obtain high compressive strength values at 7 days, it is indicated that the percentage of marble dust (MD) to be reduced. At the age of more than 7 days, an evolution of this parameter cannot be assessed by a simple percentage analysis.

2.2. General waste and recycled materials used as raw materials (GWRM)
General waste or recycled materials are produced all over the world. Although these types of raw materials represent a much smaller quantity than industrial waste, the conversion of added value is of great research interest. Some of the general waste used in the production of alkali-activated materials are ash from waste paper sludge (WPSA) and construction wastes.
2.2.1. Construction waste (CDW). Construction and demolition waste are limited in use, mainly used as fillings, as well as separately in recyclable materials. One of the promising technologies, geopolymerization, can be used to fully use construction and demolition waste.

Increased attention is required when re-using of CDW as materials for alkali-activated geopolymer production. Cities are becoming larger, therefore almost no land is available for disposal / storage in urban areas. Most of the CDW (Figure 5), about 70% is fly ash, sand cement mixture from demolition (includes hydrated cement and sand), cement concrete and sand mixture (includes hydrated cement, sand, aggregate, etc.), brick, sandstone, etc. and the remaining 30% includes wood, metals, plastics, glass, etc. [4].

Figure 5. CDW before and after milling a)concrete and b)fired clay [4].

2.3. Raw natural materials (NM)
Natural source raw materials can be also used in the production of alkali-activated geopolymer materials: kaolin (kaolinite), metakaolin (calcined kaolin), feldspace, sandy clay, diatomite, puzzolanic volcanic rocks, etc.. The main barrier when using these types of materials is the limitation to certain geological areas and also limited quantities.
2.3.1. Metakaolin (MK). MK is obtained by calcination of kaolinous clays in furnaces at temperatures of about 500-800°C. Kaolin can be obtained directly by excavating natural deposits, or it can be obtained as a component of decanting waste from the mining industry [21].
2.3.2. Other clays. In addition to metakaolin, heat-treated clays rich in *m*uscovite, obtained from tungsten mine waste (TMW), were used as alkaline activators in Portugal [22, 23].
2.3.3. Feldspar. Given that it is mostly the Earth's crust, being rich in silicon and aluminum, in approximately equal proportions, it is an important source of alkaline activator that can be used to produce geopolymer materials.

2.4. Summary of waste materials with potential use as raw materials for geopolymer production
Some of the raw materials used in literature in order to produce alkali-activated geopolymer materials are presented in Table 3, by analyzing their chemical composition and studying their capacity to provide sources of silicon (Si) and aluminium (Al), which later generate the geopolymerization process [24-41].

IOP Conf. Series: Materials Science and Engineering **877** (2020) 012033 doi:10.1088/1757-899X/877/1/012033

Table 3. Oxidic composition of waste with potential use as raw materials for the development of geopolymers, reported by different authors.

Materials	SiO_2	Al_2O_3	Fe_2O_3	CaO	MgO	Na_2O	K_2O	SO_3	FeO
Industrial waste									
Low calcium fly ash (LFA)	50.00	28.25	13.50	1.79	0.89	0.32	0.46	0.38	-
High calcium fly ash(HFA)	39.70	20.00	14.10	17.30	1.40	1.40	2.70	2.60	-
Granulated Blast Furnace Slag (GBFS)	32.46	14.30	0.61	43.10	3.94	0.24	0.33	4.58	-
Silica Fume (SF)	92.00	0.46	1.60	0.29	0.28	0.51	0.61	0.19	-
Steel Slag (SS)	15.00	6.70	15.40	44.20	10.9	0.20	0.10	0.70	-
Vanadium mine tailing (VMT)	64.2	10.27	4.98	4.46	-	5.27	2.06	-	-
Ferronickel Slag (FS)	32.74	8.32	0.76	-	2.76	-	-	-	38.8
Tungsten mine waste (TMW)	53.48	16.66	12.33	-	1.27	0.62	7.65	-	-
Cement Kiln Dust(CKD)	11.00	3.90	2.00	42.00	3.60	-	0.60	-	-
Waste paper sludge ash (WPSA)	26.25	17.50	4.40	23.40	0.90	0.10	0.20	4.63	-
Water Sludge (WS)	70.40	15.40	5.30	1.53	0.96	0.90	3.66	0.31	-
General waste and recycling materials									
Demolished Wall (DW)	76.42	1.88	1.28	9.84	0.26	0.22	0.08	2.09	-
Waste concrete (WC)	71.53	2.14	2.43	12.76	0.39	1.04	1.13	0.33	-
Waste glass (WG)	72.5	0.40	0.20	9.70	3.30	13.7	0.1	-	-
Ceramic waste-red clay brick (CW-RCB)	50.95	16.92	6.68	9.92	5.62	0.55	4.33	3.37	-
Natural materials									
Metakaolin (MK)	54.78	40.42	0.76	0.10	0.41	0.07	2.72	-	-
Kaolin (K)	48.10	36.90	0.26	0.20	0.17	0.20	1.90	-	-
Silty Clay (SK)	20.10	7.55	32.89	26.15	0.47	-	3.17	4.92	-
Diatomite (D)	59.30	10.00	18.50	1.20	-	-	-	2.74	-
Volcanic Mud (VM)	38.50	14.20	23.76	5.62	-	-	4.31	0.78	-

IOP Conf. Series: Materials Science and Engineering **877** (2020) 012033 doi:10.1088/1757-899X/877/1/012033

2.5. Summary of compressive strength values of geopolymer by prime materials

Table 4. Geopolymer compressive strength by prime materials [27-47].

Type	Prime materials (%wt)	Additives (%wt)	Si/Al Ratio	Type	Compression MPa/Aged	Alkaline Materials
IW	FA class F (90)	-	2.12	Paste	95.0/28	$NaOH+Na_2SiO_3$
IW	TMW(90)	$CaOH_2$(10)	3.20	Paste	75.0/56	$NaOH+Na_2SiO_3$
IW	FA(80)	Water Sludge(20)	2.53	Paste	70.6/90	-
GW	Ceramic waste-red clay brick	-	-	Paste	71.1/28	NaOH+KOH
IW	FA class F (90)	BA(10)	2.32	Paste	70.0/28	$KOH+K_2SiO_3$
IW	GBFS(100)	-	3.70	Paste	67.0/28	NaOH
NM	MK(100)	-	1.40	Paste	65.0/28	$NaOH+Na_2SiO_3$
NM	Natural Pozzolan(100)	-	3.44	Paste	63.0/28	$NaOH+Na_2SiO_3$
IW	Vanadium(70)	FA(30)	-	Paste	55.0/7	NaOH
NM	MK(80)	Steel Slag(20)	1.71	Paste	44.5/28	$NaOH+Na_2SiO_3$
IW	FA class F (95)	Silica Fume(5)	1.87	Mortar	35.0/28	$NaOH+Na_2SiO_3$
NM	Diatomite(100)	-	5.93	Paste	28.4/28	$NaOH+Na_2SiO_3$
NM	Kaolinit(100)	-	1.30	Paste	28.0/28	$NaOH+Na_2SiO_3$
GW	Construction waste (70)	MK(20)+ $CaOH_2$(10)	38.00	Paste	26.1/7	$NaOH+Na_2SiO_3$
IW	GBFS(75)	CKD(25)	3.70	Paste	24.0/28	NaOH
GW	Glass waste(95)	Metakaolin (5)	-	Mortar	22.0/28	NaOH
GW	Waste Paper Sludge(100)	-	1.84	Concr.	17.5/28	$NaOH+Na_2SiO_3$
IW	Ferronickel Slug (100)	-	4.13	Paste	15.8/28	$NaOH+Na_2SiO_3$
IW	Cooper mine tailing (100)	-	9.29	Paste	15.0/7	$NaOH+Na_2SiO_3$
NM	Silty Clay(75)	FA(25)	2.50	Paste	14.0/28	$NaOH+Na_2SiO_3$

3. Conclusions

Recycling waste and reducing greenhouse emissions are two of the main goals industry has to improve in order to obtain sustainable and durable development and one viable method can be provided by producing alkali-activated geopolymer materials [48]. As presented above, there is a wide range of waste or by-products that can become raw materials in the geopolymer production through alkali-activation [49].

Replacing Ordinary Portland Cement, in creating binders without clinker, by using industrial raw materials, such as fly ash or granulated furnace slag could reduce or even stop associated environmental problems.

However, careful attention is needed when mineral materials are used for geopolymer production. The variability in terms of physical and chemical properties of these materials should be very well studied. On the other hand, their local availability is also an important parameter in terms of potential environmental benefits, with possible additional costs if the source materials are transported over long distances [50-54].

4. References

[1] Abdul Aleem M I and Arumairaj P D 2012 *IJESET* **1(2)** 118–22
[2] Davidovits J 1994 *ACI SP* **144** 383–97
[3] Kartam N, Al-Mutairi N, Al-Ghusain I and Al-Humoud J 2004 *J. Waste Manag.* **24(10)** 1049–59
[4] Patak A 2016 *A laboratory scale synthesis of geopolymer from construction wastes* (Tribhuvan: Tribhuvan University, Nepal)
[5] Davidovits J 1991 *J. Therm. Anal. Calorim.* **37(8)** 1633–56
[6] Davidovits J 2005 *Geopolymer, Green Chemistry and Sustainable Development Solutions: Proceedings of the World Congress Geopolymer 2005* (France: Geopolymer Institute)
[7] Lăzărescu A, Szilagyi H, Baeră C, Ioani A 2017 *IOP Conf. Ser.: Mater. Sci. Eng.* **209** 012064
[8] Nematollahi B 2017 *Alternative binder for fiber-reinforced Strain Hardening Composites* (Hawthorn: University of Technology Hawthorn, Australia)
[9] Lăzărescu A, Szilagyi H, Ioani A, Baeră C 2018 *IOP Conf. Ser.: Mater. Sci. Eng.* **374** 012035
[10] Suwan T 2016 *Development of self-cured geopolymer cemen*t (London: Brunel University London, UK)
[11] Sujatha T, Kannapiran K and Nagan S 2012 *Asian J. Civ. Eng.* **13**(5) 635–46
[12] Komnitsas K, Zaharaki D and Perdikatsis V 2004 *J. Mater. Sci.* **42**(9) 3073–82
[13] Nuruddin M, Qazi S, Kusbiantoro A and Shafiq N 2011 *Proc. Inst. of Civ. Eng.-Constr. Mat.* **164**(6) 315–27
[14] Joshi R C and Lohtia R P 1997 *Fly Ash in Concrete: Production, Properties and Uses* (Amsterdam: Overseas Publishers Association)
[15] Dhadse S, Kumari P and Bhagia L J 2008 *J. Sci. Ind. Res. India* **67** 11–18
[16] Lăzărescu A, Szilagyi H, Baeră C and Hegyi A 2020 *Alternative concrete – Geopolymer Concrete. Emerging Research and Opportunities* (Cluj-Napoca: Napoca Star Press)
[17] Bilodeau A and Malhotra V M 2000 *ACI Mater. J.* **97**(1) 41–48
[18] Christy F C and Tensing D 2010 *Asian J. Civ. Eng.* **12** 87–105
[19] Lăzărescu A, Mircea C, Szilagyi H and Baeră C 2019 *MATEC Web of Conf.* **289** 11001
[20] Cheng T W, Ding Y C, Lee W H and Lu P C 2017 *MATEC Web of Conf.* **97** 01004
[21] Lăzărescu A, Szilagyi H, Baeră C, Ioani A and Mircea A. C. 2018 *Constr.* **19**(1/2) 19–25
[22] Pacheco-Torgal F, Castro-Gomes J and Jalali S 2008 *Constr. Build. Mater.* **22**(9) 1939–49
[23] Pacheco-Torgal F, Castro-Gomes J and Jalali S 2009 *Constr. Build. Mater.* **23**(1) 200–09
[24] Hooton R D 2008 *Cem. Concr. Res.* **38**(2) 247–258
[25] Rattanasak U, Pankhet K and Chindaprasirt P 2011 *Int. J. Min. Met. Mater.* **18**(3) 364–69
[26] Dutta D, Thokchom S, Ghosh P and Ghosh S 2010 *J. Eng. Appl. Sci.* **5**(10) 74–79
[27] Hu S, Wang H, Zhang G and Ding Q. 2008 *Cem. Concr. Compos.* **30**(3) 239–44
[28] Jiao X, Zhang Y and Chen T 2013 *Constr. Build. Mater.* **38** 43–47
[29] Komnitsas K and Zaharaki D 2007 *Miner. Eng.* **20**(14) 1261–77
[30] Pacheco-Torgal F, Castro-Gomes J and Jalali S 2007 *Int. Conf. Alkali Activated Materials-Research, Production and Utilization* 693–710
[31] Khater H 2012 *Int. J. Civ. Struct. Eng. Res.* **2**(3) 740

[32] Anuar K, Ridzuan A and Ismail S 2011 *IJCEE* **11**(1) 59–62
[33] Kongkaew B 2007 *Sludge-Based Geopolymer* (Bangkok: Kasetsart University)
[34] Khater H 2011 *J. Mater. Civil. Eng.* **24**(1) 92–101
[35] Pascual A B, Tohoue Tognonvi M and Tagnit-Hamou A 2014 *Int. J. Res. Appl. Sci. Eng. Technol.* **3**(13) 15–19
[36] Reig L, Tashima M M, Soriano L, Borrachero M V, Monzó J, and Payá J 2013 *Waste Biomass Valori.* **4** 729–36
[37] Yip C, Lukey G. and Van Deventer J 2005 *Cem. Con. Res.* **35**(9) 1688–97
[38] Hounsi A, Lecomte-Nana G, Djétéli G and Blanchart P 2013 *Constr. Build. Mater.* **42** 105–13
[39] Sukmak P, Horpibulsuk S and Shen S 2013 *Constr. Build. Mater.* **40** 566–74
[40] Phoo-Ngernkham T and Sinsiri T 2011 *KKU Eng. J.* **38**(1) 11–26
[41] Al Bakri A, Rafiza A R, Hardjito D, Kamarudin H and Niza I K 2012 *Adv. Mat. Res.* **548** 82–6
[42] Fernández-Jiménez A, Garcia-Lodeiro I and Palomo A 2007 *J. Mater. Sci.* **42**(9) 3055–65
[43] Sun Z, Cui H, An H, Tao D, Xu Y, Zhai J and Li Q 2013 *Constr. Build. Mater.* **49** 281–27
[44] Hardjito D and Fung S 2010 *Mod. Appl. Sci.* **4**(1) 44–52
[45] Wang H, Li H and Yan F 2005 *Colloids Surf. A Physicochem. Eng. Asp.* **268**(1) 1–6
[46] Allahverdi A, Mehrpour K and Kani E, 2008. *Int. J. Eng. Sci.* **19**(3) 1–5
[47] Ahmari S and Zhang L 2012 *Constr. Build. Mater.* **29** 323–31
[48] Allahverdi A and Najafi Kani E 2008 *Int. J. Civ. Eng.* **7**(3) 154–60
[49] Akcaozoğlu S and Ulu C 2014 *Constr. Build. Mater.* **58** 31–7
[50] Mathew B, Sudhakar M and Natarajan C 2013 *IJCER* **3**(1) 207–12
[51] McLellan B C, Williams R P, Lay J, van Riessen A and Corder G D 2011 *J. Clean. Prod.* **19** 1080–90
[52] Burduhos Nergis D D, Abdullah M M A B, Sandu A V and Vizureanu P 2020 *Materials* **13(2)** 343
[53] Burduhos Nergis D D, Vizureanu P and Corbu O 2019 *Revista de chimie* **70(4)** 1262-1267
[54] Scripcariu L, Mătăsaru D, Diaconescu V 2017 *UPB Scientific Bulletin Series C-Engineering and Computer Science* **73**(3) 83-92

Experimental Investigation on the Development of Geopolymer Paving Blocks by Using Romanian Local Raw Materials

A Lăzărescu[1*], H Szilagyi[1], C Baeră[2], A Hegyi[1] and V Meiță[3]
[1]NIRD URBAN-INCERC Cluj-Napoca Branch, Romania
[2]NIRD URBAN-INCERC Timișoara Branch, Romania
[3]NIRD URBAN-INCERC Bucharest, Romania

E-mail: adrian.lazarescu@incerc-cluj.ro

Abstract. For the world economic system cement and concrete are indispensable elements for the construction industry. Demand for concrete, hence for cement, is constantly growing, especially in highly developed countries, which means that alternative binders are urgently required to meet the needs of millions of people, without compromising the CO_2 levels of the atmosphere. Although environmental issues are currently not sufficiently convincing to create a sufficiently high demand for the production of geopolymer technology, this is expected to increase with the imposition of rules on carbon dioxide emissions and their effects. Preliminary alkali-activated geopolymer concrete mixtures were developed based on a rigorous trial-and-error process in order to establish the mechanical properties of the geopolymer paving blocks. The aim of this paper is to present results regarding the technology of development and optimisation for the production of geopolymer paving blocks, their mechanical properties and implementation possibilities in accordance with the intended scope of use. The study results indicate that alkali-activated geopolymer paving blocks have excellent mechanical properties, by reference to OPC paving blocks, making them suitable for practical applications.

1. Introduction

Worldwide, concrete is the second most used material and it is considered the most versatile, durable and reliable building material in civil engineering industry. Portland cement production is one of the most important carbon dioxide generators and global demand in the industry for its production as binder for concrete is rising constantly, generating environmental associated problems which are well known and start on being carefully monitored. Cement production requires large quantities of liquid fuel (between 60 and 130 kg / tonne) and the electricity amount required to produce on tonne is approximately 110 KWh [1]. In terms of environmental pollution, measurements have shown that for each tonne of cement that is produced, the industry releases into the atmosphere around 1.1 tonnes of CO_2, generated by the combustion and calcareous calcination methods [2]. World cement production indicates an obvious yearly increase, with more than 4.2 billion tonnes being produced and delivered to the industry only in 2014 [3].

A particular procedure in which the solidification the fly ash powder could be achieved is represented by its alkaline activation. When it is mixed with a certain type of alkaline activator, it creates a new binding material, with mechanical properties similar to those of Ordinary Portland Cement (OPC) concrete. Based on their properties, this type of new materials can be considered as an alternative to OPC traditional concrete. Alkali-activated geopolymers are on-going developing materials, their research being strongly connected to the worldwide need of global CO_2 emissions reduction. Characterised by excellent mechanical properties and resilience in aggressive environments, these materials represent an opportunity for both the environment and the engineering fields of novelty as a reliable alternative to the traditional technology [4-6].

In order to produce alkali-activated geopolymer concrete, fly ash can be used as main raw, binding material. Therefore, careful consideration is needed when choosing this type of material to obtain the geopolymer binder. Depending on the type of coal that is burnt, the chemical properties of the fly ash can be very different and this has to be carefully considered when fly ash is used as raw material so that the geopolymerisation process can be fully generated [7]. The two main constituents of the geopolymer binders are the alumino-silicate elements, which represent the solid part, and the alkaline activator, which represents the liquid part [8].

Paving blocks have been used in the construction industry for many years, being considered appropriate for outdoor applications. These units turned out to be cheap to produce and present exceptionally high mechanical properties, making them useful for their intended purpose: commercial, industrial or residential areas, plazas, parking areas, bus stops, etc. In addition to being economical, interlocking concrete paving blocks are also broadly obtainable in water-permeable designs, which have additional ecological benefits [9].

Although the durability of alkali-activated geopolymer materials is not exactly extensively discussed and analysed, there are numerous studies regarding the performance of these materials in terms of sustainability [10-13]. Experimental results showed excellent resistance to aging, freeze-thaw, but also to carbonation [14].

The two main barriers to the introduction of new materials in the construction industry [15] are: 1) the need for standards for the testing methods, the evaluation and admissibility conditions, given that their development and introduction takes a long time and it is a gradual process; 2) the durability performance of such materials, taking into account the fact that they must meet certain requirements, they must withstand for a long, relevant period, the in situ (put in work), specific durability challenges, and for new materials this is not quantified yet.

The aim of this paper is to present results regarding the development technology and optimisation approach for the production of geopolymer paving blocks, by using Romanian local raw materials, their mechanical properties and implementation possibilities, in accordance with the intended scope of use. Based on the results obtained for alkali-activated geopolymer concrete [16], research directions on the material applicability were identified.

2. Materials and methods

This chapter summarizes the applied experimental methodology, involving the raw materials, mix design, moulding, curing and also the testing methods for the mechanical properties' specific evaluation of the developed alkali-activated geopolymer paving blocks.

By using the full information on the production of the alkali-activated geopolymer concrete, as well as on the basis of the obtained results [6, 16], a "trial and error" experimental research was built, with the aim of producing building geopolymer materials (geopolymer paving blocks). The study of technology transfer to the industry, by enhancing the legal framework, which sets out the certification of products for the construction market, was also considered Geopolymer fly ash-based paving blocks were manufactured by the use of class F fly ash, aggregates and alkaline solution (sodium silicate + sodium hydroxide).

2.1. Fly ash

Low calcium-fly ash from a power plant in Romania was used as main raw material for the production of the alkali-activated geopolymer paving blocks. Its chemical composition was established through X-ray fluorescence methods (Table 1). As seen in Table 1, the CaO content of the fly ash was less than 10%

and $SiO_2 + Al_2O_3 + Fe_2O_3 > 70\%$, $SO_3 < 5\%$ and L.O.I. < 6%. The chemical composition of the fly ash used in this study can be classified as Class F fly ash, according to ASTM requirements [17, 18].

Table 1. Fly ash chemical composition.

Oxides	%
SiO_2	46,94
Al_2O_3	23,83
Fe_2O_3	10,08
CaO	10,72
MgO	2,63
SO_3	0,45
Na_2O	0,62
K_2O	1,65
P_2O_5	0,25
TiO_2	0,92
Cr_2O_3	0,02
Mn_2O_3	0,06
ZnO	0,02
SrO	0,03
L.O.I.	2,11

2.2. Alkaline activator

A combination between sodium silicate solution (Na_2SiO_3 solution) and sodium hydroxide solution (NaOH solution) was chosen as alkaline activator for the production of the alkali-activated geopolymer paving blocks. The Na_2SiO_3 solution was commercialy purchased and had the folowing chemical composition: SiO_2=30%, Na_2O=14% and H_2O=56%. Sodium hydroxide pearls were used to produce the NaOH solution, which was set to 10M, therefore 400g were dissolved in water, for one litre of solution. Based on the molecural weight of NaOH, in order to achieve the desired molar concentration of the solution 40g x 10 = 400g of NaOH flakes were dissolved in one liter of water.

2.3. Aggregates

Natural aggregates, granular class 0/4 mm (S) and 4/8 mm (CA) (sand and coarse aggregates) were used in this study to produce alkali-activated fly ash-based geopolymer concrete paving blocks.

2.4. Mix design, moulding and curing treatment

In order to study the mechanical properties of the alkali-activated geopolymer paving blocks, the following preliminary mixtures have been produced, considering the following parameters as constants: AA/FA ratio, Na_2SiO_3/NaOH solution ratio and S/CA ratio (Table 2).

The technology of producing alkali-activated geopolymer paving blocks (Figure 1) was based on the principles already set out for the production of alkali-activated geopolymer concrete [6, 15]. Mixing of the materials was performed at constant temperature of (20±2)°C and the following the technological steps were applied:

(1) Mixing of the sand and the coarse aggregates (0/4 mm and 4/8 mm) at low speed for 30 seconds;

(2) The addition of the necessary amount of fly ash and then the dry raw materials (fly ash + aggregates) were mixing together, for 30 seconds;

(3) The alkaline activator addition, for90 seconds to the homogenous dry mixture;

(4) After adding the total alkaline activator quantity, the mixing was continued at low speed for another 3 minutes;

(5) Casting into polypropylene moulds, with the corresponding 5 minutes vibration and starting of the heat treatment (70°C / 24 hours);

(6) After demoulding, the geopolymer paving blocks were stored in the climatic chamber at the temperature T (20±1)°C and relative humidity RH (60±5)%, and the tests for their mechanical properties were determined at the age of 7 days.

Table 2. Alkali-activated geopolymer paving blocks mix-design.

Material / Parameter	%
Fly ash	1,0
AA/FA	0,9
Na_2SiO_3	0,64
NaOH	0,26
NaOH conc.	10M
Na_2SiO_3 / NaOH	2,5
S 0/4 mm	0,93
CA 4/8 mm	0,93
S : CA	0,5 : 0,5
Binder / Aggregate	1: 1,85

Figure 1. Alkali-activated geopolymer paving blocks production.

2.5. Testing methods

Based on the results regarding the possibility of developing such construction elements [6, 15], the research initially focused on preliminary evaluation of the physical, mechanical and durability characteristics of the obtained products, for the study pf their potential, different areas of applicability:

(1) Weathering resistance – total water absorption;

(2) Weathering resistance – freeze-thaw resistance with de-icing salt;

(3) Tensile splitting strength;

(4) Abrasion resistance;

(5) Slip/skid resistance;

All the test were performed according to norm *EN 1388:2004 / AC:2006: Concrete paving blocks. Requirements and test methods*, which specifies the materials, characteristics, conditions and methods of testing the paving blocks, and it is applicable for the use of pedestrians, vehicles, bicycle lanes, parking lots, roads, highways, industrial areas, etc.. To be used in specific applications, they must comply with certain conditions at the time of their declaration as fit for use by the manufacturer.

3. Results and discussions

As mentioned in the literature, alkali-activated fly ash-based geopolymer materials properties are notably influenced by many important parameters. The physical and chemical properties of the main raw material (fly ash) used as precursor, the chemical properties of the solution used as alkaline activator and the mix-design ratios, are the most important parameters that affect the mechanical properties of the material [1, 4, 6, 10]. The 7 days compressive strength of the alkali-activated geopolymer paste (32,6 N/mm^2) [16], produced using the same materials and the same mix-design ratios, was consider relevant as testing age for the evaluation of the above nominated parameters of the alkali-activated geopolymer paving blocks.

3.1. Weathering resistance

Weathering resistance for paving blocks is represented by the total water absorption test (according to Annex E from EN 1338:2004 / AC:2006) or the freeze-thaw resistance (according to Annex D from EN 1338:2004 / AC:2006). The recommendations related to the weathering resistance considered necessary to ensure the durability of the element, are determined by well-established criteria.

Total water absorption. Total water absorption of the alkali-activated geopolymer paving blocks has been determined by submerging the test samples in water, at constant temperature of (20±5)°C, until they reached constant mass. The minimum immersion period of the samples was 3 days. After reaching the saturated state constant mass, they were weighed (M_1). Subsequently, the paving blocks were placed in the oven and dried at temperature of (105±5)°C, to reach the constant mass (M_2). The W_a (water absorption) of each paving block was expressed according to the standard, as a percentage by mass, acc. to the equation (1) [19].

$$Wa = \frac{M1 - M2}{M2} x100 \quad (1)$$

3.1.2. Freeze-thaw resistance with de-icing salt. The geopolymer paving blocks test specimens were preconditioned and then subjected to 28 freeze-thaw cycles, while their surface was covered with a 3% NaCl solution. After carrying out the 28 cycles, the exfoliated material was collected from the surface of the specimens (M) and the mass lost per unit area was calculated, acc to the eqation (2) [18].

$$L = \frac{M}{A} \; [kg/m2] \quad (2)$$

3.2. Böhme abrasion test

For the determination of this parameter 70 x 70 mm cubes were used, (Figure 2), cut from the geopolymer paving blocks and placed on the Böhme abrasive disc (Figure 3). Before the test and after every four cycles, the test specimens were weighted. Test specimens were tested for 16 cycles, each consisting of 22 full rotations of the abrasion track. After the end of the 16 cycles (Figure 4), abrasion was calculated [18].

3.3. Unpolished slip resistance (USVR)

Unpolished slip resistance of the geopolymer paving bocks has been determined by using a test equipment consisting of a friction pendulum (Figure 5), (Figure 6). At the pendulum swing (Figure 6), the force between the cursor and the test surface was measured by using a graduated scale. Prior the start of the test the geopolymer paving blocks were immersed in water at (20±2)°C for 30 minutes, according to the standard methodology requirements.

3.4. Tensile splitting strength

In order to determine the tensile splitting strength, the alkali-activated geopolymer paving bocks were placed in the testing equipment, in a special device for the tensile splitting strength test (Figure 7), according to the specifications of the standard. The splitting load was applied progressively, until the paving blocks were broken (Figure 8).

According to EN 1338:2004 / AC:2006, the tensile splitting strength T of the paving blocks should not be lower than 3,6 MPa, and the breaking load per length unit should not be lower than 250 N/mm [18]. Results obtained on geopolymer paving blocks show that the breaking load per length unit was 410 N/mm, and the tensile splitting strength was 3,6 MPa, which qualifies them as fulfilling the admissibility conditions specified by the assimilated standard.

Figure 2. Cubic specimen for the abrasion resistance evaluation.

Figure 3. Placing the cubic specimen in the testing abrasive machine.

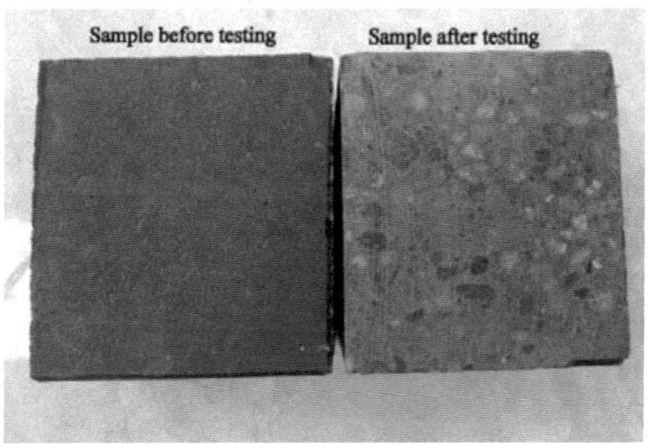

Figure 4. Before and after, Bohme abrasion test specimen.

Figure 5. Slip resistance measuring equipment.

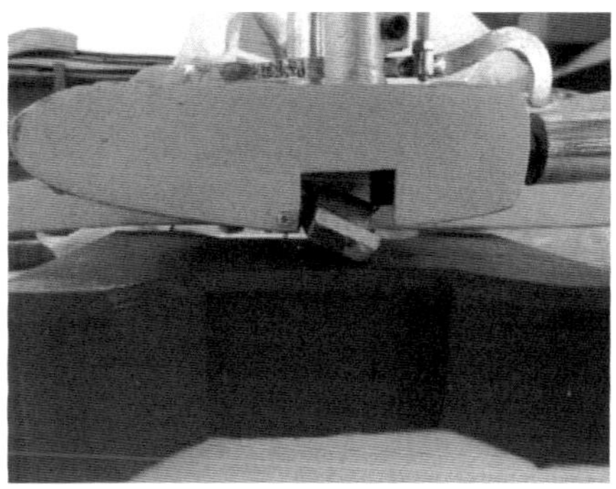

Figure 6. Unpolished Slip Resistance (USVR).

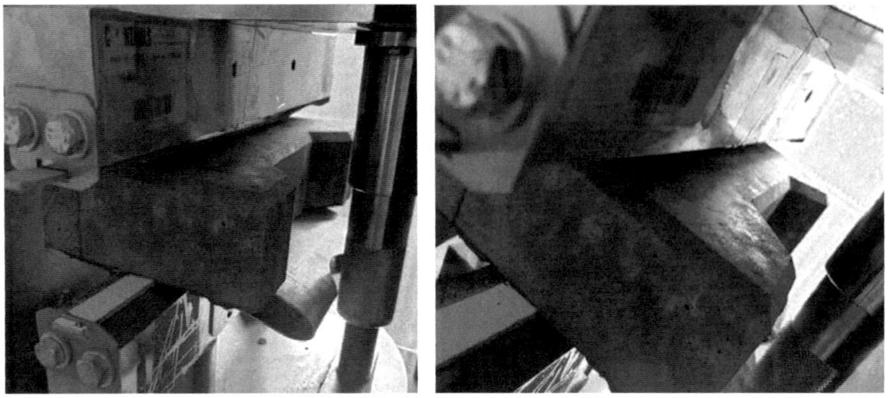

Figure 7. Loading conditions for the geopolymer paving blocks during the tensile splitting test.

Figure 8. Geopolymer paving block – failure mode after tensile splitting loading.

3.5. Mechanical properties of the alkali-activated geopolymer paving blocks

Alkali-activated geopolymer paving blocks have been subjected to tests for mechanical and durability performance evaluation by assimilating the methodology and admissibility conditions as well specified by standard EN 1338:2004 / AC: 2006. The obtained results on the nominated parameters confirmed their ability of the products to comply with this standard (Table 3), demonstrating the possibility of their use as paving elements in the construction materials industry Moreover, by optimizing the mixtures, improved performance can be achieved, and consequently the extension of the applicability interval within the specification of the EN 1338:2004 / AC: 2006.

Table 3. Geopolymer paving block characteristics.

Mechanical property	EN 1388 limit	Results
Total water absorbtion	≤ 6 %	5,6 %
Freeze-thaw resistance	≤ 1,0 kg/m^3	0,6 kg/m^3
Abrasion resistance	≤ 18000 mm^3 / 5000 mm^2	16993 mm3 / 5000 mm^2
Slip resistance	-	39 (moderate)
Tensile splitting strength	Min. 3,6 MPa	3,8 MPa

4. Conclusion and perspectives

Research regarding the production of alkali-activated geopolymer paving blocks using Romanian local raw materials could be definitely a possible answer to the need of greener concrete for sustainable development. Through the optimization of the alkali-activated geopolymer concrete mixtures, there was possible to demonstrate the viability of the material by producing geopolymer paving blocks and evaluating their mechanical and durability properties, thus opening up new opportunities for the future development of this type of material. One possible direction for the alkali-activated geopolymer paving blocks optimising process is related to psychical, mechanical and durability performance improvement by the addition of the natural fibres as disperse reinforcement [20].

Good mechanical properties, reached at very young ages (7 days) of the alkali-activated geopolymer binder, demonstrate the applicability of this material in the production of prefabricated construction products. Due to both, the raw material characteristics and the final properties of the material, the alkali-activated geopolymer paving blocks produced by using Romanian local raw materials showed all the test results within the admissibility limits imposed by EN 1388, hence demonstrating their possible use in residential driveways, light vehicles/public paths or other similar domain of use.

Acknowledgments

This paper is supported by the Programme: Research for sustainable and ecological integrated solutions for space development and safety of the built environment, with advanced potential for open innovation – "ECOSMARTCONS", Programme code: PN 19 33 04 02: "Sustainable solutions for ensuring the population health and safety within the concept of open innovation and environmental preservation", financed by Romanian Government.

5. References

[1] Hooton R D 2008 *Cem. Concr. Res.* **38**(2) 247–58
[2] Bilodeau A and Malhotra V M 2000 *ACI Mater. J.* **97**(1) 41–48
[3] Komnitsas K, Zaharaki D and Perdikatsis V 2004 *J. Mater. Sci.* **42**(9) 3073–82
[4] Palomo A, Grutzek M V and Blanco M T 1999 *Cem. Concr. Res.* **29** 1323–29
[5] Duxon P, Provis J L, Lukey G C and Van Deventer J S J 2007 *Cem. Concr. Res.* **37** 1590–97
[6] Lăzărescu A, Szilagyi H, Baeră C and Ioani A 2017 *IOP Conf. Ser.: Mater. Sci. Eng.* **209** 012064
[7] Pachenco-Torgal F, Castro-Gomez J and Jalali S 2008 *Constr. Build. Mater.* **22** 1305–14
[8] Somna K, Jaturapitakkul C, Kajitvichyanukul, P and Chindaprasirt P 2011 *Fuel* **90** 2118–24
[9] Jumutha R, Aswini A, Ellakkiya M, Karthika T and Vijai K 2017 *IOSR-JMCE* **14**(2) 6–12
[10] Sofi M, Van Deventer J S J, Mendis P and Lukey G C 2007 *Cem. Concr. Res.* **37**(2) 251–57
[11] Komnitsas K, Zaharaki D and Perdikatsis V 2004 *J. Mater. Sci.* **42**(9) 3073–82
[12] Xu H, Provis J L, Van Deventer J S J and Krivenko P V 2008 *ACI Mater. J.* **105**(2) 131–39
[13] Burduhos Nergis D D, Abdullah M M A B, Sandu A V and Vizureanu P 2020 *Materials* **13(2)** 343
[14] Medhi B 2009 *Mater. Technol.* **24**(2) 79–87
[15] Salas D A, Ramirez A D, Ulloa N, Baykara H and Boero A J 2018 *Constr. Build. Mat.* **190** 170–77
[16] Lăzărescu A, Szilagyi H, Baeră C and Hegyi A 2020 *Alternative concrete – Geopolymer Concrete. Emerging Research and Opportunities* (Cluj-Napoca: Napoca Star Press) p 152
[17] Fernandez-Jimenez A and Palomo A 2003 *Fuel* **84** 2048–53
[18] Hardjito D, Wallah S E, Sumajouw D M J and Rangan B V 2004 *ACI Mater. J.* **101** 467–72
[19] EN 1338 (2006) Concrete paving blocks. Requirements and test methods, *Romanian Standard Association, Bucharest, Romania*
[20] Silva G, Kim S, Aguilar R and Nakamatsu J 2019 *SM&T* **23** 00132

Designing Concrete with Self-healing Properties Using Engineered Cementitious Composites as a Model

A C Mircea[1,2*] and T P Toader[1,2]
[1]NIRD URBAN-INCERC Cluj-Napoca Branch, 117 Calea Florești, 400524, Cluj-Napoca, Romania.
[2]Technical University of Cluj-Napoca, 28 Memorandumului Street, 400114, Cluj-Napoca, Romania.

*E-mail: anamaria.mircea@incerc-cluj.ro

Abstract. Engineered Cementitious Composites (ECC), also known as Strain Hardening Cement-based Composites, are an easily molded mortar-based composites reinforced with polymer fibers. ECCs are designed based on micromechanics and fracture mechanics theory, to feature large tensile ductility and a variety of unique properties, including tensile properties, superior to other fiber-reinforced composites. The properties of ECCs can be custom-tailored through micromechanics design due to the interaction between the fibers and cement matrix. A structural deterioration of ECC is avoided because the fibers do not allow cracks with large widths to form, unlike conventional concrete. ECCs have the capacity to bend, generating a flexible material. Ductile properties rather than brittle had increased, unlike ordinary concrete, leading to a wide variety of applications. Obtaining superior characteristics for ECC, both in fresh and hardened state, a transition from ECC paste to concrete with self-healing properties was made. To obtain self-healing concrete, with same ECC paste behavior and characteristics, the mix-design of the paste was optimized. The aim of this article is to present the experimental results regarding the transition from ECC paste to self-healing concrete and to analyze the results in order to establish a mix-design pattern for concrete with self-healing properties.

1. Introduction

Since, both nationally and internationally, the quantity of products considered waste is very high, their removal from the list of products that pollute the environment leads to their use/re-use by incorporating them into new, innovative materials, and tending to obtain environmentally friendly / green materials [1].

The experiential study included the identification of raw materials (cement, additions, aggregates, fibers, additives, etc.) and their compatibility in order to produce materials with special performances. Performance evaluation was carried out experimentally, using standardized but also non-standardized testing methodologies.

2. Materials and methods

Engineered cementitious composite materials are an assembly between two or more distinct materials. Their combination generates different and significantly superior properties and characteristics than those of each individual component material [2]. To the matrix or also-called basic component is added a reinforcement component that alters and improves its mechanical and self-healing properties.

The reinforcement component is evenly distributed throughout the volume of the composite material matrix [3].

In designing the compositions of the engineered cementitious composite materials, the following constituents were used:

- Fine aggregate (Sand): Fine silica sand (FSS);
- Portland Cement: CEM 42.5 R;
- Polymer fibers: PVA and hybrid PP;
- Superplasticizer additive, water reducer, polycarboxylate type (HRWR);
- Additions (P): limestone slurry (SL);
- Water (A).

Several parameters vary in the mixtures as follows:

- The amount of water/additive, with the implicit effect on the ratios-water/cement, water/link; liquid/cement, liquid/link;
- The type of fly ash used – in terms of physical and chemical composition;
- The type of fibers used;

The mixing technology used is specific to this type of material and involves the use of a pallet mixer with a mixing vessel capacity of approximately 5 liters, in accordance with EN 196-1 specifications.

The technological sequences followed in order to produce this material are:

- conditioning the raw materials at a temperature of (20±2)°C and (50±5) RH;
- dry mixing of the aggregates (sand), fly ash and cement in the mixing container;
- the quantity of limestone slurry is added and the low-speed mix-up continues;
- the quantity of limestone slavery is added and the low-speed mix-up continues;
- approximately 1/3 of the amount of water is added to moisten the components in order to stimulate the reactivity of the additive and continue the mix-up;
- the additive and the rest of the water are added and the alternative mix-up continues;
- a break sequence is adopted to give the superplasticizer additive the ability to act fully on the mixture, improving its workability;
- continue mixing, while the fibers are gradually added, to ensure uniform distribution of them in the mass of the mixture.

The materials used in the mixtures have the following characteristics:

- Aggregates- fine aggregates of fine silica sand type, from a local source, with a granulation of not more than 0,3 mm;
- The binder material used in the compositions is Portland cement CEM I 42.5 R;
- Fly ash with the following chemical composition (Table 1);
- Superplasticizing additive, water reducer, allowed considerable reduction of specific ratios (A/C; L/C; A/B; L/B), while maintaining high workability;
- Fibers that satisfy the following conditions [4]:
    - tensile strength about 2 to 3 times higher than concrete;
    - adhesion to the concrete matrix at least equal to or even greater than the stretch resistance of the concrete;
    - Elasticity module of the fibers is significantly higher than those of concrete;
- Additions- limestone slurry (SL) considered chemically inert and without pozzolanic activity, added in order to optimize mixtures and to attenuate segregation and the tendency of bleeding of the mixture.

Polyvinyl alcohol (VPA) fibers used have high elasticity mode and high stretching resistance; size of 8mm and superior mechanical properties, non-corrosive, non-magnetic, chemically inert and with 100% resistance to alkalis [5].

Table 1. Fly ash chemical composition.

Oxides	%
SiO_2	53,75
Al_2O_3	26,02
Fe_2O_3	7,91
CaO	2,54
MgO	1,54
SO_3	0,35
Na_2O	0,59
K_2O	2,57
P_2O_5	0,12
TiO_2	1,02
Cr_2O_3	0,05
Mn_2O_3	0,09
ZnO	0,04
SrO	0,02
L.O.I.	3,14

Several studies were evaluated in the research for different mixtures. The hardened characteristics of the composite materials have been determined, as follows [6]:

- the density in hardened state both after taking demolding the material and at the age of 28 days;
- mechanical resistance: bending and compressive strength;
- evaluation of the repair potential of the material at different ages:28, 56, 90 days.

Tensile strength of the material was determined on 40 x 40 x 160 mm prismatic test specimens, using the three-point bending test (3PB), in accordance with EN 196-1:2016. Compressive strength was also determined in accordance with EN 196-1:2016 *Cement test methods. Part 1: Determination of resistance* [7], using the compression test of half prismatic specimens resulting from the three-point bending test (3PB). The compression load speed used was 50 N/s (0.12 MPa/s). For each sample, all data were recorded and the results obtained represent the average of the values read/calculated on a minimum of three samples [8].

The self-healing potential of the compositions was determined on coupons as follows [9]:

- R samples, which have been tested to breakage at the specific age;
- SH samples, at the specific age of 28 days until the initial cracking in the material was recorded, they were exposed to 28 SH cycles, (self-healing cycles), after which were tested under the same conditions as the original ones, at the age of 28+28 days;
- VS samples, which were kept intact until the age of retesting the SH samples, when tested under the same conditions as the SH samples, at the specific age of 28+28 days.

Exposure to SH cycles involved an alternation in a period of 24 hours between exposure of test specimens in wet environments, or immersed in water at T of (20±2)°C, for 12 h, and exposure of test specimens in air at T: (21 ± 3)°C and (50 ± 5)% RH for 12 h.

The repair potential of microcracks is presented by microscopic visual evaluation carried out before and after exposure of test specimens (coupons) to SH cycles (28 cycles).

The network of microcracks generated is characterized by:

- Number of cracks / microcracks (n);
- Open cracks: mean value (wmed), minimum value (wmin) maximum value (wmax).

In Figure 1, the coupons for assessing the repair potential of the compositions before testing are presented.

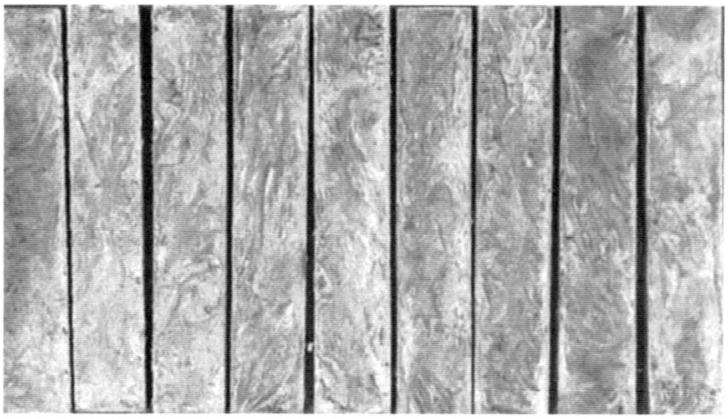

Figure 1. Test specimens (coupons) before testing.

3. Results and discussions

The evolution of composite materials at different ages following conditionings at alternating wet/dry cycles is shown in Table 2.

Table 2. Microcracks evaluation.

Samples	n	w_{max} (μm)	w_{max} (μm)	w_{med} (μm)	w_{min} (μm)
R	18	186	117	52	18
		Self-Sealing (SS)			
SH	15	134	69	19	3
		Self-Sealing (SS)			
Partially closed cracks				8	
Total closed cracks				25	

The results of the mechanical tests of the analyzed composition are presented in Figure 2 and Figure 3 respectively.

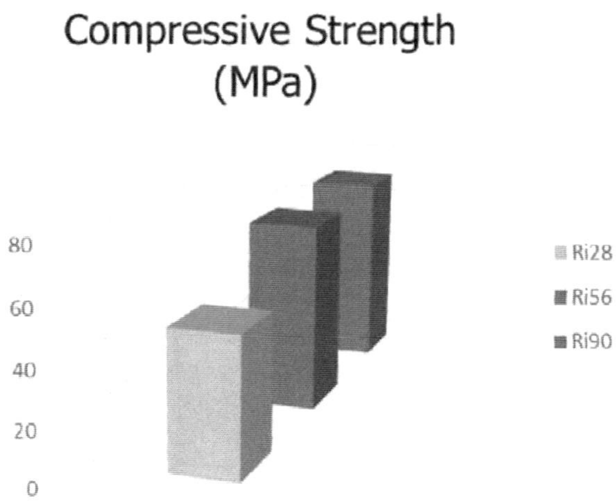

Figure 2. Compressive strength of the analyzed composition.

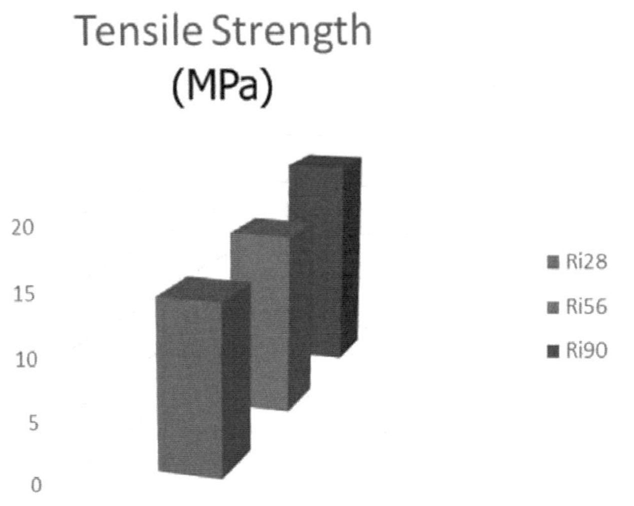

Figure 3. Tensile Strength of the analyzed composition.

The samples were tested at ages of 28, 56 and 90 days [10] and an increase in strength can be observed with testing at a later age, both at compression and at the tensile strength.

We can observe how the strengths increase from 28 days to 56 days and from 56 days to 90 days almost constantly but with no more than 50% of their value.

Figure 4 includes the comparative evaluation (before/after) for the analyzed microcracks.

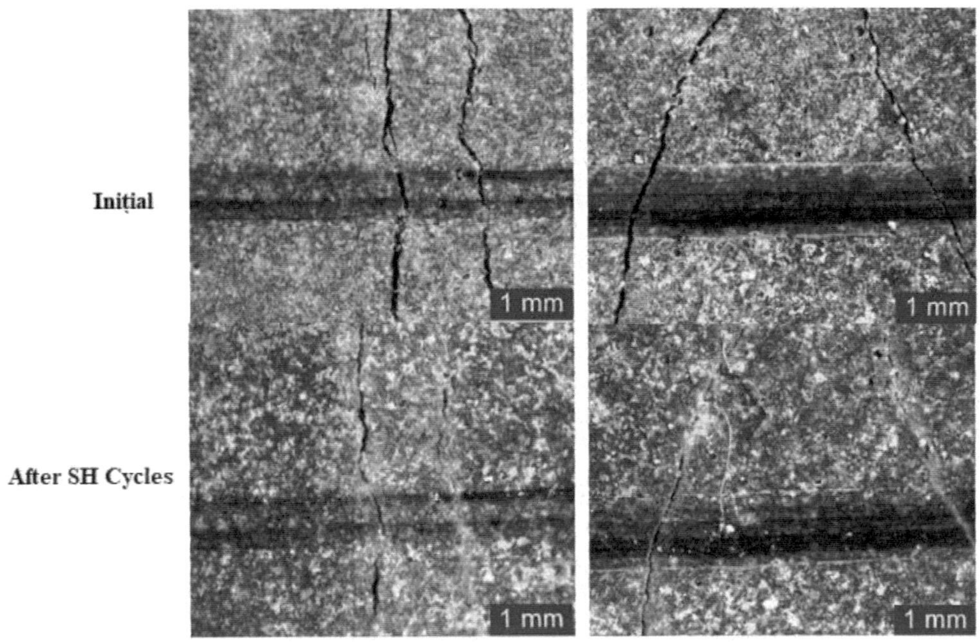

Figure 4. Microscopic analysis- specimens evaluated after microcracking and re-evaluated after supposing them at SH cycles.

4. Conclusion and perspectives

- After conditioning the samples, tested at 28 days, at 28 wet/dry cycles, 75% of the total microcracks closed completely, with only 25% partially closed;
- Following mechanical tests at different ages, a significant increase of strength from 28 to 90 days is observed, both at compression and in tensile strength;
- Since the results obtained from the testing of the compositions of cementitious composites materials with self-healing properties were very promising, the transition from small mortar-type castings to large castings – to concretes was made, trying to preserve the same compositional proportions. It has been observed that in large scale, the amount of water and additive used for small castings is too high and requires a compositional reassessment by reducing their percentages;
- The design of cementitious composite materials continues towards the creation of concrete compositions with self-healing properties, with characteristics at least the same as the cementitious composite materials studied.

5. References

[1] Szilagyi H, Baeră C, Mircea A C and Lăzărescu A 2019 *Lucrările Conferinţei de Cercetare în Construcţii, Economia Construcţiilor, Urbanism, Amenajarea Teritoriului* **15** 147–54
[2] Lepech M D and Li V C 2008 *ACI Mat. J.* **105**(4) 358–66
[3] Mircea A C, Toader T P and Dico C 2020 *Catalogul oficial al salonului "CadetINOVA"* vol 5 (Sibiu: Academia Forţelor Terestre "Nicolae Bălcescu") pp 328–331
[4] Li M and Li V C 2012 *Mater. Struct.* **46** 405–420
[5] Baeră C, Szilagyi H and Lăzărescu A 2019 *Proc. of the fib Symposium* pp 387–394
[6] Yang E-H 2008 *Designing added functions in Engineered Cementitious Composites* (Ann Arbor: University of Michigan)
[7] EN 196 (2016) Methods of testing cement. Determination of strength, *Romanian Standard Association, Bucharest, Romania*
[8] Mircea A C, Lăzărescu A, Toader T and Ionescu B 2019 *Catalogul oficial al salonului "CadetINOVA"* vol 4 (Sibiu: Academia Forţelor Terestre "Nicolae Bălcescu") pp 226–229
[9] Mircea A C and Toader T P 2020 *Catalogul oficial al salonului "CadetINOVA"* vol 5 (Sibiu: Academia Forţelor Terestre "Nicolae Bălcescu") pp 324–327
[10] Li V C 2008 Engineered Cementitious Composites (ECC) Material, Structural, and Durability Performance" *Concrete Construction Engineering Handbook, Chapter 24*, Ed. E. Nawy (Boca Raton: CRC Press)

IOP Conf. Series: Materials Science and Engineering **877** (2020) 012036 doi:10.1088/1757-899X/877/1/012036

Comparative Study on the Characteristics of Silicone Elastomers used in Dental Impression Techniques

I Gradinaru[1*], B-I Ciubotaru[2], M-F Zaltariov[2] and M Cazacu[2]
[1]Department of Implantology, Removable Dentures, Technology, "Grigore T.Popa" University of Medicine and Pharmacy, 16 Universitatii Street , 700115, Iasi, Romania.
[2]Department of Inorganic Polymers, "Petru Poni" Institute of Macromolecular Chemistry, Aleea Grigore Ghica-Voda, 41A, 700487, Iasi, Romania.

*E-mail: irina.gradinaru@umfiasi.ro; ciubotaru.bianca@icmpp.ro

Abstract. Even through nowadays technology and robotics seem to have taken a front seat on the medical field, there are still conventional methods being used in dental medicine, especially related to impression techniques. Elastomers such as silicone rubbers are widely used as impression materials for their suitable properties and behavior in dentistry. As the long-term success of the prosthetic therapy relies on the impression stage, it is important to know both the advantages and disadvantages of such materials. Nowadays *in vitro* and *in vivo* studies are being conducted to establish the precision and accuracy of these materials. In this study we report comparatively a number of characteristics determined by us under identical conditions for four such materials. Structural studies (through IR and XRF), morphology (through AFM and SEM) and surface (water contact angle) were performed. Based on the results obtained we can identify more or less significant differences between some characteristics, which could help to indicate the right material depending on the case requirements.

1. Introduction

Nowadays impression materials represent a useful tool in dental medical practice, being a method which does not require invasiveness for the patient and is easier and easier to use by dentists. These materials have been proved to be helpful and have, with the improvement of the obtaining technology in the last years, better qualities and characteristics which have been observed in both manipulation and results after usage. As the highest rate of usage is represented by elastic materials among impression materials, it is of high interest to know what requirements must these have and which are their eloquent traits for their medical application. Currently, in terms of classification of elastic materials or rubbery materials, two categories are underlined: synthetic elastomers represented by silicones, polyethers and polysulphydes as well as elastic impression materials as reversible hydrocolloids (agars) and irreversible hydrocolloid (alginates). The silicone materials can also be differentiated by the method of polymerization on setting in condensation silicone impression materials and addition silicone impression materials [1-3]. A classification of above-mentioned materials is presented in Figure 1.

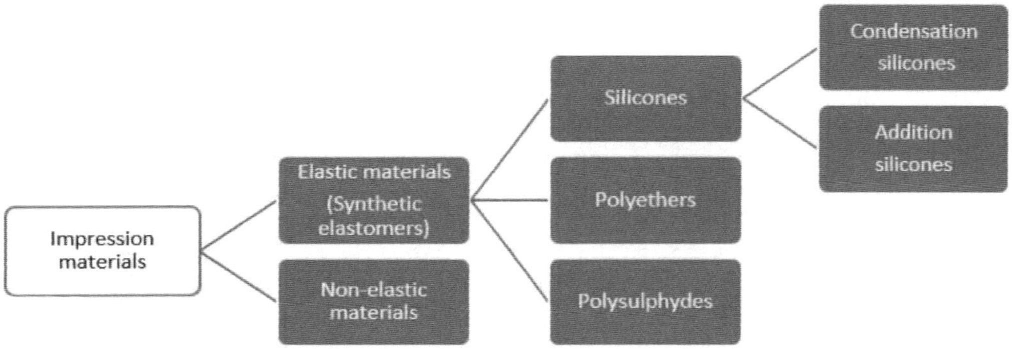

Figure 1. Classification of impression materials [1, 2].

Impression is a mandatory step in the dental prosthesis algorithm. Synthetic elastomers are irreversible elastic materials, the most commonly used category for impression taking. They are indicated to record the impression of the prosthetic fields for the fixed prosthesis as well as for the functional impression of the partially/total edentulous prosthetic field.

When an impression material is used, dentists expect that material to poses a series of requirements such as accuracy, dimensional stability which is bound to accuracy, good manipulative and handling ease and good setting characteristics. To these factors, clinical considerations can also be factors involved in selecting and using an impression material.

By definition the role of impression materials is to record intraoral structures in detail (of both teeth and soft tissues) with good accuracy which are then used for the prostheses construction [1].

In medical practice, dentists choose materials like these based on a process of association between the chemical groups contained and the properties of materials that recommend them for the ultimate purpose. Even so, it is sometimes difficult to evaluate a product based on the data provided by the producers. In this study, we have evaluated in terms of structure, morphology and surface analysis four different commercially available impression materials used in medical practice, as clinical behavior depends on materials' properties [4, 5]. For this, we selected and prepared as instructed by the producers' indications four different materials represented by Coltene Speedex putty, Coltene Speedex light body, Lascod Silaxil light body and Zhermack Oranwash L light body. The light body and putty property expresses the materials viscosity [6-9].

As chemical composition, silicone condensation materials contain polydimethylsiloxane (PDMS) as the principal component and inert inorganic mass (usually pyrolytic silica and titanium dioxide) to increase the viscosity of the liquid precursor and improve the mechanical strength of the elastomer formed by crosslinking, for the main body paste. The accelerator paste can consist of tin octoat, tetraethyl orthosilicate and sometimes also of chromium oxide or palladium metal particles with the role of capturing hydrogen, which is not beneficial for the surface of the impression [10].

For the proposed studies, all the samples were pre-prepared in the same way as 10 mm thick films, cut at the same dimensions of 2 cm^2.

2. Structural study

2.1. FTIR-ATR analysis

Attenuated Total Reflectance Infrared (ATR-IR) spectra have been measured by using a Bruker Vertex 70 spectrometer (Bruker Optics, Ettlingen, Germany) equipped with a ZnSe crystal. The registrations were carried out in the 600-4000 cm^{-1} spectral range, at room temperature with a resolution 4 cm^{-1} and accumulation of 32 scans. All samples were cut at the same dimensions of 1 cm^2, with a thickness of 10 mm for this analysis.

The main bands in the 4000-2500 cm^{-1} spectral range (Figure 2(a)) are assigned to the O-H stretching vibrations from alcohol intermolecular bonded in the structure of the materials. The C-H asymmetric and symmetric stretches in methyl and methylene groups of silicone components and alcohol can be observed at 2960-2920 cm^{-1} and 1860 cm^{-1}, respectively. The 2000-1500 cm^{-1} spectral region is characteristic for O-H deformation vibrations (1650 cm^{-1}) and vinyl stretching vibrations at 1608 cm^{-1} (from catalyst paste and vinyl silicones). The bands at 1730, 1744 and 1780 cm^{-1} are characteristic for esters present in the structure of the activators, while the bands at 1950-1870 cm^{-1} are specific for H-bonding interactions between the OH groups appeared during the hydrolysis reaction of the alkyl silicate and carbonyl groups (C=O) from activator (stannous octoate) (Figure 2(b)). In the 1600-1300 cm^{-1} spectral region there are a number of absorption bands assigned to the C-O asymmetric stretches (1554 cm^{-1}) of stannous activator, C-H bending in methylene and methyl groups at 1460 cm^{-1} and strong S=O stretching vibrations in sulphate groups (gypsum filler) at 1410 cm^{-1} and 1378 cm^{-1} (Figure 2(c)). In the 1300-600 cm^{-1} spectral range the symmetric deformation vibrations of the Si-CH_3 groups and the stretches of Si-O-Si and Si-C groups can be observed at 1260 cm^{-1}, 1080-1010 cm^{-1} and 790 cm^{-1}. The band at 868 cm^{-1} is assigned to the Si-O-C stretches in the silicate cross-linking agent, while the bands at 962, 934 and 902 cm^{-1} are due to the Si-O-C deformation vibrations (Figure 2(d)). The bands at 800-600 cm^{-1} are assigned to the C-H out-of-plane deformation vibrations and sulphate ions [11].

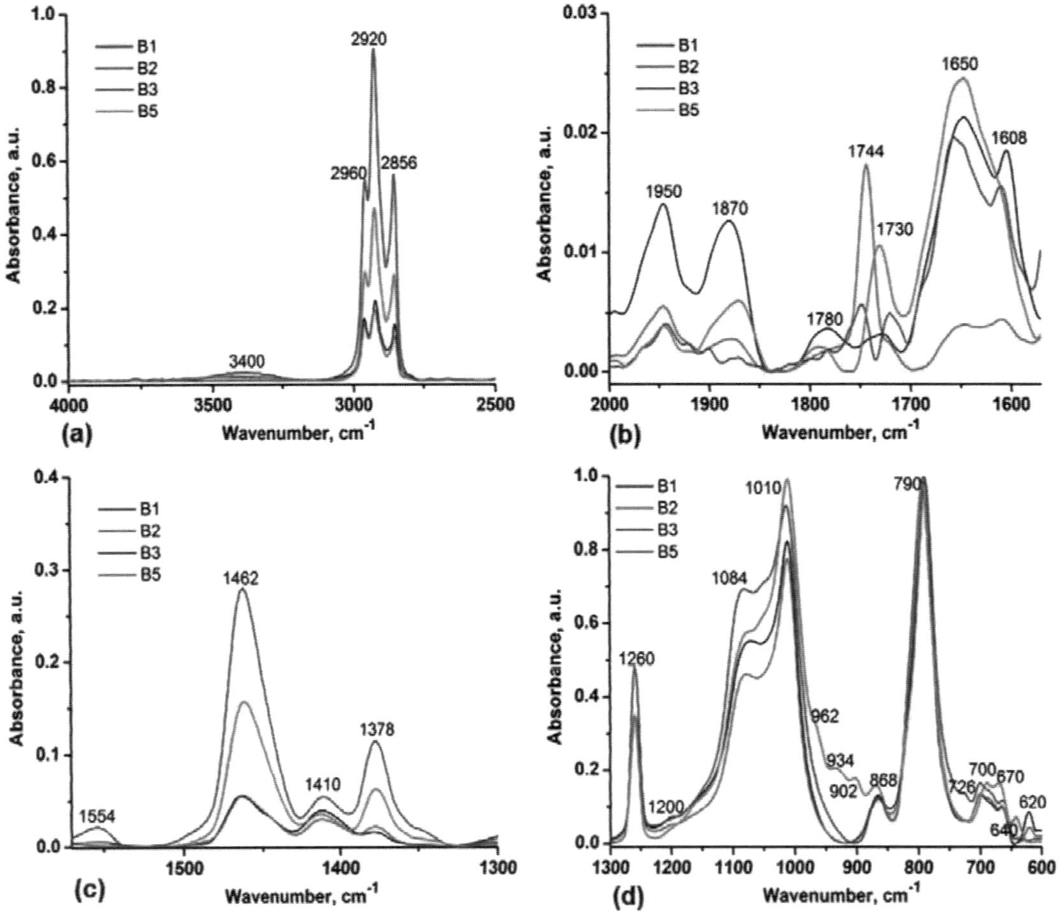

Figure 2. ATR-IR spectra of the four impression materials for Lascod Silaxil light body (B1), Zhermack Oranwash L light body (B2), Coltene Speedex light body (B3), Coltene Speedex putty (B5).

2.2. XRF analysis

Elemental composition of the samples was estimated by using X-Ray Fluorescence spectroscopy. XRF spectra were recorded on a X-Calibur Xenemetrix device equipped with a high-resolution detector and an X-ray tube, a chamber and a sample table that can accommodate various sample types and sizes and assisted with an Analytix software program. The tests were performed at ambient temperature of 32 C° and the results can be seen on Figure 3.

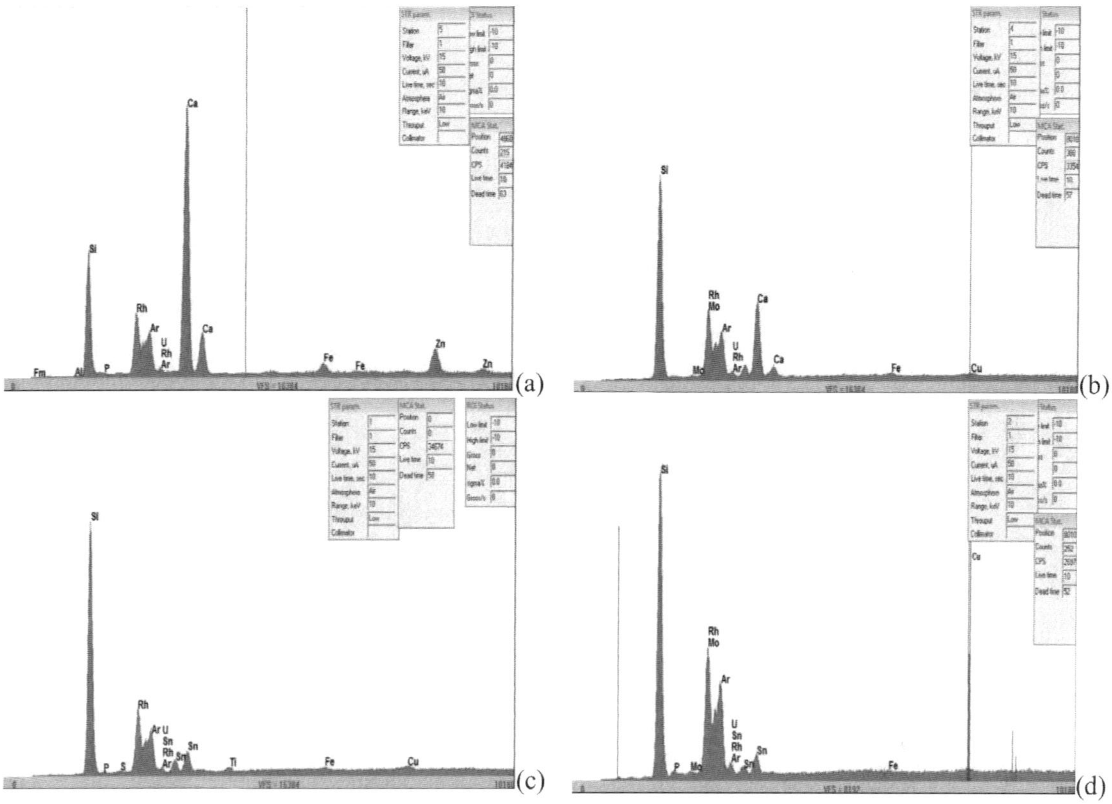

Figure 3. XRF spectra for (a) Coltene Speedex putty, (b) Coltene Speedex light body, (c) Lascod Silaxil light body and (d) Zhermack Oranwash L light body.

The XRF spectra confirm, as well as the IR spectra, that the impression materials are in fact silicone materials, the presence of Si being shown on all the results, as an element as such or in a chemical group, respectively. The presence of Cu and Fe on the XRF spectra is attributed to any added dye and fillers and/or manipulation tools used for the preparation of the materials [12].

Ferric impurities are often present in the citrine, while trace amounts of iron, copper or zinc are present in rose quartz. Additionally, phosphate or aluminium can be found in the pink quartz.

Table 1. Atomic composition and ratio for Coltene Speedex putty and light body.

Material	Atomic composition				Atomic ratio		
	Si	Ca	Cr	Zn	Si/Ca	Si/Cr	Si/Zn
Coltene Speedex putty	40120	17081	840	1323	3,35	88,7	70,4
Coltene Speedex light body	60438	4848			17,8		

Beside quartz, calcium phosphate is usually added as filler in base paste (putty), while the stannous catalyst is present in catalyst paste, so that traces of these elements can been observed in the XRF spectra.

Atomic ratio calculated for the most representative elements present in the Coltene Speedex putty and light body materials is shown in Table 1.

3. Morphology study

3.1. AFM

As impression materials used in dental medicine have to be precise, accurate for the fidelity of the desired result in the prosthetic therapy, it is of high interest for them to be smooth and have as less discontinuities as possible with a low roughness to them as a main characteristic.

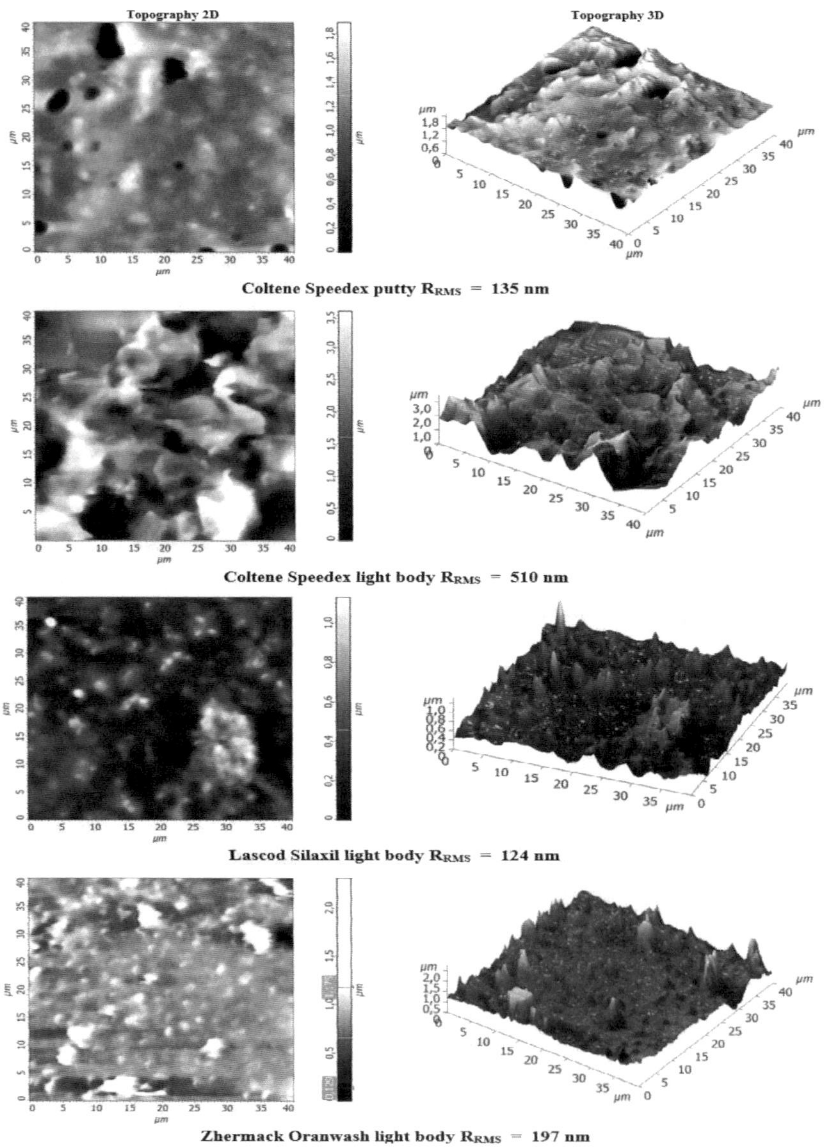

Figure 4. 2D and 3D topography of the four impression materials.

The less rough the surface of the impression material is, the more accurate will be the impression obtained with it, as even the smallest details are extremely significant [13-15].

With the purpose of obtaining fidelity characteristics, an Atomic Force Microscope Park SYSTEMS XE-100 with liquid cell and thermal control (ILX Lightwave LDT-5948 Precision Temperature Controller) was used for the morphology study of the impression materials selected and both 2D and 3D topography images can be observed in Figure 4. The estimated roughness values based on these images are between 124 nm for the Lascol Silaxil sample and 510 nm for the Coltene Speedex light body sample, the values for the other samples being in the first part of this field.

3.2. SEM analysys

As impression materials based on crosslinked silicone are solid inorganic materials, Scanning Electron Microscopy - SEM, was also conducted on film surface using an ESEM Quanta 200 Scanning Electron Microscope (Netherlands) operating at 20 kV. In dental practice, impressions taken from patients often have to maintain their shape for longer periods of time [16, 17]. As such, morphology characteristics were obtained for the materials tested after 14 days after their preparation.
SEM images captured for our samples are presented are in Figure 5.

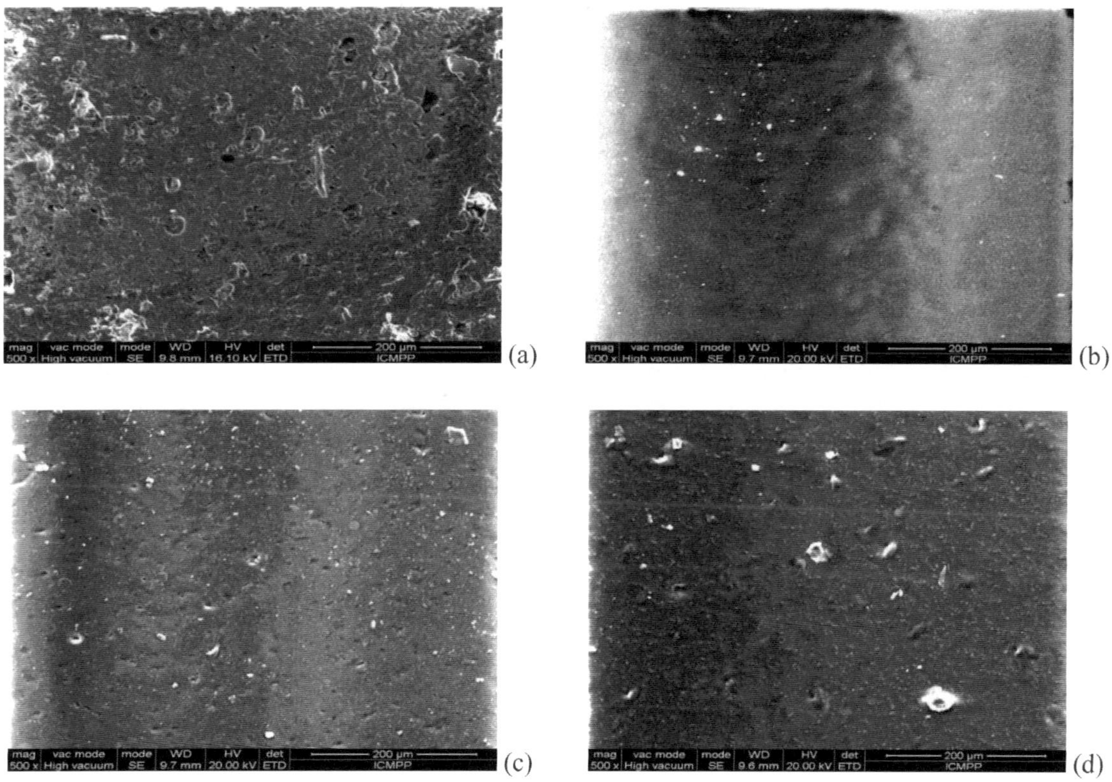

Figure 5. SEM images of impression materials (a) Coltene Speedex putty, (b) Coltene Speedex light body, (c) Lascod Silaxil light body, (d) Zhermack Oranwash L light body.

Based on the surface images obtained through SEM analysis the three samples of light body impression materials Coltene's Speedex, Lascod Silaxil and Zhermack Oranwash L seem to have close morphologies. However, unlike AFM images indicating roughness on small areas (40x40 µm), SEM images taken on larger areas indicate surface heterogeneities of materials created by the presence of additions. According to them, contrary to the AFM results, the Coltene Speedex light body sample

IOP Conf. Series: Materials Science and Engineering **877** (2020) 012036 doi:10.1088/1757-899X/877/1/012036

appears to have the most uniform surface, followed by Lascod Silaxil light body Zhermack Oranwash L light body and Coltene Speedex putty.

4. Surface study

The hydrophilic surface with a smaller static contact angle and larger work of adhesion values are those characteristics sought for the impression materials [2].

For the hydrophilic or hydrophobic characterization of the materials measurements of contact angle and work of adhesion were conducted using a CAM 101 (KSV Instruments, Helsinki, Finland) system, equipped with a liquid dispenser, video camera and drop shape analysis software. In this regard, experimental liquids were prepared and used (distilled water and a 6.8 pH buffer solution for the physiological simulation of saliva), while their surface tension was determined by using a Sigma 700 Force Tensiometer. The obtained results 72.8 N/m for distilled water and 59.9 N/m for the simulated saliva, were used in the calculations for the adhesion work $1+\cos\theta*72.8$ for water and $1+\cos\theta*59.9$ for simulated saliva. All measurements were made at room temperature.

Taking under consideration the nature of the materials used – biomaterials – one might expect that impression materials would have to be hydrophobic in order to achieve the best results when used. In this case the adhesion and cohesion forces which occur in surface tension when the materials come in contact with water would not make the material adhere to the water, which is the expected outcome [18-20].

In Figure 6 and Figure 7 are represented the static contact angle and the work of adhesion.

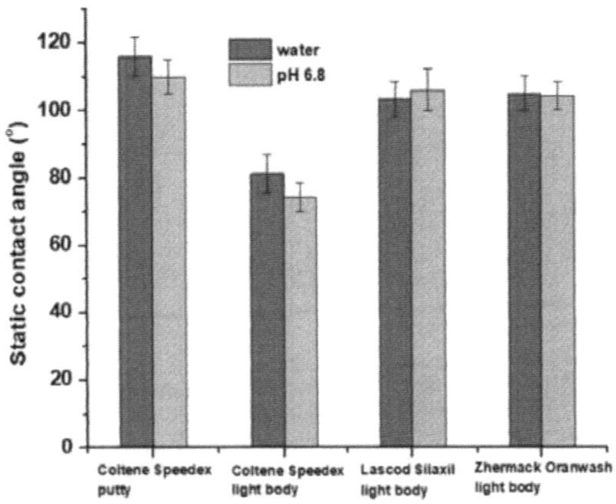

Figure 6. Static contact angle.

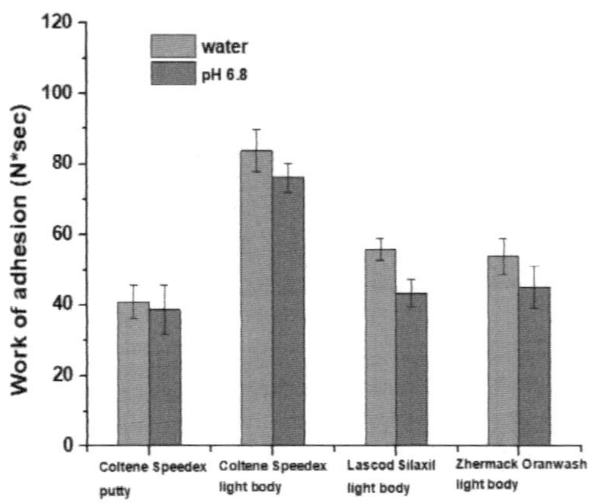

Figure 7. Work of adhesion.

As expected, the Coltene Speedex light body sample with the highest roughness has the smallest contact angle and therefore the largest adhesion work.

Correlating the contact angle values with AFM and SEM data, it could be said that the AFM provides more relevant data, as measurement technique. The hydrophilic surface and a small contact angle represent those characteristics sought for the impression materials.

5. Conclusion and perspectives

Silicone elastomer impression are among the most used impression materials in dental medicine. Four impression materials, commercially available, in light body form and putty have been tested in terms of structure, morphology and surface.

The structural study using Fourier Transform Infrared in Attenuated Total Reflected mode and X-Ray Florescence proves that these are in fact silicone-based materials, Si being the main element present on all spectra obtained, in different atomic ratios to other elements in all four materials.

The morphological studies were conducted by AFM and SEM. The AFM results that proved to be the most relevant from this perspective, indicate the Lascod Silaxil material as having the least roughness and as a result potential for rendering the details with the highest accuracy. However, roughness influences wettability, meaning that higher roughness will decrease the contact angle, as found in the Coltene Speedex light body sample, which is desirable for impression materials.

Thus, it can be concluded that none of these materials is ideal, each having its performance and limitations, which can be preferred depending on the situation.

For a better understanding of the structural part we intend to also study the materials by EDAX microanalysis (Energy Dispersive X-ray Analysis).

Acknowledgements
The authors would like to thank dr Carmen Racles for the XRF analysis, dr. Vasile Tiron for the AFM results, dr. Florica Doroftei for the SEM results and dr, Luiza Gradinaru for contact angle measurements.

6. References

[1] McCabe J F and Walls A W G 2018 *Applied Dental Materials – 9th ed.* Blackwell Publishing Ltd
[2] Tanculescu O, Łâmpu D andronache, M, Vieriu R and Doloca A 2011 Characterization of some condensation silicone impression materials. *Bulletin of the Polytechnic Institute of Iaşi, Section Chemistry and Chemical Engineering.* 57-33
[3] Chen S Y, Liang W M and Chen F N 2004 *J Dent* **32** 603-609
[4] Stober T, Johnson G H and Schmitter M 2010 *J Prosthet Dent* **103** 228-239
[5] Rubel B S 2007 *Dent Clin North Am* **51** 629-646
[6] https://www.coltene.com/products/prosthetics/c-silicones/speedex/speedex-putty/
[7] https://www.coltene.com/products/prosthetics/c-silicones/speedex/speedex-light-body/
[8] https://www.lascod.it/en/product/silaxil-body/
[9] https://www.zhermack.com/en/product/Oranwash L-l/
[10] Patrascu I 2002 *Materiale dentare* Horanda Press p 79-80
[11] Stuart B 2004 *Infrared Spectroscopy: Fundamentals and Applications* John Wiley & Sons, Ltd
[12] Motohiro Uo, Takahiro W and Tomoko S 2014 *Japanese Dental Science Review.* **10** 1016-1014
[13] Giordano R 2000 Impression materials: basic properties. *Gen Dent.* **48** 510-516
[14] Chee W L and Donovan T E 1992 *Journal of Prosthetic Dentistry* **68** 728–732
[15] Christensen G J 1997 *J Am Dent Assoc* **128** 1026-1028
[16] Pant R, Juszczyk A S and Clark R F K 2008 *J Dent* **36** 456-461
[17] Chen S Y, Liang W M and Chen F N 2004 *J Dent* **32** 603-609
[18] Walker M P, Petrie C S and Haj-Ali R 2005 *J Prosthodont* **14** 158-163
[19] Pratten D H and Craig R G 1989 *J Prosthet Dent* **61** 197-202
[20] McCabe J F and Carrick T E 2006 *Eur J Prosthodont Restor Dent* **14** 42-46

IOP Conf. Series: Materials Science and Engineering **877** (2020) 012037 doi:10.1088/1757-899X/877/1/012037

Influence of Primary Power Source on Heat-insulating Material Choice for Autonomous Telecom Unit Container

S Kiseleva[1], A Tarasenko[2*] and A Fedotov[3]
[1] Joint Institute for High Temperatures of the Russian Academy of Sciences, Moscow, Russian Federation
[2] Lomonosov Moscow State University, Faculty of Geography, Moscow, Russian Federation
[3] National Research Univerity Moscow Power Engineering Institute, 17/3 Krasnokazarmennaya Street, 111250 Moscow, Russia

E-mail: a.b.tarasenko@gmail.com

Abstract. The paper is devoted to the energy balance estimation for autonomous telecom unit, powering from photovoltaic panels and diesel-fuelled genset. Unit container has its own heat balance, basing on climate control system and parasite heating from telecom equipment inside it. Climate control system is also considered as power consumer from PV or genset. Obviously this power consumption depends on unit load graph, heat exchange with environment, environmental temperature. Energy and economical analysis has been conducted for several possible unit geographical locations, three types on heat-insulating materials and different share of solar energy in energy balance. It has been shown that in case of diesel fuel domination in energy balance and not very low environmental temperature heat-insulating material with lowest heat conductivity is not an optimal choice from operational expanses point of view (through high fuel consumption in summer for the container cooling).

1. Introduction
Problems of electric power supply for remote territories have been of great interest for many years. Diesel genset is a traditional solution of problem. Coupling it with photovoltaic panels usually solves ecology and financial problems [1]. Climate issues, concerned with severe temperature ranges, are also taken into account [2]. Telecom basic units are one of the most widespread remote power consumers. And they need uninterrupted power supply in 24/7 mode [3]. This research is devoted to the operation of telecom basic unit (repeater) powered from diesel genset combination with PV panels and battery. Different locations of unit in Russia are considered. The issues of climate control inside the unit container are also taken into account.

2. Materials and methods
Unit includes repeater itself, container with telecom equipment and climate control system (split) which can be operated as heater or air condition unit depending on temperature in container. It is assumed that temperature range (plus 5 – plus 25)°C is suitable for telecom equipment inside the container. Electric load graph for power supply system consist power fed to repeater and to climate-control system. Typical operation diagram for repeater is taken from [4] and is assumed to be similar for the every day of the year. It is given in Figure 1. Peak power (100%) is assumed to be 3 kW.

Container for telecom equipment consists of sandwich-type panels and has dimensions of 4000*2500*2400 mm [5]. Sandwich-type panels have thickness of 110 mm.

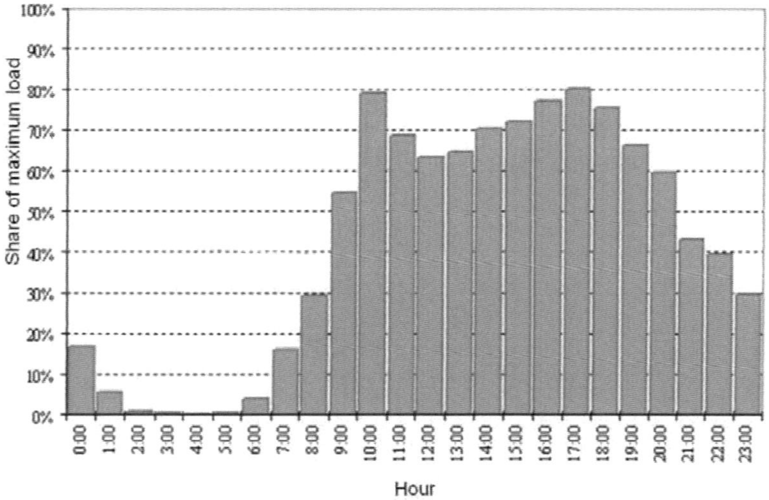

Figure 1. Basic load graph for telecom repeater.

Heat sources inside the container are:

- *repeater* with efficiency $\eta = 83\%$ [6], generated heat during Δt is

$$Q_1 = \eta * P_{tr} * \Delta t, \quad (1)$$

where P_{tr} – averaged consumed electric power during Δt, taken from Figure1 for every hour (calculation step is 1 h);

- *heat-exchange* with the environment through the container walls and roof. All heat flows are assumed to be equal and independent from the direction. Penetrated or lost heat is:

$$Q_2 = k * \frac{S}{l} * (T_{out} - T_{in}) * \Delta t, \quad (2)$$

where k stands for heat-exchange coefficient, S – total area of heat-exchange, l – heat-insulating media thickness, T_{out} – environment temperature (is taken from [7] for every day of the year and processed according to [8] to obtain hour data), T_{in}– temperature inside the container, stabilized by split:

$$k = \frac{1}{\frac{2}{\alpha} + \frac{l}{\lambda}} \quad (3)$$

Here λ stands for panel's material heat conductivity, α – for air inside the container heat transfer coefficient.

- *split operation* for heating or cooling of the air inside the container. Generated heat is equal to:

$$Q_3 = C * m * (T_2 - T_{in}), \quad (4)$$

where C is air heat capacity, m – air mass inside the container, T_2 is air temperature inside the container for the previous hour of calculation. T_{in} is derived from

$$Q_1 + Q_2 + Q_3 = 0 \quad (5)$$

Then Q_3 is calculated, taken into account condition, that plus $5^{\circ}C \leq T_{in} \leq$plus $25^{\circ}C$.

Values of Q_2 and Q_3 can be both negative and positive. Negative Q_3 value means that cooling is needed. In this case electric power consumption is estimated as 1 kWh of electric power for 3.2 kWh of heat (cooling) power. Positive Q_3 value means need heating and electric power in this case equals to needed heat.

Three approaches to telecom unit electric power supply are considered:

1) Diesel genset (G). In this case 2 gensets of equal power output are included in capital expenses due to one of them is needed to be in reserve. Here for every hour power output of genset is equal to needed electric power. Nominal power of genset used can be changed, but in case if needed power is more, than genset nominal power, load cannot be covered by such genset. Specific fuel consumption of genset is calculated for every hour depending on ratio of operated power to nominal power, according to [9].

2) Diesel genset coupled with PV panels (SG). Control system similar to SMA fuel save controller is supposed to be installed [10]. Diesel genset power cannot be lower than adjusted value (40% of nominal power), so power output of PV panels is lowered in case of low load. PV panels power output is estimated basing on [7, 8] similarly to [11]. As in [12], PV inverter efficiency depends on the translated power of PV array. PV array power also is a parameter, that can be changed for optimization.

3) Diesel genset coupled with PV panels and lead-acid battery (SGB). Battery efficiency equal to 85% [13] is taken into account in this case. Unit is fed from PV array (estimated as in case 1) and from battery, charged by excess PV output. In case if battery charge left has value of 40% of nominal energy capacity (can be optimized during calculation), genset is operated at its nominal power to feed the load and charge the battery. Such approach is supposed to decrease specific fuel consumption and CO_2 emissions at low power operation.

For every scheme total fuel consumption, number of genset service periods (1000 h of nominal operation) are estimated for 2-year period (2012-2013 actual data from [7]). Operation expenses for 10 year period are obtained multiplication of consumed fuel cost and service costs by 5. Costs for diesel fuel in Russian regions are taken from [14]. Capital costs for every scheme are estimated as sum of scheme components used. Initial prices for OPzV Lead-acid batteries, PV array mounting poles and gensets are obtained averaging data from [15]. Currency exchange course is assumed to be 70 russian roubles (RUR) for 1 US dollar (USD). For PV panels electrical and temperature parameters of Yingli-250YL were used in calculation.

3. Results and discussion

Calculation of capital and operation expenses for 3 types of power plant, feeding telecommunication unit (G – genset, SG – PV+genset, SGB – PV+genset+battery) have been taken for 4 locations in Russia. Locations included Palana (Kamchatka region with many remote consumers), Arkhangelsk (Northern port with low insolation and low temperatures), Mirny (Yakutia region with extremely low temperatures), Lago-Naki (Krasnodar region near Black Sea cost, where PV-feed telecom unit is already installed). Peak electric energy consumption was estimated as 3.4 kWe. PV array size is taken equal to 5 kWe and battery bank is of 32 kWh (8 h feed) lead-acid batteries OpZV type. Genset power output is equal to 4 kWe. Actual prices for diesel fuel for all regions were taken from [14]. Two types of commercial heat-insulating materials for telecom unit container were used – Plastmigran with specific heat conductivity of 0.08 W/m*K and Korund with 0.012 W/m*K. The results are given in Table 1.

IOP Conf. Series: Materials Science and Engineering **877** (2020) 012037 doi:10.1088/1757-899X/877/1/012037

Table 1. Technical and economic characteristics of power sources when using liquid heat-insulating media Korund (0.012 W/(m*K)) and solid heat-insulating media Plastmigran (0.08 W/(m*K)). Fuel Consumption for 10 years, tons – FC; Total Expences for 10 years, th. USD – TE-10.

	Palana (Kamchatka region)[a]		Mirny[b]		Arkhangelsk[c]		Lago-Naki[d] (Krasnodar region)	
Heat-insulating media Korund								
	FC	TE-10	FC	TE-10	FC	TE-10	FC	TE-10
SGB	28	36	26	44	28	42	29	41
G	198	106	195	180	198	147	205	145
SG	53	40	53	60	53	51	53	50
Heat-insulating media Plastmigran								
	FC	TE-10	FC	TE-10	FC	TE-10	FC	TE-10
SGB	25	34	23	41	25	38	27	39
G	184	98	177	162	182	135	193	137
SG	55	41	56	62	55	51	53	50

[a] Fuel price: 34000 rur/t [https://www.benzin-price.ru/price.php?region_id=41]
[b]Fuel price: 60700 rur/t [https://www.benzin-price.ru/price.php?region_id=14]
[c] Fuel price: 48500 rur/t [https://www.benzin-price.ru/price.php?region_id=29]
[d]Fuel price: 46270 rur/t [https://www.benzin-price.ru/price.php?region_id=23]

Calculation results showed that for all regions, due to high share of energy spent on equipment cooling, heat-insulation material with the best properties is not an optimal choice in case when G or SGB is used as energy source (Figure 2).

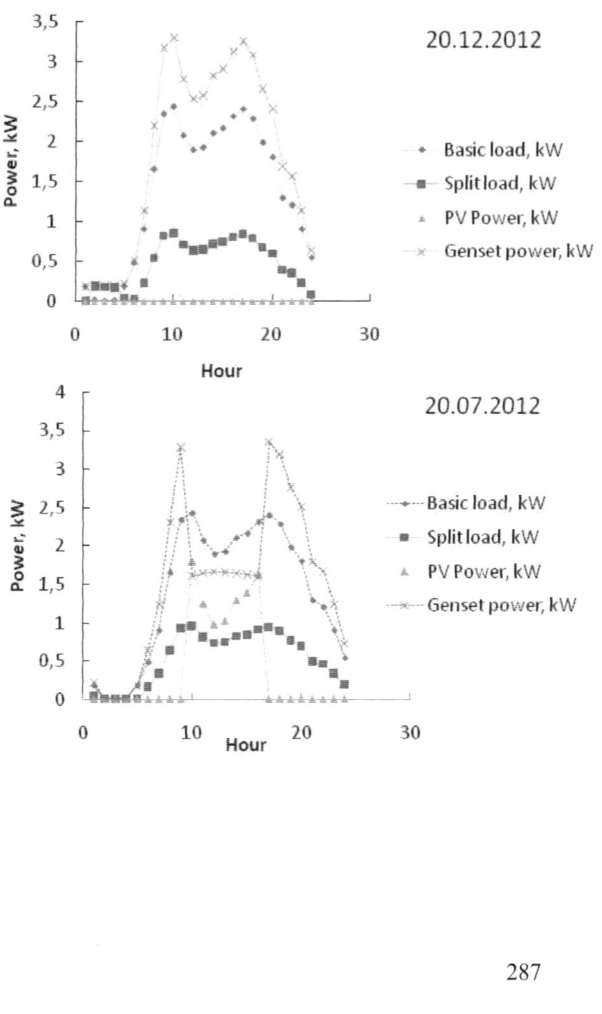

Figure 2. Power balance for Mirny during winter and summer model day.

In case of SG better heat-insulating media properties lead to equal or lower fuel consumption due to the fact that PV array can be operated only during day hours (in case of SGB significant part of solar-given energy can be transferred to night because of battery presence), when environmental temperatures are higher. Diesel fuel in SG scheme is usually consumed at night, when low environmental temperatures and low heat emission from repeater are seen. So for SG good container heat insulation is essential. For G and SGB schemes there is no significant difference for day and night energy consumption because of the constant genset operation or battery presence – solar energy during day can be absorbed into the battery and spent at night. Possibilities of SGB scheme to absorb solar energy are quite higher than for SG, because of limitations in SG scheme not to operate diesel genset at low power. Increase of fuel consumption for G and SGB scheme in case of better heat insulating media properties can be explained by domination of power consumption for cooling, not for heating during the whole year.

Results from Table 1 also showed that SGB scheme with 8-10 h possibility of battery operation, despite higher capital expenses can provide much lower fuel consumption and total expenses in 10-year period.

4. Conclusion

Several electric power feeding schemes, including diesel genset, PV array and lead-acid battery, were numerically estimated for remote telecom repeater container in 4 Russian regions with different climate conditions. Issues of climate-control inside the container due to telecom equipment heat emissions and their influence on load graph have been taken into account. Two heat-insulating medias with quite different heat conductivity have been considered in estimations. Calculation showed that application of PV array can significantly decrease diesel fuel consumption and total expenses for power supply of the remote user. Battery application leads to further fuel consumption decrease, but growth of capital expenses impedes the total costs decrease. Energy consumption for climate control appeared to be significant. For diesel genset, coupled with PV array without battery, is essential to provide good heat-insulation of the telecom equipment container because of the night diesel genset operation mainly for heating. For schemes with single genset and PV-genset-battery higher quality of heat-insulating material led to higher total diesel fuel consumption because of high energy consumption for cooling in spring, summer and autumn time. Difference between night and day consumption in this case is small, because battery increases solar energy absorption during day and allows its consumption at night.

Acknowledgments
This research has been conducted as a part of RAS Joint Institute for High Temperatures basic research program.

5. References

[1] Salameh T, Ghenai C, Merabet A and Alkasrawi M 2020 *Energy* **74** 116475
[2] Li C, Zhou D, Wang H, Cheng H and Li D 2019 *Energy* **185** 671–81
[3] Khan M J, Yadav A K and Mathew L 2017 *Ren. Sust. Energ. Rev.* **76** 577-607
[4] *Cellular radiotelephone communications as hygiene-significant electromagnetic field source.* URL: http://www.vrednost.ru/pole.php
[5] *Block-container catalogue.* URL: http://krial.nt-rt.ru
[6] *PicoCell 900 SXL repeater manual.* URL: www.shop.mmw.ru
[7] *NASA Prediction Of Worldwide Energy Resources Official Website.* URL: https://power.larc.nasa.gov/data-access-viewer/
[8] Duffie J A and Beckman W A *Solar Engineering of Thermal Processes, Fourth Edition.* Hoboken, New Jersey: John Wiley & Sons, Inc. 910
[9] Soto D 2018 *En. Sust. Dev.* **45** 180-85

[10] *SMA Solar Technologies GmbH official Web-site.* URL: https://www.sma.de /en/products/monitoring-control/sma-hybrid-controller.html

[11] Gabderakhmanova T S, Kiseleva S V, Frid S E and Tarasenko A B 2016 *J. Physics: Conf. Ser.* **774** 1-10

[12] Kolomiets Yu G, Menshikov Ya A, Tarasenko A B 2018 *Appl. Sol. En.* **54**(4) 246–50

[13] Tarasenko A B, Fedotov A A 2019 *IOP Conf. Ser.: Mater. Sci. Eng.* **564** 012137

[14] *Prices for benzene and oil products in Russian regions*. URL: https://www.benzin-price.ru/price.php

[15] *Your solar home LLC official Web-site.* URL: https://shop.solarhome.ru/

IOP Conf. Series: Materials Science and Engineering **877** (2020) 012038 doi:10.1088/1757-899X/877/1/012038

Expanded Foam Glass - an Application for Fire Resistant Multilayer Materials

O Mocanita[1], D L Chicet[2*], O Rusu[2], B Istrate[1], and C Munteanu[1]
[1]Technical University of Iasi, Mechanical Engineering Department, 61-63 D. Mangeron Blvd, 700050, Iasi, Romania
[2]Technical University of Iasi, Materials Science and Engineering Department, 41 D. Mangeron Blvd,700050, Iasi, Romania

*E-mail: dchicet@tuiasi.ro

Abstract. This study is focused on obtaining a fire-resistant multilayer material, which includes in the structure a layer with properties of high temperature resistance and low heat transfer. One of the research directions is focused on obtaining such a layer using expanded glass granules with two different dimensional ranges noted with A (1 - 2 mm) and B (0.25 - 0.5 mm). The target layer was realized with the shape of square plates of 50 x 50 mm, with a thickness of 5 and 10 mm respectively, from a mixture of quartz sand with a polymeric resin as bonding agent in which the expanded glass granules were inserted. The following samples were obtained: P5A and P5B (5 mm thick plates), respectively P10A and P10B (10 mm thickness plates). The samples were analyzed both from the morphological point of view (the emphasis being placed on the structure distribution of the expanded glass granules) and from the point of view of the resistance to high temperatures (by indirect exposure at temperatures between 400 - 600 ° C for 3 hours). The satisfactory behavior of the samples during the tests recommends the extension of the researches in this direction, in order to establish the optimal percentage of component elements.

1. Introduction

Expanded foam glass is a material that is gaining more and more ground as a construction material and also in other industrial applications, due to a positive combination of properties: lightweight and pressure-resistance, durable and dimensionally stable, 100% mineral, spherical grain shape, no hazards to health, creamy-white color, ecological - made of post-consumer recycled glass.

Currently, this type of material is used for a wide range of applications, among which can be listed: lightweight concrete [1], plaster and dry mortar, lightweight construction panels [2], fillings, 3D printing with printing beds or with SLS (selective laser sintering), cellular composites [3] or syntactic foams [4] with applications in the automotive field for shock absorber structures, soil conditioner in plant substrates [5], prefabricated sanitary parts or desalination compresses for historic masonry etc.

In addition to the low density that recommend this material for lighter structures, expanded foam glass is also characterized by a very low coefficient of thermal conductivity (λ = 0,07 W / (mK)) and a softening point at a temperature of about 700°C [5]. From the chemical composition point of view, this material belongs to the category of ceramic materials, because is created from a combination of oxides, as follows: SiO_2 (70-75%), Na_2O (10 - 15%), CaO (0, 5 - 5%), MgO (0 - 5%), K_2O (0 - 4%) [5].

Starting from these remarkable properties, in this paper was studied the possibility of obtaining fire-resistant materials, which should include in the structure expanded foam glass granules. This study is part of a larger one, focused on obtaining fire-resistant multilayer materials, in which the thermal insulation role is fulfilled by materials reinforced with expanded foam glass granules. For this purpose, two types of material were made with granules of different sizes, mixed in different percentages with quartz sand and a polymeric binder. Two properties of the samples were studied: fire resistance by exposure to open flame and heat transfer capacity. During the tests, the influence of the size of the expanded foam glass spheres on the fire resistance of the analyzed samples was also evaluated.

2. Materials and methods

The following materials werc used to obtain the samples tested in terms of fire resistance in this study:
1. quartz sand;
2. silicon-oxygen polymer resin containing ionic sodium (Na^+) components, as bonding agent (commercially available);
3. expanded foam glass in spheroidal shape (produced by Dennert Poraver GmbH), with a grain size between 1 - 2 mm (A), respectively with a grain size between 0.25 - 0.5 mm (B). The aspect of the expanded foam glass granules is shown in Figure 1 a, b, for the two categories of granulation.

(a)

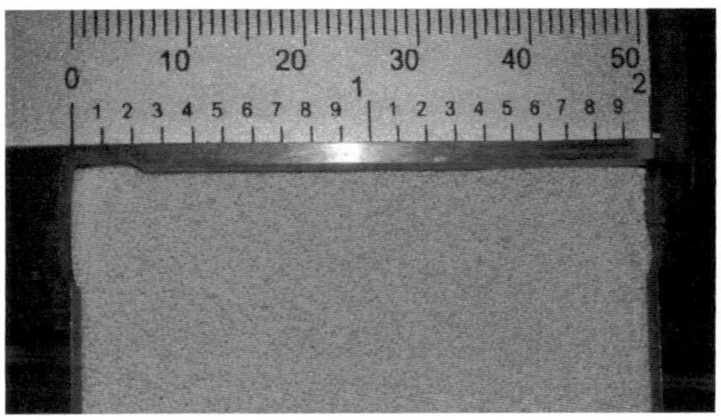

(b)

Figure 1. The aspect of the expanded foam glass granules with granulation of: a) 1 - 2 mm (A); b) 0.25 - 0.5 mm (B).

The materials were mixed in different percentages, as presented in Table 1. The mixture was modelled in specially designed shapes, in the form of plates with the following dimensions: 50 x 50 mm and thicknesses of 10 cm (P10A, P10B), respectively 5 mm (P5A, P5B). After extracting the plates from the moulds, they were left to dry in a normal environment, with a constant temperature (22 ± 2°C) and humidity between 45 - 50%. Figure 2 shows the samples aspect after drying. The homogeneous distribution of the spheroidal granules of expanded foam glass in the whole mass of the sample is observed. All the samples were rigid, with strong connections formed between the components used inside them.

Table 1. Mass ratio of the materials used to obtain the samples.

Sample	Sand ratio [wt%]	Resin ratio [wt%]	Expanded glass ratio [wt%]	
			A	B
P10A, P5A	13.4	6.2	1	-
P10B, P5B	9	3.33	-	1

IOP Conf. Series: Materials Science and Engineering **877** (2020) 012038 doi:10.1088/1757-899X/877/1/012038

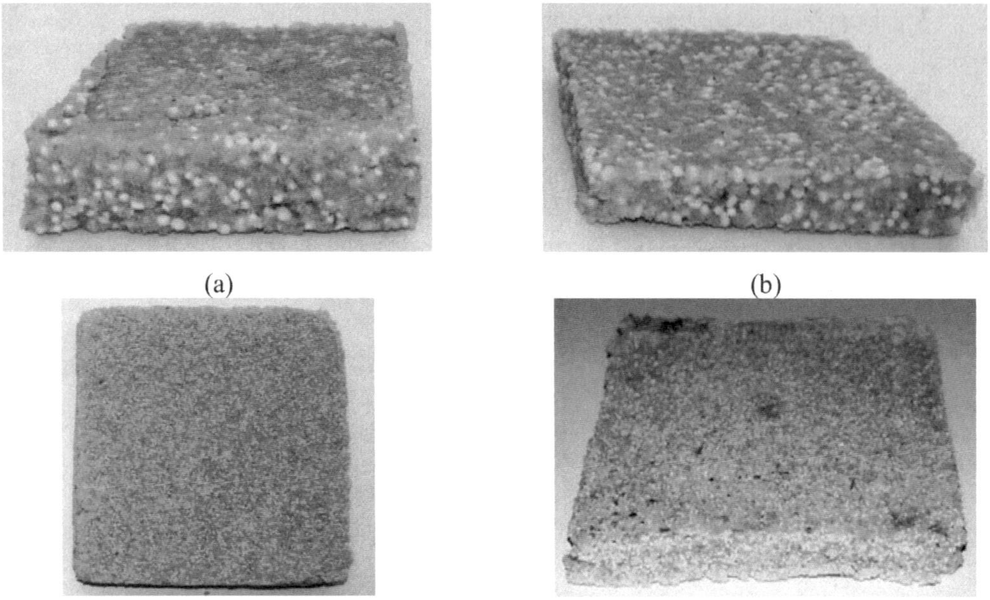

(a) (b)

(c) (d)

Figure 2. Samples aspect after drying: a) P10A; b) P5A; c) P10B; d) P5B.

For the open flame testing of the samples, a non-standardized test method was chosen, using a stand in the form of a wall, made of refractory bricks, in the middle of which a 50x50mm hole was drilled [6, 7]. In this hole, the tested samples were mounted in turn, being ensured a better sealing of the edges so that there is no passage of flames on the other side of the samples, which could lead to errors in measuring the temperature in the concerned area.

In order to obtain conditions as close as possible to reality (eg in the case of open flame fire), a flame produced by burning a gaseous fuel was used (butane - propane mixture, 87 - 13%). The burner was mounted at a fixed distance of 10 cm from the mounting surface of the samples in the test stand, as can be seen in Figure 3 a. With the help of a pressure regulator, a constant pressure of the gas was maintained throughout duration of the tests, so that the temperature recorded at the surface of the sample does not have major variations and could be maintained in the range of 600 - 650°C [8].

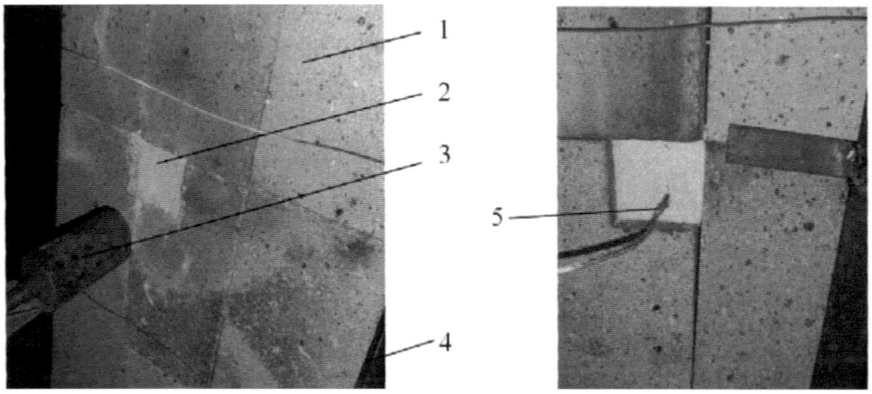

(a) (b)

Figure 3. Aspects of the test stand: a) component elements of the stand: 1 - thermo-resistant brick wall; 2 - sample; 3 - gas burner; 4 - metal frame; b) thermocouple position in case of P10A.

The recording of temperature variation during the experiments was performed using a digital thermometer TM 902C, equipped with a K-type thermocouple (-50 - 1300 ° C), which was positioned in direct contact with the surface of the sample on the side that was not exposed to direct flame, as can be seen in Figure 3 b.

The open flame exposure tests for samples P10A and P10B respectively were carried out for 60 minutes, the temperatures being recorded as follows:
- once every 30 seconds for the first 20 minutes,
- once every 1 minute, between 21 and 30 minutes,
- once every 5 minutes between 31 and 60 minutes.

The open flame exposure tests for samples P5A and P5B were carried out for 10 minutes, the temperatures being recorded once every 30 seconds.

3. Results and discussions

In this study, the two aspects necessary to characterize the behavior of a material subjected to open flame were followed: structural integrity and thermal insulation capacity. These performance criteria are regulated by SR EN 1363-1 "Fire resistance tests. General Conditions": sealing and insulation. Sealing is the time in minutes in which the specimen continues to maintain its distinct functions during the test without resulting in sustained flame. Isolation is the time in minutes in which the specimen continues to maintain its distinct functions during the test without developing on the unexposed side temperatures which: a) increase the mean temperature above the initial mean temperature by more than 140 ° C; or b) grow at any point above the initial mean temperature above 180°C [9].

If we analyze the behavior of the samples from the point of view of the first performance criterion - the sealing capacity, we can claim that the samples P10 A, respectively P10 B had a satisfactory behavior, because none of them lost their structural integrity and did not burned with flame after the tests they were subjected to. Regarding the P5A and P5B samples, respectively, they did not burn with an open flame during the tests, but they underwent geometric changes.

Figure 4 shows the aspect of the sample P10A, being observed that on the side exposed to direct flame a layer with an expanded appearance was formed, which still has rigidity and mechanical strength. The aspect of the side that has not been exposed to the flame is similar to that of the sample in its initial state, lighter in color due to dehydration and vitrification of the binder, but without major changes or destruction of the surface. These observations are supported by the cross-section aspect of the sample (Figure 4c), which show a slight damage of the expanded foam glass granules that have become brittle, but a structure of the sample still homogeneous and which has not been affected differently in thickness.

The same observations can be made in the case of sample P10B, the aspect of which is shown in Figure 5. However, a less expanded aspect of the surface exposed to direct flame is observed (Figure 5a), while the aspect of the side that was not exposed to the flame is very similar to that of the initial sample (Figure 5b). The cross section of the sample (Figure 5c) shows that no differentiated transformations occurred on the thickness of the sample, that were produced only those specific to the dehydration and vitrification of the binder, and the expanded glass granules were not affected so as to break at sectioning.

However, the geometry of the samples with a thickness of 5 mm was affected, their appearance being presented in Figure 6 a, b (P5A), respectively Figure 7 a, b (P5B). In both cases, the same expanded aspect of the surface exposed to direct flame is observed (Figure 6a, Figure 7a), which is present in this case on the sides that were not exposed to flame (Figure 6b, Figure 7b). Also, a different curvature of the samples in the central area is observed: towards the flame in the case of the P5A sample, respectively in the opposite direction to the flame in the case of the P5B sample.

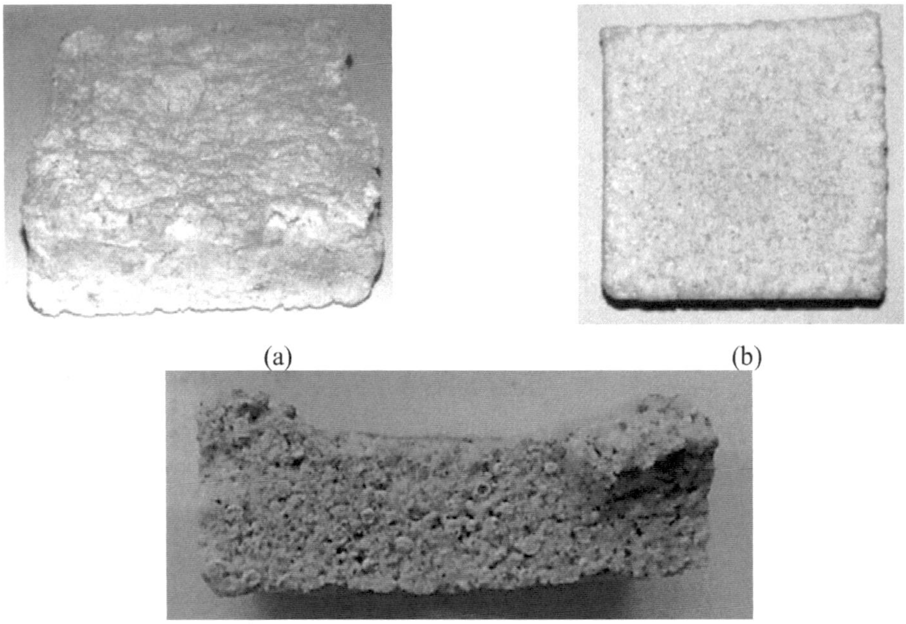

(a) (b)

(c)

Figure 4. The aspect of the P10A sample after open flame exposure for 60 minutes: a) the aspect of the side exposed to direct flame, b) the aspect of the side that has not been exposed to the direct flame, c) the aspect of the cross-section of the sample.

(a) (b)

(c)

Figure 5. The aspect of the P10B sample after open flame exposure for 60 minutes: a) the aspect of the side exposed to direct flame, b) the aspect of the side that has not been exposed to the direct flame, c) the aspect of the cross-section of the sample.

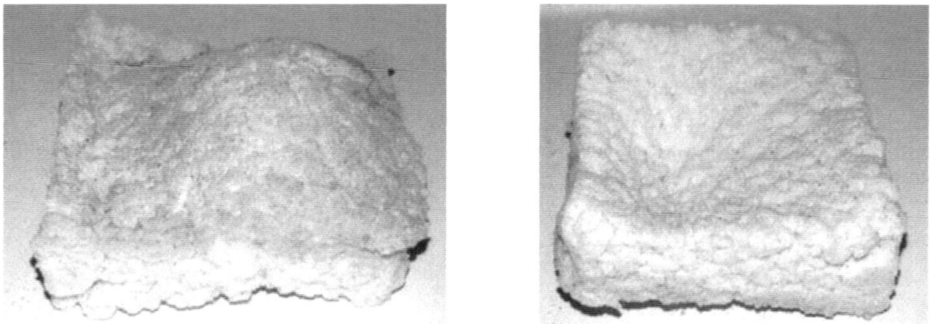

(a) (b)

Figure 6. The aspect of the P5A sample after open flame exposure for 60 minutes: a) the aspect of the side exposed to direct flame, b) the aspect of the side that has not been exposed to the direct flame.

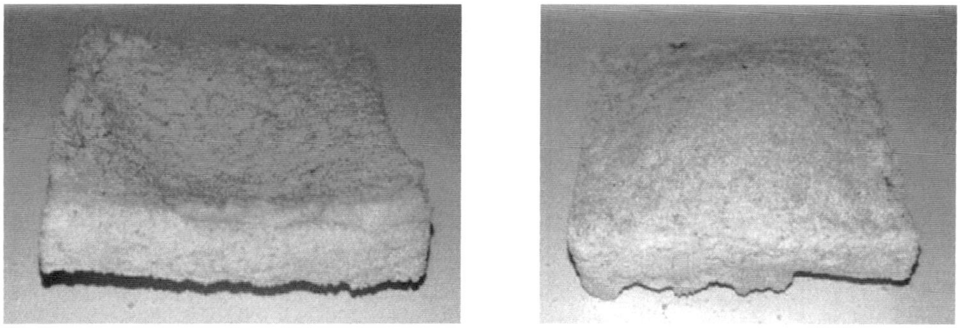

(a) (b)

Figure 7. The aspect of the P5B sample after open flame exposure for 60 minutes: a) the aspect of the side exposed to direct flame, b) the aspect of the side that has not been exposed to the direct flame.

In the case of evaluating the second performance criterion, that of thermal insulation (heat transfer), comparative graphs of the temperatures recorded during the tests were made (Figures 8 and 9), both on the sides that were not exposed to direct flame and on the side exposed to the flame.

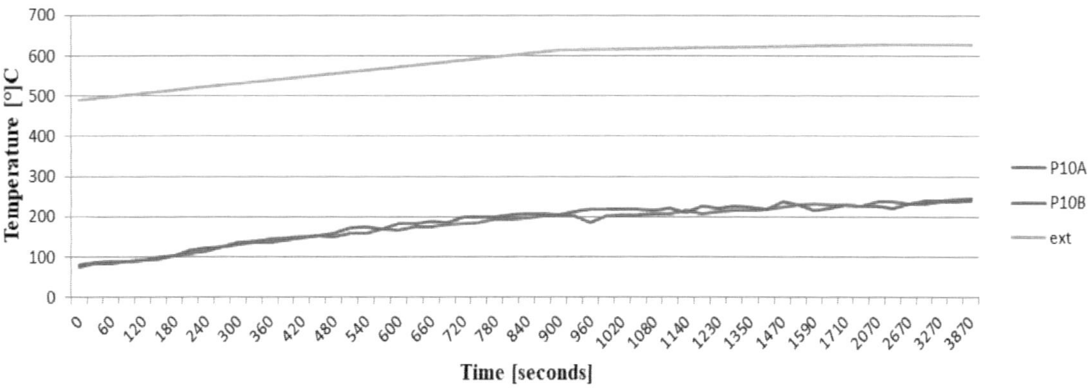

Figure 8. Temperature variation on the sides not exposed to direct flame of samples P10A and P10B in comparison with the temperature on the side exposed to direct flame.

At the analysis of the first set of samples - the one with a thickness of 10 mm - it is observed that the recorded temperatures (on the sides that were not exposed to direct flame) are similar for the two types of samples, the maximum temperatures being around 240°C (Figure 8). There was a faster increase in temperature to 210°C (in the first 30 minutes) followed by a slow increase of about 1°C / min until the end of the test. Compared to the temperature of the side exposed to the flame, a temperature difference between 350 - 400°C is observed, this being within the limit of at least 180°C imposed by SR EN 1363-1, fact for which we can say that the samples P10A, respectively P10B successfully meet criterion 2 - isolation.

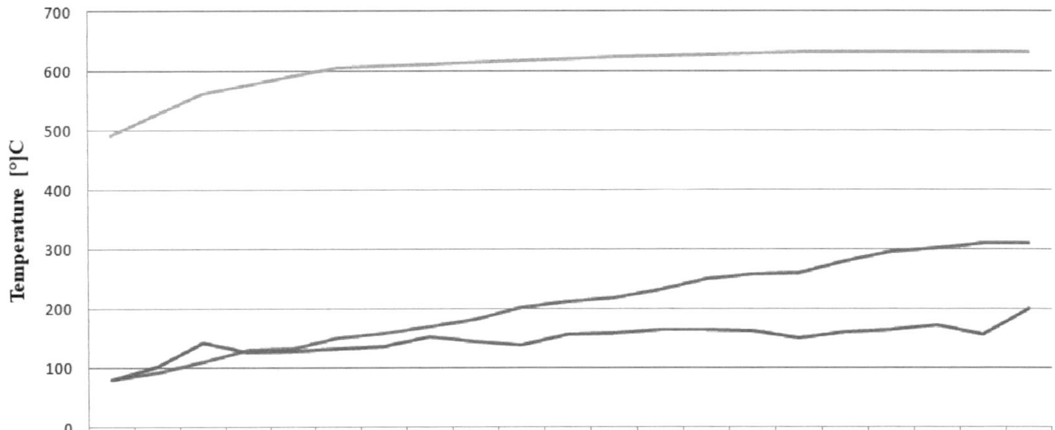

	0	30	60	90	120	150	180	210	240	270	300	330	360	390	420	450	480	510	540	570	600
P5A	80	93	110	130	132	150	158	171	183	203	213	218	232	250	258	260	281	297	303	310	310
P5B	81	102	142	127	128	132	136	152	144	138	156	158	165	164	163	151	161	164	173	157	201
ext	491		561			605						624				632					631

Figure 9. Temperature variation on the sides not exposed to direct flame of samples P5A and P5B in comparison with the temperature on the side exposed to direct flame.

Regarding the evaluation of the fulfilment of criterion 2 in the case of the thinner samples (P5A, respectively P5B), from the variation graph of the temperatures registered during the test, a much rapid increase of the temperature is observed. Values close to 200°C were reached after the first 3 minutes in the case of sample P5A, while in the case of sample P5B there was a rapid increase of the temperature in the first 2 minutes (up to about 142°C) followed by a lower speed increase to a temperature of 201°C recorded at the end of the test. However, these values must be interpreted with caution due to the bending phenomenon of the samples, produced in the first 4-5 minutes of the tests, which could have induced measurement errors, although the contact of the thermocouple with the surface was maintained throughout the test sample.

4. Conclusions

In this paper, the fulfilment of two performance criteria of two types of materials subjected to open flame tests was studied: structural integrity and thermal insulation capacity (heat transfer), regulated by SR EN 1363-1 "Fire resistance tests." These two criteria were also analysed in terms of the influence of the size of the expanded foam glass spheroidal granules used in the composition of the samples, but also in terms of the wall thickness of the samples tested at fire.

Following the performed tests, the following conclusions were drawn:

✓ both samples with a thickness of 10 mm (P10A, respectively P10B) had a satisfactory behaviour in terms of the first performance criterion ("sealing"), because they did not develop open flame and did not register major structural and geometric changes;

✓ both samples with a thickness of 10 mm had a satisfactory behaviour in terms of meeting the second performance criterion ("insulation") because, between the two sides, there were recorded temperature differences between 350 - 400°C;
✓ the samples with a thickness of 5 mm did not have a behaviour that could be considered relevant from the point of view of the isolation criterion, due to the curvature of the surface during the tests. However, the samples did not ignite and did not allow the flame to pass to the other side, behaviour that can be considered satisfactory for the evaluation from the point of view of the first criterion ("sealing");
✓ from the point of view of the influence of the size of the expanded foam glass granules on the behaviour of the two materials, we can specify that the smaller size granules (type B) ensured a better compactness of the material during the tests, the samples being less affected by the expansion phenomenon as a result of the transformation at high temperatures of the granule material and of the binder.

5. References

[1] Yu S R, van Onna D V, Spiesz P, Yu Q L and Brouwers H J H 2016 *Journal of Cleaner Production* **112** 690-701
[2] Adhikary S K, Rudžionis Ž, Vaičiukynienė D 2020 *Journal of Building Engineering* **31** 101399
[3] Gourav S 2015 *J. Civil Eng. Res.* **5**(3) 53-8
[4] Su M, Wang H, Hao H, Fiedler T 2020 *Journal of Alloys and Compounds* **821** 153233
[5] Al-Sahlani K, Broxtermann S, Lell D and Fiedler T 2018 *Materials Science & Engineering A* **728** 80–87
[6] https://www.poraver.com/us/poraver/
[7] Mocanita O, Chicet D L, Munteanu C, Istrate B and Oprișan B 2018 *IOP Conf. Series: Materials Science and Engineering* **444** 032011
[8] Mocanita O, Chicet D, Avram P, Micu C and Munteanu, C 2018 *UPB Scientific Bulletin, Series B: Chemistry and Materials Science* **81**(1) 236
[9] Mocanita O, Chicet D, Istrate B, Micu C and Munteanu C 2019 *IOP Conf. Series: Materials Science and Engineering* **572** 012052
[10] SR EN 1363-1 / 2001. Fire resistance test. Part 1: General requirements

The Effect of H_2O_2 and Lactic Acid Addition in Biological Saliva on the Corrosion Behaviour of 304L Stainless Steel

N Simionescu[1], L Benea[1*] and A Chiriac[1,2]
[1] Competences Center: Interfaces-Tribocorrosion-Electrochemical Systems (CC-ITES) Faculty of Engineering, Dunarea de Jos University of Galati, 47 Domneasca street, Galati, Romania
[2] Faculty of Medicine and Pharmacy, Dunarea de Jos University of Galati, 35 Alexandru Ioan Cuza Street, RO-800010 Galati, Romania

*E-mail: lidia.benea@ugal.ro

Abstract. The aim of this study is to assess the corrosion resistance of SS304L in biological solution (named Ringer) at body temperature 37^0 C. The addition of H_2O_2 and $C_3H_6O_3$ simulate the presence of microorganism metabolic product and inflammatory conditions. 304L stainless steel, in the oral cavity, can interact with other product capable of initiating corrosion of metals. H_2O_2 was used because is delivered by leukocytes and microscopic organism (bacteria) occurs in conditions in which the human body suffers a process of inflammation and lactic acid can be produced by a bacterial strain called *Streptococcus mitis* meeting in the dental plaque. For corrosion investigations was used in situ measurement (electrochemical) such as: OCP (open circuit potential) and EIS (electrochemical impedance spectroscopy). The electrochemical results conclude that the additions of hydrogen peroxide and lactic acid as well as these two compounds in combination have influence on the corrosion behaviour of SS304L in Ringer saliva.

1. Introduction

SS304L is progressively used within the field of biomaterials. The biocompatibility reported in various researches has led to the use of 304 type L steels in medical applications [1], especially in orthopedics for fixing fractures in the form of bone plates, intermediate nails and screws, fixed orthodontic appliances, dental crowns, tube of tracheostomy, among other non-implant medical devices and dental appliances [2, 3].

The way in which the human body responds to the phenomenon called corrosion when certain implant materials are placed in human body is one of the interesting areas [4].

The most common biomaterials implanted in the human body are derived from three systems of materials, namely: titanium alloys, cobalt chromium alloys and stainless steel [5]. After implantation of biomaterials in the human body, many problems can arise; one of them is the interaction between the material and the physiological material that can lead to implant failure.

Therefore it is important that the selected material implanted in the human body to be tested and explored in vitro in all possible situations because the human body is not a hospitable environment with metallic materials, this being a strongly oxygenated environment, with an ionic composition and concentration proteins from the fluid in the human body that complicate understanding of biomaterials and corrosion by default [6].

So corrosion remains one of the main problems facing the metal biomaterials market. Temperature and pH are other important factors that affect the behavior of metal biomaterials when inserted into the human body. So it is important that when these materials are tested in the laboratory they are tested at the internal temperature of the human body respective 37^0 C [7].

It is also important to simulate the inflammatory process, or the microorganisms, or the synergistic effect of both because they can affect the corrosion behavior due to effects such as pH modification [7].

In the literature we find few studies that have followed the synergistic effect of albumin and hydrogen peroxide on the corrosion resistance of SS316L in solution (phosphate buffered saline solution) [8], on the corrosion investigations of titanium alloy (grade 5) in FM (Fusayama Meyer) saliva [9], the effect of NaF, H_2O_2 and $C_3H_6O_3$ on the corrosion behaviour of titanium grade 4 [10], but we did not find studies to follow the synergistic effect of H_2O_2 and lactic acid on the corrosion resistance of SS304L in Ringer saliva tested at body temperature. The objective of this work is to investigate the addition of H_2O_2 and $C_3H_6O_3$ (lactic acid) in combination on the corrosion resistance of SS304L in, immersed in Ringer solution by electrochemical methods like: OCP and EIS, at body temperature, with addition of H_2O_2 (10 g/L), lactic acid ($C_3H_6O_3$) (10 g/L) and H_2O_2 (10 g/L) and lactic acid (10 g/L) in combination.

2. Experimental procedures

2.1. Materials and methods

SS304L was purchased from Direct Line Inox (Bucharest) in the form of sheet. The size of sheet was 1000 mm x 1000 mm x 1.2 mm, for corrosion experiment were cut at 25 mm x 25 mm x 1.2 mm, for electrical contact were bonded with copper electrical cable, insulated with resin to have an measurable active surface of 4.25 cm^2. The values that correspond to the SS304L passed in the quality certificate issued by the producing company are presented in Table 1.

Table 1. The chemical composition of SS304L (% wt).

Element / wt%	C	Cr	Ni	Si	Mn	Fe
SS304L	0.030	10.5-17.5	8-10.5	1.0	2.0	Balance

For the evaluation of the corrosion resistance of SS304L stainless steel, was used an electrochemical cell composed of three electrodes (SS304L used as WE (working electrode) , Platinum-Rh wire as auxiliary electrode and Silver / Silver Chloride with potassium chloride saturated solution as RE (reference electrode) connected at a equipment PGZ 301 piloted with the help of data recording software VoltaMaster 4. The tests were performed at the temperature of the human body 37^0 C ± 1^0.

The electrolyte volume of the electrochemical cell used for each experiment was 150 mL, and each experiment was triplicate to verify the data reproducibility. Prior to each experiment, 304L stainless steel samples were cleaned with NaOH (50 g / L), HCl (1:1), rinsed with distilled water dry with filter paper and placed immediately in the test solution.

Electrochemical experiments were performed in the Ringer biological solution whose chemical composition is given in Table 2 and the interest parameters of Ringer solution was measured with an equipment (analyzer multi parameter) Sension+. All chemical reagents used were by p.a, purchased from Lach:ner chemical Co (Slovakia).

The hydrogen peroxide used to prepare tested solution was purchased from Lach:ner chemical Co (Slovakia) and have a concentration of 30%. Lactic acid (85 %) was purchased from Sigma Aldrich.

Table 2. The chemical composition of Ringer solution.

Nr. Crt.	Reagents used	Ringer g / L	Ringer g / L	Ringer g / L	Ringer g / L
1.	NaCl	8.402	8.402	8.402	8.402
2.	KCl	0.302	0.302	0.302	0.302
3.	$CaCl_2$	0.298	0.298	0.298	0.298
4.	H_2O_2 (30 %)	-	-	10	10
5.	Lactic acid (LA) (85 %)	-	10	-	10
6.	Distilled water	Balance	Balance	Balance	Balance
7.	pH	6.67	1.96	5.26	1.85
8.	Conductivity [mS / cm]	14.1	16.0	13.8	15.4
9	Salinity [ppt]	8.1	9.2	8	8.9

3. Results and discussions

3.1. Open circuit potential (OCP)

The potential variation as a function in time in OCP measurement was followed for 180 minutes, with step measurement duration of 2 sec, until the equilibrium state of the potential for the SS304L tested in those four solutions is attained. This process was made to see the oxidation tendency of SS304L stainless steel in presence of Ringer solution, Ringer + 10 g/L lactic acid, Ringer + 10 g/L H_2O_2 and this two in combination.

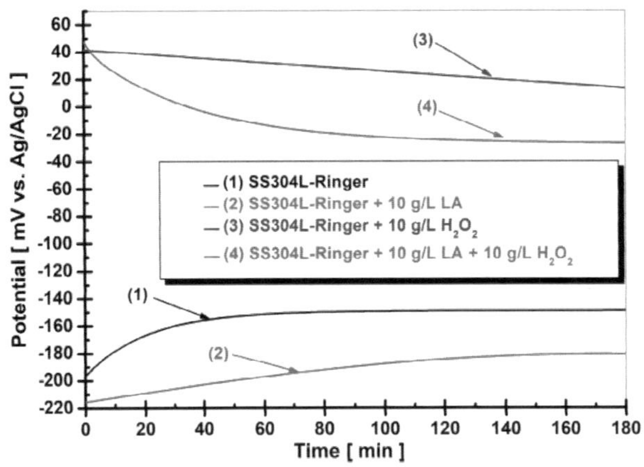

Figure 1. Open circuit potential graph of the SS304L tested in: (1) Ringer, (2) Ringer + 10 g / L LA, (3) Ringer + 10 g / L H_2O_2 and (4) Ringer + 10 g / L LA + 10 g / L H_2O_2 at 37^0 C during 180 min.

In Figure 1 can be observed, that the variation of the OCP for the SS304L tested in the biological saliva solution at temperatures which simulating the temperature of the human body, starting from a potential value of -196.7 mV vs. Reference Electrode from immersion, having a slight increase and after about 60 minutes reaching at the end of the measurement at 155.43 mV vs. Ag/AgCl.

When 10 g/L lactic acid (LA) is added to the Ringer solution, it is observed that the value of the open circuit potential decreases slightly compared to the Ringer solution, starting from a potential of -215.51 mV vs. reference electrode having a slight increase, reaching at the final of 180 min. around of potential value -204.56 mV vs. reference electrode. This behavior reveals that the amount of lactic acid added in Ringer solution affects the dissolution of SS304L.

IOP Conf. Series: Materials Science and Engineering **877** (2020) 012039 doi:10.1088/1757-899X/877/1/012039

With 10 g/L H_2O_2 added in solution it is observed that there is an increase in the potential compared to the Ringer solution and the solution containing 10 g/L lactic acid, starting from a value of 40.32 mV vs. Reference electrode at initial time, having a tendency to shift the potential towards more negative (but positive values), reaching at the ending of 180 min. of monitoring the potential at 15.28 mV vs. Ag/AgCl. This behavior is possible due to the oxidizing effect of hydrogen peroxide, thus favoring the growth of the oxide film on the surface of 304L stainless steel. At the same time leading to the attack of the oxide layer formed by exposure to oxygen. This behavior was also supported by the EIS results which showed lower R_p values for the SS304L stainless steel tested in Ringer solution with hydrogen peroxide added.

In the case of the Ringer solution with 10 g/L lactic acid + 10 g/L H_2O_2, there is an increase in the potential compared to the Ringer solution and compared with Ringer solution with lactic acid, but there is a decrease in the potential compared to the solution with 10 g/L H_2O_2. The open circuit potential starts from a potential of 47.71 mV vs. Reference electrode from immersion and reaches at the final of the measurement the potential value of -26.79 mV vs. Reference electrode. This behavior is also dependent on the fact that in the presence of lactic acid and H_2O_2 there was a decrease in pH. The results suggest the harmful effect that the synergistic effect of lactic acid and the H_2O_2 on the 304L stainless steel.

This behavior is sustained also by the lowest values obtained from the polarization resistance (EIS measurement) and microphotographs taken before and after corrosion test.

3.2. Electrochemical Impedance Spectroscopy (EIS)

EIS is powerful and precise methods which gain interest in recent decades, successfully applied to the study of corrosion systems. The EIS spectra it was made at free potential in a range of frequency started from the 100 kHz to 10 mHz, with a signal perturbation of amplitude of 10 mV. The EIS spectrum was fitted using software ZView 3.4f and the fitted quality of results was made in term of chi-square value obtained that is lower than 0.001.

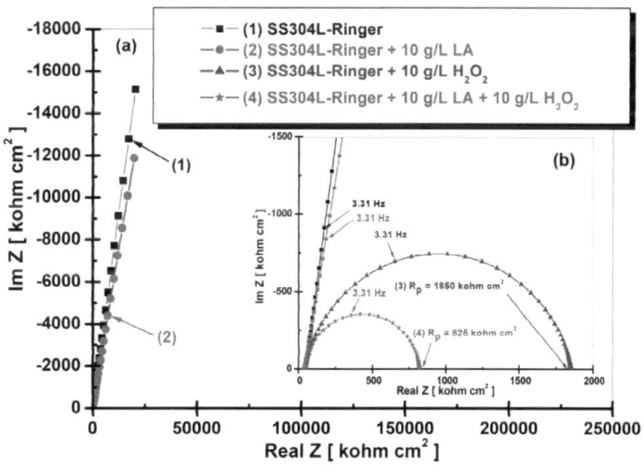

Figure 2. (a) The Nyquist representation of Electrochemical Impedance Spectroscopy diagrams recorded for SS304L immersed in (1) Ringer, (2) Ringer + 10 g/L LA and low frequency range (b) for: (3) Ringer + 10 g/L H_2O_2 and (4) Ringer + 10 g/L LA + 10 g/L H_2O_2. The experimental data are represented in the graph by the symbol and the fitted results are represented by the continuous line.

The EIS diagrams in the Nyquist representation of the SS304L tested in all solutions used are presented in Figure 2. The increase of the semicircle arc (the semicircle didn't intersect the important axis (X) named (Real Z) of Nyquist representation of Electrochemical Impedance Spectroscopy diagram for SS304L tested in solution Ringer's and Ringer + 10 g/L LA, this behaviour is attributed to the frequency dispersion) denotes an increase in the stability of the passive film and a decrease of that radius indicates a decreasing of the resistance of the passive film.

To fit the Electrochemical Impedance Spectroscopy diagrams register for SS304L tested in Ringer, Ringer + 10 g/L LA, Ringer + 10 g/L H_2O_2 and Ringer + 10 g/L LA + 10 g/L H_2O_2 it was used a circuit equivalent complex [8] and the result of R_p values from fitted spectra are given in Table 3.

Table 3. Rp (Polarization Resistance) values achieve for SS304L tested in all electrolyte used for this experiment.

Nr. crt.	Sample	Polarization resistance $[kohm \cdot cm^2]$
1.	304L stainless steel immersed in Ringer solution	4.473E11
2.	304L stainless steel immersed in Ringer + 10 g/L LA	1.473E11
3.	304L stainless steel immersed in Ringer + 10 g/L H_2O_2	1850
4.	304L stainless steel immersed in Ringer + 10 g/L LA + 10 g/L H_2O_2	825

Electrochemical impedance spectroscopy can be represented as well in Bode diagram such as figure 3a logarithm of modulus |Z| vs. logarithm of *(f)* and Figure 3b logarithm of *(f)* vs. phase angle.

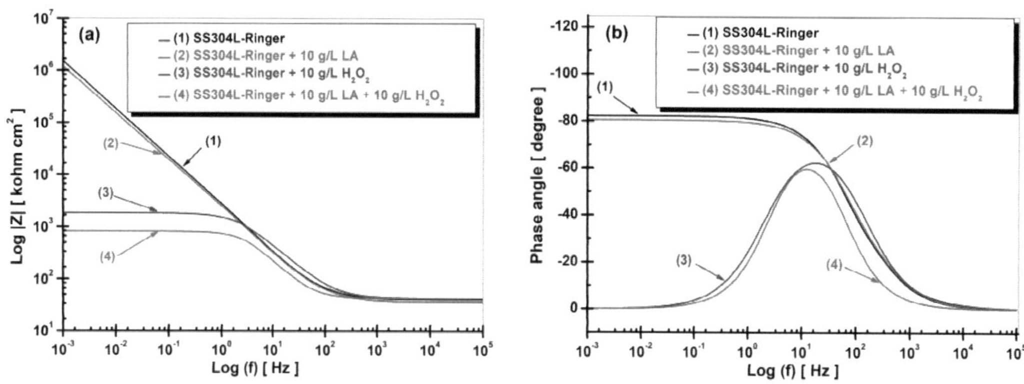

Figure 3. Development of EIS in Bode graph for: (1) Ringer, (2) Ringer + 10 g/L LA and low frequency range (b) for: (3) Ringer + 10 g/L H_2O_2 and (4) Ringer + 10 g/L LA + 10 g/L H_2O_2. (a) |Z| Modulus vs. (f) ; (b) Phase angle vs (f).

An advantage of bode diagrams is the ability to observe the impedance variation at high frequencies, which is generally omitted in the Nyquist representation. High impedance values attain at small frequency and middle frequency disclose a down reactivity or a good corrosion behavior for SS304L stainless steel indicates the presence of protective oxide film.

From Figure 3(b) it can be noted that the SS304L tested in Ringer's solution layer display an almost capacitive reply, illustrated by a phase angle near to - 82° on a wide range of frequency. This behaviour also show a distinctive behaviour of a small oxide protective film form on the surface of stainless steel immersed in Ringer solution.

3.3. Optical Microscopy investigations before and after corrosion

In Figure 4 the surfaces of SS304L was investigated ahead and after corrosion assay in Ringer solution, Ringer + 10 g/L LA, Ringer + 10 g/L H_2O_2 and Ringer + 10 g/L LA + 10 g/L H_2O_2 in order to investigate the corrosive effects.

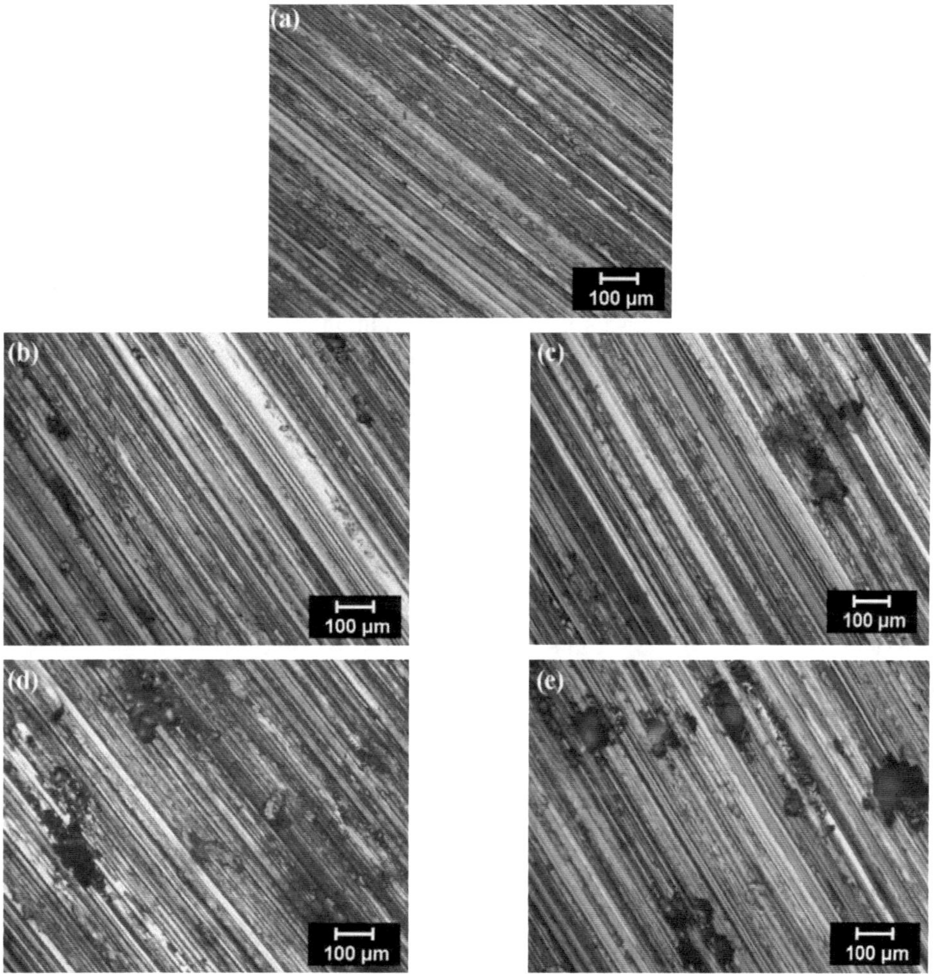

Figure 4. Optical microscopy of SS304L (a) ahead corrosion, (b) after corrosion in Ringer solution, (c) Ringer + 10 g/L LA, (d) Ringer + 10 g/L H_2O_2 and (e) Ringer + 10 g/L LA + 10 g/L H_2O_2.

In Figure 4a it can be notice that the SS304L before corrosion have a rough surface without surface defects. After corrosion in Ringer solution (Figure 4 (b)) and 4 (c) after corrosion in Ringer solution + 10 g/L LA, 304L the surface of stainless steel there are less affected by pitting corrosion.

With the addition of 10 g/L H_2O_2 in the Ringer solution it is observed that on the surface of the 304L stainless steel (Figure 4 (d)) general corrosion occurs on the surface of the material while on the surface of the SS304L immersed in Ringer saliva with 10 g/L LA + 10 g/L H_2O_2 (Figure 4 (e)) a pitting corrosion is observed which covering almost the entire surface of the material.

These results obtained by optical microscopy are in well accordance with the results obtained by the electrochemical methods respectively: OCP and EIS.

4. Conclusions

The purpose of this research study is to assess the influence of lactic acid and reactive oxygen species (H_2O_2) and the synergistic influence of $C_3H_6O_3$ and H_2O_2 in Ringer solution, on the corrosion resistance of SS304L by electrochemical investigations.

From the OCP measurements it can be observed that with the addition of 10 g/L Lactic acid the open circuit potential value slightly decreases as compared to Ringer solution. This behavior is possible due to the low pH of solution with addition of lactic acid which affects the dissolution of stainless steel. When in Ringer solution is added 10 g/L H_2O_2 it can be noted that is an increasing of the potential value as compare with Ringer solution and with the solution containing 10 g/L lactic acid. This behavior is possible due to the oxidizing effect of H_2O_2.

In the case of Ringer solution with addition of 10 g/L LA + 10 g/L H_2O_2 in combination, there is an increase in the potential compared to the Ringer solution and compared with Ringer solution with lactic acid, but there is a decrease in the potential compared to the solution with 10 g/L H_2O_2. This behavior is also dependent on the fact that in the presence of $C_3H_6O_3$ (lactic acid (LA)) and H_2O_2 there was a decrease in pH.

Through the EIS methods can be deduce that with the addition of lactic acid and hydrogen peroxide and these two compounds in combination present in Ringer solution the *Rp* value of SS304L decrease as compared with SS304L tested in Ringer electrolyte.

From the optical microscopy investigations of SS304L surfaces before and after corrosion in Ringer saliva and Ringer + 10 g/L LA (lactic acid) are the less attacked by pitting corrosion. When in Ringer solution is added 10 g/L H_2O_2 it can be seen that on the surface of the SS304L appear a general corrosion on the surface of the stainless steel while on the surface of the SS304L put in Ringer solution with 10 g/L Lactic acid + 10 g/L H_2O_2 a pitting corrosion is observed which covering almost the entire surface of the SS304L.

5. References

[1] Baila D and Tonoiu S 2019 IOP Conf. Ser.: Mater. Sci. Eng. **591** 012001
[2] Bratu F, Benea L and Celis J P 2008 *Rev. Chim. (Bucuresti)* **59** 346
[3] Davis J R 2003 *Handbook of Materials for Medical Devices* ASM International, ISBN: 087170790X, p 1
[4] Hack H and Bundy K 2005 *Corrosion Testing in In Vivo Environments* ed R Baboian (West Conshohocken, PA: ASTM International) p 500. https://doi.org/10.1520/MNL11048M
[5] Baltatu M S, Tugui C A, Perju M C, Benchea M, Spataru M C, Sandu A V and Vizureanu P 2019 *Rev. Chim. (Bucuresti)* **70** 1302
[6] Gilbert J L 2017 *Corrosion* **73** 1478
[7] Eliaz N 2019 Materials **12** 407
[8] Simionescu N, Benea L and Dumitrascu V M 2018 *IOP Conf. Ser.: Mater. Sci. Eng.* **374** 012058
[9] Dragus L, Benea L, Simionescu N, Ravoiu A and Neaga V 2019 *IOP Conf. Ser.: Mater. Sci. Eng.* **572** 012005
[10] Mabilleau G, Bourdon S, Joly-Guillou M L, Filmon R, Baslé M F and Chappard D 2006 *Acta Biomater.* **2** 121

The Influence of Blast Furnace Slag on Abrasion Resistance for Road Concretes

L M Nicula[1], O Corbu[1,2,3*] and M Iliescu[1]
[1]Technical University of Cluj-Napoca, Faculty of Civil Engineering, Romania
[2]Research Institute for Construction Equipment and Technology, Bucharest
[3]Center of Excellence Geopolymer& Green Technology (CEGeoGTech), School of Materials Engineering, University Malaysia Perlis, 01000 Kangar, Perlis, Malaysia, India

*E-mail: Liliana.nicula@infra.utcluj.ro; ofelia.corbu@staff.utcluj.ro; iliescu.mihai@staff.utcluj.ro

Abstract. In this study, the blast furnace slag was used as a mineral addition in the projected road concrete mixtures, and the blast furnace slag in the form of aggregate at 0/4 mm size was used as a percentage of 20%, 40% and 60%. Three concrete slag mixtures were compared with two other concrete mixtures made with conventional materials. Compared to the first conventional mixture, milled slag used as a 15% addition and compared to the second conventional mixture milled slag was used as a substitute for 13% of Portland cement mass. The wear / abrasion resistance was evaluated at the concrete age of 100 days, respectively 150 days. The results indicated the highest wear resistance for the mixture with 20% aggregates crushed from furnace slag, and the mixture with 60% substitution aggregates from furnace slag was observed the lowest resistance. The tendency of evolution of the wear resistance is similar to that of the compressive strength. Mixtures with the highest compression strength also recorded the lowest volume loss after the abrasion test (Bőhme). Compared to the first mixture made with conventional materials, the wear resistance obtained in blast furnace slag mixtures is lower. However, mixtures with 13% ground slag and up to 40% of aggregates crushed from blast furnace slag have higher wear resistance compared to the second mixture made with conventional materials.

1. Introduction

The use of blast-furnace slag in optimum ratio, as mineral addition to the mass of the cement and as substitute of natural sand in road concrete mixtures can be a solution both for protecting the environment and for saving the natural resources. The ground surface as well as the aquatic ecosystem are affected by the cement industry that contributes with approximately 5-8% of the carbon dioxide issued globally, followed by the Sulphur dioxide SO_2 and the nitrogen oxides NO_x, [1,2,3]. Also, the consumption of aggregates in concretes affects the natural aggregate reserves at a level of 20000 million of tons/year and an increase rate of 4.7% per year globally [4,5]. The blast-furnace slag is a secondary product obtained in the steel industry, through melting cast iron. By processing the sub-product of blast-furnace slag we obtain artificial materials that can be reused as prime materials in road constructions. The granulated blast-furnace slag is formed through the process of fast cooling in water of the melted slag, and it has a white colour, with dimensions of maximum 5 mm. The granular material resulted contains approximately 85-95% glass materials formed from calcium aluminium silicates, non-crystalline, [1,6]. The granulated and grounded blast-furnace slag has a latent hydraulic character, it is used as type II addition in concrete mixtures for the improvement of certain properties in compliance with the provisions of the SR EN 206:2002+A1:2007 [7] and SR EN 15167-1:2007 [8]. The non-granulated blast-furnace slag is obtained through the process of slow cooling in the air of the melted slag and presents a crystallin molecular structure, without any usage in the cements' manufacturing, [9].

In compliance with the ACI 201.2R-08 [10], Guide to Durable Concrete, the wear on the concrete surface can be caused by four different actions: due to human traffic on concrete floors, wear due to vehicular traffic with different types of tires and with steel wheels, abrasive materials carried by water (e.g. sand) affecting the hydrotechnical structures such as dams, and the high water velocity creating cavitation at the concrete surface [11]. The abrasion resistance is an important characteristic for the roads' exploitation. The wear from the abrasion caused by traffic appears on the concrete surface due to friction, sideslips or slipping of vehicles in motion on the tread surface. The abrasion resistance of a concrete depends on the resistance of the composing materials and the liaison between the aggregate and the cement paste [11,12]. There are two point of views regarding the type of mathematical relations between the compressive strength and the abrasion resistance. The first point of view is that the compressive strength is the most important factor for the abrasion resistance and that between the two characteristics it is developed a linear relation [1,13-15]. While Atiş showed that between the compressive strength and the abrasion resistance there is a hyperbolic mathematical relation [1,2]. Liu and collaborators studied the influence of the w/c ration on the abrasion resistance in four different water/cement ratios, between 0.26÷0.50. The increase of the abrasion resistance was registered in mixtures in which the w/c ratio decreased. These concretes had a lower porosity, a better liaison between cement paste and aggregate and an increased compressive strength [11,16]. The second point of view is that the abrasion resistance is not dependent on the compressive strength, this being influenced by the hardening conditions and the surface finishing technology [1,17-22]. Kevern et al, tested the abrasion resistance of concrete preserved in six different hardening regimes. The hardening regime was in air, the humid hardening was under plastic for 7 and 28 days, the hardening regime with oil-based protection, white pigment protection and the regime without protection for delayed evaporation. The hardening regime in humid environment with plastic protection contributed best to the densification of the concrete surface and in obtaining the best abrasion resistances due to the greenhouse effect [11,23]. Dhir showed that the long period of hardening in humid environment of 7 days, compared to 1 day and 4 days, influences the increase of the abrasion resistance [11,24]. Bakke et al, recommends the humid hardening of 5-7 days for improving the abrasion resistance [11,25].

This study considers a comparative analysis of the abrasion and compressive resistance between road concrete in which it was used the blast-furnace slag as powder and crushed aggregates at 0/4 mm and road concrete made of conventional materials. The results obtained between the two characteristics were graphically represented by a second-degree polynomial equation.

2. Materials and methods

2.1. Cement

The cement available type CEM I 42.5R was supplied by the cement factory LafargeHolcim Romania, having a Blaine surface of 4385 cm^2/g and specific weight of 3.0 t/m^3. The main characteristics of the cement are presented in Table 1, in compliance with the provisions of the standard SR EN 197-1:2011 [26], having the performances stated by the manufacturer in the declaration of performance [27].

Table 1. Cement characteristics CEM I 42.5R.

Characteristics	Obtained values	Characteristics	Declared values
Clincher-K (% of mass)	95÷100	Calcination loss, (%)	Max.5
Auxiliary components, %	0÷5	Insoluble residue, (%)	Max.5

Initial intake time	Min.60	Sulphate content (SO_3), (%)	Max.4
Stability (expansion), mm	Max.3	Chlorine content, (%),	Max.0.1
Initial compressive strength, (MPa)	Min.24	Standard compressive strength, (MPa)	Min.46

2.2. The additional constituents with cement characteristics (CSCC)

The Laplante study shows that the ultrafine silica has an important contribution to the increase of the compressive strength and to the improvement of the abrasion resistance of the concrete [11,28]. Other authors as well reported that the inclusion of the ultrafine silica (SF) increased the abrasion resistance in concrete or mortar. Turk studied the abrasion resistance of self-compacted concrete in which the cement was partially replaced with fly ash (FA) or ultrafine silica (SF). The replacement percentages with (FA) were 25%, 30%, 35%, and 40%, while for the (SF) were 5%, 10%, 15%, and 20%. The results showed an increase of the abrasion resistance in mixtures with substitution of up to 15% (SF), while in the mixtures with fly ash the abrasion resistance decreased with the increase of the (FA) content, over 30%, [1,29] studied the effect of the ultrafine silica (SF) and the ground-granulated blast-furnace slag (GGBS) on the compressive strength and abrasion resistance at the age of 7, 28, 90 and 180 days on HVFA concrete (concrete with high level of substitution of 70% fly ash (FA)). The results showed the best abrasion resistances in mixtures with 50%(FA)+20%(SF) and for 50%(FA)+10%(SF)+10%(GGBS). Naik the effects of replacing the fly ash (FA) from class C on abrasion resistance of the concrete mixtures. There were chosen five levels of cement replacement, between 15% and 70%.

Table 2. Characteristics of the granulated blast furnace slag.

Characteristics	Obtained values	Limits SR EN 15167-1:2007
Sum CaO+MgO+SiO2, %	84.05	≥ 2/3 %
Ratio $(CaO+MgO)/SiO_2$, %	1.3	≥ 1.0 %
Magnesium oxide MgO, %	5.80	≤ 18%
Sulphides, %	0.58	≤ 2.0%
Sulphate, %	0.44	≤ 2.5%
Correct calcination loss for sulphides oxidation, %	0.00	≤ 3.0%
Chlorine, %	0.004	≤ 0.10%
Humidity content, %	6.85%	≤ 1.0%
Specific surface, cm^2/g	3520	≥ 2750
Content of alkalis in cement (Na_2O equivalent), %	1.0 %	(0.5 ÷1.2) %
Activity index at 7 days, %	58.48	Min 45 %
Activity index at 28 days, %	74.94	Min 70 %
Content of tricalcium aluminate in cement C_3A, %	7.59	(6 ÷12) %

The results of the abrasion resistance could be compared to the control mixture for the replacement levels of 15% and 30%, but above these levels, the concrete did not obtain good abrasion resistances, [11,30].

In this study, it was used ground-granulated blast-furnace slag that was used as additional constituent with cement characteristics (CSCC) in road concrete mixtures. The granulated blast-furnace slag at 0/4 mm dimension was delivered by the company ArcelorMittal from Galaţi. After grounding the Blaine surface was established at a value of 5770 cm^2/g, and the specific weight of 2.77

t/m^3, testing performed through this experiment. The other granulated blast-furnace slag characteristics resulted from the testing reports [31-34] performed by the supplier. The obtained values were analysed by comparison with the recommended standard SR EN 15167-1:2007 [8], see Table 2.

2.3. Aggregates

The used aggregates proportions were 32% sand (0/4 mm dimension sand) and 68% coarse aggregate. Selecting the type of coarse aggregates for their usage in concrete mixtures resistant to wear represented interest for researchers. The aggregate resistance to fragmenting was researched by Kiliç et al. on different types of crushed coarse aggregates, respectively on magmatic rock, basalt, quartzite, chalk and sandstone. The Los Angeles coefficient (LA) was determined for each type of rock. They noticed a strong correlation between the aggregate resistance and the concrete wear resistance. The mixtures in which the aggregate had inferior resistances suffered a greater loss in wear than those with aggregates resistant to fragmenting, [11,35]. Dhir investigated the effect of the maximum dimension of the aggregate on the abrasion resistance. Four maximum aggregate dimensions were tested: 5, 10, 20, and 40 mm. the test results showed that the abrasion resistance decreased with the increase of the coarse aggregate dimension, up to 20 mm, [11,36].

The selection the coarse aggregates used in this experiment took into consideration the provisions of the regulation NE 014:2002, [37], but also the conclusions of previous studies. Coarse aggregates were supplied from two different sources, the 4/8 mm dimension sort was crushed gravel pit aggregate, while the 8/16 and 16/25 sorts were crushed rock from dacite type quarry. The source of the used quarry aggregates is famous for the characteristics complying with the concrete requirements due to very good wear resistance (the micro-Deval coefficient M_{DE} 6) and the fragmenting resistance (the Los Angeles coefficient LA_{14}), [38, 39]. The non-granulated air-cooled blast furnace slag (ABS) artificial aggregate was supplied by the company Arcelor Mittal to substitute in different proportion the natural sand. The properties of the aggregate of blast furnace slag crushed at 0/4 mm dimension resulted from the report [40], based on the lab tests performed by the supplier. The volume mass of the used aggregates was determined in our university laboratory using the method established by the SR EN 1097-6: 2002 [41], the resulted values are presented in Table 3.

Table 3. The volume mass of aggregates.

The specific weight of the aggregates ρ (t/m^3)					
Name_ sort	NA_0/4	ABS_0/4	CA_4/8	CA_8/16	CA_16/25
ρ (t/m^3)	2.7	2.59	2.68	2.66	2.67

In compliance with the standard SR EN 206+A1:2017, [7] both the natural aggregates and the blast furnace slag aggregates enter the category of current aggregates because they have the volume mass between (2.0 ÷3.0) t/m^3. The granulometric curve of the aggregate total mixture followed a framing of the interior granularity area in compliance with NE 014-2002 [37] and in NE 012/1-2007 [42], see Figure 1. The testing procedure on aggregates complied with the normative references from SR EN 12620:2003 and SR EN 12620:2003+A1:2008 [43] and SR 667:2001 [44].

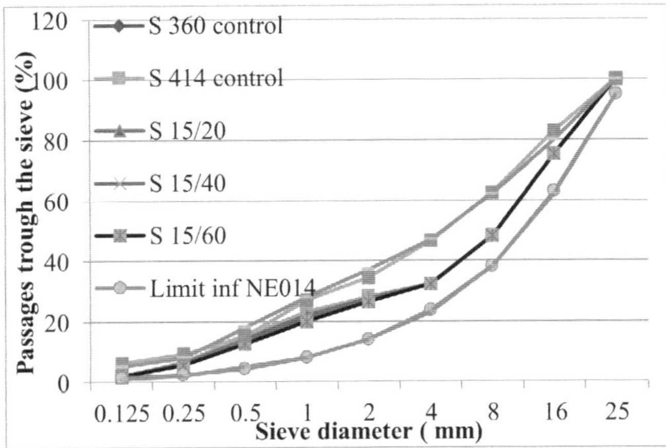

Figure 1. Aggregate granularity for total mixture.

Figure 2. Aggregates screening.

2.4. Chemical addition and water in mixture

The used additives were supplied by the company BASF Romania, the super-plasticizer additive MasterGlenium SKY 527 and the air-trainer additive Master Air 9060. The water from the concrete was taken from the water supply system of the city Cluj-Napoca, in compliance with SR EN 1008:2003 [45].

2.5. The mixtures proportion

There were a number of five mixtures in the first series of the experiment having the material quantities mentioned in Table 4. The first two control mixtures were made with Portland cement and with natural aggregates, and for the following three mixtures there were used ground-granulated blast furnace slag under 63 μm (GGBS) and crushed air-cooled blast furnace slag aggregate (ABS) with 0/4 mm dimension, in different proportions:

S 360c, control mixture with 360 kg/m^3 dosage of Portland cement and natural aggregates;
S 414c, control mixture with 360 kg/m^3 dosage of Portland cement and natural aggregates;
S 54/20, mixture with 54 kg/m^3 (GGBS) and 20% (ABS)_0/4 mm;
S 54/40, mixture with 54 kg/m^3 (GGBS) and 40% (ABS)_0/4 mm;
S 54/60, mixture with 54 kg/m^3 (GGBS) and 60% (ABS)_0/4 mm;

Table 4. Material quantities in mixtures.

Quantities	Mixtures [Kg/m^3]				
Mixture (Kg/m^3)	S 360 c	S 414 c	S 54/20	S 54/40	S 54/60
Cement (C)	360	414	360	360	360
Ground-granulated blast furnace slag (GGBS)	-	-	54	54	54
Total binder (L)	360	414	414	414	414
W /L, (water/binder)	0.44	0.41	0.41	0.43	0.43
Aggregate	1864	1831	1863	1810	1810
Density in fresh state	2380	2415	2444	2402	2402
Super-plasticizer additive	3.60	4.14	4.14	4.97	4.97
Air trainer additive	1.08	2.07	2.07	2.07	2.07

Compared with the control mixture S 360c the ground-granulated blast furnace slag (GGBS) was brought as input to the cement mass in percentage of 15%, and compared to the control mixture S 414c the ground-granulated blast furnace slag (GGBS) substituted the cement in percentage of 13%. The natural sand was substituted with crushed air-cooled blast furnace slag (ABS) aggregate_0/4 mm with a dimension of 0/4 mm.

2.6. Methods

The wear resistance was tested on abrasive disk of the Böhme device, the brand Matest, presented in Figure 4, in compliance with the standard SR EN 1338:2004/AC:2006 [46], Annex H, through the determination of the volume loss. This method was used and by Orhan 2017 [47] in his study performed on road concrete. In each projected mixture there were performed 6 cubic specimens with a side of (71 ±1,5) mm for the wear test and 6 cubic specimens with a side of 150 mm for compressive strength. The specimens were kept for 24 h in air, afterwards they were removed from forms and then kept in water at a temperature of (20±2) °C. At the age of 7 days, the 71 mm cubes and the 150 mm cubes were removed from water and they were kept in air, in a climatic chamber, at a temperature of (20±2) °C and a humidity of (65±5) %. From the age of 50 days, until the testing age of 100 days and 150 days, the specimens were placed in water at a temperature of (20±5) °C. Afterwards, at the age of 100 days, a set of 3 specimens for each mixture with the 71 mm dimension were removed from water and prepared for the wear testing. The specimens were dried up to the constant mass at a temperature of (105±5) ºC, 20 gr of standard abrasive material was scattered on the testing track, the specimen was fixed in the device with the testing side on the track and a force of (294±3) N was applied centrally. Each specimen was tested for 16 cycles, a cycle containing 22 rotations. After each cycle the disk was cleaned, the abrasive material replaced and the contact side was progressively rotated at 90º. The wear after 16 cycles, as an average of the lost volume ΔV, see the equation (1), the results fitted in the abrasion resistance class as in Table 5.

$$\Delta V = \frac{\Delta m}{\rho} \tag{1}$$

Where ΔV (mm^3) represents the volume loss after 16 cycles, Δm (g) represents the mass loss after 16 cycles, and ρ (g/mm^3) represents the specimen density. The wear test was repeated on other three specimen sets at the age of 150 days, using the same procedure.

Table 5. The abrasion resistance classes.

Class	Marking	Conditions
1	F	No measured performance
3	H	$\leq$20000 mm^3/5000 mm^2
4	I	$\leq$18000 mm^3/5000 mm^2

Figure 3. Böhme device.

The compressive testing was performed with the type Advantest 9 hydraulic digital press, of 300tf in compliance with SR EN 12390-3:2002 [48], on 3 cubes of 150 mm for each mixture, with an age of 100 and 150 days. The 150 mm specimens were tested for compression and kept in water from the age of 50 days until the age of 100 days, respectively 150 days, were used as witness samples at 150 and 300 repeated freeze-thaw cycles.

3. Results and discussions

The volume loss results of the specimen tested at 100 and 150 days and the compressive strength are presented in Table 6 and in Figures 4 and 5.

Table 6. The volume losses from abrasion testing and the compressive strength at the age of 100 days, respectively 150 days.

Mixture	Δm (g) 100 d/150d	ρ_R (g/mm^3) 100d/150d	ΔV (mm^3) 100d/150d	f_{cm} (MPa) 100d/150d
S 360c	27.73/ 25.80	0.002379/ 0.002300	11660/ 10833	85.83/87.78
S 414c	32.85/ 29.63	0.002324/ 0.002323	14138/ 12860	81.46/ 80.96
S 54/20	30.85/ 27.25	0.002375/ 0.002377	12986/ 11577	87.07/ 84.90
S 54/40	31.65/ 28.75	0.002323/ 0.002323	13627/ 12376	80.85/ 79.38
S 54/60	39.13/ 36.25	0.002378/ 0.002380	16454/ 15039	76.79/ 74.77

The volume loss evolution at the age of 100 and 150 days of the specimens tested for wear resistance registers a tendency inversely proportional with the results of the compressive strength, see Figures 4 and 5. The higher the compressive strengths, the smaller the volume losses of the tested specimens, as highlighted in the mixtures S 360c and S 54/20. The reverse situation can be seen in the mixtures S 414c, S 54/40 and S 54/60, smaller the compressive strengths led to higher volume losses for the tested specimens. However, it can be noticed a slight increase of the wear resistance (the volume loss decreases) in all mixtures once the age increases from 100 days to 150 days. This is attributed to the prolonged period of hydration of the binding agent in the made mixtures.

The volume loss after the abrasion test at the age of 100 days was smaller with 8.14% and with 9.98 % at 150 days in the S 54/20 mixture, compared with the control mixture S 414c. The mixture S 54/40 has a good behaviour regarding the wear resistance, where it can be noticed a decrease of the volume loss with 3.76% at 150 days compared to the control mixture S 414c. While the volume loss at the age of 150 days increased significantly with 16.94% in the mixture S 54/60 compared with the control mixture S 414c.

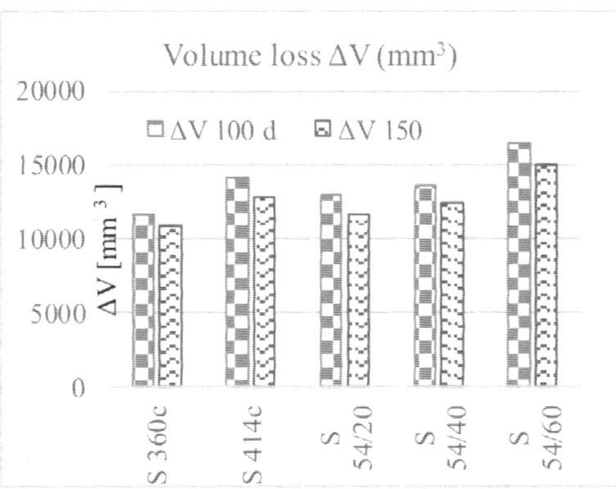

Figure 4. Volume loss through wear,

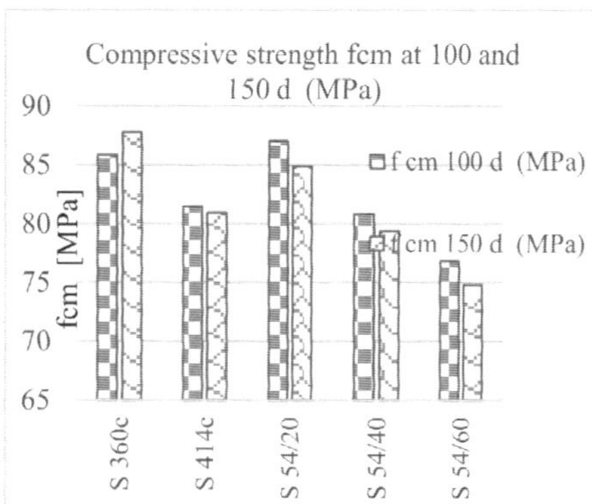

Figure 5. Compressive strength.

Figures 6 a) and b) show the relationship between the compressive strength and the volume loss after abrasion testing at 100 and 150 days. It can be noticed that between the two characteristics a second-degree polynomial equation is developed, derived through regression, having a correlation coefficient (R value) very good.

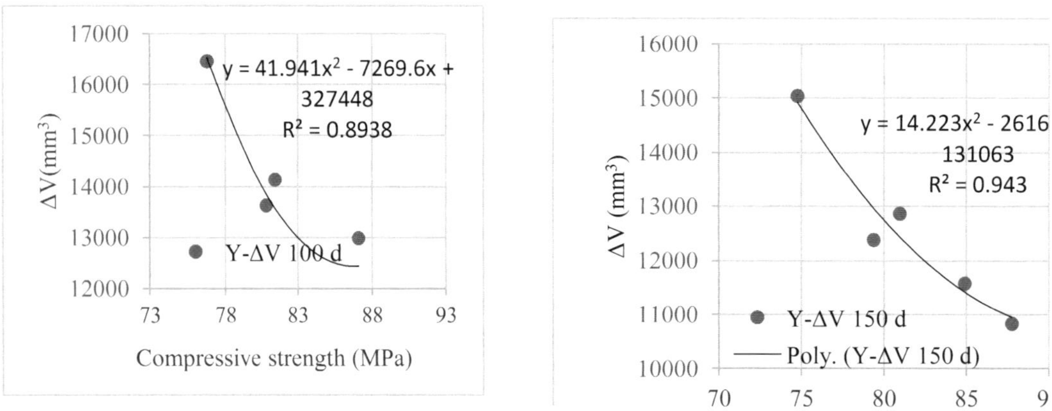

a) b)

Figure 6. The relationship between compressive strength and volume loss a) at 100 days, b) at 150 days.

The obtained results confirm those offered by Alaa M. R. [1] who also established a polynomial relationship and a correlation coefficient close to 0.9298 between the compressive strength and the depth loss.

4. Conclusions

In this study we tested the resistance to wear on three mixtures in which it was included the same quantity of 54 kg/m³ of grounded-granulated blast furnace slag (GGBS) as binding agent and three different percentages (20%, 40% and 60%) of air-cooled blast furnace slag aggregate (ABS) with a dimension of 0/4 mm that substituted the natural sand. The results were compared with two control mixtures made of different dosage of Portland cement. For the mixtures with blast furnace slag the quantity of (GGBS) became supplement of 15% of the mass of the cement compared with the first control mixture S 360c, but compared with the second control mixture, the cement quantity was substituted with 13% (GGBS).

A general conclusion from this performed study is that an increase of the compressive strength leads to an increase of the resistance to wear of the road concrete and, implicitly, a decrease of the mass loss after wear. Also, important factors positively influencing the resistance to wear are: aggregates resistant to fragmentation and wear, a low water/cement ratio and a hardening regime in the humid environment of at least 7 days.

The compressive strength and implicitly, the abrasion resistance for the two control mixtures were higher in the S 360c mixture compared with S 414c. these results are due to the fact that by increasing the cement dosage from 360 to 414 kg/m³ also increased the necessary quantity of water, but the quantity of aggregate was diminished in the S 414c mixture.

The best results were obtained for the mixtures S 54/20 and S 54/40, with 13% cement substitution with ground slag and with 20%, respectively 40% blast furnace slag aggregate, in which the abrasion resistance, at all ages, were higher than that of the control mixture S 414c. But, in the mixture in which it was substituted 60% blast furnace slag aggregate the wear resistance was lower than that obtained in the control mixture S 414c.

Between the two studied characteristics was developed a second-degree polynomial mathematical equation which shows a regression of the volume loss from the wear testing in compliance with the increase of the compressive strength.

The concrete mixtures made registered, at all ages, a volume loss smaller than 18000 mm^3/5000 mm^2, values that allowed the fitting in the abrasion resistance class 4 with marking I, in compliance with the classification in Table 5.

5. References

[1] Rashad A M, Seeem H E-D H and Shaheen A F 2014 *Internationl Journal of Concrete Structures and Materials* **8**(1) 69-81
[2] Rashad A M and Zeedan S R 2011 *Construction and Building Materials* **25** 3098–3107
[3] National Air Quality Monitoring Network *http://www.calitateaer.ro/public/assessment-page/pollutants-page/oxid-azot-page/index.html.*
[4] Chendeș R 2017 Experimental determinations regarding the reuse of concrete resulting from the demolition of buildings *http://www.upt.ro/img/files/2016-2017/doctorat/teze/rezumat/Chendes_Remus_rezumatul_tezei_ro.pdf*
[5] Bleishwitz R and Bahn-Walkowiak 2011 *Minerals Engineering,* **22**
[6] Swamy R N 1986 Concrete technology and design, *Cement replacement materials. London, U.K.: Surrey University Press, First publication* **3**
[7] xxx SR EN 206:2002+A1:2017 Concrete-Part 1: Specification, performance, production and conformity
[8] xxx SR EN 15167-1:2007 Ground granulated blast furnace slag for use in concrete, mortar and grout. Part 1: Definitions, specifications and conformity criteria
[9] User Guidelines for Waste and Byproduct Materials in Pavement Construction, Publication Number:*https://www.fhwa.dot.gov/publications/research/infrastructure/structures/97148/bfs1.cfm*
[10] American Concrete Institute ACI201.2R 2008 Guide to Durable Concrete, Michigan, USA
[11] Scott B D and Safiuddin Md 2015 *Concrete Research Letters, www.crl.issres.net* **6**(3)
[12] Papenfus N 2003 Applying Concrete Technology to Abrasion Resistance, *Proceedings of the 7th International Conference on Concrete Block Paving, Sun City, South Africa*
[13] Naik T R, Sing, S S & Hossain M M 1995 *ACI Materials Journal* **92**(6) 649–650
[14] Naik T R, Singh S S & Ramme B W 2002 *Journal of Materials in Civil Engineering* 417–426
[15] Yen T, Hsu T, Liu Y W & Chen S H 2007 *Construction and Building Materials* **21** 458–463
[16] Liu Y-W, Yen T and Hsu T-H 2005 *Cement and Concrete Research* 1814-1820
[17] Langan B W, Joshi, R C & Ward M A 1990 *Canadian Journal of Civil Engineering* **17** 19–27
[18] Nanni A 1989 *ACI Materials Journal* **86**(53) 559–565
[19] Nazari A and Riahi S 2011 *Energy and Buildings* **43** 2939–2946
[20] Riahi S and Nazari A 2011 *Science China, Technological Sciences* **54**(9) 2349–2357
[21] Ytterburg R F 1971 *Civil Engineering, ASCE* **41**(1) 68–71
[22] Nanni A 1989 *Journal of Transportation Engineering, ASCE* **114**(6) 684–694
[23] Kevern J T, Schaefer V R and Wang K 2009 *Journal of Testing and Evaluation* **37**(4) JTE101761
[24] Dhir R K, Hewlett P C and Chan Y N 1991 *Materials and Structures* **24** 122-128
[25] Bakke K J 2006 Abrasion Resistance, Significance of Tests and Properties of Concrete and Concrete-Making Materials, *ASTM STP 169D, Bridgeport,* pp. 184-193
[26] xxx SR EN 197-1:2011 Cement- Part 1: Composition, specification and conformity cririteria common cements.
[27] Declaration of Performance no. CPR-043-AE, Portland Cement, Holcim SA.
[28] Laplante P, Aïtcin P-C and Vezina D 1991 *Journal of Materials in Civil Engineering* **3**(1) 19-28
[29] Turk K, & Karatas M 2011 *Indian Journal of Engineering and Materials Sciences* **18** 49–60
[30] Naik T R, Singh S S and Hossain, M M 1995 *ACI Materials Journals* **92**(6) 649-659

[31] Laboratory Test Report no. 848/ 25.08.2016, ArcelorMittal for granular slag.
[32] Laboratory Test Report ref. AHKRO 11755-6M/01.08.2018, ArcelorMittal for granular slag.
[33] Laboratory Test Report ref. AHKRO 11802-Z/25.09.2018, ArcelorMittal for granular slag.
[34] Laboratory Test Report ref. AHKRO 11806-A/10.0802018, ArcelorMittal for granular slag.
[35] Kiliç A, Atiş C D, Teymen A, Karahan O, Özcan F, Bilim C and Özdemir M 2008 *Cement and Concrete Composites* **30**(4) 290-296
[36] Dhir R K, Hewlett P C and Chan Y N 1991 *Materials and Structures* **24** 122-128
[37] xxx NE 014-2002, The norm for the execution of cement concrete road pavements in a fixed and sliding formwork system, of Romania.
[38] Laboratory Test Report no.545/03.04.2018, SC Grandemar SA.
[39] Laboratory Test Report no.956/ 30.06.2017, SC Grandemar SA.
[40] Laboratory Test Report No.706/13.07.2016, ArcelorMittal for artificial aggregates.
[41] xxx SR EN 1097-6:2002, Test for mechanical and physical properties of aggregates -Part 6: Determination of particle density and water absorption.
[42] xxx NE 012-1-2007, Code of practice for the manufacturing of concrete, of Romania.
[43] xxx SR EN 12620:2003 and SR EN 12620+A1:2008, Aggregates for concrete.
[44] xxx SR 667:2001, Natural aggregates and worked stone for roads.
[45] xxx SR EN 1008:2003, Mixing water for concrete.
[46] xxx SR EN 1338:2004/AC:2006, Concret paving blocks. Requirements and test methods.
[47] Orhan K, Muhammet Vefa A, Metin Murtlu A 2017 Effect of fine aggregate abrasion resistance and its finess module on wear resistance of Portland cement pavements, *https://www.researchgate.net/profile/Metin_Aydin2/publication/316636335/links/5908f027458515ebb495c5da/Effects-of-fine-aggregate-abrasion-resistance-and-its-fineness-module-on-wear-resistance-of-Portland-cement-concrete-pavements.pdf*.
[48] SR EN 12390-3:2002 Thesting hardened concrete-Part 3: Compressive strength of test specimens.

IOP Conf. Series: Materials Science and Engineering **877** (2020) 012041 doi:10.1088/1757-899X/877/1/012041

Contributions to the Development of a Low-Speed Gadolinium Actuator System

V E Toader[1*], **C Ungureanu**[1], **L D Milici**[1]
[1]Faculty of Electrical Engineering and Computer Science, Ştefan cel Mare University of Suceava, RO

*E-mails: eusebiu.toader@gmail.com; costel@eed.usv.ro; dam@eed.usv.ro

Abstract. Within the EMAD Research Centre of the USV, a low-speed actuator system has been developed which is based on the properties of the intelligent material gadolinium, i.e. the modification of the magnetic properties of the material according to temperature, magnetic at a temperature of less than 22.5 degrees and paramagnetic at a temperature greater than 22.5 degrees. The speed adjustment is made by the temperature variation (cold - warm). The paper presents the constructive solution, experimental data from which the characteristics of the equipment designed, carried out and tested and the general conclusions on its implementation in industrial actuator systems.

1. Introduction

Gadolinium was discovered in 1880 by Jean Charles de Marignac, who detected his oxide using spectroscopy. Gadolinium is a chemical element with the symbol Gd and atomic number 64. Complex organic and gadolini compound solutions are used as an intravenous MRI contrast agent to improve imaging in medical nuclear magnetic resonance imaging and magnetic resonance angiography (MRA) procedures. Gadolini compounds are also used to make green phosphorus for color TV tubes. Gadolinium is used in many quality assurance applications, such as line sources and calibration ghosts, to ensure that nuclear medicine imaging systems work correctly and produce useful images of radioisotope distribution in the patient's body. It is also used as a source of gamma rays in X-ray absorption measurements or in bone density assessment for osteoporosis screening, as well as in the portable lixiscope X-ray imaging system [1].

The realization of the first engines based on the conversion of helio-thermo-magneto-motors, became a reality, only in the second half of the 20th century [2] and was favored, as already mentioned, by the enhancement of magnetic materials, with Curie temperature, possible to reach even at ambient temperature (gadolinium). This advantage, combined with the possibilities of investigation in theory, opened by the theory of the finished element and the use of the calculation technique, made it possible to obtain promising results [3, 4],

The fact that gadolinium is ferromagnetic with a Curie temperature of only 294 K (21 °C) has been used to develop simple thermomagnetic motors for laboratory demonstrations [4]. They can be operated by an electric lamp or sunlight and will not work on a warm day as the material would always be above Curie temperature. Today there is a great interest in the development of non-conventional thermomagnetic systems that can use considerable heat from the exhaust gases of the internal combustion engine to generate energy. They use materials with curia temperatures much higher than gadolinium, such as iron-nickel-chrome alloys [5]. In the case of the model studied, the

IOP Conf. Series: Materials Science and Engineering **877** (2020) 012041 doi:10.1088/1757-899X/877/1/012041

ferromagnetic material is a rotor in the form of a ring or cylinder, which moves in a magnetic field created by permanent magnets. It is heated at a time so that it passes into the state of paramagnetism when the magnetic field decreases. There is, therefore, a constant torque on the rotor that causes it to rotate. The rotor is then cooled and regains its feromagnetism before re-entering the magnetic field [4]. In Figure 1a is presented the operating principle of an engine that can generate mechanical power 100W at 1.5 rpm.

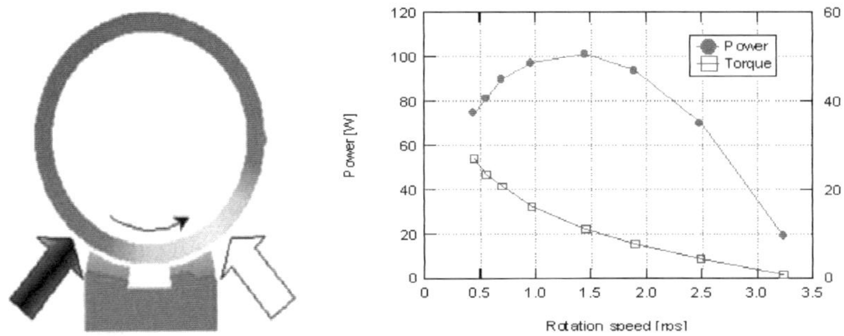

(a) (b)

Figure 1. (a) Experimental model of a thermomagnetic motor; (b) Power and torque for thermomagnetic motor [4].

The problems that occur with this variant are related to the speed of heating and cooling as it cannot vary very quickly so that the speed of rotation is low (Figure 1b), usually from 0.5 to 1.5 speeds per second [4].

2. Description of the experimental stand

The operation of this gadolinium engine (Figure 2) is possible thanks to the system made of an assembly of two permanent magnets 6 placed in front of each other, among which passes the disc with gadolinium pills, mounted in a fixed position on the support of the aluminum radiator 1.

Figure 2. Engine with gadolinium1 - aluminum radiator; 2 - pinion that transmits engine movements; 3 - disc of electroisolating material; 4 - gadolinium pills; 5 – the motor body; 6 - permanent magnets; 7 - permanent magnet support; 8 - electrical resistors.

The rotor, in the form of disc 3 (Figure 3), made of electro-isolating material, presents on its circumference the pills of gadolinium 4, arranged equidistantly and which are located in the magnetic field of the permanent magnets 6. The disc rotor begins to rotate when a temperature difference occurs between the cold source and the warm source, a temperature range that will include the Curie temperature of the gadolin. We will notice an increase in the speed of the gadolinium disc when the temperature varies to one of the outer or inner environments. Once the engine disc starts to rotate, the gadolinium pills enter the cooled enclosure and become ferromagnetic. From the rotational inertia of the disc and under the action of the magnetic field of the permanent magnets that attract the ferromagnetic gadolinium pills, the pills then reach the area heated by the two electrical resistors 8, which causes the pills to become paramagnetic, at which point the disc is printed a rotational impulse produced by the following gadolinium pills that have transited the cold source, making his movement continuous.

(a) (b) (c)

Figure 3. Experimental model of the experimental stand for the gadolinium engine: a) the rotor with gadolinium pills; b) he refrigeration system; c) general view.

For the designed, made and tested version of the gadolinium engine, an experimental stand has been developed that allows the transition of gadolinium pills on the rotor between a cold source made with a freon aggregate (refrigerator) and a warm source made with resistance powered by an adjustable source of continuous voltage.

The engine rotor (figure 3a) has gadolinium pills distributed on the circumference of the rotor to reduce the time required for the thermal exchange in the active material by decreasing the amount of heat needed to switch between the magnetic and paramagnetic states of the gadolinium pills depending on the temperature of the two media [7].

The stand produced highlights the principle of operation of engines based on thermo-magneto-mechanical conversion. It consists of the basic elements necessary for the operation of an engine in this category: the cold source, represented by the cooling system (Figure 3b), the hot source and the magnetic field source.

3. Experimental data

The built gadolinium engine was tested on the stand designed to identify the main characteristics related to speed and active mechanical torque [8].

The temperature of the environment of the two heat sources as well as the gadolinium pills on the rotor was carried out with contactless, infrared thermometers, the speed was recorded with a non-contact numerical tachogenerator and the engine torque with a torque sensor. The hot source was controlled by a variable voltage that powers electrical resistors.

Here are some experimental data relevant to the operation and operation of the gadolinium engine as follows:

• If both the cold and warm sources change their temperatures in order to widen the difference between them, a variation is obtained within wide limits of the speed, with a small slope in the first part after which the slope changes until a maximum value of 520 rpm is reached, a value which can no longer be exceeded by the change in the temperatures of the two sources (Figure 4);

• If the temperature of the hot source is the temperature of the medium (24 °C), the engine speed value has been measured at the change in the temperature of the cold source between 18 and -16 °C, with a significant increase in speed followed by a landing (Figure 5);

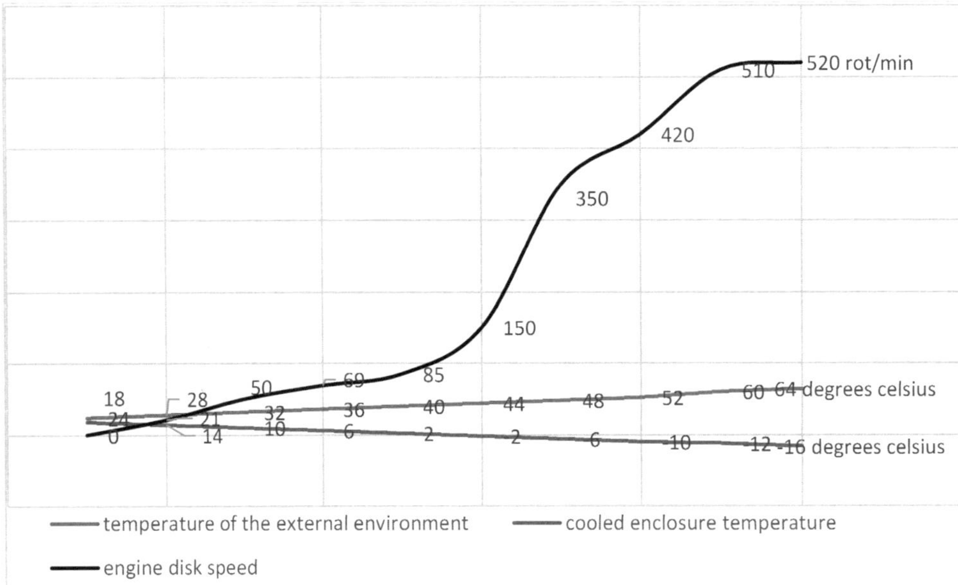

Figure 4. Change in speed when the temperature of the cold and hot source changes.

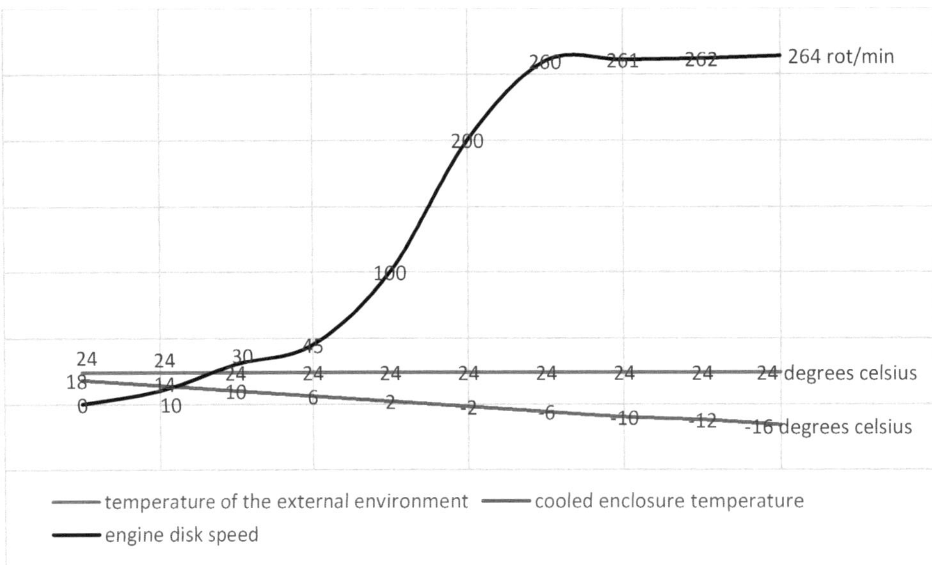

Figure 5. Change in speed when the temperature of the cold source changes.

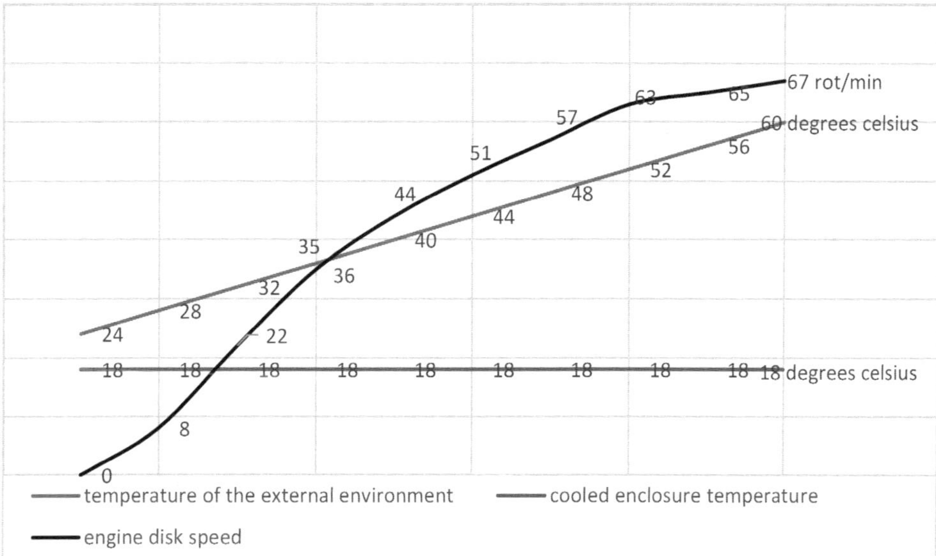

Figure 6. Change in speed when the temperature of the hot source changes.

If the temperature of the cold source is kept constant at 18 °C and the temperature of the hot source varies in the range (24 ÷ 56) °C, a slow variation in speed tends asymptotically towards a maximum value (Figure 6).

4. Conclusions

The gadolinium drive system designed, developed and tested can be used successfully in low-speed electrical operations where two optimal heat sources are available for operation, the warm source allowing heat recovery from combustion gases or technological steam.

The adjustment of the engine speed achieved can be done by changing the temperature of one of the heat sources and ensuring an optimal transfer of heat to the rotor. The speed increase can be done by changing the temperature of the heat sources or by increasing the transfer of heat by ventilating the environment at their level.

The engine has a small size, which allows its use in narrow, high thermal flux spaces to facilitate heat transfer.

Acknowledgment

This paper has been financially supported within the project entitled „DECIDE - Development through entrepreneurial education and innovative doctoral and postdoctoral research, project code POCU / 380/6/13/125031, project co-financed from the European Social Fund through the 2014 – 2020 Operational Program Human Capital"

5. References

[1] https://b2b.partcommunity.com/community/knowledge/en/detail/3253/Gadolinium, accessed on the date 04.2020
[2] douglas-self.com/MUSEUM/POWER/thermomagnetic/thermag.htm, Thermo-Magnetic Motors, accessed on the date 04.2020
[3] Duburg G J, Coumat F J and Coumat J C 1984 Moteurs magnéto-solaire. Int. Cl.3: H02 N 11/00. FR Brevet, 2538184. 1984-06-22
[4] Fang C H, Huang W N and Chen C H 2005 Reciprocating and rotary magnetic refrigeration apparatus. Int. Cl.7: F25 B 21/00. Brevet US, 6935121 B2. 2005-08-30

[5] Milici D, Milici M, Keersebilck P, Bouters A, Flore W and Ungureanu C 2011 Design a Guidance System for Solar Panels Using Virtual Instrumentation, in: Acta Electrotehnica, Proceedings of the 4th International Conference on Modern Power Systems, p. 281-283

[6] Cernuşcă D, Milici D L and Peienar M 2016 Development of special actuators using smart materials, *International Conference and Exposition on Electrical and Power Engineering*, p. 567-570

[7] Sorea N, Ungureanu C and Cernomazu D 2010 Considerations regarding the solar engines based on the thermo-magneto-mechanical conversion. *8th World Energy System Conference WESC 2010, The Scientific Bulletin of Electrical Engineering Faculty, Tîrgovişte* p. 84-90

[8] Milici M R, Milici L D, Pentiuc R D and Atanasoae P 2016 The mathematical model of a stand measuring the torque of the termobimetal actuators, *International Conference and Exposition on Electrical and Power Engineering* p. 563-566

IOP Conf. Series: Materials Science and Engineering **877** (2020) 012042 doi:10.1088/1757-899X/877/1/012042

Possibilities for Use of Vermiculite in Recultivation of Embankments Obtained as a Result of the Deposit of Mining Waste from the Extraction of Copper Ores

V Stefanova[1*], P Petrov[1] and E Zheleva[1]
[1]Faculty of ecology and landscape architecture, Univertity of Forestry, Kliment Ohridski 10 Blv., 1756 Sofia, Bulgaria

*E-mail: venistefanova3@gmail.com

Abstract. The present study examines the possibilities for vermiculite usage for recultvation of copper mining waste embankment. For this purpose, a laboratory study was carried out. The study which included the mixing of mining waste with different ratios of vermiculite and adds of other soil improvers (fertilizers and lime) with different norms. As a result is, is observed that the vermiculite addition in larger quantities did not affect the pH of the substrate. The content of major pollutants in the waste - Cu and As, remains relatively high - over the maximum permissible concentrations. Independently of the added vermiculite, it is necessary to apply calcium materials and fertilizers to improve substrate properties.

1. Introduction

In the process of mining, spaces disturbed by direct extraction and occupied by embankments are formed and rapidly increase [1]. Embankments are characterized by heavy metals and metalloids content above the maximum concentration limits for the respective zonal soils, which poses a serious risk to the environment and human health due to the pollutant properties - toxicity, mobility, soil and plant accumulation ability and migration into the food chain [2-4]. In these cases it is necessary to introduce soil improvers to immobilize the pollutants.

In recent years, vermiculite has been a widely used mineral for the adsorption of metal cations [5-7]. Vermiculite is a three-layer clay mineral - hydrated magnesium-ferrous aluminosilicate. Its structure consists of two tetrahedral silicate layers and an interlayer space of an octahedral layer containing aluminum and iron. Vermiculite is characterized by: porous structure, high sorption capacity, high ion exchange capacity, low thermal conductivity and neutral pH. Many researchers believe that due to the content of elements such as calcium, magnesium, potassium, iron and silicon, vermiculite is a natural biostimulant for plant growth. Vermiculitis exhibits good adsorption capacity for heavy metal divalent ions [8], especially suitable for the adsorption and removal of potentially toxic cations such as Cu^{2+}, Pb^{2+}, Cd^{2+}, Zn^{2+} и Ni^{2+} [9,10] due to its high cation exchange ability (120-150 meq/100g). The adsorption of metal ions from vermiculite decreases with decreasing pH and increasing ionic strength. An exception is arsenic, for which there is no evidence that it is adsorbed by vermiculite [11].

This study aims to determine the possibility of using vemiculite - alone and in combination with other ameliorants (NPK fertilizers and CaO) in biological recultivators of copper mining storage facilities for the immobilization of copper and arsenic in substrates from embankments and changes in their fertility.

2. Materials and methods

The study materials and methods are closely aligned with the stated objectives. Those elements that we aim to eliminate or neutralize in soil-forming materials are investigated.

2.1. Materials

To achieve the purpose of investigation samples were taken from embankments for the storage of mining waste from the extraction of copper ores, which examined the elements and their properties, creating a harmful environment for flora and fauna.

2.2. Methods

The methods used in the study are analytical methods and methods of vegetation experience. Chemical characterization was investigated to characterize the mine waste and determine the state of the substrate after the vermiculite treatment. The analyzes include: potentiometric determination of soil acidity (ISO 10390); total nitrogen by the Kjeldahl method. (ISO 11261); K_2O and P_2O_5 - by the methods of Petko Ivanov; determination of copper and arsenic content by combustion in aqua regia and ICP-OES.

After clarifying the agrochemical properties of the substrates and vermiculite, a vegetation experiment was conducted with different amounts of vermiculite - 5, 10 and 20%, mixed with mineral fertilizers and CaO. In plastic cups with a capacity of 1.5 l for carrying out the vegetation experiment, 1 kg of substrate was filled with the appropriate amount of vermiculite (5%, 10% and 20%). For each vermiculite amount, 4 different variants were prepared - substrate + corresponding amount of vermiculite (%V), substrate + corresponding amount of vermiculite + NPK fertilizer (%V + F), substrate + corresponding amount of vermiculite + CaO (%V + CaO), substrate + corresponding amount of vermiculite, + NPK fertilizer + CaO (%V + F + CaO). A control sample was set up without the addition of vermiculite and other ameliorants. In each of them, *Lolium perenne* L. was sown and cultivated for 5 months, so as to allow sufficient time for contact and grass growth in each variant needed for the adsorption of heavy metals.

3. Results

3.1. Characteristics of vermiculite

The acidity data and the main elements characterizing the vermiculite were provided by the company owner of the vermiculite deposit as part of the internal control of the raw material. The silicate analysis of the raw material is presented in Table 1.

Table 1. Silicate analysis of vermiculite.

Indicators	Value, %	Indicators	Value, %	Indicators	Value, %
pH	6.26	TiO_2	0.35	K_20	0.92
SiO_2	41.14	Na_20	0.23	P_20_5	8.34
$A1_20_3$	11.62	MnO	0.10	Ni0	0.15
CaO	2.56	ZnO	0.14	Cr_20_3	0.20
Mg0	27.79	CuO	0.01		

Data from analyzes of vermiculite show that it does not contain dangerous substances in quantities above the limit values. Vermiculitis has a slightly acid reaction, and the value of the elements characterizing the cation-exchange capacity indicates that the vermiculite has a high levels of base saturation. The element of silicon is in the highest quantities, followed by magnesium, aluminum and phosphorus. The amount of calcium is much smaller than magnesium, while in natural soils, the amount of calcium is usually several times higher than magnesium. However, from a chemical point of view, it should be noted that imported into the soil or soil-forming substrate, it will not adversely affect them. In addition, it can be noted that the phosphorus content is much higher than in natural soils. The potassium content is also high in comparison to the average in the soil. The content of potassium absorbed by the plants is high, which would be beneficial for the plants. All other elements are in quantities that are relatively indifferent to soil formation and, through their chemical properties, do not adversely affect soils, vegetation and the environment. These data indicate that phosphorus and potassium fertilization may not be required when vermiculite is introduced into the soil. The data from the analyzes of the vermiculite raw material indicate the potential for its use in the reclamation of landfills.

3.2. *Characteristics of mine waste from the embankments*

Data on the chemical properties of the mine waste intended for vegetation experiment: acidity, electrical conductivity, nutrient content, content of copper and arsenic are presented in Table 2. The classification for salinity and content of nutrient and content of alkaline earth elements is made by Penkov (1996) [12].

Table 2. Acidity and content of nutrients, copper and arsenic in mining waste.

pH (H_2O)	Conductivity	N	P_2O_5	K_2O	Mg	Ca	Na	Cu	As
-	μS/cm	%	mg/100g	mg/100g	mg/kg	mg/kg	mg/kg	mg/kg	mg/kg
3.6	136	0.089	0.51	2.0	734	4	384	599	19.20

As can be seen from the data in Table 2, the mining waste is characterized by an exterior acidic pH - 3.6. Low pH is a prerequisite for the mobilization of heavy metals contained in mining waste and can lead to a number of environmental problems [13]. Moreover, at these pH values, plant nutrition is characterized by an excess of micronutrient uptake [14]. Additional ameliorants are required to increase the pH value.

According to the value of the electrical conductivity, no salinity was observed in the samples tested, this is because, as the data in the table shows, the sodium content is scarce. The magnesium content is higher than that of calcium. However, the waste is poorly stored in calcium and very poorly stored in magnesium. The waste is poor in nitrogen, poorly stored in potassium and phosphorus. The analysis of the nutrient content implies the need for additional fertilization, since in vermiculite the content of P_2O_5 is 8.34% and the content of potassium is only 0.92%. The content of copper and arsenic exceeds the maximum permissible values (MPVs for industrial territories Cu - 500 mg/kg, As - 10 mg/kg according to Regulation 3 [15]). High levels of metals suggest toxicity to some microorganisms [16,17], as a result of which the mineralization of the soil and its ability for self-cleaning are impaired, the conditions for the development of vegetation also deteriorate.

3.3. *Vegetation experience*

The vegetation experience was conducted to identify the possibility of using vermiculite as a soil improver in the reclamation of embankments and the quantities required for the purpose, and more precisely, the most appropriate ratios of ameliorants to the basic soil-forming material and the corresponding ameliorations. At the end of the vegetation experience, analyzes of the substrates in the root habitat of the ryegrass were made. The vegetation experiment was carried out in closed containers, from which the water with which they are watered does not flow. Liming rate was determined by adding an aqueous solution of pure CaO to neutralization of the substrate - 0.5 g/ml. Based on the nutrient analysis, a fertilization rate of 0.05 g/ml was determined. The vegetation experiment is carried out for 5 months so as to allow sufficient time for contact between the vermiculite and the substrates.

3.4. Analyzes of soil-forming substrates at the end of the vegetation experiment

Figure 1 shows that the addition of vermiculite does not affect the pH value, despite the content of alkaline earth elements in it. The experiment also proves that the substrates need liming. The graph shows that in each of the variants, only in the case of CaO addition, an increase in pH from slightly acidic to neutral is observed, under which optimal conditions for plant nutrition are created.

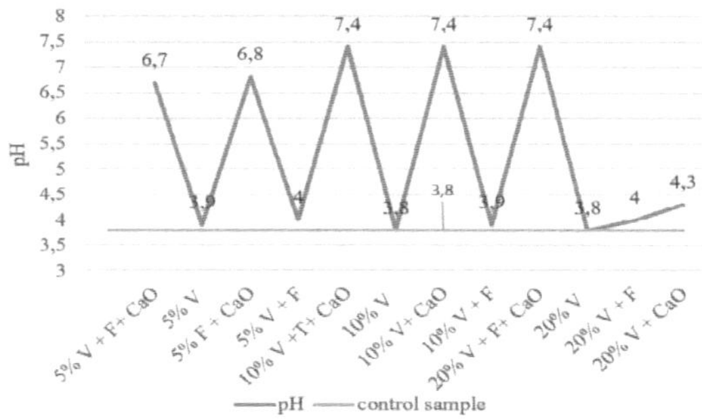

Figura 1. pH of the substrates at the end of the vegetation experiment.

Figure 2 and Figure 3 show the contents of the basic nutrients in the studied substrates in the various applications of ameliorants. Figure 2 shows that, despite fertilization, the nitrogen stock is very small to small. The introduction of more vermiculite does not increase the amount of nitrogen. Compared to the control sample a decrease in the amount of total nitrogen was observed. The data (Figure 3) show that the uptake of nutrients from mineral fertilizers is not always dependent on the acidity of the substrates and is almost unaffected by the presence of vermiculite. Potassium remains in the highest amounts in the substrate, in most cases where the pH is completely in the neutral range. In a few variants, the phosphorus content is in greater quantities - probably the element is in a water-soluble form and is absorbed by the grass root system. Only in the samples with 20% vermiculite, fertilizers and liming concentrations of potassium and phosphorus exceed those in the control sample.

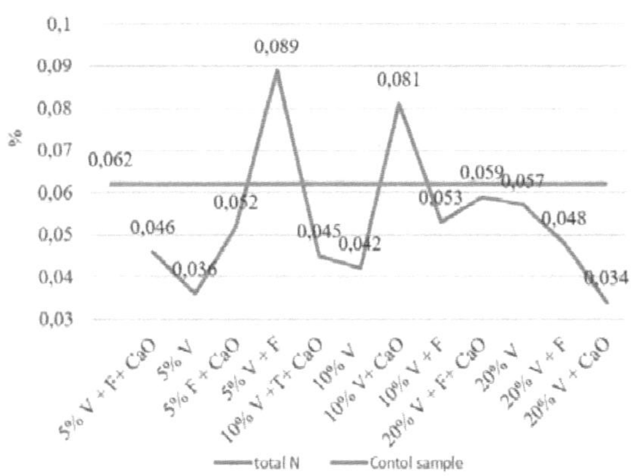

Figure 2. Total nitrogen content of the substrates at the end of the vegetation experiment,%.

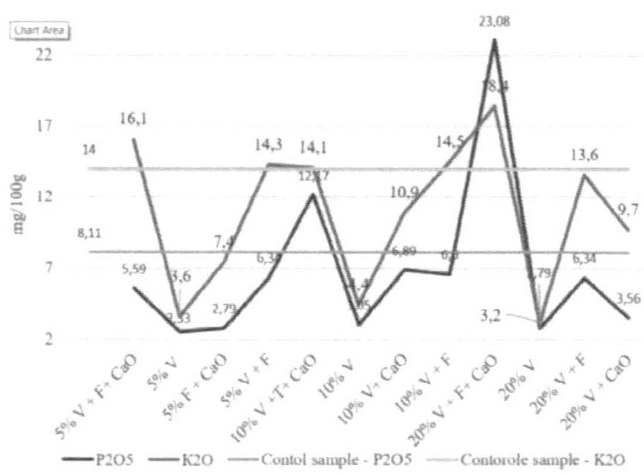

Figure 3. Content of P_2O_5 and K_2O substrates at the end of the vegetation experiment, mg/kg.

Of the metals and metalloids, only the content of copper and arsenic was examined as indicators of the deposit. Data on their content at the end of the vegetation experiment are presented in Figure 4 and Figure 5. The amount of copper is reduced compared to the control sample. The lowest concentrations were observed in the samples treated with 20% vermiculite, however, the values were close to the

MPV – 500 mg/kg. In all substrates – control, with or without vermiculite, copper is above the maximum levels, except for samples with 10% vermiculite and liming. There is no direct correlation between copper content and meliorations. The likely decrease in Cu content is due to its extraction and accumulation in the plant mass. It is very likely that in samples that have created better conditions for the development of vegetation, the amount of Cu in the soil has decreased due to the development of more biomass.

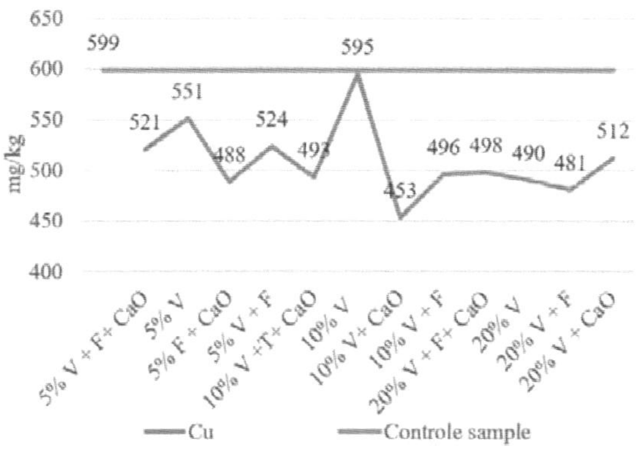

Figure 4. Copper content of the substrates end of the vegetation experiment, mg/kg.

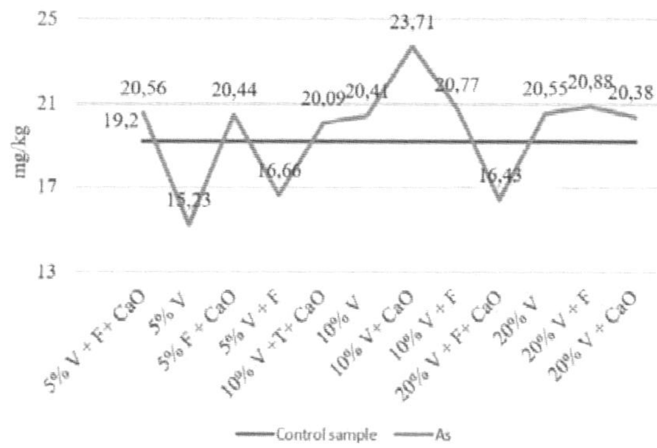

Figure 5. Arcenic content of the substrates end of the vegetation experiment, mg/kg.

The content of arsenic in all samples from the mine is between 15.23 and 23.71 mg/kg, with the highest amounts in the samples with 20% vermiculite and the smallest without vermiculite or 5% vermiculite. Whereas, that arsenic is absorbed by plants regardless of the acidity of the soil substrate, we can assume that larger amounts of vermiculite retain larger amounts of arsenic and prevent for arsenic absorption from the herbs. Although part of the samples shows a decrease in value compared to the control sample in all samples has an arsenic concentration exceeding the MPV – 10 mg/kg.

4. Conclusion

In conclusion from the vegetation experiment conducted, we can say that vermiculite does not affect the acidity of the substrates.

In limed and fertilized variants, the potassium and phosphorus content of the substrates is higher, while the nitrogen content is reduced. This suggests that from these variants the elements are absorbed more slowly by the plants. Regardless of the amount of vermiculite imported, it is necessary to bring in neutralizing calcium materials to raise the pH and immobilize the heavy metals. Fertilization with mineral fertilizers is also a necessary reclamation activity in biological reclamation.

The present study confirms that vermiculite has no effect on As content, despite additional ameliorations applied.There is a decrease in Cu content, which is probably related to the development of vegetation in each of the samples. Further investigations are needed to determine the causes for the reduction of the copper content.

Through the vegetation experience conducted under controlled conditions, the overall role of vermiculite in the remediation of embankments with the acid reaction of bulk materials cannot be demonstrated. It is necessary to continue the experiments by testing the vermiculite and the reclamation applied to the vegetation experience through.

5. References

[1] Zheleva E 2010 *Rehabilitation of damaged terrains*, PSSE, Sofia
[2] T E A Kostadinova 2019 *Journal of environmental protection and ecology* **20**(1) 397-403
[3] Kostadinova A and Todorova E 2014 *Journal of environmental protection and ecology* **15**(2) 558

[4] Zheleva E, Petrov P and Ivanova S 2016 *Journal of Environmental Protection and Ecology* **17**(4) 1324–1333
[5] Sis H and Uysal T 2014 *Appl. Clay Sci.* **95** 1-8
[6] Kebabi B, Terchi S, Bougherara H, Reinert L and Duclaux L 2017 *Appl. Clay Sci.* **139** 92-98
[7] De Freitas E D, de Almeida H J, Vieira M G A 2017 *Appl. Clay Sci.* **146** 503-509
[8] Malandrino M, Abollino O, Giacomino A, Aceto M, Mentasti E 2006 *Journal of Colloid and Interface Science* **299** 537–546
[9] Malla P B 2002 Vermiculites. Pp. 501-530 in: J. B. Dixon and D. G. Schulze (Eds.), *Soil Mineralogy with Environmental Applications*, Vol. 7. Soil Science Society of America, Inc., Madison, WI, USA
[10] Panuccio M, Sorgonà A, Rizzo M, Cacco G 2009 *Journal of Environmental Management* **90(1)** 364-374
[11] Malandrino M, Abollino O, Buoso S, Giacomino A, La Gioia C, Mentasti 2011 *Chemosphere* **181** (1–4) 149–160
[12] Пенков М 1996 Почвознание. София: Агроинженеринг- ЕООД
[13] Kostadinova A, Todorova E, Krumov T 2016 *Physics and materials* **8**(8) 25-27
[14] Малинова Л 2010 *Почвознание и замърсяване на почвите*. София: Издателска къща при ЛТУ
[15] Regulation № 3 of 1 August 2008 on the norms for the permissible content of harmful substances in soils (published in State Gazette № 71/12.08.2008)
[16] Наскова П, М. Консулова Д. Пламенов, Малчева Б 2017 Математически модел за определяне на степента на влияние на различни физико-химични фактори върху числеността на общата микрофлора в антропогенни почви (Списание за наука „Ново знание" 6-4
[17] Yankova P, Naskova P, Malcheva B, Plamenov D 2016 *International Journal of Current Research* **8**(10) 39681-39686

IOP Conf. Series: Materials Science and Engineering **877** (2020) 012043 doi:10.1088/1757-899X/877/1/012043

Progress in Electrospun Nanofibres for Air Filtration

L R Manea[1], A P Bertea[1*] and A Bertea[1]

[1] Gheorghe Asachi Technical University of Iasi, Faculty of Industrial Design and Business Management, Prof. Dr. Doc. Dimitrie Mangeron Street, No. 29, RO-70005047, Iași, Romania

*E-mail: andrei_bertea@yahoo.co.uk

Abstract. Air pollution is a growing problem worldwide, reason for seeking more efficient methods to prevent exposure to contaminated air. Air filters must have the ability to effectively remove extremely small quantities of hazardous materials from the air, without causing excessive pressure drops, all at acceptable price conditions. A modern nanotechnology for obtaining polymeric nanofibres is electrospinning, capable of producing nanofibres of various materials, with controlled structures and porosity and with small pores, which makes it a favorable candidate for efficient air filtration. Electrospun nanofibres can be used directly or incorporated into various devices to reduce the negative impact of air pollution. The article briefly reviews the latest electrospinning technologies, highlighting the structural and performance advantages in producing air filters in what regards thickness, porosity, pressure drop, permeability antibacterial properties. The types of material used to obtain the nanofibres used as filtration medium are analyzed and the structural properties and performances of the filters are discussed. Finally, the challenges of electrospinning in producing air filters and future perspectives were summarized.

1. Introduction
Air pollutants are divided into two main categories: particulate matter (PM) and gaseous contaminants [1]. The air pollution is caused by the increase of the industrial emissions in the air and the exhaust gases of cars [2, 3], and has a growing impact on public health, as well as the environment, particularly fog and haze pollution [4-9]. The environmental impact refers to the influence of atmospheric particulate matter on Earth's climate by scattering and absorbing solar radiation and by influencing cloud formation [10].

In terms of human health, it is known that mainly particles smaller than 2.5 μm in diameter, which cannot be blocked by the nasal cavity, have a strong negative effect on respiratory organs [11], as they penetrate into the lungs and cardiovascular system causing from irritation to chronic respiratory and lung cancer and worsening preexisting diseases [12, 13]. One of the most efficient methods to remove suspended particles from the air is air filtration, and electrospun nanofibre filters have proven to be amongst the most effective, due to their exceptional properties such as high specific surface area and high porosity with fine pores [14, 15].

2. Electrospinning principle
The 21st century began under the auspices of nanotechnology, with hopes of obtaining nano-materials with innovative properties and the exploitation of unique phenomena occurring at the nanometric scale [16]. All of these are expected to have a considerable impact in many fields, environmental protection

included [17]. As a result, nanotechnologies developed at the turn of the century are the foundation of future technological platforms [18].

Nanotechnologies aim to study and develop phenomena from nano-level materials in order to achieve effects and performances on a macro scale. The ability to manipulate individual molecules and perceive their characteristics at the atomic level has made this field very up-to-date.

Of all the technologies for obtaining nanofibres, the most efficient in terms of productivity and efficiency is electrospinning, a technology that evolved from electrospray technology [19]. This process is capable of producing long, continuous, controlled-oriented nanofibres with varying degrees of complexity and architectures, adaptable to different types of polymers. Electrospinning is identified as one of the key, leading technologies of the future [20, 21].

In the electrospinning process, to create an electrically charged jet of polymeric solution or melt, a high voltage current is used [22-27], respectively, one electrode is placed in the solution / melt, and the other is attached to the collector (figure 1). The electric field is applied to the capillary tube containing the polymeric fluid, maintained by its own surface tension. Thus, a charge is induced on the surface of the liquid and the mutual rejection of the electrical charges acts as a force directly opposite to the surface tension. When the intensity of the electric field increases, the hemispherical surface of the fluid at the tip of the capillary is elongated, forming a cone known as the Taylor cone. When the intensity of the electric field, in its growth, reaches a critical value, respectively when the electrostatic force of rejection exceeds the surface tension, the jet of electrically charged fluid is ejected from the tip of the Taylor cone. The jet thus formed begins a twisting process (in which the solvent evaporates, respectively in the case of melting, solidification takes place), leaving behind a polymer nanofibre with electric charge, which is randomly deposited on the collector.

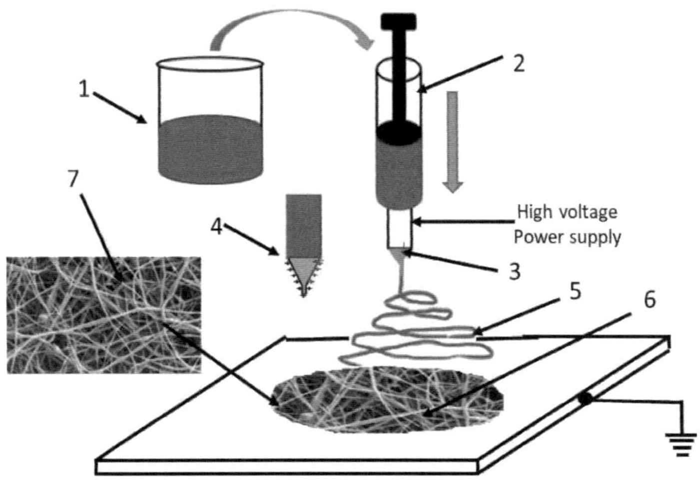

Figure 1. Scheme of the principle of electrospinning from polymeric solution 1. polymeric solution; 2. syringe; 3. needle; 4. Taylor Cone; 5. jet; 6. collector screen (rotary or stationary - the collector surface); 7. Electrospun fibre arrangement.

Electrospinning results in a network made up of submicron fibres. Larger fibrous deposits can be obtained through the use of multiple capillaries. These fibrous deposits have many special properties, such as: low weight, high permeability and small pore size, which represent advantages in filtration applications, as well as numerous other areas of application such as: sensors, protective materials, drug controlled release, sound absorbers, composite materials, antimicrobial products, biological dressings, tissue repair, enzyme immobilization, etc [19, 21, 24, 28-36].

Electrospun fibre networks offer the advantage of advanced physico-mechanical properties. In addition, the amount of electrospun fibres required to obtain a product is low, and the electrospinning process itself is characterized by high versatility, allowing spinning of fibres with different sections, under various conditions, using a wide range of polymers [26, 27, 36 -38].

3. Applications of electrospun nanofibres in filtration

Nanofibrous materials are suitable for filtration applications, as a result of their special properties [11, 39]. Especially electrospun nano-fibrous filters showed high filtration efficiency, due to their exceptional properties in what concerns high specific surface area and high porosity [14, 15]. The use of the fibres obtained by electrospinning in the field of filtration covers two domains: air filtration, respectively water filtration (both in water and wastewater treatment).

The importance of using electrospun nanofibres in the field of air filtration is attested by the numerous patents [40]. In 2007 Ngiam observed that in the first decade of the new millennium 75% of existing patents in the field of nanofibres refer to the manufacture or use of nanofibres in filter systems and a decade later the patents in this field continued to be very numerous [41, 42]. The situation is similar in terms of scientific articles dedicated to electrospun nanofibres [43].

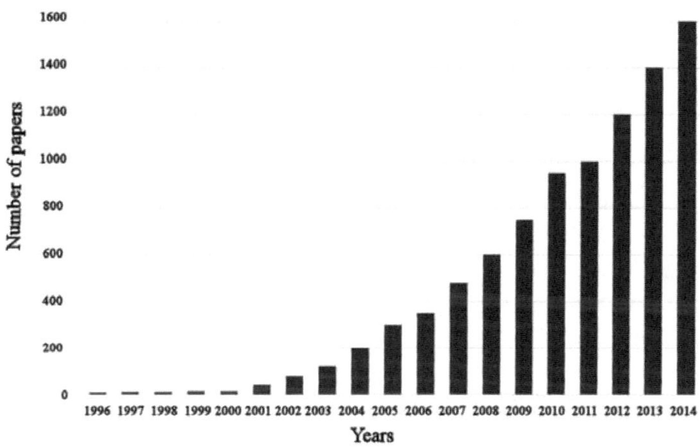

Figure 2. Evolution of the number of papers dedicated to electrospinning and electrospun polymer nanofibres [43].

Such a situation is normal if one takes into account the fact that conventional filters have certain flaws, particularly related to their efficiency in removing small particles from the air such as pathogenic agents like viruses. According to Barhate and Ramakrishna [44], high-efficiency classical filters are viable for particles up to 0.3 pm in size.

Atmospheric aerosols contain particles of different sizes ranging from ultrafine range (diameter less than 100 nm) to a gross range (over 2.5 pm); the size of viruses, for example, varies from 10 nm to 200 nm, bacteria from hundreds of nm to tens of pm, and pollen granules from tens to hundreds of pm. The fine particle range (less than 2.5 µm) can be divided into two categories: the core module (less than 40 nm) and the accumulation module (over 40 nm). Smoke and smog particles have diameters in these ranges [45, 46].

In order to protect from nanoparticles, which can have a detrimental effect on health, it is imperious necessary to increase the filter retention capacity and this can be achieved using electrospun nanofibre filters, which found use in domains as individual protection, building air filters and dust collection and vehicle cabin [47].

4. The main characteristics of air filters

The effectiveness of the filter materials is given by their capability to separate the aerosol particles from the air flow. In addition, the filtration efficiency must remain unchanged over time, under conditions of low airflow resistance and high loading capacity, without affecting the pressure drop. For this reason, the porosity of the filter materials used for this purpose is essential [48].

Air filters can be classified according to their ability to hold fine particles, respectively the smallest size of retained particles. According to the British classification system (ISO 16890) filters can be divided in 4 classes conformable to their efficiency: 1. ePM1 Filter efficiency for particle size range 0.3-1 µm; 2. ePM2.5 Filter efficiency for particle size range 0.3-2.5 µm; 3. ePM10 Filter efficiency for particle size range 0.3-10.0 µm, ISO coarse Gravimetric filter arrestance for filters not able to achieve minimum 50% efficiency at ePM10.

For high efficiency filtration HEPA (High-Efficiency Particulate Air) and ULPA (ultra low penetration air) filters are used [49]. These filters are usually made of glass and polyolefin fibres and show over 99.97% filtration efficiency for particles with a diameter greater than 300 nm [50].

Efficiency of fibrous filter commonly depends on the size of the particles, the airflow velocity and filter characteristics (fibre size and packing density). Filtration efficiency rates a filter by the percentage of contaminant removed as expressed in equation [51, 52]:

$$\text{Filtration efficiency, \%} = 100 \times \frac{(C_1 - C_2)}{C_1} \tag{1}$$

where C_1 = upstream particle concentration;
C_2 = downstream particle concentration.

The filter efficiency is expressed as a percentage (between 50% and 95% in 5% increases) for each of the ePM ranges indicated above.

Another important factor used to evaluate the filtration performance is the pressure drop, namely the differential pressure between upstream and downstream of the filter medium [51].

The aerosol particles are retained by filters by means of five mechanisms. The most predominant mechanisms are interception, inertial impact and diffusion [52-54], but gravitational deposition and electrostatic attraction can also occur [48].

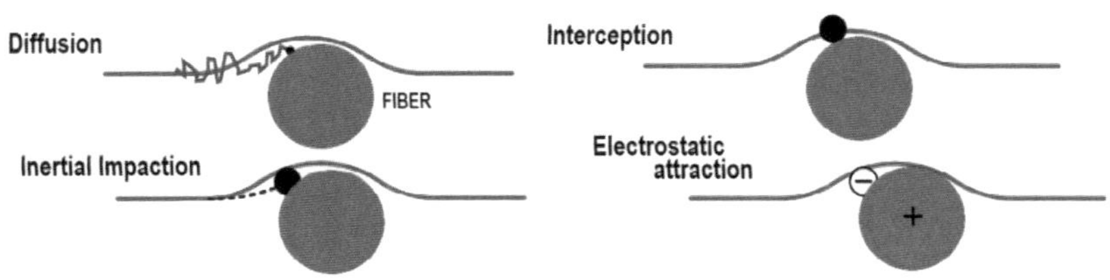

Figure 3. Mechanisms of particle capture [53].

The prevailing filtration mechanism depends largely on the size of the particles, and the filtration efficiency is the result of all the present mechanisms. An estimated range of particles in which the aforementioned mechanisms manifest and the filtration efficiency depending on the particle size is shown in Figure 4.

It can be seen that diffusion is the prevailing mechanism for small particles having a diameter of less than 0.3 µm [48], similar to the electrostatic attraction, which only takes place when the particles and / or fibres have an electrical charge.

The particle size has a great importance on the filtering efficiency, as it can be seen from Figure 5, where uniform fibre with 5 μm in diameter, a packing density of 0.05 and a thickness of 2 mm have been used at an air flow velocity of 10 cm/s with 1 g/cm^3 particle density.

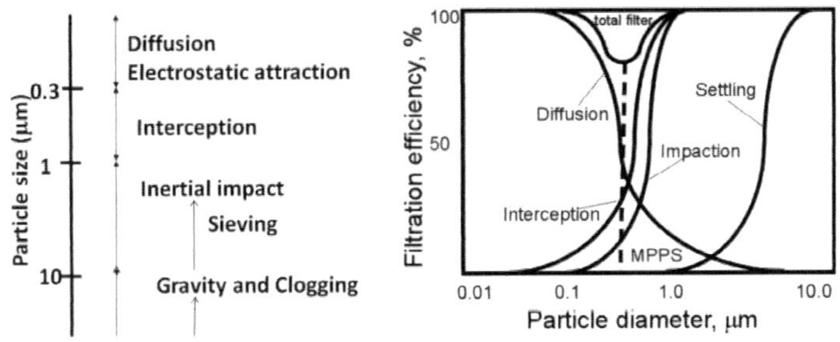

(a) (b)

Figure 4. Main filtering mechanism (a) and filtration efficiency (b) depending on particle size [44].

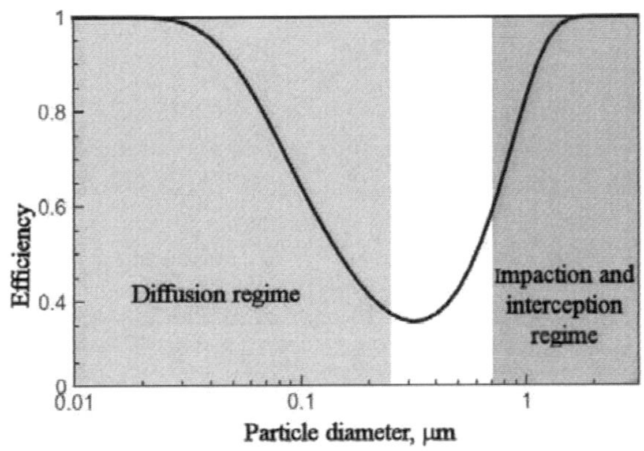

Figure 5. Filter efficiency as a function of particle diameter [54].

It can be seen that the filtration efficiency registers a minimum in the area of diameters between 300 and 700 nm, a zone that is called the most penetrating particle size (MPPS), and theoretical calculations forecast that the efficiency may be significantly increased using nanofibrous media [55].

The size of the electrospun nanofibres and the packing density can be controlled by varying the solution parameters (viscosity, conductivity and surface tension), ambient parameters (temperature and humidity) and the process parameters (fluid flow rate, spinneret to collector distance and electric field strength having a most significant influence on the properties of the obtained nanofibres) [56, 57], while the particle size as well as air flow rate depends on the application.

5. Characteristics of nanofibres that are recommended for use in air filtration

Conventional nonwoven fibre-based air filters have been used from many years for air filtration [35], as they are capable to effectively retain PM particles while letting air to pass through the filter, but these kind of filters frequently suffer from the drawbacks of too large pore size and low porosity due to their microsized diameters [58]. A solution to these disadvantages is the use of electrospun nanofibres based filters, which integrate the structural essential quality of small pore size, open stacking pores and highly adjustable porosity [7, 59]. Furthermore, they display superior service life and great dust-holding capacity in various air filtration applications [60].

The use of nanofibres in the field of air filtration is not recent, as attempts have been made in this direction for over twenty years. In order to achieve an effective filtration, it is necessary to exist an appropriate correspondence between the channels and the structural elements of the filter and the particle sizes or small droplets to be retained by the filter. This is why electrospun nanofibres are very suitable for the creation of filters, because of their very small diameters. Under these conditions, it becomes possible to remove unwanted particles smaller than 1 micron in diameter.

Nanofibres have very high surface area per unit mass, offering a outstanding capacity for the attachment of molecules, ions and different nanometer scale particles: when diameters range from 5 to 500 nm, the surface area per unit mass is varies between 10, 000 to 1,000,000 m^2/kg, as it can be seen from figure 6 [61]. Due to the very large specific surface and the high surface cohesion, it is possible to retain particles with dimensions below 0.5 μm, the efficiency of the filtration under these conditions being very high.

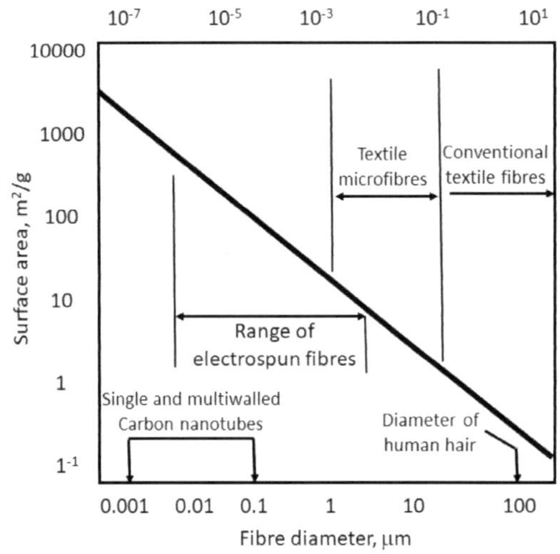

Figure 6. Variation of surface area with fibre diameter [61].

One of the decisive characteristics for the performance of the filters is their filtration efficiency, which is closely related to the fineness of the fibres, because, in general, the filtration efficiency increases linearly as the filter membrane compactness decreases and the pressure applied increases. It has been suggested that the size of the fibres in the electrospun filtration layers should be between 50 nm and 500 nm, in order to maintain both filtering efficiency and filter resistance [62].

Another reason for nanofibres filters higher air filtration performance is the difference in the flow of air across the material. Because of the slip flow at the nanofibre surface, the drag force is smaller than for the nonslip flow, causing lower pressure drop. As a result, the fraction of the air flowing near to the surface of fibres is larger, and consequently more particles will transit near the fibre, causing more efficient inertial impaction, interception, and diffusion [63].

6. Polymers used to obtain nanofibres used in air filtration

The durability of the nanofilters in the operational environment depends on their thermal, mechanical, and chemical properties, that depend on the polymer from which the filter is produced [64]. In the last years, numerous polymeric substrates were electrospun to produce nanofibres for air filtration applications. Its is considered that in general the filtration efficiency is improved when the polarity of the polymer is bigger, as the dipole–dipole or dipole–induced dipole force can help the binding of PMs to the surface of nanofibres. This is why polymers such as polyacrylonitrile, polyvinil alcohol, poly(vinylpyrrolidone), polystyrene are commonly used to make nanofilters [37].

Polyacrylonitrile, a versatile polymer with good physical properties, reasonable price and simple to electrospin [65], was used to produce electrospun nanofibres with diameters in the range of 270-400 nm that have been used as filters that showed high efficiency [50, 66]. Other research showed that electrospun PAN nanofibrous membranes exhibited high air-dust filtration efficiency of more than 99.99% in between PM0.3 and PM2.5 [67]. Electrospun PAN fibres with diameters in 270–400 nm range were used to filter NaCl nanoparticles with diameters below 80 nm and they proved to need less filter mass and showed filtration efficiency compared to commercial filters [68]. Porous bead-on-string filters produced by optimizing the polyacrylonitrile concentration of electrospun dopes and ambient humidity condition in the electrospinning process achieve an efficiency above 99% with a low pressure drop of 27 Pa at an airflow rate of 4.2 cm/s [69]. A thin coat of electrospun PAN nanofibres over a metal mesh showed over 95 % filtration efficiency (PM2.5 in a polluted city environment) with 90 % transparency [70]. TiO_2/PAN, ZnO/PAN and Ag/PAN electrospun nanofibres were used as filter media and they maintained high filtration efficiency even after the addition of nanoparticles, and the Ag/PAN nanofibre media had bactericidal activity [71, 72].

Polyvinil alcohol (PVA), a water-soluble synthetic polymer which has outstanding mechanical properties and chemical resistance, was used to produce air filters [73]. Crosslinked electrospun PVA nanofibre membranes increase with about 30% the filtration efficiency of melt-blown substrate layers and realize high removal efficiency for large particles [74]. Nonwoven polypropylene material covered by electrospun PVA nanofibre proved to be capable to eliminate organic polar compounds, ethers and compounds containing carbonyl groups from gas [75].

High-impact polystyrene (HIPS) waste was used to produce nanofibre membranes using electrospinning method and then applied as air filtration media, and the membranes with weight of 12.22 gm^{-2} displayed an efficiency greater than 99.999%, equivalent to that of the HEPA filter [76].

Nanofibre membranes made by electrospinning waste of acrylonitrile butadiene styrene (ABS), a polymer for widely industrial applications due to its good mechanical and physical properties [77], were used in air filter applications, and the results indicate that ABS waste are good as air filter media for daily needs, but the mechanical property of the membranes can be enhanced placing the membrane between two strong nonwoven membrane [78].

Another most electrospun polymer is polyamide, both 6 and 6.6. Nano-fibres obtained from polyamide 6 via electrospinning were frequently used as filtration media [79], and a very significant parameter in what concerns the filtration efficiency proved to be the solution concentration [80]. The best conditions of electrospinning process were obtained when using a 12.5% polymer solution. The pressure drop increased by increasing the electrospinning time and the thickness of the filter, and the maximum filtration efficiency was 96% for PA nanofibrous filter obtained after 240 min of electrospinning time [11].

Other studies showed that electrospun PA-6 nanofibres have good mechanical properties, and prove good filtration ability for microparticles with up to 0.3 μm diameter [81] and the PA 6 nanofilters had less pressure drop performance compared to HEPA filter [82]. PA 6 filters with 100 nm fibre diameter removed 99.9 % of aerosol particles with 1 μm - 5 μm in diameter from the air flow of 1 m/min [83]. Air filter membranes made from fibres obtained by electrospinning a 21% PA6 solution in formic acid proved to have a filtration efficiency ranging between 93-95% [84]. When using electrospun PA 6 nanofibres of 80–200 nm in diameter, the filtration efficiency was 99.993%, superior to the commercialized HEPA filter at the velocity of 5 cm/s using 0.3 μm test particles [85, 86]. Similar results were, obtained for polyamide 6.6, which showed a wider polymer solution electrospinnable range and a wider distribution of the fibre diameter (60–511 nm) compared to polyamide 6 (90–236 nm). In this case a filtration efficiency of 84.9–90.9% was obtained when an 8% PA6.6 % (w/vol) solution has been used [87]. PA-66 filter consisting in traditional nonwoven polypropylene 2D PA-66 spider-web-like nanonets supplied high filtration efficiency (up to 99.9%)

and low pressure drop [88]. Polyamide-56 nanofibre membrane with bimodal structure obtained by electrospinning proved to have good air filtration efficiency [89].

Introducing functional materials into nanofibres may lead to electrospun nanofibres with antibacterial and detoxifying properties. Polyamide 6 electrospun nanofibres were electrosprayed with ZnO nanoparticles, and the ZnO/Polyamide nanofibre mats had outstanding antibacterial efficiency against both the Gram-negative Escherichia coli and Gram-positive Bacillus cereus bacteria [90].

Electrospun polyimide nanofibre are applicable for high-temperature air and fume applications, as they have exceptional heat resistance and good flame retardant properties. These filters developed a high-efficiency air filtering, as they captured particulate matter that have a diameter of less than 2.5 micrometers at up to 370°C temperature with over 120 h working time [47].

Another polymer used to manufacture air filters is polylactic acid, which has the additional advantage that it is biodegradable and is derived from renewable resources. Polylactic acid/TiO_2 electrospun fibrous membranes showed a filtration efficiency of 99.996% with a relatively low pressure drop (128.7 Pa) and good antibacterial activity [91].

Electrospun polylactide air filter membranes captured large particle size (12.07 μm) with efficiencies over 99%, while the smallest (0.095 μm) particles resulted in the lowest capture efficiency of 92.97% [92]. Polyethylene oxide (PEO) was used to obtain nanofibre filters by coating electrospun PEO nanofibres onto a nonwoven microfibre substrate, showing that the most penetrating particle size decreased from 140 to 90 nm and the removal efficiency decreased as the face velocity increased from 5 to 10 cm/s [93].

Fibrous membranes made of electrospun polyvinylidene fluoride (PVDF) with fibre diameter from nanoscale to microscale were used as air filters, and nanofibrous membranes from 10% solution concentration proved to have best air filtration properties [94]. Electrospun cellulose acetate fibres with 0.1–1 μm have been used as filters which demonstrated a slightly better quality factor compared to commercial high efficiency glass fibre at a much lower thickness [95].

Air filters made of electrospun lignin fibres showed unfavorable mechanical properties, but lignin-based composite filters developed with polyethylene oxide (PEO) fibre filters exhibited good filtration efficiency and appropriate mechanical properties [96]. Silk fibroin solution has been electrospun to produce fibroin nanofibre membranes that exhibited a high filtration efficiency of 99.99% with a rather low air resistance of only 75 Pa [97].

Frequently two or more polymers with complementary properties are combined to produce composite polymeric nanofibre membranes which benefit from the merged properties of these polymers [9, 38, 65, 96, 108-119]. These dual/multi component nanofibre filters exhibit many characteristics such as excellent PM2.5 removal efficiency, durability and multi-functionality within a single filter [120-125].

7. Conclusions and further research directions

Nanofibres made by electrospinning are frequently applied in the area of air filtration, as they possess significant advantages such as small and controllable diameter, high porous structure and surface area to volume ratio with internal connectivity, and good mechanical properties. To improve the performance of nanofilters for air filtration, it is necessary to have in-depth knowledge of the mechanisms underlying the separation processes that takes place, namely the relation of the entire set of characteristics of the filter with the specificity of the impurities to be removed. It is essential to know exactly the relevant data that describes the relationship between the structure of the nanofiltration membrane and its transport properties. Although important development has been made in the past decades in the field of electrospinning, further improvements have to be done in what regards the nanofibre properties, such as the specific surface area, the interconnectivity of the fibres, and their thermal and mechanical properties, and this is why more fundamental experimental and theoretical analysis of the process is needed. In the future, the research will also focus on the production of functional filters, capable of selectively capture targets of interest, such as VOCs, PM2.5, viruses and many other contaminants.

8. References

[1] Al-Attabi R, Morsi YS, Schütz J and Dumée L 2019 Electrospun Membranes for Airborne Contaminants Capture, in Barhoum A, Bechelany M and Makhlouf AS (Eds) *Handbook of Nanofibres* (Switzerland: Springer Nature) p. 961

[2] Givehchi R and Tan Z 2014 *Aerosol Air Qual Res* **14** 45

[3] Muralikrishnan R, Swarnalakshmi M and Nakkeeran E 2014 *Int Res J Env Sci* **3**(4) 82

[4] Liu G, Xiao M, Zhang X, Gal C, Chen X, Liu L, Pan S, Wu J, Tang L and Clements-Croome D 2017 *Sustain Cities Soc* **32** 375

[5] Mohamed E 2017 *Environ Manage Sustain Dev* **6**(2) 429

[6] Wang Y, Zhao X, Jiao X and Chen D 2018 Electrospun Filters for Air Filtration: Comparison with Existing Air Filtration Technologies, in Focarete ML, Gualandi C and Ramakrishna S (Eds) *Filtering Media by Electrospinning - Next Generation Membranes for Separation Applications* (Switzerland: Springer International Publishing) p. 47

[7] Zhang S, Rind NA, Tang N, Liu H, Yin X, Yu J and Ding B 2019 Electrospun Nanofibres For Air Filtration, in Ding B, Wang X and Yu J (Eds) *Electrospinning: Nanofabrication and Applications* (Amsterdam: Elsevier)

[8] Bortolassi ACC, Nagarajan S, de Araújo Lima B, Guerra VG, Aguiar ML, Huon V, Soussan L, Cornu D, Miele P and Bechelany M 2019 *Mater Sci Eng C Mater Biol Appl* **102** 718

[9] Cao Q, Meng X, Tan S, Xin Z, Turng L-S, Li J, Yao Z, Zhai Z and Duan R 2020 *J Polym Res* **27**(5) 1

[10] McNeill F, Sareen N and Schwier A 2014 Surface-Active Organics in Atmospheric Aerosols, in McNeill F and Ariya P (Eds) *Atmospheric and Aerosol Chemistry* (Berlin: Springer-Verlag)

[11] Aliabadi M. 2017 *Chem Ind Chem Eng Q* **23**(4) 441

[12] Kampa M and Castanas E 2008 *Environ Pollut* **151** 362

[13] Mannucci PM, Harari S, Martinelli I, Franchini M 2015 *Intern Emerg Med* **10** 657

[14] Mao X, Si Y, Chen Y, Yang L, Zhao F, Ding B and Yu J 2012 *RSC Adv* **2** 12216

[15] Shi L, Zhuang X, Tao X, Cheng B and Kang W 2013 *Fiber Polym* **14** 1485

[16] Babar A, Iqbal N, Wang X, Yu J and Ding B 2019 Introduction And Historical Overview, in Ding B, Wang X and Yu J (Eds) *Electrospinning: Nanofabrication and Applications* (Amsterdam: Elsevier)

[17] Rubina A, Rubinová O, Blasinski P 2019 *Int Rev Appl Sci Eng* **10**(1) 51

[18] Manea L R, Bertea A, Popa A and Bertea A P 2018 *IOP Conf. Ser.: Mater. Sci. Eng.* **374** 012063

[19] Sun Y, Cheng S, Lu W, Wang Y, Zhang P and Yao Q 2019 *RSC Adv* **9** 25712-25729

[20] Rozemarie M L andrei B, Liliana H, Cramariuc R and Cramariuc O 2017 *IOP Conf. Ser.: Mater. Sci. Eng.* **209** 012063

[21] Zhao L, Duan G, Zhang G, Yang H, He S and Jian S 2020 *Nanomaterial* **10** 150

[22] Feng C, Khulbe KC and Matsuura T 2010 J *Appl Polym Sci* **115** 756–776

[23] Kumar R, Khan N, Vivekanandhan S, Satyanarayana N, Mohanty AK and Misra M 2012 *J Nanosci Nanotechno* **12** 1–25

[24] Mirjalili M and Zohoori S 2016 *J Nanostruct Chem* **6** 207–213

[25] Thomas D, Charvet A, Bardin-Monnier N and Appert-Collin J-C 2016 *Aerosol Filtration* (Oxford: Elsevier Ltd)

[26] Feltz K, Growney Kalaf E, Chen C, Martin S and Sell S 2017 *Electrospinning* **1** 46–61

[27] Yalcinkaya F 2019 *J Eng Fiber Fabr* **14** 1– 12

[28] Ko E, Lee JS, Kim H, Yang SY, Yang D, Yang K, Lee J, Shin J, Yang HS and Ryu W 2018 *ACS Appl Mater Interfaces* **10** 7614–7625
[29] Khorshidi S and Karkhaneh A 2018 *J Tissue Eng Regener Med* **12** 1974–1990
[30] Yang G, Li X, He Y, Ma J, Ni G, Zhou S 2018 *Prog Polym Sci* **81** 80–113
[31] Chen S, Li R, Li X and Xie J 2018 *Adv Drug Delivery Rev* **132** 188–213
[32] Shalumon KT, Sheu C, Chen CH, Chen SH, Jose G, Kuo CY and Chen JP 2018 *Acta Biomater* **72** 121–136
[33] Gonçalves de Pinho AR, Odila I1, Leferink A, van Blitterswijk C, Camarero-Espinosa S and Moroni L 2019 *Front Bioeng Biotech* **7**(231) 1-13
[34] Pant B, Park M and Park S-J 2019 *Pharmaceutics* **11** 305 1-21
[35] Islam S, Ang BC andriyana A and Afifi AM 2019 *SN Appl Sci* **1** 1248
[36] Nagiah N, Murdock C, Bhattacharjee M, Nair L and Laurencin C 2020 *Sci Rep* **10** 609
[37] Xue J, Wu T, Dai Y and Xia Y 2019 *Chem Rev* **119**(8) 5298
[38] Hanumantharao SN and Smitha Rao 2019 *Fibers* **7** 66
[39] Rahman MM and Thakkar A 2016 *Int J Sci Res Dev* **4**(2) 1067
[40] Lv DM, Jiang ZZ, Jiang S, Zhang Q, Xiong R and Huang C 2018 *Macromol Mater Eng* 1
[41] Ngiam M, Ramakrishna S, Raghuant M and Chan CK 2007 *Recent Pat Nanotec*h **1**(2) 137
[42] Patel G and Yadav BK 2018 *Rec Pat Nanotech* **12**(3) 114
[43] Arinstein A 2018 *Electrospun Polymer Nanofibres* (Singapore: Stanford)
[44] Barhate R and Seeram R 2007 *J Membr Sci* **296**(1-2) 1
[45] Wang N, Mao X, Zhang S, Yu J and Ding B 2014 Electrospun Nanofibres for Air Filtration, in Ding B and Yu J *Electrospun Nanofibres for Energy and Environmental Applications* (Berlin: Springer-Verlag) chapter 12 p. 299
[46] Qiu H, Tian L and Ho K 2015 *Environ Pollut* **199** 192
[47] Zhang R, Liu C, Hsu PC, Zhang C, Liu N, Zhang J, Lee HR, Lu Y, Qiu Y, Chu S and Cui Y 2016 *Nano Lett* **16**(6) 3642
[48] Hinds WC 1999 *Aerosol Technology. Properties, Behaviour and Measurement of Airborne Particles* (New York: John Wiley & Sons)
[49] Dey E, Choudhary U and Ghosh SK 2017 *American J Eng Res* **6**(10) 190
[50] Yun KM, Hogan CJ, Matsubayashi JY, Kawabe M, Iskandar F and Okuyama K 2007 *Chem Eng Sci* **62** 4751
[51] Hutten, IM 2016 *Handbook of Nonwoven Filter Media* (Amsterdam: Butterworth Heinemann)
[52] Zhu, M, Han J, Wang F, Shao W, Xiong R, Zhang Q, Pan H, Yang Y, Samal SK and Zhang F 2016 *Macromol Mater Eng* **302** 1
[53] Zobel S and T. Gries 2010 The use of nonwovens as filtration materials in Chapman RA (Ed) *Applications of nonwovens in technical textiles* (Cambridge: Woodhead Publishing)
[54] Raynor P and Peters T 2016 Controlling Nanoparticle Exposures, in Ramachandran G. *Assessing Nanoparticle Risks to Human Health* (Oxford: Elsevier)
[55] Podgórski A, Bałazy A and Gradoń L 2006 *Chem Eng Sci* **61**(20) 6804
[56] Koushkbaghi S, Jafari P, Rabiei J, Irani M and Aliabadi M 2016 *Chem Eng J* **301** 42
[57] Bertea A, Manea LR, Popa A and Bertea A 2017 The Influence of Process Parameters on the Characteristics of Electrospun 3D Nanostructures, *IOP Conf. Series: Materials Science and Engineering* **209** 012074
[58] Yang Y, Zhang S, Zhao X, Yu J and Ding, B 2015 *Sep Purif Technol* **152** 14

[59] Li J, Gao F, Liu LQ and Zhang Z 2013 *eXPRESS Polym Lett* **7**(8) 683
[60] Grafe, T and Graham K 2003 *Int Nonwov Tech Conf* **12** 51
[61] Ko F 2004 Nanofibre Technology: Bridging the Gap between Nano and Macro World, in Guceri S, Gogotsi I and Kutnezsov V (Eds) *Nanoengineered Nanofibrous Materials* (Dordrecht: Kluwer Academic Publishers)
[62] Chung HY, Hall JRB, Gogins MA, Crofoot DG and Weik TM 2008 *Polymer, polymer microfibre, polymer nanofibre and applications including filter structures* European Patent Application 1 925 352 A1
[63] Graham K, Ouyang M, Raether T, Grafe T, Mcdonald B and Knauf P 2002 Polymeric Nanofibres in Air Filtration Applications in *The Fifteenth Annual Technical Conference and Expo of the American Filtration and Separations Society*, Galveston, Texas, April 9
[64] Agboola O, Jannie Maree and Richard Mbaya 2014 *Environ Chem Lett* **12** 241
[65] Dehghan SF, Golbabaei F, Maddah B, Latifi M, Pezeshk H, Hasanzadeh M and Akbar-Khanzade F 2016 *J Air Waste Manag*e **66**(9) 912
[66] Prokopchuk NR, Shashok ZS, Prishchepenko DV and Luhin VG 2016 *Int Conf Nanomater: App* Lviv, 01FNC05-1-01FNC05-3
[67] Roche R and Yalcinkaya F 2019 *ChemistryOpen* **8** 97
[68] Zhang Q, Welch J, Park H, Wu C-Y, Sigmund W and Marijnissen JCM 2010 *J Aerosol Sci* **41**(2) 230
[69] Huang JJ, Tian Y, Wang R, Tian M and Liao Y 2020 *Sep Purif Technol* **237** 11637
[70] Liu C, Hsu PC, Lee HW, Ye M, Zheng G, Liu N, Li W and Cui Y 2015 *Nat Commun* **6** 6205
[71] Su J, Yang G, Cheng C, Huang C, Xu H and Ke Q 2017 *J Colloid Interf Sci* **507** 386
[72] Bortolassi ACC, Guerra VG, Aguiar M, Soussan L, Cornu D , Miele P and Bechelany M 2019 *Nanomaterials* **9**(1740) 1
[73] Li XW, Kong HY and He JH 2015 *Indian J Phys* **89**(2) 175
[74] Qin XH and Wang SY 2008 *J Appl Polym Sci* **109** 951
[75] Daneleviciute-Vaisniene A, Katunskis J and Buika G 2009 *Fibres Text East Eur* **17**(6) 40-4
[76] Zulfi A, Munir MM, Hapidin DA, Rajak A, Edikresnha D, Iskandar F and Khairurrijal K 2018 *Mater Res Express* **5** 1
[77] Jiang S, Schmalz H, Agarwal S and Greiner A 2020 *Adv Fiber Mat* **2** 34
[78] Zulfi A, Hapidin DA, Munir MM, Iskandar F and Khairurrijal K 2019 *RSC Adv* **9** 30741
[79] Kadam V, Wang L and Padhye R 2018 *J Ind Text* **47**(8) 2253
[80] Zhang S, Shim WS and Kim J 2009 *Mater Des* **30**(9) 3659
[81] Guibo Y, Qing Z, Yahong Z, Yin Y and Yumin Y 2013 *J Appl Polym Sci* **128** 1061
[82] Kim GT, Ahn YC and Lee JK 2008 *Korean J Chem Eng* **25** 368
[83] Schreuder-Gibson H and Gibson P 2004 Proc 24th Army Sci Conf, Massachusetts https://www.researchgate.net/publication/216777621
[84] Hameed O and Jawad JJH 2018 *J Babylon Univ* **1**(26) 122
[85] Ahn YC, Park SK, Kim GT, Hwang YJ, Lee CG, Shin HS and Lee JK 2006 *Curr Appl Phys* **6**(6) 1030
[86] Balamurugan R, Sundarrajan S and Ramakrishna S 2011 *Membranes* **1** 232
[87] Matulevicius J, Kliucininkas L, Martuzevicius D, Krugly E, Tichonovas M and Baltrusaitis J 2014 J *Nanomater* 859656 1
[88] Wang N, Wang X, Ding B, Yu J and Sun G 2012 *J Mater Chem* **22**(4) 1445

[89] Liu B, Zhang S, Wang X, Yu J and Ding B 2015 J *Colloid Interf Sci* **457** 203
[90] Vitchuli N, Shi Q, Nowak J, Kay K, Caldwell J, Breidt F, Bourham M, McCord M and Zhang X 2011 *Sci Technol Adv Mater* **12** 1
[91] Wang Z, Pan Z, Wang J and Zhao R 2016 *J Nanomater* 1
[92] Selatile MK, Ray SS, Ojijoa V and Sadiku R 2018 *J Environ Chem Eng* **6** 762
[93] Leung WW-F, Hung C-H and Yuen PT 2010 *Sep Purif Technol* **71** 30
[94] Xiao X, Wen E, Sakib N, Yue Z, Wang Y, Cheng S, Militky J, Venkataraman M and Zhu G 2019 *Autex Res J* **19** 1
[95] Chattopadhyay S, Hatton A and Rutledge G 2016 *J Mater Sci* **51** 204
[96] Chang C-Y and Chang FC 2016 *BioRes* **11**(1) 2202
[97] Gao X, Gou J, Zhang L, Duan S and Li C 2018 *RSC Adv* **8** 8181
[98] Scarlet R, Manea L R Sandu I, Cramariuc B and Sandu A V 2012 *Rev. Chim.* **63**(8) 777
[99] Scarlet R, Manea L R, Sandu I, Martinova L, Cramariuc O and Sandu I G 2012 *Rev. Chim.* **63**(7) 688
[100] Manea L R, Bertea A, Nechita E, Popescu C V and Sandu I 2016 *Rev. Chim.* **67**(8) 1607
[101] Manea L R, Scarlet R, Leon A L and Sandu I 2015 *Rev. Chim. (Bucharest)* **66**(5) 640
[102] Hristian L Sandu A V, Manea L R, Tulbure E A and Earar K 2015 *Rev. Chim. (Bucharest)* **66**(3) 342
[103] Manea L R, Hristian L, Ostafe M M, Apostol L L and Sandu I 2016 *Rev. Chim. (Bucharest)* **67**(9) 1758
[104] Manea L R, Hristian L, Leon A L and Popa A 2016 *IOP Conf. Series: Mater. Sci.Eng.* **145** 032006
[105] Manea L R, Bertea A, Popa A and Bertea A P, 2018 *IOP Conf. Ser.: Mater. Sci. Eng.* **374** 012081
[106] Manea L R, Chirita M, Hristian L and Sandu I 2016 *Mat. Plast.* **53**(3) 361
[107] Manea L R, Danu M C and Sandu I 2015 *Rev. Chim. (Bucharest)* **66**(6) 868
[108] Wang N., Raza A, Si Y, Yu J, Sun G and Ding B 2013 *J Colloid Interf Sci* **398** 240-246
[109] Wang N, Zhu Z, Sheng J, Yu J and Ding B 2014 *J Colloid Interf Sci* **428** 41-48
[110] Wang Z and Pan Z 2015 *Appl Surf Sci* **356** 1168–1179
[111] Liu Y, Park M, Ding B, Kim J, Al-Deyab SS and Kim H.-Y 2015 *Fiber Polym* **16** 629-633
[112] Zhang S, Liu H, Yin X, Yu J and Ding B 2016 *ACS Appl Mater Interf* **8** 8086-8095
[113] Yildiz O, Stano K, Faraji S, Stone C, Willis C, Zhang X, Jur JS and Bradford PD 2015 *Nanoscale* **7** 16744
[114] Huang X, Jiao T, Liu Q, Zhou J, Li B and Peng Q 2019 *Sci China Mater* **62**(3) 423
[115] Lv D, Wang R, Tang G, Mou Z, Lei J, Han J, De Smedt S, Xiong R and Huan C 2019 *ACS Appl Mater Interfaces* **11**(38) 12880
[116] Mohraz MH, Golbabaei F, Yu IJ, Mansournia MA, Zadeh AS and Dehghan SF 2019 *Int J Environ Sci Technol* **16** 681
[117] Sundarrajan S, Tan KL, Lim SH and Ramakrishna S 2014 *Procedia Eng* **75** 159
[118] Liu H, Huang J, Mao J, Chen Z, Chen G and Lai Y 2019 *Matter Captur iScience* **19** 214
[119] Jiajia X, Tong W, Yunqian D and Younan X 2019 *Chem Rev* **119**(8) 5298
[120] Nam C, Lee S, Ryu M, Lee J and Lee H 2019 *Korean J Chem Eng* **36**(10) 1565

[121] Kasa SN, Omar MF, Abdullah MMA, Ismail IN, Ting SS, Vac SC and Vizureanu P 2017 *Materiale plastice* **54(1)** 91-97

[122] Berceanu C, Tarnita D and Filip D 2010 *Journal of the Solid State Phenomena, Robotics and Automation Systems* **166-167** 45-50
[123] Calin MA, Curteza A, Toma S and Agop M 2013 *Metalurgia International* **18** 19-22
[124] Botezatu C, Duceac L D, Mastalier B, Stafie L, Jitareanu C M, Luca A C, Tarca E, Mitrea G, Iordache A C and Patrascu T 2018 *International Journal Of Medical Dentistry* **22(4)** 346-350
[125] Duceac L D, Stafie L, Banu E A, Paduraru O, Calin G and Ciuhodaru M I 2017 *Rev. Chim.* **68**(11) 2542-2545

IOP Conf. Series: Materials Science and Engineering **877** (2020) 012044 doi:10.1088/1757-899X/877/1/012044

Influence of Microparticles on Setting Time and Micromorphology of Coal Ash Geopolymers

D D Burduhos Nergis[1], P Vizureanu[1,2,*], D Țopa[3], M G Minciuna[1] and M M A B Abdullah[2]

[1]"Gheorghe Asachi" Technical University of Iasi, Blvd. Mangeron, No. 51, 700050, Iasi, Romania

[2]Center of Excellence, Geopolymer & Green Technology (CeGeoGTech), School of Material Engineering, Universiti Malaysia Perlis (UniMAP), Perlis, Malaysia

[3]"Ion Lonescu" Brad Univ. Agr. Sci. & Vet. Med. Iasi, 3 Mihail Sadoveanu Alley, Iasi 700490, Romania

*E-mail: peviz2002@yahoo.com

Abstract. Geopolymers are inorganic materials with zeolites-like microstructure and mechanical properties similar to those of Ordinary Portland cement materials [1]. However, their properties are highly depending on the constituents (raw material and activator) characteristics, as well as, on the activation particularities (mixing parameters, curing time and temperature etc.). In order to explore the influence of partial replacement of coal ash with two types of fine aggregates (glass and sand microparticles) on micromorphology and setting time, four types of geopolymers were developed. The evaluations were performed by means of electronic microscopy and Vicat method. According to this study, the coal ash replacement with glass microparticles results in an increase in the initial and the final setting time, while the replacement of coal ash with sand particles show a significant decrease. Moreover, the microstructural analysis shows different behaviour, during activation, of the studied microparticles. The surface of the glass microparticles reacts in the alkaline environment, while the sand particles did not. Therefore, the increase of initial and final setting time can be correlated with the dissolution of Si-O from the glass particles, during geopolymerisation.

1. Introduction

Nowadays, the main challenge in new materials developing is strongly related to the environmental impact of raw materials exploitation and sustainability of obtaining technology. A promising class of oxidic materials, ideal to substitute Ordinary Portland Cement (OPC) materials and to eliminate the negative effects produced by those, are geopolymers [1, 2].

A geopolymer is rich in aluminium and silicon material with porous structure and characteristics similar to OPC materials [3]. However, since as source of raw material for geopolymers, any powder rich in aluminium and silicon oxides can be used, globally multiple types of wastes have been identified as compatible [4]. Furthermore, the lack of standardization and the wide range of parameters that can vary in each geopolymer has lead to multiple unknowns which are hard or impossible to predict or simulated. In order to develop specific standards for these materials, different scenarios (composition) must be designed and analysed. According to previous studies [5, 6], multiple factors influence the properties of geopolymers. As can be seen in Table 1, those are related to the raw materials characteristics, the type of the activation solution, the curing stage parameters and the parameters of the mixing stage.

Table 1. Factors that influence the properties of geopolymers [5, 7].

Raw material	Particle size distribution Humidity Chemical composition
Activator solution	Solid to liquid ratio Sodium silicate to sodium hydroxide ratio Sodium hydroxide molar concentration
Mixing stage	Mixing time Mixing speed
Curing stage	Curing time Curing temperature

In case of geopolymers with reinforcing element few more factors must be taken into consideration, such as particles pozzolanicity or particles adhesion to the matrix [6].

According to Lateef N A et al. [8] the particle size distribution has a direct effect on the compressive strength, structure and permeable void ratio of geopolymers. For example when a fly ash source with the average particle size of 42.3 μm has used a value for compressive strength of 22 MPa was obtained in case of 28-days age samples, while after the fly ash was milled and sifted (the average particle size decreases to 4.78 μm) the compressive strength increased up to 67.3 MPa. For the same type of particles, the absorption decreased up to 21% and the microcracks resulted from the curing process are fewer in number.

The quality of the final structure and its properties results mainly from the uniformity and strength of the tetrahedral structure formed by the Si-O-Al system [9]. Any impurity or chemical element (except Si, O and Al) from the raw material, especially the calcium content, can lead to the formation of defects in the structure that significantly influence the final properties of the geopolymers [10]. To evaluate the influence of calcium content on the setting time of geopolymers, P. Chindaprasirt et al. [11] replaced different percentages of ash with Portland cement type I, which contains a high content of calcium hydroxide ($Ca(OH)_2$) and calcium oxide (CaO). According to the study, the increase in calcium content in the composition causes a sharp decrease in setting time regardless of the percentage of replacement.

The chemical activator or activation solution plays a vital role in initiating the geopolymerization process. In general, a strongly alkaline environment is required to increase the hydrolysis of aluminosilicate particles contained in the raw material, therefore its concentration has a pronounced effect on the mechanical properties of the geopolymers [12]. On the other hand, the dissolution of Si^+ and Al^+ species during geopolymer synthesis depends mainly on the NaOH concentration, where the dissolution capacity of silicon and aluminum oxides is controlled by the NaOH concentration and mixing parameters [13]. Considering the fact that for geopolymers manufacturing alkaline solution with high concentration are used, adequate health and safety conditions must be introduced during the obtaining process [14-16].

Somna et al. [17] studied the compressive strength of fly ash-based geopolymers, hardened at ambient temperature by changing the NaOH concentration from 4.5 M to 16.5 M. It was observed that by increasing the NaOH concentrations from 4, 5 to 9.5 M, there is a significant increase in the compressive strength of the samples. However, the increase of NaOH concentrations from 9.5 to 14 M positively influences the mechanical properties of the final product, but in a reduced. The influence on the compressive strength of the alkaline activator concentration is significant due to the degree of dissolution of Si^+ and Al^+ which directly affects the formation of sialates bonds. Above 16.5 M a decrease is observed due to the excess of hydroxide groups which determines the precipitation of the aluminosilicate gel at early ages.

Conventional methods of manufacturing geopolymers involve mixing the solid component, the raw materials rich in aluminum and silicon with the reinforcing elements if any, with a liquid component, a strongly alkaline solution. There are several studies about the ideal mixing procedure, however, the most common method involves mixing the solid components, if there are several, as well as the liquid ones, followed by mixing the liquid component with the solid one [18]. Another method is to mix the materials rich in silicon and aluminum with the sodium silicate solution, and after a few minutes, the NaOH solution is introduced in the binder while mixing. However, the geopolymers obtained by the second method have lower mechanical strength properties, except for those based on fly ash [19]. After mixing the components, the resulting geopolymeric paste is poured into moulds, where it is to harden at room temperature or at a slightly high temperature. At the same time, according to the study conducted by Z. Yahya et al. [20], if after filling the moulds, the surface in direct contact with the atmosphere is covered with a foil, the evaporation rate of the water in the mixture is reduced, so the geopolymer will have a smaller number of cracks.

Up to know, the geopolymers successfully participated in multiple domains such as civil construction, water filtration, heavy metals encapsulation, heat resistant shields etc. due to their novel properties including corrosion resistance, compressive and flexural strength and high-temperature resistance [21-23, 25-27]. However, there are still a few challenges in developing geopolymers with different types of particles in the matrix. Therefore, this study aims to develop and analyse four types of coal-ash based geopolymers which contain one or two types of reinforcing elements.

2. Materials and methods

The obtained geopolymer was manufactured by mixing sodium silicate and sodium hydroxide solution (alkaline activator) with coal ash and reinforcing particles (solid component).

2.1. Materials

2.1.1. Coal ash. Coal ash is the main waste which results from coal combustion in power plants, these mineral powder rich in aluminium and silicon oxides ends by being deposited on vaste areas near the cities powerplants. The coal ash used in this study comes from CET II - Holboca Iasi Romania ash dumps, which occupied an area of approximately 50 hectares in 2013. Due to the fact that the performance of coal ash in geopolymers is influenced by its characteristics (chemical composition, humidity, particle size distribution etc.). In order to remove the negative effects produced by the impurities with high dimensions from the coal ash, only the particles with the diameter lower than 80 μm have been used. Moreover, the powder was dried at 120 until its mass remains constant for 30 minutes, i.e. humidity ≈ 0.

According to ASTM C618-92a, the coal ash used in this study belongs to class F fly ashes because the sum of the three main oxides (silicon, aluminum and iron) is higher than 70% (eq. 1), as can be seen in Table 2.

$$SiO_2 + Al_2O_3 + Fe_2O_3 = 47.8 \pm 0.32\ \% + 28.6 \pm 0.27\% + 10.2 \pm 0.95\ \% = 86.6 \pm 1.54\ \% \quad (1)$$

Table 2. Indigenous fly ash oxide chemical composition.

Oxide	**SiO_2**	**Al_2O_3**	**Fe_xO_y**	**CaO**	**K_2O**	**MgO**	**TiO_2**	**Na_2O**	**P_2O_5**	**Oth.[1]**
%, weight	47.80	28.60	10.20	6.40	2.40	2.00	1.30	0.60	0.40	0.30
Stat. error, %	0.32	0.27	0.95	0.77	0.71	1.09	1.81	0.63	0.24	-

[1] Sum of chemical elements lower than 0.1 %.

2.1.2. Reinforcing particles. Another mineral waste that results in large quantities, both from the food industry and from the civil construction industry is glass. This inert material does not decompose naturally, producing proven negative effects on the environment following storage in waste dumps [5]. Therefore, the use of glass waste in the manufacture of environmentally friendly materials has become a global concern [4]. Due to the ability to incorporate geopolymer paste, by the introduction of glass powder in the composition of these materials, a geopolymer with higher compressive strength can be obtained while an environmental problem is solved. As can be seen from Table 2, the glass powder, with particles smaller than 10 μm in diameter, shows higher SiO_2, CaO and Na_2O content, but lower aluminium oxide concentration.

Table 3. Glass powder oxide chemical composition.

Oxide	SiO_2	Al_2O_3	Fe_xO_y	CaO	MgO	Na_2O	Oth.[1]
%, weight	70-71	1.5-2	0.8-1	9-11	2-3	12-14	< 0.1

[1] Sum of chemical elements lower than 0.1 %.

In order to study the behaviour of natural aggregates (sand), compared with waste (glass), two types of geopolymers with different percentage of sand particles have been obtained. As natural aggregates, a class of river sand with particles lower than 4 mm in diameter has been used.

2.1.3. Activator solution. As an alkaline solution for coal ash-based geopolymers activation, a mixture of sodium hydroxide and sodium silicate solutions was used. The NaOH solution was prepared at a molar concentration of 10 by dissolving high purity NaOH flakes (98%) in distilled water 24 hours prior use. The sodium silicate solution with a density of 1.37 g/cm³ and pH lower than 11.5 has been commercially acquired (Scharlad Company, Barcelona).

2.1.4. Sample preparation. The samples mixture was prepared according to the BS EN 196-1:1995 by means of a variable speed mixer. In order to increase the homogeneity of the samples, firstly, the solid component was poured into the mixer and stirred in a dry state for 4 min. Secondly, the liquid component was added gradually and mixed for 10 min. until a homogeneous binder was obtained. The mix proportion of liquid and solid component of each sample are presented in Table 4.

Table 4. Samples components mix proportion.

Sample	Na_2SiO_3/NaOH ratio % weight	Liquid to Solid ratio % weight	Solid component, % weight		
			Coal ash (CA)	Glass powder (GP)	River sand (RS)
100CA	1.5	1.5	100	0	0
70CA_30GP			70	30	0
30CA_70RS			30	0	70
15CA_15GP_70RS			15	15	70

The micromorphology analysis was realized on samples cured at 70 °C for 8 hours and aged for 28 days in normal environmental condition.

2.2. Methods

The chemical composition of raw materials and obtained samples were measured by a Bruker S8 Tiger Wavelength Dispersive X-ray Fluorescence (Bruker, Karlsruhe, Germany) instrument. The samples micromorphology has been analysed using a scanning electron microscope FEI, model QUANTA 450 FEG (FEI company, WA, USA) in environmental SEM condition. The setting time of the geopolymer paste was evaluated by Vicat test according to ASTM C 191-04a. The test was conducted at room temperature (22 ±2 °C) with a Vicat apparatus (Figure 1.a).

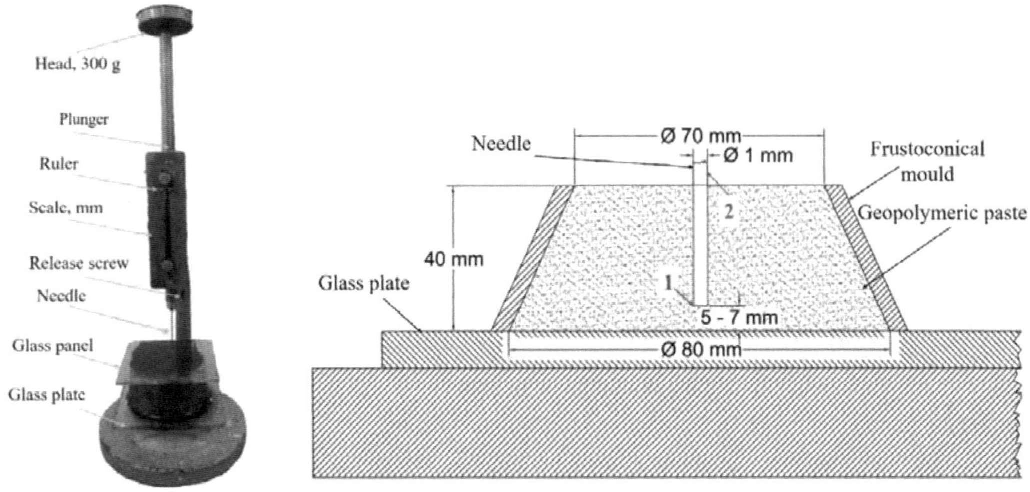

(a) (b)

Figure 1. Vicat apparatus for setting time measurement: (a) components; (b) schematic representation of measurement principle.

The Vicat method consists in testing the penetration resistance, with a flat tip needle with a diameter of 1 mm, of a geopolymer sample that was poured into a frustoconical mould specific to the device used (Figure 1. b). After placing the mould on a glass plate, it is filled with freshly obtained geopolymer paste (immediately after mixing the components) and covered with a glass panel that prevents water evaporation. Every 30 minutes, the penetration resistance shall be checked by removing the glass panel and releasing the needle using the release screw. The needle is pushed into the sample surface under the action of its weight, the weight of the plunger and that of the head (300 g). Depending on the depth of penetration, it is determined whether the two points (initial or final) specific to the setting time have been reached. After checking the depth on the scale, the needle is removed from the paste by lifting the rod, further, the sample is covered with the glass panel for the next 30 minutes.

3. Results and discussion

3.1. Chemical composition analysis

As can be seen in Table 5, the chemical composition of the obtained geopolymers shows high differences between the concentration of the main elements (Si, Na, Fe and Al). Except for oxygen, the element with the highest concentration in all samples is silicon.

Table 5. Chemical composition of the studied samples.

Sample	Element	Si	Na	Fe	Al	Ca	K	Ti	Mg	O
100CA	[%], wt.	21.46	10.98	3.14	8.86	2.25	1.05	0.20	0.73	balance
	error, [%]	0.33	0.79	0.27	0.70	0.49	1.10	1.83	1.72	
70CA_30GP	[%], wt.	22.06	12.27	2.47	6.83	3.24	0.99	0.27	1.00	
	error, [%]	0.34	0.74	0.28	0.75	0.52	1.14	1.89	1.91	
30CA_70RS	[%], wt.	25.72	10.83	3.08	4.79	2.20	1.13	0.29	0.80	
	error, [%]	0.36	0.65	0.26	0.80	0.72	1.08	1.72	1.81	
15CA_15GP_70RS	[%], wt.	26.41	11.01	2.30	4.42	2.83	0.81	0.73	0.90	
	error, [%]	0.35	0.69	0.27	0.77	0.62	1.11	1.79	1.72	

Sample 100CA contains the highest concentration of aluminum (Al) and iron (Fe), but the lowest concentration of silicon (Si). However, the sample 70CA_30GP exhibit a decrease of more than 20% in the concentration of Fe and Al, but due to the replacement of 30% of coal ash with glass powder, the calcium concentration in the sample increases up to 30%. The chemical composition of the 30CA_70RS sample shows an increase of approximately 17% in the Si concentration, but by replacing 70% of the amount of coal ash with sand particles, the Al and Ca content decreases by 46%, respectively 3%. In the case of 15CA_15GP_70RS sample, the increase of Si concentration is of approximately 17% while the decrease of aluminum content reaches 50%. Therefore, by decreasing the coal ash content from the composition of the geopolymers, the aluminum content decrease while the silicon content increases.

According to the chemical composition, the 100CA sample has better adhesion properties due to the ratio between the Si and Al concentration of approximately 2.4 which corresponds to a geopolymer with linear 2D structure [27, 32]. However, by increasing the percentage of reinforcing particles, the Si content increases, therefore, those sample will exhibit lower flexibility but higher hardness and compressive strength.

3.2. Micromorphology analysis

From the microstructural point of view, the samples show a homogeneous structure with high compactness of the matrix that depends on the type or the percentage of reinforcing particles. Moreover, all samples show unreacted coal ash particles, but the sample 100CA (Figure 2) shows the highest number, therefore, in some areas the matrix continuity is interrupted. Additionally, the SEM micrograph at 500X also highlights the cracks formed during the curing process, due to the fast evaporation of water. However, due to the self-healing capacity of geopolymers, which is related to the continued geopolymerisation reaction between unreacted particles and the activation solution from the gel pores, some cracks are repaired over time. At the same magnification, the sample 70CA_30GP (Figure 3) shows a homogeneous structure, with a high degree of ash dissolution.

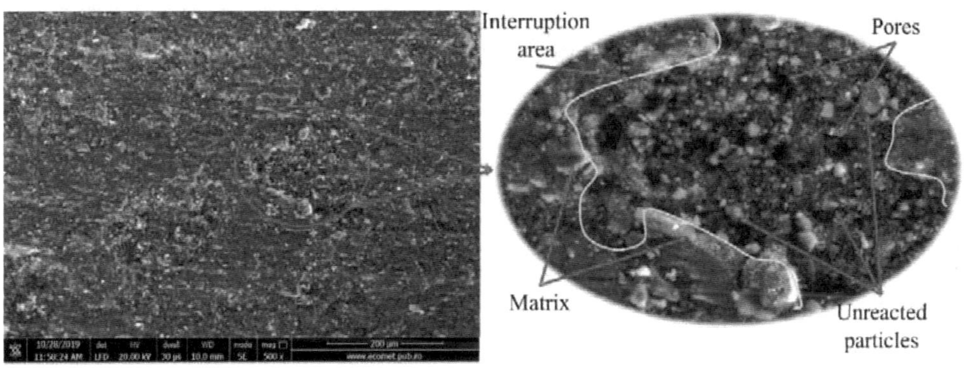

Figure 2. The micromorphology of 100CA sample.

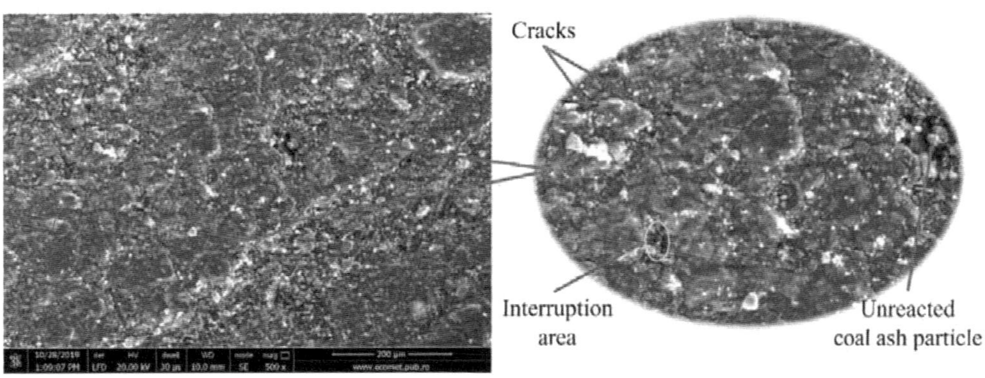

Figure 3. The micromorphology of 70CA_30GP sample.

The micromorphology of the geopolymers with river sand (Figure 4) respectively river sand and glass powder (Figure 5) exhibit a high increase in the number of large pores, besides, the number of unreacted particles and small pores significantly decrease. The effect is produced by the high dimensions particles which block the air bubbles evaporation from the geopolymeric paste, during curing.

Figure 4. The micromorphology of 30CA_70RS sample.

Figure 5. The micromorphology of 30CA_70RS sample.

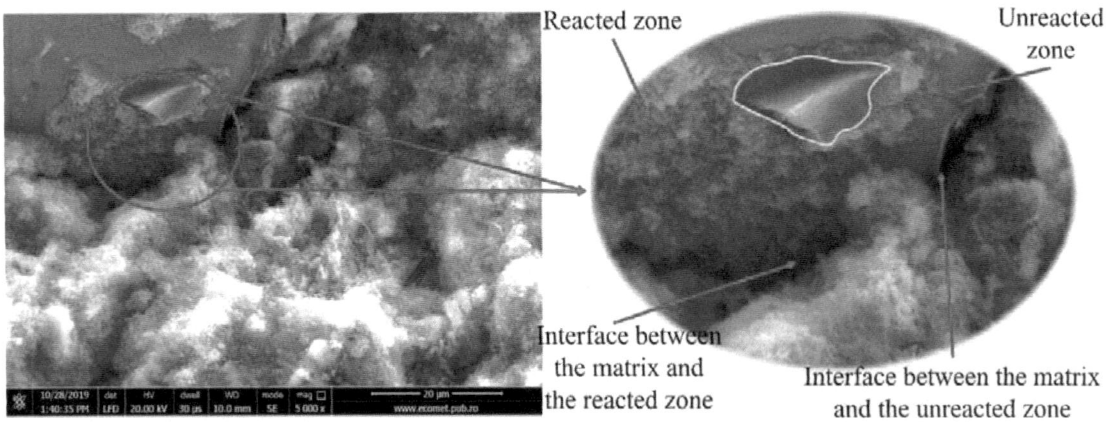

Figure 6. 70CA_30GP sample micromorphology at high magnification (10kX).

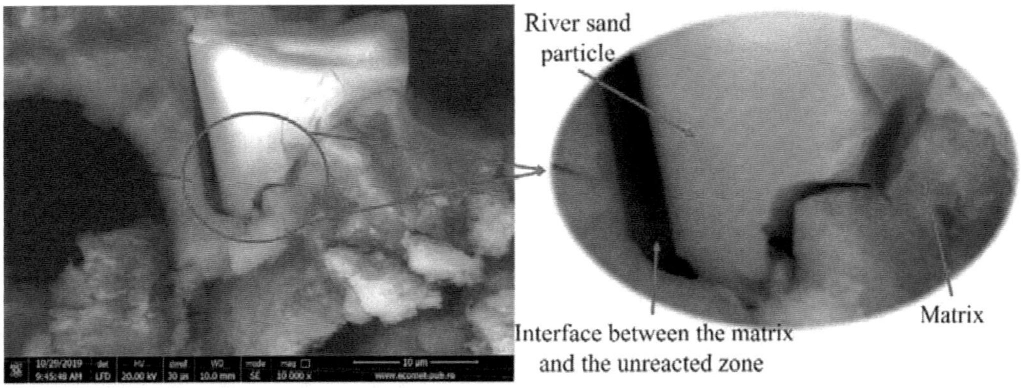

Figure 7. 30CA_70RS sample micromorphology at high magnification (10kX).

In environmental SEM condition, the interface between the reinforcing particles and the matrix can be studied at high magnification. As can be seen from Figure 6, the surface of the glass particles reacted in the alkaline medium (reacted zone), therefore, the matrix adheres strongly to their surface. In the case of the samples with river sand as reinforcing elements (Figure 7) the matrix does not adhere to the surface of the particles, thus, a clear delimitation can be observed between the matrix and the sand particles.

3.3. Setting time

The interpretation of the setting time for these materials is similar to the equilibrium diagram specific to alloy systems, where the cooling liquid line provides information about the temperature at which the first crystallization germs appear, and the solidus line highlights the temperature at which the transition to solid-state occurs for the entire volume of material [24]. In the case of geopolymers, the first point (position 1 of the needle tip) refers to the initial setting time, which is specific to the loss of plasticity of the mixture and the passage of a certain period of time after mixing the components. The second point (position 2 of the needle tip) refers to the final setting time and this is characterized by the total time elapsed from the mixing of the components to the actual hardening of the sample (Figure 8). In other words, the initial setting time is determined by the time elapsed from the pouring in the mould of the geopolymeric paste until the needle penetrates to a depth of only (5 ÷ 7) mm in the geopolymer paste (Figure 1.b). The final setting time is specific to the moment when the needle no longer penetrates the surface of the sample.

The coal ash-based geopolymers analysed in this study shows an initial setting between 4.78 and 6.88 hours and a final setting time between 22.95 ÷ 25.00 hours at ambient temperature (Table 6).

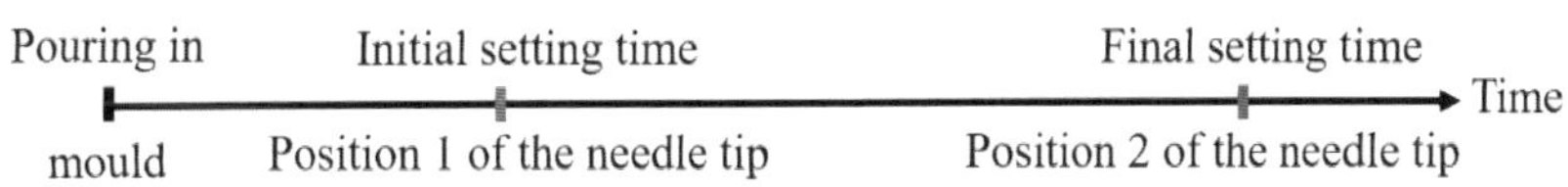

Figure 8. Time distribution of the points specific to setting time.

Table 6. Setting time of coal ash based geopolymers.

Sample	**100CA**		**70CA_30GP**		**30CA_70RS**		**15CA_15GP_70RS**	
	a	**b**	**a**	**b**	**a**	**b**	**a**	**b**
Test 1	290	1420	410	1500	280	1380	350	1410
Test 2	300	1420	410	1490	290	1370	350	1400
Test 3	290	1430	420	1510	290	1380	370	1430
Average, [min.]	**293**	**1423**	**413**	**1500**	**287**	**1377**	**357**	**1414**

a – initial setting time, [min.]; b – final setting time, [min.].

It is essential that the initial setting time to be long enough to cover all the cast-in-place operations. However, according to the obtained results, the initial and final setting time of the samples with glass powder is higher than that of 100CA and 30CA_70RS samples. In case of this materials calcium silicate hydrate (C-S-H), calcium aluminate silicate hydrate (C-A-S-H) and sodium aluminate silicate hydrate (N-A-S-H) are formed during geopolymerisation. As reported in previous studies the compounds that include calcium have lower setting time because the dissolution rate of Si^{4+} and Al^{3+} species is lower compared to the dissolution rate of Ca^{2+}. Therefore, the setting time of the samples with glass powder should be lower, but this inversion can be explained through the dissolution of high content of Si^{4+}, as a consequence of the glass particles reaction in the alkaline environment, that results in a high initial and setting time.

4. Conclusions

Four types of geopolymers have been obtained by activating the coal ash with a solution of sodium silicate and sodium hydroxide. In order to evaluate the influence of reinforcing elements of setting time and micromorphology, two types of particles have been introduced in the geopolymers matrix.

The microstructural analysis confirms a different behaviour, during geopolymerisation, between the glass powder and the river sand particles. According to this study, the glass particles react partially in the alkaline environment while the matrix adheres strongly. However, the river sand particles didn't react in the described condition showing a clear delimitation between their surface and the matrix of the samples.

Due to the fineness of the glass powder, the Si^{4+} content from their surface is dissolved in high concentration by the alkaline solution. Therefore, the setting time of the samples with glass particles (70CA_30GP and 15CA_15GP_70RS) show higher initial and final setting time, even with a higher Ca content.

5. References

[1] Burduhos Nergis D D, Abdullah M M A B, P Vizureanu and Mohd Tahir M F 2018 *IOP Conference Series: Materials Science and Engineering* **374** 012019 doi:10.1088/1757-899X/374/1/012019
[2] Luhar S and Gourav S 2015 *Journal of Civil Engineering Research* **5(3)** 53-58
[3] Jamil N H, Ibrahim W M A W, Abdullah M M A B, Sandu A V and Tahir M F M 2017 *IOP Conference Series: Materials Science and Engineering* **209** 012004
[4] Bouaissi A, Li L, Moga L, Abdullah M M A B and Sandu A V 2018 *Revista de Chimie* **69(7)**, Doi: 10.37358/RC.18.7.6390
[5] Burduhos Nergis D D, Abdullah M M A B, Sandu A V and Vizureanu P 2020 *Materials* **13(2)** 343
[6] Burduhos Nergis D D, Vizureanu P and Corbu O 2019 *Revista de chimie* **70(4)** 1262-1267
[7] Part W K, Mahyuddin Ramli C and Chee B 2015 *Construction and Building Materials* **77** 370-395
[8] Assi L N, Deaver E E and Ziehl P 2018 *Construction and Building Materials* **167** 372-380
[9] Kloušek D, Brus J, Urbanova M, Andertova J, Hulinsky V and Vorel J 2007 *Journal of Materials Science* **42(22)** 9267-9275
[10] He P, Wang M, Shuai F, Jia D, Yan S, Yuan J, Xu J, Wang P, Zhou Y 2016 *Ceramics International* **42** 14416-14422
[11] Chindaprasirt P, Chareerat T, Hatanaka S and Cao T 2011 *J. Mater. Civ. Eng.* **23** 264-70
[12] Vargas A, Dal Molin D, Vilela A, Silva F, Pavão B and Veit H 2011*Cement and Concrete Composites* **33** 653-660
[13] Rattanasak U and Chindaprasirt 2009 *Minerals Engineering* **22** 1073-1078
[14] Bejinariu C, Burduhos-Nergis D P, Cimpoesu N, Bernevig-Sava M A, Toma S L, Darabont D C and Baciu C 2019 *Qual-Access to Success* **20(1)** 71-76
[15] Burduhos Nergis D P, Baciu C, Vizureanu P, Lohan N and Bejinariu C 2019 *IOP Conference*

Series: Materials Science and Engineering **572** 012027
[16] Bejinariu C, Darabont D C, Baciu E R, Ionita I, Sava M A B and Baciu C 2017 *Environ. Eng. Manag. J.* **16(6)** 1395–1400
[17] Somna K, Jaturapitakkul C, Kajitvichyanukul P and Chindaprasirt P 2011 *Fuel* **90** 2118-24
[18] Burduhos Nergis D D, Abdullah M M A B and Vizureanu P 2017 *European Journal of Materials Science and Engineering* **2(4)** 111-118
[19] Komnitsas K and Zaharaki D 2007 *Miner. Eng.* **20** 1261-1277
[20] Yahya Z, Husin K, Abdullah M M A B, Ismail K N and Razak R A 2016 *Key Eng. Mater.* **673** 21-28
[21] Burduhos Nergis D P, Vizureanu P, A V Sandu and C Bejinariu 2020 *Applied Sciences* **10** 2753, Doi: 10.3390/app10082753.
[22] Nazari A and Sanjayan 2016 *Handbook of Low Carbon Concrete* (Elsevier: Butterworth-Heinemann Publication) ed. 1
[23] Liew Y-M, Heah C-Y, Mustafa A B M and Kamarudin H 2016 *Prog. Mater. Sci.* **83** 595-629
[24] Baltatu M S, Nabiałek M, Savin A and Vizureanu P 2019 *Materials* **12** 2973, Doi 10.3390/ma12182973
[25] Baltatu M S, Vizureanu P, Balan T, Lohan M and Tugui C A *Book Series: IOP Conference Series-Materials Science and Engineering* **374** 012023
[26] Irimiciuc S A, Agop M, Nica P, Gurlui S, Mihaileanu D, Toma S and Focsa C 2014 *Japanese Journal Of Applied Physics* **53(11)** 173-179
[27] Nejneru C, Cimpoesu N, Stanciu S, Vizureanu P and Sandu AV 2009 *Metalurgia international* **14(7)** 95-98

Recycling of Aqueous Phase from Hydrothermal Liquefaction and Municipal Wastewater by Microalgae

N Chernova[1], S Kiseleva[1*], M Vlaskin[2] and A Grigorenko[2]
[1] Lomonosov Moscow State University, Faculty of Geography, Moscow, Russian Federation.
[2] Joint Institute for High Temperatures of the Russian Academy of Sciences, Moscow, Russian Federation.

*E-mail: k_sophia_v@mail.ru

Abstract. The search for alternatives to hydrocarbon fuels remains an actual task. Microalgae (MA) as raw material for the production of biofuel remain an urgent object of research among other types of biomass, and the cultivation of MA is constantly growing. One of the promising technologies for biofuel production from MA is hydrothermal liquefaction (HTL) that allows processing wet biomass and turning all carbon-containing components (lipids, carbohydrates, proteins) into fuel. However, the hydrothermal liquefaction process leads to the formation of a significant amount of the aqueous phase, which is a by-product with low energy value and also needs to be processed. It is important to consider the possibility of using the aqueous phase after HTL, as well as municipal wastewater, for the cultivation of MA in combination with the production of biofuels. The MA strains capable to grow in a dilute aqueous phase after HTL (*Galdieria sulphuraria rsemsu G-1, Chlorella vulgaris rsemsu Chv-20/11-Ps, Arthrospira platensis rsemsu Bios)* and in wastewater (*Arthrospira platensis rsemsu Bios, Chlorella ellipsoidea rsemsu Chl-el)* were experimentally selected from collection of RSE Laboratory at LMSU. The article is devoted to the experimental study of the degree of nutrients utilization by MA from wastewater and an HTL-aqueous phase.

1. Introduction

Bioenergy remains the largest renewable energy industry. At the same time, microalgae (MA) are a promising object of research among other types of biomass and the scale of MA using for energy purposes is increasing [1]. On the other hand, the growth of global energy production and urbanization, which results in high rates of generation of municipal waste and wastewater, requires integrated waste management technologies with the complete waste utilization. Utilization of organic waste with energy production by various technologies is one of the methods for solving this complex problem.

In this work, we consider the cultivation of MA using wastewater and liquid phase from the HTL of MA. MA for liquid waste treatment are selected because:

- wastewater has a complex composition, the purification of wastewater from nitrogen and phosphorus and other nutrients, as well as from heavy metals using biological objects is optimal;
- MA are considered as a promising raw material for the biofuels production (third-generation biofuels), since they have higher growth rates than terrestrial biomass, and at the same time do not compete with the production of food and feed;

- MA cultivation in wastewater reduces the use of fresh water, nutrients (fertilizers), which generally leads to a reduction in the cost of biofuels obtained from them.

2. The use of MA for wastewater treatment: existing approaches and results

The development of wastewater treatment technologies using MA is quite an urgent task [2-4]. Real or model wastewater with various COD and BOD indicators as well as the concentrations of nitrogen, phosphorus, and dissolved carbon compounds have been used. The consumption efficiency of these substances by MA is estimated by the rate and total amount of nitrogen absorbed from wastewater (in the form of NH_4^+, NO_3^-, NO_2^-, total nitrogen), phosphorus (PO_4^{3-}, total phosphorus), dissolved and total carbon, etc. [5]. At the same time, it is taken into account that in addition to organic and inorganic chemicals, bacteria and protozoa are also present in the wastewater, which compete for nutrients with MA and even eat them. MA is cultivated on wastewater of various compositions. According to the results of experiments, strains of MA are selected that demonstrate a significant growth rate and, accordingly, the absorption of nutrients and the most resistant to toxic components of wastewater [6]. A number of studies also set the task of improving methods for the cultivation of MA in wastewater [7], studying the effectiveness of consortia of MA and bacterial organisms [5, 8], and analyzing the possibility of cultivating MA in wastewater in open cultivators in a wide range of climatic conditions [2, 9].

3. Technologies for obtaining biofuel from MA

The MA biomass can be processed into a wide range of biofuels: bioethanol, biodiesel, bio-oil, biohydrogen, etc. A traditional technology is the transesterification of MA lipids into biodiesel. The obvious disadvantages of this method are the high energy costs in the process of separation of MA from the culture medium and drying, as well as the use of unsafe organic solvents (such as methanol) for the extraction of lipids from MA biomass. In addition, in the production of biodiesel, only the lipid part is converted to fuel, while the rest part (most of the MA biomass), including proteins and carbohydrates, is not used to produce liquid biofuel.

Thermochemical conversion technologies don't have these disadvantages; they are carried out by heating and decomposing biomass in the presence or absence of air (oxygen). Types of thermochemical conversion are direct combustion (the main product is electric energy or heat), gasification (synthesis gas), pyrolysis (synthesis gas, bio-oil, coal), and liquefaction (bio-oil, synthesis gas, coal). Nowadays pyrolysis and hydrothermal liquefaction technologies of MA biomass have been actively promoted.

Under the MA pyrolysis in order to obtain solid, liquid, and combustible gaseous products, the biomass is heated to temperatures of 400–600°C; in some cases the temperature can reach 800°C. The MA pyrolysis which is aimed at producing combustible gases with a high hydrogen and methane content and high calorific value is less studied. MA cells are small in size, the dry biomass of MA is loose, and therefore it does not require additional grinding, like other types of biomass. However, for effective pyrolysis, the MA biomass must contain a minimum amount of moisture, which significantly increases the cost of technology implementing.

Hydrothermal liquefaction (HTL) is a method of thermochemical conversion of wet MA, which produces a liquid biofuel referred to as "bio-oil" or "biocrude", as well as gaseous, water-containing (aqueous phase) and solid by-products ("biochar"). The biomass processing by the HTL method does not require preliminary drying of the feedstock: MA can be fed to the hydrothermal liquefaction reactor in the wet state, for example, in the form of an aqueous suspension. In addition, when producing fuel by the HTL method, not only lipids, but also carbohydrates and proteins contribute to the bio-oil yield, which increases the total yield of the fuel. However, the hydrothermal liquefaction process leads to producing the significant amount of the aqueous phase, which is a by-product, has limited energy value and also needs to be utilized.

The cultivation of MA biomass using the municipal wastewater and its conversion by the HTL method with the subsequent utilization of the aqueous phase is the subject of this study.

4. Materials and methods

The following MA strains with a high growth rate and capability of responding to change in environmental conditions (a decrease / increase of the main nutrient elements concentration, toxic elements, etc.) from the collection of the Renewable Source Energy Laboratory at Lomonosov Moscow State University (RSEL MSU) were used for experiments:

– for recycling waste water - *Arthrospira platensis rsemsu Bios (P)*, *Gloeotila pulchra rsemsu Pz-6*, *Galdieria sulphuraria rsemsu G-1*, *Chlorella ellipsoidea rsemsu Chl-el;*

– for recycling HTL-aqueous phase - *Arthrospira platensis rsemsu Bios (P), A. platensis rsemsu 1/02, A. platensis rsemsu 1/02-T, Galdieria sulphuraria rsemsu G-1, Chlorella vulgaris rsemsu Chv-20/11-Ps.*

All of these MA strains, excepting the *Arthrospira platensis rsemsu Bios (P)* culture, are algologically pure. They are maintained in the RSEL MSU collection on sterile nutrient media (*Chlorella ellipsoidea* and *Chlorella vulgaris* - on Tamiya medium; *Galdieria sulphuraria* on Allen medium; *Gloeotila pulchra* – on BG-11 medium) [10]. The culture of *Arthrospira platensis rsemsu Bios (P)* is grown in the RSEL MSU using Zarrouk's medium [11] in open planar cultivators (volume is about 1000-500 l) by the semi-continuous method for more than 30 years and is represented by a stable consortium of MA *Arthrospira platensis* with heterotrophic bacteria-associates that were identified as representatives of the genera *Pseudomonas* and *Bacillus*. The latter are able to assimilate dissolved organic matter and toxic elements and thus provide a more efficient municipal wastewater treatment.

Municipal wastewater taken from the observation well in Moscow in October 2018 was used as a model medium for MA cultivation. The cultivation was carried out in Erlenmeyer flasks with a volume of 250-500 ml, the initial MA concentration was about 0.16 g (dry weight) / l. The cultivation was carried out in undiluted wastewater, as well as in wastewater diluted with freshly boiled tap water (dilution range used: 5 ... 50 times). The reference represents the growth of *A. platensis* biomass on 1/2 or 1/3 of Zarrouk's medium; for *Gloeotila pulchra* - on the ½ medium BG-11; *Chlorella ellipsoidea* and *Chlorella vulgaris* – Tamiya medium; for *Galdieria sulphuraria* – Allen medium. Each experiment was carried out three times. Biomass growth was monitored colorimetrically. All flasks were arranged on a shaker (rotation 120 rpm; T = 33°C; lighting $30 \mu E/m^2/s$ during 24 h). The cultivation of MA in wastewater was carried out for 10 to 47 days. The concentration of nutrients in wastewater was studied by chemical analysis of the nutrients in the culture fluid at different stages of MA cultivation. COD, BOD_5, nitrates (NO_3^-), phosphates (PO_4^{3-}), ammonium (NH_4^+) were considered as indicators. The state of MA cultures at different stages of the experiments was visually monitored by microscopy of culture fluid samples (LOMO Mikmed-2 and Axioplan 2 Imaging microscopes with an AxioCam MBm camera and AxioVision 3.1. Carl Zeiss).

Hydrothermal processing experiments were performed at a laboratory reactor (Figure 1). HTL was carried out at the installation of the Joint Institute for High Temperatures of the Russian Academy of Sciences. Hydrothermal liquefaction experiments were conducted at T = 270°C and 300°C.

The yields of bio-oil, gaseous products, solid residue and aqueous phase were 34-46%, 12-18%, 12-18% and 10-24%, respectively. The molecular formulas of the substances in the composition of the final products were determined. It has been established that compounds containing one and two nitrogen atoms dominate in bio-oil, while ON2, N3 classes are also present. Bio-oil from MA with a high content of lipids and carbohydrates is close to traditional oil [12]. The aqueous phase was used for cultivating the MA and thus for recirculation of this by-product.

5. Results and discussion

5.1. Composition of municipal wastewater

Due to the autumn period with high precipitation, municipal wastewater samples were significantly diluted (Table 1). The amount of nutrients (nitrogen ammonia, nitrates and orthophosphates) are present in a small amount - 110, 2.4 and 15 mg/l, respectively. COD (total organic matter in heavily polluted effluents) 140 mgO/l and BOD_5 (readily oxidizable organic matter concentration in effluents)

85 mgO_2 / l indicate that the sample contains a small amount of dissolved organic matter. According to the value of the hydrogen index (pH = 8.2), the wastewater is alkaline. The amount of oil products (2.7 mg/ l) and total aromatic substances (0.668 mg/l) exceeds the average for this class of wastewater. A comparison of the wastewater composition with the culture media composition for the growth of MA shows that the wastewater contains all the trace elements necessary for the growth and development of MA. An analysis of published data on the composition of wastewater showed that the wastewater used in the study is typical, but with a low COD value and high nitrogen ammonia content.

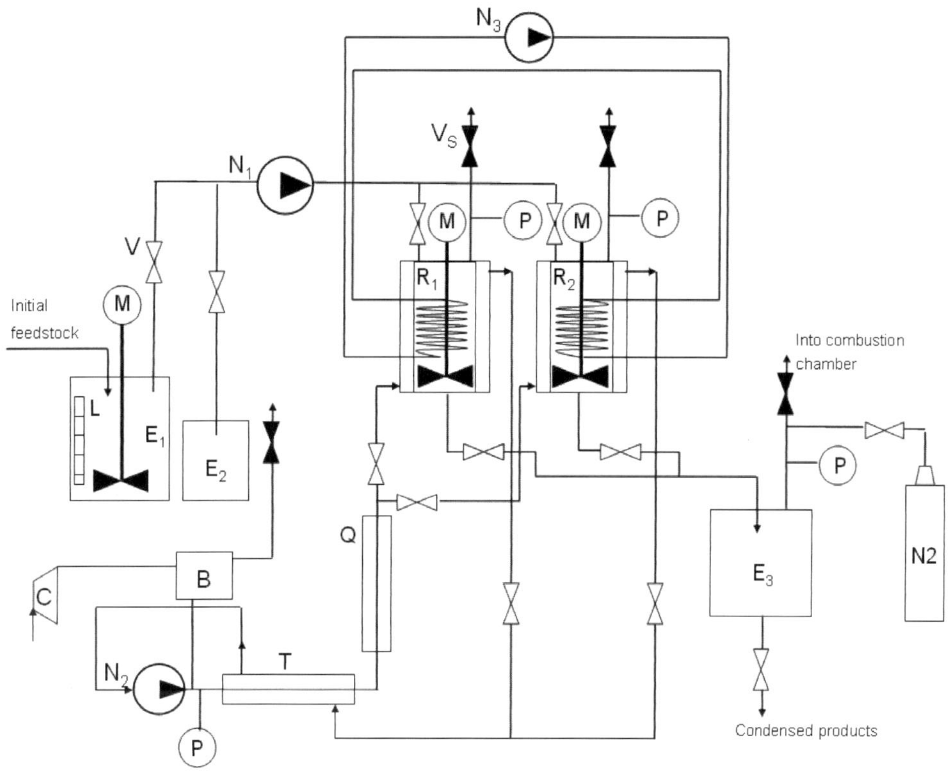

Figure 1. Scheme of the experimental plant: R_1, R_2 – reactors-autoclaves with agitators and cooling/heating coils, E_1 – tank for receiving the raw feedstock, E_2 – tank with water, E_3 – tank for receiving the products of hydrothermal liquefaction, N_1 – pump for the loading of MA suspension into the reactor, N_2 – high-pressure pump for circulating water, N_3 – reverse pump for heat recuperation, C – air compressor, B – buffer tank, T – heat exchanger, Q – heater, N2 – high-pressure tank with nitrogen, M – agitators, P – pressure sensors, V – valves, V_S – safety valves, L – level sensor.

Table 1. The chemical composition of the municipal wastewater sample from the observation well of Moscow.

Chemical Composition, units	Value	Chemical Composition, units	Value
Hydrogen indicator (pH), units pH	8,2	Hydrocarbonates, mg / l	820
COD (bichromate oxidation), mgO / l	140	Carbonates, mg / l	<6
BOD_5, mgO_2 / l	85	Ammonium, mg / l	110
Nitrates, mg / l	2,4	Phosphates, mg / l	15
Potassium, mg / l	20	Magnesium, mg /l	14

5.2. Determination of the wastewater dilution degree for microalgae cultivation

Since wastewater contains a significant amount of substances toxic to MA, an important step of the study was the selection of dilution degree. On the one hand, the toxicity of dissolved substances should be reduced, and on the other hand, the nutrients concentration should not be significantly reduced. The experimental results of the MA cultivation at different waters dilution degree are shown in Figure 2.

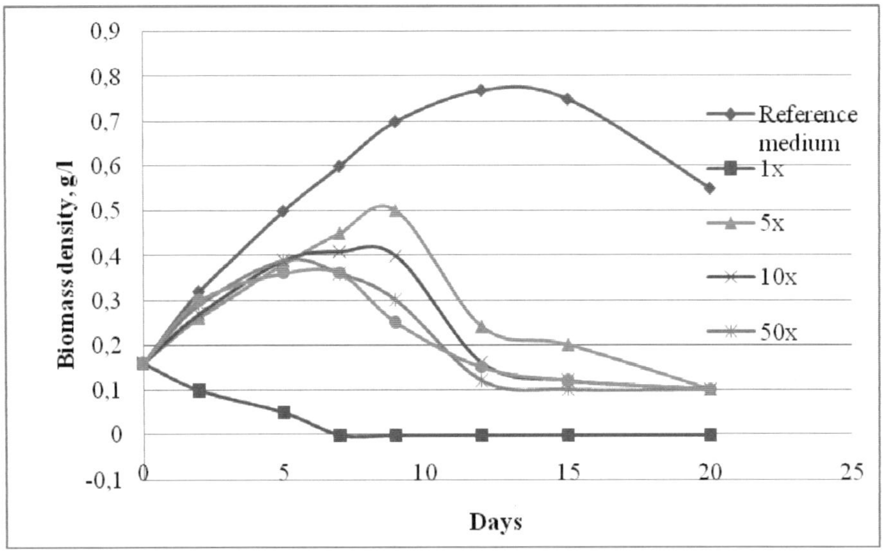

Figure 2. Dynamics of the *Arthrospira platensis Bios (P)* biomass density in wastewater with varying dilution degree.

As can be seen, a positive growth dynamic of the *A. platensis Bios* biomass in diluted wastewater was observed at all dilution degrees during the first 7–9 days. Moreover, a 2-fold increase in biomass occurred on the 5th day of the experiment. With all wastewater dilution degrees, the growth of biomass lagged behind the reference medium (the MA cultivating on the 1/3 of the Zarrouk's nutrient medium composition, without wastewater). Apparently, toxic substances (ammonium compounds, oil products and aromatic compounds) in wastewater without dilution suppressed the growth of MA. The optimal wastewater dilution degree is not higher than 5 x. And with dilutions of more than 5x, the biomass growth stopped after 5-7 days due to the complete exhaustion of nutrients. Thus, the need for wastewater dilution for sustainable growth of the *A. platensis Bios* biomass was determined and the dilution limits (no more than 5x) were found. In addition, it was found that for the arthrospira, sodium bicarbonate must be added to the wastewater, as arthrospira grows only in alkaline waters in the presence of bicarbonate anion as a carbon source.

5.3. Characteristics of microalgae growth in wastewater

Three clonal algologically pure cultures of *A.platensis* did not grow well in weakly diluted wastewater, and by 5-7 days the cultures died. On the contrary, the culture of *A.platensis Bios (P)* with associative bacteria showed good growth in wastewater diluted even 2 times. Therefore, experiments with the *A.platensis Bios (P)* culture with associative bacteria were carried out in 2-3 times diluted wastewater. The growth of this culture not only does not lag behind the reference medium, but in the period of 7–12 days, its growth in wastewater with dilution degree 2x even exceeds the reference medium (Figure 3).

After 12 days of the experiment, the growth of cultures slowed down, and all biomass accumulation curves reached a plateau. Subsequently, portions of wastewater equal to their volume at the beginning of the experiment were added to these experimental cultures. In this case, an additional dilution of wastewater (1.3x and 1.75x, respectively) occurred. The results of further cultivation of MA cultures under modified experimental conditions are presented in Figures 3, 4.

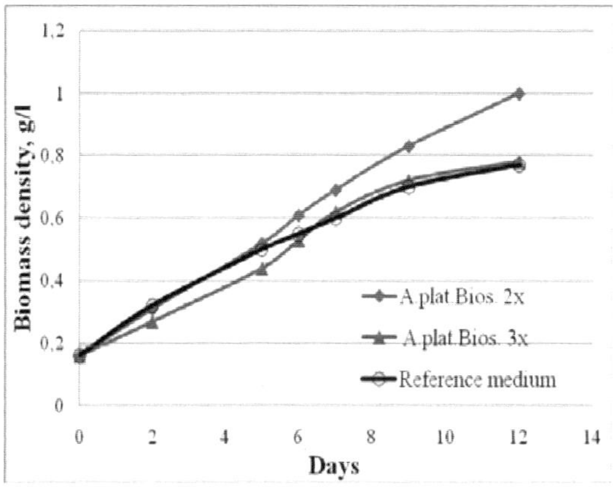

Figure 3. Dynamics of Arthrospira platensis Bios (P) biomass density in wastewater with 2x and 3x dilution.

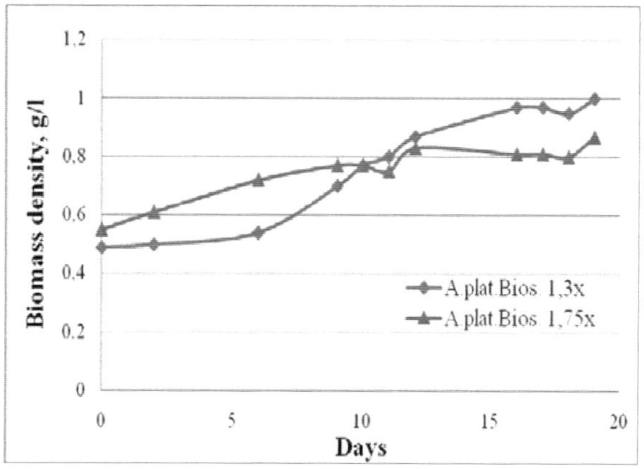

Figure 4. Continuation of the experiment: the growth of Arthrospira platensis Bios (P) biomass in wastewater with the addition of a portion of wastewater after 12 days of growth.

On the 19th day of the experiment, the cultures density in both cases reached the same values as in the previous stage of the experiment. Then, the growth of culture was continued up to 47 days with the addition of the same wastewater portions. At such conditions the growth of MA occurred at approximately constant speed. After 47 days, the growth of A. platensis Bios *(P)* slowed down and reached a plateau, and a significant increase cannot be achieved without a periodic selection of the grown biomass from the culture medium. Apparently, at a biomass density of more than 1 g / l on dry matter, the culture reaches the maximum density at which illumination and nutrients are limited, and exametabolites accumulate that adversely affect cell growth (Figure 4).

Similar experiments with the periodic wastewater and nutrient medium addition were carried out with other strains of MA (*Chlorella ellipsoidea, Gloeotila pulchra* and *Galdieria sulphuraria*). The reaction of various MA to the periodic wastewater and nutrients addition was strain-specific. For example, chlorella unlike other strains in wastewater without the addition of nutrients in the first 12 days showed a significant increase in biomass, while the galdieria and gleotila did not give biomass increase, although their cells remained in a viable state (Figure 5).

Additions of wastewater on the 12th day of experiments caused a significant growth of chlorella cells until the maximum density of 0.8 g / l on dry matter was reached on the 20th day. The next addition of Tamiya medium (25% by volume) stabilized the growth of chlorella cells at a level of 0.6 g / L dry matter for a further 20 days. But the culture of *Chlorella ellipsoidea* has significant adhesive properties to the inner surface of glass flasks, which undoubtedly reduced the real values of the culture density. MA *Gloeotila* and *Galdieria* reacted differently to the wastewater and culture media addition. *Gloeotila* increased biomass by 5 times on the 47th day of growth, adapting to growth conditions in wastewater. For the extremophilic alga *Galdieria*, the growth conditions were unsuitable (Figure 5). Thus, for the effective cultivation of specific MA in wastewater, the search for specific key factors for their growth and development is needed.

5.4. Effectiveness of wastewater treatment using microalgae

Inorganic nitrogen and phosphorus are particularly difficult to remove from wastewater. A comparative analysis of the wastewater main chemical indicators before and after growing MA showed that *A.platensis Bios (P)* is capable of efficient growth in slightly diluted wastewater. However, a prerequisite for growth is the creation of alkaline conditions in a nutrient medium by adding NaHCO3 in an amount of at least 6 g / l (Table 2). *A.platensis Bios (P)* cells completely consumed phosphorus from the wastewater in the form of orthophosphates and nitrogen in the form of ammonium and nitrates, the concentration of which decreased by 4-30, 58-73 and 24 times, respectively.

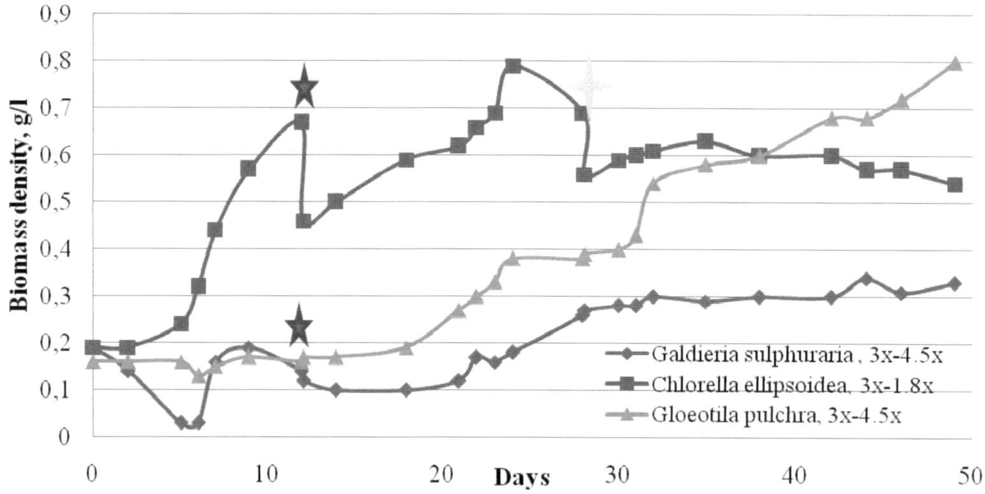

Figure 5. Dynamics of the MA *Chlorella ellipsoidea, Gloeotila pulchra* and *Galdieria sulphuraria* biomass density. The legend indicates the initial and final dilution of wastewater. The cultivation of MA in flasks with a volume of 250 ml, the initial volume of the nutrient medium is 100 ml:★ - the addition of 50 ml of wastewater for chlorella, the addition of 50 ml of the BG-11 medium for gleotila, the addition of 50 ml of the Allen medium for galdierium.✧ – addition 34 ml ¼ Tamiya medium for chlorella.

In experiments with *Chlorella ellipsoidea*, due to the low cell growth rate at the beginning of the experiment and exhaustion of nutrients from wastewater, addition of Tamiya nutrient medium was carried out on the 27th day of the experiment with a potassium nitrate salt content of 5 g/l, which increased the nitrate content in the nutrient medium to 807.7 mg/l. At the end of the experiment (47 days), nitrates were assimilated by more than 70%, and phosphates and ammonium were assimilated completely by chlorella cells.

Table 2. Comparative analysis of the wastewater main chemical indicators before and after growing MA (growing time 47 days).

Composition indicator, units	Experimental results				Relative error, %
	Source wastewater	*A. platensis rsemsu Bios-* 2x-1,2x	*A.platensis rsemsu Bios-* 3x-1,5x	*Chlorella ellipsoidea 3x-1,8x*	
COD (bichromate oxidation), mgO / l	140	140	120	140	15
BOD_5, mgO_2 / l	85	68	53	72	13
Potassium, mg / l	15	<0,5	4	0,95	10
Nitrates, mg / l	2,4	<0,5	<0,1	235	20
Ammonium, mg / l	110	1,9	1,5	1,3	12

Gloeotila pulchra adopted to growth conditions by 25 days; as a result, for 47 days of cultivation in diluted wastewater with the addition of BG-11 medium at intermediate stages, it assimilated about 67% of the main biogenic elements from wastewater. *Galdieria sulphuraria* did not adopt to cultivation on wastewater.

After growing MA in wastewater, BOD_5 values significantly decreased, which indicates a decrease in the readily soluble organics in the wastewater.

5.5. Efficiency of purification of the HTL aqueous phase using microalgae

As mentioned above, in the production of biofuels from MA by the hydrothermal liquefaction method, a significant amount of the aqueous phase is formed as a by-product that has no energy value. Therefore, the study of the biofuel production cycle including the cultivation of MA using aqueous phase is relevant.

The possibility of the nutrients utilization from an HTL aqueous phase (AP) obtained in the process of *A. platensis 1/02-P* cultivation has been studied. It was found that the aqueous phase contains a large amount of nutrients, which are necessary for the MA cultivation, as well as trace elements (58 compounds were identified). The nutrients content in AP is orders of magnitude higher than the standard nutrient media required for the MA cultivation. However, an aqueous phase is characterized by a toxic effect on MA. Growth inhibition is associated with the presence of a large number of toxic compounds (phenols, cyclic nitrogen compounds, heavy metals, ammonium ions). Intensive dilution of AP is necessary to prevent the toxic effect of growth inhibitors. It was done by using a mixture of aqueous phase and standard nutrient medium or only distilled water. MA strains that could grow for a long time (more than a month) on diluted HTL aqueous phase have been experimentally selected: *Galdieria sulphuraria*, *Chlorella vulgaris* and *A. platensis Bios (P)* (collection of Renewable Source Energy Laboratory at LMSU).

Growth analysis of candidate-strains (*Galdieria sulphuraria, Chlorella vulgaris*) in AP after the dilution with distilled water in 500 times showed that the most active growth is in case of *Galdieria*. Its growth in the first 25 days of cultivation exceeds that in the nutrient media. This result confirms the ability of *Galdieria* to mixotrophy. *Chlorella* had a steady growth also. However, it was still lower than in reference media. Consortia of *A. platensis Bios* with associate bacteria showed steady growth in 300-400 times diluted HTL aqueous solution. Cultivation of this consortia required the bicarbonate ions addition to create more favorable conditions for the growth and development of arthrospira. It allows partially recycling the by-product of bio-oil from MA (aqueous phase) [13].

6. Conclusion

The methods of MA cultivation in municipal wastewater and the aqueous phase after hydrothermal liquefaction of MA were experimentally selected. The effectiveness of wastewater and aqueous phase treatment were analyzed.

Based on experiments with seven candidate MA strains, two promising MA cultures (*Arthrospira platensis Bios (P)* and *Chlorella ellipsoidea rsemsu Chl-el*), which are capable to grow efficiently in wastewater of different dilutions for a long time, were selected.

It was shown that the consortium of *A. platensis rsemsu Bios (P)* with associative bacteria can grow efficiently in wastewater, diluted by 5 times and assimilate nutrients within 8 days. At higher dilutions, due to lack of nutrients, cell growth ceased for 7 days. The addition of 6 g/l $NaHCO_3$ to wastewater created more favorable conditions (pH>8.5; a sufficient concentration of HCO_3^- anions) for the growth of arthrospira. When the wastewater was diluted 1.2-1.5 times the addition of $NaHCO_3$ provided the effective growth of this MA consortium with associative bacteria for 47 days, as a result of which the concentration of nitrogen and phosphorus ions in the wastewater decreased to trace values. In addition, the amount of readily soluble organic matter (BOD_5) decreased from 85 to 68÷53 mgO_2/l, which indicates the ability of the MA to switch to mixotrophic growth with the consumption of short-chain organics.

Chlorella ellipsoidea in wastewater without adding nutrients in the first 12 days showed a significant increase in biomass (biomass density increased by 3.5 times). The addition of wastewater portion and ¼ Tamiya medium on the 12th day of the experiment, as in the case of arthrospira, led to a significant growth of chlorella cells until their maximum density (0.8 g/l dry matter) was reached.

It was found that for more efficient wastewater utilization and treatment by MA growth, it is necessary to select specific biomass growth key factors. The experiments duration (47 days) made it possible to identify the adaptive abilities of some MA to wastewater conditions and to obtain an algo-bacterial culture capable of "excessive" absorption of wastewater nutrients.

The aqueous phase (AP) after hydrothermal liquefaction of the MA biomass contains a wide variety and a large concentration of nutrients and trace elements necessary for the MA cultivation. Moreover, the aqueous phase can have a toxic effect on MA due to the high content of phenols, heavy metals, etc., which requires, as in the case of wastewater, intensive AP dilution for MA cultivation. MA strains that can grow for a long time in diluted HTL aqueous phase (*Galdieria sulphuraria, Chlorella vulgaris* and *Arthrospira platensis Bios (P)* from the collection of Renewable Source Energy Laboratory at LMSU) have been experimentally selected. The most active growth was demonstrated by the consortium A. platensis Bios (P) with heterotrophic associative bacteria, which grew well on less diluted AP (300x – 400x), and when sodium bicarbonate ions were added in an amount of 6 g / l, their growth was enhanced due to more favorable conditions for the arthrospira growth.

Galdieria sulphuraria also grew well for more than a month in the aqueous phase, which confirms its ability to mixotrophy, but with a dilution of 500x.

The obtained results confirm that the technology based on wastewater treatment and aqueous phase after HTL by MA is technically feasible and promising. It allows partial processing of a by-product of the bio-oil production from MA by hydrothermal liquefaction technology.

The reported study was funded by Russian Foundation for Basic Research, project number 18-58-45009.

7. References

[1] Su Y, Song K, Zhang P, Su Y, Cheng J and Chen X 2017 *Renew. Sust. Energ. Rev.* **74** 402–11

[2] Jämsä M, Lynch F, Santana-Sánchez A, Laaksonen P, Zaitsev G, Solovchenko A and Allahverdiyeva Y 2017 *Algal Res.* **26** 65–73

[3] Rui H, Zhen Y, Jinchen L, Min G, Weiling M and Yi Z 2017 *Int. J. Agric. & Biol Eng.* **10(1)** 107–13

[4] Tejido-Nuñez Y, Aymericha E, Sanchoa L and Refardt D 2019 *Ecol. Eng.* **136** 1–9

[5] Posadas E, Alcantara C, Garcia-Encina P A, Gouveia L, Guieysse B, Norvill Z, Acien F G, Markou G, Congestri R, Koreiviene J and Munoz R 2017 *Microalgae-Based Biofuels and Bioproducts: From Feedstock Cultivation to End Products.* ed R Muñoz and C Gonzalez-Fernandez UK: Woodhead Publishing 67-92

[6] Chiu S–Y, Kao C–Y, Chen T–Y, Chang Y–B, Kuo C–M and Lin C–S 2015 *Biores. Techn.* **184** 179–189

[7] Han T, Lu H F, Ma S S, Zhang Y H, Liu Z D and Duan N 2017 *N. Int J Agric & Biol. Eng.* **10(1)** 1–29

[8] Sriram S and Seenivasan R 2012 *J. Algal Biomass Utln.* **3**(2) 9–13

[9] Chernova N, Kiseleva S, Vlaskin M and Rafikova Y 2018 *MATEC Web of Conferences*, **178** 09011

[10] *Catalog of microalgae cultures in the collections of the USSR* 1991 *(Katalog cultur microvodoroslei v collektsiyakh SSSR)* Moscow: IFR RAS 48, 53-5 (in Russ.)

[11] Zarrouk C 1966 *Influence de diverse facteursphysiques et chimiques sur la croissance et la photosynthese de Spirulina maxima (Setch et Gardner)* Geitler: University of Paris

[12] Vlaskin M, Grigorenko A, Chernova N, Kiseleva S, Kumar V and Pruthi V 2018 International Conference on New Horizons in Green Chemistry & Technology (Uttarakhand. Uttaranchal University. 27-28 November 2018) *SSRN Electron. J.*

[13] Chernova N, Kiseleva S, Vlaskin M, Grigorenko A and Rafikova Y 2019 *IOP Conf. Ser.: Mater. Sc. Eng.* **564** 012101

Temperature Deformation Influence on the Locking Equipment Destruction in Water Supply Systems at High-Rise Buildings Upper Floors

A Antimonov[1] and N Pushkareva[2*]
[1] Department of Mechanical Engineering, Ural Federal University, Institute of New Materials and Technologies, Mira Street 19, 620002 Ekaterinburg, Russia
[2] Department of Physics, Ural Federal University, Institute of Fundamental Education, Mira Street 19, 620002 Ekaterinburg, Russia

*E-mail: nbpush@mail.ru

Abstract. The locking equipment in water supply systems destruction problem on the upper floors of high-rise buildings is being solved. After destruction products quality study did not reveal defects of a metallographic nature or rejects in manufacture. Therefore, the hypothesis that the destruction cause is low-cycle material fatigue under the dynamic loads influence, arising due to temperature deformations in the pipeline, was proposed. In this paper, we consider a pipeline scheme in which a temperature gradient arises in the fluid flow direction. At different temperature strains of the pipeline parts bending stresses emerge. These stresses cyclic nature is determined by the hot- and cold-water consumption frequency. The pipeline model where the locking equipment was installed is a beam with pinched ends. The bending stresses calculation for this beam was performed using the initial parameter method. Plots of moments and transverse forces in the beam section are plotted. The permissible stress calculation under the dynamic loads action was made. The pipeline section size, within which shut-off equipment can be installed without destruction, was determined.

1. Introduction

The metal structures destruction cause in many cases is material fatigue [1-7]. This phenomenon essence is the appearance and development cyclic loads fatigue crack under the action [8-14]. As the crack grows, the part cross-section gradually decreases. When the stress in this section reaches the ultimate strength, the part suddenly collapses. This process feature is the fact that the calculated static stresses are in the range of elasticity and significantly less than the permissible. Sudden destruction can lead to structural bearing capacity loss and a major technological disaster of varying severity and damage degree [15-18].

2. Problem statement

A significant amount of work has been devoted to problems of strength under the dynamic loads action [19-21]. In many respects, this concerns the high-speed machines operation, in which loads changes with a high frequency. At a low frequency for a relatively short period of time, their action is less dangerous, while with prolonged exposure, structures destruction is also possible. This phenomenon is called low-cycle fatigue [22-25]. The technical systems reliability that, at first glance, are in a static state without visible external forces acting on them, are particularly unpredictable. In

IOP Conf. Series: Materials Science and Engineering **877** (2020) 012046 doi:10.1088/1757-899X/877/1/012046

this paper, we consider just such case when it was necessary to establish the structural elements destruction cause, which service purpose were no requirements for their bearing capacity. Such objects are brass ball valves, which are used as locking equipment in the water supply systems of high-rise buildings. These faucets body destruction on the upper floors during operation was the cause of flooding the lower floors with hot- or cold-water.

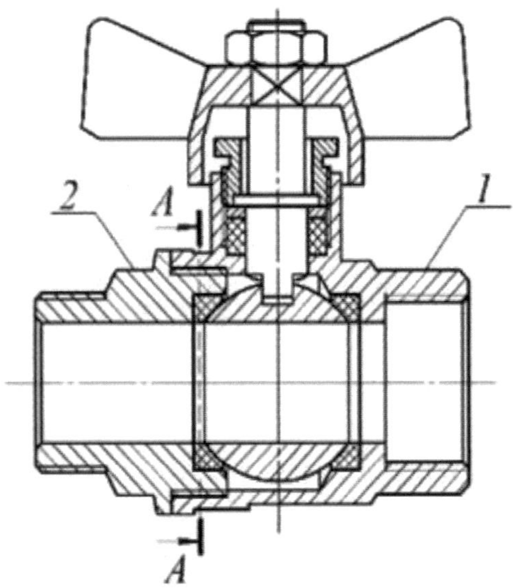

Figure 1. Faucet drawing.

Figure 2. The appearance of the destroyed faucet.

The faucet drawing is shown in 'Figure 1'. The crane consists of a body 1 and a fitting 2 connected by a thread. The analysis shows that the hazardous section A-A of the crane body is located in the threaded connection area between the body and the fitting. The body wall thickness in this section is 1.5 mm. The destroyed faucet is shown in 'Figure 2', from which it follows that the faucet breakdown occurred in this section.

3. Process dynamics

The faucet destruction occurs mainly on the pipelines of the upper-floors some time after the buildings commissioning. This allows to suggests that the failure cause is material fatigue due to dynamic loads that arise in the pipeline thermal deformations case.

A pipeline diagram with a faucet, which is installed at the upper floor level, is presented on 'Figure 3'. The pipeline consists of two racks H hight and traverse length l_0. Traverse has two sections separated by a faucet, with length l_1 and l_2, which may vary in length. The faucet is used to control the flow of hot or cold-water.

When users begin to make a water consume, a temperature gradient arises in the pipeline in the fluid flow direction. With hot-water supply, one of the racks is heated more strongly than the other. It is known that with fluctuations in temperature, body sizes change.

Therefore, one of the racks lengthens more than the other, and the pipeline cross-section undergoes bending deformation, as shown in Figure 4.

When supplying cold water, the reverse process takes place. One of the racks cools faster than the other and, accordingly, is shortened more. In this case, the pipeline traverse bending deformation also occurs. With a water flow variable mode, the pipe racks temperature periodically changes, and accordingly, deformations and stresses in the pipeline system change over time.

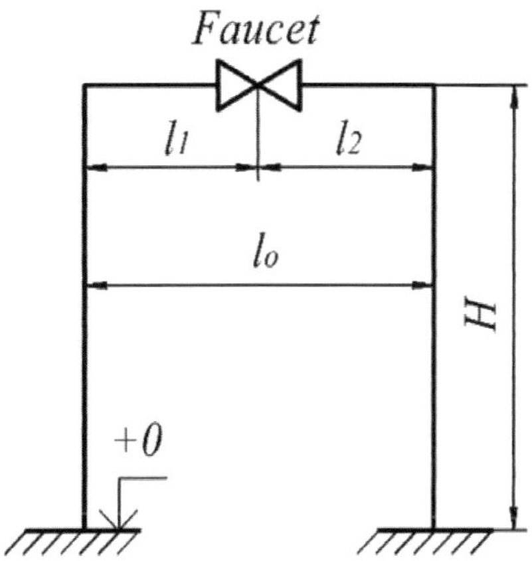

Figure 3. The pipeline scheme.

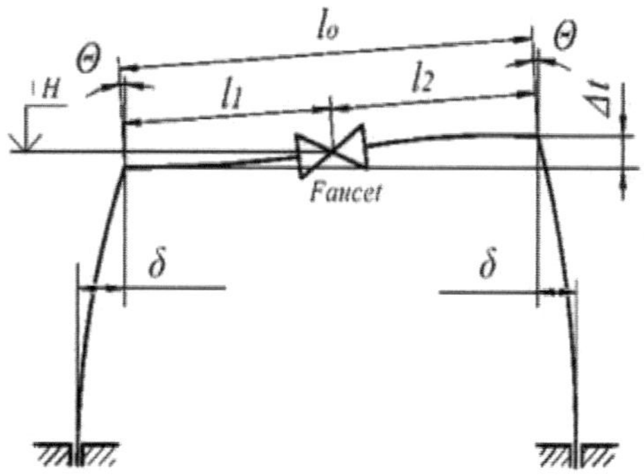

Figure 4. The pipeline deformation scheme.

4. Stress Calculation Method

The bending stresses calculation in the pipeline cross-section is performed for the pipeline without thermal compensators, when the cross-beam fully accepts the loads from the racks temperature deformations. The geometric model of the pipeline for calculating stresses is presented in Figure 5.

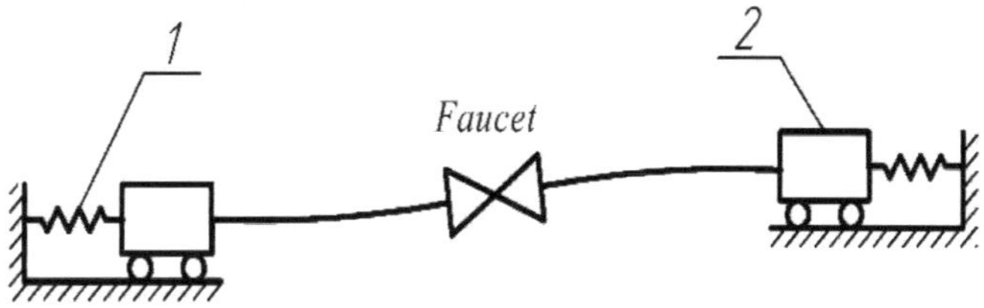

Figure 5. Pipeline model for stress analysis: 1 - elastic element, 2 – roller.

The traverse is presented as a beam with movable pinched ends. The ends mobility is ensured by the rollers movement 1 in the horizontal direction, which simulates the horizontal movement δ the pipe racks ends when they are bent (Figure 4). The elastic elements stiffness 2 is analogous to the stiffness of the uprights bending, which creates tensile stresses in the traverse. In the calculations, we will neglect the elastic elements stiffness, which eliminates the influence on the tensile stress strength in the traverse. We take this effect into account due to an increase in bending stresses, taking the rotation angles sections at the ends of the traverse Θ equal to zero (Figure 4).

Bending stress σ in traverse we define by the formula:

$$\sigma = \frac{M}{W} \tag{1}$$

Where: M – bending moment; W – cross section resistance axial moment.

For a beam with an annular cross section, calculation W produced by the formula:

$$W = \frac{\pi D^3}{32}(1 - \chi^4);\ \chi = \frac{d}{D} \tag{2}$$

Where: D and d – outer and inner diameter of the ring.

To calculate the bending moment, we use the initial parameter method [26]. The reverse diagram in the beam form with pinched ends for calculating stresses is presented in 'figure 6'. Under the external forces action of in the beam cross section, bending moments and transverse forces arise, the which distribution along the beam axis is represented by diagrams $M_x$and Q_x.

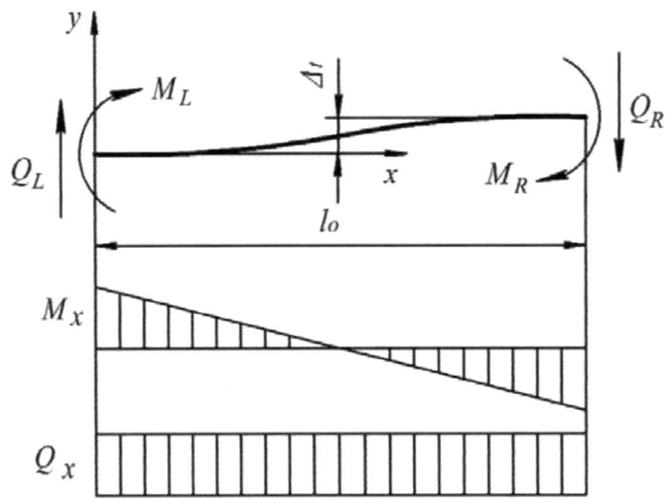

Figure 6. The traverse diagram in the beam form with pinched ends for calculating stresses.

Let's combine the origin with the left end of the traverse. The rotation angle and deflection for its right end are defined as follows:

$$\Theta_R = \Theta_L + \frac{1}{EI}\left(M_L l_0 + Q_L \frac{l_0^2}{2}\right), \quad y_R = y_L + \Theta_L l_0 + \frac{1}{EI}\left(M_L \frac{l_0^2}{2} + Q_L \frac{l_0^3}{6}\right) \tag{3}$$

Where: Θ_L, Θ_R, y_L, y_R, M_L, Q_L – accordingly, rotation angles, deflections, moments and transverse forces in sections for the left and right ends of the beam; E - elastic modulus; I - inertia axial moment

$$I = \frac{\pi D^4}{64}(1 - \chi^4); \chi = \frac{d}{D}. \tag{4}$$

Assuming from the boundary conditions in these equations $\Theta_L = \Theta_R = 0$; $y_L = 0$;$y_R = \Delta t$, where Δt – the difference in the length of the pipe racks, heated to various temperatures, we obtain for the transverse force and bending moment in the cross section at the right end of the beam:

$$Q_L = -\frac{2M_L}{l_0}, M_L = \frac{2EI\Delta t}{l_0^2} \tag{5}$$

The minus sign in the expression for Q_L means that this force has direction opposite to that chosen in 'figure 6'. Then in an arbitrary pipeline beam section with the coordinate x will haveQ_x:

$$Q_x = Q_L; M_x = M_L + Q_L \cdot x \tag{6}$$

Whence for a bending moment at the right end of the pipeline beam when $x = l_0$ we obtain

$$M_R = -\frac{2EI\Delta t}{l_0^2}. \quad (7)$$

Thermal deformation value Δt determined by the formula:

$$\Delta t = \alpha \cdot H \cdot \delta t \quad (8)$$

Where: α – linear expansion coefficient; H – pipeline rack height; δt– racks temperature difference.

Calculation resultsQ_L, M_L, σ and δt depending on the racks without heat compensators temperature difference are presented in table 1.

Table 1. Calculation results Q_L, M_L, σ and δt depending on the racks without heat compensators temperature difference.

Temperature difference in racks δt, degrees	1	5	7	15	20	25	30	35	40
Deformation Δt, mm	0.4	1.9	2.7	5.7	7.5	9.4	11.3	13.2	15
Transverse force Q, N	22	102	144	304	400	501	602	703	799
Bending moment M, Nm	11	51	72	152	200	251	301	352	400
Bending stress σ, MPa	9	40	57	120	158	198	238	278	315

When carrying out the calculations it is accepted: for steel – $\alpha = 125 \cdot 10^{-7} deg^{-1}$. Elastic modulus $E = 1 \cdot 10^5$ MPa, rack height at 10 floors $H = 30$ m; traverse length $l_0 = 1000$ mm; the outer and inner faucet body diameters in a dangerous section are respectively equal $D = 35$ mm, $d = 32$ mm.

5. Fatigue strength calculations.

Regardless of the current load type, the condition for bending strength has the form

$$\sigma \le [\sigma] \quad (9)$$

Where: σ and $[\sigma]$ – calculated and allowable stresses.

In fatigue calculations, the permissible stress is determined by the following formula [27]:

$$[\sigma] = \frac{2[\sigma_{+1}][\sigma_{-1}]}{(1-r)[\sigma_{+1}]+(1+r)[\sigma_{-1}]} \quad (10)$$

Where: $[\sigma_{+1}]$ – permissible stress under static load; $[\sigma_{-1}]$ – endurance limit for a product with a symmetrical stress cycle taking into account the stress concentration influence, scale factor and surface roughness; r – cycle characteristic defined by the formula

$$r = \frac{\sigma_{min}}{\sigma_{max}} \quad (11)$$

Where: σ_{min} and σ_{max} – minimum and maximum cycle stress.

The stresses in the pipeline traverse vary in time along a pulsating cycle, for which $\sigma_{min} = 0$. Then $r = 0$ and:

$$[\sigma] = \frac{2[\sigma_{+1}][\sigma_{-1}]}{[\sigma_{+1}]+[\sigma_{-1}]} \quad (12)$$

Values $[\sigma_{+1}]$ and $[\sigma_{-1}]$ are determined from the following relations:

$$[\sigma_{+1}] = \frac{\sigma_T}{n}; [\sigma_{-1}] = \frac{\sigma_{-1}}{k} \quad (13)$$

Where: σ_T– material yield strength; σ_{-1} – material endurance limit for a smooth reference material; n – safety factor under static load; k – coefficient that taking into account the stress concentration influence, scale factor and surface roughness on reducing the fatigue limit.
For most materials, the value of these coefficients is such that $[\sigma_{+1}]$ and $[\sigma_{-1}]$ decrease in relation to σ_T and σ_{-1} on 30-50%. For brass L68, close in chemical composition to brass faucet, we have $\sigma_T =$ 91 MPa; $\sigma_{-1} = 120$ MPa [1]. Let's accept a decrease in these values by 40%. Then, rounding off the results, we have $[\sigma_{+1}] = 55$ MPa and $[\sigma_{-1}] = 70$ MPa. As a result, for permissible stresses, we obtain $[\sigma] = 62$MPa. Comparing the calculated stresses from the table with the permissible stress, we find that even with a pipe racks heating temperature difference in 10 degrees, the stress at the beam ends can exceed the permissible.

6. Conclusions

From the bending moment analysis diagram, presented in 'figure 6', it follows that in accordance with the bending moment, the bending stresses vary linearly along the pipeline crosshead length, are equal to zero in the middle and change sign when going through zero. Thus, within the traverse there is a length section l_S with ends equally spaced from the middle where safe stress apply. Faucet installation within this area reduces the chance of its destruction. This section length is determined by the formula:

$$l_S = \frac{[\sigma]}{\sigma_{CE}} l_0 \quad (14)$$

Where: σ_{CE} - calculated stress at the ends of traverse.
The safe section length value, depending on the temperature difference between the racks for the pipe traverse length $l_0 = 1$ m presented in table 2.

Table 2. The safe section length value, depending on the temperature difference between the racks for the pipe traverse length $l_0 = 1$ m.

Rack temperature difference δt, degree	1	5	10	15	20	25	30	35	40
Safe section length l_S, mm	1000	1000	1000	1000	1000	940	775	667	591

From the table it follows that with a temperature difference in the racks up to 7 degrees, the locking equipment can be installed anywhere in the traverse. With a greater temperature difference in the racks, its installation must be done closer to the beam center, and the safest place is its center.

7. References

[1] Ekobori T 1971 *Fizika i mehanika razrushenija i prochnosti tvjordyh tel* [*Physics and mechanics of fracture and solid strength*] (Moscow: Metallurgija) p 264
[2] Ivanova V S 1979 *Razrushenie metallov* [*Metal destruction*] (Moscow: Metallurgija) p 168
[3] Meshkov J J 1981 *Fizicheskie osnovy razrushenija stal'nyh konstrukcij* [*Physical foundations of the destruction of steel structures*] (Kiev: Naukova dumka) p 238
[4] Astaf'ev V I 2001 *Nelinejnaja mehanika razrushenija* [*Nonlinear fracture mechanics*] (Samara: Samara University) p 630
[5] Terent'ev V F and Petuhov A N 2013 *Ustalost' vysokoprochnyh metallicheskih materialov* [*Fatigue of high strength metal materials*] (Moscow: IMET RAN – CIAM) p 515

[6] Cherepanov G P 2012 *Mehanika razrushenija. Serija Matematika i mehanika* [*The mechanics of destruction. Series Mathematics and Mechanics*] (Moscow-Izhevsk: Izdatel'stvo «IKI») p 872
[7] Terent'ev V F and Korableva S A 2015 *Ustalost' metallov* [*Metal fatigue*] (Moscow: Nauka) p 479
[8] Starokon' I V 2012 *J. Sovremennye problemy nauki i obrazovanija* **4** 100-103
[9] Fishgojt A V 2003 *J. Izvestija Rossijskoj akademii nauk. Serija fizicheskaja* **6(67)** 881–86
[10] Den'gaeva O V *J. Tehnika i tehnologija* **5** 34–40
[11] Vasjuta B N. 2003 *J. Izvestija vysshih uchebnyh zavedenij. Stroitel'stvo*. **10(538)** 4–13
[12] Shi K, Cai L and Chenet L al 2014 A theoretical model of semi–elliptic surface crack growth. *Chi. J. Aeronaut.* **3(27)** 730–34
[13] Fatemi A and Vang L 1998 *Int. J. Fatigue* **20(1)** 9–34
[14] Lautrou N, Thevenet D and Cognard J–Y 2009 *J. Fatigue Fract Eng Mater Struct* **32(5)** 403–17
[15] Suresh S 2003 *Fatigue of Materials*. 2nd. ed. (Cambridge: Cambridge Uniiversity Press) p 679
[16] Bathias S 2005 *Gigacycle fatigue in mechanical practice* (Yerlag: Magsel Dekker) p 304
[17] Zshbu M and Dzhons D 2010 *Konstrukcionnye materialy. Polnyj kurs. Uchebnoe posobie* [*Construction materials. Full course. Tutorial*] (Dolgoprudnyj:Izdatel'skij dom «Intellekt») p 672
[18] Gilmor R 1984 *Prikladnaja teorija katastrof* [*Applied Catastrophe Theory*] (Moscow: Mir) p 350
[19] Dubrovskij V A and Podvolockaja NI 2005 *J. Problemy mashinostroenija i nadezhnosti mashin* **4** 119-124
[20] Kolesnikov K S 2002 *Problemy dinamiki i prochnosti ispolnitel'nyh mehanizmov i mashin: Materialy nauch. konf* [*Problems of dynamics and strength of actuators and machines*] *In: Proc. of Sci. conf.* (*Astrahan*), (Astrahan': Inform.-izdat. centr AGTU) p 315
[21] Svetlickiy V A and Narajkin O S 2005 *Problemy prikladnoj mehaniki, dinamiki i prochnosti mashin : sbornik statej* [*Problems of applied mechanics, dynamics and strength of machines: collection of articles*] (Moscow: Izd -vo MGTU) p 315
[22] Collins G 1984 *Povrezhdenie materialov v konstrukcijah. Analiz, predskazanie, predotvrashhenie* [*Damage to materials in structures. Analysis, prediction, prevention*] (Moscow: Mir) p 624
[23] Troshhenko V T 2009 *Prochnost' materialov i konstrukcij*. T. 2. *Ustalost' metallov* [*Strength of materials and structures. V 2. Fatigue of metals*] (Kiev: In-t problem prochnosti) p 664
[24] Il'inyh A V 2016 *Malociklovaja ustalost' konstrukcionnyh splavov pri peremennyh parametrah cikla. V sbornike: Mehanika, resurs i diagnostika materialov i konstrukcij Sbornik materialov* [*Low-cycle fatigue of structural alloys with variable cycle parameters. In the collection: Mechanics, resources and diagnostics of materials and structures Materials collection*] (Ekaterinburg: IMASh UrO RAN) pp 292–293
[25] Vansovich K A and Ljagova AA 2015 *J. Gazovaja promyshlennost'* **12(731)** 96–99
[26] Pisarenko G S, Jakovlev A P and Matveev V V 1988 *Spravochnik po soprotivleniju materialov Izd. 2-e, pererab* [*Handbook of materials resistance*] (Kiev: Nauk dumka) p 736
[27] Beljaev N M 2014 *Soprotivlenie materialov: ucheb. posobie* [*Resistance of materials: Tutorial*] (Moscow: Al'jans) p 607

Studies on Two Icons Oklads with the Representation of St. Great Martyr George

L R Bivol[1], V G Ghica[1*], M I Petrescu[1], G Iacob[1], E Vasile[1], M Buzatu[1], D Gheorghe[1] and T A Kovács[2]
[1] University POLITEHNICA of Bucharest, 313 Spl. Independenţei, Bucharest, Romania
[2] ÓBUDA University, 1081 Budapest, Hungary

*E-mail: gvghica@yahoo.com

Abstract. The work represents a comparative study of two icon oklads representing St. George killing the dragon: the first of Romanian origin according to the stamp with the silver title "12" - examples of Romanian oklads with this silver title can be found in the works of S.S. Duicu, and the second oklad with stamps that seem to be specific to Russian silverware workshops. The Romanian icon oklad depicting St. George killing the dragon has in the central-lower part two stamps - the first stamp represent the monogram of master controller or craftsman or silver master "КЛ", and the second represents the silver title "12". The initials of the master are written in Cyrillic letters, as well as the inscription from the upper part of the oklad "C. ГЕОРГІЕ". The investigations on the two icons oklads were achieved through XRF analysis performed with a portable X-MET 5000 Oxford analyzer (Oxford X-MET 5000 Handheld XRF Analyzer) and SEM-EDX analyzes performed on a Phillips XL-30-ESEM electron microscope using a Quanta Inspect F50, with a field emission gun (FEG) with 1.2 nm resolution and an Energy Dispersive X-ray Spectrometer (EDXS) having 133 eV resolution at MnKα. Following the analyses it was observed that the silver title 84 stamped on the Russian icon oklad does not correspond to reality.

1. Introduction

Saint George was a Christian martyr during the reign of Diocletian [1], his history being recorded since the VII-VIII centuries in the early calendars and martyrologies (as for example in: the Irish Martyrology of Óengus, the Martyrology of Bede, and the Old English Martyrology) [2], its cult reaching its peak in the late Middle Ages [2]. He was born in Lydda, Syrian Palestine in 280 and was martyred on April 23, 303 in Nicomedia, Bithynia, Roman Empire [3, 4] for his beliefs, according to early accounts he came from Cappadocia, Central Turkey or even Nubia [5].

The legend of St. George killing the dragon is first recorded in an 11th-century Georgian manuscript kept at the Patriarchal Library in Jerusalem. Although the legend of St. George killing the dragon is considered to be fictitious [1], being included in the apocryphal writings by the Council of Nicaea [6] it has become very popular in both Western Europe - after being translated and printed by Claxon in The Golden Legend [2], as well as in Eastern Europe [7].

Saint Great Martyr George is considered one of the most important martyrs of the churches in Western and Eastern Europe, recognized and appreciated globally [5], he represents the personification of the ideals of the Christian chivalry [2], and the oldest representation of St. Great Martyr George dates from the 10th century, is on the facade of the Armenian Holy Cross Church on

Akdamar Island and depicts Saint George killing a man and not a dragon. The name St. George spread both in onomatology and in toponymy [8], he being also protector of cities, countries or even protector of the army [10].

From the 11th century, the Holy Great Martyr George is one of the military saints who are associated with the killing of the dragon with a spear. He is depicted on a white horse, killing the dragon that is the medieval symbol of evil [9, 11] and thus saving the princess sentenced to death. In Romanian mythology, the dragon is a snake with legs and wings [12], and in Russian and Greek iconography it is represented in small dimensions, the emphasis being obviously on the strong, victorious and savior image of St. George.

The revetment of icons with precious-metal sheets began in Constantinople, and in Russia it played a special role in the development of religious art. The decoration of the icons with oklads represents a connection between the brilliance offered by the material used in their realization (silver, gold or precious stones) and holiness [13].

2. Materials and Methods

2.1. Analysis of the two sets of icons oklads – front and back

Two sets of icons oklads were analyzed: St. George (Ro), with Romanian markings (Figure 1) and St. Martyr George (Russian), with Russian markings (Figure 2). On the surfaces the icons oklads is represented the scene of the dragon's murder by the Saint Great Martyr George. From the iconographic point of view, he is represented as a young man with curly hair, up to his ears, without a beard [10], dressed in armor specific to Romanian soldiers [5]. The image of the armors of the Romanian soldiers was used both in the case of the icon oklad that was supposed to be of Russian origin, and in that of the icon oklad that has stamped the mark "12", specific to some Romanian icons oklads [14, 15, 16].

Figure 1. Saint George (Ro), with Romanian markings, general images – front and back.

From the time of Peter the Great, Russian official religiosity has experienced a strong development based on elements specific to the Roman Empire. Both the images of saints and the portraits of tsars were attributed complex symbolic structures that had their origin in the "secondary mythology" of antiquity [17]. In turn, Romanian iconography was strongly influenced by the religious art of Kievan Rus' [18].

Figure 2. Saint Great Martyr George (Rus), with Russian markings, general images – front and back.

2.2. Significance of documents and markings

The identification of St. Great Martyr George was based on the incised inscriptions (in Slavic letters) on the two oklads and the specific iconography. Both oklads are made up of two detachable elements - the oklad and the riza. The riza has no punches or inscriptions, instead the actual oklads, in addition to the inscriptions regarding the identity of the saint represented also have a series of markings, specific to the oklads of Romanian and Russian icons, respectively.

In the upper-central part of the icon oklad of Saint George (Ro) with Romanian markings the inscription "С. ГЕОРГІЕ" (St. George) is incised, and in the upper-central part of the second icon oklad of Saint Martyr George (Russian) with Russian markings is incised the inscription "С В МУ ГЕОРГЇИ" (St. Martyr George).

In the central-lower part of the Saint George (Ro) icon oklad, two adjacent markings are stamped (Figure 3a): the first mark represents the monogram of the master controller or of the silver master "КЛ", and the second one signifies the standard silver title "12".

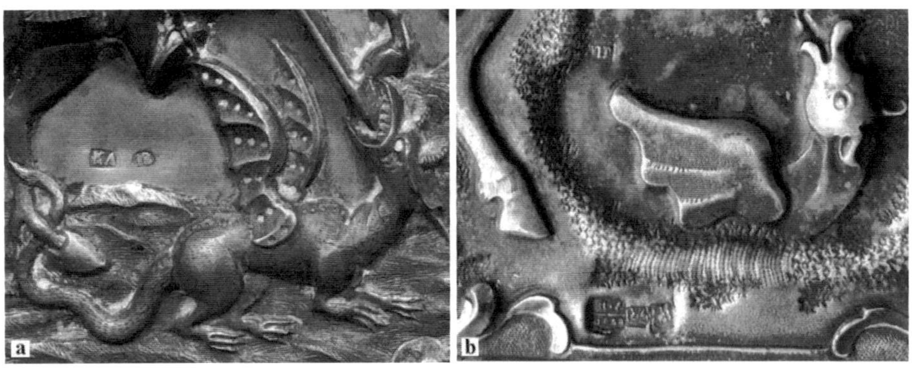

Figure 3. a) Saint George (Ro), with Romanian markings and b) Saint Martyr George (Rus), with Russian markings (detail).

The silver title "12" (12 löthige) was used in Germany between 1600 -1800 [18], and from 1809 in the Hungarian District of Austria [19]. This "12" is found stamped on icons oklads from Romania [2, 21] and corresponds to 750/1000 parts pure silver, the equivalent of 75% pure silver and 25% copper or copper and other metals.

In the central-lower area of the icon oklad of Saint Martyr George (Russian) there are four types of markings: the marking with the initials of the craftsman "ПГ" (located above the series of three adjacent marks); the verification mark with the year of verification and the initials of the master controller "И ·A / 1858"; the marking of the silver title "84" and the mark of the city (in this case the coat of arms of Moscow).

On the Russian territory, starting with May 1, 1798, the silver title of "84" (84 zolotnik Russian silver) became a mandatory standard [22], and from the 19th century the revetment of icons was made by mechanical techniques such as lamination and punching [23, 24]. These techniques were also used in the manufacturing of Russian icons oklads with the representation of The Madonna and Child, probably made at the end of the 19th century [25].

Silver title "84" represents 875/1000 parts pure silver [21] equivalent to 87.5% purity of silver, and percentage represents the equivalent of 87.5% pure silver and 12.5% copper or other metals [25]. In the 18th and 19th centuries, especially in the northern part of the country, the Russian icons oklads are common [18, 27].

The methods of analysis and investigation were used to discover the differences between the materials used to make these icons, especially since one of them raised suspicions about it.

Thus, X-ray Fluorescence Spectrometry (XRF) was used, which was performed with the Oxford X-MET 5000 Portable XRF Analyzer (Oxford X-MET 5000 Handheld XRF Analyzer). Method parameters: RX tube current 10 μA, high tube voltage 40 kV, XRF SiPIN detector, dead time 10u S, analysis time 15 S. Type of calibration used: 1-EC (Empirical Calibration) - Calibration with standards, where the matrix allowed; 2-FP (Fundamental Parameter) - General calibration without standards. The spectrometer was previously checked with standards GBW 07236 and 316, Skyray instrument co., Ltd, TRH-801 for Pb in painting and Fe, Cr, Ni, Cu in metal alloys respectively.

The SEM-EDX investigations, made on a polished sample, were performed on a Phillips XL-30-ESEM electron microscope using a Quanta Inspect F50, with a field emission gun (FEG) with 1.2 nm resolution and an Energy Dispersive X-ray Spectrometer (EDXS) having 133 eV resolution at MnKα.

3. Results and discussions

Following the XRF analyzes (Tables 1 and 2), a surprise appeared when the result for the Russian icon was displayed; the concentration of Ag was too low than that indicated by the mark inscribed on the punch.

Following some metallographic analyzes the thickness of the upper layer was determined [21]. Metallographic analyzes were performed using an optical microscope Reichert Univar, for this a sample of about 2 mm^2 was taken from an unimportant area (near the pin hole which supports the icon oklad) [21].

Investigations on the metallographic microscope reveals that the sample has a uniform layer (of about 25 μm) deposited on the base material. It was also observed that its microstructure is composed of polyhedron grains with different shades and crystallographic orientation which represent the typical structure of copper and copper alloys.

In this paper we continue the research to find out exactly the types of materials used to make icons oklads with the representation of Saint Great Martyr George. The XRF analysis in order to determine the chemical compositions was done on both sides of the icons oklads, namely the icon oklad of Saint George (Ro) and icon oklad of Saint Martyr George (Rus). The performed by means of XRF analysis are presented in Tables 1 and 2.

Table 1. Chemical composition analysis (XRF) of the icon oklad of St. George (Ro) – front and back.

Side	Chemical composition, %			
	Ag	Cu	Zn	Pb
Front	71.9175	26.5053	0.7410	0.8359
Back	83.9011	15.0388	0.3609	0.6991

Table 2. Chemical composition analysis (XRF) of the icon oklad of St. Martyr George (Rus) – front and back.

Side	Chemical composition, %			
	Cu	Ag	Zn	Co
Front	74.1969	25.5789	0.2240	-
Back	96.8998	3.0026	-	0.0389

After the XRF analyses, the sample were investigated by Scanning Electron Microscopy (Figure 4) in which the morphological aspect of the support material and the size of the deposited layer are observed and Energy-dispersive X-ray spectroscopy (Table 3) to determine more precisely the chemical composition of the two layers.

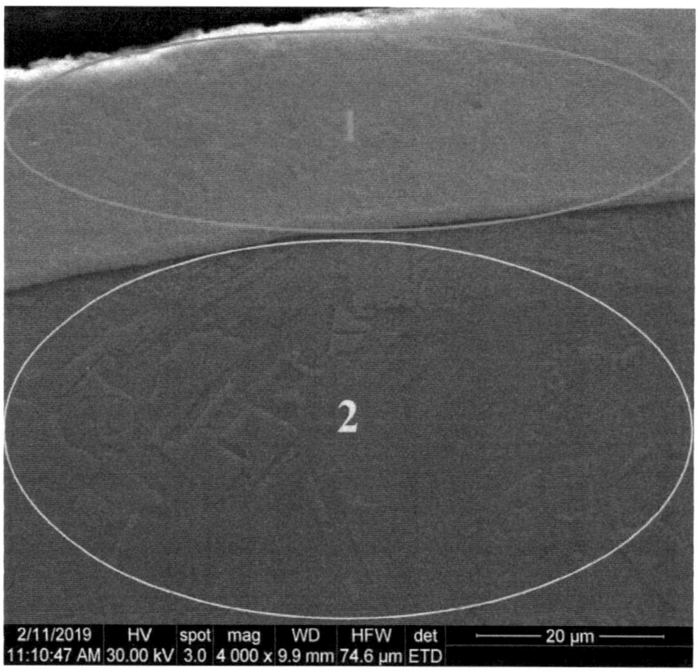

Figure 4. Electron microscopy image (x4000) for the Russian icon oklad.

The EDX analysis on the sample (Figure 4) was made for the both layer:
- EDX1 which belongs to the silver layer deposited on the copper substrate;
- EDX2 for the copper substrate.

IOP Conf. Series: Materials Science and Engineering **877** (2020) 012047 doi:10.1088/1757-899X/877/1/012047

Table 3. EDX analysis of the icon oklad of St. Martyr George (Rus).

Analysis	Elem	Wt %	At %	K-Ratio	Z	A	F
EDX 1	AgL	96.23	93.80	0.9493	0.9969	0.9897	1.0000
	CuK	2.78	4.60	0.0259	1.0649	0.8760	1.0000
	ZnK	0.99	1.60	0.0096	1.0710	0.9007	1.0000
	Total	100.00	100.00				
EDX 2	AgL	12.23	7.59	0.0907	0.9368	0.7912	1.0000
	CuK	84.96	89.53	0.8402	1.0071	0.9819	1.0000
	ZnK	2.81	2.88	0.0281	1.0110	0.9893	1.0000
	Total	100.00	100.00				

The SEM-EDX electron microscopy analyses, performed on a slightly polished sample which led to the removal of the electrolytically deposited silver layer, are captured in Figures 5 and 6.

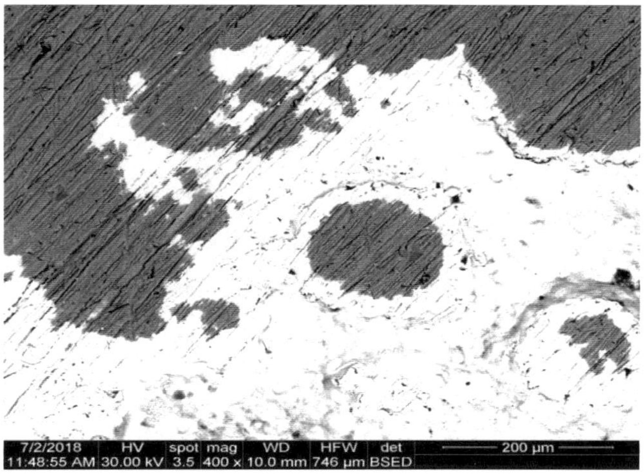

Figure 5. Electron microscopy image (x400) for a polished area of the Romanian icon oklad. The specific aspect of the polishing process is observed.

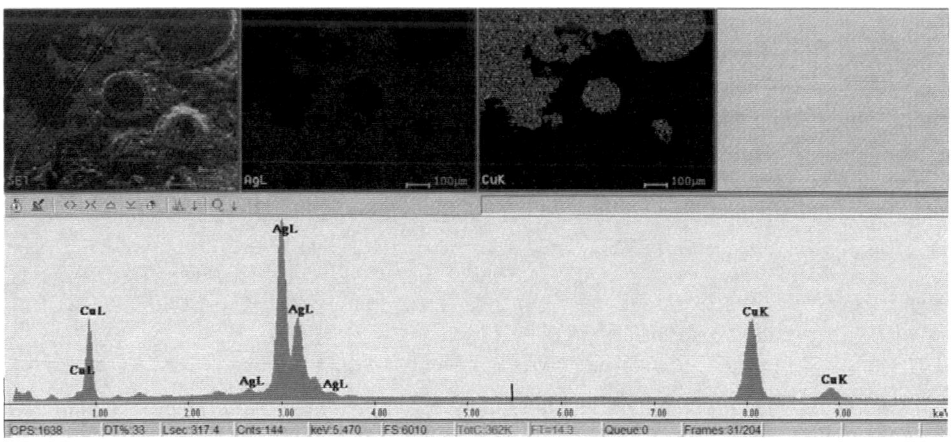

Figure 6. Sample investigation with the overall distribution of the elements per area analyzed - Romanian icon oklad.

According to the SEM-EDX analysis for the selected micro-zone, the distribution of each element is observed, with silver in high concentration. Then, the XRF analysis was repeated on a section from which the silver oxide layer (Figure 7a and Table 4) and the Ag layer (Figure 7b and Table 5) were removed.

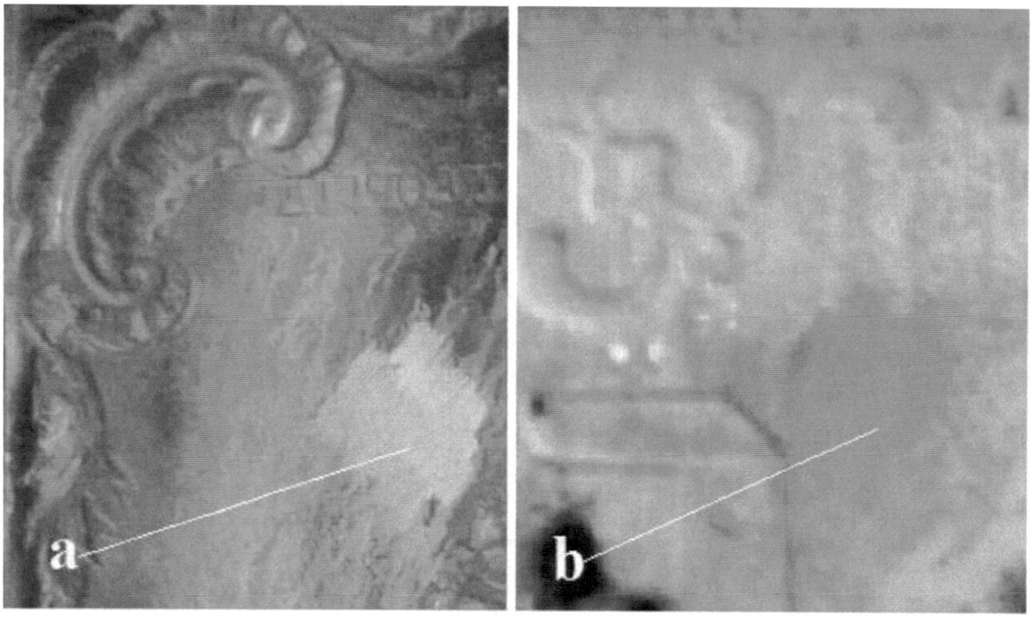

Figure 7. Images of polished areas in order to remove the surface layer: a) Romanian icon oklad; b) Russian icon oklad.

Table 4. XRF results on the icon oklad of St. George (Ro) – back.

Side	Chemical composition, %			
	Ag	Cu	Zn	Pb
Back	70.0951	28.2537	0.6939	0.9571

Table 5. XRF of the icon oklad of St. Martyr George (Rus) – back.

Side	Chemical composition, %			
	Cu	Ag	Zn	Co
Back	96.8998	3.0026	-	0.0389

The silver color remaining after polishing the layer of Romanian icon oklad leaves no doubt. The XRF analysis confirmed that it is a plastic deformed silver plate with a silver concentration between 70-80%, which corresponds to the inscribed mark.

After polishing the layer of Russian icon oklad was observed a reddish color that leaves no doubt. The XRF analysis confirmed that it was a copper substrate plate.

Following the investigations, it was found that the icon oklad is made of copper, which was then plated with silver.

4. Conclusion and perspectives

The two icons oklads have similar iconographic features (the representation of St. Great Martyr George killing the dragon with a spear).

On both oklads are silver title markings. In the first case, the silver title "12" is stamped, which is found stamped on a large part of the Romanian icon oklads, and in the second case, icon oklad with the silver title "84", specific to Russian silverware workshops, have been circulating on Romanian territory since the 18th century.

The oklad with Romanian markings approaches the value of the silver title standard "12" which represents 750/1000 purity of silver, the percentage equivalent of 75% pure silver and the remaining 25% copper and / or other metals.

In the case of Russian oklad markings - with the stamp of the silver title "84" which would correspond to the values of the standard 84 zolotnik Russian silver which represents 875/1000 silver parts and which is the percentage equivalent of 87.5% pure silver and 12.5% copper or other metals, following metallographic analyzes, XRF and SEM-EDX investigations showed that it was a copper plate on which electrolytic Ag was deposited; the presence of markings specific to the Russian school on the oklad proves the intention to be false.

Acknowledgments

The work has been funded by the Operational Programme Human Capital of the Ministry of European Funds through the Financial Agreement 51668/09.07.2019, SMIS code 124705.

5. References

[1] Watkins B 2015 *The book of saints: a comprehensive bibliographical dictionary*, Bloomsbury Academic; Bloomsbury T&T Clark US
[2] Farmer D H 2004 *The Oxford Dictionary of Saints*, Oxford University Press, USA
[3] Foakes F J 2005 *A History of the Christian Church: From the Earliest Times to A.D. 461*, Cosimo Press, USA
[4] Ball A 2005 *Encyclopedia of Catholic Devotions and Practices*, Our Sunday Visitor Inc., USA
[5] Morgan G 2006 *St. George: Knight, Martyr, Patron Saint and Dragonslayer*, Pocket Essentials; 1 edition, USA
[6] De Voragine J, 2012 *The golden legend: readings on the saints*, Princeton University Press, USA
[7] Kuehn S 2011 *The Dragon in Medieval East Christian and Islamic Art*, BRILL, USA
[8] Cartojan N 1938 *Popular books in Romanian literature. The Age of Greek Influence* **2** (Bucharest: King Carol II Foundation for Literature and Art)
[9] Guiley R E 2001 *The encyclopedia of saints*, Visionary Living Inc., USA
[10] Dionisie din Furna 2000, *The ermine of Byzantine painting* (Bucharest: Sophia Publishing House)
[11] Olinescu M 2001 *Romanian mythology* (Bucharest: Saeculum I.O. Publishing House)
[12] Pamfile T 1916 *Romanian mythology I. Enemies and friends of man* (Bucharest: Socec & Comp. Bookstores)
[13] Cormack R 2007 *Icons*, The British Museum Press, UK
[14] Duciu S S 2013 *Testimonies of Orthodox Art and Culture from the Diocese of Tulcea 17th-20th century* (Craiova: Universitaria Publishing House)
[15] Duciu S S 2014 *Testimonies of Orthodox Art and Culture from the Diocese of Alexandria and Teleorman 17th-20th century* (Craiova: Universitaria Publishing House)
[16] Bivol L R, Vasile E, Gheorghe D, Ghica V G, Buzatu M, Petrescu M I, Kovács T A and Iacob G 2019 U.P.B. Sci. Bull., Series B **81(3)** 213-224
[17] Tarasov O 2002 *Icon and Devotion: sacred spaces in Imperial Russia*, Reaktion Books, UK
[18] Nicolescu C 1973 *The art of precious metals in Romania* (Bucharest: Publishing House)

[19] Smith S L, *European Silver Quality Markings. The Beer Stein Journal*, Nov.1994. Accesed June 24, 2018 from: https://www.beerstein.net/articles/bsj-2c.chtm
[20] Markham C A 1898 *Handbookto foreign hall marks on gold and silver plate: (With the exception of those in French plate)* Containing 163 stamps (London: Gibbings and Co.)
[21] Bivol L R, Ghica V G, Buzatu M, Petrescu M I, Iacob G, Vasile E, Gheorghe D and Kovács T A 2019 *IOP Conference Series: Materials Science and Engineering* **572** 012097
[22] Postnikova-Loseva M M, Platonova N G and Ulyanova B L 1995 *Gold and silver craft in the 15-20th centuries: the territory of the USSR* (Moscow: Science Publishing)
[23] Simion V 2014 *Pro Patrimonio - Romanian old art studies* (Bucharest: Basilica Publishing House)
[24] Andryushchenko A I 1904 *The guide on gold and silver craftsmanship* (Nizhny Novgorod: Ripol Classic Publishing House)
[25] Bivol L R, Ghica V G, Buzatu M, Petrescu M I, Iacob G, Vasile E, Gheorghe D and Kovács T A 2018 *IOP Conference Series: Materials Science and Engineering* **374** 012076
[26] https://en.wikipedia.org/wiki/Silver_standards [20.03.2020]
[27] Petrescu S 1977 *Treasures from Neamțu and Secu* (Bucharest: Socec & Co. Workshops)

Landfill Waste Fire Effects Over Town Areas Under Rainwaters

I Cocean[1], M Diaconu[1], A Cocean[1], C Postolachi[1] and S Gurlui[1*]
[1]Alexandru Ioan Cuza University of Iasi, Faculty of Physics, Atmosphere Optics, Spectroscopy and Laser Laboratory (LOASL), 11 Carol I Bld. 700506 Iasi, Romania

*E-mail: sgurlui@uaic.ro

Abstract. Identifying a dark brown color of the rainwater deposited on the streets after a landfill waste fire, the color persisting few months after the fire, an analysis of water samples collected from the streets after rainfall reveled characteristic compounds as for those resulted from urban waste combustion. A simulation of burning waste materials specific to be thrown on the landfill was conducted in the laboratory and the resulting ashes have been analyzed and compared with the dry material obtained from the rain waters collected. FTIR spectra of the two samples proved noticeable similarity among them, evidencing very toxic components, both organic and inorganic, among which dioxins and furans, sulfonic compounds, imines, imides, aromatic and aliphatic nitro and nitroso compounds, HCN and HNO3. Significant fluorescence intensity was determined for the rainwaters using LIF method and peaks specific to dioxin fluorophores were obtained in the resulted spectrum. Therefore, persistent pollution generated by garbage dumps fires is proven herein to affect both air and soil with the possibility to also contaminate town water supply, raising a major threat for human's health, as well as for vegetation and all livings.

1. Introduction

Identifying a dark brown color of the rainwater deposited after a landfill waste fire, the color persisting few months after the fire, an analysis of water samples collected from the streets after rainfall reveled characteristic compounds as for those resulted from urban waste combustion. It has been proved a long time source of soil level contaminants, but not limited to that, as it has been proved before pollution circuit air – rain – soil – water – back to the air [1, 2]. Previous studies and reports warned of the danger of inducing major pollution by the burning of waste, especially landfill waste, which is a mixture of materials and substances that lead to the formation of high toxicity chemical compounds by combustion, among which, dioxin and furans being by far one of the most dangerous due to their carcinogenic effect [3-7]. Scientists have been concerned to finding methods to reduce and neutralize such emissions [8, 9]. Emissions of other very dangerous chemical species due to their toxic effects, such as cyanates and cyanides, including HCN, have been reported in forest or vegetation wildfires. Taking in consideration that landfill waste is a mixture of materials and substances that reflect the waste incineration mills, only being open fires, but also being associated to hazardous fires of vegetation, the magnitude of the damages to environment due to pollutants is highly enhanced. The study presented herein is aimed to evaluate long time damages of landfill waste fire, its persistence and extensive spread under rainwater effects in atmosphere and on soil. A landfill waste fire in Romania, which had lasted for two weeks before being extinguished, proved to be a long time source of soil level contaminants. Dark brown color of the rain waters on the streets persisted after few months from the fire episode and its strong smell obviously denoted extremely high contamination.

2. Methods and materials

Samples of rainwaters were collected from the streets after few months from the landfill waste fire and analyzed. The sample of water was dried on a glass slab and the resulted solid phase was grated out and then analyzed with FTIR spectroscopy to identify chemical groups and chemical species assessed to those. In order to receive as much accurate information, the landfill waste fire was simulated in laboratory, burning different materials together. The resulted ashes and incompletely combusted materials were washed with water and the resulted residual waters were processed for FTIR analysis after that using the same as proceeded with the rain waters collected from the streets. Three combinations of materials were used in the simulations:

S1: polystyrene, wool, cardboard, wrapping foil for aliments, cotton, plastic, paper, wood;

S2: wool, cardboard, wrapping foil for aliments, cotton, plastic, paper, wood, polyacrylonitrile, leaves, pectin, detergent, earth, which means (S1 – polystyrene) + (polyacrylonitrile, leaves, pectin, detergent, earth);

S3: wool, cardboard, wrapping foil for aliments, cotton, plastic, paper, wood, polyacrylonitrile, leaves, pectin, detergent, earth, aluminum foil, which means S2 + aluminium foil. Furthermore, the water collected from the street of Horezu (Horezu sample) was analyzed with laser induced fluorescence (LIF) as well as UV-Vis absorption.

3. Results and discussions

3.1. Fourier transform infrared spectroscopy (FTIR)

The source of pollution detected on the streets of Horezu is proven the landfill waste fire by the good similitude between the FTIR spectrum of the rainwaters collected from the streets of Horezu (Horezu sample) and the two simulation samples S2 and S3, as seen in Figure 1 (b).

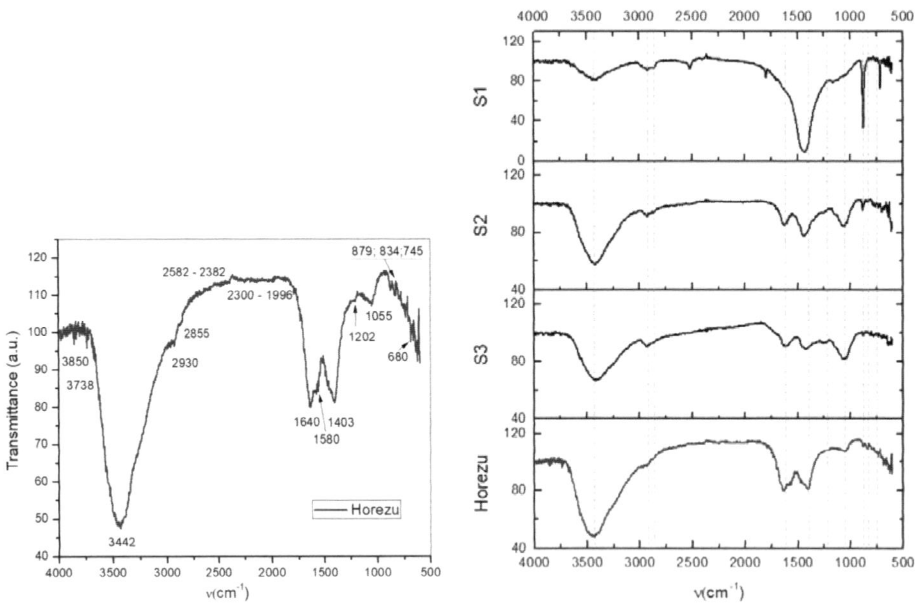

(a) (b)

Figure 1. FTIR spectrum of: (a) waters collected from the streets of Horezu after rain (*Horezu*); (b) FTIR compared spectra of waters collected from the streets of Horezu after rain (*Horezu*) and residual waters after washing the burnt mixtures as described before (*S1, S2, S3*).

IOP Conf. Series: Materials Science and Engineering **877** (2020) 012048 doi:10.1088/1757-899X/877/1/012048

Based on the FTIR spectra of the Horezu sample (Figure 1a), also compared to the S1 and S2 simulation samples (Figure 1b), aliphatic and aromatic structures were identified, as well as the functional groups attached to them. Thus, the composition of the pollutants from the Horezu sample could be evaluated, as well as the sources, respectively the waste materials from the landfill waste that generated them because of the combustion. The results are centralized in Table 1.

Table 1. FTIR spectrum bands interpretation.

Component	Spectrum bands (Figure 1) and associated groups [10, 11]	Sources in the waste combustion
aliphatic groups $-CH_2-CH_2-$ **aromatic** and **heteroaromatic** groups 	2930 cm^{-1}, 2855 cm^{-1} **C-H aliphatic stretching vibrations** **Aromatic C-H stretching** in the range 3080 – 3030 cm^{-1} overlapping with the wide band at 3442 cm^{-1}; 1580 cm^{-1} **aromatic C – C stretching, ring skeleton vibration;** 1202 cm^{-1} **aromatic C – H in plan deformation;** 680 cm^{-1} **C – H out of plan deformation**	
(aromatic) **dioxin** (2,3,7,8-tetrachlorodibenzo-p-dioxin) 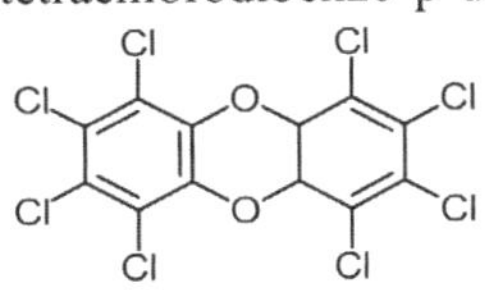 and (heteroaromatic) **furans** (oxacyclopentadiene) 	1055 cm^{-1} and 879 cm^{-1}: **C – O – C symmetric and asymmetric stretching**; 834 cm^{-1} **C-Cl stretching** 745 cm^{-1} **heteroaromatic C – H out of plan deformation**	Cardboard, paper, food wastes combustion, as well as combustion of other materials with aromatic, phenolic structures
(heteroaromatic) **pyrrole** 	3442 cm^{-1} **NH stretching vibration** free and H-bonded; 745 cm^{-1} **C – H out of plan deformation**	Plastic, dyestuffs, detergents combustion
(heteroaromatic) **pyridine** 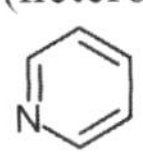	1640 cm^{-1} **C = N stretching vibration**;	Polyacrylonitrile
(aliphatic and aromatic) **imine** 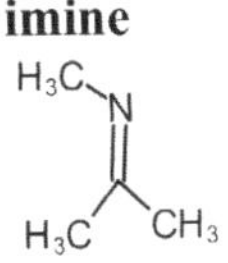 and **oxime** (aldoxime -1 and ketoxime - 2) 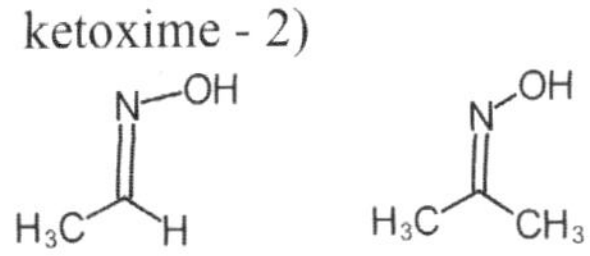	1640 cm^{-1} **C = N stretching vibration**; 3442 cm^{-1}; 3738 cm^{-1}; 3850 cm^{-1} **O – H stretching vibrations**	Leaves combustion

Component	Spectrum bands (Figure 1) and associated groups [10, 11]	Sources in the waste combustion
(1) (2)		
(aliphatic and aromatic) **Sulfoxides** 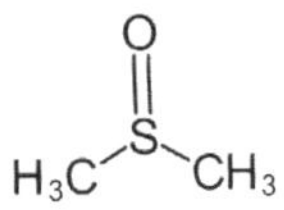 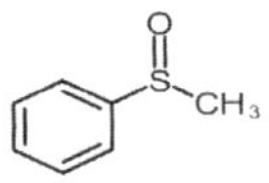 and **sulfones** 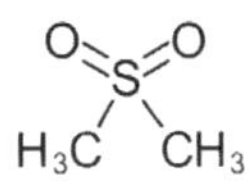 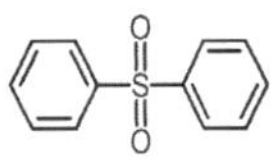	1055 cm^{-1} **S = O stretching;** 1403 cm^{-1} **- SO_2 stretching asymmetric**; in the same range **R – SO_2 – hal** 1202 cm^{-1} **- SO_2 stretching symmetric**; in the same range **R – SO_2 – hal** 879 cm^{-1} **S – O stretching**;	Detergents, bleachers combustion
(aliphatic and aromatic) **thiols** and **sulfides** 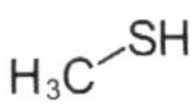	2582 cm^{-1} – 2382 cm^{-1} **S – H stretching vibration** 879 cm^{-1} **S – H deformation vibration**; 745 cm^{-1} **C – S stretching vibration**	Wool combustion
(aliphatic and aromatic) **cyanates, cyanides** and $\mathbf{OC \equiv N;\ C \equiv N}$ **dyazo**- components $\mathbf{-N^{+} \equiv N}$	2300 – 1996 cm^{-1} $\mathbf{OC \equiv N;\ C \equiv N; -N^{+} \equiv N}$ **stretching;**	Polyacrylonitrile, organic dyes
(aliphatic and aromatic) epoxides and peroxides 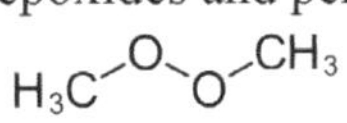	3442 cm^{-1} **O – O – H stretching**; 1202 cm^{-1} **C – O – O stretching**; 834 cm^{-1} **O – O stretching**.	Bleachers, paper
(aliphatic) **formates** 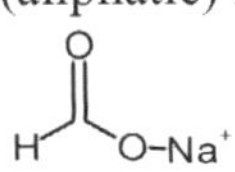 and (aromatic) **benzoates** 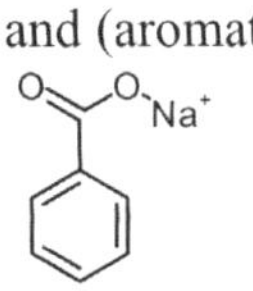	1580 cm^{-1} **$(COO)^{-}$ stretching asymmetric**; 1403 cm^{-1} **$(COO)^{-}$ stretching symmetric**; 745 cm^{-1} **$(COO)^{-}$ deformation vibration in formates**; 680 cm^{-1} **$(COO)^{-}$ deformation vibration in benzoates**	Textiles (wool, PAN, cotton), processed wood, plastics, detergents

Based on Table 1, vibrational bands of aliphatic, aromatic and heteroaromatic groups are evidenced, as well as other specific groups added to the first ones, according to which the substances that resulted from combustion are identified. Thus, the **aliphatic** groups are denoted by the bands at 2930 cm^{-1}, 2855 cm^{-1} for C-H stretching vibrations. Bands in the **aromatic** range C-H stretching vibrations may be overlapping with the 3442 cm^{-1} very broad band. Aromatic C-C stretching vibration in 1590 cm^{-1} – 1560 cm^{-1} range and aromatic C-H in plan deformation in the range 1250 cm^{-1} – 950 cm^{-1} and C – H out of plan deformation in the range 900 cm^{-1} – 650 cm^{-1} confirm the aromatic structures. Ring skeleton in the 1590 cm^{-1} - 1560 cm^{-1} range is also specific to **heteroaromatic** compounds (dioxin, furans, pyridines, pyrrole, and others) denoted by the bands in the range 990 cm^{-1} – 725 cm^{-1} of the C – H out of plan deformation in heteroaromatic groups. In addition to the ring skeleton vibration, the heteroaromatic species are also evidenced by specific functional groups vibrations, namely **pyrrole** structure is evidenced by the 3442 cm^{-1} band of NH stretching vibration,

pyridine by 1640 cm^{-1} band of C = N stretching vibration; and **dioxin** by C – Cl stretching vibration at 834 cm^{-1}. Other chemical species denoted by their specific groups are **imine** and **oximes** with C = N stretching vibrations at 1640 cm^{-1}, **oxime** being also denoted by the 3442 cm^{-1}, 3850 cm^{-1} and 3738 cm^{-1} bands of O – H group stretching. Bands at 1055 cm^{-1}, 1403 cm^{-1}, 1202 cm^{-1}, 879 cm^{-1} are evidence of **sulfoxides and sulfones**. Thiols and sulfides specific bands are indicated by the bands in the range 2582 cm^{-1} – 2382 cm^{-1} and 745 cm^{-1}, while **cyanates and cyanides** are indicated in the range 2300 cm^{-1} – 1996 cm^{-1} of weak bands. Testing with gas sensor the emissions of rainwaters collected from the streets of Horezu (Horezu sample), HCN was detected. The bands of 3442 cm^{-1}, 1202 cm^{-1} and 834 cm^{-1} may also indicate **epoxides and peroxides**. There are also evidences of carboxylates with the 1580 cm^{-1}, 1403 cm^{-1} bands, as **formates** (745 cm^{-1}) and **benzoates** (680 cm^{-1}) [10, 12, 13]. As the band at 1640 cm^{-1} is in the same range with C=O, **formaldehyde** and **urea formaldehyde** could be also in the mixture of chemicals resulted from waste combustion [1]. The difference in the FTIR spectrum of S1 simulated sample compared with the Horezu sample and S1, S2 spectra shows the influence of the added "waste materials". The missing band at 1640 cm^{-1} of C = N vibration in the spectrum of S1 sample shows that imine, oxime and pyridine didn't result from its combustion. Imine and oxime result from leaves combustion and the source of pyridine in S2 and S3 simulation samples is polyacrilonitrile combustion. Both, leaves and polyacrilonitrile, missing from the S1 initial mixture, the result reflects the composition used for each simulation and provides information on the presence of imines, oximes and pyridine in the composition of Horezu sample. In the rainwaters collected from the streets of Horezu (Horezu sample), toxic substances on which the World Health Organization warns were thus detected by the FTIR method. Among all, dioxin and dioxin-like substances are known for its effects on human's immunity system, also producing skin lesions, carcinogenicity and other effects on human health. Hydrogen cyanide and cyanides, are also of high toxicity (oral, respiratory and dermal), as well as pyridines, pyrroles, imines and oximes, sulfoxides, sulfones, thiols, sulfides, cyanates, cyanides, dyazo- components, epoxides and peroxides, formates, benzoates.

3.2. Laser induced fluorescence (LIF)

Laser induced fluorescence using pulsed laser beam of 355 nm wavelength and 10 ns pulse width, at 100mJ energy proved high intensity fluorescence for the rainwaters collected from the streets of Horezu (Horezu sample) after few months from the landfill waste fire with time life of 50 ns. The decay of intensity emission was 10 times within 20 ns, noticed between 30 ns delay and 50 ns delay measurements.

Fluorescence spectrum evidences content of aluminum in the waste material that entered into reaction during combustion. The emission at 487 nm from the spectrum of Figure 2 (a) denotes the fluorophore Al (III) that entered in bonds with organic compounds [14], but also a number of chemical compounds of biological origin, including chlorophyll, but also wood lignin, as well as organic pigments and dyes of biological and/or synthetic origin [15-20]. Secondary peaks are noticed in the emission for both 30 ns delay and 50 ns delay, the emission after 50 ns presenting shifts to lower wavelengths (hypsochromic shift or blue shift). Bands within 406 nm – 593 nm could also indicate benzofurans of different structures as per Jérémie Grolleau et al [21] fluorsecence analysis, despite slight shifts, but as measurements conditions are different, we cannot precisely tell if they are exactly the same structures in the study presented herein. Moreover, the bands shifts between 30 ns delay and 50 ns delay for the same sample show how important the conditions and parameters of the method used for fluorescence analysis are. The shift in the LIF spectra of Figure 2 could denote either tautomeric structures changing during emission, or free radicals formation and other reactions in process, as well as Swan peaks organic carbon generated during laser interaction with the organic compounds in the sample of water [22-25].

IOP Conf. Series: Materials Science and Engineering **877** (2020) 012048 doi:10.1088/1757-899X/877/1/012048

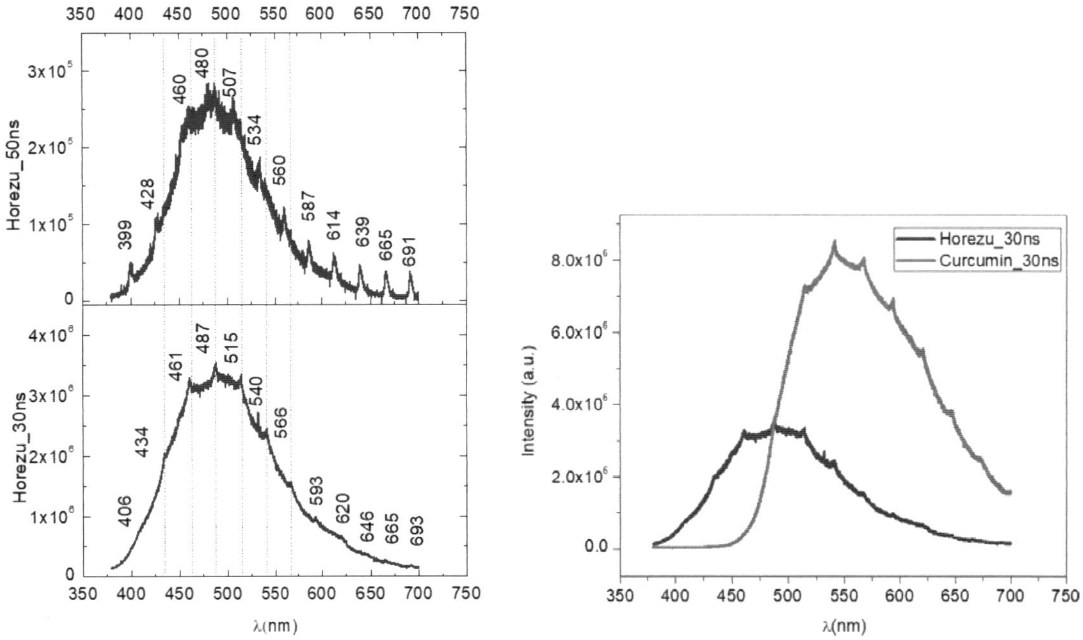

a) b)

Figure 2. Compared fluorescence spectra 30 ns and 50 ns delay of rain waters collected from the streets of Horezu few months after the landfill waste fire (a), compared fluorescence spectra 30 ns delay of rain waters collected from the streets of Horezu few months after the landfill waste fire and curcumin.

3.3. UV-VIS
Absorbance UV-Vis spectrum of the Horezu sample rainwater (Figure 3) denotes dichloromethane with the peak at 233 nm.

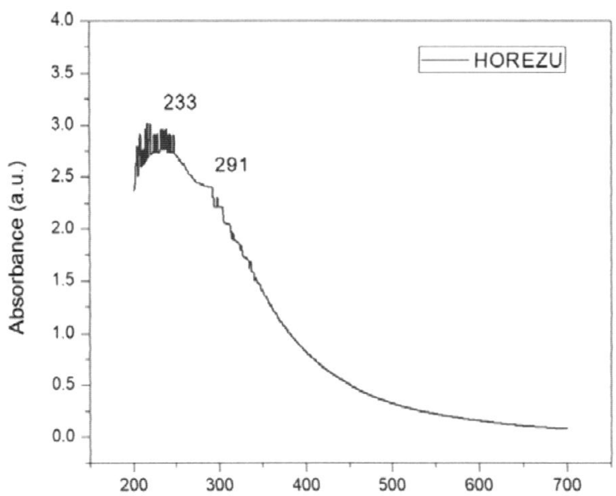

Figure 3. Absorbance UV-VIS spectrum.

3.4. Effects on human health
In the rainwaters collected from the streets of Horezu (Horezu sample), toxic substances on which the World Health Organization warns were thus detected by the FTIR method. Among all, dioxin and

dioxin-like substances are known for its effects on human's immunity system, also producing skin lesions, carcinogenicity and other effects on human health. Hydrogen cyanide and cyanides, are also of high toxicity (oral, respiratory and dermal), as well as pyridines, pyrroles, imines and oximes, sulfoxides, sulfones, thiols, sulfides, cyanates, cyanides, dyazo- components, epoxides and peroxides, formates, benzoates. UV-Vis spectra of the Horezu sample evidenced dichloromethane, a volatile liquid, producing respiratory disorders, known for inducing germ cell mutagenity and carcinogenicity. Adding the flammable properties of some of the identified substances in the Horezu sample, we have the image of the magnitude of the danger when exposing the population of Horezu to the long-term effects caused by the landfill waste fire.

The morphology meaning the shape of volcanic ashes is depended upon a plethora of diverse eruptions and kinematic processes [46, 47].

Eruptions of less viscous basaltic magmas form droplet-like particles of volcanic ashes which are partly controlled through surface tension, the friction of air, an acceleration of the droplets after leaving the vent of a volcano. The shapes are found ranging from a perfect sphere to variously twisted, elongated droplets having even fluidal surfaces [47].

While the morphology of volcanic ashes from eruptions of highly viscous magmas dependent mostly upon the shape of vesicles in the upcoming magma prior to disintegration. The vesicles are produced through the spreading out of gases embodied with magma prior to solidification of the magma. The particles of volcanic ashes can have dissimilar vesicular particles and vesicularity and also enormously high ratios of surface area to volume [46].

The broken vesicle walls are formed owing to concavities, tubes, and troughs observed on surfaces of grain [47]. Angular, vesicular fragments that are pumiceous or thin vesicle-wall fragments are found occurring in vitric volcanic ashes generated from the eruptions of highly viscous magma. Whereas, the lithic fragments of volcanic ash are found angular to sub rounded in shape. Lithic morphology in volcanic ashes is commonly depended upon the mechanical properties of the wall rock which are shattered by the explosive expansion of gases embodied in magma or spalling when it touches the surface.

The morphology of volcanic ash particles generated through phreatomagmatic types of eruptions is managed by forces present in the chilled magma that result in divisions of the glass to yield small pyramidal or blocky particles of glass ash [48].

In the case of phreatomagmatic types of eruptions, merely a trivial role in the determination of grain shape is played by the density and vesicle shape. In this case, the coming up magma is cooled swiftly when coming into contact with surface or groundwater. The principal kinds of pyroclastic shapes of fragments formed due to the forces in the "quenched" magma are (1) vesicular and irregular but having even surfaces; (2) equant and blocky (3) drop-like or sphere-shaped (3) convoluted and moss-like and, (5) plate-like.

4. Conclusion and perspectives

The mixture of materials of high variety of chemical structures led by combustion to a number of chemical species with significant danger potential to the human health, among which dioxins and furans, pyridines, pyrroles, imines and oximes, sulfoxides, sulfones, thiols, sulfides, cyanates, cyanides, dyazo- components, epoxides and peroxides, formates, benzoates. Since the tests were performed few months after the fire episode on the landfill waste of Horezu, it is proven with the analysis results presented herein the persisting toxicity in the atmosphere and on the soil, "recycled" by rain and evaporation phenomena.

5. References

[1] Cocean A, Cocean I, Cazacu M M, Bulai G, Iacomi F and Gurlui S 2018 *Applied Surface Science* **443** 83–90, Doi: 10.1016/j.apsusc.2018.02.156
[2] Cocean I, Cocean A, Iacomi F and Gurlui S 2020 *Applied Surface Science* **499** 142487, Doi: 10.1016/j.apsusc.2019.04.179

[3] Adrian M. Cunliffe, Paul T. Williams 2009 *Waste Management* **29** 739–748, Doi: 10.1016/j.wasman.2008.04.004
[4] Pan Y, Yang L, Zhou J, Liu J, Qian G, Ohtsuka N, Motegi M, Oh K, Hosono S 2013 *Chemosphere* **92** 765–771, doi: 10.1016/j.chemosphere.2013.04.003
[5] Liu H, Kong S, Liua Y and Zeng H 2012 *Procedia Environmental Sciences* **16** 661 – 668, Doi: 10.1016/j.proenv.2012.10.091
[6] Buekens A and Huang H 1998 *Journal of Hazardous Materials* **62** 1–33
[7] Villanger G D, Gabrielsen K M, Kovacs K M, Lydersen C, Lie E, Karimi M, Sørmo E G and Jenssen B M 2013 *Chemosphere* **92** 828–842, Doi: 10.1016/j.chemosphere.2013.04.036
[8] Wu H-L, Lu S-Y, Li X-D, Jiang X-G, Yan J-H, Zhou M-S, Wang H 2012 *Chemosphere* **86** 361–367, Doi:10.1016/j.chemosphere.2011.10.016
[9] Mukherjee A, Debnath B and Ghosh S K 2016 *Procedia Environmental Sciences* **35** 528 – 540, Doi: 10.1016/j.proenv.2016.07.037
[10] Pretch E, Bülmann P and Badertscher M 2009 Structure Determination Of Organic Compounds. Tables Of Spectral Data, Fourth, Revised and Enlarged Edition, *Springer – Verlag Berlin Heidelberg* ISBN **978** – 3 – 540 – 93810 – 1, Doi: 10.1007/978-3-540-93810-1
[11] Lippert T, Wokaun A and Lenoirt D 1991 *Environ. Sci. Technol.* **25**, 1485-1489
[12] Miller F A and Wilkins C H 1952 *Anal. Chem.* **24** (8) 1253–1294
[13] M. Siciliani de Cumis, F. D'Amato, S. Viciani, B. Patrizi, P. Foggi and C L Galea 2013 *Laser Phys*. **23** 025603 (5pp), Doi: 10.1088/1054-660X/23/2/025603
[14] Wang J, L Yuanyuan, L. Kai, M. Xiangru, H. Hongwei 2017 *Chemistry* **23(21)** 5081-5089, Doi: 10.1002/chem.201606024
[15] Maze C H 1995 *Mar. Ecol. Prog. Ser.* **120** 185-191
[16] Chin D P, Shiratori I, Shimizu A, Kato K, Mii M and Waga I 2018 *Scientific Reports* **8** 16556, Doi:10.1038/s41598-018-34837-2
[17] Nazir S and Khan M S 2018 *Advancements in Life Sciences – International Quarterly Journal of Biological Sciences*
[18] Stewart Jr C N, Millwood R J, Halfhill M D, Ayalew M, Cardoza V, Kooshki M, Capelle G A, Kyle K R, Piaseki D, McCrum G and Di Benedetto J 2005 *Journal of Fluorescence* **15** (5), Doi: 10.1007/s10895-005-2977-5
[19] Donaldson L, Radotic K, Kalauzi A, Djikanovic D, Jeremic M 2010 *Journal of Structural Biology* **169** 106–115, doi: 10.1016/j.jsb.2009.09.006
[20] Radotic K, Kalauzi A, Djikanovic D, Jeremic M, Leblanc R M and Cerovic Z G 2006 *Journal of Photochemistry and Photobiology B: Biology* **83** 1–10, doi: 10.1016/j.jphotobiol.2005.12.001
[21] Grolleau J, Petrov R, Allain M, Skene W G and Frère P 2018 *ACS Omega* **3** 18542−18552, Doi: 10.1021/acsomega.8b02384
[22] Cocean I, Cocean A, Postolachi C, Pohoata V, Cimpoesu N, Bulai G, Iacomi F and Gurlui S, 2019 *Applied Surface Science* **488** 418–426, Doi: 10.1016/j.apsusc.2019.05.207
[23] Tsuboi Y, Kimoto N, Kabeshita M and Itaya A 2001 *Journal of Photochemistry and Photobiology A: Chemistry* **145** 209–214
[24] Harilal S S, Lahaye N L and Phillips M C 2017 *Optics Express*, Doi: 10.1364/OE.25.002312
[25] Burns I S and Kaminski C F 2011 *Z. Phys. Chem.* **225** 1343–1366, Doi: 10.1524/zpch.2011.0182

Study of Physico-Chemical Characteristics of Some Major Urban Air Pollutants

S Garofalide[1,2], M Diaconu[1], I Cocean[1], A Cocean[1], V Pelin[1,3], S Gurlui[1*] and L Leontie[1,2]

[1]Alexandru Ioan Cuza University of Iasi, Faculty of Physics, Optical Atmosphere, Lasers and Spectroscopy Laboratory (LOA-SL), Blvd. Carol I, nr. 11, 700506, Iaşi
2Alexandru Ioan Cuza University of Iasi, Institute of Interdisciplinary Research, Integrated Center of Environmental Science Studies in the North - Eastern Development Region (CERNESIM)
[3]Alexandru Ioan Cuza University of Iasi, Geography - Geology Faculty, Geoscience Doctoral School, 20A Carol I Blvd, 700505 Iasi, Romania

*E-mail: sgurlui@uaic.ro

Abstract. Urban air/environmental pollution as a major concern of contemporary civilization, substantially affecting the health of the residents, requires identification of both physico-chemical specificity and generating sources. Particle matter (PM) monitored in the atmosphere should be associated to their chemical composition that allows identification of sources as enhancement contributor and sinks as mitigation factor but also as possible spreading agent as it has been reported in rain fall for some of the pollutants cathegories. After a number of episodes of increased amounts of measured PM 2.5 and PM 10, considerable dust deposits on cars were collected and SEM-EDX analyses were performed, showing the agglomerated morphology and elemental composition of the dust. Sodium aluminate and calcium chloride were identified as major contributors to the dust composition. Such chemical components may end-up into the water supply and tab water as previously reported affecting the quality of the water. As sources, dust-binding solutions for streets cleaning and/or coagulant agents in constructions are to be considered.

1. Introduction

Urban pollution is mainly associated to the highly intense car traffic, with emission of oil combustion chemical compounds in gas, liquid or solid state contaminating the air [1- 5], the soil and waters in a circuit that has been called as "atmosphere self- cleaning" [6] or "wet removal" [3, 7] and the connection of the atmosphere composition with the components deposited on soil has been proven when rain water was analyzed after an abundant rainfall episode following to a drought period of time [6, 8]. Pollutants of organic origin specific to result from fuel combustion, including soot that has being intensly studied due to its negative impact on the environment quality, proved to be a serious matter for the environmental issue as contributors to the particles inducing respiratory disorder [9-14], including from the perspective of particles PM 10 that are sourced both in smoke and dust [15-17].

Assimilating an existing technology of using coagulants or flocculants for cleaning the waters of organic dispersed contaminants [18-21], but also in dust control applications [22], some municipalities have started to apply the procedure for a dry – cleaning process of the streets [23]. The method is

IOP Conf. Series: Materials Science and Engineering **877** (2020) 012049 doi:10.1088/1757-899X/877/1/012049

known as "dust – binding" and it consists in using "calcium chloride liquid solutions" to reduce streets dust emission. Helsinki municipality, after a study in the REDUST Life + project in 2011-2014 as to the cost – effectiveness impact, has used the cleaning procedure, but also Seoul and others. However, problems that may arise from the use of one or the other of the coagulants or flocculants require careful approach and this is the aim of the study presented herein.

During winter time, within 13 – 15 January 2020, the quantity of dust on the streets of Iasi was obviously more than usually. On the cars driving in the traffic of Iasi City, a consistent layer of dust was noticed too. The situation was unusual, such episodes of dusty environment being specific only for when construction activities are carried out in a certain area, or when antiderapants such as calcium chloride mixed with sand are used. In the same period of time, IS1 Monitoring Station of Romanian Environmental Protection Agency (APM – Agentia de Protectia Mediului) measured high levels of PM10.

2. Methods of investigation

After collecting dust from the surface of the car, the sample was analyzed using Scanning electron microscopy coupled with energy dispersive spectroscopy (SEM-EDS) as morphology and elemental composition. Evaluation of the dust origin was performed based on the PM10 and benzene levels registered in the atmosphere in the period of time between 13 – 15 January 2020, and their compared variation at different times.

3. Results and discussions

3.1. Scanning electron microscopy coupled with energy dispersive spectroscopy (SEM-EDS)

Morphological aspect observed in Figure 1(a) denotes agglomerated structures consisting in granular/spherical, fusiform, fibrous and acylar shapes with dimensions varying between less than 150 μm to over 1400 μm (1.4 mm) for granular structures while fibrous/fusiform structures show dimensions over 2000 μm (2 mm) length, as seen in SEM image presented in Figure 1 (a - d).

(a) (45 x) magnitude

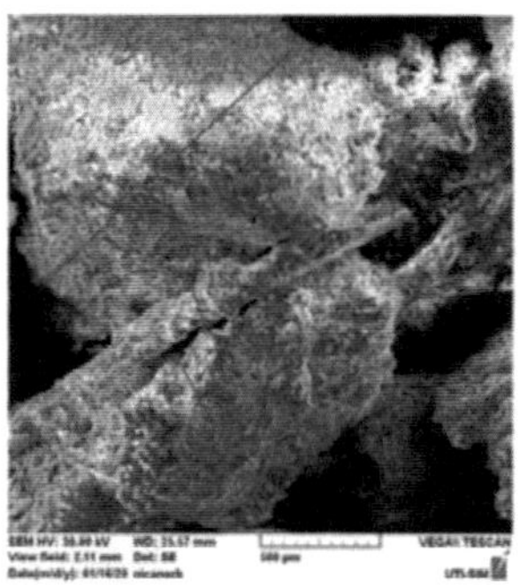
(b) (100 x) magnitude

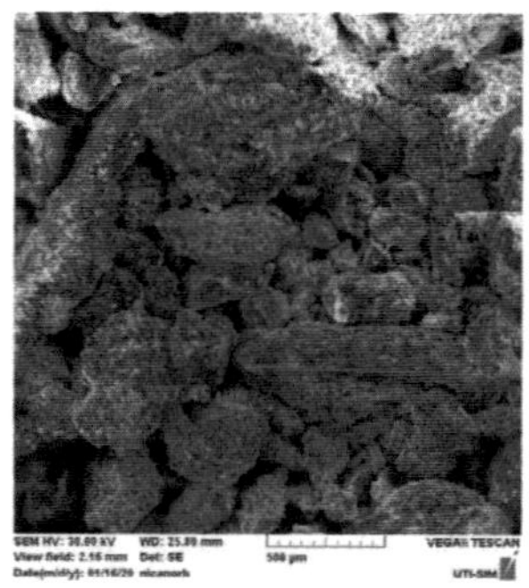
(c) (100 x) magnitude

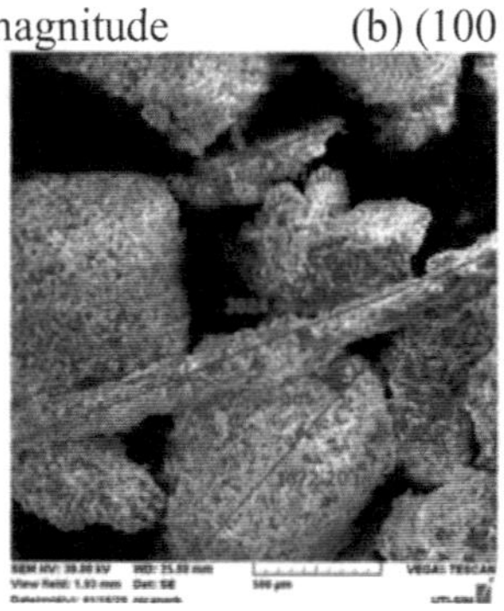
(d) (100x) magnitude

(e) (400 x) magnitude

Figure 1. SEM images of dust sample of different magnitude: 45 x (a); 100 x (b), (c), (d).

Even from the images with magnification orders of 45 x and 100 x, an aggregation and / or flocculation structure had been observed (Figure 1 (a - d)).

In the SEM image of magnifying order 400 (Figure 1 e), the micro-composite structure of each of the granular and fibrous elements seen in the images of Figure 1 (a – d) is clearly identified. The aggregate particles in the composite structures are spheroids and flakes of average sizes from 10 μm to 40 μm in diameter. Based on elemental composition (Table 1), there is evidence about the presence of so called "dust –binders", known as coagulants or flocculants in physico – chemical processes. Such binders provide a micro – continuous phase that is used in water cleaning and recently on dust – cleaning to enhance the particles size the desired effect being to make the particles filterable when cleaning the water or to keep them on soil when it is about dust cleaning.

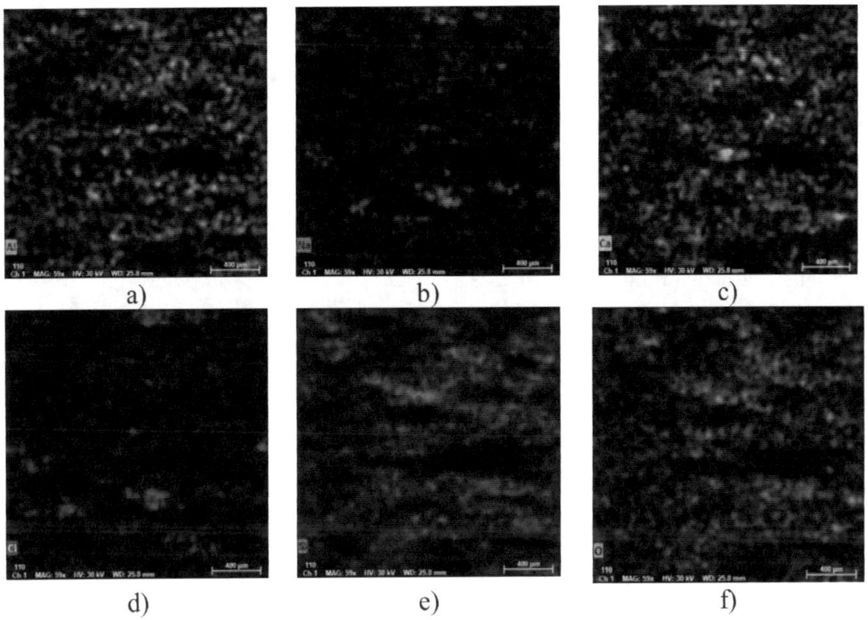

Figure 2. Maps with elemental distribution on a segment of sample of 400 μm: a) Al; b) Na; c) Ca; d) Cl; e) Si; f) O.

Analyzing the elemental composition (Table 1) and elements distribution (Figure 2) performed with energy dispersive spectroscopy (EDS), the flocculent agents, namely sodium aluminates, $NaAlO_2$, $Na_2O{\cdot}Al_2O_3$, or $Na_2Al_2O_4$ or their alkali form in solution, $NaAl(OH)_4$ are denoted, while as for coagulant agents there are indications of calcium chloride, $CaCl_2$, as well as sodium chloride, NaCl. These binding agents develop continuous phase in solution contributing to binding the organic and inorganic particles. Flocculation develops fibrous and acylar formations that are usually attributed to organic components denoted by carbon detected in elemental composition bound with sodium aluminates [6, 24]. Coagulation develops the granular shapes and represents the binding with $CaCl_2$ of the sand particles denoted by the silicon from the elemental composition [22].

Table 1. Elemental composition of the dust sample.

Element	Mass [%]			Atom [%]		
	(1)	(2)	(3)	(1)	(2)	(3)
Oxygen	48.91558	42.10785	52.24393	64.19317	50.53882	62.26211
Carbon	-	13.17617	7.802855	-	21.06569	12.387
Silicon	18.13641	11.64905	18.28044	13.55859	7.964786	12.4107

IOP Conf. Series: Materials Science and Engineering **877** (2020) 012049 doi:10.1088/1757-899X/877/1/012049

Element	Mass [%]			Atom [%]		
	(1)	(2)	(3)	(1)	(2)	(3)
Chlorine	8.98456	12.24241	3.448827	5.320994	6.631064	1.854871
Sodium	8.772009	9.623185	4.196467	8.011425	8.038035	3.480494
Calcium	6.21738	4.808191	5.390304	3.257217	2.303782	2.564478
Aluminium	4.006958	3.121396	3.965323	3.118124	2.221508	2.802229
Iron	2.803692	2.515865	2.498967	1.054085	0.865074	0.853203
Potassium	1.170298	0.755887	1.077129	0.628469	0.371249	0.525293
Magnesium	0.993114	-	1.095764	0.857924	-	0.859633
	100	100	100	100	100	100

3.2. PM10 and benzene analysis

Following the data of the measurements registered by IS1 Monitoring Station of Romanian Environmental Protection Agency (APM), during a period of time when the Municipality of Iasi announced the dust – binding procedure to be used for streets cleaning, and when the most dust was deposited on the cars during traffic or while parking, few aspects are to be noticed. As previously discussed, EDS analysis proved presence of the "binders" as sodium aluminate and calcium chloride, and their effect was notticed in the SEM images as agglomerates and micro-composite structures. The data from IS1 Monitoring Station represented in the plot of Figure 3 show 5 μg/m^3 PM10 and 9 μg/m^3 benzene on January 13, 2020 at 10 am hours, while starting with 12 am hours of the same day, the PM10 concentration in air versus benzene changes, meaning that there is a sudden growth in PM10 measured (8 μg/m^3) compared to benzene (6 μg/m^3).

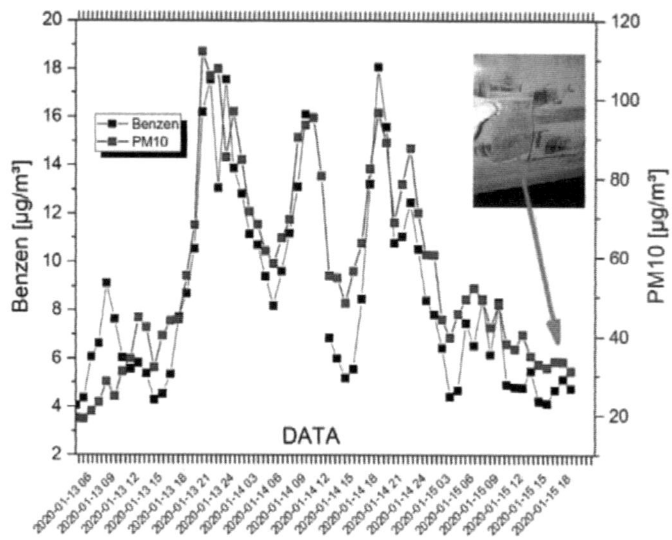

Figure 3. PM10 and benzene in the periode of time 13 – 15 January 2020 as registered by APM (Environmental Protection Agency), IS1 Monitoring Station.

From that time untill January 15, 2020 at 6 pm hours the concentration in PM10 is the predominant one. It could be in relation with the dust – binding process during cleaning performed on the streets of Iasi, meaning that solution of dust – binders had been administrated on the streets on January 13, 2020. Another source of sodium aluminate on the streets could be from constructions activity where the flocculent agent is used in cements for a fast binding of the composite. However, such activity was not in the area. As for calcium chloride, it is also an anti-skid agent, used during winter when mixed with sand. However, the explanation for the IS1 Monitoring Station data measurements may be in absorbing and/or binding the benzen with sodium aluminate, more than the measured values are very close for PM10 and benzene and identical in some points like on January 13, at 6 pm hours, or on January 14, at 9 am hours.

As SEM images show, the 10 μm particles are bound in micro-composite structures due to flocculation and/or coagulation and are kept on the soil but if they are not removed, the micro-composites are crushed due to their low mechanical strength under compression during traffic. Advanced drying of the dust formed composite could be also a cause for the crushing because water is one of the key elements in the flocculation and coagulation process.

4. Conclusion and perspectives

The streets pollution with flocculent and coagulant agents either used with the aim of "keeping the dust at soil", has been proven with the study presented herein as being inefficient due to the low strength of the bindings within the micro – composite structures which are crushed during mechanical actions or by dehydration of the aggregate components involved. That resulted in a high level of the PM10 dispersed in the air and the particles carry organic components such as benzene detected at very close level with PM10 by IS1 Monitoring Station which denotes benzene binding on the particles.

It is obvious that the dust – binding without proceeding with vacuum cleaning after that is not a reliable environmentally safe process to be used on the streets of the cities.

If the binding agents arrived on the streets either accidentally from construction or from use as anti – skid agents, the same immediate vacuum cleaning is required to prevent grinding the micro – composite back into dust - like particles that are trained in the air.

5. References

[1] Cocean A, Cocean I, Cazacu M M, Bulai G, Iacomi F and Gurlui S 2018 *Applied Surface Science* **443** 83–90, DOI: 10.1016/j.apsusc.2018.02.156

[2] Schlager H, Grewe V and Roiger A 2012 Chemical Composition of the Atmosphere, U. Schumann (ed.), *Atmospheric Physics, Research Topics in Aerospace*, (Berlin Heidelberg: Springer-Verlag), Doi: 10.1007/978-3-642-30183-4_2

[3] Smith J N, Dunn M J, VanReken T M, Iida K, Stolzenburg M R, McMurry P H and Huey L G 2008 *Geophysical Research Letters* **35** L04808, Doi:10.1029/2007GL032523

[4] Zhao F and Zeng N 2014 *Earth Syst. Dynam.* **5** 423–439, Doi:10.5194/esd-5-423-2014

[5] Xia X, Che H, Zhu J, Chen H, Cong Z, Deng X, Fan X, Fu Y, Goloub P, Jiang H, Liu Q, Mai B, Wang P, Wu Y, Zhang J, Zhang R and Zhang X 2016 *Atmospheric Environment* **124** 243 – 251, doi.org/10.1016/j.atmosenv.2015.05.071

[6] Cocean I, Cocean A, Iacomi F and Gurlui S 2020 *Applied Surface Science* **499** 142487, Doi: 10.1016/j.apsusc.2019.04.179

[7] Thompson A M and Cicerone R J 1982 *Journal of Geophysical Research* **87** C11 8811-8826

[8] Chameides W L and Davis D D 1982 *Journal Of Geophysical Research* **87** C7 4863-4877

[9] Kumar P and Bohidar H B 2010 *Indian Journal of Experimental Biology* **48** 1037 – 1042

[10] Li Z, Song C, Song J, Lv G, Dong S and Zhao Z 2011 *Combustion and Flame* **58** 1624–1630

[11] Lapuerta M, Oliva F, Agudelo J R and Boehman A L 2012 *Combustion and Flame* **159** 844–853

[12] Wang C, Huddle T, Huang C-H, Zhu W, Vander Wal R L, Lester E H and Mathews J P 2017 *Carbon* **117** 174-181, Doi: 10.1016/j.carbon.2017.02.059

[13] Cain J P, Gassman P L, Wang H and Laskin A 2010 *Phys. Chem. Chem. Phys.* **12** 5189-5198, Doi: 10.1039/B927518E

[14] Vander Wala R L, Bryg V M and Haysc M D 2010 *Aerosol Science* **41** 108 – 117

[15] Colls D J and Tiwary A 2010 Air Pollution: Measurement, Modelling and Mitigation, 3rd edn., (New York: *Routledge*)
[16] Phalen R F and Phalen R N 2013 *Introduction to Air Pollution Science: A Public Health Perspective*, (Burlington: Jones & Barlett Learning)
[17] Soroiu C, Gurlui S O and Bujoreanu C 2019 *Buletinul Institutului Politehnic Din Iaşi* **65** (69) 2
[18] Dominguez J R, Gonzalez Teresa, Garcia H M, Sanchez-Lavado F and de Heredia J B 2007 *Journal of Hazardous Materials* **148**(1-2) 15-21, Doi: 10.1016/j.jhazmat.2007.05.003
[19] Matilainen A, Vepsäläinen M and Sillanpää M 2010 *Advances in Colloid and Interface Science* **159** 189–197
[20] Jiang J-Q 2015 The role of coagulation in water treatment *Elsevier Ltd.*
[21] Contreras C A, Sugita S and Ramos E 2006 Preparation of Sodium Aluminate From Basic Aluminium Sulfate, 2 (August 2006), doi: 10.2240/azojomo0220 *AZojomo* (ISSN **1833**-122X)
[22] Reyier J A 1972 *Comparison between Calcium Chloride and Magnesium Chloride as Dust-binding Agents on Gravel Roads; Examination Project* (Stockholm: The Institute for Road-building, Royal Technical College)
[23] Männikkö J-P, Niemi J, Ritola R, Kupiainen K, Pirjola L, Väkevä O and Virtanen T 2014 *REDUST: Best practices in winter maintenance to reduce respirable street dust* (Report based on reports published during the REDUST (EU LIFE+) project during years 2011–2014), Helsinki, December 2014, ISBN **978**-952-272-794-7; ISBN (PDF) **978**-952-272-795-4
[24] Sweetman A M, Jarvis S P, Sang H, Lekkas I, Rahe P, Wang Y, Wang J, Champness N R, Kantorovich L and Moriarty P 2014 *Nature Communications* **5** 3931, Doi: 10.1038/ncomms4931

Instrumental Analytical Techniques Applied to Old Gate Tower from Corvins' Castle

R M Ion[1,2*], S Tincu[3], I Minca[1], I D Dulama[4], I A Bucurica[4], M L Ion[5] and A I Gheboianu[4]

1 National R&D Institute for Chemistry and Petrochemistry – ICECHIM, Research Group „Evaluation and Conservation of Cultural Heritage", Bucharest, Romania.
2 Valahia University of Târgoviște, Doctoral School of Materials Engineering, Târgoviște, Romania;
3 Corvin's Castle, Hunedoara, Romania.
4 "Atelierul de Creatie" NGO, Bucharest, Romania.
5 Multidisciplinary Science and Technology Research Institute of Valahia University of Târgoviște - ICSTM-UVT, Târgoviște, Romania.

*E-mail: rodica_ion2000@yahoo.co.uk

Abstract. Nowadays, a better understanding of the factors (including environmental) that influence the degradation of materials from different buildings and monuments, is absolutely necessary. In this paper some analytical techniques, as Fourier Transformed IR Spectroscopy (FTIR), X-ray diffraction (XRD), X-ray fluorescence (XRF), optical microscopy (OM) and SEM-EDS, have been used for evaluation the structure and chemical composition and the characterization of the degradation products, causes and mechanisms of disintegration / degradation of materials and aging processes. Few samples from the Old Gate Tower of the Corvins'Castle, Hunedoara, have been collected and subjected to be analysed and fruitful conclusions have been obtained referring to the raw materials provenance and weathering / deterioration processes which they supported. Similar minerals with those identified at the other towers from Corvins'Castle have been evidenced: quartz, calcite and dolomite, being majority, and additional apatite minerals. Also, ion chromatography could be used to identify the presence of different ions, responsible for the accelerated process of the monument's walls.

1. Introduction

Known as one of 10 most beautiful castles in Europe, the Corvins'Castle was built in 1446, over the site of an older fortification, the orders of John Hunyadi (in Hungarian: *Hunyadi János*, in Romanian: *Iancu or Ioan de Hunedoara*). Built in a Renaissance-Gothic style Corvin Castle, the castle is recognized as an imposing structure with eight towers, colourful roofs and many balconies adorned with sculptures and embroidery in stone [1]. The castle is flanked by 4 rectangular towers and 3 circular towers, which is considered an innovation for the military architecture of the 15th century Transylvania [2-4]. The castle has two bridges, two platforms, 42 chambers and is built on an area of 7000 square meters. The main entrance into the castle contains a wooden bridge, sustained by 4 massive piers built of dolomite limestone blocks, with crenels at the top. Also, the external yard was added during the 17th century used for administration and storage [5]. The castle contains 8 towers: 2

of rectangular shape and 6 with circular shape. The rectangular towers (the old gate tower, the new gate tower), are placed in the northwest and southeast areas of the castle.

The Old Gate Tower is located in the south-eastern part and was built during the time of Ioan de Hunedoara. With a rectangular form, with two levels of defense that superimpose the access corridor, it had the last floor provided but battlements and its exterior was covered by a painting with geometric motifs, of German origin. The entrance to the tower was blocked by the construction of the White Tower in the days of Prince Gabriel Bethlen.

The castle in the XVIth century was one of the few civil constructions, built from this part of Europe. Now, the Old Gate Tower has brick and brick vaults with missing bricks, dislocated and the stone floor is damaged.

Consideration is given to sanitation, cleaning, filling, roasting, undoing of cement plaster and repairing them with breathable plaster and paint based on lime, painting restorations, carpentry repairs, wood ceiling restoration works, and overhaul.

Whereas the New gate tower has been partially characterized in our previous studies [6], the Old Gate tower has never been characterized and evaluated from compositional, structural or morphological point of view. From this period there are few civil constructions, identified in the old tower area of the gate on the occasion of the last restoration works. Consideration is given to sanitation, cleaning, filling, roasting, undoing of cement plaster and repairing them with breathable plaster and paint based on lime, painting restorations, carpentry repairs, wood ceiling restoration works, and overhaul. total installations, renouncing gas heating and switching to electric heating.

The purpose of this work is to study some samples taken from the Poarta Veche Tower, Figure 1, in order to identify the origin of the raw materials, possibly from the exploitation of natural resources in this area and to highlight the processes and mechanisms of alteration / deterioration of this tower. For this purpose, sensitive analytical techniques were used such as: FTIR, WDXRF, XRD, OM SEM-EDS and, ion chromatography, and their results will be discussed in detail.

2. Experimental part

2.1. Materials

The samples received from archaeologists are only stone samples, they are degraded, as shown in Figure 3. These samples are samples detached from the tower, without heritage value, they can no longer be reused.

2.2. Equipments

For X-ray diffraction (XRD) analysis, a X-ray diffractometer Rigaku Ultima IV (Cu anode (2 kW) as X-ray tube, detector of NaI) has been used. A PDXL 2.2. (processing) software has been used for data processing.

For X-ray Fluorescence Spectrometry (WDXRF) has been used a Supermini200 - Rigaku as a Benchtop sequential spectrometer, wavelength dispersive with a PC-controlled system and equipped with solid and liquid sample accessories.

For Fourier-transformed infrared spectroscopy (ATR-FTIR) a Vertex 80 spectrometer (Bruker Optik GMBH, Germania) has been used, with DRIFT technique. All the measurements have been recorded over the range 4000-200 cm^{-1}.

For microscopic analysis, the optical microscopy (Primo Star ZEISS, with transmitted light and magnification 4X - 100X) and Scanning Electron Microscopy with Energy Dispersive Spectroscopy (SEM-EDS) (SU-70 (Hitachi, Japan), have been used.

These microscopic techniques allowed both a clear and sensitive image of the samples, and an elementary analysis of them was allowed.

Ion chromatography was performed using a Dionex ICS-1000 ion chromatograph, with an Ion Pac AS14A Dionex anion column, equipped with a Dionex ASRS-ULTRA suppressor. The analyses were performed on a solid sample in water solution.

Figure 1. The Old Gate tower (up) and the aspect of the investigated samples (down).

3. Results and discussion

The Hunedoara region is located in the Southern Carpathian area, and from a geological point of view it is composed of sediment-volcanic units and sedimentary formations such as: limestone, marl, clay shale, conglomerate, sandstone, magmatite (basalt), chert, andesite and pyroclasts [7 8]. From a mineralogical point of view, it contains mostly quartz and dolomite, with K-feldspar, plagioclase, biotite, muscovite, garnet and various opaque minerals. Apart from these, the presence of the apatite is noted, which could be an explanation for the sustainability of this monument as a whole. The apatites are present in this area as minerals that accompany gold deposits and are recognized as good stabilizers and antimicrobial agents [9-11].

The stone from the Old Gate Tower, is differently coloured with aspect that varies from different shades of grey to reddish, due to the deposits of ferrous minerals; also, are visible the alteration changes caused by the weather or possible restoration processes carried out over time and about which not much details are known.

The whole composition of the stone present in this tower, should be evaluated by using different analytical techniques, in order to offer to the restorers, the solution of the new compatible that should be used.

In order to determine the main chemical components from the investigated samples, FTIR analysis have been recorded (Figure 2), and led to the following composition, Table 1.

Table 1. The FTIR bands and assignments for the Old Gate Tower samples.

FTIR band, cm^{-1}	Assignment
1088-1092	quartz
796	Si-O (stretching mode)
777-783	Si-O (symmetrical stretching)
692-694	Si-O (bending modes)
752-758	Feldspars
646	
465-471	
1400	calcite
900–1000	SO_4^{2-} (most probably from gypsum, that could overlap each other in the region)
560 - 600	PO_4^{3}-
1000–1100	(apatite [12])
630	
868	CO_3^{-2}
397	(apatite [12])
1799	

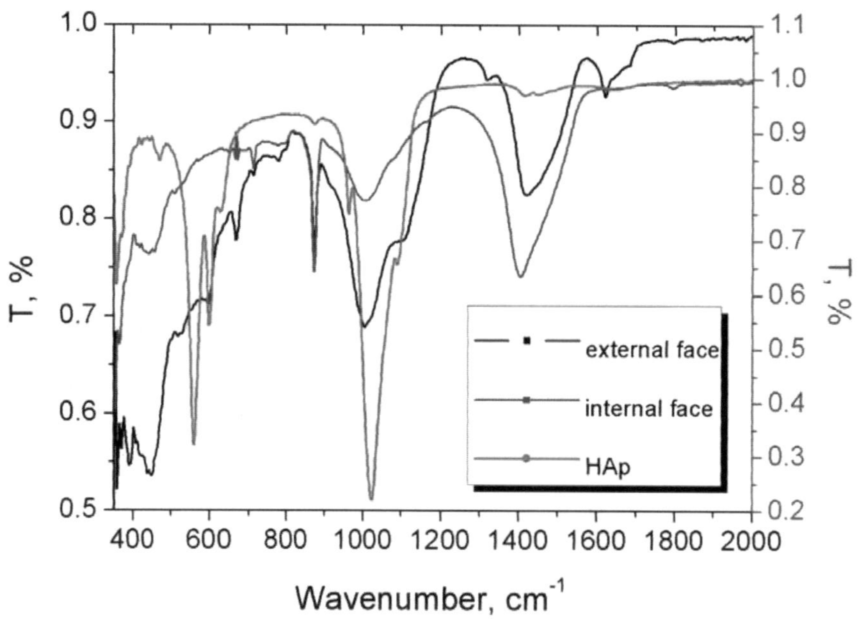

Figure 2. The FTIR spectra of Old Gater Tower samples.

XRD indicates mostly the presence of dolomite, calcite and quartz, with small amount of illite, muscovite, paragonite, montmorillonite, wonesite, feldspars, chlorite and some clayey raw material:

whitmoreite, kornelite, micas and other heavy minerals. Also, hydroxyapatite was identified, too and recognized by the used analytical techniques (Figure 3).

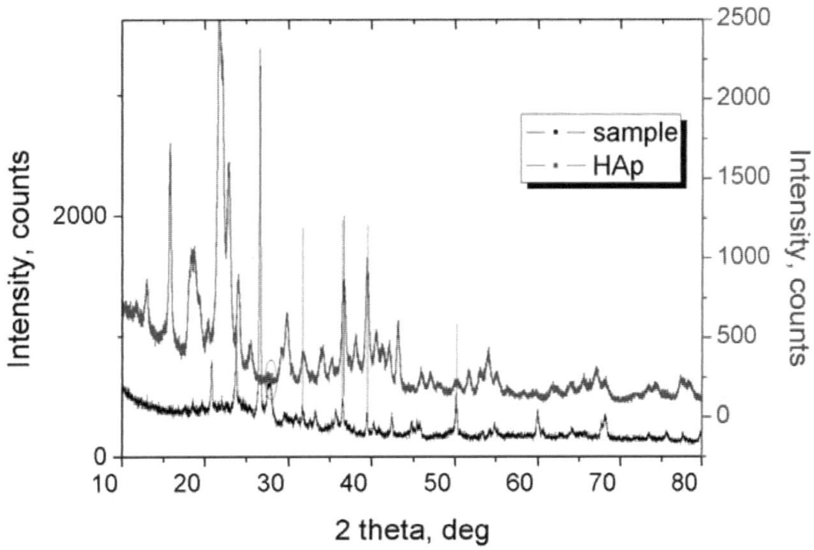

Figure 3. XRD diagram for the stone collected from the Old Gate Tower.

Except the above-mentioned minerals, mica, zircon and titanite could be identified by the XRD analysis, but in very low concentration, such result being in agreement with those expected to Gneiss stones mentioned by the literature [13], known as a common and widely distributed type of metamorphic rock. Gneiss is formed by high temperature and high-pressure metamorphic processes acting on formations composed of igneous or sedimentary rocks. Gneiss nearly always shows a banded texture characterized by alternating darker and lighter colored bands and without a distinct foliation.

XRD indicates mostly the presence of dolomite (25%), calcite (23.6%) and quartz (42.8%), with minor amount of other minerals (ankerite, illite, orthoclase, halloysite, celestine, anglesite, cerussite, zircon, rutile, ilmenite) and some clayey raw material (whitmoreite, kornelite, micas). Furthermore, this sample also contained feldspars - orthoclase visible by XRD. The presence of apatite is visible and marked in the area 2 theta = 28-30 degree), Figure 3.

The XRF of this sample (Table 2) coincides with the FTIR analysis, showing high amounts of CaO and SiO2, present in many of the minerals presented above.

Table 2. XRF analysis of the sample from Old Gate Tower.

Component	Result (mass%)
MgO	13.995
Al_2O_3	7.0008
SiO_2	37.7993
P_2O_5	0.4317
SO_3	1.9603
Cl	0.0688
K_2O	1.1202

IOP Conf. Series: Materials Science and Engineering **877** (2020) 012050 doi:10.1088/1757-899X/877/1/012050

CaO	33.3002
TiO_2	0.4591
MnO	0.1333
Fe_2O_3	3.1323
SrO	0.0401
ZrO_2	0.0295
PbO	0.5294

The WDXRF results put into evidence the existence of P2O5 and Cl– SO3 and K2O most probably related to the hanksite and gypsum (as basanite, which is generated during the transformation of anhydrite into gypsum, possibly due to poor baking and the climatic factors that occured in time). Also, a high amount of calcite is obtained, followed by halloysite and celestine and sand, due to its use in some restoration works.

Also, some information regarding not only the internal morphology that developed in antiquity, but also the extent of the vitrification (the glassy phase) has been obtained by optical microscopy. The presence of porous structure is visible by OM and SEM-EDS (Figure 4) [14, 15].

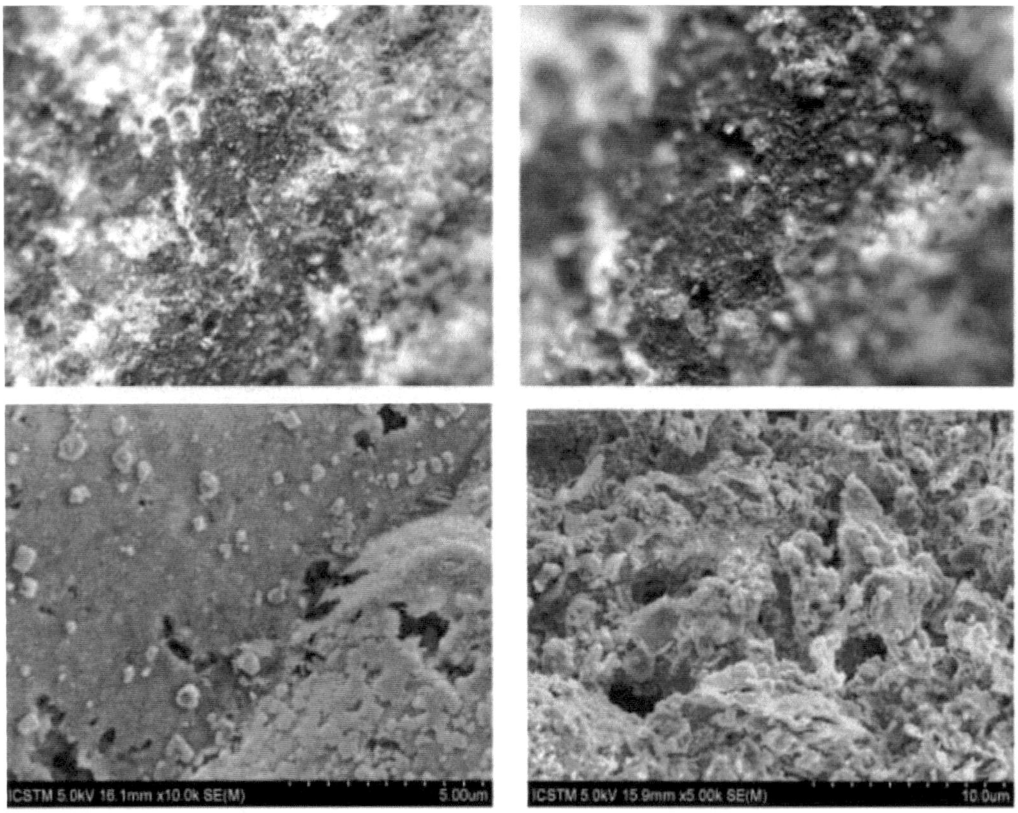

Figure 4. OM and SEM images of the samples collected from Old Gate Tower.

In the obtained images, it is visible a whitish structure, with a roughly round shape, with multiple openings, porous that seems detached from the basic structure. These can be attributed to hydroxyapatite, also identified by the disordered structure as agglomeration of particles of different sizes and morphologies, visible in SEM images.

Also, the gypsum is visible in images, as an advanced state of deterioration, by the dark areas, and also the calcite, through the hexagonal structures present near the gypsum areas.

From all these analytical techniques, different minerals present in the samples taken from this tower could be identified, and even though each technique partially identifies different minerals, depending on its performance and sensitivity, it is important that all these data be corroborated in order to provide solutions for an optimum and viable conservation procedure.

Thus, the EDS technique (Figure 5) revealed high iron concentrations, which leads to the conclusion that there are ferrous mineral reserves in this area (a fact well known and reported in the literature [7]), and these can be corroborated with the reddish aspect observed by the OM technique. The stone samples from Old Gate Tower are in open areas and are vulnerable to physical, chemical, mechanical and biological weathering, irreversibly affecting their structural properties and aesthetic aspects. Atmospheric conditions and atmospheric pollutants have led to a complex process of aging of the stone. Different forms of stone degradation, such as black crusts, efflorescence and discoloration could be identified by the presence of salts in the form of sodium sulphate, which is recognized as one of the most destructive agents in porous stones [16]. The gypsum and halite crystals generated inside the porous stones of the building lead to the destruction of the basic structure of the stone, on the one hand due to the internal pressure induced by the newly formed compounds, and because of the shrinking of the pore space it is [17]. The qualitative and quantitative identification of the different anions responsible for the alteration and degradation of the stone was obtained by ion chromatography, Table 3.

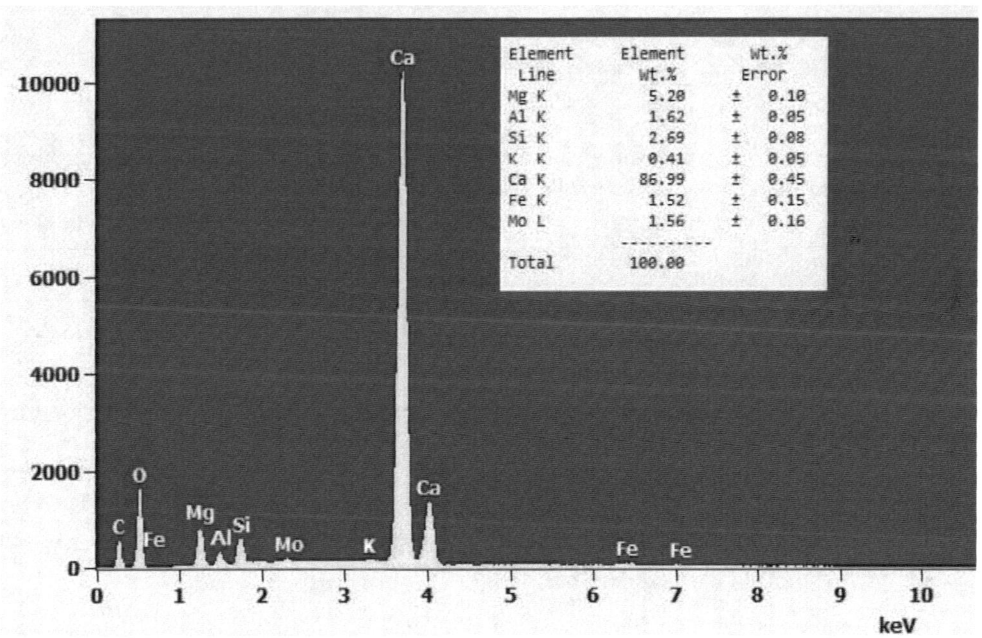

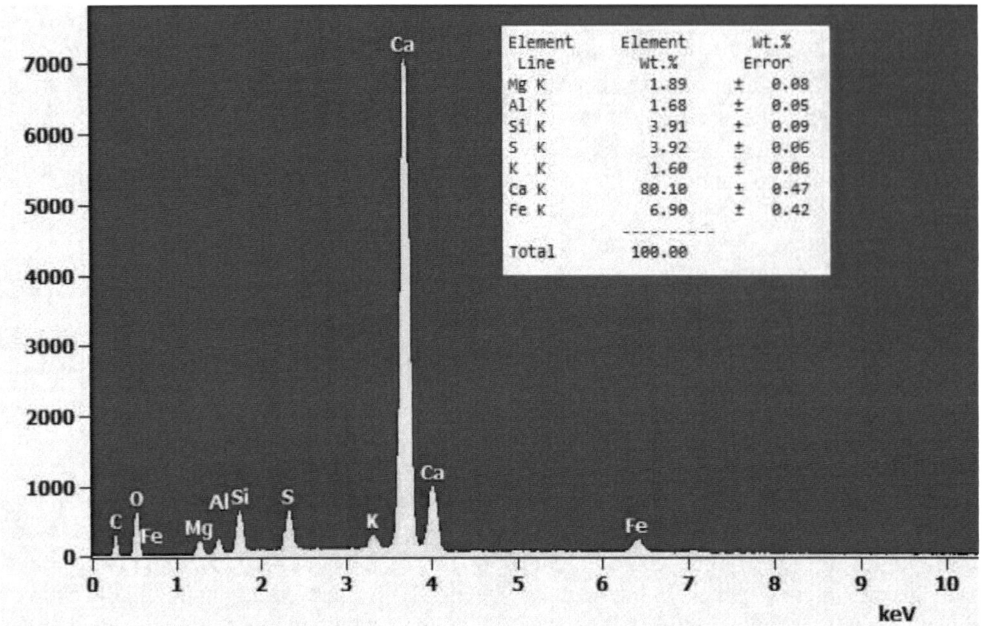

Figure 5. EDS analysis for the Old Gate Tower.

Table 3. The anions concentration of the investigated soil (in mg/l).

F^-	Cl^-	NO_3^{2-}	PO_4^{3-}	SO_4^{2-}
0.014	**0.137**	**0.134**	**0.08**	**0.847**

The white spots found on some samples indicated efflorescence, an alteration caused by NaCl (high concentration of chlorine anion), identified by the ion chromatography technique, in agreement with the literature [18]. Also, the highest concentration of sulphate anions, is a good indicator for gypsum present here, also responsible for degradation processes occurred at this monument.

4. Conclusions

In this paper, an archaeometry study of some samples from the Old Gate Tower, Corvins'Castle was analyzed. Through the used analytical techniques were identified: the origin of the raw materials, of the natural resources and highlighted the processes of alteration and deterioration of these samples, as well as their causative factors. Sensitive and performant analytical techniques, such as FTIR, WDXRF, XRD, microscopic techniques such as OM and SEM-EDS, in conjunction with ion chromatography data, have led to the elucidation of mineralogical composition, structure, but also degradation forms, which affect stability and durability of the monument. All these data will optimize the restoration and preservation of the monuments, preserving our cultural heritage.

Acknowledgments

The authors are grateful to the financial support of the projects: PN-III-51PCCDI/ 2018 from UEFISCDI-MEC, Romania.

5. References

[1] Velescu O 1961 *Castelul de la Hunedoara* (Bucharest: Editura Meridiane: Monumentele Patriei Noastre)
[2] Bogdan A 1970 *Bul Monum Istor* **2** 18
[3] Roman C C and Tincu S 2012 *Sargetia Acta Musei Devensis* **3** 243

[4] Vatasianu V 1933 *Boabe de Grau* **6** 420

[5] Pataki I *1973 Hunedoara Domain at the Beginning of the 16th Century: Study and Documents vol 39* (Bucharest: Publishing House of the Academy of the Socialist Republic of Romania)

[6] Ion R M Tincu S, Iancu L, Grigorescu R M, Radulescu C, Teodorescu S, Dulama I D, Stirbescu R M, Bucurica I A, Minca I, Ion M L and Gheboianu A I 2019 *IOP Conf Ser: Mater Sci Eng* **572** 012088

[7] Balintoni I 1997 *Geotectonica terenurilor metamorfice din România* (Cluj-Napoca: Editura Carpatica) p. 176

[8] Balintoni I and Iancu V 1986 *Studii şi Cercetări de Geologie,* Geo*fizică şi Geografie, seria Geologie, Bucureşti* **31** 51

[9] Ion R M, Iancu L, Grigorescu R, Carutiu-Turcanu D, Tincu S, Ion N, Bucurica I, Teodorescu S, Dulama I and Stirbescu R M 2018 *IOP Publishing Ltd Vol 374 (Bristol, UK)*

[10] Ion R M, Iancu L, Carutiu D T, Schroder V, Tincu S, Roman C, Ion N, Bucurica I A, Teodorescu S and Dulama I D 2018 *Proc Geophys Res Abstr* **20** 5198

[11] Ion R M, Iancu L, Grigorescu R M, Tincu S, Minca I, Teodorescu S, Dulamă I D, Bucurica A I, Ion M L, Stirbescu R M and Grigore M E 2018 *J Arch Urb Herit* **2** 35

[12] Akyuz S, Akyuz T, Basaran S, Bolcal C and Gulec A 2007 *Journal of Molecular Structure* **834** 150

[13] Blatt H and Tracy R J 1996 *Petrology: Igneous, Sedimentary and Metamorphic*, 2nd ed Freeman 359–65

[14] Adriano P, Silva A S, Veiga R, Mirao J and Candeias A 2009 *Mat Charact* **60** 610

[15] Cardell C, Guerra I, Romero-Pastor J, Cultrone G and Rodriguez-Navarro A 2009 *Anal. Chem.* **81** 604

[16] Gobbi G, Zappia G and Sarbbioni C 1998 *Atmospheric Environment* **32** 783

[17] Flatt R J, Caruso F, Sanchez A M A and Scherer G W 2014 *Nat Commun* **5** 4823

[18] Turrión M-B, Gallardo J-F and González M-I 1999 *Communications in Soil Science and Plant Analysis* **30** 1137

IOP Conf. Series: Materials Science and Engineering **877** (2020) 012051 doi:10.1088/1757-899X/877/1/012051

On the Security of a Smart City SDN-Based Network

L Scripcariu[1*], P D Mătăsaru[1] and F Mocanu[1]
[1] "Gheorghe Asachi" Technical University of Iași, 67 Prof. Dr. D. Mangeron Bd., Iași, Romania

*E-mail: lscripca@etti.tuiasi.ro

Abstract. Our research work focuses on the security techniques applied on a software-defined network (SDN), designed for a smart city. Information security is essential for the good functionality of the applications used in a smart city. Information should be correct and complete otherwise the 'life' of the city can be compromised by various cyberattacks targeting, for example, the public transportation system, the traffic information on highways or location services for emergency crews and many others. Securing all devices connected to the SDN, including end-user devices, wireless sensors, smart home equipment is essential because various cyberattacks may be launched and the smart city may be disturbed. This paper presents a risk analysis, the ways to choose the security solutions according to the smart city security policy and our proposal for the SDN security system. We compare different security techniques and present some recommended solutions.

1. Introduction

The smart city concept represents the future of the high-density urban areas [1]. It is the best way to offer many services to a lot of people and institutions, with real-time information and high-accuracy prediction. It implies using information technology all over the place, in order to collect data using large sensor network and to distribute information to interested users via different smart devices connected to a dense metropolitan network.

Nowadays, Intelligent Transportation System (ITS) is a real fact in some countries and it will be implemented all over the world [2]. This is only one application of smart cities. Many others can be developed for the economical field, health and emergency services, environmental monitoring, commerce, fashion, advertising campaigns, entertainment, sport events, smart homes, people safety, waste disposal services etc.

A high-density wide area network is needed to implement the smart city concept and a lot of energy is consumed to maintain functional all its components. Alternative energy sources must be installed in the city with backup energy solutions to sustain the entire system 24/7 with no interruption. Planning a smart city is not an easy task because it has to deal with traditional city structures, historical sites, old buildings and maybe a low communication capacity [3].

A smart city needs a large number of devices connected to the metropolitan network by a lot of communication equipment. The amount of data is big so cloud storage is required and a high capacity of data processing in large data centers is needed. A traditional metropolitan area network (MAN) has a mesh physical topology and it is composed by many interconnected routers, grouped in the autonomous systems of different Internet service providers.

A lot of data is uploaded and downloaded in the same time so broadband communication systems are needed. Usually, the smart city network is created in a traditional city, with an existing communication infrastructure and all existing networks are used. The communication infrastructure has to be extended systematically in order to offer the network services all over the city. All communication and network service providers are involved in developing the demanding cooperative project of a smart city. External services can also be used, for example cloud storage.

Software-Defined Networking (SDN) is a performant solution to centralize the control of MAN [4]. The network architecture is simplified by SDN and the network management is also improved. The routers are replaced by less complex and faster forwarding devices that are clients of the network controllers. The SDN concepts applied to a smart city are described in Section 2.

The overall network security and the security of data transferred over the network are critical because the risks are high and the consequences of lost or false information can be serious. Therefore the security risk analysis for the SDN of a smart city, presented in Section 3, is important in order to decide which security measures should be applied.

Section 4 describes the security techniques that must be adopted in order to design a robust communication network for a smart city.

2. Concepts of SDN

Software-Defined Networking (SDN) concepts separates the logical control plane responsible for decisions made in a network from the data transfer plane that includes all forwarding devices (FD) such as switches and routers (R). Above all of these is the management plane (MP) that includes different applications (A) for routing, network virtualization or security services like access control, network monitoring or attack detection.

The network controllers (NC), also named network operating systems (NOS), operate the forwarding devices by application programming interfaces (API) that are easy to reconfigure when it is needed. The network controller can be implemented in a centralized or a distributed way. The distributed NC avoid the collapse of the entire SDN layered structure produced by the centralized NC falling down. The software-defined network is more flexible and more reliable than the traditional network. The new way of networking is characterized by centralized control and management (Figure 1).

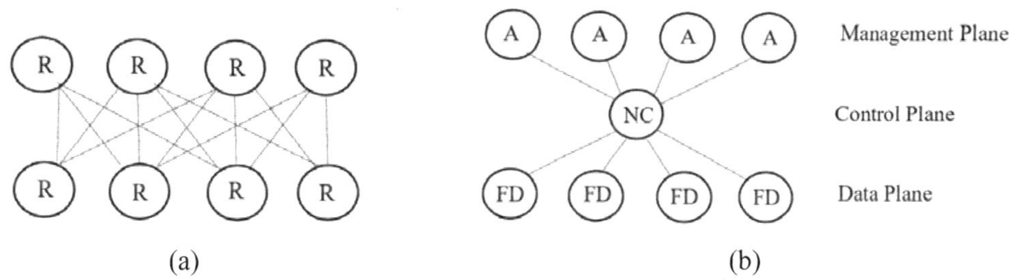

Figure 1. (a) Mesh Topology; (b) SDN Topology.

The traditional metropolitan network, with a mesh topology, is composed by a lot of routers and L3-switches that embed hardware and software capabilities. Routing, switching and forwarding are all embedded in one device with no separation between the control capability and the data transfer function. Routing information is transferred from each router to other routers by using routing protocols.

Network service providers (NSP) manage their routers grouped in autonomous systems (AS), eventually partitioned into areas of routers. Core routers from one area have Internet access through the boundary routers that communicates with each other by a network backbone connected to the public network through a border router. Such a packet transport network is complex and it is hard to configure and to secure.

The router processors does a lot of operational tasks producing high transmission delay. Each NSP has a lot of network servers for different network services (web, file transfer, e-mail, database management, DNS, logging history, authentication, etc.) that can be in-house servers or hosted in data centers with maintenance administration employees or external IT services. There are pros and cons for all these solutions regarding their effectiveness and security. It is a better choice to use cloud servers in order to reduce the costs of software license and computing devices purchasing, of maintenance and of IT services.

Using cloud services is an improvement but the packet-switching network topology is still complicated and difficult to manage, to extent and to reconfigure. Therefore, SDN must be considered as a good solution for the network sustaining a smart city.

In order to design the network of a smart city, a **macro-map** has to be created. This macro-map contains different small networks interconnected by the city transport network. These small networks can be sensor networks, vehicular networks, wireless communication networks, private networks, etc. The cooperation between these networks is essential for the success of the project but it is critical regarding the security. Securing the borders of each network as part of the smart city network is very important.

Implementing the smart city network based on SDN concepts is a good strategy because the network control is centralized. An SDN controller can be programmed for each component network, all these controllers being supervised by the city management stations from the Network Operating Center (NOC). The centralized control is the key for a unitary robust wide-spread system.

3. Security Risk Analysis for a Smart City Network

The smart city network is vulnerable to a lot of attacks intended to block the communication, to steel information, to insert false data, to take control on a service or on the entire network.

The smart city offers a lot of services in many fields such as traffic control, health, emergency services, environmental monitoring, commerce, event advertising, smart homes, people safety, etc. All these services need large databases to manage the customers and a wide high-density network to communicate with them. These databases must secure personnel information. Public information must also be secured in order to avoid misinformation of people by the city network. Database applications have to be developed to manage people, devices, institutes, locations, events, facts, and all data transferred by the smart city network. Cloud servers can be used in order to have access to services from any location, on all the devices connected to the SDN-enabled smart city network (Figure 2).

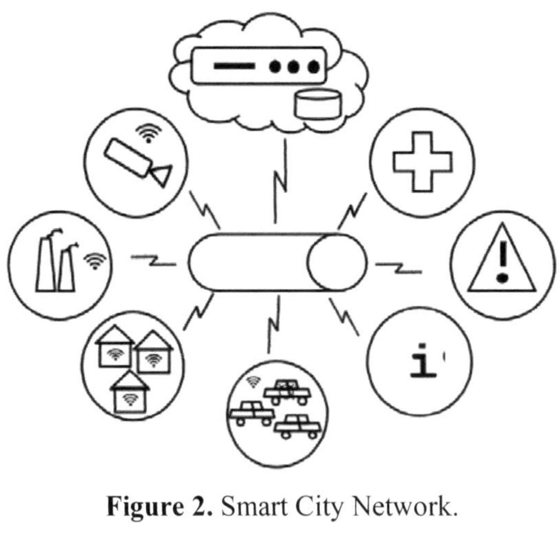

Figure 2. Smart City Network.

Web applications must be developed to send information to people, on their personal smart devices or on public devices located throughout the city. Large public displays can offer visual information to people, on the street. Audio information can also be delivered on different places from the city.

Many sensors and cameras installed in the city send data to the city servers using the SDN-enabled metropolitan network [5].

All devices can get information from the network, for example, emergency applications can shut down energy or natural gas sources when an earth quake occurs.

The actuator network that is a component of the smart city network is used to turn off or on different equipment from the city area.

Traffic lights can be configured to keep opened the traffic on one street, without stopping for a period of time, when congestion is imminent or emergency vehicles have to pass.

The risks of malfunction exists in any point of the network, on all the layers of the 'onion' security model: physical access, logical access, service access, application and information layers (Figure 3).

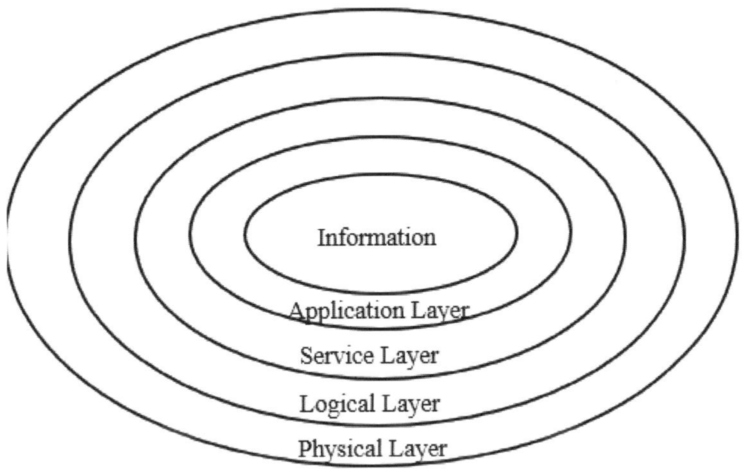

Figure 3. 'Onion' Security Model.

Protection of people, institutions, devices, applications, information and of the entire city is the goal of the security system designed for the smart city network.

Physical risks are numerous and can involve power outage, communication link failure, equipment failure or steel, including personal devices with authorized access to restricted 'areas' of the network, such as access cards, smart phones, notebooks, etc.

Logical risks concern the steel of authorized user account or devices credentials. Attackers get access to various network equipment and devices and can produce many damages in the city, like traffic events, energy supply interruption, surveillance cameras shut down, servers blocking, disrupting of normal operation of a smart house or of different institutions.

Network services vulnerabilities can be exploited by attackers in order to steel information, to use them without authorization and to misinform people or officials in order to create disorder in the city, traffic congestion, false emergency calls and so on.

Financial consequences can occur by attacking the automatic toll collection network of the city. On the one hand, payment is not made and the service is used free of charge or toll collection can be redirected to other bank accounts, but on the other hand users' banking information can be stolen if this data is not protected.

Application servers can be attacked in various ways by web attacks, SQL insert attacks, Denial-of-Service (DoS), DNS attacks, etc.

People can be targeted by cyberattacks based on social engineering methods, like phishing, that exploits the human nature weaknesses and its lack of awareness or knowledge about cybersecurity.

The real problem is that the smart city network is accessible to so many people and devices, with more or less computing resources and various security mechanisms, that it is very difficult to manage and secure all of them. Attackers will always exploit the weakest link of the system chain and thus they can get access to other more important components of the network and produce serious damages in the system.

All people connected to the network and using network services should be aware of the risk they are exposed and apply all recommended security measures with high responsability.

4. Security Measures for a Smart City Network

Risk analysis leads to a systematic security plan that includes security measures for all the components of the network [7], including in terms of Quality of Service, as the principles of trade-off between different levels for security and QoS go hand in hand [8].

An overall view of the network security shows its complexity. The division of tasks between all actors acting on the smart city 'scene' is a crucial choice for security efficiency.

A smart city network includes many users that access home, public or and private enterprise networks. Personal device security is usually the responsibility of the owner. It is important to update periodically the device in order to run the best security software modules, antivirus, anti-spyware and anti-malware defense programs and keep it safe from today's cybersecurity threats.

The software developers should provide updates to their application programs in order to offer the strongest security options. Automatic software update must be compulsory and not optional for all devices connected to the smart city network. Devices that do not run the latest software and have not enabled all security options should not gain access to the network. Devices must be certified in order to access a network service. Messages sent via this network should be digitally signed and message authentication codes must be used to check their authenticity by the destination node.

The network administrators have to secure all devices, users' accounts and communication links. Basic principles of cybersecurity must be applied to users' account configuration [6]. Principle of Least Privilege (POLP) is essential in avoiding potential damages produced by people with or without intention. Account configuration should be based on roles, seniority, place and time of the day criteria. The risk analysis conducts to a particular risk grade associated to a user or a device that is also a criterion of granting or revoking privileges. Mutual authentication is recommended in order to avoid the masquerade attack and to protect users' credentials.

Public information and periodical training of the employed staff is very important for them to be awareness of the risks to which the entire network is exposed. Public users and employees should be aware of the risk of social engineering based-attacks like phishing and to be trained to have a responsible attitude that counteracts these malicious actions.

Backup solution must be created for each component of the network like connections, forwarding devices, SDN controllers, servers, databases and human personnel.

For the energy supply sources of the city, backup solutions, including alternative energy sources, must be ready at any moment to become active in order to avoid service interruption.

Virtualization is an option for the network security. Configuring Virtual Private Networks (VPN) is a good security solution for enterprises to keep their network traffic secure, with data encryption and authorized access only [9]. VPNs can be offered as a service by the network providers.

Information protection can be ensured applying the CIA principle that means Confidentiality-Integrity-Availability. Different encryption protocols can be applied but it depends on the device's resources and the importance of data [10]. Information redundancy can be ensured by sending it on multiple path and check its integrity after the transmission.

Servers can be 'separated' by installing Virtual Machines (VM) and configuring multiple virtual servers. Virtual containers can be configured to share the virtual space to many services and users.

Using cloud servers, the security of them is the responsibility of the cloud service provider. Robust security can be offered by a third part authority as a professional service with some financial costs but with the guaranty of no security incidents.

SDN allows centralizing the security measures. In this case, the number of nodes that requires higher security levels is reduced and these are the network controllers and the servers. Service orchestration is supervised by the city security staff. Network monitoring and logging history analysis is important to identify possible attacks, to detect intrusion and prevent them. For such a large network, this must be done automatically, using specialized software.

A Security Incident and Event Management (SIEM) system should be used to monitor the security of the smart city network as a whole. Being such a complex task, Machine Learning (ML) based security mechanisms can be used to secure all network components by the mean of the SDN controllers. Artificial intelligence (AI) algorithms could create behavior patterns for all the entities involved in the communication process that can be used to do automatic recognition of security events. Security scripts run by the SDN controllers can be written in different programming languages. Network programmers are responsible for the applications developed by them for software-defined networking with guarantee of security. A monitoring, alert and interventional security system have to be designed for the smart city SDN-based network in order to launch automatic countermeasures against imminent or ongoing network attacks, to quarantine risky objects from the network, to block access of intruders and to alert network administration staff to act properly on each situation.

5. Conclusions

The design of a smart city communication network is a very complex task with a lot of responsibility regarding the possible consequences of malfunction or disruption. SDN-networking is a good option for the metropolitan network design with a centralized topology. Cyberattacks are a reality of nowadays Internet and various security solutions must be adopted in order to avoid or to minimize the effects of any possible attack. A security risk analysis for the smart city network is presented. The security has to be ensured on all levels, starting with the physical one, continuing with the logical access control, users' accounts, devices' certificates, network services usage and, finally, ensuring the information integrity. The availability of services is an overall task of network administration that can be accomplished by continuously monitoring the network, real-time alerts, prevention mechanisms and interventional techniques. Different security techniques are recommended in order to ensure the security of the SDN-based network for a smart city.

6. References

[1] Albino V, Berardi U, Dangelico R 2015 *Journal of Urban Technology* **22**(1) 3-21
[2] Stawiarska E and Sobczak P 2018 *Sustainability* **10**(1318) 1-32
[3] Greco I, Cresta A 2015 *International Conference on Computational Science and Its Applications* Proceedings Book 563-576
[4] Kreutz D, Ramos F M V, Verisimo P E, Rothenberg C E, Azodolmolky S and Uhlig S 2015 *Proceedings of the IEEE* **103**(1) 14-76
[5] Ceken C and Al-Hubaishi M 2019 Integrating SDN – Enabled Wireless Sensor Networks into the Internet, *10th IEEE International Conference on Intelligent Data Acquisition and Advanced Computing Systems: Technology and Applications 18-21 September 2019, Metz, France* Proceedings Book 1090-1094
[6] Craigen D, Diakun-Thibault N and Purse R 2014 *Management Review* 13-21
[7] Diaconescu V D, Scripcariu L, Matasaru P D, Diaconescu M R, Ignat C A 2018 *IOP Conference Series-Materials Science and Engineering* **374** 1-10
[8] Mătăsaru P D, Scripcariu L and Diaconu F 2017 An up-to-date overview on QoS for Satellite IP networks *Proceedings of the International Symposium on Signals, Circuits and Systems Iasi, Romania (ISSCS 2017)*
[9] Scripcariu L and Bogdan I 2009 *Revista de Telecomunicatii, Bucuresti* **2** 32-37
[10] Scripcariu L 2015 *International Symposium on Signals, Circuits and Systems, Iasi Proceedings Book* 1-4

Intelligent Communication, Specific to the Engineering of Materials, under the Conditions of Ensuring the Circular Economy and Technological Digitization

M M Vescan[1], V F Soporan[1*], S Pădurețu[2], T R Lehene[1] and M D Crișan[1]
[1]Department of Environmental Engineering Entrepreneurship and Sustainable Development, Technical University of Cluj-Napoca,103-105, Muncii Boulevard, RO 400641, Cluj-Napoca, Romania.
[2]Department of Modern Languages and Communication, Technical University of Cluj-Napoca, 103-105, Muncii Boulevard, RO 400641, Cluj-Napoca, Romania

*E-mail: vfsoporan@gmail.com

Abstract. The paper presents the problem of meeting the conditions imposed by the circular economy with direct references to a field of materials engineering, the manufacture of castings. The study has several motivations, one related to the analysis of the ways in which the development of the circular economy can be achieved by a company producing castings. The second one is to establish the partnerships that are necessary in order to materialize the conditions in the castings industry. The third motivation, based on the analyzes carried out at the level of the first two, aims at calibrating the communication for the actors that contribute to the realization of the circularity desideratum in the industrialization of the cast products in the new conditionings of the digitization instruments. To structure an action methodology in the direction of intelligent communication, specific to materials engineering, the paper is structured as follows: the conditions and values to which the circular economy is reported, the engineering design elements of solutions for a circular economy, communication partnerships, analysis of development measures of the circular economy in partnerships and the results of the analysis of the expected activities and partnerships.

1. Introduction

From the perspective of socio-economic behaviors, often imposed institutionally, there is an increasing need for communication for many areas served by engineering. One of the most interesting is that of natural resources, especially that of materials that support the materialization of functional coverage by designing products and services. Their transformation into industrial developments of great diversity, at the level of an increasingly complex economy, requires a fundamental change in the way traditional engineering is practiced. Therefore, one of the most dynamic changes has been in the fields of materials engineering, making a necessary distinction between what it means and what materials science represents.

Therefore, engineering is related to the design of the truth established in a certain space of developments, defined by existing conditions and existing partnerships in the design, operation and use of products and services necessary to achieve functionalities very well defined by market requirements and regulated by competent authorities.

In order to understand more deeply and exactly the involvement of engineering in the fields it serves, it is important to carry out an analysis on the following levels: assessment of the existing situation, analysis of public policies, knowledge of the regulatory framework, deepening and detailing technical and procedural solutions, responsible involvement in understanding financial instruments and mechanisms, establishing the impact of developed solutions on socio-economic developments. Therefore, it is essential to follow - at the level of communication - the aspects analyzed in order for the solutions offered to be in line with the requirements imposed at multidisciplinary level.

To properly calibrate the necessary communication in materials engineering, it is important to make an analysis on the following aspects: the conditions to which materials are subjected, the requirements of the fields served, the actors of the activities carried out or the beneficiaries of specific activities. The reasoning to be made, given the specifics of the paper, is achieved by imposing two conditions, one related to ensuring the participation of materials engineering in the particular development of the circular economy and the second which aims to use the tools of the most dynamic field of moment, that of the digitization of the activities included at the level of the present industries in the current socio-economic framework.

Therefore, even if there is a very well defined specialization at the level of university training, that of materials engineering, it is appreciated that the practice in this field requires having a good communication with the representatives of the categories involved in existing multidisciplinary processes.

The present paper, in addition to its scientific objective, is intended to be a plea for an opening of materials engineering to new socio-economic developments and a recognition of a profound differentiation between materials engineering and materials science. At least in Romania, materials engineering has emerged as a transformation of the old metallurgical specializations, which in turn have emerged as a differentiator, required by new industrial developments, by the fundamental specializations particularly to chemistry and physics. Reducing engineering only to the requirements of chemistry or other fundamental disciplines is not appreciated as favorable in today's course of technological developments and the requirements of transforming products and services into industrial objects.

2. The conditions and values at which the circular economy is reported

For the practical enforcement of the principles of the circular economy, it is important that materials engineering should find ways to communicate the conditions considered in the design of a shaped material, a product or a service, namely:

- safety conditions in ensuring the functionalities, which ensure the behavior of the material at the demands imposed by the accomplishment of the required functionalities or the safety in operation at the imposed demands;
- processability conditions, ensuring a certain geometric shape at a certain dimensional accuracy and a morphological structure of the material;
- conditions for increasing the duration of exploitation, by ensuring a sustainability corresponding to the socio-economic requirements;
- environmentally friendly conditions, by ensuring measures concerning the manufacturing, processing or operation part in an environmentally friendly relationship;
- disposal conditions at the end of the life cycle, by ensuring a material, energetic and environmental recovery of the materials from the products to be disposed of or by ensuring an intelligent degradability at the end of the life cycle.

If the first three conditionings were present at the level of the general-practicing engineer until the 1990s, the references being made regarding the existing situation in Romania, the last three decades imposed a reformulation of engineering, by adding the other three groups of conditionings, private, as a major responsibility for sustainable development. This maturity manifests itself today in the form of the circular economy, which is represented as "*a principle of economic organization that aims to systematically reduce the amount of raw materials and energy throughout a life cycle of a product or service and at all levels of the organization of society, in order to ensure the protection of biodiversity and the proper development of individuals*" [1].

For the exact understanding of the conditions, with the motivation to build a tailor-made communication, the following are the values to which the circular economy refers through the material value of products and services, related by the overall consumption of materials in the atmosphere or from the stock of inorganic materials.

Given the need for efficient communication on the potential of materials included in the manufacturing process of castings, in the desire to use them efficiently, the indicator of "material intensity" is used, which is the total mass of raw materials needed to make a product. The concept was developed by the Wuppertal Institute, which, in approaching the "material intensity" indicator, defined five categories of flows: "abiotic raw materials, which are non-renewable raw materials such as minerals (aggregates, sand, ore of various metals, etc.) or fossil fuels, which are the result of drilling or extraction activities and are consumed indirectly in the manufacturing process of a product (e.g. for one tonne of steel not only one tonne of minerals is consumed); biotic raw materials, which are renewable and which represent the biomass consumed, whether or not obtained from agricultural or forestry activities; earthworks, determined by the activities of agricultural production, forestry, construction works, etc.; water consumed in the manufacture of a product; the air used, used either for combustion or for chemical transformation" [2].

Using a synthesis of the intensity of the materials made by Nicholas Roussat and adopting it for materials that enter the process of elaboration of a ton of cast iron, we have the situation presented in Table 1 [3]. The calculation is approximate, based on a calculation of the average annual consumption in a cast iron foundry for a period of seven years, between 2012-2018.

Table 1. The material strength of the load components for making a ton of cast iron, good parts.

Component	Consumption at ton of cast iron, [t]	Abiotic material	Water	Air
		t/ amount of component		
Old cast iron	0.593	4.744	44.475	0.385
Scrap metal	0.394	3.677	32.264	0.304
Turnings	0.049	0.456	4.013	0.038
Sand	0.325	0.715	0.682	0.013
Recirculated material	0.195	1.817	15.97	0.151
Chromite sand	0.009	0.031	0.029	0.001
Hardener	0.049	0.412	0.352	0.137
Refractory paint	0.019	0.133	0.129	0.002
Metal shot blasters	0.011	0.136	1.485	0.030
Total/ton of good parts	1.644	12.174	99.399	2.061

From the point of view of communication and awareness of the actors present in the manufacturing process of castings, the data calculated on the basis of "material intensity" are significant. Thus, it can be estimated that for a ton of cast parts delivered, 1,644 tons are consumed directly, and at the level of biotic materials we have an extraction of resources in an equivalent of 12,174 tons of raw materials, a specific consumption of 99,399 tons of water and a use of 2,061 tons of air. Basically, if the principles of recirculation and recycling were not used as instruments of the circular economy, this would be the real consumption in relation to the use of natural potential.

At the same time, at the level of partnerships that can be created through industrial symbiosis, these consumptions can be the basis for the use of the non-useful part that results in the manufacturing process, especially used sand, which is generated in a quantity of 324 kg/t and slag in an amount of 34 kg/t Their use would substantially reduce the quantities that are extracted from nature for use for various lucrative purposes.

Therefore, the example given and the definition used show that materials engineering provides the tools and reasoning necessary to materialize the circular economy.

3. Engineering design elements of solutions for the circular economy

The following is a methodology for determining the elements of intelligent construction of the circular economy, with a careful focus on a specific field of materials engineering. In a particular approach, we called this process "*circular economy engineering*" [4], which came to be introduced as a specialty in the master studies. Starting from the structure of the mentioned discipline, we adapt a methodology for building solutions for the circular economy at the level of a field of materials engineering, that of the manufacture of castings. It presents the following stages: the selection of favorable public policy elements, their transposition in the construction of the circular economy and the determination of the necessary solutions to ensure circularity.

Following the proposed methodology, that of following the institutional formulations, we notice that in the European Union there was a serious analysis on the mechanisms to support the circular economy. At the level of a public policy document, "the notion is defined by ensuring an efficient economy in the use of resources, imposing an objective of maximizing economic growth and limiting the pressure on resources" [5].

Without going into the details of the institutional presentations, moving to the second stage of the proposed methodology, the favorable policy elements are presented grouped on the following categories of analysis, Figure 1: technical-scientific, technical-productive, technical-economic and innovative construction .

In the first category, that of technical-scientific analyzes, the defining elements of engineering are taken into account, those of design in order to ensure the principles of ensuring the circular economy, of which we mention the following, with a schematic presentation in Figure 2: light loading, ensuring sustainability, efficiency in the use of resources, substitution of hazardous or deficient materials, ecological conception from the perspective of exploitation, intelligent management of the neutral parts generated throughout the life of materials, products or services.

In the second category, that of technical-productive analyzes, the following specific actions were retained: audit of productive activities from the perspective of how to ensure waste management activities; imposing technological infrastructure efficiently in terms of consumption of material and energy resources; choosing environmentally friendly materials at the level of technological processes; organization of technological flows with a view to ensuring good waste management.

In the third category, that of technical-economic analysis, the following specific actions were retained: ensuring an open market for the sale of raw materials generated following recycling; repositioning of maintenance and repair services on the market, development of end-of-life product collection systems; supporting industrial symbiosis to make the by-products market a reality in a significant way; the development of new methods of consumption, which stimulate the rental, lending, exchange of services or sharing of products, as alternatives to their individual ownership; unlocking specific investments; elimination of environmentally harmful subsidies; progressive taxation of the consumption of material and energy resources.

In the fourth category, that of innovative solutions, it is emphasized that action must be taken at the following levels: technological, organizational, social and vocational training.

4. Communication partnerships

Starting from the necessary conditions, from the values to which the reporting is made and from the elements of the engineering design of the solutions for the circular economy, we notice that in order to establish the communication modalities it is necessary to establish the necessary partnerships to transform a field or activities according to the principles of circular economy.

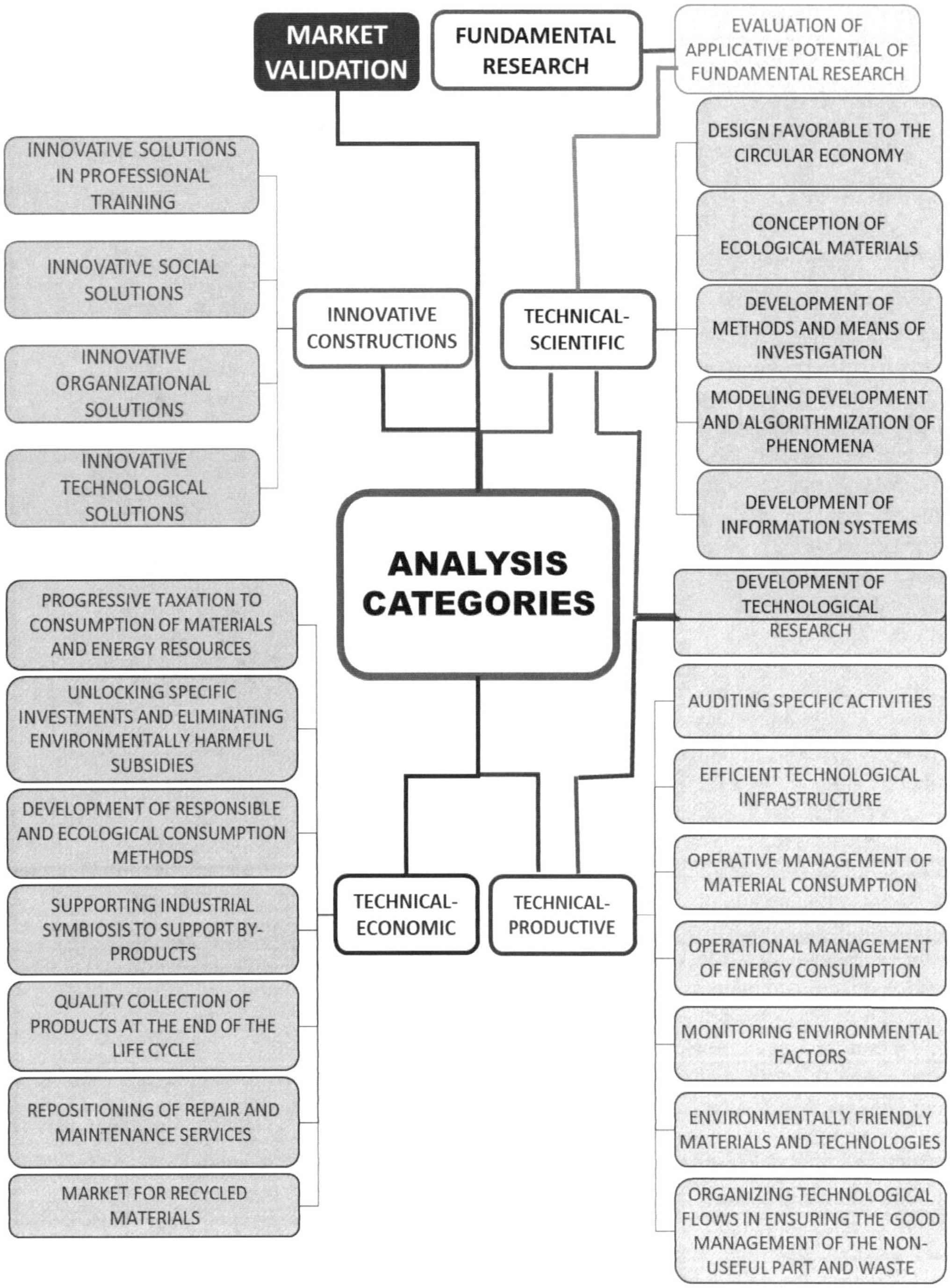

Figure 1. Categories of analysis in order to promote the circular economy.

From the analysis made, we establish that the process of transforming materials engineering, especially the manufacture of castings, to the requirements of the circular economy can not be achieved only by the participation of specialists in one field, requiring the involvement of several specializations. This can be a valid assessment, first of all, for those who organize engineering training systems - at the level of industrial developments the process is much faster - with the desire to streamline activities in the global market. The castings industry can be an example in this direction, by the fact that at the "world foundry congresses", which take place every two years, they addressed issues of multidisciplinary action, in which partnerships took place between parts manufacturers, foundries, suppliers of raw materials, beneficiaries from various fields, representatives of the scientific environment, technological research and academia. As a justification for the diversity of partnerships, but also for imposing the circular economy in the manufacture of castings, there are some aspects related to the congresses in Bilbao, in 2014, and in Krakow, in 2018. Thus, for the first congress mentioned the keywords made reference to energy, environment, market trends, demands and innovation, and the following issues were addressed: advanced engineering, robotics and automation, environmental protection and sustainable development, energy efficiency and rapid prototyping. One of the works of the authors, in the aforementioned congress, constitutes the support of a "method of evaluation of the tendencies of the manufacture of the castings in relation to the principles of development of the circular economy." [6].

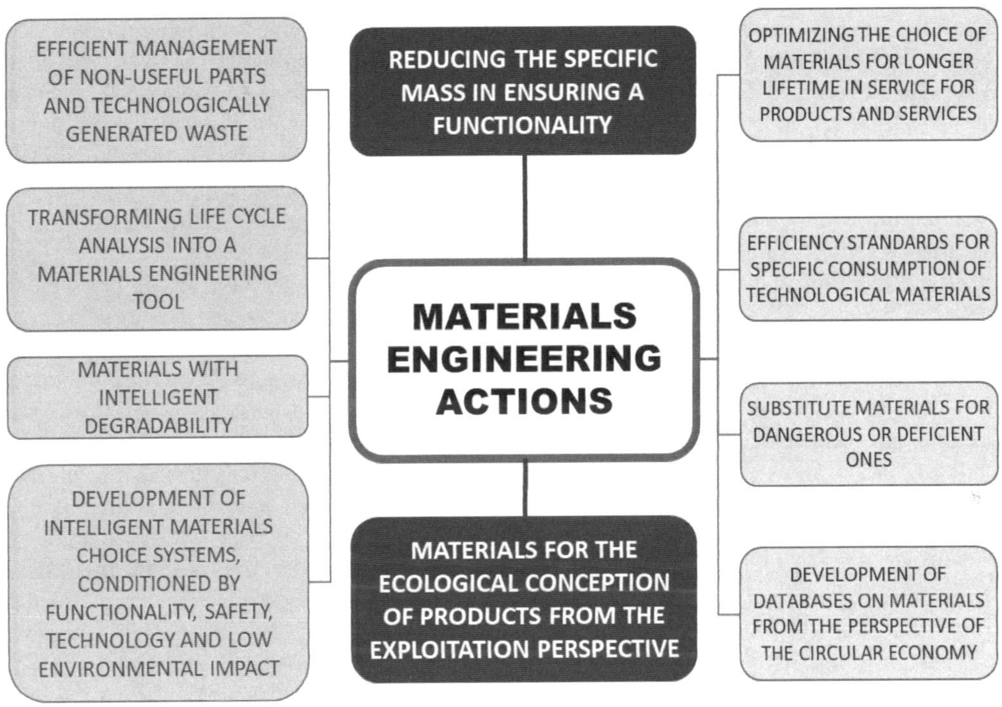

Figure 2. Materials engineering actions to support the circular economy.

5. The analysis of measures for the development of circular economy in partnership system

In order to establish the measures and actions to be undertaken at the level of a technological process for obtaining the castings from the perspective of achieving the objectives of the circular economy, the following procedure was performed: dividing the process into phases, stages and activities; establishing the necessary measures; intuition of the necessary partnerships.

At the level of the process division, the following phases were taken into account: 1. Analysis of specific information; 2. Analysis of manufacturing readiness in terms of available technologies and technological infrastructure; 3. Analysis of the casting process; 4. Analysis of the results of the manufacturing process; 5. Analysis of exploitation and disposal at the end of the life cycle; 6. Analysis of other phases that contribute to the realization of castings; 7. Analysis of the whole process from the perspective of fulfilling the conditions for ensuring the circular economy.

Each phase is divided into several stages, take for example the structure of the second, third and seventh phase. Thus, the "analysis of the preparation of the manufacture from the point of view of the necessary technologies and infrastructure" (phase 2) presents the following stages: 2.1. analysis of the constructive conception; 2.2. technological design analysis; 2.3. manufacturing preparation analysis; and the "analysis of the manufacturing process" (phase 3) presents the following stages: 3.1. direct activities; 3.2. indirect activities, and for the "analysis of the whole process from the perspective of fulfilling the conditions for ensuring the circular economy" (phase 7) the following stages are presented: 7.1. pollution monitoring; 7.2. recording and interpretation of specific material and energy consumption; 7.3. recording specific energy consumption per unit of product; 7.4. recording other consumptions per unit of product; 7.5. waste management in relation to the product unit; 7.6. management of recirculated material introduced in the production process and reporting to the product unit; 7.7. the management of the recirculated material introduced within the technological process by reference to the product unit; 7.8. the management of the by-products generated within the technological process in relation to the product unit; 7.9. energy management recovered at the level of the technological process in relation to the product unit; 7.10. the management of other resources consumed at the level of the technological process and their reporting on the product unit.

The present stages are divided into main activities of the technological process, for example, several details are presented. Thus, the "analysis of technological design" (stage 2.2) presents the following specific activities: 2.2.1. general analysis of the casting technology depending on the nature of the alloy, the type of technology and the constructive features of the required functionalities (wall thickness, overall dimensions and dimensional differences to be ensured); 2.2.2. power system analysis and optimization; 2.2.3. analysis and optimization of solidification by correlating the masonry and / or cooling system with the constructive design typologies; 2.2.4. analysis of the cooling process and constructive optimization from the point of view of the stresses that form in the castings; 2.2.5. constructive-technological optimization with the composition of the alloy, the characteristics of the shape and the casting conditions; and "indirect activities" (stage 3.2) have the following specific activities: 3.2.1. the flow of transport and handling of materials entering the production circuit; 3.2.2. the flow of use of protective equipment and materials; 3.2.3. the flow necessary to ensure the utilities of the technological process; 3.2.4. the flow of activities for ensuring the maintenance of technological equipment; 3.2.5. the flow of current and general technological cleaning activities; 3.2.6. the flow of integrated and sustainable management activities of productive waste and other similar household waste; 3.2.7. other indirect flows present within the technological flow; so that at the level of "waste management by reference to the product unit" (stage 7.5) the following activities are present: 7.5.1. waste management generated during elaboration; 7.5.2. waste management generated at the realization of forms; 7.5.3. waste management generated at casting; 7.5.4. waste management generated at cleaning and finishing, 7.5.5. waste management generated in the heat treatment phase; 7.5.6. waste management for painting and surface coating operations; 7.5.7. waste management for roughing and processing; 7.5.8. waste management in operations other than those specified.

Starting from the presentation of the activities, in Table 2 are presented, in a selective way, intuitively, the necessary partnerships in order to fulfill the proposed objectives and to find the best ways to communicate.

Table 2. Activities supporting the circular economy together with the partners required for the specified activities (selective presentation).

Code	Activity	Measures	Partnerships
2.2.1.	General analysis of the casting technology depending on the nature of the alloy, the type of technology and the constructive features of	Development of computer programs for technological design in order to optimize material and energy consumption, adoption of technologies and infrastructure necessary for the established objectives. They can be	Specialists in the field of informatics, mathematics, materials physics, chemistry, circular economy, modeling of physico-chemical processes and those of a

Code	Activity	Measures	Partnerships
	the required functionalities (wall thickness, overall dimensions and dimensional differences to be ensured)	positioned on various levels of use, covering the typology of processes, alloys and the level of technological equipment. Within the mentioned instruments, it is necessary, taking into account the objectives of ensuring the circular economy, the development of IT structures aimed at managing the life cycle of the process, environmental protection and sustainable development measures.	metallurgical nature. They will work with specialists in various fields for which cast parts are delivered and with those specialized in the practices of casting various alloys, etc.
2.2.2.	Power system analysis and optimization	Supporting computer programs for computer-aided design of power systems, which, based on the models created, are to achieve an optimization of specific consumption. At the same time, it will be followed from the beginning of the technological design activity of the management modalities in order to recirculate the detached power supply networks from the part level. The transition from traditional to computer-aided design aims to achieve effects on the mentioned optimizations, even if often an interaction between the two systems is necessary.	Specialists in mathematics, hydraulics, thermotechnics, especially in the phenomena of heat transfer in liquid, solid and gaseous phase, in mass transfer. Specialists in the field of modeling physical phenomena specific to the development of technological processes. Specialists in the field of informatics, with experience in the field of industrial activities, etc.
2.2.5.	Constructive technological optimization with alloy composition, shape characteristics and casting conditions	Promoting computer-aided operational design software, the process of obtaining castings, usable in the offices of study and current technological analysis.	Specialists in computer-aided programming, adapted to the design of technological processes, etc.
3.1.1.	Alloy processing flow	Designing, developing and following the elaboration processes on the principles of the circular economy; reduction of material and energy consumption; calculation of rechargeable or recycled material with a high share of rechargeable material; avoiding materials that alter the heritage of recycled material; avoidance of the use of dangerous substances; establishing flows for obtaining by-products from the general useless structure within the manufacturing process; generating flows for reuse and recycling of used materials or products, etc.	Specialists in the field of metallurgy, logistics, physics, chemistry, use of hazardous substances, obtaining construction materials, making shaped products used in the area of public works, etc.
3.2.1.	The flow of transport and handling of materials entering the production circuit	Optimizing transport and handling flows, in order to reduce specific consumption. Materialization directly at the level of production costs of these activities. Development of recycling activities for infrastructure used at the end of the life cycle, and for consumables regeneration and recycling activities.	Specialists in the field of logistics, organization of material flows, means of transport, life cycle analysis, management of consumables used in transport, etc.
4.2.2.	The non-useful part turned	Collaborative realization of studies for	Specialists in construction

Code	Activity	Measures	Partnerships
	into a by-product	the inclusion of a large part of the non-useful part in the by-product category, by respecting the existing institutional regulations.	materials, public works, chemistry, materials testing, case law on the use of material resources, etc.
6.2.2.	Analysis of transport activities	Management on the principles of the circular economy of internal and external transport activities: internalization of transport activities at the level of manufacturing costs of castings; optimization of circuits and raw materials and finished products; use of ecological means of transport; regeneration and recycling of consumables used in the means of transport; reduction of material and energy costs of transport and handling per unit of product.	Specialists in the organization of industrial transport, in the logistics of materials specific to obtaining castings, maintenance of internal or external means of transport, management of specific consumables in transport, etc.
7.1.1.	Air pollution	Complex management of atmospheric polluting activities specific to manufacturing processes: activity monitoring measures; adoption of environmentally friendly technologies; introduction of depollution equipment; seeking solutions based on the principles of circular economy of pulverulent materials retained in depollution activities.	Specialists in physics, chemistry, meteorology, air pollution management, infrastructure specific to depollution facilities, etc.
7.5.1.	Management of waste generated during processing	Approaching the elaboration process on the principle of inputs and outputs, paying special attention to reducing losses through handling, reducing the neutral part generated, looking for solutions to ensure recirculation and those related to the imposition of the resulting by-products.	Specialists in the field of metallurgy, electrical installations, automation, physico-chemical analysis, chemistry of refractory materials, etc.

6. The results of the analysis of the expected activities and partnerships

In order to establish the specificity of the communication, it is important to evaluate the process partners, namely the actions that are subsumed to the functionalities of the market economy from the perspective of materials engineering, by respecting the conditions related to ensuring circularities. The analysis of the partnerships shows that the process of ensuring circularity is carried out together with another very important one, that of digitalization within the fourth industrial revolution "Industry 4.0". Therefore, it can be appreciated that the engineering of the manufacture of castings is and will be strongly marked by powerful knowledge in terms of achieving an environmentally friendly production through the use of computer systems and infrastructure.

Given this trend, it is appreciated the importance of creating an IT platform for industries served by materials engineering, as a tool for positioning towards the socio-economic environment, i.e. producers, suppliers and beneficiaries to the requirements of the circular economy. To these components are added, given the existence and need for institutionalized conditioning, also the public authorities and global structures acting at the financial, economic and sustainable development level.

Being aware that the problem posed is an interdisciplinary one, or more precisely a multidisciplinary one, it can be seen that at the level of communication an engineering approach is needed to provide the essential data in having a proper picture of the goal. With this motivation in mind, we consider that the engineer specialized in materials engineering is necessary to penetrate the constitutive mechanisms and use of the virtual image, which gives him the possibility to perform an analysis of the manufacturing process from this perspective. This can offer the possibility of establishing a strong interpretive basis, of solid positioning for the specific actions of the "smart economy" and for the analysis of a phenomenon, present at the moment, that of defining a multidimensional truth and its use in identifying "false information" as such.

In order to achieve the proposed objective, it is necessary for the communication to emphasize the importance of the following activities:

- critical analysis of the conditions and indicators specific to the circular economy from the perspective of technological solutions;
- evaluation of the systemic modalities of virtual communication, in order to adopt the variant usable within the materials engineering;
- critical engineering analysis of manufacturing systems from the perspective of the circular economy and the use of its constituent elements.

7. Structuring the architecture of a collaborative platform by transposing the circular economy at the level of the manufacture of castings

The design of the structure of the collaborative platform architecture took into account the ways and conditions that virtual communication, through its specific tools, can integrate the principles of the circular economy, through the mutations and effects produced in the castings industry. For this, through the analysis made to the actors present in Romania at the level of cast parts, the following elements were pursued: the elements of interest of the circular economy that transpire from the virtual image of organizations, the ways to migrate to new business models based on information access and economic circularity; the influence of the principles of the circular economy in the design and manufacture of castings, which involve the use of tools specific to virtual communication.

Given the objective of the paper, the studies focused on the analysis of the major elements that determine the new economic models in the casting industry, models that correlate virtual communication, specific manufacturing and the circular economy. In this sense, a detailed analysis was performed on the following levels: external communication through presentation sites, the structure of manufacturing processes, the presence of elements subscribed to the circular economy.

In the analysis of the first level, that of communication, starting from the fact that the main objective of IT platforms is to facilitate the interconnectivity of economic actors, we noticed, for the analyzed situation, that of Romanian producers' sites, the need to improve communication between the protagonists of this field, especially the facilitation of the relationship between them, according to the interests they have, with priority in meeting the demand with the supply. Such an approach involves, however, planning strategic paths by developing a "communication plan" for this purpose, a document focused on messages and communication objectives, materialized in the virtual image.

Thus, from this perspective, the following elements were proposed at the level of restructuring the virtual image: elaboration of a general picture on the virtual communication mode in the economic sector of castings, analysis and validation of the types of virtual messages conveyed, capturing the ways in which the communication objectives of the economic actors in the manufacture of castings are affected in the communication process. They materialize and receive quantifiable specificity in image indicators and sub-indicators such as: content, expression, meaning, identity, visibility, interactivity, dissemination, etc. These were presented in detail in an "*analysis of the manufacture of castings and the perspective of virtual communication*" [7].

Following the analysis of the position of manufacturers, with reference to the characteristics of business models, considered essential in promoting the circular economy in the manufacture of castings under the current IT conditions, four types of characteristics of these business models were identified: access and co-use (sharing) of services, products or raw materials, within communities, or value chains in a certain market; advanced use of state-of-the-art IT platforms, networks and mobile communication technologies for monitoring and management, use of aggregates, customers and information; management and optimization of material assets (tangible or intangible), which are co-used, for the purpose of a valuable and relevant delivery and recovery; communication and relationship with users through social networks, exchanging offers, news, recommendations, appreciations or information.

Without going into detail on the development of the architecture of the IT platform (static platform and dynamic platform), starting from the analysis made and the need to adapt engineering behaviors to the requirements of economic models and virtualities of the moment, the main functions are presented:

the main functions are presented: general presentation of the fields (manufacture of castings, circular economy, informatics applied to industrial environments), socio-economic data of the fields; platform competences (animation of relations between members, communication - media communication and events, information, publication of specific materials, partnerships, technical coordination, legal services, professional information services and human resource provision, environmental protection, hygiene and security services, eco-responsible activities, domain development considerations); platform partners (professional associations, public authorities, licensing and certification institutions, research institutions, vocational and university training organizations, casting manufacturers, suppliers - technological infrastructure, raw materials, design, consulting, etc.), the technological potential of the platform (technologies, alloys terminology, computer aided manufacturing, etc.), innovation, research and development (technological innovation, organizational innovation, social innovation, fundamental research with application potential, technological research, innovative solutions in vocational training, etc.), eco-responsible activities environmental, social responsibility, etc.), considerations on the development of the fields, jobs and professional training, the news of the field, the management of the adherents of the platform, etc.

At the level of structuring the dynamic platform, the following components were taken into account: the platform management unit, the permanent virtual exhibition of the manufacture of castings at the conditions of the circular economy, the virtual conference of technical-scientific communications, multilingual institutional and technical dictionaries, general offer, general demand, specific commercial actions and component units. Starting from the analysis categories in the promotion of the circular economy, presented in Figure 1, we established the correlation which exists between the analysis categories and the defined functions of the IT platform for the promotion of the circular economy.

8. Conclusions

The construction of the paper presents a methodology for analyzing the problem of applying the circular economy in the manufacture of castings, with practical utility in implementing specific actions in a field of materials engineering, namely the manufacture of castings. At the same time, according to the analysis made, a procedural variant is offered for the choice of the partnerships necessary for the realization of the specific measures and the modalities of communication at their level.

One of the novelty elements used in the established procedure is the introduction of the notion of "material intensity" in the analysis. According to it, for the manufacturer's attention and for creating a suggestive image at the communication level, it was calculated that a ton of good castings means the direct management of 1,644 tons of materials, which transformed from the point of view of biotic consumption represents 12,174 tons of materials, from that of water 99,399 tons, and in terms of air consumption 2,061 tons. From the analysis of the results of the casting process, it was found that the used sand has the potential to be capitalized in partnerships with the beneficiaries from the cement industry. Thus, through industrial symbiosis measures, the option of reducing the pressure on the environment was reached by the fact that out of the 325 kg of used sand generated per ton of good cast piece, 295 are taken over in the cement manufacturing process, for which the industrial landfill storage has been replaced with an intermediate platform for preparation in view of dispatching.

Considering the fact that the problems of applying the circular economy at the level of the castings industry is a multidisciplinary one, a fact that requires the realization of highly complex partnerships in a regime of maximum operability. Starting from this reality, demonstrated procedurally, the paper suggests a possible architecture of an action platform for the circular economy. Considering the specificity of the existing situation at the level of the production of castings in Romania, with a value of 125,000 tons, from the discussions carried on at the level of producers, it was found that another development variant would not be possible, lacking large producers that could set the tone in this area. Therefore, the awareness communication becomes an absolutely necessary tool, in the given situation of Romania, which justifies the need for such an approach in introducing the principles of the circular economy in the manufacture of castings.

At the same time, the data provided in this paper can support the reform of training programs in engineering for the manufacture of castings, by introducing notions that support the acquisition of skills necessary for good management of the circular economy in materials engineering.

9. References

[1] Vincez A and Georgeault *L Economie circulaire. Systeme economique et finitude de ressources* 2016 (Paris : De Boeck Superieur) p 115

[2] *** *Buletin de studii Industria Sârmei S.A.* 1938 (București: Socec & Co S.A.)

[3] Rtthoff M, Rohn H and Liedt E 2003 Calculting MIPS – Resourse productivity of products and services *Wuppertal Institute for Climate, Environment and Energy*

[4] Roussat N 2007 *Optimisation des strategies de gestion du contenu matieres premieres et energie des dechets produits et utilises sur un territoire urbanise, These pour obtenir le grade de docteur* (Lyon : Institut National des Sciences Appliques de Lyon)

[5] Soporan V F 2017 "Circular economy engineering", lecture notes specialization of Master studies "Management and engineering of natural resources and waste" Technical University of Cluj-Napoca

[6] *** Communication of the Commission "Towards a circular economy: a zero waste program in Europe", COM (2014) 398 final, European Union

[7] Soporan V F, Crișan M D, Lehene T R and Pop A L 2016 *IOP Conf. Ser.: Mater. Sci. Eng.* **133** 012063

[8] Vescan M M, Soporan V F and Pop A L 2016 Castings manufacture from the perspective of digital communication in virtual space *Proceedings of International Conference on Innovative Technologies, Publisher: Faculty of Engineering University of Rijeka* pp 147-150

Considerations Regarding the Recalibration of the Manufacture of Castings from the Perspective of the Development of the Circular Economy

V F Soporan[1*], M M Vescan[1], T R Lehene[1], S Păduretu[2] and T Gabor[1]
[1]Department of Environmental Engineering Entrepreneurship and Sustainable Development, Technical University of Cluj-Napoca,103-105, Muncii Boulevard, RO 400641, Cluj-Napoca, Romania.
[2]Department of Modern Languages and Communication, Technical University of Cluj-Napoca, 103-105, Muncii Boulevard, RO 400641, Cluj-Napoca, Romania

*E-mail: vfsoporan@gmail.com

Abstract. The paper presents the issue of recalibration of cast parts from the perspective of the circular economy, by presenting the specific issue, by setting the main objective of the study, that of identifying areas that require interventions in the technological process and ensuring socio-economic functionality. The analysis, in general, includes the presentation of the specifics of the analyzed process, the construction of assessment indicators of the circular economy and the quantification and hierarchy of the influencing factors. Particular attention is paid to the quantitative indicators and value appreciation of the manufacture of castings to the rigors of the circular economy. At the level of interpretive analysis, are presented the critical areas that require a multidisciplinary intervention to ensure the specific conditionalities of the circular economy.

1. Considerations for addressing the issue of the circular economy in the manufacture of castings

In tackling the proposed theme, the following considerations were taken into account: the importance of the circular economy in the context of current economic and social developments, the influence that materials engineering has on the specific developments of circularity, the interest of Romanian specialists, both theoretical and practical, including among these the achievements of the authors of this paper as well.

For the first consideration, one where engineering must play a key role, are taken into account, by proper management of the manufacturing processes of products and services, in order to achieve the proposed functionalities, energy and material consumption in a circuit that considers their entire life cycle.

In this process, materials engineering has a special influence, a topic which includes the issue of elaboration, processing into utility products for various industries, shaping by various processes in functional geometric structures suitable for various components of machines and equipment, tracking operational safety and recovery, at the end of the life cycle, in order to reintroduce in the different manufacturing processes the materials included in the various functional structures.

By defining "materials engineering" in this way, we consider that "castings manufacturing" also falls into this category, with the specification that it is one of the best technological examples with adaptability to the circular economy, which is why we conducted a study on recalibration of the castings industry to the new requirements of the circular economy.

The third consideration is related to the fact that among Romanian specialists there were significant concerns regarding the promotion of the circular economy, both theoretically and practically. At a theoretical level, the work "*La croissance: Entropie – Ecologie – Economie*" [1] by Nicholas Georgescu-Roegen stands out, in which an extension of the entropy law to economic problems is made, the theoretical bases of the circular economy in which we have to deal with a decrease in terms of the potential of resources needed in the conduct of various technological processes.

At a pragmatic level, the "first metallurgy magazine" in Romania stands out, published in 1938, which justifies, on the principles of the circular economy, the construction of a steelworks in a company specialized in plastic deformation (lamination, wiredrawing, obtaining nails and other metallurgical products). Some of the measures taken for the rational use of metallic materials, revolutionary at the level of the interwar period, are present in the mentioned paper, among which we mention the following: "*reuse of scrap metal; methods suitable for the execution of a thing with a minimum of material cost; constructive forms adapted to the demands and qualities of the materials; ennobling steel by adding other materials to increase its strength; protection against acid and moisture attack.*" [2] At the same time, the presence at the "*Dusseldorf Exhibition of 1938*" [2] highlights the concerns in the 1930s regarding the rational use of resources, giving examples of the production of synthetic petrol from coal or in obtaining synthetic rubber from coal and lime. In the same context, it is emphasized the existence of a stand for "used materials", which aims "*to draw everyone's attention to all those materials, which, not being immediately useful, are discarded. The exhibition aims to show that old things have an appreciable economic value, they can be used to manufacture new objects, thus having the character of raw material. And, in order for the visitor to keep in mind the importance of used raw materials, it is shown in detail how they are sorted, what operations are then subjected to, with which machines are processed and what manufactured products they can give rise to.*" [2].

Regarding the concerns of the authors of this paper, they have developed numerous studies on the application of the circular economy in the manufacture of castings, including: "Rudology of the manufacture of castings", "Methodology for assessing the manufacture of castings from the perspective of the circular economy" [3], "Engineering of the institutionalization of the circular economy in the manufacture of castings" [4] or "Engineering of the circular economy and good management of material resources" [5].

In structuring the analysis, two fundamental works were used, one related to "Circular economy. Economic systems and resource limitation" [6] and the other related to specifying the relations between "engineering and the environment in the castings industry" [7].

2. The objective of the study and the way of performing the analysis

The proposed study on recalibration of castings manufacturing, as a component of wrought materials engineering, aims to identify areas that require interventions in technological processes and socio-economic operations from the perspective of institutional development of the circular economy.

The analysis is performed by taking into account the following constituent elements: the specifics of the technological process; use and adaptation of assessment indicators; quantification and ranking of influencing factors.

The first element, that of the specificity of the cast products industry from the perspective of the circular economy, aims at an evaluation of the technological process, the evaluation of the materials entered in the technological process, the evaluation of the environmental impact and the evaluation of the circularities that are ensured.

The second element, that of loading with institutional value the indicators adapted and used in the quantitative and value assessment of the manufacture of castings to the rigors of the circular economy, considers the proposal of the following evaluation structure:

- **Indicator**:
  - of mass recovery at the level of castings (IVM), calculated as a percentage ratio between the mass of good castings and the mass of materials entered in the technological process of obtaining castings;
  - the non-useful part generated in the technological process (IPNU), calculated as a percentage ratio between the mass of the non-useful part and the mass of the good castings;
  - the part with recirculation potential generated in the process of obtaining the castings (IPPRECIRCULAR), calculated as a percentage ratio between the mass generated with recirculation potential and the mass of the good castings;
  - the part with potential for recycling generated in the technological process of obtaining castings (IPRECYCLING), calculated as a percentage ratio between the mass generated with potential for recycling and the mass of good castings;
  - the part requiring special disposal treatment (IPNTSE), calculated as a percentage ratio between the mass of the generated material requiring special disposal treatment and the mass of the good castings;
- **Unit expenses**:
  - with the electricity consumed in the filtration installations, expressed in Euro / 1000 tons of castings;
  - intended for household waste management, expressed in Euro / 1000 tons of castings;
  - intended for the management of hazardous waste, expressed in Euro / 1000 tons of castings;
  - intended for wastewater monitoring, expressed in Euro / 1000 tons of castings;
  - intended for monitoring environmental factors, expressed in Euro / 1000 tons of castings;
  - intended for domestic water management, expressed in Euro / 1000 tons of castings;
  - intended for other environmental expenses, expressed in Euro / 1000 tons of castings;
  - total unit average, expressed in Euro / 1000 tons of castings;
- **Unitary investments** for investments in the field of environmental protection, expressed in Euro / 1000 tons of castings.

The third element, that of the factors and measures influencing the circular economy, was constructed taking into account the model offered by Aurez V and Georgeault L.

3. Materialization of the procedure for substantiating the study

The materialization of the substantiation procedure of the study was achieved by quantifying the production in a cast iron foundry over a period of seven years and reporting it to the level of environmental costs and investments made in that period, identifying areas of intervention to ensure economic development in the calibration process.

In order to have a basis on the total account of the materials entered in the technological process, the scheme from Figure 1 is used, which shows the inputs of the materials in the technological process and the output of the delivered cast parts and the useless part generated in the technological process, and in Table 1 their quantitative values are shown, as an average of the 7-year analysis performed.

The values identified on environmental expenditures and environmental investments for the same period are presented in Table 2.

For the situation taken into account in the analysis, Table 3 presents the calculated values of the reporting indicators according to the rigors of circular economy, the reporting addressing the average of the data obtained over the analyzed interval and presented in Table 2.

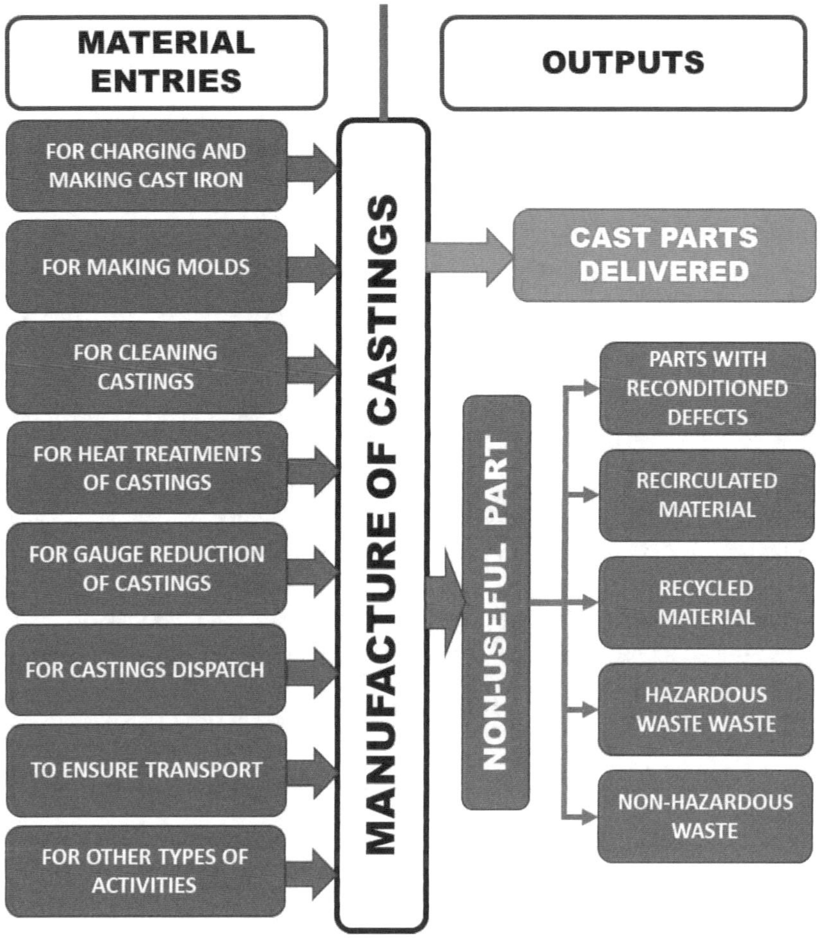

Figure 1. Inputs and outputs at the level of casting manufacturing.

Table 1. Quantities of materials included in the technological process and production of castings and the neutral part generated.

Input / generated material	UM	Value
MATERIAL ENTRIES IDENTIFIED IN THE TECHNOLOGICAL PROCESS		
For charging and making cast iron	t	17,416
For the preparation of forming mixtures	t	7,096
For making molds	t	76
For cleaning and sandblasting molds	t	204
For priming castings	t	27
For castings dispatch	m^3	49
For making wooden models	m^3	192
Used to make metal models	t	102
For making spare parts	t	27
Used in equipment maintenance activities	t	28
Used for transport	t	139
TOTAL MATERIALS USED	**t**	**25,356**

Input / generated material	UM	Value
OUTPUTS IDENTIFIED FOLLOWING THE TECHNOLOGICAL PROCESS		
Cast parts delivered	**t**	**15,713**
Recirculating cast iron (burrs, casting networks, scrap parts, etc.)	t	4,700
Metallic pellet waste	t	1
Ferrous cast iron chip	t	839
Ferrous steel iron chip	t	10.0
Aluminum waste	t	0.5
Cables	t	0.2
TOTAL MATERIAL WITH RECIRCULATION POTENTIAL	**t**	**5,550.7**
Used foundry sand	t	5,544
Lining and refractory materials from metallurgical processes	t	10
Waste from refractory bricks	t	2
Shredded grinding parts	t	0.1
Used oil	t	1.0
Sawdust and wood shavings	t	7.0
Rubber waste	t	1.0
TOTAL MATERIALS WITH RECYCLING POTENTIAL	**t**	**5,565.1**
Waste paints containing organic solvents	t	3.1
Metal packaging contaminated with dangerous substances	t	15.2
Plastic packaging contaminated with dangerous substances	t	9.4
TOTAL MATERIALS WITH NEUTRALIZATION TREATMENT	**t**	**27.7**
TOTAL NON-USEFUL PART GENERATED		**11,136.8**

Table 2. Data on the production achieved and its reporting regarding environmental expenditures and investments made in the 2018 interval.

Indicator	Value of expenses, Euro							
	2012	**2013**	**2014**	**2015**	**2016**	**2017**	**2018**	**Annual average**
Electricity filtration installations	101,886	188,088	162,350	152,681	138,526	124,361	153,961	145,979
Household waste	103,744	2,763	3,347	5,892	6,213	8,188	7,110	19,608
Hazardous waste	5,767	0	3,638	4,285	3,235	9,805	5,518	4,607
Wastewater monitoring	806	1,914	2,330	11,054	2,459	1,067	286	2,845
Monitoring environmental factors, Air	36,668	5,456	5,283	4,954	4,631	4,408	4,520	8,709
Sewage, household water	36,753	39,850	40,554	44,41	34,536	28,823	33,066	36,846
Other environmental expenses	81	52	48	41	39	9,192	38	1,356
Total	**158,095**	**238,073**	**217,550**	**214,652**	**189,641**	**185,846**	**204,501**	**201,194**
Environmental investments	**61,904**	**35,346**	**116,518**	**266,031**	**0**	**122,146**	**42,525**	**92,067**

Table 3. Calculated values of the indicators related to the rigors of the circular economy.

Indicator of reporting to the rigors of the circular economy	UM	Value
Mass recovery index at the level of castings (IVM)	%	68.68
The index of the non-useful part generated in the technological process (IPNU)	%	64.14
Index of the part with recirculation potential generated in the process of obtaining castings (IPPRECIRCULAR)	%	31.86
	%	31.95
Index of the part requiring special disposal treatment (IPNTSE)	%	0.16
Unit costs of electricity consumed in filtration plants	€/1000 t of castings	26,231.62
Unit costs on household waste management	€/1000 t of castings	3,523.45
Unit costs for hazardous waste management	€/1000 t of castings	827.85
Unit costs for wastewater monitoring	€/1000 t of castings	511.23
Unit costs for monitoring environmental factors	€/1000 t of castings	1,564.95
Unit costs for wastewater management	€/1000 t of castings	6,621.02
Unit costs for other environmental costs	€/1000 t of castings	243.66
Total unit environmental costs	€/1000 t of castings	36,153.45
Unitary investment for investments in the field of environmental protection	€/1000 t of castings	16,543.93

4. Interpretation of data obtained at the level of indicators from the perspective of the circular economy

The use of indicators in relation to the rigors of the circular economy gives us the possibility of some interpretations, which offers the possibility to determine the critical areas regarding the recalibration of the castings industry to the new requirements.

Thus, from the point of view of annual inputs and corresponding outputs, it is estimated that a ton of good castings is made - from a manufacturing point of view – out of 1.68 tons of materials specific to the technological process, which is a good capitalization of the resources introduced.

From the recirculation perspective, a first-rate process in building the circular economy, the value of the indicator is favorable to the process, in the sense that the manufacture of a ton of castings generates a quantity of 318 kilograms of materials that are recirculated in the technological process. Another interpretation, equally useful in ensuring the principles of the circular economy, is that there is a significant consumption of energy and material resources that is embedded in the networks of castings, processing additives and masses. Therefore, it is important that this indicator is as small as possible in order to reduce the consumptions mentioned. A drastic, technologically unprepared reduction can lead to a substantial increase in scrap, which requires a thorough increase in IT tools to ensure a secure design of manufacturing technology. From this perspective, it results that one of the measures to ensure the circular economy is related to the digitization of the castings industry.

Another element of ensuring the conditions of the circular economy is related to the knowledge of the non-useful part that has the potential to capitalize on recycling. Thus, the adopted indicator, which measures the weight of the process mentioned, at the level of the analysis it is found that for one ton of cast parts a quantity of 319.5 kg of materials is generated that have the potential to be capitalized at the level of recycling. In order to bring the process as close as possible to the circular economy it is important to have a reduction of the mentioned quantity, and for the entry in the recycling area to meet the related conditions it is important to refer to the specific by-product conditions. Therefore, in the entry into the recycling area of the non-useful part generated in the manufacture of castings, it is important to cooperate with other areas, especially with the area of building materials, public works and other sectors of the chemical industry and metallurgy. A relevant example in this field is the study, carried out by two large French centers - CERIB and CTIF, which presents the results of the recovery of waste from the castings industry at the level of construction materials [8].

The fifth indicator gives us an image of the non-useful part, generated within the technological process, which requires a special neutralization treatment. In the case analyzed, it is found, as an average of the results processed in seven years, that for a ton of good cast parts it is necessary to neutralize a quantity of 16 kilograms, usually the materials that enter the hazardous zone. From the perspective of the circular economy, it is important that this amount is as small as possible.

At the level of indicators, which make an assessment of the expenditures made for ensuring the conditions of the circular economy, the importance of their knowledge is underlined. Given the fact that there are not enough data neither in the literature nor in the data which companies operate with, it is currently difficult to make a comparative analysis.

5. Conclusions

From the analysis of the existing situation for an average productive situation, in terms of ensuring technological developments, it is estimated that the success of including the manufacture of castings in the new rigors of the circular economy is not possible without establishing appreciation indicators, without cooperation with new fields of advanced engineering, which includes elements of intelligent informatics in ensuring the development and monitoring of the technological process.

The research carried out for a particular situation can be a model for analyzing the manufacturing process of castings, by referring to the requirements of the circular economy.

At the same time, there is a need to institutionalize the indicators of reporting to the circular economy and to stimulate the market economy of companies operating in the field of casting, from the perspective of efforts to ensure circularity for an industry that globally ensures a production of over 100 million tons.

6. References

[1] Georgescu-Roegen N 1979 *La decroissance: Entropie – Ecologie – Economie* (Lausanne : Pierre-Marcel Favre)

[2] *** *Buletin de studii Industria Sârmei S.A.* 1938 (București : Socec & Co S.A.)

[3] Soporan V F, Crișan M D, Lehene T R and Pop A L, 2016 *IOP Conf. Ser.: Mater. Sci. Eng.* **133** 012063

[4] Vescan M M, Soporan V F, D M Crișan, Lehene T R, Pădurețu S and Samuila V 2017 *IOP Conf. Ser.: Mater. Sci. Eng.* **209** 012096

[5] Soporan VF, Lehene TR, Pădurețu S and Gabor T 2019 *IOP Conf. Ser.: Mater. Sci. Eng.* **572** 012086

[6] Aurez V and Georgeault L 2016 *Economie circulaire. Systeme economique et finitude des resources* (Paris : De Boeck Superieure)

[7] Duquet B, Laplaiche N, Le Deaut A and Rousiere D 2015 *Inginerie et environnement en fonderie* (Paris : Techniques de l'ingenier)

[8] Petitpain M 2016 *Economie circulaire: Projet valorification croisee de co-produits industriels, Etudes et recherches* (Epernon: CERIB)

IOP Conf. Series: Materials Science and Engineering **877** (2020) 012054 doi:10.1088/1757-899X/877/1/012054

Innovation for Trap Particle and Eliminate Germs in Air By Corona Electrostatic System

S Ketkaew[1*]
[1]Faculty of Engineering, Ramkhamhaeng University, Ramkhamhaeng Road, Huamark, Bangkapi, Bangkok Thailand 10240.

*E-mail: siseerot@hotmail.com

Abstract. This article presents an innovation for trap particle and eliminate germs in air by corona electrostatic system. The air circulation rate is 165 cubic meters per hour by designing the power supply with IC#TL494. The IC#TLP250 uses separate ground and drive Power MOSFET#IRFP460 by bringing DC high voltage from the transformer#TLF14511 to the dust collector unit and the electron separation unit. By designing and building a machine capable of trapping dust particles smaller in size from 0.3 microns up to the results of the tests showed that the dust collecting efficiency of 95.37% from the experiment by measuring the amount of dust using Personal pump and the dust collection efficiency of 95.86% from measuring dust using Handy laser partical counter and part of electron separation can produce corona for producing ozone gas up to 3 ppm to eliminate bacteria in the air. The UV lamps for ozone emissions are reduced to 0.1 ppm for disaster prevention to users.

1. Introduction

In the modern era, there is an increase in air pollution. Which is the effect of various technology development has affected the living conditions and health of the people Therefore, it has invented the method of dust trapping by using high voltage, direct current electricity to reduce air pollution which has been applied by using high voltage switching transformers to produce high voltage to create a dust collector set, where air pollutants are whatever substances are present in the air that affects human health or other living. The air is beings that are undesirable to humans or is a substance that directly or indirectly affects human well-being. There are many types of toxins or contaminants in the air, including dust, odors, smoke, soot and radioactivity, such as carbon monoxide (CO), sulfur oxides (SOX), nitrogen oxides (NOX), hydrocarbons (HC), mercury (Hg), lead (Pb) Radioactive aerosols, etc. These substances, if exceeded in the air, will cause danger to humans, animals and plants. Acute effect can lead to death. Some types can cause chronic poisoning (Chronic Effect) and later death. Toxins may enter the body by contact breathing or indirectly, such as contaminating food with clothing attached and then transmitted into the body for another.

From previous research on dust and cyclone dust collectors for the rice mills of Surin, Ngarmamngam, Sasiwawan Intharakong, Kittisak Chaisuwan, Saran Yu Suksawat, Anawin Kannkaew and Proud Laosong have taken the Abstract: The dust that floats in the air affects the health of living things. Both humans and animals Small particles can penetrate deep into the respiratory system causing inflammation or allergies. This research presents Cyclone and electrostatic dust collectors for rice mills the purpose is to reduce dust in rice mills. Cyclone and Electrostatic Dust Trap Designed to have a rough surface finish panel in order to easily corona increasing the ability to trap dust using DC voltage is not very high. From the test results by measuring the amount of dust coming out of the cyclone found an average amount of dust of 7.3 mg/m^3. The remaining dust will enter the set Trap dust with static electricity when measuring dust at the exit of the electrostatic dust collector, the average dust content is 0.909 mg/m^3. From the dust measurement results, it can be seen that cyclone and electrostatic dust collectors can help reduce dust problems droplets in the rice mill down [1].

For research on innovations in dust trapping and airborne germs elimination by using an electrostatic system together with a corona discharge focus on design studies and creating an innovation to trap dust to remove contaminants in the air by focusing on solid parts such as dust, smoke, incense, cigarette smoke and particles between 200 microns and 0.01 microns in size. By creating an electrostatic suite consist of stainless steel plates that are placed at an appropriate distance and alternately feed the positive and negative voltage. To create an electric field between the stainless sheets when various dust and smoke passed through will be captured by stainless steel sheet. And the corona set produces ozone gas for removal of airborne germs With UV lamps to help reduce the amount of ozone gas that is released from the dust trap innovation to the amount of ozone gas suitable for the atmosphere in the room.

2. Equipment and methods

The working principle of the innovation to trap dust and eliminate airborne germs as in the Figure 1. Innovation to trap dust and eliminate airborne germs is to cause the air around the ionized wire to break down into ionizing air (by supplying high voltage direct current to the corona wire) Which will cause the electric field at the surface of the conductor wire to have a high electric field intensity and when the atoms or molecules of the air pass near the wire will cause the air to become ionized.

The dirty air, such as dust, smoke and pollution When passing through the dust trap innovation, this air will circulate through the pre-filter, which filters dust or contaminated particles in the front. From then, this air will be given energy from the high intensity electric field until ionization and the charging of dust to various particles will move to be stuck on a dust collector (Collector Plates) with the opposite electric charge. And air free from dust with UV lamps to reduce the amount of ozone gas to approximately 0.1 ppm for the safety of users, therefore will get clean air free from dust, germs and smells.

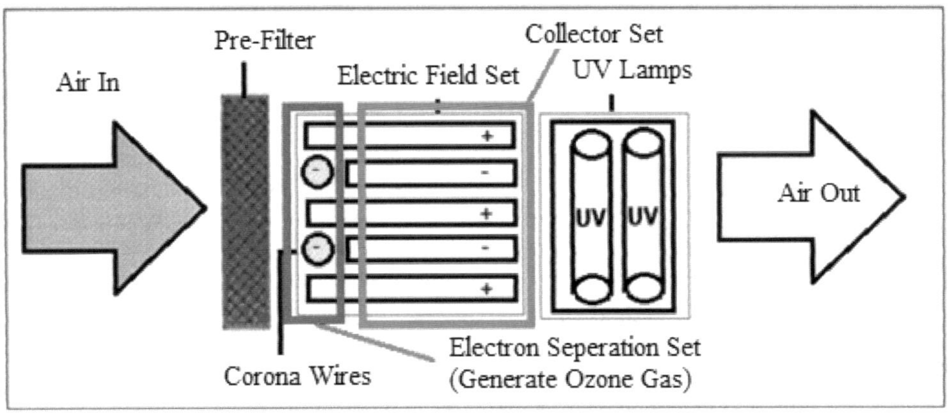

Figure 1. The internal structure of the innovation to trap dust and eliminate airborne germs.

In general air, dust, cigarette smoke and other germs are mixed in the air. Therefore, if we pass through the air to be ionized by high voltage circuit dust will be charged and will move to adhere to the set of dust traps. Therefore, making the air we breathe clean Free from dust and germs, improving physical health.

3. The dust trap innovation is divided into 2 major types

3.1. Dust trap innovation by using mechanical principles

This dust trap innovation relies on the use of air filters as filters to remove dust or contaminants from the air by using an exhaust fan is a vacuum that circulates through the air filter. The exhaust fan used must be large in order to overcome the friction or flow resistance of the filter which is high. And the filter used cannot be cleaned.

3.2 Dust trap innovation by using electrical principles

This dust trap innovation relies on air disintegration. (Ionization) and then the dust is charged and then moved to the dust trap. The fan used is smaller than the mechanical type due to the small friction between the dust collectors and the dust collection sheet can be washed and cleaned.

4. Build and design

4.1. Design of an electrostatic system and corona discharge (Electric field cell sets) for trapping dust and producing ozone.

The electrostatic and corona discharge systems consist of a set of stainless steel plates that alternately feed and subtract electric potential. For the dust collection sheet to be used, the thickness should be small in order not to have a strong resistance to the air flow which causes turbulence and will reduce the efficiency of dust trapping. Therefore, the thickness of the stainless steel sheet for dust trapping is set to be equal to 0.09 centimeters. Of the electron separation set and the dust collection set, as in Figure 2.

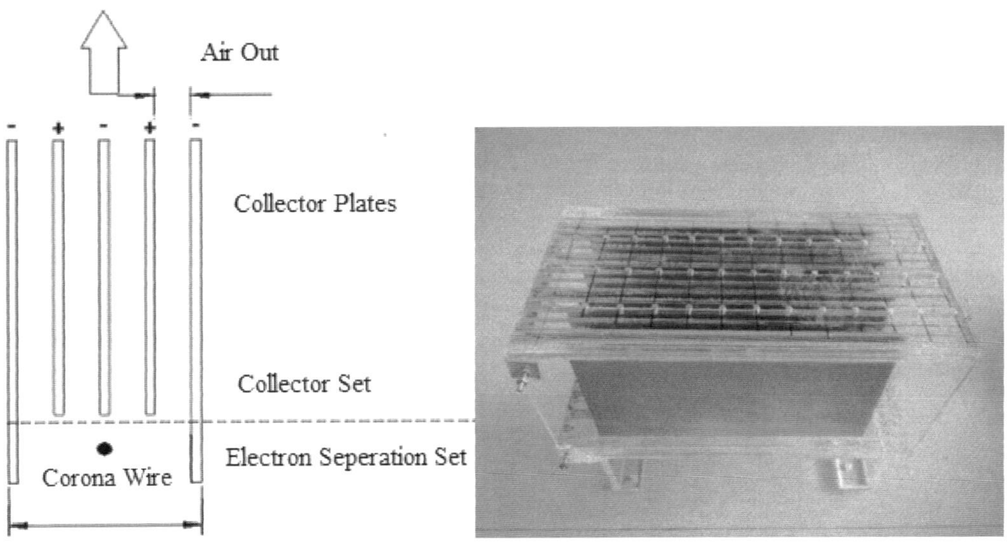

Figure 2. Electrostatic and Corona Discharge Kit (Electric field cell sets).

Electrostatic system is a series of electrodes. The uniform electric field means the electrode that has every electric field. The points between them are equal which can be calculated from the relationship.

$$E_{max} = E = \frac{V}{d} \tag{1}$$

Where: V is the voltage that is fed between the electrodes;
d is the distance between the electrodes;
E is the electric field stress at any point between the electrodes;
E_{max} is the maximum electric field stress.

Corona Discharge Kit Consists of corona wire placed between stainless steel sheets which has the opposite voltage If considering the electric field stress of the corona set will be according to the equation.

$$E = \frac{V}{d \times \eta^*} \tag{2}$$

Where: E is the electric field intensity;
V is the high voltage;
d is the distance between the corona wire and stainless steel sheet;
$\eta^*$is an electric field factor.

4.2 Design and construction of a high-voltage switching power supply for static and corona discharge charging systems.

Switching high-voltage power supply design introduced the converter circuit principle By using IC No. TL494 as a pulse width modulation generator as a switching control circuit. Which uses mosfet as a device for conducting current which has a switching frequency of approximately 20 kHz. Finally, it will pass through high frequency transformers (Figure 3). To achieve the desired high voltage.

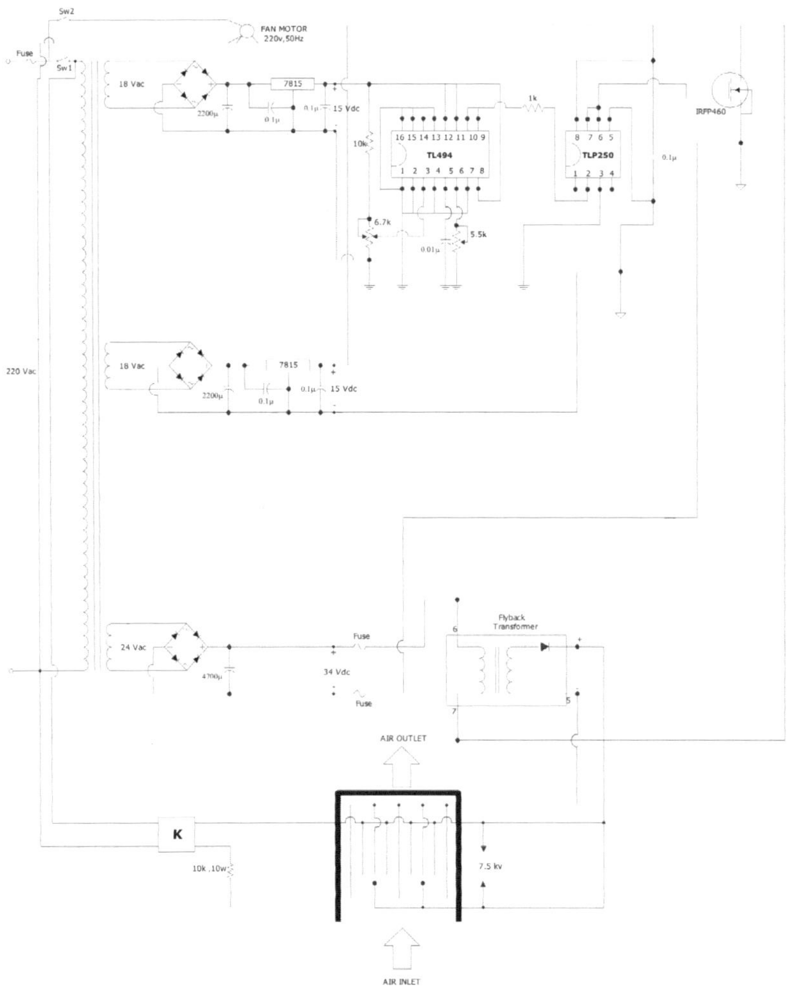

Figure 3. High-voltage DC power supply circuit.

5. Results and criticism of experimental results.

Electric field cell test results with the instrument for analyzing the amount of dust as in Figure 7(a) and the Handy laser particle counter for measuring the dust as in Figure 7 (b) tested at a room size of 25 square meters) (testing time 45 minutes)

The test results of measuring dust trapping efficiency with the tools for analyzing the amount of dust, as in Figure 7(a), it is equal to 95.37%.

The test results for measuring dust collection efficiency with the Handy laser particle counter, as shown in Figure 7(b), are equal to 95.86%.

Analyze the test results of dust collection efficiency with an instrument for analyzing dust quantity as shown in Figure 7(a) and Handy laser particle counter for dust measurement as in Figure 7(b). Both types of dust gauges had no different percentage of dust measurement efficiency. (Dust measurement has other components and factors. Get involved May cause the measurement of dust to be inaccurate as well)

The test results measure the amount of ozone that the electric field cells produce (Test time 45 minutes) and the amount of microorganisms before and after using the innovation as in the Figure 4.

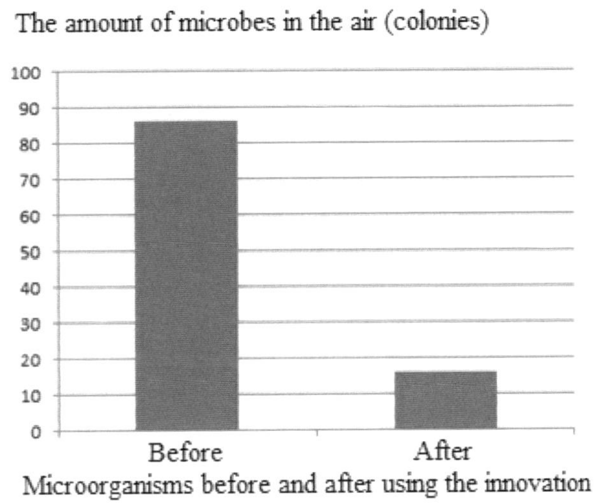

Figure 4. Graph of microbes measured in the air before and after using the innovation to trap dust and eliminate germs in the air.

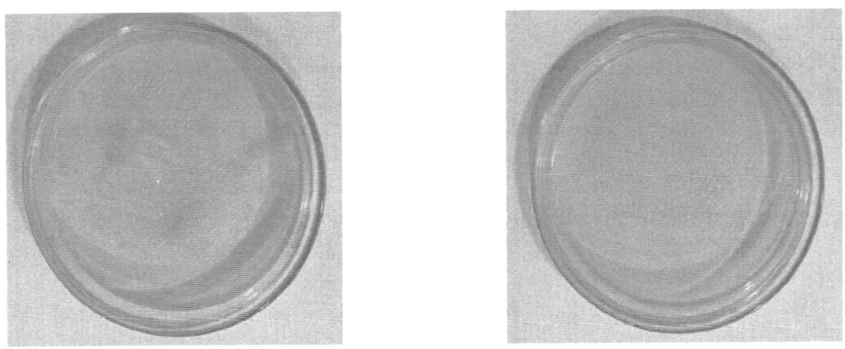

(a) (b)

Figure 5. (a) The amount of microbes measured in the air before using the innovation to trap dust and eliminate germs in the air; (b) The amount of microbes measured in the air after using the innovation to trap dust and eliminate germs in the air.

IOP Conf. Series: Materials Science and Engineering **877** (2020) 012054 doi:10.1088/1757-899X/877/1/012054

The test result is that the amount of ozone that the electric field cells produce is equal to 3 ppm. The amount of ozone gas is only 0.1 ppm and the amount of microbes in the air before using the innovation to trap dust and eliminate airborne germs is equal to 86.3 colonies and the amount of microbes in the air after using the innovation to trap dust and get rid of airborne pathogens the value of 15.9 colonies. The result of measuring the amount of microbes in the air will find that the amount of microbes in the air after using the innovation to trap SRGB and spray disinfectant in the air has less volume.

6. Conclusion

Innovative dust trapping and air germ elimination by using an electrostatic system together with a corona discharge designed and built electric field cell sets (Electron separator and dust collector set) in which the dust collector plates are placed in parallel by stainless steel plates and feed high voltage DC to create the force of the electric field to collect small dust particles. And the electron separation kit used the corona wire to cause the electric field intensity to produce ozone gas by testing the efficiency of dust trapping using equipment for analyzing the amount of dust, as in Figure 7(a), it is equal to 95.37% and the test of measuring the efficiency of dust collecting with Handy laser particle counter as in Figure 7(b) is equal to 95.86%.

Figure 6. Innovation, dust trapping and eliminate airborne germs. By using an electrostatic system together with a corona discharge that has been designed and built.

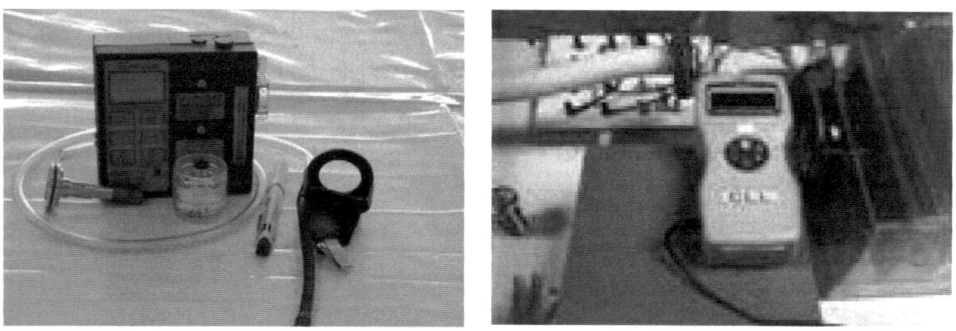

(a) (b)

Figure 7. (a) Equipment used to collect dust samples. In order to analyze the amount of dust; (b) Handy laser particle counter for measuring dust.

The ozone produced by the electric field cells is equal to 3 ppm and the results of the measurement of microbes in the air will find that the amount of microbes in the air after using the innovation to trap dust and eliminate germs in the air will be less. And after passing the UV lamp the amount of ozone gas will be only 0.1 ppm. In the future, this innovation can be further developed commercially in response to and drive the research policy of Thailand as well.

7. References

[1] Surin Haengmamngam, Sasiwawan Intharawong, Kittisak Chaisuwan, Saran Yu Suksawat, Anawin Karnkaew, Phumjai Laopong 2015 Cyclone and Electrostatic dust collectors for rice mills, *The 8th Thailand Renewable Energy for Community Conference*, organized by the Faculty of Engineering Rajamangala University of Technology Thanyaburi, 4-6 November 164-167

[2] Ketkaew S 2002 *Proceeding of the IEEE International Conference on Power System Technology* 1611-1614

[3] Suwat D 1994 Technique & Design of Switching Power Supply, *https://www.electronics-tutorials.ws/power/switch-mode-power-supply.html*

[4] Suwat D 1995 *IEEE Transactions on Broadcast and Television Receivers* **BTR-16 (3)** 203-212

[5] Mohan N, Undeland T M, William R and Robbins P 1995 Power Electronics. (New Yourk: *John Wiley & Sons*)

[6] Louis S T 1989 *High Frequency Switching Power Supply* (London: McGraw-Hill)

[7] Ketkaew S 2012 *Procedia Engineering* **32** 148-154

[8] TL494 Data sheet 2002 *Texas instruments* 1-10

IOP Conf. Series: Materials Science and Engineering **877** (2020) 012055 doi:10.1088/1757-899X/877/1/012055

Toxic Gas Eliminator from Print Ink by Applied High Frequency High Ripple Pulse Corona

S Ketkaew[1*]
[1] Faculty of Engineering, Ramkhamhaeng University, Huamark, Bangkapi, Bangkok Thailand 10240.

*E-mail: siseerot@hotmail.com

Abstract. This paper presents the design and development of toxic gas eliminator from print ink using high frequency high ripple pulse corona technique for decrease VOCs gas in the suction gas of printing ink using high voltage AC switched mode power supply. Using mini inverter principle and control of the switching by IC No.MB3579. The inverter is designed to operate at 10 kHz to 20 kHz through a high voltage switching transformer at output voltage of 1.5 kV, and at the input voltage of 18 V_{AC} 50 Hz, By adapting the non-uniform electric field to the electrode design 3 tubes, one - hour operating yields the ozone gas (O_3) generating capacity of 6.3 ppm enables decrease VOCs gas as well.

1. Introduction

Nowadays, ozone gas is utilized, for example, using ozone gas to wash fruits instead of pomegranate. To eliminate germs and reduce the amount of chlorine in the water-pool. But ozone has disadvantages as well. If used in too much doses, it can cause irritation to the body. But if we use the right amount and suitable for the work that uses ozone gas in the application above, it will bring good health benefits. In this article, we will create an ozone generator with high-frequency switching high-voltage circuits by using IC No.MB3579 as the PWM generator. By sending the high frequency square wave waveform to the signal amplifier IC in order to drive the Power MOSFET high-voltage switching transformers work to send a voltage of 1.5 kV to the electrode in which ozone is produced by the principle of oxygen molecules breaking down which will produce ozone from the equation $O_2 + O = O_3$, which ozone has the ability to withstand a certain voltage. And heat affects the amount of ozone produced. Therefore, it has to be controlled amount of high voltage and the appropriate high frequency for ozone production. The chemical energy needed is 493 kJ/mol to 762 kJ/mol. To convert the unit, the required energy area is 5.583 kWh/m^3 to 8.631 kWh/m^3. There is approximately 21% oxygen. Therefore, the required energy range is 1.172 kWh/m^3 to 1.812 kWh/m^3 is enough to produce ozone in the air gap of the electrode rods.

2. Materials and methods

Therefore, this research requires experimentation by spraying ozone gas into VOCs in the gas suction pipe from the printing ink. To see the amount of VOCs that has decreased or not.

The principles of ozone production are as follows.

a) Materials used to generate ozone gas are 2 types which are

- Use dry air that is everywhere.
- Use pure oxygen

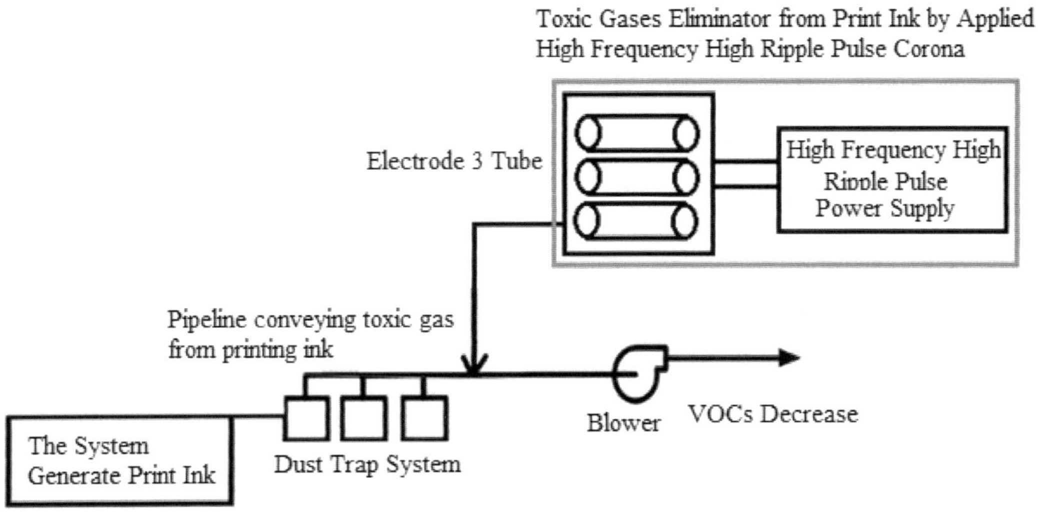

Figure 1. Block diagram of the VOCs reduction system in the gas suction pipe from the printing platform by using a toxic gas remover from the created ink.

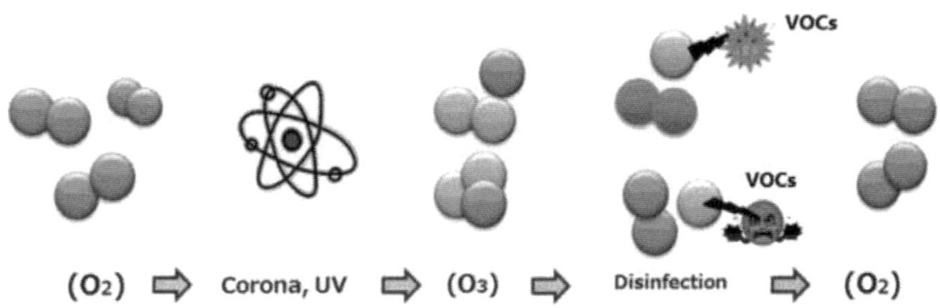

Figure 2. Reaction to destroy pollution (VOCs) of ozone.

b) Ozone generator can be separated according to the following principles

- use electrical principles which we call corona discharge process which is divided into 2 types according to frequency which are low frequency type and high frequency type. For hot weather and moisture, high frequency type is more suitable because it can produce ozone concentration is higher.

- using UV lamps (Ultraviolet: UV), this type will generate ozone. Low concentration will be used in air treatment for water treatment is not suitable because the solubility is very low and therefore not effective.

3. Research objectives

a) To study the design and construction of high-ripple high-frequency pulse power supplies and co-axial cylindrical electrode cells In order to produce ozone gas

b) To study ozone production methods with the use of highly uneven electric fields

c) To study the use of ozone gas to eliminate toxic gas from printing ink.

4. Method of research

The research design is divided into 2 main parts as follows

a) Design of an electrode for ozone production

b) The high ripple pulsed power supply design

5. The design of toxic gases eliminator.

5.1 Electrode stick design (Ozone production rods)

In the design of a 2-layer co-axial cylindrical rod, the design principle is quantity.

Ozone gas is formed very well in the electric field stress which is very uneven. Therefore, the design of the electrode is used in double layers insulation because the use of different values of insulation is suitable for uneven electrical fields. Which will cause the electric field stress in each valuable layer having different values are similar. Therefore, the design of the 2-axis co-ordinated cylindrical rod design, in which the 2-axis co-ordinated cylindrical rod consists of:

a) Insulation class 1 uses glass as an insulator because glass has an effect on the formation of ozone gas with a coefficient (1) of 8 with a diameter of 2.6 centimeters and a length of 30 centimeters;

b) Insulation layer 2 will use air as insulation with a coefficient (2) equal to 1;

c) Cathode terminals are used as copper coils. Is inside of insulating glass;

d) Anode terminal is used as a cylindrical stainless steel rod. With a diameter of 2.9 centimeters and a length of 30 centimeters.

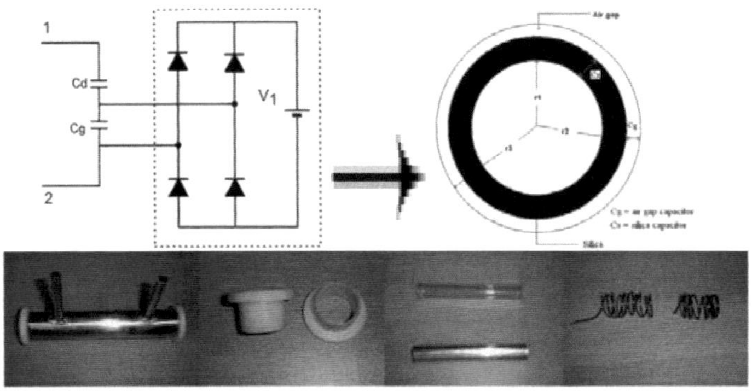

Figure 3. A set of double-layered cylindrical rod electrode structure.

At electrical energy area of 5.58 kWh/m^3 to 7.73 kWh/m^3, in the air there is 21% (O_2) oxygen using at electrical energy range 1.172 kWh/m^3 to 1.620 kWh/m^3

Therefore, the electric power equation is equal to equation 1.

$$W = \frac{1}{2}\int_{Vol} \varepsilon E^2 dv \tag{1}$$

Therefore, the electric field intensity is equation 2:

$$E = \sqrt{\frac{2 \times W}{\varepsilon \; vol}} \tag{2}$$

When the glass is valued and precious weather is Therefore the voltage when fed to the electrode unit is then done creating ozone gas equal to equation 3:

$$V = \frac{E.r_2.\varepsilon_2\left(\varepsilon_1 \ln\left(\frac{r_3}{r_2}\right) + \varepsilon_2 \ln\left(\frac{r_2}{r_1}\right)\right)}{\varepsilon_1 \times \varepsilon_2} \tag{3}$$

5.2 High voltage pulse switched mode power supply design

Switched mode power supply, high voltage, high frequency, switching type using IC No.MB3579 as a (PWM) pulse width modulation generator Which is the control circuit of the Power MOSFET with a

IOP Conf. Series: Materials Science and Engineering **877** (2020) 012055 doi:10.1088/1757-899X/877/1/012055

switching frequency of about 10 kHz to 20 kHz and then send the driving signal to the IC No.TLP250 to separate the ground and amplify the signal between the control unit and the converter set. After that, drive it to power final Power MOSFET going through the high voltage switching transformer to get the high voltage, high frequency as required by creating a high voltage of 1.5 kV to supply the 3 tubes electrode as in the Figure 4.

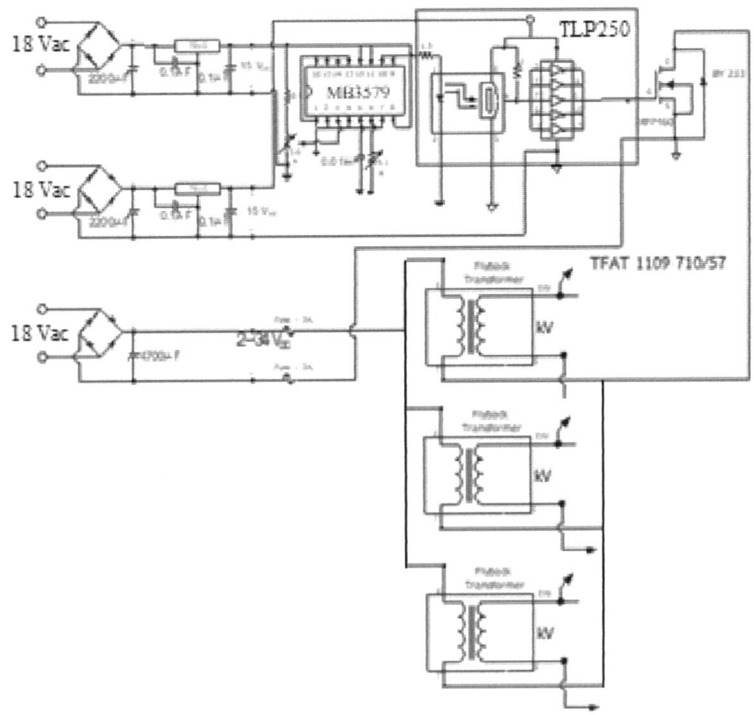

Figure 4. High voltage pulse switched mode power supply.

6. Research results

6.1. The measurement results of the Power MOSFET driver and high-voltage high-frequency (HVHF) signal on the output side.

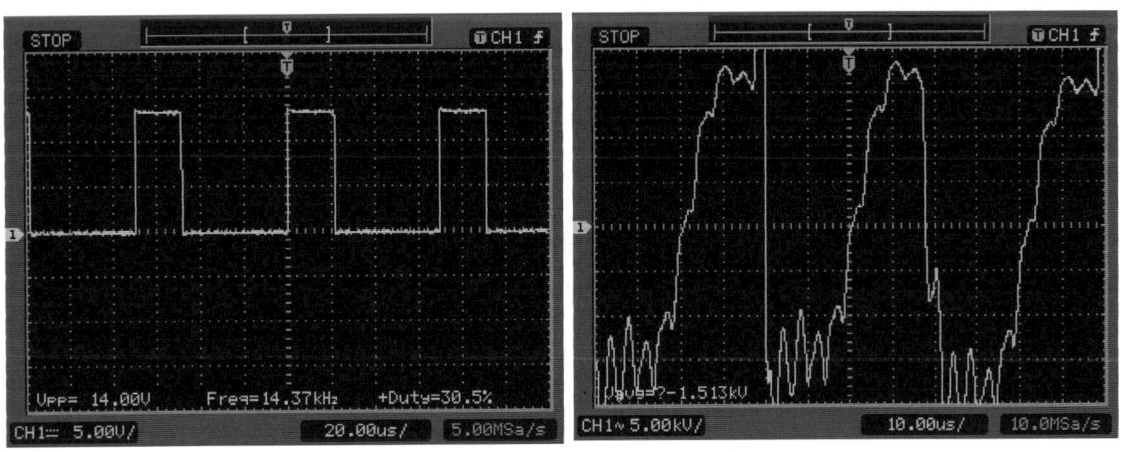

a) b)

Figure 5. a) Power MOSFET driver signal using the probe x 10; b) high frequency high voltage signal output using the probe x 1000.

6.2. Electrical effects and the amount of ozone gas produced in Table 1.

Table 1. Relationship between electrical parameters and the amount of ozone produced.

f_{SW} (kHz)	$V_{in(rms)}$ (V)	$I_{in(rms)}$ (A)	P_{in} (W)	PF	V_{out} (kV)	Ozone Gas Quantity (ppm), (3 tubes)
14.37	220	1.5	346.94	0.83	1.5	6.3

f_{SW} (kHz) is the switching frequency of the mini converter.
V_{in} (rms) (V) is the input voltage of the mini converter.
I_{in} (rms) (A) is the input current of the mini converter.
P_{in} (W) is the input power of the mini converter.
PF is the input power of the mini converter.
V_{OUT} (kV_p) is the output voltage of the high voltage switching transformer.
Ozone Gas Quantity (ppm) is the amount of ozone gas.

6.3. Effect of VOCs (before and after ozone spraying) in accordance with Table 2.

Table 2. The amount of VOCs (before and after ozone spraying) within 45 minutes.

Before spraying the ozone gas into the suction pipe, the smell of printing ink	119.35 ppm
After spraying the ozone gas into the suction pipe, the smell of printing ink	12.78 ppm

7. Conclusion and discussion of research findings

This research has designed and built a toxic gas elimination machine from printing ink. By the application of high frequency pulse spark plasma in the test, which is divided into 3 main parts: Part 1 tested the high voltage pulse spark power supply that was created of which the power supply can be adjusted to 1.5 kV at 14.37 kHz frequency. In the second part, the co-axial cylindrical rod test is performed or the ozone generator that is created. The electrode can produce ozone by 6.3 ppm and in the third part, the test by using the produced ozone gas to reduce the amount of VOCs in the gas suction tube from the printing ink. The result is a gas remover. Poisoning from this created ink can decrease the amount of VOCs gas to only 12.78 ppm (with the amount of VOCs beginning at 119.35 ppm) according to Table 2, in the future; the researcher will apply ozone gas to eliminate other air pollution which is one part that helps reduce global warming.

8. Suggestions

In this research there should be a test to reduce and increase the number of electrodes (ozone rods) to see the change in the amount of VOCs at the gas suction tube. In the 45 minute testing period in order to use the information to develop a toxic gas scrubber from the ink.

9. References

[1] Ketkaew S 2012 *Academic conference Network engineering The 4th Rajamangala University of Technology* **4** 617-620
[2] Alonso J M, Cardesín J, Corominas E L, Rico-Secades M, García J 2004 *IEEE Transactions on Industry Applications* **40**(2)
[3] Ketkaew S 2012 *Procedia Engineering* **32** 148-154

[4] Ketkaew S 2007 *AU Journal of Technology. Published by Assumption University (ABAC)* **11(1)** 42-47
[5] Brow J D 1998 *TECH Talk* **2(1)** 1-2
[6] Kuff E E, Zaengl W S and Kuffel J 2000 *High-Voltage Engineering Fundamentals* (Oxford: Butterworth-Heinemann)
[7] Alston L L 1968 *High-Voltage Technology* (London: Oxford University Press)
[8] Dun S 2016 Technique & Design of Switching Power Supply (Thailand: Intelthai)
[9] TL494 Data sheet 2002 *Texas instruments* 1-10

IOP Conf. Series: Materials Science and Engineering **877** (2020) 012056 doi:10.1088/1757-899X/877/1/012056

Cellulosic and Tannins Containing Wastewater Treatment Using MBBR Technology and Fungal Strain

I C Moga[1*], O G Iordache[2], G Petrescu[1], E C Mitran[2], A G Tanasa[1], I Sandulache[2], G A Pantazi[1], L O Secareanu[2], G Anghelache[1] and C Lite[2]
[1]DFR Systems SRL, Bucuresti, Romania
[2]National R&D Institute for Textile and Leather, Bucharest, Romania

*E-mail: corinamoga@yahoo.com

Abstract. Since the beginning, Mobile Bed Biofilm Reactor (MBBR) technology has been extensively used, both at the level of small on-site treatment units and at industrial scale. Moreover, this technology represents a starting point for many researches aimed at improving performance, such as the use of microorganisms, enrichment with anammox bacteria to accelerate nitrogen removal and more. Within the present paper, a new generation of carriers (consisting of a mix of high-density polyethylene + talcum + cellulose) was bio-augmented with a WRF (White Rot Fungi) strain, namely *Cerioporus squamosus*, in static conditions (data not shown in this paper). The wastewater, targeted for treatment, originated from National R&D Institute for Textile and Leather, INCDTP Bucharest, leather subsidiary, Leather and Footwear Research Institute, technological flux, characterized by high tannins concentration, and cellulosic content. Wastewater treatment aimed the reduction of COD value, as a water quality parameter, with satisfactory results, obtaining a percentage reduction rate of 48.53%. Also, GC-MS chromatography analysis was carried out on five vegetal tannins, used in leather treatment, highlighting main compounds for Mimosa, Chestnut, Gambier, Myrobalan and Quebracho natural tannins.

1. Introduction

Wastewaters containing cellulose fibers and various sources of tannins can lead to significant changes to the physicochemical properties of these sources of wastewater, rendering them more and more difficult to treat. Within National R&D Institute for Textile and Leather, along with its subsidiary, Leather and Footwear Research Institute, the generated wastewaters have high content of cellulose (due to the treatment and processing of textile materials) and tannins (due to tanning of leather-based materials).

Cellulose is a very abundant solid material as it is the main component in wood (40-45) % and other plant-based materials (till 90%) [1]. Cellulose is a raw material that is used in many industries like, pulp and paper, textile, tannery, wood, biotechnological industry, pharmaceutical. In 1839 this raw material was first segregated from plants by Anselme Payen [2]. Since then, the cellulose extraction has been developed, and now it can be extracted from different sources (plants, wood, tunicates, algae, bacteria etc. [3]). The global cellulose market size was 191.74 billion Euro in 2018 (Figure 1) and is projected to reach 266.47 billion Euro by 2026 [1].

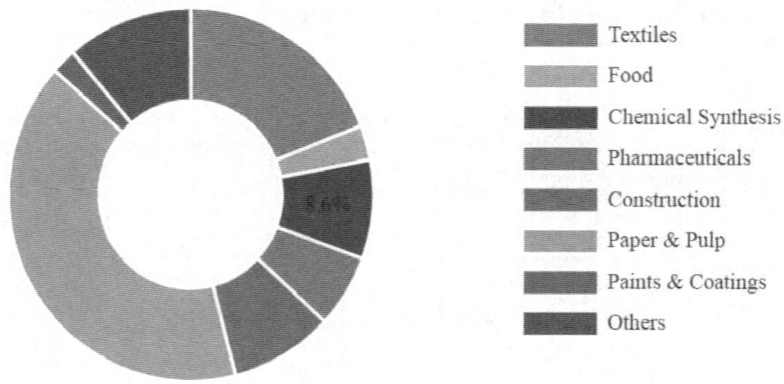

Figure 1. Global cellulose market for 2018 [1].

Cellulose has a linear structure similar with a chain made from hundreds to thousands of linked glucose units. The polymerization degree varies for different sources such as 10000 for cellulose chains from nature or approx. 15000 for cellulose found in cotton [4]. In Figure 2 is presented the chemical structure of cellulose.

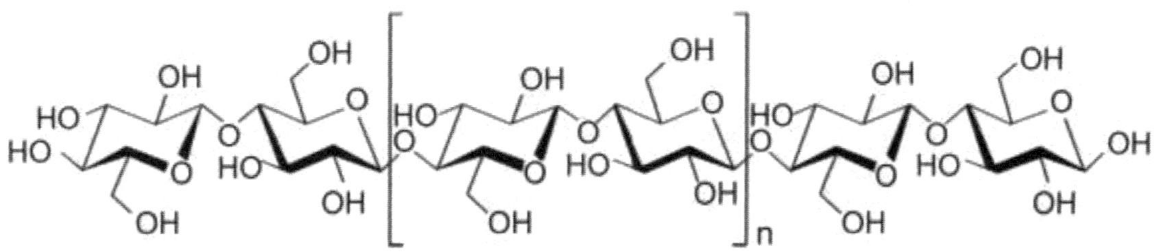

Figure 2. Structure of cellulose.

Vegetable tannins are widely used in the treatment of leather materials, representing an important stage in leather treatment, and for inducement of leather characteristic properties. Quebracho extracts are used to produce all types of leather (bags, saddles, belts, upper shoes, garment, car seats), but its qualities are highlighted when used to produce natural, full vegetable leather, like harnesses and vacchetta [5]. Chestnut extract results in: gross weight yield, excellent physical and chemical properties, even re-tanning and attractive and stable colour. In re-tanning of chrome leather, up to 30% of chestnut extract can be added to obtain smooth, compact and full leather with a good light fastness [6]. The application of Mimosa is very common worldwide and it can be modified by mixing with other types of tanning materials to adopt its characteristics to the specific leather type in re-tanning. Properties are natural soluble extracts, uniform in quality, pale in colour, low in salts and sludge formation [7]. Myrobalans are marketed in the form of whole nuts crushed nuts or solid and spray dried extracts wich are used in tanning of hides. Powdered myrobalans extracted with cold water are used for dyeing red [8]. Gambier is widely used as an industrial raw material in the textile industry, cosmetic industry, pharmaceutical industry, and as an additive to food products. In traditional societies, gambier is used as a dye for textiles and rubber and for leather tanning. Gambier contains natural polyphenol compounds including tannins. Complex phenolic compounds such as tannins dissolve in polar solvents such as water (especially in hot water), methanol, ethanol, and acetone [9].

Treatment of tannins containing wastewaters is a difficult process, due to the high solubility of these chemicals in water, and to the fact that they inhibith growth of microorganisms in activated sludge [10-13].

2. Materials and Methods

2.1. GC-MS analysis

GC-MS analysis was carried out on pure solutions of each vegetal tannin, in order to assess the main compounds, on an Agilent Technologies 6890N spectrophotometer, with 7694E Headspace module (Agilent Technologies) and MS 5973N MS detector. The solutions were sonicated for 10 minutes, after which they were filtered and analyzed on gas chromatography with Mass Spectroscopy detection. The injection of the samples was done by the automatic liquid injection mode. The method and system parameters were as follows: Capillary column: DB-35 MS, length: 30 m, internal diameter 0,25 mm; layer thickness: 0,25 µm; injection system: splitless; injector temperature: 300°C; constant flow: 1,8 mL/min; carrier gas Helium; temperature program: 70 °C (2 min) at 310°C with 10°C/min, 310°C (5 min); injection volume: 1,0 µl; auxiliary: 300°C; MS detector: scan mode; scanning interval: 30-500 amu.

2.2. COD analysis

COD analysis was carried out according to SR ISO 6060, which implies the boiling with reflux for a certain duration, of the water samples mixed with mercury sulphate (III), with a known volume of potassium dichromate, in the presence of a silver catalyst in a strongly acidic environment (sulfuric acid), so that part of the potassium dichromate is reduced by the oxidizable materials present. The excess potassium dichromate is titrated with iron (III) sulphate and ammonium solution. The COD value is calculated from the reduced amount of potassium dichromate.

For the boiling stage with reflux, a C.O.D. thermoreactor was used, namely ECO6, Velp Scientifica type, with a temperature set to 200°C.

All reagents used to determine the chemical oxygen content were of a known analytical quality:

a) Sulfuric acid (ρ=1,84 g/mL). $c(H_2SO_4)$=4 mol/L;

b) Silver sulphate (Ag_2SO_4);

c) Potassium dichromate, reference standard solution. $c(K_2Cr_2O_7)$=0.040 mol/L;

d) Iron (II) sulphate and ammonium, titrated solution. $c[(NH_4)_2Fe(SO_4)_2 * 6H_20]$=0.12 mol/L;

e) Feroin, indicator solution.

The chemical oxygen consumption (COD) expressed in milligrams oxygen per liter is calculated according to the formula (ec. 1):

$$\text{COD (mg/L)} = \frac{800\, c\,(V1-V2)}{V0} \quad (1)$$

Where: c = concentration of the amount of substance of iron (II) sulphate and ammonium solution;

V_0 = the volume of the sample to be analyzed, before dilution (if performed), in milliliters;

V_1 = volume of iron (II) sulphate and ammonium solution, used for titration of the control sample, in milliliters;

V_2 = volume of iron (II) sulphate and ammonium solution, used for titration of the sample to be analyzed, in milliliters;

8000 = molar mass of ½ O_2, in milligrams per liter.

2.3. Cerioporus squamosus functionalized HDPE carriers

HDPE carriers (with addition of talcum and cellulose) were functionalized, in static conditions, with a WRF representative, namely Cerioporus squamosus. First, the strain was grown in fresh culture, on Sabouraud nutritive broth, for 7 days, at 28°C. Afterwards, the borth, containing the fungal biomass, was poured over HDPE carriers (which were previously sterilized at 121°C, for 15 min.) enough for the liquid media to cover the carriers with just a few mm. The inoculated carriers were then incubated for 10 days, at 28°C. Treatment was carried out by extraction of bio-augmented carriers, and placement inside the testing aliquots, containing the wastewater.

3. Results and Discussions

Within FunCell project, textile and leather industry originated wastewater sample was subjected to COD reduction, with the help of *Cerioporus squamosus* functinoalized HDPE carriers. Also, GC-MS analysis was conducted on the sample, in order to asses tha main constituents of five tannins currently used in leather tanning: Mimosa, Quebracho, Chestnut, Gambier and Myrobalan.

Mimosa is a vegetable tannin, which is generally extracted from Acacia mollissima bark grown in Australia, Southeast Asia and South Africa. The bark of the tree is used for the tanning process, because it has a very high tannin content, up to 30 percent. The skin tanned with mimosa tannin has a reddish color. Mimosa extract is also used in natural medicine to treat headaches and diarrhea. The general structure of the main compound is presented in Figure 3.

Quebracho is a common name in Spanish for describing the species of very hard wood trees (wood density 0.9–1.3). The etymology of the name derives from quiebrahacha or quebrar hacha. Quebracho produces tannins that can be extracted from the heart of both red Quebracho (Schinopsis lorentzii) [17] and white (Aspidosperma quebracho-blanco). The general structure of the main compound is presented in Figure 4.

Chestnut extract is obtained from the heart and the sapwood (all the young layers located between the bark and the heart of a tree trunk, through which water and mineral salts pass) of the species of Castanea sativa and dentata. The general structure of the main compound is presented in Figure 5.

Gambier or gambir is an extract derived from the leaves of Uncaria gambir, a shrub native to Southeast Asia. Gambier is produced in Indonesia and Malaysia where it was an important commercial product at the end of the 19th century. It can be used as a tanning agent, food additive brown dye [16-17] and as a herbal medicine. Also known as pale catechu [17] or white catechu, it is often confused with other forms of catechu (which is an extract of Acacia species, especially Acacia catechu). The general structure of the main compound is presented in Figure 6.

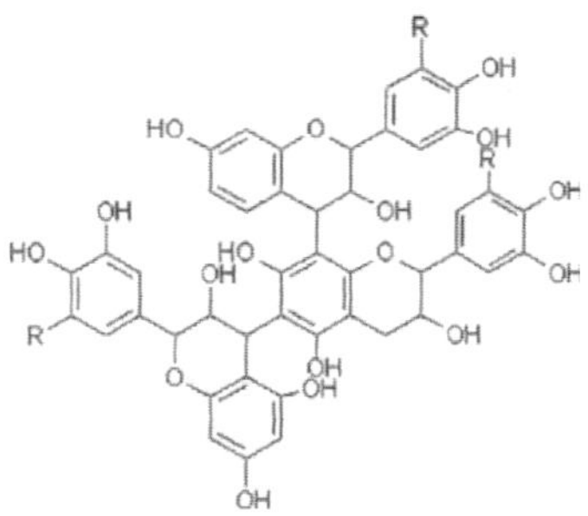

Figure 3. Molecular structure of main compound found in Mimosa tannin.

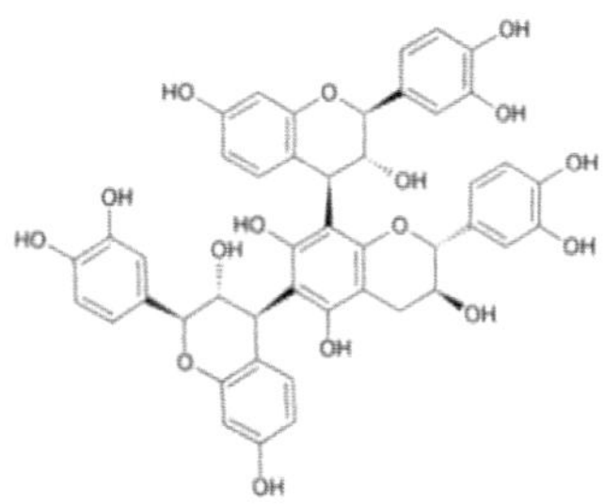

Figure 4. Molecular structure of main compound found in Quebracho tannin.

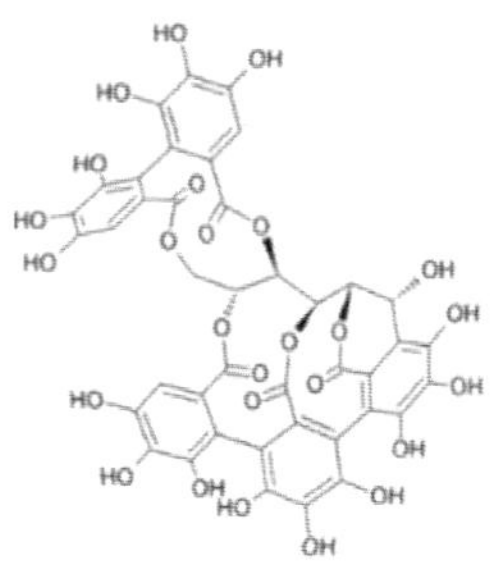

Figure 5. Molecular structure of main compound found in Chestnut tannin.

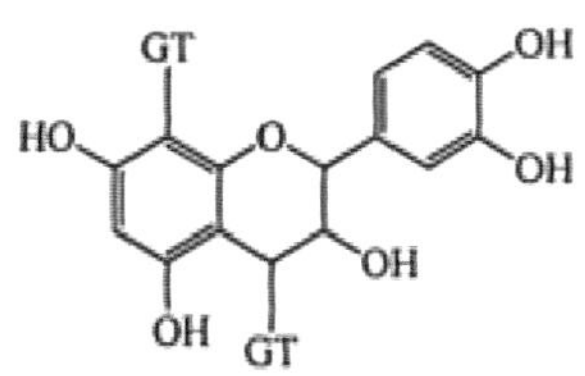

Figure 6. Molecular structure of main compound found in Gambier tannin.

IOP Conf. Series: Materials Science and Engineering **877** (2020) 012056 doi:10.1088/1757-899X/877/1/012056

Myrobalan is the dry extract of ripe fruits of Terminalia chebula (belonging to the Combretaceae family) and related species. These fruits are harvested from January to April and are dried by settling in thin layers, most of them in hue.

Characteristics:

- Color: yellow-brown to brown;
- Smell: very weak;
- Taste: astringent and slightly pungent;
- Form: mucilaginous.The general structure of the main compound is presented in Figure 7.

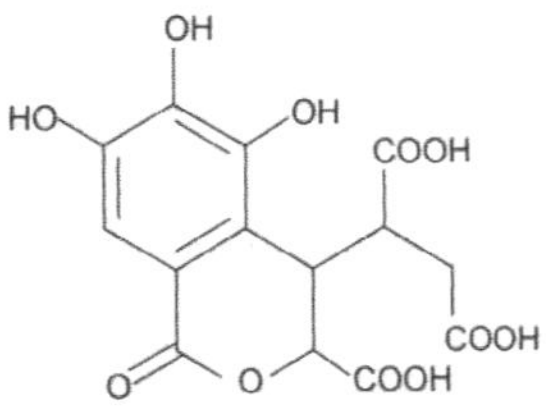

Chebulic acid

Figure 7. Molecular structure of main compound found in Myrobalan tannin.

Myrobalan contains about 30% of the hydrolyzable tannins, which consist of chebulinic acid, chebulagic acid and D-galloyl glucose. Contains free tannic acid, gallic acid, ellagic acid and mirobalanine from the resin. Due to its antiseptic and healing properties, it is used in India as a medicine for different types of conditions such as: chronic ulcers or different types of wounds. Commercially, it is used in the dyeing and tanning industry.

GC-MS analysis results are shown in Tables 1-5, which highlight the presence of the five tannins used, together with their main compounds. This analysis will be the basis of future experiments on tannin degradation in aqueous solutions.

Table 1. Chestnut tannin main compounds.

No.	Rt	CAS	Compound name	Area
1.	3.283	98-01-1	Furfural	51476
2.	5.258	97-69-8	N-Acetyl-L-alanine	5344
3.	6.046	6843-45-4	3-Methylimidazolidine-2,4-dione	22109
4.	9.180	52485-92-4	Methyl α-D-ribofuranoside	23898
5.	9.774	4007-18-5	3,4-dimethyl-sydnone	12370
6.	9.871	3201-20-5	2,4-dihydro-4,4,5-trimethyl-3H-pyrazol-3-one	4207
7.	10.027	125425-35-6	Triethylene glycol monomethyl ether	8867
8.	10.427	87-66-1	Pyrogallol	23157
9.	11.420	2595-97-3	D-Allose	47297
10.	13.837	149-91-7	Gallic Acid	16914

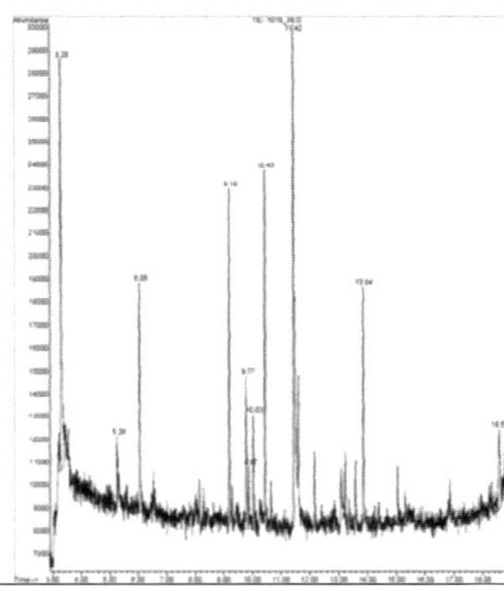

Table 2. Gambier tannin main compounds.

No.	Rt	CAS	Compound name	Area
1	8.571	120-80-9	Catechol	52202
2	10.238	108-46-3	Resorcinol	137335
3	11.419	87-66-1	Pyrogallol	29530
4	14.932	18979-60-7	4-Propylresorcinol	64141

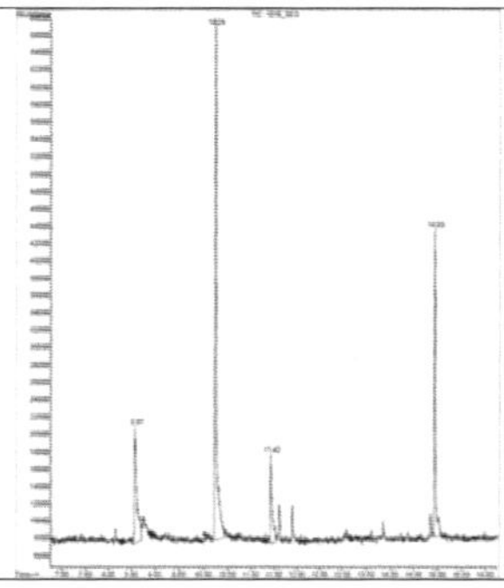

Table 3. Quebracho tannin main compounds.

No.	Rt	CAS	Compound name	Area
1	8.565	120-80-9	Catechol	189877
2	10.243	108-46-3	Resorcinol	704221
3	11.414	87-66-1	Pyrogallol	41921
4	14.932	18979-60-7	4-Propylresorcinol	302142

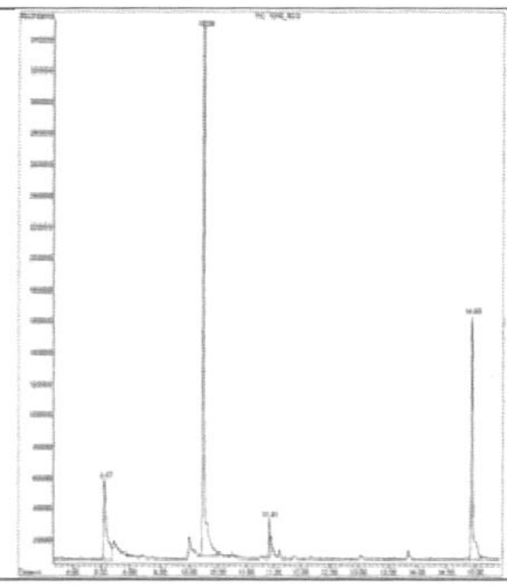

Table 4. Mimosa tannin main compounds.

No.	Rt	CAS	Compound name	Area
1.	10.243	108-46-3	Resorcinol	639603
2.	11.419	87-66-1	Pyrogallol	109404
3.	13.070	57-50-1	Sucrose	44375
4.	13.842	498-07-7	Levoglucosan	20389
5.	14.932	451-13-8	Homogentisic Acid	63595
6.	15.655	146-72-5	3-O-methyl-D-glucose	939664

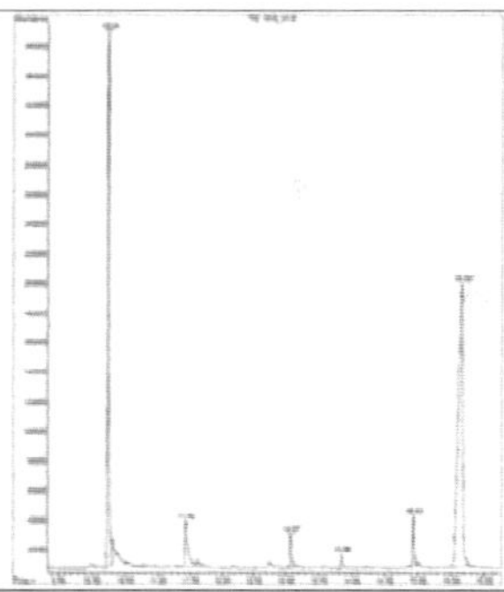

Table 5. Myrobalan tannin main compounds.

No.	Rt	CAS	Compound name	Area
1.	3.277	98-01-1	Furfural	25302
2.	10.017	67-47-0	5-hydroxymethyl furfural	90084
3.	11.414	87-66-1	Pyrogallol	251925
4.	13.076	57-50-1	Sucrose	26676
5.	13.837	498-07-7	Levoglucosan	31507
6.	15.040	33818-21-2	1,6-Anhydro-α-D-galactofuranose	10391
7.	15.579	99-16-1	Allantoic acid	11708
8.	18.666	149-91-7	Gallic Acid	564637

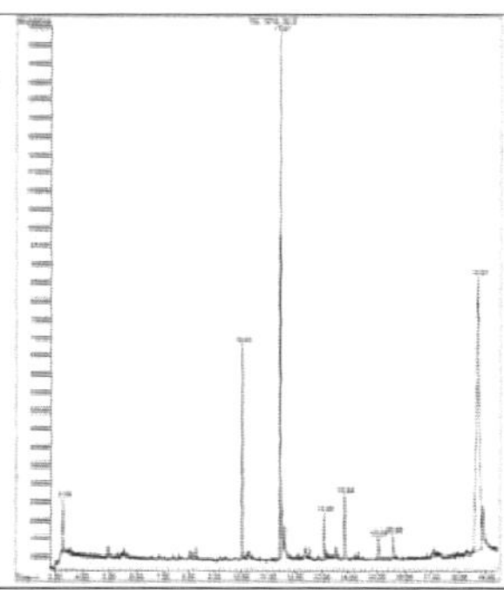

COD analysis is a general indicator of the water quality, measuring the capacity of dissolved oxygen depletion, in the samples contaminated with organic matter. Specifically, the analysis determines the equivalent amount of oxygen required for chemical oxidation of organic compounds in water. Often, COD analysis is used to estimate BOD (Biological Oxygen Demand) values, between these 2 indicators existing strong correlations.

The most common method of determining chemical oxygen demand is the method that involves digestion of the sample for two hours at high temperature, under acidic conditions, in which potassium dichromate serves as an oxidant for the organic matter present in the water samples. Silver sulphate is present in the environment, as a catalyst, and mercuric sulfate is used to complex chlorinated compounds, possibly present in samples. Following digestion, the degree of oxidation is determined by indirectly measuring the oxygen requirement through the electrons consumed in reducing Cr^{6+} to Cr^{3+}, by titration techniques or spectrophotometry. The level at which the determined experimental results are close as values to the theoretical values, first of all depends on the water load subjected to the analyzes and on the degree of oxidation, the COD analysis strongly depending on the composition of the analyzed water. According to the standard, a large number of organic compounds are oxidized in the proportion of 90% -100%, and if the samples contain a large amount of such compounds (the case of municipal effluents), the COD value is a good approximation of the theoretical consumption of oxygen.

Figure 8. Acidic water sample digestion in ECO6 thermoreactor.

The result of the analysis revealed a COD value, for the tested wastewater, of 55,849 mg/L COD (Figure 8). Following biological treatment, with *Cerioporus squamosus* bio-augmented HDPE carriers (data not shown here), it was observed a percentage reduction of 48.53% of COD content.

4. Conclusion and perspectives

MBBR technology is not currently emerging as a novel treatment technology, for conventional wastewater, as the validates efficiency of this technology is already settled throughout the years. However, new demands, regarding treatment of industrial wastewater, also dictates necessity of novelty within the MBBR systems. FunCell project took a stab at obtaining novel generations of carriers, combining HDPE with talcum and cellulose.

Furthermore, bio-agumentation of these structures was possible with fungi strains, which indeed represents a novelty in this thematic area. Within this research paper, the authors carried out identification of main compounds from five main vegetal tannins widely used in the leather industry. Also, COD reduction was possible, with almost 50% reduction efficiency, from a source of natural wastewater sample, with bio-augmented HDPE carriers, thus making a step forward validation of this novel technology in the field of industrial wastewater treatment.

Acknowledgements
This work was supported by a grant of the Romanian National Authority for Scientific Research and Innovation, CCCDI – UEFISCDI, project number COFUND-MANUNET III-FUNCELL, within PNCDI III.

5. References

[1] *** 2020, *Cellulose market size, share & industry analysis, by derivate type (commodity cellulose pulp, cellulose fibers, cellulose ethers, cellulose esters, microcrystalline cellulose, nanocellulose, and others), by end-use industry (textile, food, chemical synthesis, pharmaceuticals, construction, pulp & paper, paint & coating and others) and regional forecast 2019-2026*, Fortune Business Insights Publishing
[2] Castro R I and Morales-Quintana L 2019 *Cellulose* **26(5)** 3009-3020
[3] Dong Y, Bi J, Zhu D, Meng D, Ming S, Guo W, Chen Z, Liu Q, Guo L and Li T 2019 *Cellulose* **26(12)** 7355-7370
[4] Jiang Y, Zhang Y, Ding L, Joshua A., Wang B, Feng X, Chen Z, Mao Z and Sui X 2019 *Carbohydrate polymers* **223** 115079
[5] Gurreiro O, Soldado D, Fialho L, Cachuco L, Garrido A, Francisco A and Jeronimo E 2019 *Inclusion of rockrose and quebracho condensed tannins extracts in dairy goat diets–influence on the production, chemical composition and fatty acid profile of milk*
[6] Quave C L, Lyles J and Horswill A R 2019 *U.S. Patent Application no.* **16/225,281**
[7] El-Sayed A, El-Sakhaw M, Kamel S, El-Gendy A and Abouzeid R 2019 *Egyptian Journal of Chemistry* **62(5)** 777-787
[8] Khan M A, Khan M, Srivastava P K and Mohammad F 2005 *Colourage* **52(12)** 53-60
[9] Sy S and Kasman M 2019 *IOP Conference Series: Materials Science and Engineering* **546(2)** 022032
[10] He Q, Yao K, Sun D and Shi B 2007 *Biodegradation* **18(4)** 465-72
[11] Botezatu C, Duceac L D, Mastalier B, Stafie L, Jitareanu C M, Luca A C, Tarca E, Mitrea G, Iordache A C and Patrascu T 2018 *International Journal Of Medical Dentistry* **22(4)** 346-350
[12] Tarnita D, Tarnita D.N, Popa D, Grecu D and Niculescu D 2010 *Romanian Journal of Morphology and embryology* **51(1)** 145-150
[13] Berceanu C, Tarnita D and Filip D 2010 *Journal of the Solid State Phenomena, Robotics and Automation Systems* **166-167** 45-50
[14] Hobbs C E 2019 *Polymers* **11**(2) 224
[15] Gironi F and Piemonte V 2011 *Chem. Eng. Res. Des.* **89(7)** 857–862
[16] Sakti A S, Saputri F C and Mun'im A 2019 *Pharmacognosy Journal* **11(1)** 119-123
[17] Manalu D T and Armyanti T 2019 *Agribusiness and Agricultural Economics Journal* **2(1)** 46-67

Assessing the Contamination of the Dambovita River Through Heavy Metal Indices

F D Dumitru[1], Gy Deák[1], M A Moncea[1], A G Baraitaru[1], P I Gheorghe[1] and I E Ciobotaru[1*]

[1] National Institute for Research and Development in Environmental Protection (INCDPM Bucharest), Bucharest, Romania

*E-mail: irinaelenaciobotaru@yahoo.com

Abstract. The continuous population growth, which leads to increased social and economic development, affects the quality and quantity of the freshwater resources, both directly through changes in landscapes and pollution caused by industrialization, urbanization, and agriculture, as well as indirectly through climate change pressures (such as higher air temperature, drought, frequent extreme precipitation events). Within this study, the heavy metal contamination of Dambovita River was assessed, as it is the main drinking water source for the city of Bucharest. Thus, surface water samples and sediments were collected from four sampling sites along the Dambovita River within two campaigns, in order to assess the seasonal fluctuations. The heavy metal pollution index (HPI) together with the metal index (MI) of the collected water samples were calculated for cadmium, nickel and lead, for each sampling site. Additionally, in order to establish the potential ecological risk index, the contamination factors and potential ecological risk factors were assessed considering each of the monitored heavy metals (Cd, Ni and Pb). The results showed that the HPI values were below the critical pollution index value of 100, along with MI values below 1. Also, the sediment analyses further confirmed that all the sampling sites can be classified with low to medium levels of potential ecological risk.

1. Introduction

Water, one of the most important natural resource, is under continuous pressures generated by the combination of both naturally occurring conditions and anthropogenic activities. This leads to the continuous variation of the chemical composition hindering the water quality monitoring [1, 2]. One of the most frequent issues in water management is the contamination with hazardous substances, among which heavy metals can affect the aquatic organisms and also human health due to their capacity to accumulate in quantities exceeding the allowable thresholds [3, 4]. Therefore, the identification and quantification of heavy metal contamination sources, extent and persistence (highlighted by seasonal variations) are relevant in establishing proper amelioration solutions. A useful tool in assessing the long term water quality trends are the quality indices.

In this context, several indices were used by researchers to analyze both water quality and long-term sediments contamination, such as heavy metal pollution index, metal quality index, arithmetic weighted water quality index, multi-parametric quality index, and potential ecological risk index [1, 5-8].

Among these, the heavy metal pollution index (HPI) is the most frequently used by researchers, as it rates the total quality of water with respect to the cumulated influence of individual heavy metals, providing an overview on the overall pollution generated by heavy metals [9]. Another index used for establishing the general quality of surface waters is the metal index (MI). This index analyzes the potential cumulative impact of hazardous substances, namely heavy metals, on the human health [10].

Additionally, as heavy metals are integrated in the organic matter of the sediments, the potential ecological risk index (RI) can assess the combined pollution risk to an ecological system [11]. RI considers various factors, such as synergy between analyzed heavy metals, toxicity level, heavy metals concentration and ecological sensitivity of heavy metals [12].

Although the assessment of each individual index is relevant for highlighting the water quality status, it should be conducted together with other analysis tools (indices) in order to comprehend the overall status of the analyzed water bodies.

Therefore, in the context of increased industrialization and urbanization in Bucharest, the purpose of the present study was to analyze the Dambovita River water quality using heavy metal indices, highlighting the possible water contamination which could provide further support to competent authorities in managing the valuable water resources.

2. Materials and methods

2.1. Sampling

Water and sediment samples were collected in two campaigns (summer - C1 and autumn - C2) from four locations on the Dambovita River, as it can be seen in Figure 1. The samples were collected in polyethylene containers, and subsequently transported in the laboratory for further processing. Water samples were preserved by using high purity nitric acid (69%) at a pH < 2 until further analysis. The obtained sediment samples were initially dried at room temperature, and subsequently crushed, sieved, digested and analysed.

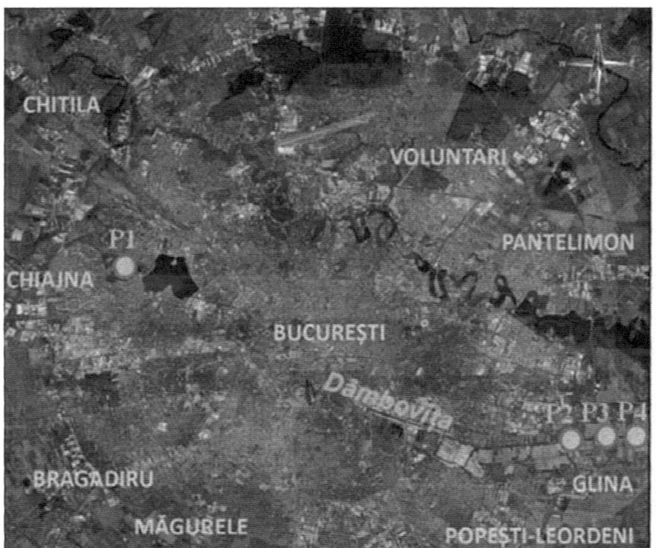

Figure 1. Sampling locations on the Dambovita River.

The concentrations of Cd, Ni and Pb were determined using a HighResolution Continuum Source atomic absorption spectrometer (ContrAA 700, Analytik Jena, Germany). Water samples were analysed using the graphite furnace technique, applying a method similar to ISO 15586/2003, and the sediment samples were analysed using the flame technique.

2.2. Methods

2.2.1. Heavy Metal Pollution Index. HPI was calculated using the weighted arithmetic quality mean method (13), which implies to first establish a rating scale for each selected metal giving it a weight and subsequent selection of the pollution parameter on which the index is to be based.

The rating scale ranges between zero and one, being equal to an inversely proportional value to the recommended standard of each analyzed heavy metal [5].

The HPI model is calculated using the following equation [5]:

$$HPI = \frac{\sum_{i=1}^{n} W_i Q_i}{\sum_{i=1}^{n} W_i} \quad (1)$$

Where: W_i represents the unit weight for each ith heavy metal, Q_i represents the sub-index for each ith heavy metal, and n represents the number of analyzed heavy metals [5]. This sub-index of the heavy metals is calculated by equation (2).

$$Q_i = \sum_{i=1}^{n} \frac{|M_i - I_i|}{(S_i - I_i)} \cdot 100 \quad (2)$$

Where: M_i represents the monitored value for each ith heavy metal, I_i represents the ideal value for each ith heavy metal, and S_i represents the recommended standard value for each ith heavy metal, in ppb [5].

Within this study, the recommended standard (S_i), and the desirable maximum value (I_i) considering each analyzed heavy metal were identified in the national legislation (Ministerial Order No. 161/2006), as quality Class III for the recommended standard (S_i) and quality Class I for the desirable maximum value (I_i).

Although Prasad and Bose have determined the critical pollution index value as being equal to 100 [13], within this study the obtained HPI values will be analyzed using the three classes scale established by Edet and Offiong (low for HPI values <15, medium for HPI values ranging between 15-30 and high for HPI values >30) [14].

2.2.2. Metal Index. This indicator estimates the water quality trend, by correlating the heavy metal concentration (C_i) to the maximum allowable concentration (*MAC*) of each analyzed heavy metal, as it can be determined from the following equation [10]:

$$MI = \sum_{i=1}^{n} \frac{C_i}{(MAC)_i} \quad (3)$$

The MAC values were selected from the regulatory document regarding the classification of surface water quality (Ministerial Order No. 161/2006). For MAC values lower than the analyzed heavy metal concentration, the water quality can be considered degraded. Thus, the obtained results will characterize the water quality as a MI value higher than 1 is a threshold of warning, while values lower than 0.3 highlight a very pure water [15].

2.2.3. Potential Ecological Risk Index. RI characterizes the potential ecological risks determined by the accumulation of heavy metals in sediments, being determined by the potential ecological risk factor (E_R) through the degree of contamination (C_D) and toxic-response factor (T_R) [7].

$$RI = \sum_{i=1}^{n} E_R^i \quad (4)$$

$$E_R^i = T_R^i \cdot C_F^i \quad (5)$$

$$C_F^i = \frac{C_D^i}{C_R^i} \qquad (6)$$

Where: E_R^i represents the potential ecological risk index for each ith heavy metal, T_R^i represents the toxic response factor for each ith heavy metal, C_F^i represents the pollution coefficient for each ith heavy metal, C_D^i represents the measured concentration for each ith heavy metal in each sampling point; C_R^i represents the reference value, which can be considered the background value of each heavy metal in sediments [7]. The toxic response factors determined for Cd, Ni and Pb are 30, 5 and respectively 5 [8].

There are five grades of E_R^i and four grades of RI, as it can be seen in Table 1.

Table 1. Grades of potential ecological metal contamination [7].

E_R^i	Grades of E_R^i (risk)	RI	Grades of RI (risk)
<40	Low	<150	Low
40-80	Moderate	150-300	Moderate
80-160	Considerable	300-600	Considerable
160-320	High	≥600	Very high
≥320	Very high		

3. Results and discussion

3.1. HPI determination

In order to determine HPI of the sampling points, the concentrations of Cd, Ni and Pb were determined, taking also into account the temporal variations of heavy metals in the Dambovita River with the sampling being performed in two campaigns. Table 2 presents the parameters used in this study to calculate HPI.

Table 2. Sub-indices values for HPI determination, based on surface water quality standard (M.O. 161/2006).

Heavy metal	I_i, µg/L	S_i, µg/L	W_i
Cd	0.50	2.00	0.50
Ni	10.00	50.00	0.02
Pb	5.00	25.00	0.04

Figure 2 presents the HPI variation obtained for both campaigns in the sampling points. Although the individual concentration of heavy metals do not highlight a contamination risk, the HPI values place the analyzed water samples in the high class area of critical pollution index. The lowest HPI value (61.38) was observed at P3 sampling point (near the wastewater treatment plant from Glina) in the second campaign, while the highest HPI (80.60) was observed in the first campaign at P2 (approximately 400 m upstream the wastewater treatment plant from Glina). Although the overall water quality with respect to the heavy metals is alarming, all HPI values are lower than the critical value of 100. Following the analysis it can be highlighted that the HPI recorded the following variation in sampling locations: P3 < P4 < P2 < P1.

Besides the spatial variation, from figure 2 it can be noticed a temporal variation, as in the second campaign there were lower HPI values than in the first sampling campaign. This may be due to the diluting effect owing to the increase in precipitations.

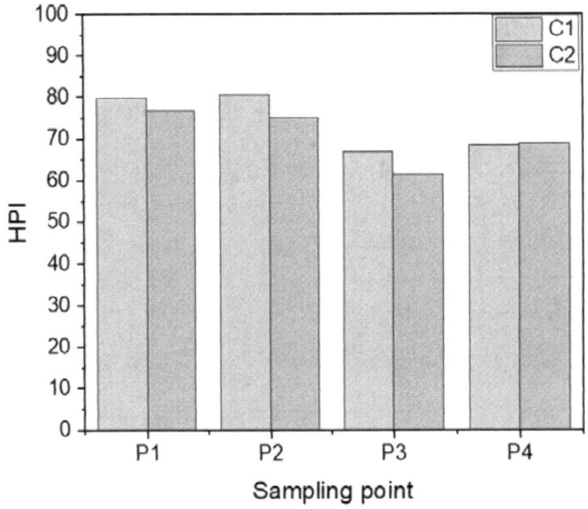

Figure 2. HPI variation in the sampling points.

The seasonal behavior of heavy metal persistence was also observed by Mondol et al. [16].

3.2. MI determination

In order to determine the MI values, first the MAC values were established for each heavy metal analyzed within this study. Accordingly, the maximum allowable concentration for Cd was 3 μg/L, for Ni was 20 μg/L and for Pb was 10 μg/L.

Figure 3 presents the MI for all sampling points. It can be seen that all selected sampling points along the Dambovita River have MI values well below the warning level, indicating pure and very pure water. A similar behavior was observed by Ionescu et al. when investigating the heavy metal contamination of two lakes from Bucharest [17]. Accordingly, the Metal Index spatial variation in the study area showed the following order: P1 < P2 < P4 < P3.

In terms of temporal variation, a slight increase of the MI values in the second campaign compared to the first one can be explained by many factors, such as bioaccumulation, dissolution of sediments, and complexes with organic matter [18].

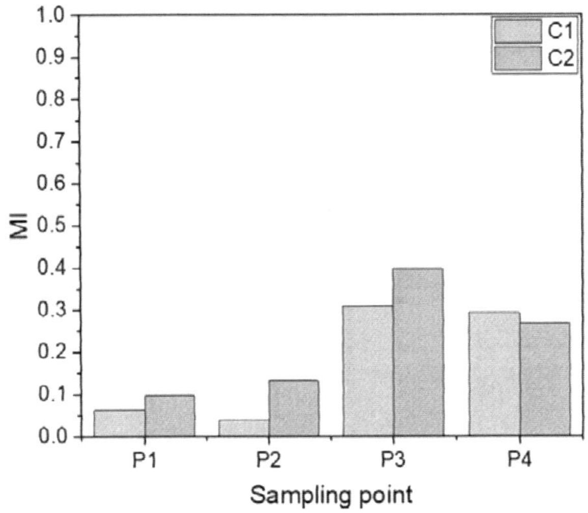

Figure 3. MI variation in the sampling points.

3.3. RI determination

In order to establish the comprehensive potential ecological risk index (RI), the potential ecological risk index for Cd, Ni and Pb (E_R^i) were initially calculated. The obtained results are presented in Table 3.

Table 3. RI computation in the four sampling points.

Campaign	Sampling point	E_R^{Cd}	E_R^{Ni}	E_R^{Pb}	RI
C1	P1	7.17	5.76	1.20	14.12
	P2	144.14	7.37	15.73	167.23
	P3	59.28	3.97	6.27	69.51
	P4	14.92	7.58	25.38	47.88
C2	P1	11.24	6.01	1.75	19.00
	P2	118.46	7.68	10.97	137.12
	P3	76.40	11.96	16.71	105.07
	P4	25.04	6.94	6.47	38.44

From Table 3 it can be noticed that the ecological risk index values of each analyzed heavy metal fluctuates with the sampling point. Accordingly, Cd analysis highlights a considerable potential ecological risk in the sampling point P2 and a moderate risk at P3, while the other heavy metals have low potential ecological risk in all sampling points. Therefore, the potential ecological risk arrays in the following order: $E_R^{Cd} > E_R^{Pb} > E_R^{Ni}$.

The RI values highlight the potential ecological risk levels along the Dambovita River. The corresponding ecological risk grades were generally low, except for the sampling point P2 which had a slight moderate grade.

4. Conclusions

The present paper established the baseline characterization of heavy metals along the Dambovita River. The overall status of water quality was evaluated considering three indices, heavy metal pollution index, metal index, and potential ecological risk index.

The HPI values showed spatial and temporal variations, highlighting the influence of both natural and anthropogenic factors on the heavy metal pollution (such as discharge from industrial and urban activities, increased precipitations). Although none of the samples presented higher values than the critical pollution index value of 100, all the samples had HPI values categorized in the high class, accentuating the need for further monitoring.

The MI values obtained in all the sampling points showed that the total trend assessment indicates the existence of pure water.

However, considering the sediments potential to accumulate heavy metals and the possibility of a subsequent release of heavy metals back into the water, the paper showed a considerable potential ecological risk of Cd upstream of the wastewater treatment plant, through the analysis of the potential ecological risk index.

Therefore, a continuous monitoring of the heavy metal indices should be performed for the Dambovita River, and if necessary suitable solutions should be identified in order to ameliorate any temporary excess.

Acknowledgements

This work was supported by a grant of the Romanian Ministry of Research and Innovation, CCCDI-UEFISCDI, project number 26PCCDI/01.03.2018, "Integrated and sustainable processes for environmental clean-up, wastewater reuse and waste valorization" (SUSTENVPRO), within PNCDI III.

5. References

[1] Radu V, Ivanov A, Ionescu P, Deak Gy and Tudor M 2016 *Environ. Eng. Manag. J.* **15** 1069
[2] Ionescu P, Radu V and Diacu E 2015 *Rev. Chim.* **66** 1725
[3] Monroy M, Maceda Veiga A and de Sostoa A 2014 *Sci. Total. Environ.* **487** 233
[4] Chen C, Ward D, Williams J and Fisher N 2016 *J. Mar. Sci. Eng.* **4** 41

[5] Mohan S, Nithila P and Reddy S 1996 *J. Environ. Sci. Healrh.* **A31** 283
[6] Balan I, Shivakumar M and Kumar P 2012 *Chron. Young Sci.* **3** 146
[7] Hakanson L 1980 *Water Res.* **14** 975
[8] Soliman N, Nasr S and Okbah M 2015 *J. Environ. Health Sci.* **13** 70
[9] Mahato M, Singh G, Singh P, Singh A and Tiwari A 2017 *Bull. Environ. Contam. Toxicol.* **99** 54
[10] Caerio S, Costa M, Ramos T, Fernandes F, Silveira N, Coimbra A, Medeiros G and Painho M 2005 *Ecolog. Indicators* **5** 151
[11] Yang Z, Wang Y, Shen Z, Niu J and Tang Z 2009 *J. Hazard. Mater.* **166** 1186
[12] Jiao X, Teng Y, Zhan Y, Wu J, Lin X 2015 *PLoS One.* **10** e0127736
[13] Prasad B and Bose J 2001 *Environ. Geol.* **41** 183
[14] Edet A and Offiong O 2002 *GeoJurnal* **57** 295
[15] Bakan G, Boke-Ozkoc H, Tulek S and Cuce H 2010 *Turk. J. Fisheries Aquat. Sci.* **10** 453
[16] Mondol M, Chamon A, Faiz B and Elahi S 2011 *J. Bangladesh Acad. Sci.* **35** 19
[17] Ionescu P, Marcu E, Radu V, Ciobotaru I and Tociu C 2017 *Ecoterra* **14** 19
[18] Abdullah E 2013 *Environ. Earth Sci.* **3** 114

IOP Conf. Series: Materials Science and Engineering **877** (2020) 012058 doi:10.1088/1757-899X/877/1/012058

Comparison of Total Phenolic Content, Antioxidant Activity and Extraction Yield from Apium Graveolens Waste Using Unconventional Extraction Methods

S Spînu[1,2], A Orțan[1*], D Ionescu[1,3], I Moraru[4] and M Drugulescu[1]
[1]University of Agronomic Sciences and Veterinary Medicine of Bucharest, Romania.
[2]University of Bucharest, Faculty of Physics, Doctoral School of Biophysics and Medical Physics, Romania.
[3]Hofigal Export Import S.A., Romania.
[4]SC Laboratoarele Medica SA, Romania.

*E-mail: alina_ortan@hotmail.com

Abstract. Apium graveolens is a medicinal and aromatic plant used in pharmaceutical industry, cosmetic industry, gastronomy and traditional medicine. Aerial parts of Apium graveolens (stems and leaves) are rich in bioactive compounds (flavonoids, phenols, furanocoumarins, volatile oils, sesquiterpene alcohols, fatty acids, etc.). It is known for its hepatoprotective, antioxidant, anticancer, antidiabetic, anti-inflammatory, antimicrobial, analgesic, anti-spasmolytic, anti-infertility, cardiotonic activities. This is why studies in this direction have not yet ceased and are increasingly pursuing the discovery of new, unconventional methods for identifying, quantifying and isolating bioactive compounds from Apium graveolens waste. The aim of this study was to determine and correlate the total phenolics, total flavonoids, antioxidant activity and extraction yield from ethanolic extracts (70%) of stems and leaves waste with different unconventional extraction methods. Apium graveolens aerial parts waste were obtained after aqueous extraction of the leaves and stems. The extraction yield was expressed as function of freeze-dried extracts, and the extraction methods were ultrasound assisted extraction, microwave assisted extraction and accelerated solvent extraction. Regarding the chosen extraction methods, numerous studies have shown, especially in the case of microwave assisted extraction, the presence of bioactive compounds in plant waste, in a reasonable percentage, therefore this study aimed to identify an optimal extraction method also in the case of Apium graveolens aerial parts waste. Total phenols content was determined by Folin-Ciocalteu spectrophotometric method, the flavonoids content waste determined by aluminium chloride spectrophotometric method and the antioxidant activity was studied using DPPH (2,2-Diphenyl-1-picrylhydrazyl) spectrophotometric method. The findings of this work indicated a positive dependence between the extraction yield of the waste material and the considerable content of bioactive compounds still present (phenols and flavonoids) for all considered extraction methods.

1. Introduction

Plants represent a rich source of various bioactive compounds with different biological properties. Thus, plants are still present in the researchers' studies in order to improve the recovery methods of active compounds. Using different extraction methods, identification and purification methods in the quantitative and qualitative analysis, many targeted studies reported important results, therefore in the last time the interest for studies regarding the valorization of plants wastes has also increased.

Apium graveolens L., celery, is a medicinal and aromatic plant with many properties and many applications in traditional medicine, gastronomy, but also in pharmacology. Apium graveolens L. belongs to Apiaceae family, to the genus Apium, originally from the Mediterranean area, and is spread all over Romania territory. From a medicinal point of view, the leaves, the root and the seeds of the plant are used.

Celery is an excellent source of vitamins, phenolic compounds, volatile oil and other nutrients. The leaves contain essential oil having as major components: limonene, myrcene, 1,2 ethanediol, γ-terpinene, 2-(2-propenyl), Z-β-ocimene, trans-β-guaiene, 2,5 pyrrolidinedione, 1-phenyl, 6-butyl-1,4-cycloheptadiene, furan [1], phenolic compounds (gallic acid, vanillic acid, chlorogenic acid, caffeic acid, rutin, epicatechin, quercetin, myricetin), ascorbic acid, carotens and minerals. Celery also contains furanocoumarins, including xanthotoxin, bergapten and 5-methoxypsoralen, which are chemicals that react to sunlight [2].

Extracts or the essential oils obtained from the aerial parts of Apium graveolens have been investigated in animal or cellular experimental settings for a variety of benefits and pharmacological activities such as: antioxidant, antimicrobial, hepatoprotective, anti-spasmolytic, hypocholesterolemic activities or anxiolytic potential, etc. [3-8].

The aim of this study was to determinate the total phenolics content, total flavonoids content, antioxidant activity and extraction yield from Apium graveolens aerial parts, as leaves and stems waste. The extraction of the bioactive compounds from waste plant material was an important step, therefore we adopted three extraction methods as: microwave-assisted extraction, ultrasound-assisted extraction and accelerated solvent extraction. These three extraction technologies were chosen due to the fact that it has been demonstrated by many studies as being appropriate in obtaining useful compounds from plant material.

2. Material and Methods

The plant material consisted of dried Apium graveolens leaves and stems, unprocessed and waste. The waste material was obtained by two aqueous extraction of Apium graveolens leaves and stems, at 1000C, for 3 hours the first one, and 1.5 hours the second one.

The plant material (unprocessed and waste) was powdered (Figure 1) and was extracted by three different extraction methods: microwave-assisted extraction and ultrasound-assisted extraction with 70% bioethanol (1:10), and accelerated solvent extraction. Microwave-assisted extraction (MAE) was performed using Ethos Easy Milestone equipment, at 1000C, for 55 min, pulsatory microwaves. Ultrasound-assisted extraction (UAE) was performed in an ultrasound bath for 30 min, followed by oven for 3 hours at 680C. Accelerated solvent extraction (ASE) was performed using Dionex ASE 350, all extractions were performed in 22 ml extraction cells, containing 7 g of sample. The extraction conditions are presented in Table 1. All the experiments were carried out in triplicate.

Table 1. ASE Extraction conditions.

Extraction Solvent	Bio-ethanol 70%
Temperature	100°C
Pressure	1500 psi
Heat-up Time	5 min
Static Time	10 min
Flush Volume	100%
Purge Time	90 s
Static Cycles	3

(a) (b)

Figure 1. (a) The powdered unprocessed plant material; (b) the powdered waste material.

Regarding extraction yield determination, the solvents from the extracts were removed in vacuo and freeze-dried for 45 hours, at -550C using Lyo Cube 4-8 Alpha 2-4 LSC plus Christ Freeze Dryier and the extraction yield was expressed as function of freeze-dried extracts weight.

Determination of total phenolics content (TPC) from unprocessed and waste Apium graveolens leaves and stems extracts was performed using Folin-Ciocalteu method adapted after [9]. The spectrophotometric method was used to measure the absorbance of samples at 765 nm, using UV-VIS Spectrophotometer Specord 210 Plus, Analitik Jena. The total phenolics content was expressed as mg of gallic acid equivalent by using the equation obtained from gallic acid calibration curve. The obtained concentration of phenolics content was averaged from triplicate samples and the samples were diluted in a ratio 1:100.

Determination of total flavonoids content (TFC) from unprocessed and waste Apium graveolens plant material extracts was carried out by aluminium chloride method as described by [9], using 0.5 ml of sample. The absorbance was measured at 415 nm with the UV-VIS spectrophotometer. The total flavonoids content was expressed as mg of rutin equivalent by using the equation obtained from rutin calibration curve. The obtained concentration of flavonoids content was averaged from the triplicate samples and the samples were diluted with a dilution factor of 100.

Determination of antioxidant activity using α-diphenyl-β-picrylhydrazyl (DPPH) radical scavenging of the extracts was performed according to [10]. Over time, the determination of antioxidant activity by DPPH assay was often applied for various plants extracts [11], therefore it was considered a traditional method. The absorbance of the samples was read at 517 nm on the UV-VIS spectrophotometer. The determinations were carried out in triplicate and the results were averaged. The antioxidant activity was calculated using the formula:

$$AA=\frac{A_{DPPH}-A_{sample}}{A_{DPPH}}*100 \tag{1}$$

Where: AA – antioxidant activity, $\mathbf{A_{DPPH}}$ - the absorbance of the DPPH solution without sample, $\mathbf{A_{sample}}$ - the absorbance of the sample (which contain the mixture of extract and DPPH solution).

The extraction yield was determined and the highest percentage in the case of aerial parts waste was found in the extracts obtained by UAE. The lowest extraction yield of the waste material was found in the extracts obtained by ASE method. The extraction yield values are presented in Table 2. As it can be observed from the table, extraction yields varied from 58.84 to 80.45 % from the dried unprocessed leaves and stems and from 54.41 to 65.16% from the waste leaves and stems. As the results indicated, there were significant differences between the yield obtained from waste material and from unprocessed material, by MAE method (about 17.5%), and by UAE method (about 9.8%), but there was not a substantial difference in case of ASE method (about 4.4%).

Table 2. The extraction yield of Apium graveolens aerial parts (leaves and stem).

Extraction method		Extraction Yield (%)
Microwave assisted extraction	Unprocessed plant material	80.45
	Waste plant material	62.92
Ultrasounds assisted extraction	Unprocessed plant material	75.04
	Waste plant material	65.16
Accelerated solvent extraction	Unprocessed plant material	58.84
	Waste plant material	54.41

The total phenolics content present in the extracts from aerial parts of Apium graveolens unprocessed and waste was obtained by Folin-Ciocalteu method and their concentrations were expressed as equivalents of gallic acid, calculated as [12] described, adapted and multiplied with the dilution factor of 100. The resulted concentrations are presented in Table 3. The values indicate that in case of waste plant material, the highest total phenolics content was obtained using ASE method, followed by UAE method and MAE method.

Table 3. Total phenolics content and total flavonoids content of Apium graveolens aerial parts (leaves and stem).

Plant type/ Extraction method	Total phenolics content (mgGAE/g DW)	Total flavonoids content (mgRUE/g DW)
Apium graveolens unprocessed / Microwave assisted extraction	11.34	11.71
Apium graveolens waste / Microwave assisted extraction	9.47	10.37
Apium graveolens unprocessed / Ultrasounds assisted extraction	14.83	23.57
Apium graveolens waste / Ultrasounds assisted extraction	10.80	11.94
Apium graveolens unprocessed / Accelerate solvent extraction	12.46	17.21
Apium graveolens waste / Accelerate solvent extraction	11.38	12.01

The analytical data of the total flavonoids content of the extracts from unprocessed and waste celery aerial parts material is presented in Table 3. The concentrations were calculated as [12] described, adapted and multiplied with the dilution factor of 100, and were expressed in rutin as mg RUE/g. The obtained results varied, and the greatest total flavonoids content for the waste material was obtained using ASE method. Hence, it was observed that in this case, for both unprocessed and waste material, the ASE method offered a better total content, but even the others extraction methods offered good results, in the same range.

Due to the fact that majority of plants contain important contents of phenols and flavonoids, the scientists studied and reported the wide spectra of biochemical and pharmacological effects of these natural compounds. The concentrations of TFC and TPC obtained in this study could be compared with other similar studies results [13-18] present in literature.

More, as the results indicated, the hydroalcoholics extracts of Apium graveolens waste material contained considerable values of TFC and TPC. There are no results in the scientific literature until now regarding the quantitative determination of TFC and TPC from Apium graveolens waste material, up to our knowledge.

A substantial difference was found between TFC in waste and unprocessed plant material extracts, revealed by UAE method (23.57 mgGAE/g vs. 11.94 mgGAE/g), at close TPC values (14.83 mgGAE/g vs. 10.80 mgGAE/g). Regarding the TPC, there were no significant differences between the unprocessed plant material extracts (less then 3.5 mgGAE/g), or between the waste material extracts (less then 2 mgGAE/g). But, the variation of TPC and TFC in the extracts depend on many factors, such as extraction methods, drying process, cultivation, growth and harvesting process of the plant, soil nature, the preserving process, etc, as is well known.

As expected, the results revealed a high determination coefficient ($y = 0.0106x - 0.0657$, R^2=0.9971) in the case of TPC determination, but also in the case of TFC ($y = 0.0027x - 0.0035$, R^2= 0.9993).

The free radical-scavenging activity of the Apium graveolens aerial parts extracts was performed through DPPH method and the results are presented in Figure 2. The obtained extracts were able to reduce the stable radical DPPH to the yellow-colored diphenylpicrylhydrazine. The highest value of the antioxidant activity percentage from the waste material was found in the case of MAE method. The high antioxidant activities of the extracts could explain the values of the total phenolics content.

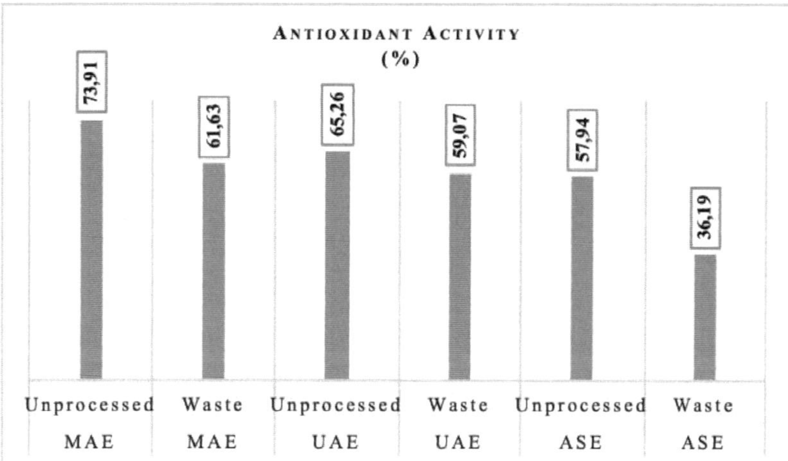

Figure 2. Antioxidant activity of Apium graveolens aerial parts (leaves and stem).

The MAE of antioxidants from Apium graveolens L. dried aerial parts was demonstrated to be effective, fast and practical, for unprocessed and waste material.

The results from the analysis of aerial parts of Apium greveolens waste showed important content of total phenols and flavonoids and good antioxidant activity, facts that confirmed the importance and the abundant presence of celery in the Romanian diet and cuisine.

3. Conclusions

In the present study, the decreasing values of antioxidant activity among the Apium graveolens aerial parts waste extracts assayed through all the described methods were found to be MAE> UAE > ASE. The opposite order was identified for the total phenolics contents and the total flavonoids content of the aerial parts waste extracts.

This study emphasizes that the waste material of Apium graveolens still contains bioactive compounds as phenols and flavonoids, making it a valuable and valorization-worthy source nowadays. More, it was found a positive dependence between the extraction yield of the waste material and the considerable content of bioactive compounds still present (phenols and flavonoids) for all considered extraction methods.

Aknowledgement
The authors gratefully acknowledge the support obtained through the project SusMAPWaste, SMIS 104323, Contract No. 89/09.09.2016, from the Operational Program Competitiveness 2014-2020, project co-financed from the European Regional Development Fund.

4. References

[1] Rożek E, Nurzyńska-Wierdak R, Sałata A and Gumiela P 2016 *Acta Sci. Pol. Hortorum Cultus* **15** 147-157
[2] Lombaert G A, Siemens K H, Pellaers P, Mankotia M and Ng W 2001 *J AOAC Int* **84** 1135-43
[3] Popovic M, Kaurinovi, B, Trivic S, Mimica-Dukic N and Bursa M 2006 *Phytother Res.* **20** 531–537
[4] Abd El-Mageed N M 2011 *Pharmacogn Mag.* **7** 151–156
[5] Branković S, Gočmanac-Ignjatović M, Kostić M, Veljković M, Miladinović B, Milutinović M and Radenković M 2015 *Acta med. Median.* **54** 11-16
[6] Hassanen N H M, Eissal A M F, Hafez S A M and Mosa E A M 2015 *Int.J.Microbiol.App.Sci* **4** 284-296
[7] Dianat M, Veisi A, Ahangarpour A and Moghaddam H F 2015 *Avicenna j. Phytomed* **5** 203-209
[8] Tanasawet S, Boonruamkaew P, Sukketsiri W and Chonpathompikunlert P 2017 *Asian Pac.J.Trop.Biomed* **7** 20-26
[9] Fierascu I, Ungureanu C, Avramescu S M, Fierascu R C, Ortan A, Soare L C and Paunescu A 2015 *Rom Biotech Lett* **20** 10626-36
[10] Ionescu D, Spînu S, Orţan A, Moraru I, Fîntîneru G, Fierăscu RC, Fierăscu I and Drugulescu M 2017 *AgroLife Scientific Journal* **6** 141-145
[11] Tonguc Yayintas O, Sogut O, Konyalioglu S, Yilmaz S and Tepeli B 2017 *AgroLife Scientific Journal* **6** 205-213
[12] Abdelhady M I S, Motaal A A and Beerhues L 2011 *Am. J. Plant Sci.* **2** 847-850
[13] Nour V, Trandafir I and Cosmulescu S 2017 *Not Bot Horti Agrobo* **45** 179-184
[14] Chiriac V, Stratulat D N, Calin G, Nichitus S, Burlui V, Stadoleanu C, Popa M and Popa I M 2016 *IOP Conference Series-Materials Science and Engineering* 133 012055
[15] Calin G, Mitrea G, Marcu C, Burlui A, Trusca R, Vasile E and Iacomi F 2019 *Rev. Chim.* **70**(9) 3444-3447
[16] Duceac L D, Marcu C, Ichim D L, Ciomaga I M, Tarca E, Iordache A C, Ciuhodaru M I, Florescu L, Tutunaru D, Luca A C and Stafie L 2019 *Rev. Chim.* **70(7)** 2622-2626
[17] Tarnita D, Tarnita D.N, Popa D, Grecu D and Niculescu D 2010 *Romanian Journal of Morphology and embryology* **51(1)** 145-150.
[18] Baltatu S, Vizureanu P, Mareci D, Burtan LC, Chiruta C and Trinca LC 2016 *Materials and corrosion-werkstoffe und corrosion* 67(12) 1314-20

Experimental Research on the Behaviour of the Systems for Ventilated Facades, Exposed to the Fire Action

H Dragnc[1*], A Simion[2] and D Stoica[3]
[1] INCERC laboratory of applied research and construction tests from National Research and Development Institute URBAN-INCERC, Bucharest, Romania
[2] INCERC laboratory of applied research and construction tests from National Research and Development Institute URBAN-INCERC, Bucharest, Romania
[3] INCERC laboratory of applied research and construction tests from National Research and Development Institute URBAN-INCERC, Bucharest, Romania

*E-mail: horatiudragne@yahoo.com

Abstract. When establishing the constructive structures that must reach the required levels of fire performance, one must start from the fact that the fire is an accident. The structures must remain viable for the normal fire time considered. The behaviour of the construction to the fire is dependent, besides other factors, on the contribution to the fire of the elements and parts of the construction, of the materials and products for the construction, as well as on the fire resistance of some of them. Due to the fact that the initiation and evolution of the fires is different, it was imposed that the determination of the fire behaviour of the constructions and the materials that enter their composition should be done under specified conditions, by exposing them to a specified ignition source, in a good context.

1. Introduction

The requirement of the built space regarding the "Fire safety" of the constructions requires, in order to be satisfied, that they should be designed and executed so that, in the case of a fire, the following objectives are ensured:

- maintaining the carrying capacity of the construction for a certain period of time;
- limiting the production and propagation of fire and smoke inside the construction;
- limiting the spread of fire to the neighboring constructions;
- intervention for evacuation, respectively rescue of the people caught inside the construction;
- security of the intervention teams.

Meeting the criteria and levels of performance regarding the "fire safety" requirement must be ensured by a set of interdependent measures, regarding the conditions of placement, configuration and fire conformation of the constructions, the performance conditions for the structures, for products, utilities and installations, including for fire safety installations. These measures are subject to specific technical regulations regarding the location, design, execution and operation of constructions, developed under the authority of the line ministries.

Construction products represent, through their fire performance, one of the determining factors for:

- assessing the development of a fire in a room, including the emission of smoke, the spread of fire and smoke in the building and in the neighboring buildings [6];

- evaluating the fire performance of the construction and establishing the design rules for the different parts of it, relevant for the fire safety.

The general purpose of fire safety legislation [1] is to provide fire safety. To achieve this goal, the requirements for structures (load bearing capacity), building materials (fire prevention and limitation of fire propagation, limitation of fire size), modalities of evacuation the buildings (including the prevention of fire propagation to other neighboring buildings) are established to define how they should be designed and executed according to their purpose. The structural design rules, including those regarding fire, are laid down by Eurocodes, in force in the territory of the EU member countries. The aim of the Eurocode program initiated by the European Commission is to establish a harmonized set of technical rules for the design of constructions and for civil engineering works. These rules initially served as an alternative to the national provisions of EU member countries, replacing them later (SR EN 1991-1-2: 2004/AC:2009). The European classification system does not include requirements for the toxicity of the resulting combustion products. The toxicity of the combustion products and the environmental aspects of the industrial products are of increasing interest. Also, an increasing importance is given to the type and concentrations of the products resulting from the burning of construction materials. These aspects must also be subject to legislative action. The spread of flame on the surface of a solid body and the existence of combustible materials are important factors in the issue of fire safety because it influences the initial development of fire and the rate of heat emission. In an enclosure fire, the first object that ignites is usually singular. If it is located near a wall covered with combustible material, the wall cladding can ignite and maintain fire. In this case, the vertical flame spread is significant and rapid, especially in the presence of air in the case of a ventilated façade [4]. The propagation and growth of the fire were extensively studied during the years 1980-1990 [2]. For example, the research program EUREFIC (European Reaction to Fire Classification) provided valuable information on the evaluation of surface coatings. Within this program, large-scale tests were carried out by three laboratories: MTB-Finland, NBL-Norway and SP-Sweden [7] in order to collect data for the development of a prediction model for fire growth in fire propagation tests in the test rooms (ISO/DIS 9705) based on the results of calorimetry tests (ISO/DIS 5660). The patterns of fire propagation and fire growth vary in complexity, from simple empirical correlations to approaches that solve the conservation equations of gas and solid phase integrated into numerical models [5].

2. Experimental procedures

Researchers from the *INCERC Fire Safety Research and Testing Laboratory* conducted several experimental tests to assess structural behaviour of buildings with ventilated façades exposed to the action of fire (Figure 1). Specifically, it refers to an approach in designing and evaluating the structural components of buildings with ventilated façades, which are efficient in their fire performance.

Researchers found important to understand what performance criteria is chosen when selecting a particular material or method [3], including exposure to fire, duration, aesthetics, cost and maintenance. The application of protective and insulating materials is one of the most common means of protection of structural metallic elements in fire. The purpose of the data collection is to evaluate the position of all metallic structural elements in relation to the position of the exterior facade of the brick and to promote the construction materials that are effective in both a fire and have good thermophysical and hydrothermal properties. The parameters evaluated in the fire test were:

- The effective bending resistance of steel exposed to high temperature;
- Tilt of the linear elastic zone of the steel exposed to high temperature;
- The factor of reducing the effective flow resistance of steel exposed to high temperature;
- Reduction factor for the effective slope of the linear elastic range of steel exposed to high temperature;
- Structural temperature of steel.

Figure 1. Ventilated facade system before fire resistance testing.

Temperature variations in materials lead to changes in their properties and to changes in structural strength characteristics, a fact that becomes evident when the temperature rises significantly. The unique possibility of obtaining a passive control of the resistance of the steel section during the fire, is its protection against fire by a properly chosen specific insulating materials. In addition, the passive protection of the system is taken into consideration when designing the cover and must be properly maintained in operation. The sustainability indicators of a metal structure exposed to fire can be grouped as follows: the first indicator indicates the safety and strength of the structure, the other refers to the impact on society and at last, the passive design of a protection against fires and the environment, including the life cycle aspects of the materials used. This is an assessment that leads to limiting the costs of protection that would be disproportionately large compared to the costs of the structure itself. In the Eurocode, fire resistance is the capacity of a structure to perform its function required for a certain fire exposure and a period of time. For a certain load level, the temperature at which a failure of a structural steel element is expected is characterized by a critical temperature where an effective flow force is large enough to generate plastic stresses, according to Eurocode 3.1. The first limit of temperature is 1400 °C, where both structural safety and damage limitation resistance are taken into account; the second is 2600 °C, where, result that structural failure and collapse are inevitable. An image that shows structural damage during the experimental test is shown in Figure 2.

Figure 2. Fire affecting the ventilated façade system. Breaking the windows under the action of the compartment fire and propagation the fire outside from compartment and on the façade.

IOP Conf. Series: Materials Science and Engineering **877** (2020) 012059 doi:10.1088/1757-899X/877/1/012059

At this point the temperature of steel was higher than 700 ° C and the bending of the steel structure can be seen.

3. Results and discussions

The mechanisms of flame propagation on the vertical surface of the facades depend on many factors, of which the most important are:

- the presence of combustible materials;
- the existence of cavities as part of the exterior plating system or as defects / local degradation of the products used in plating.

Under the conditions of a facade fire, the flames can extend five to ten times higher than the initial height of the flame, regardless of the nature of the materials from which the facade system is made.

A real fire produced in a room grows, evolves and extinguishes in direct correlation with the mass and energy balance in the compartment in which it is produced. The energy developed depends on the type and quantity of available fuel and the ventilation conditions. In the pre-flashover phase the room temperature is reduced and the fire is still local (minimal). This period is usually particularly important for fire escape and fighting. It usually does not produce effects on structures. If the fire evolves towards stage 3 (generalized fire), the structure of the ventilated façade is affected. Following the largescale tests, during fire reaction tests carried out by the INCERC laboratory for applied research and construction tests, similar results were obtained.

The materials tested by our laboratory are presented in table 1 and figure 3 and for them, the maximum heat release rate (HRR) and total heat dissipated was found.

Table 1. Index of materials

Acronym	Description	Maximum HRR (kW/m^2)	Total heat (kW)
ACM_PE1	Higher density Polyethylene	1364	106
ACM_PE2	Lower density Polyethylene	1123	107
ACM_FR1	Polyethylene with mineral filling (medium size)	123	62
ACM_FR2	Polyethylene with mineral filling (maximum size)	195	71
ACM_FR3	Polyethylene with mineral filling (minimum size)	144	66
ACM_NC1	Polyethylene with aluminum	14	3
ACM_NC2	Polyethylene with corrugated aluminum	30	1
HPL_PF	High pressure laminate	530	172
HPL_FR	High pressure laminate with mineral filling	263	67
MWB_1	Mineral wood board (low density)	150	38
MWB_2	Mineral wood board (high density)	194	28
PF1	Phenolic foam (medium density)	64	19
PF2	Phenolic foam (low density)	62	18
PF3	Phenolic foam (high density)	65	20
PIR1	Polyisocyanurate (high density)	116	16
PIR2	Polyisocyanurate (medium density)	107	14
PIR3	Polyisocyanurate (low density)	104	13
SW	Stone wool	6	0,07
GW	Glass wool	9	0,8

The materials chosen are similar to the materials used in the study of the fire at the Greenfell tower (only the density could not be corelated exactly) because it was desired to make a comparison between the results. They also cover the largest range of construction products commonly used in ventilated façade systems. Special materials and nano-materials have not been studied because they represent emerging technologies with a very small percentage of the total materials used in construction in 2019. If we compare same products of the generic type (like the ACM_FR, or PIR), there is a small difference in combustion and combustion behavior for the respective product type, as well as for insulation materials, in smoke toxicity. The HPL_PF is identified as problematic during the tests. The combustion behavior under laboratory conditions for PIR and PF the more resistant character of PIR material (which has a higher initial peak followed by a lower HRR) compared to PF material (with a lower peak HRR, but with higher constant HRR). This difference is also found in toxicity tests. No heat release from mineral wool and glass wool is to be expected, but it is crucial to also demonstrate the existence of non-flammable alternative insulating materials (Figure 3).

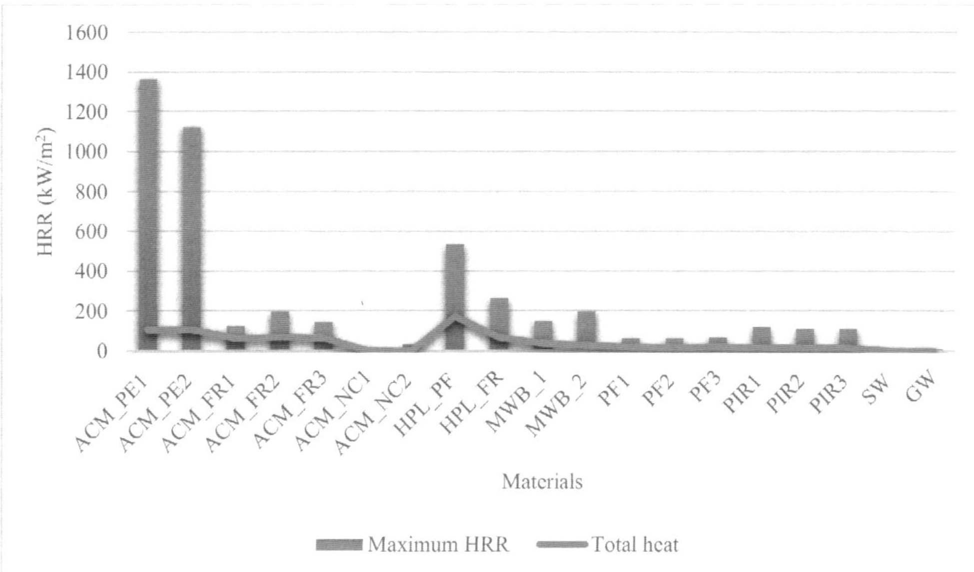

Figure 3. Fire reaction parameters of materials tested.

Smoke toxicity data showed a higher growth factor for PIR materials compared to the PF materials. The non-combustible GW and SW products indicate a smoke toxicity much lower than PIR product.

The graphic shown in figure 3 that show the total heat release and the peak of HRR show large differences in the fire behavior of the materials and highlight the most dangerous combinations of materials used in building the facades. Preventing the spread of fire on the ventilated façade must be done along the length of the construction, not just the vertical it, regardless of the fire stability level / degree fire resistance insured. Thus, every 20 linear meters of the facade, or dilatation, coughing or anti-seismic, which occurs first will provide fire resistant barriers E30. In the situation where a façade it has a length of less than 40 m and more than 20 m, then the barrier fire resistant will be mounted at half distance. Ventilated facades adjacent to the escape routes used in the case fire will be made of non-combustible materials, according to the provisions of the specific technical regulations regarding security at fire. This determines adjacent means at distances less than 3.00 m in front of the exterior staircase. Preventing the spread of fire on the ventilated facade can be integrated into the architectural achievement.

In the situation of the facades ventilated with glass windows, the fallowing protection of the facades must be achieved:

- through the protection and finishing component / the finishing material the façade system;
- through the rhythmic interruption of the vertical void inside the façade system;

- by the thermal insulation component / the thermal insulation material;
- by protecting the contour of the glazed or unprotected hollow;
- by parapets with a height of at least 1,20 m, fireproof E30 near floors or screens with a height of at least 0.50 m, fireproof E30.

4. Conclusions

Architectural features of ventilated facades are proposed for environmentally friendly or sustainable buildings. In such cases there are potential risks of rapid spread of fire through the cavity the facade due to the "horn effect" and such design normally does not can be accepted if all the measures are not fully respected fire protection. Currently, buildings with ventilated facades are not on fully covered by the legal regulations in the field of construction, especially when the combination of normal ventilation is desired building with ventilation in front due to the advantage of the exchange of the heat. Scenario for the capture and spread of smoke and heat in the cavity of the facade is identified as quite dangerous and should be better understood. Studying the spread of fire is a necessity, considering the fire as a complex action on the constructions and the effects it produces. Therefore, the experimental determinations were made within the Laboratory, with the purpose of determining the optimal test parameters, the optimal mounting positions of the temperature sensors and the other measuring and control devices, as well as the flame propagation mechanisms on the glass front, under natural ventilation conditions. The fire measures have proved successful in the case of buildings completely protected by sprinkler systems or those where the space between the curtain wall and the floor was properly sealed by the perimeter barrier. These seem to be the conditions under which architects can put their innovative design ideas into practice. The results of the tests that show the heat release and the emission of toxic gases show that the polyisocyanurate and the mineral insulation behave in a similar way during the fire, which can be summarized that insulation products do not constitute a fire safety hazard. The tests done in the laboratory show that using the fire reaction only, we can estimate the fire safety of a building façade by testing the materials and the combination of the materials that form it. Real scale test for the ventilated facades got similar results as those predicted by the fire reaction tests. This lowers the cost to improve the fire safety in building by conducting smaller and efficient test. However, in order for the results to be accurate, specialist in the fire safety with experience must determine the dangerous combinations between different materials and create a new methodology that must be approved by local legislation. This will drastically improve the fire safety of ventilated facades which right now represent an uncontrolled fire hazard that can be dangerous in certain conditions. This is mainly because in a high building with a ventilated façade there is a danger that smoke and hot gases from a space set on fire to spread through the existing gaps in the facade and to can accumulate and spread horizontally or vertically in other spaces they have openings related to the intermediate space between the facades for the purpose of natural ventilation. Fire behaviour conditions and security measures, materials and equipment used in the design and construction ventilated façade systems are to be determined very accurate and presented in the technical documentation by the specialized fire designers.

The fire resistance can be ensured by the solid support, concrete type reinforced or masonry. In the case of light walls, multilayer fire resistance will be found by testing the particular type of wall together with the façade ventilated. Continuity of the fuel components of the facades ventilated should stop at least next to the joints clamping, expansion or seismic constructions, through systems with non-combustible products of at least 1 m wide, limiting the fire propagation. Permanent consideration must be given to setting the risks in a high-rise building, such as the design and implementation of automatic fire extinguishing systems, the fire department's response capacity, the occupancy of the building and the associated loading, the concept of evacuation, compartmentalization, and not in lastly, the scenarios for evaluating security threats.

5. References

[1] BS 8414-1 2002 *Fire performance of external cladding systems applied to the face of the building* BSI 4-6 (London: British Standards Institution)

[2] ISO 13785-1 2002 *Reaction-to-fire tests for façades. Part 1: Intermediate-scale test, Part 2: Large-scale test* (London: British Standards Institution)
[3] Klopovic S and Turan O F 2001 *Fire Safety Journal* **36(2)** 99-133
[4] Klopovic S and Turan O F 2001 *Fire Safety Journal* **36(2)**135-172
[5] Simion A and Dragne H 2017 *IOP Conference Series: Materials Science and Engineering* **209** 012102
[6] Suzuki T, Sekizawa A, Yamada T, Yanai E, Satoh H, Kurioka H, Kimura Y 2000 *Proceedings of the Fourth Asia-Oceania Symposium of Fire Science and Technology, Tokyo, Japan* 363-373
[7] Swedish National Testing and Research Institute 1994 *SP FIRE 105. External wall assemblies and façade claddings – reaction to fire* **5**

Reducing Power Consumption in Smart Monitoring Systems with BLE Wireless Technology

Ș C Stoica[1], P D Mătăsaru[1*] and L Scripcariu[1]
[1] "Gheorghe Asachi" Technical University of Iasi, Faculty of Electronics, Telecommunications and Information Technology, Romania

*E-mail: dmatasaru@etti.tuiasi.ro

Abstract. This research focuses on alternative low power consumption in smart monitoring systems and on various techniques to facilitate access to different monitoring sensors connected via Bluetooth Low Energy (BLE). Due to high complexity and power demands that new smart monitoring equipment have, the need for reduced power consumption in the context of new regulations/architectures for the Internet of Things has occurred. Sensors can have dimensions of a few centimeters, are not required to have cabling or power supply and can transmit data for a long period of time. A network of sensors connected through BLE wireless technology with a corresponding architecture can provide a simple and versatile solution to multiple environments that need to be monitored. Data provided by the smart monitoring systems must be carried and analyzed either in upper layers of the network or close to the edge. Dedicated applications allow users to access data via smart-devices and data can be downloaded on any device, such as mobile phones. All these require communication technologies with optimized and reduced power consumption, also increasing battery lifetime and becoming more eco-friendly.

1. Introduction

As the world population rapidly increases, inversely proportional the resources of the planet also continue to fast diminishing, therefore more and more researchers become concerned about the waste management, as well as recycling. In an era dominated by computers, technology and artificial intelligence one can refer to waste in the form of energy consumption, limited data storage, raw processing power and so on, the need to reduce consumption turning into one of the main actual challenges.

Among the factors that might lead to this type of resources waste can be the diversity of existing regulations, rules or standards that define the modern equipment used in either households or in enterprise environment. Potential solution to such problems can be achieved by interconnecting all this equipment with commodity sensors, measuring set-up and embedded systems throughout machine learning techniques. This can generate a reliable communication between connected devices and all the real-time analytics that is being generated, thus creating a global connected network of connected networks also known as the Internet of Things (IoT) [1].

As mainstream IoT has become part of daily life, the need to store and process all this ever-increasing number of raw data (also known as Big Data) has been strongly accelerating in the last decade. International Data Corporation (IDC) has estimated that the total load generated by all IoT connected devices and users, from business to healthcare, will be reaching 175 Zettabytes of data by the year 2025, compared to only 33 Zettabytes of data in 2018 and 10 times the data generated in 2016 [2].

Wireless sensor networks (WSN) has proved to be a reliable way of real time data acquisition from a group of spatially dispersed and dedicated sensors for monitoring and recording the physical conditions of the environment and organizing the collected data at a central location. WSNs measure environmental conditions like temperature, sound, pollution levels, humidity, wind, and so on and can help in taking autonomous decision making in various applications ranging from agricultural related activities (i.e. automating irrigation, fertilizer or pesticide spreading) [3, 4] to controlling and maintaining a constant temperature and humidity in strict environments such as museums and art galleries [5]. In combination with low-cost and power efficient communication modules that can be controlled and connected via Bluetooth Low Energy (BLE), WSNs may be the answer to the challenge to further reduce the power demands in smart monitoring systems, allowing scalability and extending life time in battery powered devices, as well as reducing the overall environmental impact..

2. Analysis of Bluetooth Low Energy power consumption

With the accelerated and ongoing switching to the Internet of Things, where all devices are constantly collecting and exchanging data from edge to the cloud, Bluetooth Low Energy (BLE) represents a convenient way of communication between network nodes which are wirelessly transmitting data in any given WSN [6]. The sensors can transmit data to a BLE-enabled device that is responsible for processing information, analyze it and make decisions close to the edge if necessary (e.g. life support systems or any other systems that require immediate response in case of emergency). Also, some of the data may be forwarded up-network for additional analysis and eventually storage. The BLE system can operate in low-power mode, where most of the time sensors are in a low-power state with reduced power consumption (e.g. during deep sleep mode the current consumption can be reduced as low as 1.3 – 1.4 μA), making this solution energy efficient and suitable for complete systems with strict power requirements (constrained devices). However, a BLE system can distinguish five power modes: active, sleep, deep sleep, hibernate and stop, power consumption variating depending on the mode. A lot of technologies used in IoT, such as BLE or Zigbee, have an activity cycle in which the sleep mode power consumption dominates the overall current drawn by the System on Chip (SoC). A usual power consumption in the 5 modes in a Programmable-System-on-Chip (PSoC) produced by Cypress Semiconductor [7] is presented in Table 1.

Table 1. Power Mode and typical Current Range (PSoC produced by Cypress Semiconductor) [7].

Active	Sleep	DeepSleep	Hibernate	Stop
1.3 mA to 14 mA	1.0 mA to 3 mA	1.3 μA to 15 μA	150 nA to 1 μA	20 nA to 80 nA

An in depth analysis of the power consumption of Bluetooth Low Energy, ZigBee and ANT Sensor Nodes in a Cyclic Sleep Scenario and the experimental results has been done in [8], proving that BLE is a better choice for diminishing power consumption and extending battery duty cycle (Table 2).

Table 2. Power consumption of Bluetooth Low Energy, ZigBee and ANT Sensor Nodes in a Cyclic Sleep Scenario [8].

	BLE	Zigbee	ANT
Sleep current	0.78 μA	4.18 μA	3.1 μA
Awake current	4.5 mA	9.3 mA	2.9 mA
Min current (at 120 sec interval)	10.1 μA	15.7 μA	28.2 μA

Another study [9] pointed out the impact of sleep current consumption on an application powered by a coin-cell battery which might be in active mode 1 minute per day (0.07% active, 99.93% in sleep mode – as could be the case in a museum) comparing BLE and Wi-Fi.

Table 3. Impact of sleep current consumption on an application powered by a coin-cell battery which might be in active mode 1 minute per day [9].

	Active current	Sleep current	Average current	225 mAh coin cell battery autonomy
WiFi SoC	100 mA	10 μA	80 μA	117 days
		1 μA	71 μA	132 days
		0.2 μA	70.2 μA	133 days
BLE SoC	7 mA	10 μA	14.9 μA	629 days
		1 μA	5.9 μA	1588 days
		0.2 μA	5.1 μA	1838 days

As seen in Table 3, the duty cycle of a battery lasts for several years due to low power consumption when using Bluetooth Low Energy.

3. Analysis of system architectures and impact on power consumption

A smart monitoring solution for a museum [10] was analyzed considering a low-power BLE in order to achieve the power constrains required by the network of sensors in such an environment. Considering that in spaces such as museums, internet connections and a stable power source may be limited, the proposed solution involved that several nodes are responsible for monitoring quantities of interest in different sites.

These nodes must have small dimensions and are to be powered by a small size battery, using a low power low range wireless connection in order to send the collected data. Due to their small size (usually lower than 3 cm in diameter), the sensors may be easily mounted, however this reduction in size may also become challenging when it comes to wirelessly transmit data. Although this small-size battery may bring some important challenges (e.g. limited capacity, low voltage, required power output, weight, size, shape), a short-range wireless connection between nodes may be achieved to distances up to 10 m, providing an average life expectancy of a few years without battery replacement. Depending of their complexity, some edge nodes can partially store information in a local memory buffer, increasing the reliability of the system if an error occurs in the radio transmission.

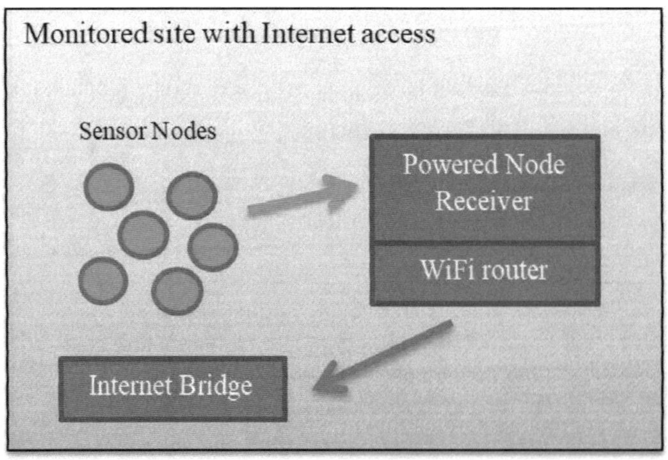

Figure 1. Simplified measuring solution for a site with no Internet access and power source (worst case scenario) [10].

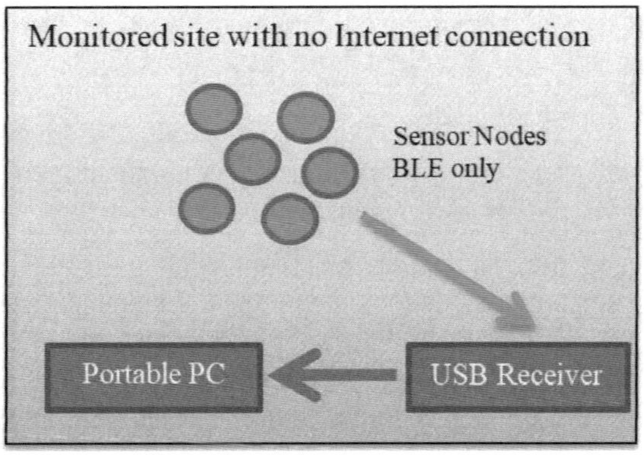

Figure 2. Measuring solution for a site with reliable internet access and power source [10].

In the worst-case scenario presented in Figure 1, the monitoring system relies only on the internal memory of the transmitting nodes to manage collected data. Depending on capacity, power constrains or specific user needs, a given amount of data is stored locally to be later transferred via USB or other portable receiving devices whose role is to make the data available to users and larger storage facilities in the cloud.

In the scenario presented in Figure 2, the sensors (measuring nodes in the network) are constantly sending data via cabling (not always possible and not always desired due to esthetic reasons) or via a Wi-Fi connection to a powered receiver, in real time.

If a problem with the power distribution or with the Internet link arises, the data still may be collected and stored locally for a limited period of time, until the connection is re-established. This requires an additional battery for back-up, with limited capabilities depending on the power consumption, the range of transmission and frequency of reporting. A lower power consumption increases the service period for back-up if the problem with the main power-supply persists.

The architecture of the monitoring system proposed in [10] is a button architecture designed starting from the measurement sensor, which is responsible for collecting the museum environmental data, an important issue in artifact and exhibit conservation.

Based on user's specific needs, sensors are connected to a controller and wireless transmitter equipped with an antenna using a non-volatile serial memory capable of locally storing data, the measuring node being powered with a Lithium battery (Figure 3). Therefore, power consumption at all parts becomes a critical aspect which needs to be addressed, impacting on size, lifetime and performance of the system.

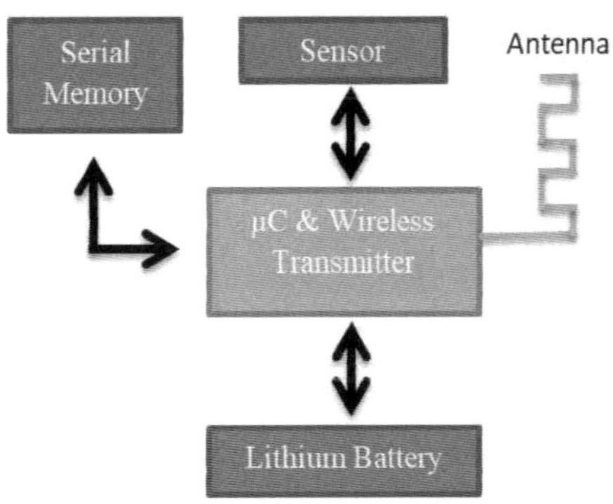

Figure 3. Block diagram for a measuring node [10].

In the end, the collected data is available to the user who can access the network-based storage depicted in Figure 4 using a PC or a smartphone in order to process, eventually store all the accumulated data.

Another example of the advantages of WSN is represented in [11], where sensors are used in order to provide constant measurements of humidity without impacting the esthetics of the room. In this case due to the spherical layout of the room the sensors were placed circularly, which made battery-powered nodes the valid choice.

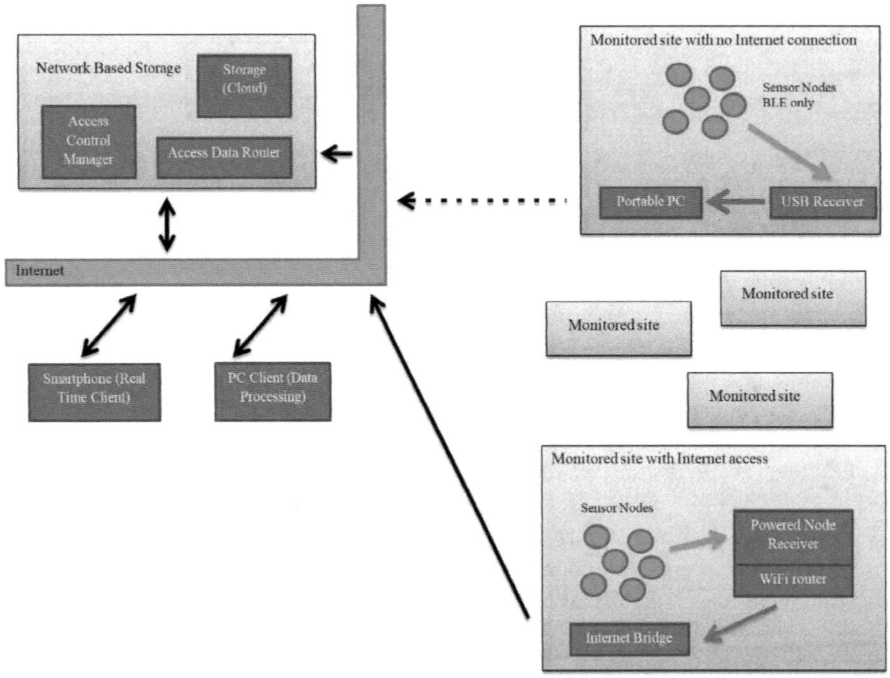

Figure 4. Architecture of a monitoring system [10].

Considering the small dimensions of the sensors and their light weight they can be mounted everywhere with ease and provide accurate measurements throughout an increased life cycle, if power consumption is optimized and diminished per system using BLE. Another important issue concerning the measurements values that were collected during an entire year was revealed, that in the museum room the variation of humidity is greatly varying during every season. By efficiently controlling of the air conditioner, the dehumidifier and even the central heating of the room in order to maintain constant room humidity, not only the preservation of the museum exhibits was achieved, but also an optimal energy consumption in the room which decreased overall costs and helped the battery life additionally increase, due to small temperature variations.

4. Conclusions

This article presents some of the benefits of using Bluetooth Low Energy (BLE) combined with wireless sensor networks as a feasible alternative to lower power consumption, increase coverage and preserve accuracy of data collected in different environments with or without a stable Internet connection. It is estimated that a rapid increase in wireless sensor networks (WSNs) that might benefit from this technology, considering the accelerated development in the Internet of Things. Equipment in wider areas will be inter-connected and could be capable of working in scenarios in which the lack of a power source and a reliable internet connection could pose a problem. Not only the power consumption of the measurement nodes in a WSN by using BLE is optimized and minimized, but also the overall slow variable environmental conditions also helped in preserving battery life and decrease consumption for low range low power devices. Another important result would be the fact that extending the duty-cycle of batteries, the number of batteries that need disposal would be small, thus the hole process becomes more eco-friendly.

5. References

[1] Al-Fuqaha A, Guizani M, Mohammadi M, Reinsel D, Aledhari M and Ayyash M 2015 *IEE Communications Surveys and Tutorials* **17(4)** 2347-2376

[2] Reinsel D, Gantz J and Rydning J 2017 *Data Age 2025: The Evolution of Data to Life-Critical. 2017 Available online: https://www.seagate.com/www-content/our-story/trends/files/Seagate-WP-DataAge2025- March-2017.pdf*

[3] Yoo S, Kim J, Kim T, Ahn S, Sung J and Kim D 2007 *A2S Automated Agriculture System Based on WSN,* IEEE International Symposium on Consumer Electronics, Doi: 10.1109/ISCE.2007.4382216

[4] Kurmi S K, Verma R. and Sharma A K 2013 *International Journal of Emerging Technology and Advanced Engineering* **3** 663-667

[5] Diaconescu V D, Scripcariu L, Matasaru P D, Diaconescu M R and Ignat C A 2018 *IOP Conference Series: Materials Science and Engineering* **374** 012074

[6] Jung K K and Kim Y 2018 *International Conference on Electronics, Information, and Communication (ICEIC)* **1-3**

[7] Sadasivam K and Kingsburyhttps M 2018 *PSoC 4 Low-Power Modes and Power Reduction Techniques* (San Jose: Cypress)

[8] Dementyev A, Hodges S, Taylor and Joshua S 2013 *In IEEE International Wireless Symposium* **1-4**, Doi: 10.1109/IEEE-IWS.2013.6616827

[9] Genevey S and Dominguez C *Improving Battery-Powered Device Operation Time Thanks To Power Efficient Sleep Mode. Available online: https://www.design-reuse.com/articles/42405/improving-battery-powered-device-operation-time-thanks-to-power-efficient-sleep-mode.html*

[10] Lombardo L, Corbellini S, Elsayed E, Angelini E, Parvis M, Grassini 2017 *IEEE International Workshop on Measurement and Networking*, Doi: 10.1109/IWMN.2017.8078394

[11] Diaconescu V D, Scripcariu L, Matasaru P D, Diaconescu M R and Ignat C A 2018 *IOP Conference Series Materials Science and Engineering* **572** 012099

Simulation of Pedestrian Walking Through Angled-Corridors for Evacuation Behaviour Study

N A Rahman[1*], N A Alias[1], N S A Sukor[1], H Halim[1], H Gotoh[2] and F H Hassan[3]
[1]School of Civil Engineering, Engineering Campus, Universiti Sains Malaysia, 14300, Penang, MALAYSIA
[2]Graduate School of Civil and Earth Resources Engineering, Kyoto University, Katsura Campus, Nishikyo-ku, 615-8540, Kyoto, JAPAN
[3]School of Computer Sciences, Universiti Sains Malaysia, 11800 Penang, MALAYSIA

E-mail: celindarahman@usm.my

Abstract. The effect of acute- and obtuse- angled-corridors on the evacuation behaviour is rarely studied. In the previous work, we have confirmed that corridor with less than 90° turning angle has negative impact to the pedestrians walking behaviour due to the tendency of the pedestrians to slow down their motion when approaching the turning of the corridor. This restrict the flow of walkers during high-density situation and cause congestion near that turning of the corridor. In this contribution, empirical data collected from previous experimental work is utilised to further study the effect of turning angle on the evacuation behaviour. The simulations of walking through angled-corridor were reproduced which reflects evacuation scenarios. Three different types of angled-corridor are taken into consideration (a 60°-, a 90°-, and a 135°- angled-corridors) with different sets of number of pedestrians (NOP: 60, 70, 80, 90 and 100). Three metrics of evacuation behaviour measurement are reported: (1) inflexion points, (2) escape time, and (3) interaction force. Inflexion point is used to evaluate the restriction flow occurred during the evacuation. From the results, the 60° angled-corridor contained the highest number of inflexion points. With the highest number of inflexion points, the escape time for the 60° angled-corridor is 14.8% to 24% longer compared to the other two types of angled-corridors. This shows that this kind of angled-corridor need to be avoided in the future design of walkways. On top of that, the 60° angled-corridor also gives the maximum interaction force with 213.75 N. This reconfirmed that the corridor with less than 90° turning angle is not suitable to be built as a walkway. This work is relevant to be studied due to its immediate applications in assessment of crowd safety for building egress.

1. Introduction

Many complex geometries of walkways can be witnessed in public buildings (like train stations, shopping malls, universities and parks) and one of their kinds is an angled-corridor. In our previous experimental study on pedestrians walking through angled-corridor [1], we have confirmed that corridor with less than 90° turning angle (i.e. 60° angled-corridor) has negative impact to the pedestrians walking behaviour due to the tendency of the pedestrians to slow down their motion when approaching the turning of the corridor. This could restrict the flow of walkers during high-density situation and cause congestion near that turning angle of corridor. As the turning angle of corridor increases (i.e. 135° angled-corridor), the impacts to pedestrian walking velocity become less.

The findings in the previous study is crucial to be further studied in postulating the effect of angled-corridor on evacuation behaviour. From current literature review conducted, the effects of acute and obtuse angled-corridors on evacuation behaviour is still limited. Shiwakoti et al. (2011) [2] highlighted the importance aspect of collective egress under emergency conditions is the turning movement when a sudden change in the direction or the layout of the escape area occurs. Dias *et al.* (2012) [3] suggested that right angles in the egress path are 20% to 25% less effective compared with the straight egress paths. As further examination with microsimulation suggests, this figure could be as significant as 65% for higher densities.

In this contribution, empirical data collected from previous experimental work [1] is utilised to further study the effect of turning angle on the evacuation behaviour. The simulations of walking through angled-corridor were performed which reflects evacuation scenarios. Three different types of angled-corridor are taken into consideration (a 60°-, a 90°-, and a 135°- angled-corridors) and three metrics of evacuation behaviour measurement are reported. This work is relevant to be studied due to its immediate applications in assessment of crowd safety for building egress.

2. Methods

This study is the continuation of the previous study [1]. In the previous study, we reported experimental findings of pedestrians walking behavior while walking through angled-corridors. The setting of the angled-corridor is as shown in Figure 1. From that study, we came to the conclusion that pedestrians tend to slow down their walking pace when approaching the turning angle of the corridor (β). In high density situation, restriction to the flow of walkers could be triggered. Hence, these findings are worth to be further studied in postulating the effect of turning angle on evacuation behavior.

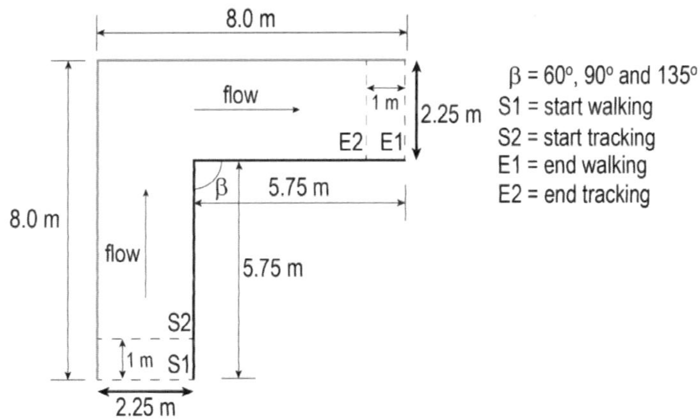

Figure 1. Schematic of angled-corridor.

To facilitate this study, a microscopic Distinct Element Method (DEM)-based model, which has been used previously in simulating hypothetical evacuation scenarios [4–6] is employed. DEM-based model is capable to calculate the interaction force (between humans and boundaries) and egress time, in which the important metrics of evacuation behavior measurement in this study.

2.1. Validation of the model

For determination of the degree of validity of the model, pedestrians walking through angled-corridor was reproduced first by using DEM-based model. This validation work is crucial to confirm the reliability of DEM-based model. From the previous experimental work, three walking scenarios were selected to be reproduced: a 60° angled corridors with number of persons (NOP) 60, a 90° angled-corridors with NOP 60 and a 135° angled-corridors with NOP 60. The reliability of the model is based on the comparison of 'time-in' and 'time-out' near turning angle of corridor between simulation and empirical data from [1]. If the difference between simulation and empirical data is less than 5%, the model is considered reliable.

2.2. Simulation of evacuation scenarios

In studying the turning angle effect on evacuation behaviour, simulations were performed for three different angled-corridors: a 60°-, a 90°-, and a 135°- angled-corridors and five different number of pedestrians (NOP) were simulated: NOP: 60, 70, 80, 90, and 100. Note that the NOP is the variable used in this simulation works. It starts with NOP 60 and increase by 10 until achieve NOP 100. The NOP starts from 60 because of the maximum participation in the previous experimental work is up to 60 pedestrians. A total of 15 simulations is conducted. Three metrics of evacuation behavior are collected from these simulations: (1) inflexion points, (2) escape time, and (3) interaction force.

2.3. Safe level of interaction force

The interaction force acting on each pedestrian is a significant output from the current simulations. This metric can be used to confirm the interaction between pedestrians are in the safe level. Previously, the investigations of injury on human have been reported. The experimental works on the maximum endurable force during earthquake evacuation had been conducted by [7]. They found that the maximum endurable contact force that can be withstood by a fit young female and a middle aged male weighing 60 kg and 74 kg is 162 N minimum and 242 N maximum. Meanwhile, study conducted by [8] on the level of comfortable loads for twenty-one pedestrians with the range of age 20 to 25 years old had been tested on three barrier types loaded on upper and lower chest and abdomen. They found that the comfortable limit of loads was ranged from 175 N to 247 N. Hence, the benchmark for the safe level of interaction forces used in this study is in the range of 162 N and 242 N as discovered from the most recent studies [7].

3. Results and discussion

The results are discussed according to the following criteria: (1) reliability of DEM-based model; (2) restriction flow due congested situation; and (3) the safe evacuation.

3.1. Reliability of DEM-based model

During the experimental work [1], the 'time-in' and 'time-out' near the turning angle of corridor has been collected. The 'time-in' is measured when the first pedestrian reached the red line (1 m before the turning angle) and the 'time-out' is measured when the last pedestrian reach the yellow line (1 m after the turning angle) (shown in Figure 2). Similarly, in the simulation works, the 'time-in' and 'time-out' of the pedestrians are also collected. Then, the comparison is made between findings in [1] and simulation works.

Task	Angled-corridor	Experiment	Simulation
37A (NOP 60)	60°	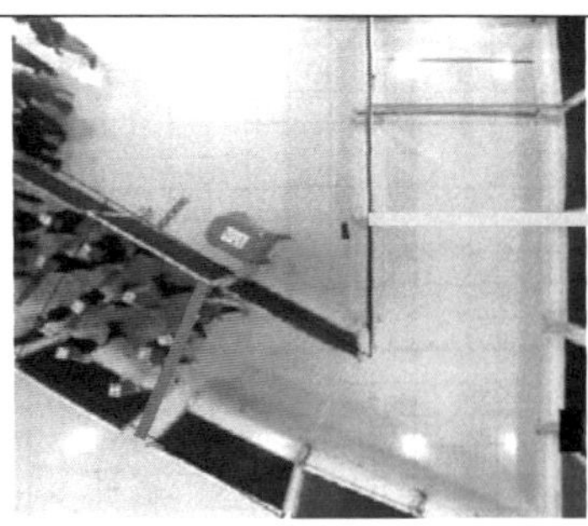 Time in: 3.14 s	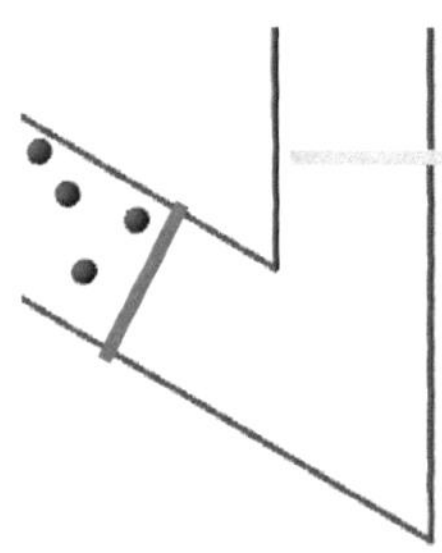 Time in: 4.00 s

IOP Conf. Series: Materials Science and Engineering **877** (2020) 012061 doi:10.1088/1757-899X/877/1/012061

37A (NOP 60)	60°	 Time out: 47.04 s	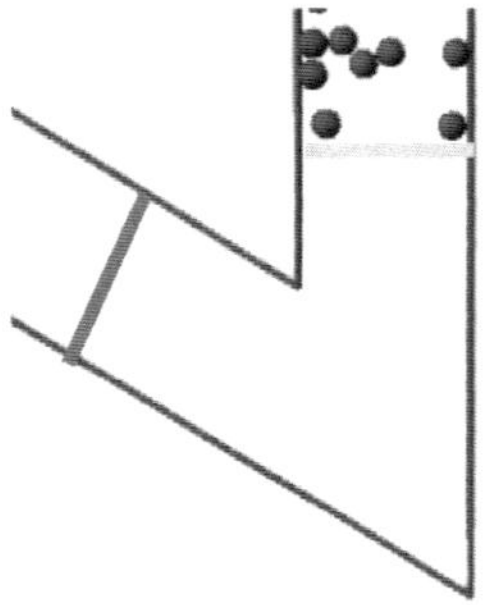 Time out: 49.00 s
13B (NOP 60)	90°	 Time in: 4.18 s	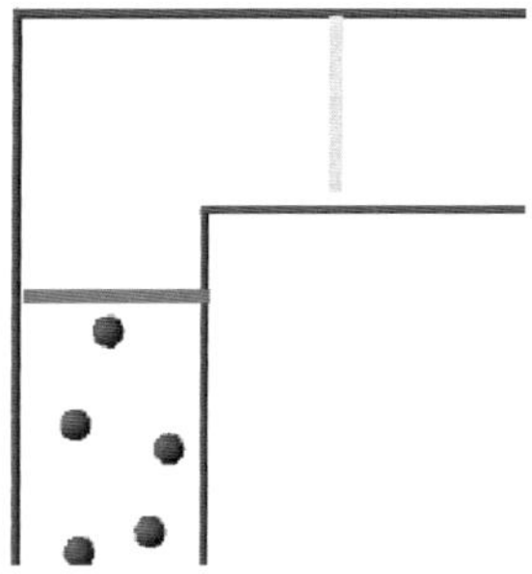 Time in: 5.00 s
		 Time out: 44.12 s	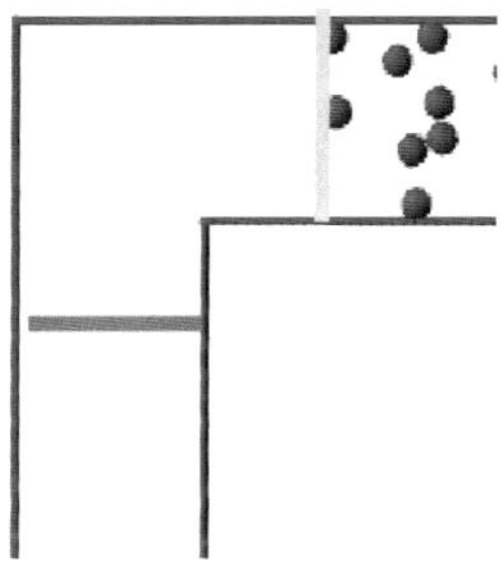 Time out: 45.00 s
29B (NOP 60)	135°	 Time in: 4.13 s	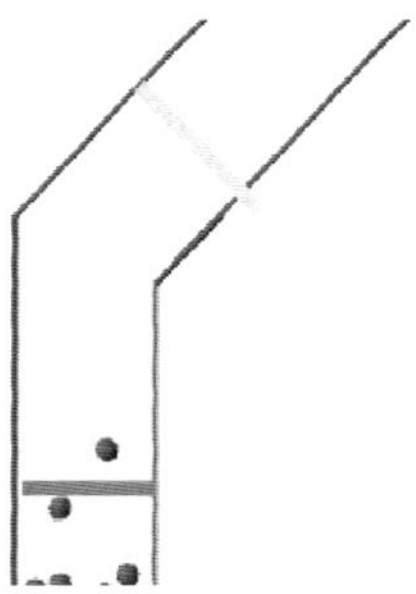 Time in: 5.00 s

29B (NOP 60) 135°

Time out: 37.28 s

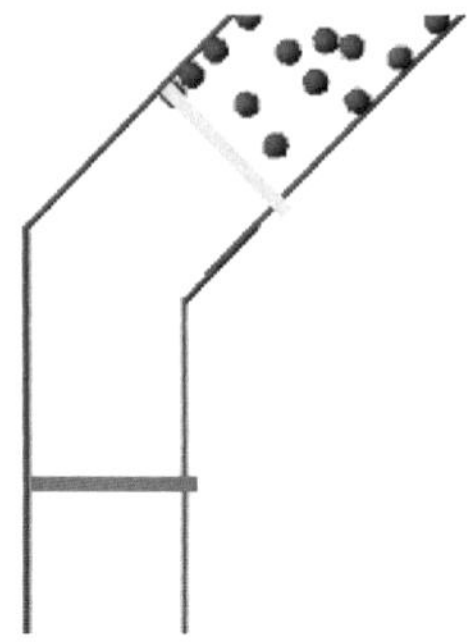

Time out: 38.00 s

Figure 2. The time-in and time-out for experimental and simulation works.

From Table 1, it shows that the difference of 'time-in' and 'time-out' at angled-corridors between experiment and simulation works are acceptable. The percentage range of the differences are between 0.48% to 2.51% and the time differences between experimental and simulation are ranged around 0.06 s to 1.10 s. Hence, the DEM-based model is considered reliable and suitable for this study. Meanwhile, Figure 2 shows visually the 'time-in' and 'time-out' for both experimental and simulation works.

Table 1. Difference of time-in and time-out at angled-corridor between experiment and simulation works.

Angled-corridor	Experiment			Simulation			(A) – (B) (s)	%
	Time-in (s)	Time-out (s)	Difference (s) (A)	Time-in (s)	Time-out (s)	Difference (s) (B)		
60°	3.14	47.04	43.90	4.00	49.00	45.00	1.10	2.51
90°	4.18	44.12	39.94	5.00	45.00	40.00	0.06	0.15
135°	4.13	37.29	33.16	5.00	38.00	33.00	-0.16	-0.48

*Can be referred in [1]

3.2. Restriction flow due to congested situation

In measuring the restriction flow that may occur at the turning angle of the corridor, the time taken to complete the task of walking through angled-corridor (escape time) is plotted versus the number of escaped pedestrian (NEP). Figure 3(a)–(e) show the graph of the escape time versus NEP for different number of pedestrian (NOP). From the graphs, it can be witnessed that, the 60° angled-corridor takes longer time than the other two types of angled-corridor.

Table 2. Number of inflexion point counted.

Angled-corridor	Number of Inflexion point				
	NOP 60 (SIM1-SIM3)	NOP 70 (SIM4-SIM6)	NOP 80 (SIM7-SIM9)	NOP 90 (SIM10-SIM12)	NOP 100 (SIM13-SIM15)
60°	10	11	12	9	5
90°	10	8	7	9	5
135°	4	6	5	7	5

IOP Conf. Series: Materials Science and Engineering **877** (2020) 012061 doi:10.1088/1757-899X/877/1/012061

To confirm further the reason of longer time taken in 60° angled-corridor, the inflexion points are observed and counted (as shown in circle in Figure 3). The inflexion points are the uniform lines in the graphs. These inflexion points indicate that the flow is restricted at certain points along the corridor. Table 2 shows the frequency of the occurrence of the inflexions points for every type of angled-corridor. The highest inflexion points is observed at 60° angled-corridor. This confirmed that the restrictions of flow mostly occurred at 60° angled-corridor.

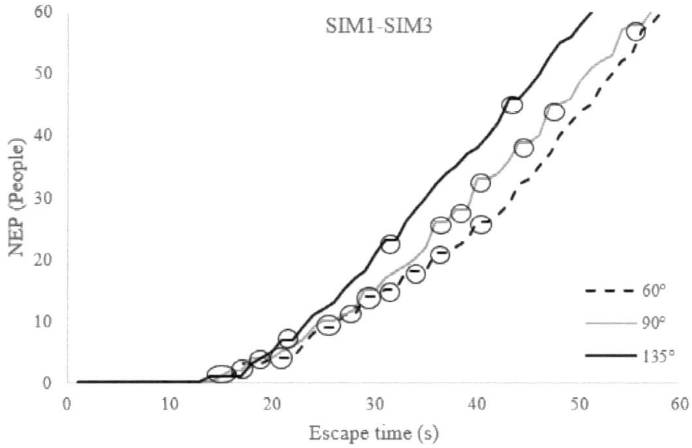

(a) NOP 6;

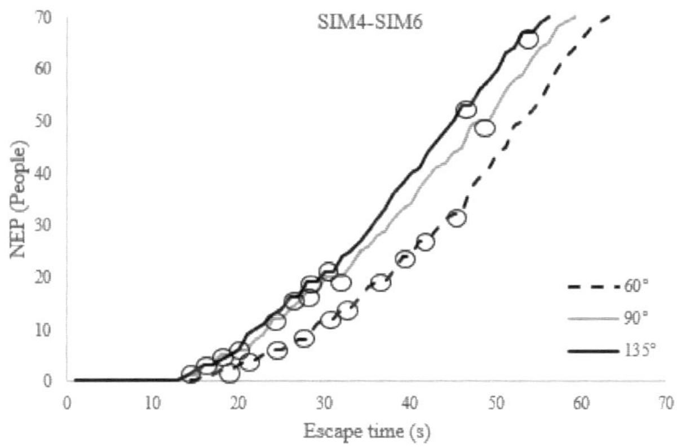

(b) NOP 70;

Figure 3. The escape time versus the number of escaped pedestrian (NEP) for all types of angled-corridor (from NOP 60 to 100).

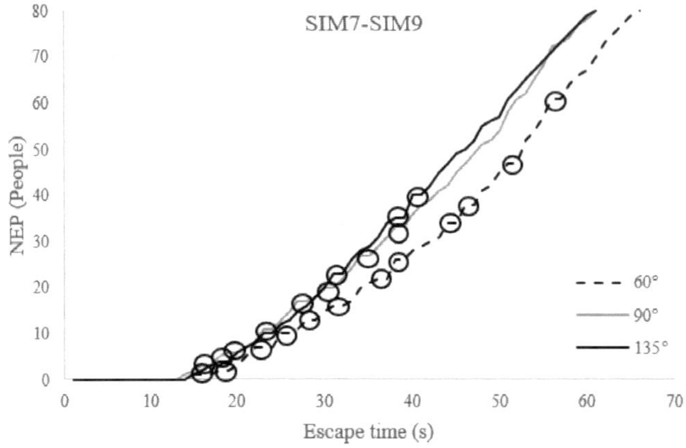

(c) NOP 80;

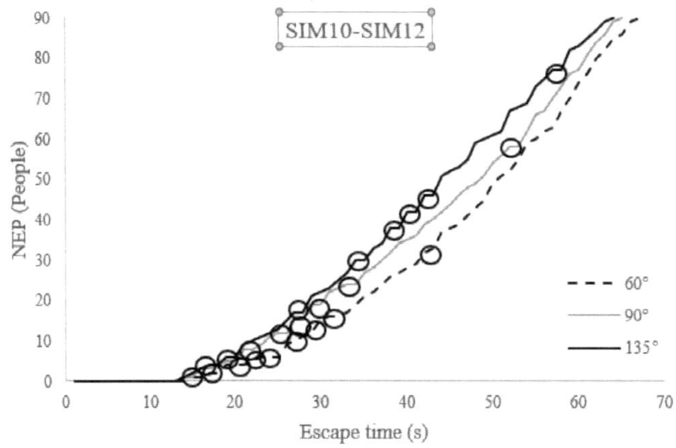

(d) NOP 90;

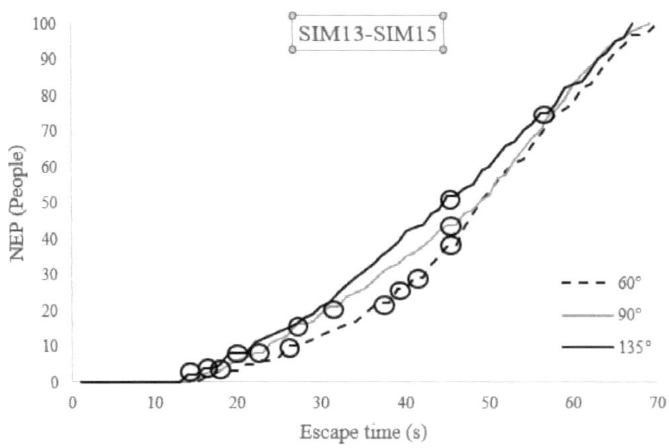

(e) NOP 100;

Figure 3. (continue) The escape time versus the number of escaped pedestrian (NEP) for all types of angled-corridor (from NOP 60 to 100).

3.3. The safe evacuation

As mentioned earlier, the interaction force acting on each pedestrian is a significant output from the current simulations. This force can be used to affirm the safest evacuation scenario simulated so that the findings from this study are more valuable.

A time series of maximum interaction force is plotted (Figure 4). The interaction force analysis was conducted to ensure the maximum interaction force acting on each pedestrian is within the safe level of interaction force (162 N – 242 N). Meanwhile, Table 3 shows the maximum interaction force with the snapshots for NOP 80 for all three types of angled-corridor. The snapshots visualise the situation at time the maximum interaction force occurred for each types of angled-corridor. As can be seen, during t = 38 s, at 60° angled-corridor, the pedestrians accumulate near the turning angle of corridor. This shows that a 60° angled-corridor has a strong tendency for a congestion to occur. Similar situation occurs at the 90° angled-corridor at t = 32 s. Meanwhile, in 135° angled-corridor, less pedestrians are seen accumulated near the turning angle. This confirmed our previous findings which pedestrians prefer to occupy space available to suit their own speed and comfort since the corridor is nearly similar to the straight corridor.

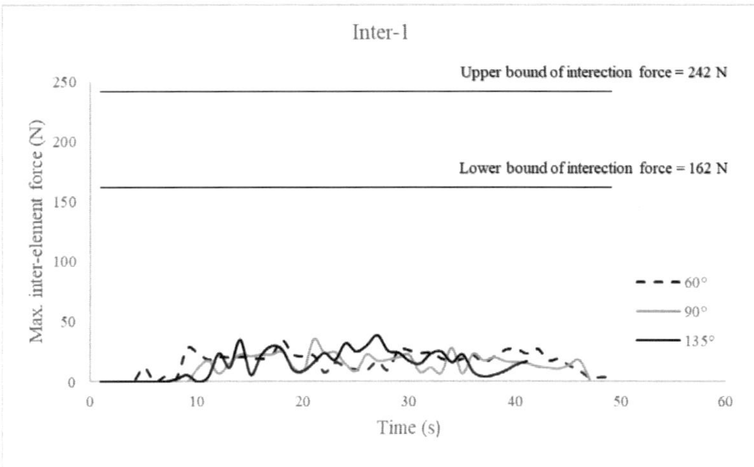

(a) NOP 60

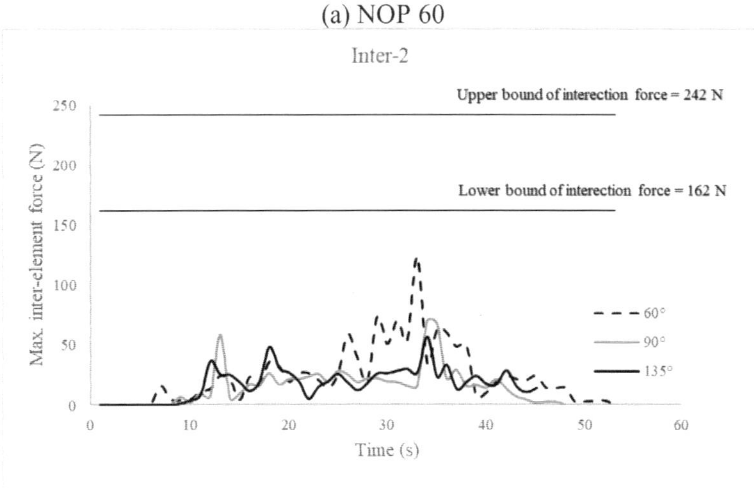

(b) NOP 70

Figure 4. The time series of the maximum interaction force.

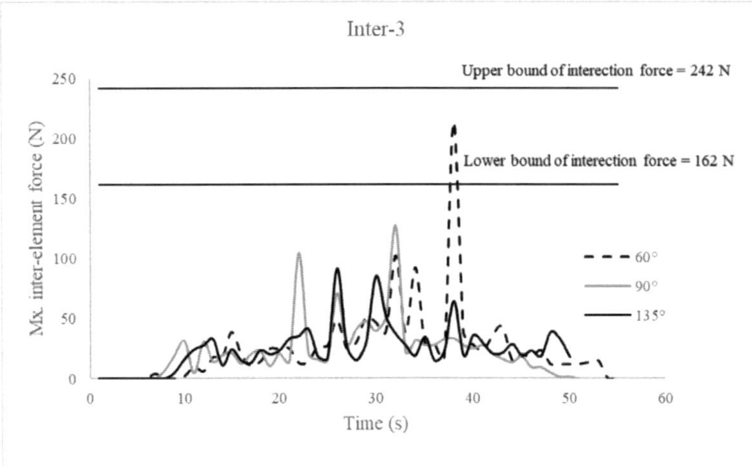

(c) NOP 80

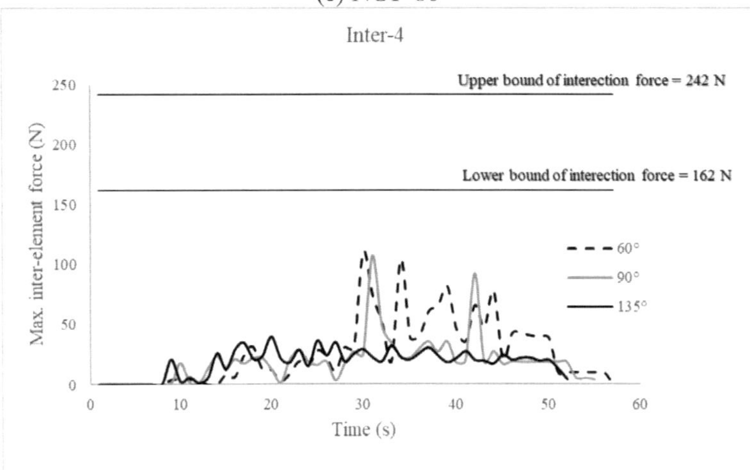

(d) NOP 90

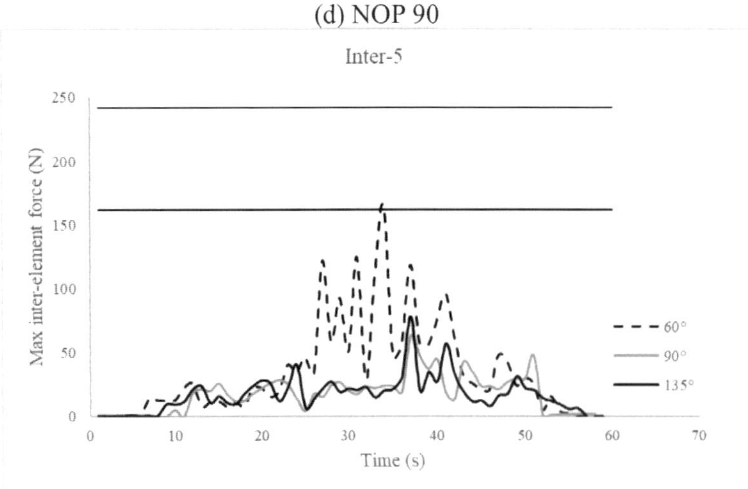

(e) NOP 100

Figure 4. The time series of the maximum interaction force.

Table 3. Maximum interaction force for Inter-3 with NOP 80.

Number of pedestrian	Angled-corridor	Max. inter-element force (N)	Snapshot from simulation work
Inter-3 (NOP 80)	60°	213.75	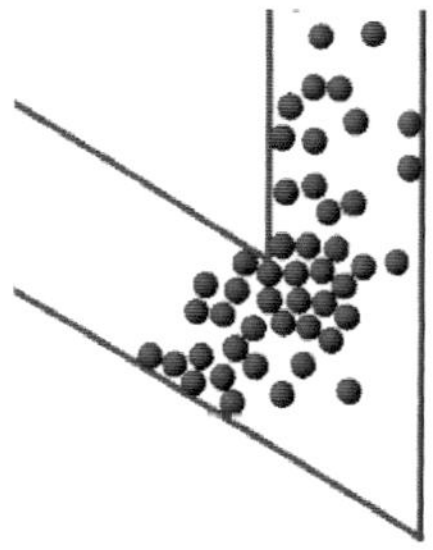 t = 38 s
	90°	126.98	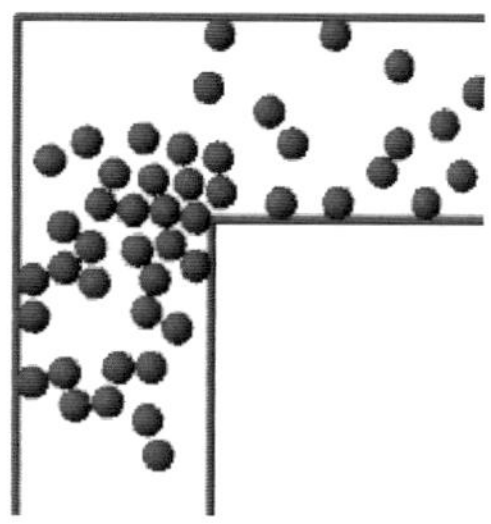 t = 32 s
	135°	91.29	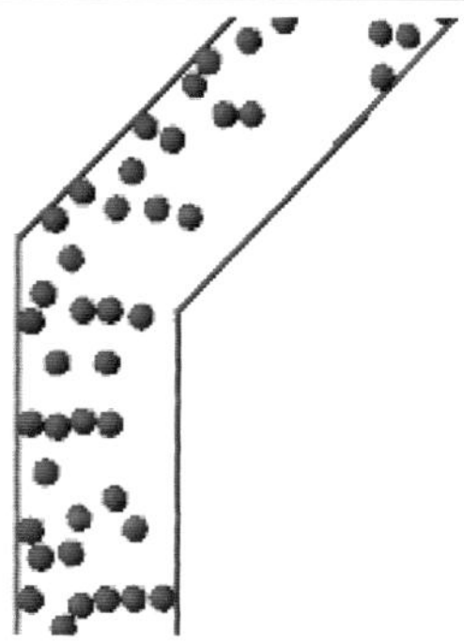 t = 26 s

4. Conclusion

In this contribution, the simulation works are performed to further study the effect of angled-corridor on evacuation behaviour. Three metric of measurements were collected: (1) inflexion points, (2) escape time, and (3) interaction force. Inflexion point is used to evaluate the restriction flow that may occur during the evacuation. As can be witnessed, the 60° angled-corridor contained the highest number of inflexion points. With the highest number of inflexion points, the escape time for the 60° angled-corridor is longer. This shows that this kind of angled-corridor need to be avoided in the future design of walkways. On top of that, the same corridor also gives the maximum interaction force with 213.75 N. This reconfirmed that the corridor with less than 90° turning angle is not suitable from the perspective of safety issue.

Acknowledgement
The authors would like to thank the sponsorship of this study, Universiti Sains Malaysia, under Short Term Grant 2016 (304/PAWAM/60313040).

5. References

[1] Rahman N A, Alias N A, Sukor N S A, Halim H, Gotoh H and Hassan F H 2019 *IOP Conference Series: Materials Science and Engineering* **572** 012114
[2] Shiwakoti N, Sarvi M, Rose G and Burd M 2011 *Transportation research record* **2234(1)** 97-104
[3] Dias C, Sarvi M, Shiwakoti N and Burd M 2012 *Transportation research record* **2312(1)** 120-127
[4] Gotoh H, Harada E and Andoh E 2012 *Safety science* **50(2)** 226-332
[5] Harada E, Gotoh H and Rahman N B A 2015 *Safety science* **79** 105-115
[6] Rahman N A, Abustan M S, Mat N C, Gotoh H and Harada E 2015 *Applied Mechanics & Materials* **802**
[7] Langston P A, Masling R and Asmar B N 2006 *Safety Science* **44(5)** 395-417
[8] Smith R A and Lim L B 1995 *Safety science* **18(4)** 329-335

Polyester Dyeing and its Environmental Impact

LR Manea[1], A-P Bertea[1*] and A Bertea[1]
[1] Gheorghe Asachi Technical University of Iasi, Faculty of Industrial Design and Business Management, Prof. Dr. Doc. Dimitrie Mangeron Street, No. 29, RO-70005047, Iași, Romania

*E-mail: andrei_bertea@yahoo.co.uk

Abstract. Dyeing and printing of textiles are responsible for 30% of the pollutants contained in the effluents. The impact of textile dyeing processes on environment is highly variable, most important issues being the colour, the organic charge, and the solids in suspension. The present study investigates the characteristics of the polyester dyeing process in what concerns the environmental issues. Disperse dyes are non ionic, have very limited solubility in water at room temperature and have substantivity for polyester fibre. Two disperse dyes were used, and dyeing were performed at different concentrations. The exhaustion degree, the volume of wastewater used in the process, the specific flow, pH, temperature and chemical oxygen demand of the wastewater, the total residuum, volatile substances and the biological oxygen demand were determined in all the cases. The data obtained showed that the global pollutant content of the wastewater from disperse dyes dyeing is high (more than two times higher than those obtained in the case of dyeing with acid dyes). Absorption spectra of the initial and exhausted dyebath were determined, in order to evaluate the dilution degree needed to make wastewater colour acceptable.

1. Introduction

Most of the pollutants contained in the effluents from textile chemical processing are caused by preparation processes (about 70%). For the remaining 30% dyeing and printing are responsible [1-10]. Complete dye exhaustion in the dye bath is not possible for most dyeing systems. This is because there is a balance in dyeing and washing, which always leaves some of the dye in the wastewater. In commercial processes, there can be from almost none to more than 50% of the total dye left in the residual dyeing bath. In the textile sector, about 10-20% of the used dyes are lost in the wastewater through incomplete exhausting and washing process [11-18]. In addition to dyes, wastewater from textile dyeing contains significant amounts of auxiliaries, such as salt, acids, equalizing agents, etc., and this is why the textile dyeing is considered as one of the most environmentally unfriendly industrial processes [19-25].

The impact of dyes on environment is highly variable, but the colour they generate is of great interest. Even though colour is one of the key attraction of any textile material, the dyes may cause threats to environment and living beings [26-28]. While coloured organic substances generally impart only a small degree of the total organic load in a wastewater, the high degree of colour is easily detected (human eye can detect some dyes at concentrations as low as 0.005 mg/L) and detracts from the aesthetic value of streams, rivers, etc. These small quantities of dyes that are visible in water give rise to worries. Most dyestuffs have a mammalian toxicity low enough so they do not cause problems, but, as long as the public is concerned, the presence of coloured wastewater is often more important than the presence of the soluble colourless organic substances which usually contribute to the major fraction of the biochemical oxygen demand.

Polyester fibre is currently the most used fibre in the textile industry, due to its physical properties, price, recyclability and versatility, advantages unmatched by any other fibre. Also polyester can be successfully blended with cotton with various proportions. Since 1990, the consumption of polyester fibres has increased at a sustained rate of almost 7% per year globally.

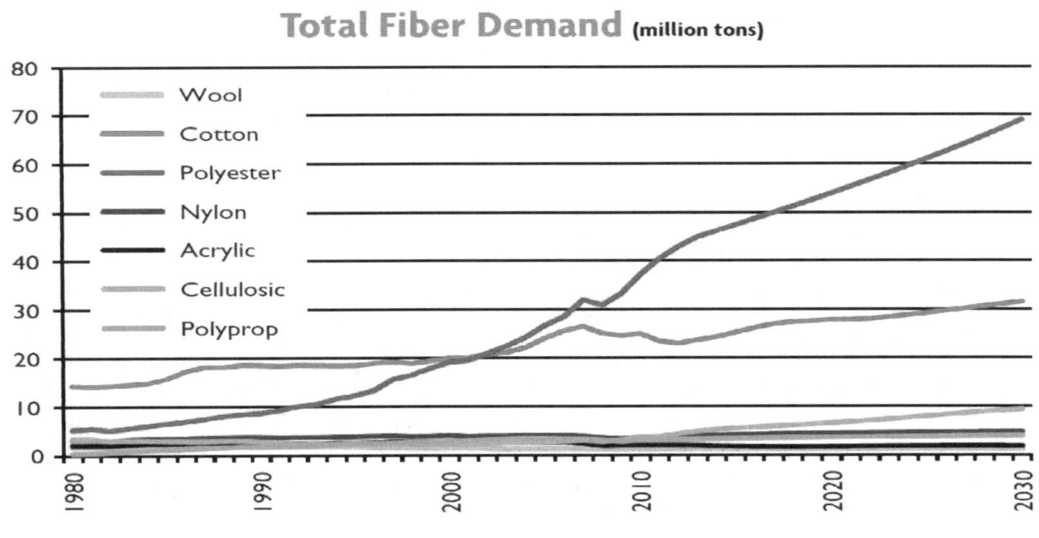

Figure 1. Evolution of fibre demand [29].

Because the polyester fibres are characterized by a very compact structure, with a high crystallinity and without defined dyeing areas, only disperse dyes, applied from a stable aqueous dispersion, are used for their dyeing. Besides polyester, disperse dyes are the main water-insoluble dyes for cellulose acetate fibres [4-5].

Disperse dyes are non ionic, have very limited solubility in water at room temperature and have substantivity for polyester fibre [30-37]. As disperse dyes have very limited solubility in water, some particulate disperse dye may still be on fibre surface after the dyeing phase has completed, and these dyes particles have to be removed from the surface as it will considerably reduce the wet fastness, wash fastness, sublimation and dry cleaning fastness, as well as making the shade become dull. The usual procedure for removing this unwanted dye is called reduction clearing, and it generates supplemental pollution to the wastewater [38-41].

2. Experimentals

The paper investigates the characteristics of the polyester dyeing process in what concerns the environmental issues.

We have used commercial samples of two disperse dyes: Disperse Blue 19 and Disperse Black 9. The dyeings were made on 100% polyester fabric under HT conditions. Different concentration dyeings were produced in sealed dye pots of 300 cm^3 capacity, housed in a laboratory-scale dyeing machine, using a 20:1 L:R, at a maximal temperature of 130°C. The dyeing recipes are showed in Table 1.

When dyeing polyester with disperse dyes the levelness of the dyeing is determined by the dispersion stability, rate of dyeing and migration. An increase in migration demands longer dyeing times. The rate of dyeing and therefore the heating rate must be controlled in such a way that the dyeing is level from the beginning by means of controlled adsorption.

Table 1. Dyeing recipes.

No.	Dyestuff	Dye conc. %	Dispersant, g/l	pH*	Liquor: ratio	T°C	Time
1	Disperse Blue19	1	2	≅4,5	1:20	130	1h
2		2					
3		3					
4	Disperse Black 9	8	2	≅4,5	1:20	130	1h
5		12					
6		16					

* with CH_3COOH

The HT dyeing process comprise the next phases:

- dyebath set at 50°C
- dyebath auxiliaries adding and running for 10 min
- dye adding and running for 10 min
- heating to 130°C at 1.5°C/min
- fixing for 60 min at the dyeing temperature
- rinsing and aftertreating.

The method used is shown in the figure below (Figure 2).

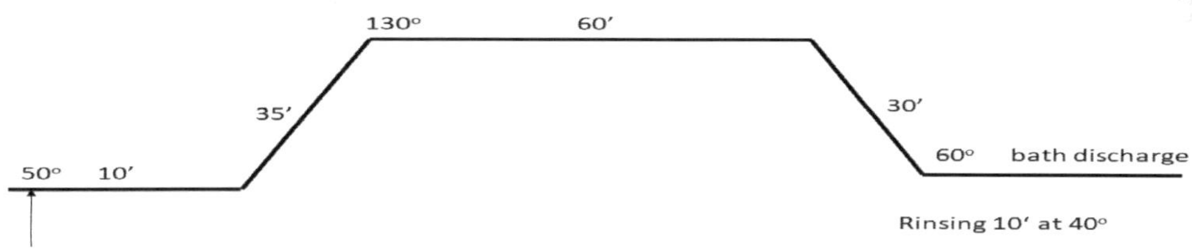

Figure 2. Polyester dyeing process.

After 40°C rinsing follows a reducing washing conforming to the recipe:

- soda ash: 6 ml/l, 36°Be
- sodium dithionite: 3g/L
- surfactant: 1g/L
- Liquor ratio = 1:20
- t° = 80°C
- time: 30'

The reductive washing process is schematised in the next diagram:

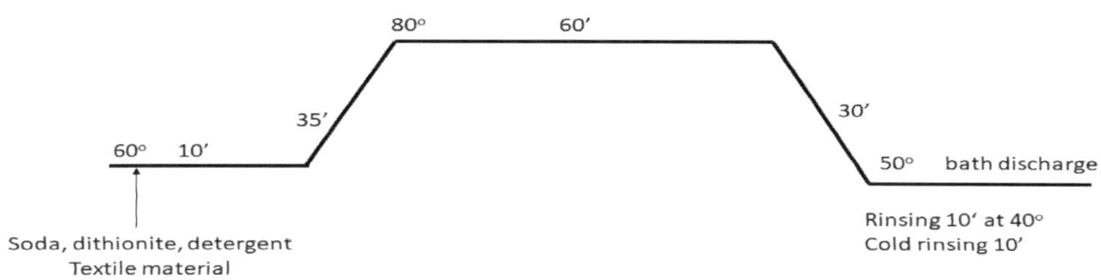

Figure 3. Reductive washing parameters.

IOP Conf. Series: Materials Science and Engineering **877** (2020) 012062 doi:10.1088/1757-899X/877/1/012062

The exhaustion degree, the volume of wastewater used in the process, the specific flow, pH, temperature and chemical oxygen demand of the wastewater, the total residuum, volatile substances and the biological oxygen demand were determined in all the cases.

We have determined the exhaustion degree for the analysed dyes by determining the absorbance of the dyestuff solutions with a spectrophotometer before and after the dyeing process. The cuvettes for the spectrophotometer were clear plastic and held one millilitre of sample. The spectrophotometer compared each sample to blank containing de-ionised water. The wavelength was selected so as to obtain maximum absorbance. The absorbance was converted into concentration using a calibration line.

The exhaustion degree was calculated using the relation:

$$Exhaustion\ degree = \frac{C_{initial} - C_{final}}{C_{initial}} \times 100 \qquad (1)$$

Where: $C_{initial}$ represents the concentration of the initial dye solution;

C_{final} represents the concentration of the dye in the exhausted bath.

The exhaustion degree for each dye and dyeing concentration is indicated in Table 2.

Table 2. Dye exhaustion.

Dye	Variant	Dye concentration (%)	Exhaustion degree, (%)
Disperse Blue19	1	1	93.7
	2	2	90.7
	3	3	91.6
Disperse Black 9	4	8	89.5
	5	12	87.5
	6	16	87.3

The process of industrial dyeing of polyester with disperse dyes is a pretty long process (over 8 hours), takes place in enclosed system, at 130°C and uses in the final stage inorganic reductive products. During the dyeing and the washing stages there have been determined bath losses that reached about 10-12 % in the dyeing process and about 5% in the reductive washing. Although the reductive wash uses sodium dithionite, it is improbable that sulphur compound are relieved in the atmosphere, especially because of the strong alkaline medium of the washing process.

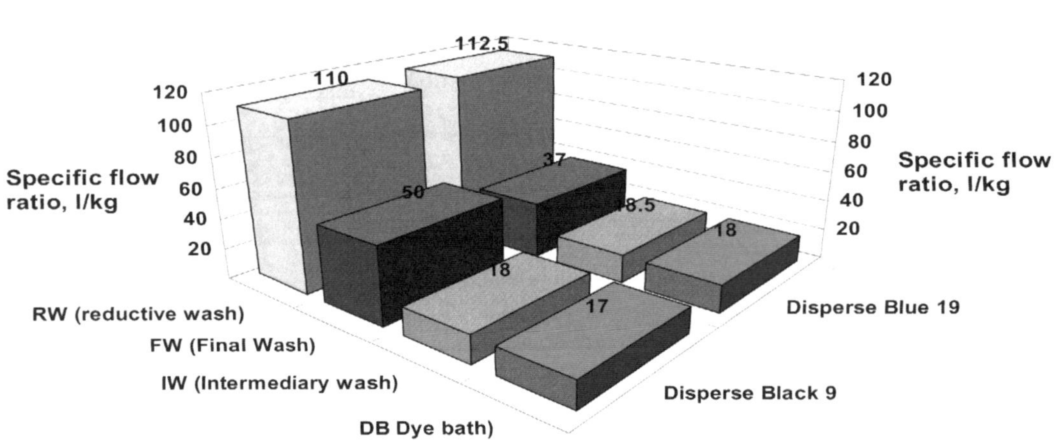

Figure 4. Global water consume.

The wastewater is formed by the water that is used in the four stages of the process: the dyeing, first washing, reductive washing and final washing. Because of the significant duration between the dyeing process and the last washing (over 5 hours), we have studied the four categories of wastewater separately. The global water consume exceeds 110 L/kg, as it can be seen in Figure 4.

We have determined some of the most environmental relevant parameters of the wastewater. The Chemical Oxygen Demand (COD) value was determined according to the Romanian standard method: a sample of the wastewater (treated and untreated) is refluxed in strongly acidic solution with a known excess of potassium dichromate in the presence of silver sulphate. The organic matter that exists in the sample is oxidized by the potassium dichromate. The dichromate that exceeds the stoichiometric value was titrated with ferrous ammonium sulphate, using ferroin as an indicator. At equivalence point the colour of the solution changes from green blue to reddish blue. The amount of $K_2Cr_20_7$ consumed is converted in the oxidable matter that is calculated in terms of equivalent oxygen [42].

The 5 Days Biochemical Oxygen Demand (BOD5) value was determined according to the Romanian standard method as well: filling with sample, to overflowing, an airtight bottle (300-mL bottles having a ground-glass stopper and a flared mouth) and incubating it at 20°C + 1°C for 5 days. Dissolved oxygen is measured initially and after incubation, and the BOD is computed from the difference between initial and final DO. Because the initial DO is determined shortly after the dilution is made, all oxygen uptake occurring after this measurement is included in the BOD measurement. [43].

To determine the Total Suspended Solids content, samples of 200 mL have been vacuum filtered using Glass microfiber filters. Finally the filters were dried one hour at 103-105°C, cooled in a desiccator and weighed [44]. The same sample was used to determine the Volatile organic compounds (VOCs) content. The filter have been introduced in a furnace and kept at 550° C ± 50° C for 30 minutes. In the end the samples were air cooled, desiccated and weighed [45].

The reducer content in the wastewater has been analysed by means of an iodometric method.

The environmental relevant characteristics of the wastewater produced in the dyeing process with the two studied disperse dyes are shown in Tables 3 and 4.

The physical and chemical characteristics of the wastewater from the dyeing and washing process lead to the following conclusions:

- wastewater is intensely coloured and has a high suspension content (especially for Disperse Black 9).
- even if the dyeing concentration differs significantly, most of the parameters of the wastewater are comparable, probably because of the higher exhaustion degree of the black dye.

Table 3. Environmental characteristics of the Disperse Blue 19 dyeings.

	Variant 1			Variant 2			Variant 3		
	D	RW	FW	D	RW	FW	D	RW	FW
1	18	18.5	112.5	18	18.5	112.5	18	18.5	112.5
2	4.5	13.6	10.8	4.8	13.8	10.9	4.7	13.7	11
3	60	50	35-40	60	50	35-40	60	50	35-40
4	2560	1420*	400[1]	2680	1430[1]	400[1]	2698	1425[1]	400*
5	2.9	7.2	0.3	3.1	7.6	0.3	3.1	7.6	0.3
6	55	4.7	26	61	7.4	30	62	6.9	34
7	358	32.7	41	353	31.9	32.0	351	29.7	33.6
8	101.3	41.7	4.24	100.7	42.2	4.45	100.9	43.8	4.44

* values distorted due to reductive substances presence.

Where 1: Specific flow ratio (l/Kg material) 2: pH; 3: t°C; 4: COD - Cr, mg/l; 5: Total residuum, mg/l; 6: VOC in the total residuum, %; 7: COD_5, mg/l; 8: Total reductive matter mval/l

The wastewater produced in the reductive wash process, with a strong alkaline reaction, shows a high degree or organical content. The values of COD were distorted by the presence of the ions S^{2-},

$(S_2O_3)^{2-}$, $(S_2O_4)^{2-}$, SO_3^{2-} as follows: $(S_2O_4)^{2-}$: 32 mg O_2/mval; $(SO_3)^{2-}$: 8 mg O_2/mval; $(S_2O_3)^{2-}$: 40 mg O_2/mval; S^{2-} : 32 mg O2 /mval.

Correlating these values with the COD values that we have determined, we can conclude that the reductive system is complex, containing a mixture of the indicated ions.

Table 4. Environmental characteristics of the Disperse Black 9 dyeings.

	Variant 1			Variant 2			Variant 3		
	D	RW	FW	D	RW	FW	D	RW	FW
1	117	18	110	117	18	110	117	18	110
2	7.9	14.0	10.2	7.8	13.9	10.4	7.9	14.1	10.2
3	60	50	35-40	60	50	35-40	60	50	35-40
4	1621	864[1]	185[1]	3920	1270[1]	240[1]	4685	1400	264
5	2.6	6.8	1.2	3.2	7.4	1.9	5.3	9.6	3.1
6	65	75	9.9	77	51	11	81	65	12.3
7	842	90.4	82.8	1800	99.8	100.1	1985	102.7	100.9
8	102.2	51.3	3.11	189.2	53.9	3.96	199.2	56.7	4.51

Where: 1 – 8 have the same signification. In what concerns the biodegradability of the two dyes, important differences have been found. The BOD5/COD indicates a much better biodegradability in the case of the black dye, the blue dye having only modest biological degradation capabilities.

The values of the spectral absorbance quotient for the initial dyebath and the exhausted one are shown in Figures 5 and 6.

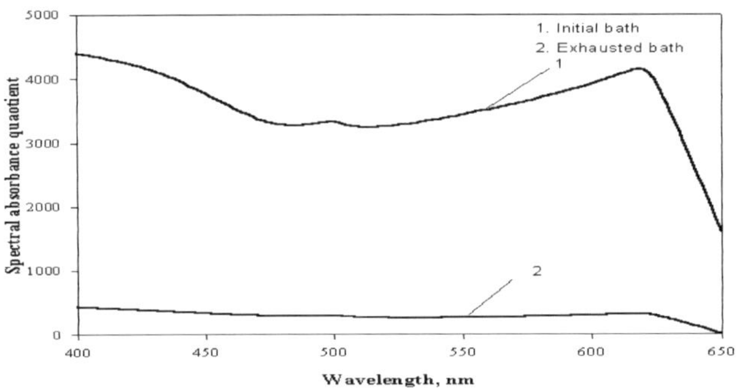

Figure 5. Spectral absorbance quotient for Disperse Blue 19.

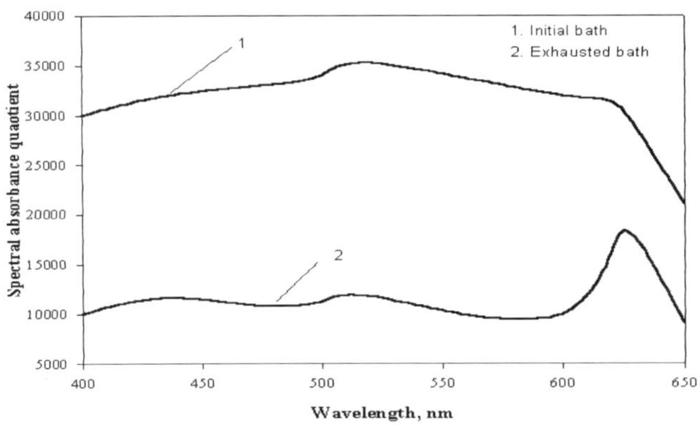

Figure 6. Spectral absorbance quotient for Disperse Black 9.

It can be seen that important reduction of the spectral absorbance quotient is obtained in both cases, even if the blue dye proves to exhaust better.

3. Conclusions

The results that we have obtained indicate that the dyeing of polyester with disperse dyes in the HT process produces wastewater with high pollutant content, and large quantities of water are required. The alkaline reaction and the high content of sulphur salts of the reducing dyeing wastewater are other important environmental significant characteristics.

The washing process implies the use of reductive substances and consumes more than three quarters of the total water needed in the process, which means that the potential for environmental improvements in the rinsing procedures is in this way considerable.

4. References

[1] Bertea AF, Bertea AP, Butnaru R and Bertea A 2013 *Poluare şi ecotehnologii în chimia textilă* (Iaşi: Performantica)
[2] Holkar C, Jadhav A, Pinjari D, Mahamuni N and Pandit A 2016 *J. Env. Manag.* **182** 351
[3] Mahapatra N 2015 *Textiles and environment* (New Delhi: Woodhead)
[4] Muthu S 2018 *Detox Fashion - Sustainable Chemistry and Wet Processing* (Singapore: Springer)
[5] Muthu S 2018 *Sustainable Innovations in Textile Chemical Processes* (Singapore: Springer)
[6] Muthu S 2018 *Sustainable Innovations in Textile Chemistry and Dyes* (Singapore: Springer)
[7] Mondal S, Kumar M and De P 2019 *Advances in Dye Removal Technologies* (Singapore: Springer)
[8] Agrawal B 2016 Res. *J. Rec. Sciences* **5**(10) 20
[9] Bertea A, Manea L R, Bertea A P and Sandu I 2016 *Rev. Chim-Bucharest* **67(12)** 2446
[10] Bertea AF, Butnaru R and Bertea AP 2013 *Environ. Eng. Manag. J* **12(5)** 1085
[11] Saini R 2017 *Int. J. Chem. Eng. Res.* **9(1)** 121
[12] Manea LR, Bertea A and Bertea AP 2019 *IOP Conf. Ser.: Mater. Sci. Eng.* **572** 012111
[13] Choudhury A 2018 *Adv. Mat. Tech. Env.* **2(1)** 145
[14] Stănescu I, Butnaru R, Bertea A and Bertea AP 2014 *Ind. Textila* **65(6)** 318
[15] Jasim N 2020 *IOP Conf. Ser.: Mater. Sci. Eng.* **745** 012138
[16] Kumbasar P and Körlü A 2016 *Textile Wastewater Treatment* (ExLi4EvA)
[17] Manea LR, Bertea A and Bertea AP 2019 *IOP Conf. Ser.: Mater. Sci. Eng.* **572** 012077
[18] Islam MR and Mostafa MG 2018 *J. Environ. Sci. & Natural Resources* **11(1&2)** 131
[19] Mudnoor V and Chakraborty J 2020 *Indian J. Fibre Text.* **45** 109
[20] Muthu S 2017 *Textiles and Clothing Sustainability - Sustainable Textile Chemical Processes* (Singapore: Springer)
[21] Shahin A, Mohammadi L, Rahdar A, Rahdar S, Dehghani R, Igwegbe C5 and Kyzas G 2020 *Nanomaterials* **10(556)** 1
[22] Shaid A, Osman S, Hannan A and Bhuiyan R 2013 *Int. J. Eng. Res. Dev.* **5(8)** 45
[23] Stanescu I, Manea LR, Bertea A, Bertea AP and Sandu I 2016 *Rev. Chim-Bucharest* **67(11)**, 2276
[24] Hassaan M and El Nemr A 2017 *Am. J. Env. Sc. Eng.* **1(3)** 64
[25] Stănescu I, Manea LR, Bertea A, Bertea AP and Sandu IC 2016 *Rev. Chim-Bucharest* **67(10)** 2082
[26] Mia R, Selim M, Shamim A, Chowdhury M, Sultana S, Armin M, Hossain M, Akter R, Dey S and Hasnun A 2019 *J. Textile Eng. Fashion Technol.* **5(4)** 220
[27] Tusief M, Malik M, Mohsin M, Asghar H, Iqbal M and Mahmood N 2020 *Pak. J. Ssci. Ind. Res. Ser. A: phys. sci.* **63A(1)** 40
[28] Yusuf M 2018 *Handbook of Textile Effluent Remediation* (Singapore: Pan Stanford Publishing)

[29] Textile World [Internet]. Man-Made Fibers Continue To Grow; 2015 [updated 2015 Feb 5; cited 2020 Feb 15]. Available from: https://www.textileworld.com/textile-world/fiber-world/2015/02/man-made-fibers-continue-to-grow/

[30] Maftei D, Asaftei I V, Sandu I, Manea L R, Birsa L M and Earar K 2015 *Rev. Chim (Bucharest)* **65(5)** 673

[31] Asaftei I V, Sandu I G, Birsa L M, Manea L R and Earar K 2015 *Rev. Chim (Bucharest)* **65(3)** 336

[32] Sarbu L G, Apostu M O, Sandu I G, Manea L R and Bahrin L G 2014 *Rev. Chim (Bucharest)* **65(11)** 1327

[33] Secula M S, Cretescu I, Cagnon B, Manea L R, Stan C S and Breaban I G 2013 *Materials* **6(7)** 2723

[34] Cailean D, Barjoveanu G, Musteret C P, Sulitanu N, Manea L R and Teodosiu C 2009 *Environmental Engineering and Management Journal* **8(3)** 503

[35] Gherasimescu C, Leva M, Butnru R, Muresan A and Manea L R 2011 *Industria Textila (Bucharest)* **62(1)** 19

[36] Amariei N, Manea L R, Bertea A P, Bertea A and Popa A 2017 *IOP Conf. Ser.: Mater. Sci. Eng.* **209** 012092, doi: 10.1088/1757-899X/209/1/ 012092

[37] Amariei N, Manea L R, Bertea A P, Cramariuc R, Bertea A and Cramariuc O 2017 *IOP Conf. Ser.: Mater. Sci. Eng.* **209** 012091, doi: 10.1088/1757-899X/209/1/012091

[38] Manea L R, Scarlet R, Amariei N, Nechita E and Sandu I G 2015 *Rev. Chim.* **66** (4) 542

[39] Calin M A, Khenoussi N, Schacher L, Adolphe D and Manea L R, Gradinaru I, Zetu I and Stratulat S 2013 *Mat. Plast. (Bucharest)* **50 (4)** 257

[40] Manea L R, Popa A and Bertea A 2017 *IOP Conf. Ser.: Mater. Sci. Eng.* **209** 012074

[41] Bertea A, Manea L R, Bertea A and Hristian L 2017 *IOP Conf. Ser.: Mater. Sci. Eng.* **209** 012073 doi:10.1088/1757-899X/209/1/ 012073

[42] SR ISO 6060 1996 Calitatea apei. Determinarea consumului chimic de oxygen (Bucharest: Romanian Standards Association)

[43] SR EN 1899-1 2003 Calitatea apei. Determinarea consumului biochimic de oxigen după n zile (CBOn). Partea 1: Metoda prin diluare şi însămânţare cu aport de aliltiouree (Bucharest: Romanian Standards Association)

[44] STAS 6953 1981 Determinarea substantelor extractibile cu solvent (Bucharest: Romanian National Standardization Body)

[45] STAS 12586 1987 Nămoluri rezultate de la tratarea apelor de suprafaţă şi epurarea apelor uzate. Determinarea umidităţii, a substanţei uscate, a pierderilor la calcinare (substanţe volatile) şi a substanţelor minerale (Bucharest: Romanian National Standardization Body)

AUTHOR INDEX